T0202285

Springer Series in Statistics

Advisors:
P. Bickel, P. Diggle, S. Fienberg, U. Gather,
I. Olkin, S. Zeger

Springer Series in Statistics

For other titles published in this series, go to
http://www.springer.com/series/692

Trevor Hastie
Robert Tibshirani
Jerome Friedman

The Elements of Statistical Learning

Data Mining, Inference, and Prediction

Second Edition

 Springer

Trevor Hastie
Stanford University
Dept. of Statistics
Stanford CA 94305
USA
hastie@stanford.edu

Robert Tibshirani
Stanford University
Dept. of Statistics
Stanford CA 94305
USA
tibs@stanford.edu

Jerome Friedman
Stanford University
Dept. of Statistics
Stanford CA 94305
USA
jhf@stanford.edu

ISSN: 0172 7397
ISBN: 978-0-387-84857-0   e-ISBN: 978-0-387-84858-7
DOI: 10.1007/b94608

Library of Congress Control Number: 2008941148

Printed on acid-free paper

springer.com

To our parents:

Valerie and Patrick Hastie

Vera and Sami Tibshirani

Florence and Harry Friedman

and to our families:

Samantha, Timothy, and Lynda

Charlie, Ryan, Julie, and Cheryl

Melanie, Dora, Monika, and Ildiko

Preface to the Second Edition

In God we trust, all others bring data.

–William Edwards Deming (1900-1993)[1]

We have been gratified by the popularity of the first edition of *The Elements of Statistical Learning*. This, along with the fast pace of research in the statistical learning field, motivated us to update our book with a second edition.

We have added four new chapters and updated some of the existing chapters. Because many readers are familiar with the layout of the first edition, we have tried to change it as little as possible. Here is a summary of the main changes:

---

[1] On the Web, this quote has been widely attributed to both Deming and Robert W. Hayden; however Professor Hayden told us that he can claim no credit for this quote, and ironically we could find no "data" confirming that Deming actually said this.

Chapter	What's new
1. Introduction	
2. Overview of Supervised Learning	
3. Linear Methods for Regression	LAR algorithm and generalizations of the lasso
4. Linear Methods for Classification	Lasso path for logistic regression
5. Basis Expansions and Regularization	Additional illustrations of RKHS
6. Kernel Smoothing Methods	
7. Model Assessment and Selection	Strengths and pitfalls of cross-validation
8. Model Inference and Averaging	
9. Additive Models, Trees, and Related Methods	
10. Boosting and Additive Trees	New example from ecology; some material split off to Chapter 16.
11. Neural Networks	Bayesian neural nets and the NIPS 2003 challenge
12. Support Vector Machines and Flexible Discriminants	Path algorithm for SVM classifier
13. Prototype Methods and Nearest-Neighbors	
14. Unsupervised Learning	Spectral clustering, kernel PCA, sparse PCA, non-negative matrix factorization archetypal analysis, nonlinear dimension reduction, Google page rank algorithm, a direct approach to ICA
15. Random Forests	New
16. Ensemble Learning	New
17. Undirected Graphical Models	New
18. High-Dimensional Problems	New

Some further notes:

- Our first edition was unfriendly to colorblind readers; in particular, we tended to favor red/green contrasts which are particularly troublesome. We have changed the color palette in this edition to a large extent, replacing the above with an orange/blue contrast.

- We have changed the name of Chapter 6 from "Kernel Methods" to "Kernel Smoothing Methods", to avoid confusion with the machine-learning kernel method that is discussed in the context of support vector machines (Chapter 12) and more generally in Chapters 5 and 14.

- In the first edition, the discussion of error-rate estimation in Chapter 7 was sloppy, as we did not clearly differentiate the notions of conditional error rates (conditional on the training set) and unconditional rates. We have fixed this in the new edition.

- Chapters 15 and 16 follow naturally from Chapter 10, and the chapters are probably best read in that order.

- In Chapter 17, we have not attempted a comprehensive treatment of graphical models, and discuss only undirected models and some new methods for their estimation. Due to a lack of space, we have specifically omitted coverage of directed graphical models.

- Chapter 18 explores the "$p \gg N$" problem, which is learning in high-dimensional feature spaces. These problems arise in many areas, including genomic and proteomic studies, and document classification.

We thank the many readers who have found the (too numerous) errors in the first edition. We apologize for those and have done our best to avoid errors in this new edition. We thank Mark Segal, Bala Rajaratnam, and Larry Wasserman for comments on some of the new chapters, and many Stanford graduate and post-doctoral students who offered comments, in particular Mohammed AlQuraishi, John Boik, Holger Hoefling, Arian Maleki, Donal McMahon, Saharon Rosset, Babak Shababa, Daniela Witten, Ji Zhu and Hui Zou. We thank John Kimmel for his patience in guiding us through this new edition. RT dedicates this edition to the memory of Anna McPhee.

Trevor Hastie
Robert Tibshirani
Jerome Friedman

Stanford, California
August 2008

Preface to the First Edition

We are drowning in information and starving for knowledge.

Rutherford D. Roger

The field of Statistics is constantly challenged by the problems that science and industry brings to its door. In the early days, these problems often came from agricultural and industrial experiments and were relatively small in scope. With the advent of computers and the information age, statistical problems have exploded both in size and complexity. Challenges in the areas of data storage, organization and searching have led to the new field of "data mining"; statistical and computational problems in biology and medicine have created "bioinformatics." Vast amounts of data are being generated in many fields, and the statistician's job is to make sense of it all: to extract important patterns and trends, and understand "what the data says." We call this *learning from data*.

The challenges in learning from data have led to a revolution in the statistical sciences. Since computation plays such a key role, it is not surprising that much of this new development has been done by researchers in other fields such as computer science and engineering.

The learning problems that we consider can be roughly categorized as either *supervised* or *unsupervised*. In supervised learning, the goal is to predict the value of an outcome measure based on a number of input measures; in unsupervised learning, there is no outcome measure, and the goal is to describe the associations and patterns among a set of input measures.

This book is our attempt to bring together many of the important new ideas in learning, and explain them in a statistical framework. While some mathematical details are needed, we emphasize the methods and their conceptual underpinnings rather than their theoretical properties. As a result, we hope that this book will appeal not just to statisticians but also to researchers and practitioners in a wide variety of fields.

Just as we have learned a great deal from researchers outside of the field of statistics, our statistical viewpoint may help others to better understand different aspects of learning:

> There is no true interpretation of anything; interpretation is a vehicle in the service of human comprehension. The value of interpretation is in enabling others to fruitfully think about an idea.

–Andreas Buja

We would like to acknowledge the contribution of many people to the conception and completion of this book. David Andrews, Leo Breiman, Andreas Buja, John Chambers, Bradley Efron, Geoffrey Hinton, Werner Stuetzle, and John Tukey have greatly influenced our careers. Balasubramanian Narasimhan gave us advice and help on many computational problems, and maintained an excellent computing environment. Shin-Ho Bang helped in the production of a number of the figures. Lee Wilkinson gave valuable tips on color production. Ilana Belitskaya, Eva Cantoni, Maya Gupta, Michael Jordan, Shanti Gopatam, Radford Neal, Jorge Picazo, Bogdan Popescu, Olivier Renaud, Saharon Rosset, John Storey, Ji Zhu, Mu Zhu, two reviewers and many students read parts of the manuscript and offered helpful suggestions. John Kimmel was supportive, patient and helpful at every phase; MaryAnn Brickner and Frank Ganz headed a superb production team at Springer. Trevor Hastie would like to thank the statistics department at the University of Cape Town for their hospitality during the final stages of this book. We gratefully acknowledge NSF and NIH for their support of this work. Finally, we would like to thank our families and our parents for their love and support.

Trevor Hastie
Robert Tibshirani
Jerome Friedman

Stanford, California
May 2001

> The quiet statisticians have changed our world; not by discovering new facts or technical developments, but by changing the ways that we reason, experiment and form our opinions ....

–Ian Hacking

Contents

1
Introduction

Statistical learning plays a key role in many areas of science, finance and industry. Here are some examples of learning problems:

- Predict whether a patient, hospitalized due to a heart attack, will have a second heart attack. The prediction is to be based on demographic, diet and clinical measurements for that patient.

- Predict the price of a stock in 6 months from now, on the basis of company performance measures and economic data.

- Identify the numbers in a handwritten ZIP code, from a digitized image.

- Estimate the amount of glucose in the blood of a diabetic person, from the infrared absorption spectrum of that person's blood.

- Identify the risk factors for prostate cancer, based on clinical and demographic variables.

The science of learning plays a key role in the fields of statistics, data mining and artificial intelligence, intersecting with areas of engineering and other disciplines.

This book is about learning from data. In a typical scenario, we have an outcome measurement, usually quantitative (such as a stock price) or categorical (such as heart attack/no heart attack), that we wish to predict based on a set of *features* (such as diet and clinical measurements). We have a *training set* of data, in which we observe the outcome and feature

T. Hastie et al., *The Elements of Statistical Learning, Second Edition,*
DOI: 10.1007/b94608_1,
© Springer Science+Business Media, LLC 2009

TABLE 1.1. *Average percentage of words or characters in an email message equal to the indicated word or character. We have chosen the words and characters showing the largest difference between* spam *and* email.

	george	you	your	hp	free	hpl	!	our	re	edu	remove
spam	0.00	2.26	1.38	0.02	0.52	0.01	0.51	0.51	0.13	0.01	0.28
email	1.27	1.27	0.44	0.90	0.07	0.43	0.11	0.18	0.42	0.29	0.01

measurements for a set of objects (such as people). Using this data we build a prediction model, or *learner*, which will enable us to predict the outcome for new unseen objects. A good learner is one that accurately predicts such an outcome.

The examples above describe what is called the *supervised learning* problem. It is called "supervised" because of the presence of the outcome variable to guide the learning process. In the *unsupervised learning problem*, we observe only the features and have no measurements of the outcome. Our task is rather to describe how the data are organized or clustered. We devote most of this book to supervised learning; the unsupervised problem is less developed in the literature, and is the focus of Chapter 14.

Here are some examples of real learning problems that are discussed in this book.

Example 1: Email Spam

The data for this example consists of information from 4601 email messages, in a study to try to predict whether the email was junk email, or "spam." The objective was to design an automatic spam detector that could filter out spam before clogging the users' mailboxes. For all 4601 email messages, the true outcome (email type) email or spam is available, along with the relative frequencies of 57 of the most commonly occurring words and punctuation marks in the email message. This is a supervised learning problem, with the outcome the class variable email/spam. It is also called a *classification* problem.

Table 1.1 lists the words and characters showing the largest average difference between spam and email.

Our learning method has to decide which features to use and how: for example, we might use a rule such as

$$\text{if } (\text{\%george} < 0.6) \text{ \& } (\text{\%you} > 1.5) \quad \text{then spam}$$
$$\text{else email.}$$

Another form of a rule might be:

$$\text{if } (0.2 \cdot \text{\%you} - 0.3 \cdot \text{\%george}) > 0 \quad \text{then spam}$$
$$\text{else email.}$$

FIGURE 1.1. *Scatterplot matrix of the prostate cancer data. The first row shows the response against each of the predictors in turn. Two of the predictors,* svi *and* gleason, *are categorical.*

For this problem not all errors are equal; we want to avoid filtering out good email, while letting spam get through is not desirable but less serious in its consequences. We discuss a number of different methods for tackling this learning problem in the book.

Example 2: Prostate Cancer

The data for this example, displayed in Figure 1.1[1], come from a study by Stamey et al. (1989) that examined the correlation between the level of

---

[1]There was an error in these data in the first edition of this book. Subject 32 had a value of 6.1 for lweight, which translates to a 449 gm prostate! The correct value is 44.9 gm. We are grateful to Prof. Stephen W. Link for alerting us to this error.

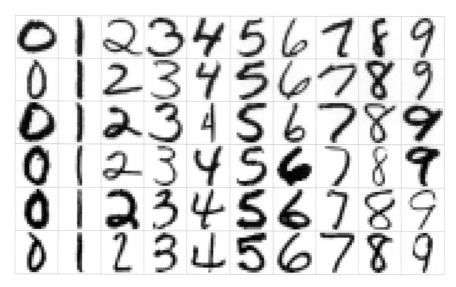

FIGURE 1.2. *Examples of handwritten digits from U.S. postal envelopes.*

prostate specific antigen (PSA) and a number of clinical measures, in 97 men who were about to receive a radical prostatectomy.

The goal is to predict the log of PSA (lpsa) from a number of measurements including log cancer volume (lcavol), log prostate weight lweight, age, log of benign prostatic hyperplasia amount lbph, seminal vesicle invasion svi, log of capsular penetration lcp, Gleason score gleason, and percent of Gleason scores 4 or 5 pgg45. Figure 1.1 is a scatterplot matrix of the variables. Some correlations with lpsa are evident, but a good predictive model is difficult to construct by eye.

This is a supervised learning problem, known as a *regression problem*, because the outcome measurement is quantitative.

Example 3: Handwritten Digit Recognition

The data from this example come from the handwritten ZIP codes on envelopes from U.S. postal mail. Each image is a segment from a five digit ZIP code, isolating a single digit. The images are 16×16 eight-bit grayscale maps, with each pixel ranging in intensity from 0 to 255. Some sample images are shown in Figure 1.2.

The images have been normalized to have approximately the same size and orientation. The task is to predict, from the 16×16 matrix of pixel intensities, the identity of each image $(0, 1, \ldots, 9)$ quickly and accurately. If it is accurate enough, the resulting algorithm would be used as part of an automatic sorting procedure for envelopes. This is a classification problem for which the error rate needs to be kept very low to avoid misdirection of mail. In order to achieve this low error rate, some objects can be assigned to a "don't know" category, and sorted instead by hand.

Example 4: DNA Expression Microarrays

DNA stands for deoxyribonucleic acid, and is the basic material that makes up human chromosomes. DNA microarrays measure the expression of a gene in a cell by measuring the amount of mRNA (messenger ribonucleic acid) present for that gene. Microarrays are considered a breakthrough technology in biology, facilitating the quantitative study of thousands of genes simultaneously from a single sample of cells.

Here is how a DNA microarray works. The nucleotide sequences for a few thousand genes are printed on a glass slide. A target sample and a reference sample are labeled with red and green dyes, and each are hybridized with the DNA on the slide. Through fluoroscopy, the log (red/green) intensities of RNA hybridizing at each site is measured. The result is a few thousand numbers, typically ranging from say −6 to 6, measuring the expression level of each gene in the target relative to the reference sample. Positive values indicate higher expression in the target versus the reference, and vice versa for negative values.

A gene expression dataset collects together the expression values from a series of DNA microarray experiments, with each column representing an experiment. There are therefore several thousand rows representing individual genes, and tens of columns representing samples: in the particular example of Figure 1.3 there are 6830 genes (rows) and 64 samples (columns), although for clarity only a random sample of 100 rows are shown. The figure displays the data set as a heat map, ranging from green (negative) to red (positive). The samples are 64 cancer tumors from different patients.

The challenge here is to understand how the genes and samples are organized. Typical questions include the following:

(a) which samples are most similar to each other, in terms of their expression profiles across genes?

(b) which genes are most similar to each other, in terms of their expression profiles across samples?

(c) do certain genes show very high (or low) expression for certain cancer samples?

We could view this task as a regression problem, with two categorical predictor variables—genes and samples—with the response variable being the level of expression. However, it is probably more useful to view it as *unsupervised learning* problem. For example, for question (a) above, we think of the samples as points in 6830–dimensional space, which we want to *cluster* together in some way.

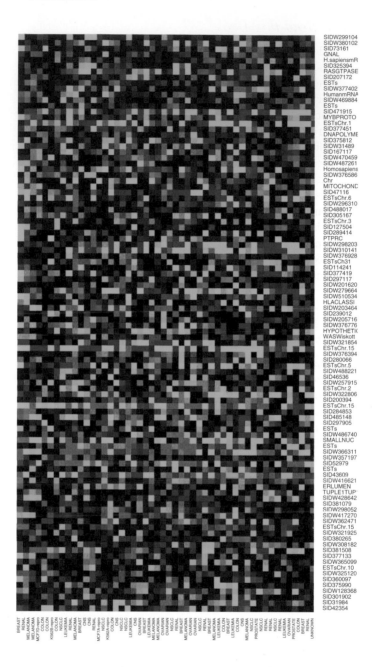

FIGURE 1.3. *DNA microarray data: expression matrix of* 6830 *genes (rows) and* 64 *samples (columns), for the human tumor data. Only a random sample of* 100 *rows are shown. The display is a heat map, ranging from bright green (negative, under expressed) to bright red (positive, over expressed). Missing values are gray. The rows and columns are displayed in a randomly chosen order.*

Who Should Read this Book

This book is designed for researchers and students in a broad variety of fields: statistics, artificial intelligence, engineering, finance and others. We expect that the reader will have had at least one elementary course in statistics, covering basic topics including linear regression.

We have not attempted to write a comprehensive catalog of learning methods, but rather to describe some of the most important techniques. Equally notable, we describe the underlying concepts and considerations by which a researcher can judge a learning method. We have tried to write this book in an intuitive fashion, emphasizing concepts rather than mathematical details.

As statisticians, our exposition will naturally reflect our backgrounds and areas of expertise. However in the past eight years we have been attending conferences in neural networks, data mining and machine learning, and our thinking has been heavily influenced by these exciting fields. This influence is evident in our current research, and in this book.

How This Book is Organized

Our view is that one must understand simple methods before trying to grasp more complex ones. Hence, after giving an overview of the supervising learning problem in Chapter 2, we discuss linear methods for regression and classification in Chapters 3 and 4. In Chapter 5 we describe splines, wavelets and regularization/penalization methods for a single predictor, while Chapter 6 covers kernel methods and local regression. Both of these sets of methods are important building blocks for high-dimensional learning techniques. Model assessment and selection is the topic of Chapter 7, covering the concepts of bias and variance, overfitting and methods such as cross-validation for choosing models. Chapter 8 discusses model inference and averaging, including an overview of maximum likelihood, Bayesian inference and the bootstrap, the EM algorithm, Gibbs sampling and bagging. A related procedure called boosting is the focus of Chapter 10.

In Chapters 9–13 we describe a series of structured methods for supervised learning, with Chapters 9 and 11 covering regression and Chapters 12 and 13 focusing on classification. Chapter 14 describes methods for unsupervised learning. Two recently proposed techniques, random forests and ensemble learning, are discussed in Chapters 15 and 16. We describe undirected graphical models in Chapter 17 and finally we study high-dimensional problems in Chapter 18.

At the end of each chapter we discuss computational considerations important for data mining applications, including how the computations scale with the number of observations and predictors. Each chapter ends with Bibliographic Notes giving background references for the material.

We recommend that Chapters 1–4 be first read in sequence. Chapter 7 should also be considered mandatory, as it covers central concepts that pertain to all learning methods. With this in mind, the rest of the book can be read sequentially, or sampled, depending on the reader's interest.

The symbol indicates a technically difficult section, one that can be skipped without interrupting the flow of the discussion.

Book Website

The website for this book is located at

<code>http://www-stat.stanford.edu/ElemStatLearn</code>

It contains a number of resources, including many of the datasets used in this book.

Note for Instructors

We have successively used the first edition of this book as the basis for a two-quarter course, and with the additional materials in this second edition, it could even be used for a three-quarter sequence. Exercises are provided at the end of each chapter. It is important for students to have access to good software tools for these topics. We used the R and S-PLUS programming languages in our courses.

2
Overview of Supervised Learning

2.1 Introduction

The first three examples described in Chapter 1 have several components in common. For each there is a set of variables that might be denoted as *inputs*, which are measured or preset. These have some influence on one or more *outputs*. For each example the goal is to use the inputs to predict the values of the outputs. This exercise is called *supervised learning*.

We have used the more modern language of machine learning. In the statistical literature the inputs are often called the *predictors*, a term we will use interchangeably with inputs, and more classically the *independent variables*. In the pattern recognition literature the term *features* is preferred, which we use as well. The outputs are called the *responses*, or classically the *dependent variables*.

2.2 Variable Types and Terminology

The outputs vary in nature among the examples. In the glucose prediction example, the output is a *quantitative* measurement, where some measurements are bigger than others, and measurements close in value are close in nature. In the famous Iris discrimination example due to R. A. Fisher, the output is *qualitative* (species of Iris) and assumes values in a finite set $\mathcal{G} = \{ \textit{Virginica}, \textit{Setosa} \text{ and } \textit{Versicolor} \}$. In the handwritten digit example the output is one of 10 different digit *classes*: $\mathcal{G} = \{0, 1, \ldots, 9\}$. In both of

T. Hastie et al., *The Elements of Statistical Learning, Second Edition*,
DOI: 10.1007/b94608_2,
© Springer Science+Business Media, LLC 2009

these there is no explicit ordering in the classes, and in fact often descriptive labels rather than numbers are used to denote the classes. Qualitative variables are also referred to as *categorical* or *discrete* variables as well as *factors*.

For both types of outputs it makes sense to think of using the inputs to predict the output. Given some specific atmospheric measurements today and yesterday, we want to predict the ozone level tomorrow. Given the grayscale values for the pixels of the digitized image of the handwritten digit, we want to predict its class label.

This distinction in output type has led to a naming convention for the prediction tasks: *regression* when we predict quantitative outputs, and *classification* when we predict qualitative outputs. We will see that these two tasks have a lot in common, and in particular both can be viewed as a task in function approximation.

Inputs also vary in measurement type; we can have some of each of qualitative and quantitative input variables. These have also led to distinctions in the types of methods that are used for prediction: some methods are defined most naturally for quantitative inputs, some most naturally for qualitative and some for both.

A third variable type is *ordered categorical*, such as *small, medium* and *large*, where there is an ordering between the values, but no metric notion is appropriate (the difference between medium and small need not be the same as that between large and medium). These are discussed further in Chapter 4.

Qualitative variables are typically represented numerically by codes. The easiest case is when there are only two classes or categories, such as "success" or "failure," "survived" or "died." These are often represented by a single binary digit or bit as 0 or 1, or else by -1 and 1. For reasons that will become apparent, such numeric codes are sometimes referred to as *targets*. When there are more than two categories, several alternatives are available. The most useful and commonly used coding is via *dummy variables*. Here a K-level qualitative variable is represented by a vector of K binary variables or bits, only one of which is "on" at a time. Although more compact coding schemes are possible, dummy variables are symmetric in the levels of the factor.

We will typically denote an input variable by the symbol X. If X is a vector, its components can be accessed by subscripts X_j. Quantitative outputs will be denoted by Y, and qualitative outputs by G (for group). We use uppercase letters such as X, Y or G when referring to the generic aspects of a variable. Observed values are written in lowercase; hence the ith observed value of X is written as x_i (where x_i is again a scalar or vector). Matrices are represented by bold uppercase letters; for example, a set of N input p-vectors x_i, $i = 1, \ldots, N$ would be represented by the $N \times p$ matrix **X**. In general, vectors will not be bold, except when they have N components; this convention distinguishes a p-vector of inputs x_i for the

ith observation from the N-vector $\mathbf{x}_j$ consisting of all the observations on variable X_j. Since all vectors are assumed to be column vectors, the ith row of $\mathbf{X}$ is x_i^T, the vector transpose of x_i.

For the moment we can loosely state the learning task as follows: given the value of an input vector X, make a good prediction of the output Y, denoted by $\hat{Y}$ (pronounced "y-hat"). If Y takes values in $\mathbb{R}$ then so should $\hat{Y}$; likewise for categorical outputs, $\hat{G}$ should take values in the same set $\mathcal{G}$ associated with G.

For a two-class G, one approach is to denote the binary coded target as Y, and then treat it as a quantitative output. The predictions $\hat{Y}$ will typically lie in $[0, 1]$, and we can assign to $\hat{G}$ the class label according to whether $\hat{y} > 0.5$. This approach generalizes to K-level qualitative outputs as well.

We need data to construct prediction rules, often a lot of it. We thus suppose we have available a set of measurements (x_i, y_i) or (x_i, g_i), $i = 1, \dots, N$, known as the *training data*, with which to construct our prediction rule.

2.3   Two Simple Approaches to Prediction: Least Squares and Nearest Neighbors

In this section we develop two simple but powerful prediction methods: the linear model fit by least squares and the k-nearest-neighbor prediction rule. The linear model makes huge assumptions about structure and yields stable but possibly inaccurate predictions. The method of k-nearest neighbors makes very mild structural assumptions: its predictions are often accurate but can be unstable.

2.3.1   Linear Models and Least Squares

The linear model has been a mainstay of statistics for the past 30 years and remains one of our most important tools. Given a vector of inputs $X^T = (X_1, X_2, \dots, X_p)$, we predict the output Y via the model

$$\hat{Y} = \hat{\beta}_0 + \sum_{j=1}^{p} X_j \hat{\beta}_j. \tag{2.1}$$

The term $\hat{\beta}_0$ is the intercept, also known as the *bias* in machine learning. Often it is convenient to include the constant variable 1 in X, include $\hat{\beta}_0$ in the vector of coefficients $\hat{\beta}$, and then write the linear model in vector form as an inner product

$$\hat{Y} = X^T \hat{\beta}, \tag{2.2}$$

where X^T denotes vector or matrix transpose (X being a column vector). Here we are modeling a single output, so $\hat{Y}$ is a scalar; in general $\hat{Y}$ can be a K–vector, in which case β would be a $p \times K$ matrix of coefficients. In the $(p + 1)$-dimensional input–output space, $(X, \hat{Y})$ represents a hyperplane. If the constant is included in X, then the hyperplane includes the origin and is a subspace; if not, it is an affine set cutting the Y-axis at the point $(0, \hat{\beta}_0)$. From now on we assume that the intercept is included in $\hat{\beta}$.

Viewed as a function over the p-dimensional input space, $f(X) = X^T \beta$ is linear, and the gradient $f'(X) = \beta$ is a vector in input space that points in the steepest uphill direction.

How do we fit the linear model to a set of training data? There are many different methods, but by far the most popular is the method of *least squares*. In this approach, we pick the coefficients β to minimize the residual sum of squares

$$\text{RSS}(\beta) = \sum_{i=1}^{N} (y_i - x_i^T \beta)^2. \tag{2.3}$$

$\text{RSS}(\beta)$ is a quadratic function of the parameters, and hence its minimum always exists, but may not be unique. The solution is easiest to characterize in matrix notation. We can write

$$\text{RSS}(\beta) = (\mathbf{y} - \mathbf{X}\beta)^T (\mathbf{y} - \mathbf{X}\beta), \tag{2.4}$$

where $\mathbf{X}$ is an $N \times p$ matrix with each row an input vector, and $\mathbf{y}$ is an N-vector of the outputs in the training set. Differentiating w.r.t. β we get the *normal equations*

$$\mathbf{X}^T (\mathbf{y} - \mathbf{X}\beta) = 0. \tag{2.5}$$

If $\mathbf{X}^T\mathbf{X}$ is nonsingular, then the unique solution is given by

$$\hat{\beta} = (\mathbf{X}^T\mathbf{X})^{-1}\mathbf{X}^T\mathbf{y}, \tag{2.6}$$

and the fitted value at the ith input x_i is $\hat{y}_i = \hat{y}(x_i) = x_i^T \hat{\beta}$. At an arbitrary input x_0 the prediction is $\hat{y}(x_0) = x_0^T \hat{\beta}$. The entire fitted surface is characterized by the p parameters $\hat{\beta}$. Intuitively, it seems that we do not need a very large data set to fit such a model.

Let's look at an example of the linear model in a classification context. Figure 2.1 shows a scatterplot of training data on a pair of inputs X_1 and X_2. The data are simulated, and for the present the simulation model is not important. The output class variable G has the values BLUE or ORANGE, and is represented as such in the scatterplot. There are 100 points in each of the two classes. The linear regression model was fit to these data, with the response Y coded as 0 for BLUE and 1 for ORANGE. The fitted values $\hat{Y}$ are converted to a fitted class variable $\hat{G}$ according to the rule

$$\hat{G} = \begin{cases} \text{ORANGE} & \text{if } \hat{Y} > 0.5, \\ \text{BLUE} & \text{if } \hat{Y} \leq 0.5. \end{cases} \tag{2.7}$$

Linear Regression of 0/1 Response

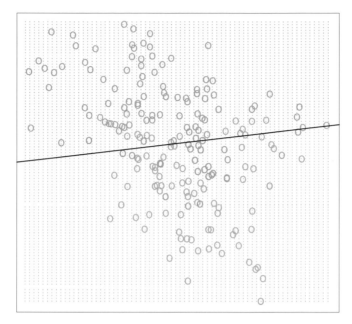

FIGURE 2.1. *A classification example in two dimensions. The classes are coded as a binary variable (*BLUE = 0, ORANGE = 1*), and then fit by linear regression. The line is the decision boundary defined by* $x^T \hat{\beta} = 0.5$*. The orange shaded region denotes that part of input space classified as* ORANGE*, while the blue region is classified as* BLUE*.*

The set of points in $\mathbb{R}^2$ classified as ORANGE corresponds to $\{x : x^T \hat{\beta} > 0.5\}$, indicated in Figure 2.1, and the two predicted classes are separated by the *decision boundary* $\{x : x^T \hat{\beta} = 0.5\}$, which is linear in this case. We see that for these data there are several misclassifications on both sides of the decision boundary. Perhaps our linear model is too rigid— or are such errors unavoidable? Remember that these are errors on the training data itself, and we have not said where the constructed data came from. Consider the two possible scenarios:

Scenario 1: The training data in each class were generated from bivariate Gaussian distributions with uncorrelated components and different means.

Scenario 2: The training data in each class came from a mixture of 10 low-variance Gaussian distributions, with individual means themselves distributed as Gaussian.

A mixture of Gaussians is best described in terms of the generative model. One first generates a discrete variable that determines which of the component Gaussians to use, and then generates an observation from the chosen density. In the case of one Gaussian per class, we will see in Chapter 4 that a linear decision boundary is the best one can do, and that our estimate is almost optimal. The region of overlap is inevitable, and future data to be predicted will be plagued by this overlap as well.

In the case of mixtures of tightly clustered Gaussians the story is different. A linear decision boundary is unlikely to be optimal, and in fact is not. The optimal decision boundary is nonlinear and disjoint, and as such will be much more difficult to obtain.

We now look at another classification and regression procedure that is in some sense at the opposite end of the spectrum to the linear model, and far better suited to the second scenario.

2.3.2   Nearest-Neighbor Methods

Nearest-neighbor methods use those observations in the training set $\mathcal{T}$ closest in input space to x to form $\hat{Y}$. Specifically, the k-nearest neighbor fit for $\hat{Y}$ is defined as follows:

$$\hat{Y}(x) = \frac{1}{k} \sum_{x_i \in N_k(x)} y_i, \tag{2.8}$$

where $N_k(x)$ is the neighborhood of x defined by the k closest points x_i in the training sample. Closeness implies a metric, which for the moment we assume is Euclidean distance. So, in words, we find the k observations with x_i closest to x in input space, and average their responses.

In Figure 2.2 we use the same training data as in Figure 2.1, and use 15-nearest-neighbor averaging of the binary coded response as the method of fitting. Thus $\hat{Y}$ is the proportion of ORANGE's in the neighborhood, and so assigning class ORANGE to $\hat{G}$ if $\hat{Y} > 0.5$ amounts to a majority vote in the neighborhood. The colored regions indicate all those points in input space classified as BLUE or ORANGE by such a rule, in this case found by evaluating the procedure on a fine grid in input space. We see that the decision boundaries that separate the BLUE from the ORANGE regions are far more irregular, and respond to local clusters where one class dominates.

Figure 2.3 shows the results for 1-nearest-neighbor classification: $\hat{Y}$ is assigned the value y_ℓ of the closest point x_ℓ to x in the training data. In this case the regions of classification can be computed relatively easily, and correspond to a *Voronoi tessellation* of the training data. Each point x_i has an associated tile bounding the region for which it is the closest input point. For all points x in the tile, $\hat{G}(x) = g_i$. The decision boundary is even more irregular than before.

The method of k-nearest-neighbor averaging is defined in exactly the same way for regression of a quantitative output Y, although $k = 1$ would be an unlikely choice.

15-Nearest Neighbor Classifier

FIGURE 2.2. *The same classification example in two dimensions as in Figure 2.1. The classes are coded as a binary variable* (BLUE $= 0$, ORANGE $= 1$) *and then fit by 15-nearest-neighbor averaging as in (2.8). The predicted class is hence chosen by majority vote amongst the 15-nearest neighbors.*

In Figure 2.2 we see that far fewer training observations are misclassified than in Figure 2.1. This should not give us too much comfort, though, since in Figure 2.3 *none* of the training data are misclassified. A little thought suggests that for k-nearest-neighbor fits, the error on the training data should be approximately an increasing function of k, and will always be 0 for $k = 1$. An independent test set would give us a more satisfactory means for comparing the different methods.

It appears that k-nearest-neighbor fits have a single parameter, the number of neighbors k, compared to the p parameters in least-squares fits. Although this is the case, we will see that the *effective* number of parameters of k-nearest neighbors is N/k and is generally bigger than p, and decreases with increasing k. To get an idea of why, note that if the neighborhoods were nonoverlapping, there would be N/k neighborhoods and we would fit one parameter (a mean) in each neighborhood.

It is also clear that we cannot use sum-of-squared errors on the training set as a criterion for picking k, since we would always pick $k = 1$! It would seem that k-nearest-neighbor methods would be more appropriate for the mixture Scenario 2 described above, while for Gaussian data the decision boundaries of k-nearest neighbors would be unnecessarily noisy.

1-Nearest Neighbor Classifier

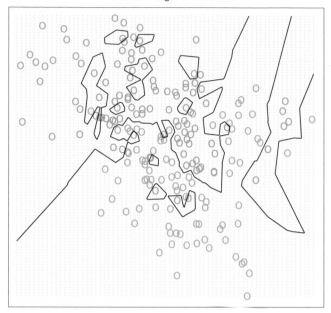

FIGURE 2.3. *The same classification example in two dimensions as in Figure 2.1. The classes are coded as a binary variable (BLUE = 0, ORANGE = 1), and then predicted by 1-nearest-neighbor classification.*

2.3.3 From Least Squares to Nearest Neighbors

The linear decision boundary from least squares is very smooth, and apparently stable to fit. It does appear to rely heavily on the assumption that a linear decision boundary is appropriate. In language we will develop shortly, it has low variance and potentially high bias.

On the other hand, the k-nearest-neighbor procedures do not appear to rely on any stringent assumptions about the underlying data, and can adapt to any situation. However, any particular subregion of the decision boundary depends on a handful of input points and their particular positions, and is thus wiggly and unstable—high variance and low bias.

Each method has its own situations for which it works best; in particular linear regression is more appropriate for Scenario 1 above, while nearest neighbors are more suitable for Scenario 2. The time has come to expose the oracle! The data in fact were simulated from a model somewhere between the two, but closer to Scenario 2. First we generated 10 means m_k from a bivariate Gaussian distribution $N((1,0)^T, \mathbf{I})$ and labeled this class BLUE. Similarly, 10 more were drawn from $N((0,1)^T, \mathbf{I})$ and labeled class ORANGE. Then for each class we generated 100 observations as follows: for each observation, we picked an m_k at random with probability 1/10, and

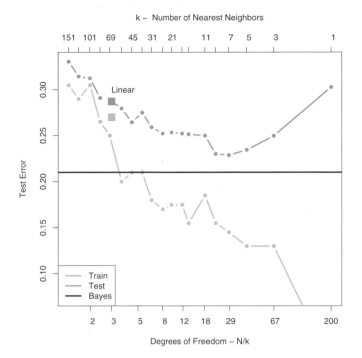

FIGURE 2.4. *Misclassification curves for the simulation example used in Figures 2.1, 2.2 and 2.3. A single training sample of size 200 was used, and a test sample of size 10,000. The orange curves are test and the blue are training error for k-nearest-neighbor classification. The results for linear regression are the bigger orange and blue squares at three degrees of freedom. The purple line is the optimal Bayes error rate.*

then generated a $N(m_k, \mathbf{I}/5)$, thus leading to a mixture of Gaussian clusters for each class. Figure 2.4 shows the results of classifying 10,000 new observations generated from the model. We compare the results for least squares and those for k-nearest neighbors for a range of values of k.

A large subset of the most popular techniques in use today are variants of these two simple procedures. In fact 1-nearest-neighbor, the simplest of all, captures a large percentage of the market for low-dimensional problems. The following list describes some ways in which these simple procedures have been enhanced:

- Kernel methods use weights that decrease smoothly to zero with distance from the target point, rather than the effective 0/1 weights used by k-nearest neighbors.

- In high-dimensional spaces the distance kernels are modified to emphasize some variable more than others.

- Local regression fits linear models by locally weighted least squares, rather than fitting constants locally.

- Linear models fit to a basis expansion of the original inputs allow arbitrarily complex models.

- Projection pursuit and neural network models consist of sums of non-linearly transformed linear models.

2.4 Statistical Decision Theory

In this section we develop a small amount of theory that provides a framework for developing models such as those discussed informally so far. We first consider the case of a quantitative output, and place ourselves in the world of random variables and probability spaces. Let $X \in \mathbb{R}^p$ denote a real valued random input vector, and $Y \in \mathbb{R}$ a real valued random output variable, with joint distribution $\Pr(X, Y)$. We seek a function $f(X)$ for predicting Y given values of the input X. This theory requires a *loss function* $L(Y, f(X))$ for penalizing errors in prediction, and by far the most common and convenient is *squared error loss*: $L(Y, f(X)) = (Y - f(X))^2$. This leads us to a criterion for choosing f,

$$\text{EPE}(f) \quad = \quad \text{E}(Y - f(X))^2 \qquad (2.9)$$

$$= \quad \int [y - f(x)]^2 \Pr(dx, dy), \qquad (2.10)$$

the expected (squared) prediction error . By conditioning[1] on X, we can write EPE as

$$\text{EPE}(f) = \text{E}_X \text{E}_{Y|X} \left([Y - f(X)]^2 | X \right) \qquad (2.11)$$

and we see that it suffices to minimize EPE pointwise:

$$f(x) = \text{argmin}_c \text{E}_{Y|X} \left([Y - c]^2 | X = x \right). \qquad (2.12)$$

The solution is

$$f(x) = \text{E}(Y|X = x), \qquad (2.13)$$

the conditional expectation, also known as the *regression* function. Thus the best prediction of Y at any point $X = x$ is the conditional mean, when best is measured by average squared error.

The nearest-neighbor methods attempt to directly implement this recipe using the training data. At each point x, we might ask for the average of all

---

[1]Conditioning here amounts to factoring the joint density $\Pr(X, Y) = \Pr(Y|X)\Pr(X)$ where $\Pr(Y|X) = \Pr(Y, X)/\Pr(X)$, and splitting up the bivariate integral accordingly.

those y_is with input $x_i = x$. Since there is typically at most one observation at any point x, we settle for

$$\hat{f}(x) = \text{Ave}(y_i | x_i \in N_k(x)), \qquad (2.14)$$

where "Ave" denotes average, and $N_k(x)$ is the neighborhood containing the k points in $\mathcal{T}$ closest to x. Two approximations are happening here:

- expectation is approximated by averaging over sample data;

- conditioning at a point is relaxed to conditioning on some region "close" to the target point.

For large training sample size N, the points in the neighborhood are likely to be close to x, and as k gets large the average will get more stable. In fact, under mild regularity conditions on the joint probability distribution $\Pr(X, Y)$, one can show that as $N, k \to \infty$ such that $k/N \to 0$, $\hat{f}(x) \to \text{E}(Y|X = x)$. In light of this, why look further, since it seems we have a universal approximator? We often do not have very large samples. If the linear or some more structured model is appropriate, then we can usually get a more stable estimate than k-nearest neighbors, although such knowledge has to be learned from the data as well. There are other problems though, sometimes disastrous. In Section 2.5 we see that as the dimension p gets large, so does the metric size of the k-nearest neighborhood. So settling for nearest neighborhood as a surrogate for conditioning will fail us miserably. The convergence above still holds, but the *rate* of convergence decreases as the dimension increases.

How does linear regression fit into this framework? The simplest explanation is that one assumes that the regression function $f(x)$ is approximately linear in its arguments:

$$f(x) \approx x^T \beta. \qquad (2.15)$$

This is a model-based approach—we specify a model for the regression function. Plugging this linear model for $f(x)$ into EPE (2.9) and differentiating we can solve for β theoretically:

$$\beta = [\text{E}(XX^T)]^{-1}\text{E}(XY). \qquad (2.16)$$

Note we have *not* conditioned on X; rather we have used our knowledge of the functional relationship to *pool* over values of X. The least squares solution (2.6) amounts to replacing the expectation in (2.16) by averages over the training data.

So both k-nearest neighbors and least squares end up approximating conditional expectations by averages. But they differ dramatically in terms of model assumptions:

- Least squares assumes $f(x)$ is well approximated by a globally linear function.

- k-nearest neighbors assumes $f(x)$ is well approximated by a locally constant function.

Although the latter seems more palatable, we have already seen that we may pay a price for this flexibility.

Many of the more modern techniques described in this book are model based, although far more flexible than the rigid linear model. For example, additive models assume that

$$f(X) = \sum_{j=1}^{p} f_j(X_j). \tag{2.17}$$

This retains the additivity of the linear model, but each coordinate function f_j is arbitrary. It turns out that the optimal estimate for the additive model uses techniques such as k-nearest neighbors to approximate *univariate* conditional expectations *simultaneously* for each of the coordinate functions. Thus the problems of estimating a conditional expectation in high dimensions are swept away in this case by imposing some (often unrealistic) model assumptions, in this case additivity.

Are we happy with the criterion (2.11)? What happens if we replace the L_2 loss function with the L_1: $\mathrm{E}|Y - f(X)|$? The solution in this case is the conditional median,

$$\hat{f}(x) = \mathrm{median}(Y|X = x), \tag{2.18}$$

which is a different measure of location, and its estimates are more robust than those for the conditional mean. L_1 criteria have discontinuities in their derivatives, which have hindered their widespread use. Other more resistant loss functions will be mentioned in later chapters, but squared error is analytically convenient and the most popular.

What do we do when the output is a categorical variable G? The same paradigm works here, except we need a different loss function for penalizing prediction errors. An estimate $\hat{G}$ will assume values in $\mathcal{G}$, the set of possible classes. Our loss function can be represented by a $K \times K$ matrix $\mathbf{L}$, where $K = \mathrm{card}(\mathcal{G})$. $\mathbf{L}$ will be zero on the diagonal and nonnegative elsewhere, where $L(k, \ell)$ is the price paid for classifying an observation belonging to class $\mathcal{G}_k$ as $\mathcal{G}_\ell$. Most often we use the *zero–one* loss function, where all misclassifications are charged a single unit. The expected prediction error is

$$\mathrm{EPE} = \mathrm{E}[L(G, \hat{G}(X))], \tag{2.19}$$

where again the expectation is taken with respect to the joint distribution $\Pr(G, X)$. Again we condition, and can write EPE as

$$\mathrm{EPE} = \mathrm{E}_X \sum_{k=1}^{K} L[\mathcal{G}_k, \hat{G}(X)]\Pr(\mathcal{G}_k|X) \tag{2.20}$$

Bayes Optimal Classifier

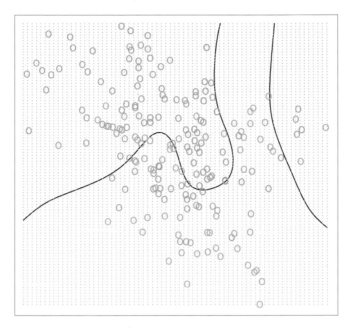

FIGURE 2.5. *The optimal Bayes decision boundary for the simulation example of Figures 2.1, 2.2 and 2.3. Since the generating density is known for each class, this boundary can be calculated exactly (Exercise 2.2).*

and again it suffices to minimize EPE pointwise.

$$\hat{G}(x) = \operatorname{argmin}_{g \in \mathcal{G}} \sum_{k=1}^{K} L(\mathcal{G}_k, g) \Pr(\mathcal{G}_k | X = x). \qquad (2.21)$$

With the 0–1 loss function this simplifies to

$$\hat{G}(x) = \operatorname{argmin}_{g \in \mathcal{G}} [1 - \Pr(g | X = x)] \qquad (2.22)$$

or simply

$$\hat{G}(x) = \mathcal{G}_k \text{ if } \Pr(\mathcal{G}_k | X = x) = \max_{g \in \mathcal{G}} \Pr(g | X = x). \qquad (2.23)$$

This reasonable solution is known as the *Bayes classifier*, and says that we classify to the most probable class, using the conditional (discrete) distribution $\Pr(G | X)$. Figure 2.5 shows the Bayes-optimal decision boundary for our simulation example. The error rate of the Bayes classifier is called the *Bayes rate*.

Again we see that the k-nearest neighbor classifier directly approximates this solution—a majority vote in a nearest neighborhood amounts to exactly this, except that conditional probability at a point is relaxed to conditional probability within a neighborhood of a point, and probabilities are estimated by training-sample proportions.

Suppose for a two-class problem we had taken the dummy-variable approach and coded G via a binary Y, followed by squared error loss estimation. Then $\hat{f}(X) = \mathrm{E}(Y|X) = \Pr(G = \mathcal{G}_1|X)$ if $\mathcal{G}_1$ corresponded to $Y = 1$. Likewise for a K-class problem, $\mathrm{E}(Y_k|X) = \Pr(G = \mathcal{G}_k|X)$. This shows that our dummy-variable regression procedure, followed by classification to the largest fitted value, is another way of representing the Bayes classifier. Although this theory is exact, in practice problems can occur, depending on the regression model used. For example, when linear regression is used, $\hat{f}(X)$ need not be positive, and we might be suspicious about using it as an estimate of a probability. We will discuss a variety of approaches to modeling $\Pr(G|X)$ in Chapter 4.

2.5   Local Methods in High Dimensions

We have examined two learning techniques for prediction so far: the stable but biased linear model and the less stable but apparently less biased class of k-nearest-neighbor estimates. It would seem that with a reasonably large set of training data, we could always approximate the theoretically optimal conditional expectation by k-nearest-neighbor averaging, since we should be able to find a fairly large neighborhood of observations close to any x and average them. This approach and our intuition breaks down in high dimensions, and the phenomenon is commonly referred to as the *curse of dimensionality* (Bellman, 1961). There are many manifestations of this problem, and we will examine a few here.

Consider the nearest-neighbor procedure for inputs uniformly distributed in a p-dimensional unit hypercube, as in Figure 2.6. Suppose we send out a hypercubical neighborhood about a target point to capture a fraction r of the observations. Since this corresponds to a fraction r of the unit volume, the expected edge length will be $e_p(r) = r^{1/p}$. In ten dimensions $e_{10}(0.01) = 0.63$ and $e_{10}(0.1) = 0.80$, while the entire range for each input is only 1.0. So to capture 1% or 10% of the data to form a local average, we must cover 63% or 80% of the range of each input variable. Such neighborhoods are no longer "local." Reducing r dramatically does not help much either, since the fewer observations we average, the higher is the variance of our fit.

Another consequence of the sparse sampling in high dimensions is that all sample points are close to an edge of the sample. Consider N data points uniformly distributed in a p-dimensional unit ball centered at the origin. Suppose we consider a nearest-neighbor estimate at the origin. The median

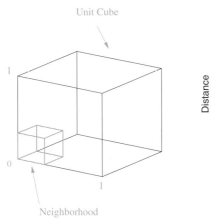

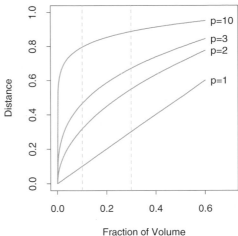

FIGURE 2.6. *The curse of dimensionality is well illustrated by a subcubical neighborhood for uniform data in a unit cube. The figure on the right shows the side-length of the subcube needed to capture a fraction r of the volume of the data, for different dimensions p. In ten dimensions we need to cover 80% of the range of each coordinate to capture 10% of the data.*

distance from the origin to the closest data point is given by the expression

$$d(p, N) = \left(1 - \frac{1}{2}^{1/N}\right)^{1/p} \qquad (2.24)$$

(Exercise 2.3). A more complicated expression exists for the mean distance to the closest point. For $N = 500$, $p = 10$, $d(p, N) \approx 0.52$, more than halfway to the boundary. Hence most data points are closer to the boundary of the sample space than to any other data point. The reason that this presents a problem is that prediction is much more difficult near the edges of the training sample. One must extrapolate from neighboring sample points rather than interpolate between them.

Another manifestation of the curse is that the sampling density is proportional to $N^{1/p}$, where p is the dimension of the input space and N is the sample size. Thus, if $N_1 = 100$ represents a dense sample for a single input problem, then $N_{10} = 100^{10}$ is the sample size required for the same sampling density with 10 inputs. Thus in high dimensions all feasible training samples sparsely populate the input space.

Let us construct another uniform example. Suppose we have 1000 training examples x_i generated uniformly on $[-1, 1]^p$. Assume that the true relationship between X and Y is

$$Y = f(X) = e^{-8||X||^2},$$

without any measurement error. We use the 1-nearest-neighbor rule to predict y_0 at the test-point $x_0 = 0$. Denote the training set by $\mathcal{T}$. We can compute the expected prediction error at x_0 for our procedure, averaging over all such samples of size 1000. Since the problem is deterministic, this is the mean squared error (MSE) for estimating $f(0)$:

$$
\begin{aligned}
\mathrm{MSE}(x_0) &= \mathrm{E}_{\mathcal{T}}[f(x_0) - \hat{y}_0]^2 \\
&= \mathrm{E}_{\mathcal{T}}[\hat{y}_0 - \mathrm{E}_{\mathcal{T}}(\hat{y}_0)]^2 + [\mathrm{E}_{\mathcal{T}}(\hat{y}_0) - f(x_0)]^2 \\
&= \mathrm{Var}_{\mathcal{T}}(\hat{y}_0) + \mathrm{Bias}^2(\hat{y}_0).
\end{aligned}
\tag{2.25}
$$

Figure 2.7 illustrates the setup. We have broken down the MSE into two components that will become familiar as we proceed: variance and squared bias. Such a decomposition is always possible and often useful, and is known as the *bias–variance decomposition*. Unless the nearest neighbor is at 0, $\hat{y}_0$ will be smaller than $f(0)$ in this example, and so the average estimate will be biased downward. The variance is due to the sampling variance of the 1-nearest neighbor. In low dimensions and with $N = 1000$, the nearest neighbor is very close to 0, and so both the bias and variance are small. As the dimension increases, the nearest neighbor tends to stray further from the target point, and both bias and variance are incurred. By $p = 10$, for more than 99% of the samples the nearest neighbor is a distance greater than 0.5 from the origin. Thus as p increases, the estimate tends to be 0 more often than not, and hence the MSE levels off at 1.0, as does the bias, and the variance starts dropping (an artifact of this example).

Although this is a highly contrived example, similar phenomena occur more generally. The complexity of functions of many variables can grow exponentially with the dimension, and if we wish to be able to estimate such functions with the same accuracy as function in low dimensions, then we need the size of our training set to grow exponentially as well. In this example, the function is a complex interaction of all p variables involved.

The dependence of the bias term on distance depends on the truth, and it need not always dominate with 1-nearest neighbor. For example, if the function always involves only a few dimensions as in Figure 2.8, then the variance can dominate instead.

Suppose, on the other hand, that we know that the relationship between Y and X is linear,

$$
Y = X^T \beta + \varepsilon,
\tag{2.26}
$$

where $\varepsilon \sim N(0, \sigma^2)$ and we fit the model by least squares to the training data. For an arbitrary test point x_0, we have $\hat{y}_0 = x_0^T \hat{\beta}$, which can be written as $\hat{y}_0 = x_0^T \beta + \sum_{i=1}^{N} \ell_i(x_0)\varepsilon_i$, where $\ell_i(x_0)$ is the ith element of $\mathbf{X}(\mathbf{X}^T\mathbf{X})^{-1}x_0$. Since under this model the least squares estimates are

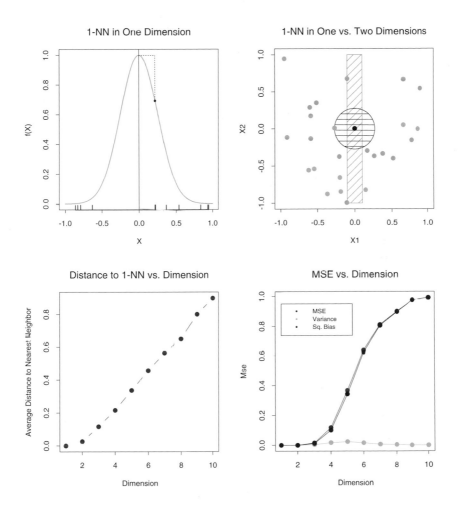

FIGURE 2.7. *A simulation example, demonstrating the curse of dimensionality and its effect on MSE, bias and variance. The input features are uniformly distributed in $[-1,1]^p$ for $p = 1, \ldots, 10$ The top left panel shows the target function (no noise) in $\mathbb{R}$: $f(X) = e^{-8||X||^2}$, and demonstrates the error that 1-nearest neighbor makes in estimating $f(0)$. The training point is indicated by the blue tick mark. The top right panel illustrates why the radius of the 1-nearest neighborhood increases with dimension p. The lower left panel shows the average radius of the 1-nearest neighborhoods. The lower-right panel shows the MSE, squared bias and variance curves as a function of dimension p.*

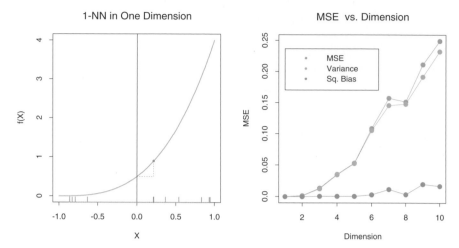

FIGURE 2.8. *A simulation example with the same setup as in Figure 2.7. Here the function is constant in all but one dimension:* $f(X) = \frac{1}{2}(X_1 + 1)^3$. *The variance dominates.*

unbiased, we find that

$$
\begin{aligned}
\mathrm{EPE}(x_0) &= \mathrm{E}_{y_0|x_0}\mathrm{E}_{\mathcal{T}}(y_0 - \hat{y}_0)^2 \\
&= \mathrm{Var}(y_0|x_0) + \mathrm{E}_{\mathcal{T}}[\hat{y}_0 - \mathrm{E}_{\mathcal{T}}\hat{y}_0]^2 + [\mathrm{E}_{\mathcal{T}}\hat{y}_0 - x_0^T\beta]^2 \\
&= \mathrm{Var}(y_0|x_0) + \mathrm{Var}_{\mathcal{T}}(\hat{y}_0) + \mathrm{Bias}^2(\hat{y}_0) \\
&= \sigma^2 + \mathrm{E}_{\mathcal{T}}x_0^T(\mathbf{X}^T\mathbf{X})^{-1}x_0\sigma^2 + 0^2.
\end{aligned}
\tag{2.27}
$$

Here we have incurred an additional variance σ^2 in the prediction error, since our target is not deterministic. There is no bias, and the variance depends on x_0. If N is large and $\mathcal{T}$ were selected at random, and assuming $\mathrm{E}(X) = 0$, then $\mathbf{X}^T\mathbf{X} \to N\mathrm{Cov}(X)$ and

$$
\begin{aligned}
\mathrm{E}_{x_0}\mathrm{EPE}(x_0) &\sim \mathrm{E}_{x_0}x_0^T\mathrm{Cov}(X)^{-1}x_0\sigma^2/N + \sigma^2 \\
&= \mathrm{trace}[\mathrm{Cov}(X)^{-1}\mathrm{Cov}(x_0)]\sigma^2/N + \sigma^2 \\
&= \sigma^2(p/N) + \sigma^2.
\end{aligned}
\tag{2.28}
$$

Here we see that the expected EPE increases linearly as a function of p, with slope σ^2/N. If N is large and/or σ^2 is small, this growth in variance is negligible (0 in the deterministic case). By imposing some heavy restrictions on the class of models being fitted, we have avoided the curse of dimensionality. Some of the technical details in (2.27) and (2.28) are derived in Exercise 2.5.

Figure 2.9 compares 1-nearest neighbor vs. least squares in two situations, both of which have the form $Y = f(X) + \varepsilon$, X uniform as before, and $\varepsilon \sim N(0, 1)$. The sample size is $N = 500$. For the orange curve, $f(x)$

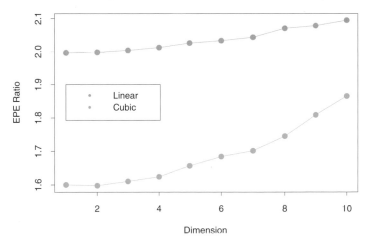

FIGURE 2.9. *The curves show the expected prediction error (at $x_0 = 0$) for 1-nearest neighbor relative to least squares for the model $Y = f(X) + \varepsilon$. For the orange curve, $f(x) = x_1$, while for the blue curve $f(x) = \frac{1}{2}(x_1 + 1)^3$.*

is linear in the first coordinate, for the blue curve, cubic as in Figure 2.8. Shown is the relative EPE of 1-nearest neighbor to least squares, which appears to start at around 2 for the linear case. Least squares is unbiased in this case, and as discussed above the EPE is slightly above $\sigma^2 = 1$. The EPE for 1-nearest neighbor is always above 2, since the variance of $\hat{f}(x_0)$ in this case is at least σ^2, and the ratio increases with dimension as the nearest neighbor strays from the target point. For the cubic case, least squares is biased, which moderates the ratio. Clearly we could manufacture examples where the bias of least squares would dominate the variance, and the 1-nearest neighbor would come out the winner.

By relying on rigid assumptions, the linear model has no bias at all and negligible variance, while the error in 1-nearest neighbor is substantially larger. However, if the assumptions are wrong, all bets are off and the 1-nearest neighbor may dominate. We will see that there is a whole spectrum of models between the rigid linear models and the extremely flexible 1-nearest-neighbor models, each with their own assumptions and biases, which have been proposed specifically to avoid the exponential growth in complexity of functions in high dimensions by drawing heavily on these assumptions.

Before we delve more deeply, let us elaborate a bit on the concept of *statistical models* and see how they fit into the prediction framework.

2.6   Statistical Models, Supervised Learning and Function Approximation

Our goal is to find a useful approximation $\hat{f}(x)$ to the function $f(x)$ that underlies the predictive relationship between the inputs and outputs. In the theoretical setting of Section 2.4, we saw that squared error loss lead us to the regression function $f(x) = \mathrm{E}(Y|X = x)$ for a quantitative response. The class of nearest-neighbor methods can be viewed as direct estimates of this conditional expectation, but we have seen that they can fail in at least two ways:

- if the dimension of the input space is high, the nearest neighbors need not be close to the target point, and can result in large errors;

- if special structure is known to exist, this can be used to reduce both the bias and the variance of the estimates.

We anticipate using other classes of models for $f(x)$, in many cases specifically designed to overcome the dimensionality problems, and here we discuss a framework for incorporating them into the prediction problem.

2.6.1   A Statistical Model for the Joint Distribution $\Pr(X, Y)$

Suppose in fact that our data arose from a statistical model

$$Y = f(X) + \varepsilon, \tag{2.29}$$

where the random error ε has $\mathrm{E}(\varepsilon) = 0$ and is independent of X. Note that for this model, $f(x) = \mathrm{E}(Y|X = x)$, and in fact the conditional distribution $\Pr(Y|X)$ depends on X *only* through the conditional mean $f(x)$.

The additive error model is a useful approximation to the truth. For most systems the input–output pairs (X, Y) will not have a deterministic relationship $Y = f(X)$. Generally there will be other unmeasured variables that also contribute to Y, including measurement error. The additive model assumes that we can capture all these departures from a deterministic relationship via the error ε.

For some problems a deterministic relationship does hold. Many of the classification problems studied in machine learning are of this form, where the response surface can be thought of as a colored map defined in $\mathbb{R}^p$. The training data consist of colored examples from the map $\{x_i, g_i\}$, and the goal is to be able to color any point. Here the function is deterministic, and the randomness enters through the x location of the training points. For the moment we will not pursue such problems, but will see that they can be handled by techniques appropriate for the error-based models.

The assumption in (2.29) that the errors are independent and identically distributed is not strictly necessary, but seems to be at the back of our mind when we average squared errors uniformly in our EPE criterion. With such a model it becomes natural to use least squares as a data criterion for model estimation as in (2.1). Simple modifications can be made to avoid the independence assumption; for example, we can have $\text{Var}(Y|X = x) = \sigma(x)$, and now both the mean and variance depend on X. In general the conditional distribution $\text{Pr}(Y|X)$ can depend on X in complicated ways, but the additive error model precludes these.

So far we have concentrated on the quantitative response. Additive error models are typically not used for qualitative outputs G; in this case the target function $p(X)$ *is* the conditional density $\text{Pr}(G|X)$, and this is modeled directly. For example, for two-class data, it is often reasonable to assume that the data arise from independent binary trials, with the probability of one particular outcome being $p(X)$, and the other $1 - p(X)$. Thus if Y is the 0–1 coded version of G, then $\text{E}(Y|X = x) = p(x)$, but the variance depends on x as well: $\text{Var}(Y|X = x) = p(x)[1 - p(x)]$.

2.6.2 Supervised Learning

Before we launch into more statistically oriented jargon, we present the function-fitting paradigm from a machine learning point of view. Suppose for simplicity that the errors are additive and that the model $Y = f(X) + \varepsilon$ is a reasonable assumption. Supervised learning attempts to learn f by example through a *teacher*. One observes the system under study, both the inputs and outputs, and assembles a *training* set of observations $\mathcal{T} = (x_i, y_i)$, $i = 1, \ldots, N$. The observed input values to the system x_i are also fed into an artificial system, known as a learning algorithm (usually a computer program), which also produces outputs $\hat{f}(x_i)$ in response to the inputs. The learning algorithm has the property that it can modify its input/output relationship $\hat{f}$ in response to differences $y_i - \hat{f}(x_i)$ between the original and generated outputs. This process is known as *learning by example*. Upon completion of the learning process the hope is that the artificial and real outputs will be close enough to be useful for all sets of inputs likely to be encountered in practice.

2.6.3 Function Approximation

The learning paradigm of the previous section has been the motivation for research into the supervised learning problem in the fields of machine learning (with analogies to human reasoning) and neural networks (with biological analogies to the brain). The approach taken in applied mathematics and statistics has been from the perspective of function approximation and estimation. Here the data pairs $\{x_i, y_i\}$ are viewed as points in a $(p + 1)$-dimensional Euclidean space. The function $f(x)$ has domain equal to the p-dimensional input subspace, and is related to the data via a model such as $y_i = f(x_i) + \varepsilon_i$. For convenience in this chapter we will assume the domain is $\mathbb{R}^p$, a p-dimensional Euclidean space, although in general the inputs can be of mixed type. The goal is to obtain a useful approximation to $f(x)$ for all x in some region of $\mathbb{R}^p$, given the representations in $\mathcal{T}$. Although somewhat less glamorous than the learning paradigm, treating supervised learning as a problem in function approximation encourages the geometrical concepts of Euclidean spaces and mathematical concepts of probabilistic inference to be applied to the problem. This is the approach taken in this book.

Many of the approximations we will encounter have associated a set of parameters θ that can be modified to suit the data at hand. For example, the linear model $f(x) = x^T\beta$ has $\theta = \beta$. Another class of useful approximators can be expressed as *linear basis expansions*

$$f_\theta(x) = \sum_{k=1}^{K} h_k(x)\theta_k, \tag{2.30}$$

where the h_k are a suitable set of functions or transformations of the input vector x. Traditional examples are polynomial and trigonometric expansions, where for example h_k might be x_1^2, $x_1 x_2^2$, $\cos(x_1)$ and so on. We also encounter nonlinear expansions, such as the sigmoid transformation common to neural network models,

$$h_k(x) = \frac{1}{1 + \exp(-x^T\beta_k)}. \tag{2.31}$$

We can use least squares to estimate the parameters θ in f_θ as we did for the linear model, by minimizing the residual sum-of-squares

$$\text{RSS}(\theta) = \sum_{i=1}^{N} (y_i - f_\theta(x_i))^2 \tag{2.32}$$

as a function of θ. This seems a reasonable criterion for an additive error model. In terms of function approximation, we imagine our parameterized function as a surface in $p + 1$ space, and what we observe are noisy realizations from it. This is easy to visualize when $p = 2$ and the vertical coordinate is the output y, as in Figure 2.10. The noise is in the output coordinate, so we find the set of parameters such that the fitted surface gets as close to the observed points as possible, where close is measured by the sum of squared vertical errors in $\text{RSS}(\theta)$.

For the linear model we get a simple closed form solution to the minimization problem. This is also true for the basis function methods, if the basis functions themselves do not have any hidden parameters. Otherwise the solution requires either iterative methods or numerical optimization.

While least squares is generally very convenient, it is not the only criterion used and in some cases would not make much sense. A more general

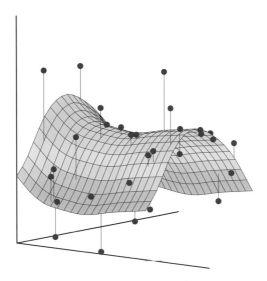

FIGURE 2.10. *Least squares fitting of a function of two inputs. The parameters of $f_\theta(x)$ are chosen so as to minimize the sum-of-squared vertical errors.*

principle for estimation is *maximum likelihood estimation.* Suppose we have a random sample y_i, $i = 1, \ldots, N$ from a density $\text{Pr}_\theta(y)$ indexed by some parameters θ. The log-probability of the observed sample is

$$L(\theta) = \sum_{i=1}^{N} \log \text{Pr}_\theta(y_i). \tag{2.33}$$

The principle of maximum likelihood assumes that the most reasonable values for θ are those for which the probability of the observed sample is largest. Least squares for the additive error model $Y = f_\theta(X) + \varepsilon$, with $\varepsilon \sim N(0, \sigma^2)$, is equivalent to maximum likelihood using the conditional likelihood

$$\text{Pr}(Y|X, \theta) = N(f_\theta(X), \sigma^2). \tag{2.34}$$

So although the additional assumption of normality seems more restrictive, the results are the same. The log-likelihood of the data is

$$L(\theta) = -\frac{N}{2} \log(2\pi) - N \log \sigma - \frac{1}{2\sigma^2} \sum_{i=1}^{N} (y_i - f_\theta(x_i))^2, \tag{2.35}$$

and the only term involving θ is the last, which is $\text{RSS}(\theta)$ up to a scalar negative multiplier.

A more interesting example is the multinomial likelihood for the regression function $\text{Pr}(G|X)$ for a qualitative output G. Suppose we have a model $\text{Pr}(G = \mathcal{G}_k|X = x) = p_{k,\theta}(x)$, $k = 1, \ldots, K$ for the conditional probability of each class given X, indexed by the parameter vector θ. Then the log-likelihood (also referred to as the cross-entropy) is

$$L(\theta) = \sum_{i=1}^{N} \log p_{g_i,\theta}(x_i), \qquad (2.36)$$

and when maximized it delivers values of θ that best conform with the data in this likelihood sense.

2.7   Structured Regression Models

We have seen that although nearest-neighbor and other local methods focus directly on estimating the function at a point, they face problems in high dimensions. They may also be inappropriate even in low dimensions in cases where more structured approaches can make more efficient use of the data. This section introduces classes of such structured approaches. Before we proceed, though, we discuss further the need for such classes.

2.7.1   Difficulty of the Problem

Consider the RSS criterion for an arbitrary function f,

$$\text{RSS}(f) = \sum_{i=1}^{N} (y_i - f(x_i))^2. \qquad (2.37)$$

Minimizing (2.37) leads to infinitely many solutions: any function $\hat{f}$ passing through the training points (x_i, y_i) is a solution. Any particular solution chosen might be a poor predictor at test points different from the training points. If there are multiple observation pairs $x_i, y_{i\ell}$, $\ell = 1, \ldots, N_i$ at each value of x_i, the risk is limited. In this case, the solutions pass through the average values of the $y_{i\ell}$ at each x_i; see Exercise 2.6. The situation is similar to the one we have already visited in Section 2.4; indeed, (2.37) is the finite sample version of (2.11) on page 18. If the sample size N were sufficiently large such that repeats were guaranteed and densely arranged, it would seem that these solutions might all tend to the limiting conditional expectation.

In order to obtain useful results for finite N, we must restrict the eligible solutions to (2.37) to a smaller set of functions. How to decide on the nature of the restrictions is based on considerations outside of the data. These restrictions are sometimes encoded via the parametric representation of f_θ, or may be built into the learning method itself, either implicitly or explicitly. These restricted classes of solutions are the major topic of this book. One thing should be clear, though. Any restrictions imposed on f that lead to a unique solution to (2.37) do not really remove the ambiguity caused by the multiplicity of solutions. There are infinitely many possible restrictions, each leading to a unique solution, so the ambiguity has simply been transferred to the choice of constraint.

In general the constraints imposed by most learning methods can be described as *complexity* restrictions of one kind or another. This usually means some kind of regular behavior in small neighborhoods of the input space. That is, for all input points x sufficiently close to each other in some metric, $\hat{f}$ exhibits some special structure such as nearly constant, linear or low-order polynomial behavior. The estimator is then obtained by averaging or polynomial fitting in that neighborhood.

The strength of the constraint is dictated by the neighborhood size. The larger the size of the neighborhood, the stronger the constraint, and the more sensitive the solution is to the particular choice of constraint. For example, local constant fits in infinitesimally small neighborhoods is no constraint at all; local linear fits in very large neighborhoods is almost a globally linear model, and is very restrictive.

The nature of the constraint depends on the metric used. Some methods, such as kernel and local regression and tree-based methods, directly specify the metric and size of the neighborhood. The nearest-neighbor methods discussed so far are based on the assumption that locally the function is constant; close to a target input x_0, the function does not change much, and so close outputs can be averaged to produce $\hat{f}(x_0)$. Other methods such as splines, neural networks and basis-function methods implicitly define neighborhoods of local behavior. In Section 5.4.1 we discuss the concept of an *equivalent kernel* (see Figure 5.8 on page 157), which describes this local dependence for any method linear in the outputs. These equivalent kernels in many cases look just like the explicitly defined weighting kernels discussed above—peaked at the target point and falling away smoothly away from it.

One fact should be clear by now. Any method that attempts to produce locally varying functions in small isotropic neighborhoods will run into problems in high dimensions—again the curse of dimensionality. And conversely, all methods that overcome the dimensionality problems have an associated—and often implicit or adaptive—metric for measuring neighborhoods, which basically does not allow the neighborhood to be simultaneously small in all directions.

2.8 Classes of Restricted Estimators

The variety of nonparametric regression techniques or learning methods fall into a number of different classes depending on the nature of the restrictions imposed. These classes are not distinct, and indeed some methods fall in several classes. Here we give a brief summary, since detailed descriptions are given in later chapters. Each of the classes has associated with it one or more parameters, sometimes appropriately called *smoothing* parameters, that control the effective size of the local neighborhood. Here we describe three broad classes.

2.8.1 Roughness Penalty and Bayesian Methods

Here the class of functions is controlled by explicitly penalizing RSS(f) with a roughness penalty

$$\text{PRSS}(f; \lambda) = \text{RSS}(f) + \lambda J(f). \tag{2.38}$$

The user-selected functional $J(f)$ will be large for functions f that vary too rapidly over small regions of input space. For example, the popular *cubic smoothing spline* for one-dimensional inputs is the solution to the penalized least-squares criterion

$$\text{PRSS}(f; \lambda) = \sum_{i=1}^{N} (y_i - f(x_i))^2 + \lambda \int [f''(x)]^2 dx. \tag{2.39}$$

The roughness penalty here controls large values of the second derivative of f, and the amount of penalty is dictated by $\lambda \geq 0$. For $\lambda = 0$ no penalty is imposed, and any interpolating function will do, while for $\lambda = \infty$ only functions linear in x are permitted.

Penalty functionals J can be constructed for functions in any dimension, and special versions can be created to impose special structure. For example, additive penalties $J(f) = \sum_{j=1}^{p} J(f_j)$ are used in conjunction with additive functions $f(X) = \sum_{j=1}^{p} f_j(X_j)$ to create additive models with smooth coordinate functions. Similarly, *projection pursuit regression* models have $f(X) = \sum_{m=1}^{M} g_m(\alpha_m^T X)$ for adaptively chosen directions α_m, and the functions g_m can each have an associated roughness penalty.

Penalty function, or *regularization* methods, express our prior belief that the type of functions we seek exhibit a certain type of smooth behavior, and indeed can usually be cast in a Bayesian framework. The penalty J corresponds to a log-prior, and PRSS($f; \lambda$) the log-posterior distribution, and minimizing PRSS($f; \lambda$) amounts to finding the posterior mode. We discuss roughness-penalty approaches in Chapter 5 and the Bayesian paradigm in Chapter 8.

2.8.2 Kernel Methods and Local Regression

These methods can be thought of as explicitly providing estimates of the regression function or conditional expectation by specifying the nature of the local neighborhood, and of the class of regular functions fitted locally. The local neighborhood is specified by a *kernel function* $K_\lambda(x_0, x)$ which assigns weights to points x in a region around x_0 (see Figure 6.1 on page 192). For example, the Gaussian kernel has a weight function based on the Gaussian density function

$$K_\lambda(x_0, x) = \frac{1}{\lambda} \exp\left[-\frac{||x - x_0||^2}{2\lambda}\right] \tag{2.40}$$

and assigns weights to points that die exponentially with their squared Euclidean distance from x_0. The parameter λ corresponds to the variance of the Gaussian density, and controls the width of the neighborhood. The simplest form of kernel estimate is the Nadaraya–Watson weighted average

$$\hat{f}(x_0) = \frac{\sum_{i=1}^{N} K_\lambda(x_0, x_i) y_i}{\sum_{i=1}^{N} K_\lambda(x_0, x_i)}. \tag{2.41}$$

In general we can define a local regression estimate of $f(x_0)$ as $f_{\hat{\theta}}(x_0)$, where $\hat{\theta}$ minimizes

$$\text{RSS}(f_\theta, x_0) = \sum_{i=1}^{N} K_\lambda(x_0, x_i)(y_i - f_\theta(x_i))^2, \tag{2.42}$$

and f_θ is some parameterized function, such as a low-order polynomial. Some examples are:

- $f_\theta(x) = \theta_0$, the constant function; this results in the Nadaraya–Watson estimate in (2.41) above.

- $f_\theta(x) = \theta_0 + \theta_1 x$ gives the popular local linear regression model.

Nearest-neighbor methods can be thought of as kernel methods having a more data-dependent metric. Indeed, the metric for k-nearest neighbors is

$$K_k(x, x_0) = I(||x - x_0|| \leq ||x_{(k)} - x_0||),$$

where $x_{(k)}$ is the training observation ranked kth in distance from x_0, and $I(S)$ is the indicator of the set S.

These methods of course need to be modified in high dimensions, to avoid the curse of dimensionality. Various adaptations are discussed in Chapter 6.

2.8.3 Basis Functions and Dictionary Methods

This class of methods includes the familiar linear and polynomial expansions, but more importantly a wide variety of more flexible models. The model for f is a linear expansion of basis functions

$$f_\theta(x) = \sum_{m=1}^{M} \theta_m h_m(x), \tag{2.43}$$

where each of the h_m is a function of the input x, and the term linear here refers to the action of the parameters θ. This class covers a wide variety of methods. In some cases the sequence of basis functions is prescribed, such as a basis for polynomials in x of total degree M.

For one-dimensional x, polynomial splines of degree K can be represented by an appropriate sequence of M spline basis functions, determined in turn by $M - K - 1$ *knots*. These produce functions that are piecewise polynomials of degree K between the knots, and joined up with continuity of degree $K - 1$ at the knots. As an example consider linear splines, or piecewise linear functions. One intuitively satisfying basis consists of the functions $b_1(x) = 1$, $b_2(x) = x$, and $b_{m+2}(x) = (x - t_m)_+$, $m = 1, \ldots, M - 2$, where t_m is the mth knot, and z_+ denotes positive part. Tensor products of spline bases can be used for inputs with dimensions larger than one (see Section 5.2, and the CART and MARS models in Chapter 9.) The parameter M controls the degree of the polynomial or the number of knots in the case of splines.

Radial basis functions are symmetric p-dimensional kernels located at particular centroids,

$$f_\theta(x) = \sum_{m=1}^{M} K_{\lambda_m}(\mu_m, x)\theta_m; \qquad (2.44)$$

for example, the Gaussian kernel $K_\lambda(\mu, x) = e^{-||x - \mu||^2/2\lambda}$ is popular.

Radial basis functions have centroids μ_m and scales λ_m that have to be determined. The spline basis functions have knots. In general we would like the data to dictate them as well. Including these as parameters changes the regression problem from a straightforward linear problem to a combinatorially hard nonlinear problem. In practice, shortcuts such as greedy algorithms or two stage processes are used. Section 6.7 describes some such approaches.

A single-layer feed-forward neural network model with linear output weights can be thought of as an adaptive basis function method. The model has the form

$$f_\theta(x) = \sum_{m=1}^{M} \beta_m \sigma(\alpha_m^T x + b_m), \qquad (2.45)$$

where $\sigma(x) = 1/(1 + e^{-x})$ is known as the *activation* function. Here, as in the projection pursuit model, the directions α_m and the *bias* terms b_m have to be determined, and their estimation is the meat of the computation. Details are given in Chapter 11.

These adaptively chosen basis function methods are also known as *dictionary* methods, where one has available a possibly infinite set or dictionary $\mathcal{D}$ of candidate basis functions from which to choose, and models are built up by employing some kind of search mechanism.

2.9   Model Selection and the Bias–Variance Tradeoff

All the models described above and many others discussed in later chapters have a *smoothing* or *complexity* parameter that has to be determined:

- the multiplier of the penalty term;

- the width of the kernel;

- or the number of basis functions.

In the case of the smoothing spline, the parameter λ indexes models ranging from a straight line fit to the interpolating model. Similarly a local degree-m polynomial model ranges between a degree-m global polynomial when the window size is infinitely large, to an interpolating fit when the window size shrinks to zero. This means that we cannot use residual sum-of-squares on the training data to determine these parameters as well, since we would always pick those that gave interpolating fits and hence zero residuals. Such a model is unlikely to predict future data well at all.

The k-nearest-neighbor regression fit $\hat{f}_k(x_0)$ usefully illustrates the competing forces that affect the predictive ability of such approximations. Suppose the data arise from a model $Y = f(X) + \varepsilon$, with $\mathrm{E}(\varepsilon) = 0$ and $\mathrm{Var}(\varepsilon) = \sigma^2$. For simplicity here we assume that the values of x_i in the sample are fixed in advance (nonrandom). The expected prediction error at x_0, also known as *test* or *generalization* error, can be decomposed:

$$
\begin{aligned}
\mathrm{EPE}_k(x_0) &= \mathrm{E}[(Y - \hat{f}_k(x_0))^2 | X = x_0] \\
&= \sigma^2 + [\mathrm{Bias}^2(\hat{f}_k(x_0)) + \mathrm{Var}_\mathcal{T}(\hat{f}_k(x_0))] \qquad (2.46) \\
&= \sigma^2 + \left[f(x_0) - \frac{1}{k} \sum_{\ell=1}^{k} f(x_{(\ell)}) \right]^2 + \frac{\sigma^2}{k}. \qquad (2.47)
\end{aligned}
$$

The subscripts in parentheses (ℓ) indicate the sequence of nearest neighbors to x_0.

There are three terms in this expression. The first term σ^2 is the *irreducible* error—the variance of the new test target—and is beyond our control, even if we know the true $f(x_0)$.

The second and third terms are under our control, and make up the *mean squared error* of $\hat{f}_k(x_0)$ in estimating $f(x_0)$, which is broken down into a bias component and a variance component. The bias term is the squared difference between the true mean $f(x_0)$ and the expected value of the estimate—$[\mathrm{E}_\mathcal{T}(\hat{f}_k(x_0)) - f(x_0)]^2$—where the expectation averages the randomness in the training data. This term will most likely increase with k, if the true function is reasonably smooth. For small k the few closest neighbors will have values $f(x_{(\ell)})$ close to $f(x_0)$, so their average should

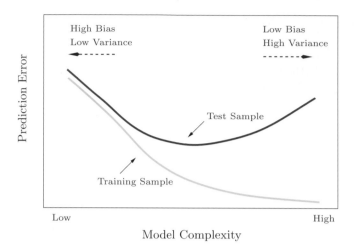

FIGURE 2.11. *Test and training error as a function of model complexity.*

be close to $f(x_0)$. As k grows, the neighbors are further away, and then anything can happen.

The variance term is simply the variance of an average here, and decreases as the inverse of k. So as k varies, there is a *bias–variance tradeoff*.

More generally, as the *model complexity* of our procedure is increased, the variance tends to increase and the squared bias tends to decrease. The opposite behavior occurs as the model complexity is decreased. For k-nearest neighbors, the model complexity is controlled by k.

Typically we would like to choose our model complexity to trade bias off with variance in such a way as to minimize the test error. An obvious estimate of test error is the *training error* $\frac{1}{N}\sum_i(y_i - \hat{y}_i)^2$. Unfortunately training error is not a good estimate of test error, as it does not properly account for model complexity.

Figure 2.11 shows the typical behavior of the test and training error, as model complexity is varied. The training error tends to decrease whenever we increase the model complexity, that is, whenever we fit the data harder. However with too much fitting, the model adapts itself too closely to the training data, and will not generalize well (i.e., have large test error). In that case the predictions $\hat{f}(x_0)$ will have large variance, as reflected in the last term of expression (2.46). In contrast, if the model is not complex enough, it will *underfit* and may have large bias, again resulting in poor generalization. In Chapter 7 we discuss methods for estimating the test error of a prediction method, and hence estimating the optimal amount of model complexity for a given prediction method and training set.

Bibliographic Notes

Some good general books on the learning problem are Duda et al. (2000), Bishop (1995),(Bishop, 2006), Ripley (1996), Cherkassky and Mulier (2007) and Vapnik (1996). Parts of this chapter are based on Friedman (1994b).

Exercises

Ex. 2.1 Suppose each of K-classes has an associated target t_k, which is a vector of all zeros, except a one in the kth position. Show that classifying to the largest element of $\hat{y}$ amounts to choosing the closest target, $\min_k ||t_k - \hat{y}||$, if the elements of $\hat{y}$ sum to one.

Ex. 2.2 Show how to compute the Bayes decision boundary for the simulation example in Figure 2.5.

Ex. 2.3 Derive equation (2.24).

Ex. 2.4 The edge effect problem discussed on page 23 is not peculiar to uniform sampling from bounded domains. Consider inputs drawn from a spherical multinormal distribution $X \sim N(0, \mathbf{I}_p)$. The squared distance from any sample point to the origin has a χ_p^2 distribution with mean p. Consider a prediction point x_0 drawn from this distribution, and let $a = x_0/||x_0||$ be an associated unit vector. Let $z_i = a^T x_i$ be the projection of each of the training points on this direction.

Show that the z_i are distributed $N(0, 1)$ with expected squared distance from the origin 1, while the target point has expected squared distance p from the origin.

Hence for $p = 10$, a randomly drawn test point is about 3.1 standard deviations from the origin, while all the training points are on average one standard deviation along direction a. So most prediction points see themselves as lying on the edge of the training set.

Ex. 2.5

(a) Derive equation (2.27). The last line makes use of (3.8) through a conditioning argument.

(b) Derive equation (2.28), making use of the *cyclic* property of the trace operator [trace$(AB) =$ trace(BA)], and its linearity (which allows us to interchange the order of trace and expectation).

Ex. 2.6 Consider a regression problem with inputs x_i and outputs y_i, and a parameterized model $f_\theta(x)$ to be fit by least squares. Show that if there are observations with *tied* or *identical* values of x, then the fit can be obtained from a reduced weighted least squares problem.

Ex. 2.7 Suppose we have a sample of N pairs x_i, y_i drawn i.i.d. from the distribution characterized as follows:

$$x_i \sim h(x), \text{ the design density}$$
$$y_i = f(x_i) + \varepsilon_i, \; f \text{ is the regression function}$$
$$\varepsilon_i \sim (0, \sigma^2) \text{ (mean zero, variance } \sigma^2)$$

We construct an estimator for f *linear* in the y_i,

$$\hat{f}(x_0) = \sum_{i=1}^{N} \ell_i(x_0; \mathcal{X}) y_i,$$

where the weights $\ell_i(x_0; \mathcal{X})$ do not depend on the y_i, but do depend on the entire training sequence of x_i, denoted here by $\mathcal{X}$.

(a) Show that linear regression and k-nearest-neighbor regression are members of this class of estimators. Describe explicitly the weights $\ell_i(x_0; \mathcal{X})$ in each of these cases.

(b) Decompose the conditional mean-squared error

$$E_{\mathcal{Y}|\mathcal{X}}(f(x_0) - \hat{f}(x_0))^2$$

into a conditional squared bias and a conditional variance component. Like $\mathcal{X}$, $\mathcal{Y}$ represents the entire training sequence of y_i.

(c) Decompose the (unconditional) mean-squared error

$$E_{\mathcal{Y},\mathcal{X}}(f(x_0) - \hat{f}(x_0))^2$$

into a squared bias and a variance component.

(d) Establish a relationship between the squared biases and variances in the above two cases.

Ex. 2.8 Compare the classification performance of linear regression and k-nearest neighbor classification on the zipcode data. In particular, consider only the 2's and 3's, and $k = 1, 3, 5, 7$ and 15. Show both the training and test error for each choice. The zipcode data are available from the book website www-stat.stanford.edu/ElemStatLearn.

Ex. 2.9 Consider a linear regression model with p parameters, fit by least squares to a set of training data $(x_1, y_1), \ldots, (x_N, y_N)$ drawn at random from a population. Let $\hat{\beta}$ be the least squares estimate. Suppose we have some test data $(\tilde{x}_1, \tilde{y}_1), \ldots, (\tilde{x}_M, \tilde{y}_M)$ drawn at random from the same population as the training data. If $R_{tr}(\beta) = \frac{1}{N} \sum_1^N (y_i - \beta^T x_i)^2$ and $R_{te}(\beta) = \frac{1}{M} \sum_1^M (\tilde{y}_i - \beta^T \tilde{x}_i)^2$, prove that

$$E[R_{tr}(\hat{\beta})] \leq E[R_{te}(\hat{\beta})],$$

where the expectations are over all that is random in each expression. [This exercise was brought to our attention by Ryan Tibshirani, from a homework assignment given by Andrew Ng.]

3
Linear Methods for Regression

3.1 Introduction

A linear regression model assumes that the regression function $E(Y|X)$ is linear in the inputs $X_1, \ldots, X_p$. Linear models were largely developed in the precomputer age of statistics, but even in today's computer era there are still good reasons to study and use them. They are simple and often provide an adequate and interpretable description of how the inputs affect the output. For prediction purposes they can sometimes outperform fancier nonlinear models, especially in situations with small numbers of training cases, low signal-to-noise ratio or sparse data. Finally, linear methods can be applied to transformations of the inputs and this considerably expands their scope. These generalizations are sometimes called basis-function methods, and are discussed in Chapter 5.

In this chapter we describe linear methods for regression, while in the next chapter we discuss linear methods for classification. On some topics we go into considerable detail, as it is our firm belief that an understanding of linear methods is essential for understanding nonlinear ones. In fact, many nonlinear techniques are direct generalizations of the linear methods discussed here.

T. Hastie et al., *The Elements of Statistical Learning, Second Edition*,
DOI: 10.1007/b94608_3,
© Springer Science+Business Media, LLC 2009

3.2 Linear Regression Models and Least Squares

As introduced in Chapter 2, we have an input vector $X^T = (X_1, X_2, \ldots, X_p)$, and want to predict a real-valued output Y. The linear regression model has the form

$$f(X) = \beta_0 + \sum_{j=1}^{p} X_j \beta_j. \tag{3.1}$$

The linear model either assumes that the regression function $\mathrm{E}(Y|X)$ is linear, or that the linear model is a reasonable approximation. Here the β_j's are unknown parameters or coefficients, and the variables X_j can come from different sources:

- quantitative inputs;

- transformations of quantitative inputs, such as log, square-root or square;

- basis expansions, such as $X_2 = X_1^2$, $X_3 = X_1^3$, leading to a polynomial representation;

- numeric or "dummy" coding of the levels of qualitative inputs. For example, if G is a five-level factor input, we might create X_j, $j = 1, \ldots, 5$, such that $X_j = I(G = j)$. Together this group of X_j represents the effect of G by a set of level-dependent constants, since in $\sum_{j=1}^{5} X_j \beta_j$, one of the X_js is one, and the others are zero.

- interactions between variables, for example, $X_3 = X_1 \cdot X_2$.

No matter the source of the X_j, the model is linear in the parameters.

Typically we have a set of training data $(x_1, y_1) \ldots (x_N, y_N)$ from which to estimate the parameters β. Each $x_i = (x_{i1}, x_{i2}, \ldots, x_{ip})^T$ is a vector of feature measurements for the ith case. The most popular estimation method is *least squares*, in which we pick the coefficients $\beta = (\beta_0, \beta_1, \ldots, \beta_p)^T$ to minimize the residual sum of squares

$$
\begin{aligned}
\mathrm{RSS}(\beta) &= \sum_{i=1}^{N} (y_i - f(x_i))^2 \\
&= \sum_{i=1}^{N} \left(y_i - \beta_0 - \sum_{j=1}^{p} x_{ij} \beta_j \right)^2.
\end{aligned}
\tag{3.2}
$$

From a statistical point of view, this criterion is reasonable if the training observations (x_i, y_i) represent independent random draws from their population. Even if the x_i's were not drawn randomly, the criterion is still valid if the y_i's are conditionally independent given the inputs x_i. Figure 3.1 illustrates the geometry of least-squares fitting in the $\mathbb{R}^{p+1}$-dimensional

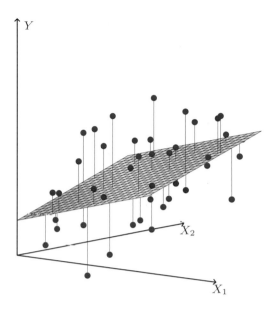

FIGURE 3.1. *Linear least squares fitting with* $X \in \mathbb{R}^2$. *We seek the linear function of* X *that minimizes the sum of squared residuals from* Y.

space occupied by the pairs (X, Y). Note that (3.2) makes no assumptions about the validity of model (3.1); it simply finds the best linear fit to the data. Least squares fitting is intuitively satisfying no matter how the data arise; the criterion measures the average lack of fit.

How do we minimize (3.2)? Denote by $\mathbf{X}$ the $N \times (p+1)$ matrix with each row an input vector (with a 1 in the first position), and similarly let $\mathbf{y}$ be the N-vector of outputs in the training set. Then we can write the residual sum-of-squares as

$$\text{RSS}(\beta) = (\mathbf{y} - \mathbf{X}\beta)^T (\mathbf{y} - \mathbf{X}\beta). \tag{3.3}$$

This is a quadratic function in the $p+1$ parameters. Differentiating with respect to β we obtain

$$\frac{\partial \text{RSS}}{\partial \beta} = -2\mathbf{X}^T (\mathbf{y} - \mathbf{X}\beta)$$

$$\frac{\partial^2 \text{RSS}}{\partial \beta \partial \beta^T} = 2\mathbf{X}^T \mathbf{X}. \tag{3.4}$$

Assuming (for the moment) that $\mathbf{X}$ has full column rank, and hence $\mathbf{X}^T\mathbf{X}$ is positive definite, we set the first derivative to zero

$$\mathbf{X}^T (\mathbf{y} - \mathbf{X}\beta) = 0 \tag{3.5}$$

to obtain the unique solution

$$\hat{\beta} = (\mathbf{X}^T \mathbf{X})^{-1} \mathbf{X}^T \mathbf{y}. \tag{3.6}$$

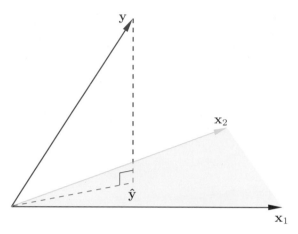

FIGURE 3.2. *The N-dimensional geometry of least squares regression with two predictors. The outcome vector* **y** *is orthogonally projected onto the hyperplane spanned by the input vectors* **x**₁ *and* **x**₂. *The projection* **ŷ** *represents the vector of the least squares predictions*

The predicted values at an input vector x_0 are given by $\hat{f}(x_0) = (1 : x_0)^T \hat{\beta}$; the fitted values at the training inputs are

$$\hat{\mathbf{y}} = \mathbf{X}\hat{\beta} = \mathbf{X}(\mathbf{X}^T\mathbf{X})^{-1}\mathbf{X}^T\mathbf{y}, \tag{3.7}$$

where $\hat{y}_i = \hat{f}(x_i)$. The matrix $\mathbf{H} = \mathbf{X}(\mathbf{X}^T\mathbf{X})^{-1}\mathbf{X}^T$ appearing in equation (3.7) is sometimes called the "hat" matrix because it puts the hat on $\mathbf{y}$.

Figure 3.2 shows a different geometrical representation of the least squares estimate, this time in $\mathbb{R}^N$. We denote the column vectors of $\mathbf{X}$ by $\mathbf{x}_0, \mathbf{x}_1, \ldots, \mathbf{x}_p$, with $\mathbf{x}_0 \equiv 1$. For much of what follows, this first column is treated like any other. These vectors span a subspace of $\mathbb{R}^N$, also referred to as the column space of $\mathbf{X}$. We minimize $\mathrm{RSS}(\beta) = \|\mathbf{y} - \mathbf{X}\beta\|^2$ by choosing $\hat{\beta}$ so that the residual vector $\mathbf{y} - \hat{\mathbf{y}}$ is orthogonal to this subspace. This orthogonality is expressed in (3.5), and the resulting estimate $\hat{\mathbf{y}}$ is hence the *orthogonal projection* of $\mathbf{y}$ onto this subspace. The hat matrix $\mathbf{H}$ computes the orthogonal projection, and hence it is also known as a projection matrix.

It might happen that the columns of $\mathbf{X}$ are not linearly independent, so that $\mathbf{X}$ is not of full rank. This would occur, for example, if two of the inputs were perfectly correlated, (e.g., $\mathbf{x}_2 = 3\mathbf{x}_1$). Then $\mathbf{X}^T\mathbf{X}$ is singular and the least squares coefficients $\hat{\beta}$ are not uniquely defined. However, the fitted values $\hat{\mathbf{y}} = \mathbf{X}\hat{\beta}$ are still the projection of $\mathbf{y}$ onto the column space of $\mathbf{X}$; there is just more than one way to express that projection in terms of the column vectors of $\mathbf{X}$. The non-full-rank case occurs most often when one or more qualitative inputs are coded in a redundant fashion. There is usually a natural way to resolve the non-unique representation, by recoding and/or dropping redundant columns in $\mathbf{X}$. Most regression software packages detect these redundancies and automatically implement some strategy for removing them. Rank deficiencies can also occur in signal and image analysis, where the number of inputs p can exceed the number of training cases N. In this case, the features are typically reduced by filtering or else the fitting is controlled by regularization (Section 5.2.3 and Chapter 18).

Up to now we have made minimal assumptions about the true distribution of the data. In order to pin down the sampling properties of $\hat{\beta}$, we now assume that the observations y_i are uncorrelated and have constant variance σ^2, and that the x_i are fixed (non random). The variance–covariance matrix of the least squares parameter estimates is easily derived from (3.6) and is given by

$$\text{Var}(\hat{\beta}) = (\mathbf{X}^T\mathbf{X})^{-1}\sigma^2. \tag{3.8}$$

Typically one estimates the variance σ^2 by

$$\hat{\sigma}^2 = \frac{1}{N-p-1}\sum_{i=1}^{N}(y_i - \hat{y}_i)^2.$$

The $N - p - 1$ rather than N in the denominator makes $\hat{\sigma}^2$ an unbiased estimate of σ^2: $\text{E}(\hat{\sigma}^2) = \sigma^2$.

To draw inferences about the parameters and the model, additional assumptions are needed. We now assume that (3.1) is the correct model for the mean; that is, the conditional expectation of Y is linear in $X_1, \ldots, X_p$. We also assume that the deviations of Y around its expectation are additive and Gaussian. Hence

$$
\begin{aligned}
Y &= \text{E}(Y|X_1, \ldots, X_p) + \varepsilon \\
&= \beta_0 + \sum_{j=1}^{p} X_j\beta_j + \varepsilon,
\end{aligned} \tag{3.9}
$$

where the error ε is a Gaussian random variable with expectation zero and variance σ^2, written $\varepsilon \sim N(0, \sigma^2)$.

Under (3.9), it is easy to show that

$$\hat{\beta} \sim N(\beta, (\mathbf{X}^T\mathbf{X})^{-1}\sigma^2). \tag{3.10}$$

This is a multivariate normal distribution with mean vector and variance–covariance matrix as shown. Also

$$(N - p - 1)\hat{\sigma}^2 \sim \sigma^2\chi^2_{N-p-1}, \tag{3.11}$$

a chi-squared distribution with $N - p - 1$ degrees of freedom. In addition $\hat{\beta}$ and $\hat{\sigma}^2$ are statistically independent. We use these distributional properties to form tests of hypothesis and confidence intervals for the parameters β_j.

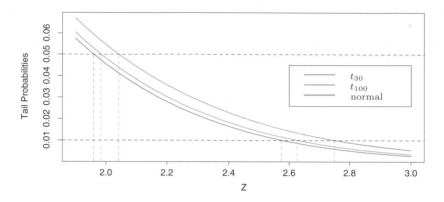

FIGURE 3.3. *The tail probabilities* $\Pr(|Z| > z)$ *for three distributions,* t_{30}, t_{100} *and standard normal. Shown are the appropriate quantiles for testing significance at the* $p = 0.05$ *and* 0.01 *levels. The difference between* t *and the standard normal becomes negligible for* N *bigger than about* 100.

To test the hypothesis that a particular coefficient $\beta_j = 0$, we form the standardized coefficient or *Z-score*

$$z_j = \frac{\hat{\beta}_j}{\hat{\sigma}\sqrt{v_j}}, \tag{3.12}$$

where v_j is the jth diagonal element of $(\mathbf{X}^T\mathbf{X})^{-1}$. Under the null hypothesis that $\beta_j = 0$, z_j is distributed as t_{N-p-1} (a t distribution with $N - p - 1$ degrees of freedom), and hence a large (absolute) value of z_j will lead to rejection of this null hypothesis. If $\hat{\sigma}$ is replaced by a known value σ, then z_j would have a standard normal distribution. The difference between the tail quantiles of a t-distribution and a standard normal become negligible as the sample size increases, and so we typically use the normal quantiles (see Figure 3.3).

Often we need to test for the significance of groups of coefficients simultaneously. For example, to test if a categorical variable with k levels can be excluded from a model, we need to test whether the coefficients of the dummy variables used to represent the levels can all be set to zero. Here we use the F statistic,

$$F = \frac{(\text{RSS}_0 - \text{RSS}_1)/(p_1 - p_0)}{\text{RSS}_1/(N - p_1 - 1)}, \tag{3.13}$$

where RSS_1 is the residual sum-of-squares for the least squares fit of the bigger model with $p_1 + 1$ parameters, and RSS_0 the same for the nested smaller model with $p_0 + 1$ parameters, having $p_1 - p_0$ parameters constrained to be zero. The F statistic measures the change in residual sum-of-squares per additional parameter in the bigger model, and it is normalized by an estimate of σ^2. Under the Gaussian assumptions, and the null hypothesis that the smaller model is correct, the F statistic will have a $F_{p_1-p_0,N-p_1-1}$ distribution. It can be shown (Exercise 3.1) that the z_j in (3.12) are equivalent to the F statistic for dropping the single coefficient β_j from the model. For large N, the quantiles of $F_{p_1-p_0,N-p_1-1}$ approach those of $\chi^2_{p_1-p_0}/(p_1-p_0)$.

Similarly, we can isolate β_j in (3.10) to obtain a $1-2\alpha$ confidence interval for β_j:

$$(\hat{\beta}_j - z^{(1-\alpha)}v_j^{\frac{1}{2}}\hat{\sigma}, \ \ \hat{\beta}_j + z^{(1-\alpha)}v_j^{\frac{1}{2}}\hat{\sigma}). \tag{3.14}$$

Here $z^{(1-\alpha)}$ is the $1-\alpha$ percentile of the normal distribution:

$$
\begin{aligned}
z^{(1-0.025)} &= 1.96, \\
z^{(1-.05)} &= 1.645, \ \ \text{etc.}
\end{aligned}
$$

Hence the standard practice of reporting $\hat{\beta} \pm 2 \cdot \text{se}(\hat{\beta})$ amounts to an approximate 95% confidence interval. Even if the Gaussian error assumption does not hold, this interval will be approximately correct, with its coverage approaching $1 - 2\alpha$ as the sample size $N \to \infty$.

In a similar fashion we can obtain an approximate confidence set for the entire parameter vector β, namely

$$C_\beta = \{\beta|(\hat{\beta} - \beta)^T \mathbf{X}^T \mathbf{X}(\hat{\beta} - \beta) \leq \hat{\sigma}^2 \chi^2_{p+1}{}^{(1-\alpha)}\}, \tag{3.15}$$

where $\chi^{2(1-\alpha)}_\ell$ is the $1-\alpha$ percentile of the chi-squared distribution on ℓ degrees of freedom: for example, $\chi^{2(1-0.05)}_5 = 11.1$, $\chi^{2(1-0.1)}_5 = 9.2$. This confidence set for β generates a corresponding confidence set for the true function $f(x) = x^T\beta$, namely $\{x^T\beta|\beta \in C_\beta\}$ (Exercise 3.2; see also Figure 5.4 in Section 5.2.2 for examples of confidence bands for functions).

3.2.1 Example: Prostate Cancer

The data for this example come from a study by Stamey et al. (1989). They examined the correlation between the level of prostate-specific antigen and a number of clinical measures in men who were about to receive a radical prostatectomy. The variables are log cancer volume (lcavol), log prostate weight (lweight), age, log of the amount of benign prostatic hyperplasia (lbph), seminal vesicle invasion (svi), log of capsular penetration (lcp), Gleason score (gleason), and percent of Gleason scores 4 or 5 (pgg45). The correlation matrix of the predictors given in Table 3.1 shows many strong correlations. Figure 1.1 (page 3) of Chapter 1 is a scatterplot matrix showing every pairwise plot between the variables. We see that svi is a binary variable, and gleason is an ordered categorical variable. We see, for

TABLE 3.1. *Correlations of predictors in the prostate cancer data.*

	lcavol	lweight	age	lbph	svi	lcp	gleason
lweight	0.300						
age	0.286	0.317					
lbph	0.063	0.437	0.287				
svi	0.593	0.181	0.129	−0.139			
lcp	0.692	0.157	0.173	−0.089	0.671		
gleason	0.426	0.024	0.366	0.033	0.307	0.476	
pgg45	0.483	0.074	0.276	−0.030	0.481	0.663	0.757

TABLE 3.2. *Linear model fit to the prostate cancer data. The Z score is the coefficient divided by its standard error (3.12). Roughly a Z score larger than two in absolute value is significantly nonzero at the $p = 0.05$ level.*

Term	Coefficient	Std. Error	Z Score
Intercept	2.46	0.09	27.60
lcavol	0.68	0.13	5.37
lweight	0.26	0.10	2.75
age	−0.14	0.10	−1.40
lbph	0.21	0.10	2.06
svi	0.31	0.12	2.47
lcp	−0.29	0.15	−1.87
gleason	−0.02	0.15	−0.15
pgg45	0.27	0.15	1.74

example, that both lcavol and lcp show a strong relationship with the response lpsa, and with each other. We need to fit the effects jointly to untangle the relationships between the predictors and the response.

We fit a linear model to the log of prostate-specific antigen, lpsa, after first standardizing the predictors to have unit variance. We randomly split the dataset into a training set of size 67 and a test set of size 30. We applied least squares estimation to the training set, producing the estimates, standard errors and Z-scores shown in Table 3.2. The Z-scores are defined in (3.12), and measure the effect of dropping that variable from the model. A Z-score greater than 2 in absolute value is approximately significant at the 5% level. (For our example, we have nine parameters, and the 0.025 tail quantiles of the t_{67-9} distribution are ±2.002!) The predictor lcavol shows the strongest effect, with lweight and svi also strong. Notice that lcp is not significant, once lcavol is in the model (when used in a model without lcavol, lcp is strongly significant). We can also test for the exclusion of a number of terms at once, using the F-statistic (3.13). For example, we consider dropping all the non-significant terms in Table 3.2, namely age, lcp, gleason, and pgg45. We get

$$F = \frac{(32.81 - 29.43)/(9 - 5)}{29.43/(67 - 9)} = 1.67, \tag{3.16}$$

which has a p-value of 0.17 ($\Pr(F_{4,58} > 1.67) = 0.17$), and hence is not significant.

The mean prediction error on the test data is 0.521. In contrast, prediction using the mean training value of lpsa has a test error of 1.057, which is called the "base error rate." Hence the linear model reduces the base error rate by about 50%. We will return to this example later to compare various selection and shrinkage methods.

3.2.2 The Gauss–Markov Theorem

One of the most famous results in statistics asserts that the least squares estimates of the parameters β have the smallest variance among all linear unbiased estimates. We will make this precise here, and also make clear that the restriction to unbiased estimates is not necessarily a wise one. This observation will lead us to consider biased estimates such as ridge regression later in the chapter. We focus on estimation of any linear combination of the parameters $\theta = a^T \beta$; for example, predictions $f(x_0) = x_0^T \beta$ are of this form. The least squares estimate of $a^T \beta$ is

$$\hat{\theta} = a^T \hat{\beta} = a^T (\mathbf{X}^T \mathbf{X})^{-1} \mathbf{X}^T \mathbf{y}. \tag{3.17}$$

Considering $\mathbf{X}$ to be fixed, this is a linear function $\mathbf{c}_0^T \mathbf{y}$ of the response vector $\mathbf{y}$. If we assume that the linear model is correct, $a^T \hat{\beta}$ is unbiased since

$$
\begin{aligned}
\mathrm{E}(a^T \hat{\beta}) &= \mathrm{E}(a^T (\mathbf{X}^T \mathbf{X})^{-1} \mathbf{X}^T \mathbf{y}) \\
&= a^T (\mathbf{X}^T \mathbf{X})^{-1} \mathbf{X}^T \mathbf{X} \beta \\
&= a^T \beta. \tag{3.18}
\end{aligned}
$$

The Gauss–Markov theorem states that if we have any other linear estimator $\tilde{\theta} = \mathbf{c}^T \mathbf{y}$ that is unbiased for $a^T \beta$, that is, $\mathrm{E}(\mathbf{c}^T \mathbf{y}) = a^T \beta$, then

$$\mathrm{Var}(a^T \hat{\beta}) \leq \mathrm{Var}(\mathbf{c}^T \mathbf{y}). \tag{3.19}$$

The proof (Exercise 3.3) uses the triangle inequality. For simplicity we have stated the result in terms of estimation of a single parameter $a^T \beta$, but with a few more definitions one can state it in terms of the entire parameter vector β (Exercise 3.3).

Consider the mean squared error of an estimator $\tilde{\theta}$ in estimating θ:

$$
\begin{aligned}
\mathrm{MSE}(\tilde{\theta}) &= \mathrm{E}(\tilde{\theta} - \theta)^2 \\
&= \mathrm{Var}(\tilde{\theta}) + [\mathrm{E}(\tilde{\theta}) - \theta]^2. \tag{3.20}
\end{aligned}
$$

The first term is the variance, while the second term is the squared bias. The Gauss-Markov theorem implies that the least squares estimator has the smallest mean squared error of all linear estimators with no bias. However, there may well exist a biased estimator with smaller mean squared error. Such an estimator would trade a little bias for a larger reduction in variance. Biased estimates are commonly used. Any method that shrinks or sets to zero some of the least squares coefficients may result in a biased estimate. We discuss many examples, including variable subset selection and ridge regression, later in this chapter. From a more pragmatic point of view, most models are distortions of the truth, and hence are biased; picking the right model amounts to creating the right balance between bias and variance. We go into these issues in more detail in Chapter 7.

Mean squared error is intimately related to prediction accuracy, as discussed in Chapter 2. Consider the prediction of the new response at input x_0,

$$Y_0 = f(x_0) + \varepsilon_0. \tag{3.21}$$

Then the expected prediction error of an estimate $\tilde{f}(x_0) = x_0^T \tilde{\beta}$ is

$$\begin{aligned} E(Y_0 - \tilde{f}(x_0))^2 &= \sigma^2 + E(x_0^T \tilde{\beta} - f(x_0))^2 \\ &= \sigma^2 + \text{MSE}(\tilde{f}(x_0)). \end{aligned} \tag{3.22}$$

Therefore, expected prediction error and mean squared error differ only by the constant σ^2, representing the variance of the new observation y_0.

3.2.3 Multiple Regression from Simple Univariate Regression

The linear model (3.1) with $p > 1$ inputs is called the *multiple linear regression model*. The least squares estimates (3.6) for this model are best understood in terms of the estimates for the *univariate* ($p = 1$) linear model, as we indicate in this section.

Suppose first that we have a univariate model with no intercept, that is,

$$Y = X\beta + \varepsilon. \tag{3.23}$$

The least squares estimate and residuals are

$$\hat{\beta} = \frac{\sum_1^N x_i y_i}{\sum_1^N x_i^2}, \tag{3.24}$$

$$r_i = y_i - x_i \hat{\beta}.$$

In convenient vector notation, we let $\mathbf{y} = (y_1, \dots, y_N)^T$, $\mathbf{x} = (x_1, \dots, x_N)^T$ and define

$$\begin{aligned} \langle \mathbf{x}, \mathbf{y} \rangle &= \sum_{i=1}^N x_i y_i, \\ &= \mathbf{x}^T \mathbf{y}, \end{aligned} \tag{3.25}$$

the *inner product* between $\mathbf{x}$ and $\mathbf{y}$[1]. Then we can write

$$\hat{\beta} = \frac{\langle \mathbf{x}, \mathbf{y} \rangle}{\langle \mathbf{x}, \mathbf{x} \rangle},$$

$$\mathbf{r} = \mathbf{y} - \mathbf{x}\hat{\beta}. \tag{3.26}$$

As we will see, this simple univariate regression provides the building block for multiple linear regression. Suppose next that the inputs $\mathbf{x}_1, \mathbf{x}_2, \ldots, \mathbf{x}_p$ (the columns of the data matrix $\mathbf{X}$) are orthogonal; that is $\langle \mathbf{x}_j, \mathbf{x}_k \rangle = 0$ for all $j \neq k$. Then it is easy to check that the multiple least squares estimates $\hat{\beta}_j$ are equal to $\langle \mathbf{x}_j, \mathbf{y} \rangle / \langle \mathbf{x}_j, \mathbf{x}_j \rangle$—the univariate estimates. In other words, when the inputs are orthogonal, they have no effect on each other's parameter estimates in the model.

Orthogonal inputs occur most often with balanced, designed experiments (where orthogonality is enforced), but almost never with observational data. Hence we will have to orthogonalize them in order to carry this idea further. Suppose next that we have an intercept and a single input $\mathbf{x}$. Then the least squares coefficient of $\mathbf{x}$ has the form

$$\hat{\beta}_1 = \frac{\langle \mathbf{x} - \bar{x}\mathbf{1}, \mathbf{y} \rangle}{\langle \mathbf{x} - \bar{x}\mathbf{1}, \mathbf{x} - \bar{x}\mathbf{1} \rangle}, \tag{3.27}$$

where $\bar{x} = \sum_i x_i / N$, and $\mathbf{1} = \mathbf{x}_0$, the vector of N ones. We can view the estimate (3.27) as the result of two applications of the simple regression (3.26). The steps are:

1. regress $\mathbf{x}$ on $\mathbf{1}$ to produce the residual $\mathbf{z} = \mathbf{x} - \bar{x}\mathbf{1}$;

2. regress $\mathbf{y}$ on the residual $\mathbf{z}$ to give the coefficient $\hat{\beta}_1$.

In this procedure, "regress $\mathbf{b}$ on $\mathbf{a}$" means a simple univariate regression of $\mathbf{b}$ on $\mathbf{a}$ with no intercept, producing coefficient $\hat{\gamma} = \langle \mathbf{a}, \mathbf{b} \rangle / \langle \mathbf{a}, \mathbf{a} \rangle$ and residual vector $\mathbf{b} - \hat{\gamma}\mathbf{a}$. We say that $\mathbf{b}$ is adjusted for $\mathbf{a}$, or is "orthogonalized" with respect to $\mathbf{a}$.

Step 1 orthogonalizes $\mathbf{x}$ with respect to $\mathbf{x}_0 = \mathbf{1}$. Step 2 is just a simple univariate regression, using the orthogonal predictors $\mathbf{1}$ and $\mathbf{z}$. Figure 3.4 shows this process for two general inputs $\mathbf{x}_1$ and $\mathbf{x}_2$. The orthogonalization does not change the subspace spanned by $\mathbf{x}_1$ and $\mathbf{x}_2$, it simply produces an orthogonal basis for representing it.

This recipe generalizes to the case of p inputs, as shown in Algorithm 3.1. Note that the inputs $\mathbf{z}_0, \ldots, \mathbf{z}_{j-1}$ in step 2 are orthogonal, hence the simple regression coefficients computed there are in fact also the multiple regression coefficients.

---

[1]The inner-product notation is suggestive of generalizations of linear regression to different metric spaces, as well as to probability spaces.

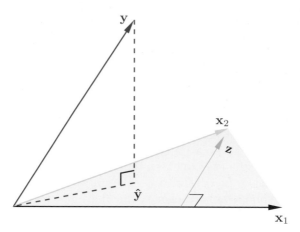

FIGURE 3.4. *Least squares regression by orthogonalization of the inputs. The vector $\mathbf{x}_2$ is regressed on the vector $\mathbf{x}_1$, leaving the residual vector $\mathbf{z}$. The regression of $\mathbf{y}$ on $\mathbf{z}$ gives the multiple regression coefficient of $\mathbf{x}_2$. Adding together the projections of $\mathbf{y}$ on each of $\mathbf{x}_1$ and $\mathbf{z}$ gives the least squares fit $\hat{\mathbf{y}}$.*

Algorithm 3.1 *Regression by Successive Orthogonalization.*

1. Initialize $\mathbf{z}_0 = \mathbf{x}_0 = \mathbf{1}$.

2. For $j = 1, 2, \ldots, p$

   Regress $\mathbf{x}_j$ on $\mathbf{z}_0, \mathbf{z}_1, \ldots, \mathbf{z}_{j-1}$ to produce coefficients $\hat{\gamma}_{\ell j} = \langle \mathbf{z}_\ell, \mathbf{x}_j \rangle / \langle \mathbf{z}_\ell, \mathbf{z}_\ell \rangle$, $\ell = 0, \ldots, j - 1$ and residual vector $\mathbf{z}_j = \mathbf{x}_j - \sum_{k=0}^{j-1} \hat{\gamma}_{kj} \mathbf{z}_k$.

3. Regress $\mathbf{y}$ on the residual $\mathbf{z}_p$ to give the estimate $\hat{\beta}_p$.

The result of this algorithm is

$$\hat{\beta}_p = \frac{\langle \mathbf{z}_p, \mathbf{y} \rangle}{\langle \mathbf{z}_p, \mathbf{z}_p \rangle}. \tag{3.28}$$

Re-arranging the residual in step 2, we can see that each of the $\mathbf{x}_j$ is a linear combination of the $\mathbf{z}_k$, $k \leq j$. Since the $\mathbf{z}_j$ are all orthogonal, they form a basis for the column space of $\mathbf{X}$, and hence the least squares projection onto this subspace is $\hat{\mathbf{y}}$. Since $\mathbf{z}_p$ alone involves $\mathbf{x}_p$ (with coefficient 1), we see that the coefficient (3.28) is indeed the multiple regression coefficient of $\mathbf{y}$ on $\mathbf{x}_p$. This key result exposes the effect of correlated inputs in multiple regression. Note also that by rearranging the $\mathbf{x}_j$, any one of them could be in the last position, and a similar results holds. Hence stated more generally, we have shown that the jth multiple regression coefficient is the univariate regression coefficient of $\mathbf{y}$ on $\mathbf{x}_{j \cdot 012 \ldots (j-1)(j+1) \ldots, p}$, the residual after regressing $\mathbf{x}_j$ on $\mathbf{x}_0, \mathbf{x}_1, \ldots, \mathbf{x}_{j-1}, \mathbf{x}_{j+1}, \ldots, \mathbf{x}_p$:

The multiple regression coefficient $\hat{\beta}_j$ represents the additional contribution of $\mathbf{x}_j$ on $\mathbf{y}$, after $\mathbf{x}_j$ has been adjusted for $\mathbf{x}_0, \mathbf{x}_1, \ldots, \mathbf{x}_{j-1}$, $\mathbf{x}_{j+1}, \ldots, \mathbf{x}_p$.

If $\mathbf{x}_p$ is highly correlated with some of the other $\mathbf{x}_k$'s, the residual vector $\mathbf{z}_p$ will be close to zero, and from (3.28) the coefficient $\hat{\beta}_p$ will be very unstable. This will be true for all the variables in the correlated set. In such situations, we might have all the Z-scores (as in Table 3.2) be small— any one of the set can be deleted—yet we cannot delete them all. From (3.28) we also obtain an alternate formula for the variance estimates (3.8),

$$\mathrm{Var}(\hat{\beta}_p) = \frac{\sigma^2}{\langle \mathbf{z}_p, \mathbf{z}_p \rangle} = \frac{\sigma^2}{\|\mathbf{z}_p\|^2}. \tag{3.29}$$

In other words, the precision with which we can estimate $\hat{\beta}_p$ depends on the length of the residual vector $\mathbf{z}_p$; this represents how much of $\mathbf{x}_p$ is unexplained by the other $\mathbf{x}_k$'s.

Algorithm 3.1 is known as the *Gram–Schmidt* procedure for multiple regression, and is also a useful numerical strategy for computing the estimates. We can obtain from it not just $\hat{\beta}_p$, but also the entire multiple least squares fit, as shown in Exercise 3.4.

We can represent step 2 of Algorithm 3.1 in matrix form:

$$\mathbf{X} = \mathbf{Z}\boldsymbol{\Gamma}, \tag{3.30}$$

where $\mathbf{Z}$ has as columns the $\mathbf{z}_j$ (in order), and $\boldsymbol{\Gamma}$ is the upper triangular matrix with entries $\hat{\gamma}_{kj}$. Introducing the diagonal matrix $\mathbf{D}$ with jth diagonal entry $D_{jj} = \|\mathbf{z}_j\|$, we get

$$\begin{aligned} \mathbf{X} &= \mathbf{Z}\mathbf{D}^{-1}\mathbf{D}\boldsymbol{\Gamma} \\ &= \mathbf{Q}\mathbf{R}, \end{aligned} \tag{3.31}$$

the so-called QR decomposition of $\mathbf{X}$. Here $\mathbf{Q}$ is an $N \times (p+1)$ orthogonal matrix, $\mathbf{Q}^T\mathbf{Q} = \mathbf{I}$, and $\mathbf{R}$ is a $(p+1) \times (p+1)$ upper triangular matrix.

The $\mathbf{QR}$ decomposition represents a convenient orthogonal basis for the column space of $\mathbf{X}$. It is easy to see, for example, that the least squares solution is given by

$$\begin{aligned} \hat{\beta} &= \mathbf{R}^{-1}\mathbf{Q}^T\mathbf{y}, \tag{3.32} \\ \hat{\mathbf{y}} &= \mathbf{Q}\mathbf{Q}^T\mathbf{y}. \tag{3.33} \end{aligned}$$

Equation (3.32) is easy to solve because $\mathbf{R}$ is upper triangular (Exercise 3.4).

3.2.4 *Multiple Outputs*

Suppose we have multiple outputs $Y_1, Y_2, \ldots, Y_K$ that we wish to predict from our inputs $X_0, X_1, X_2, \ldots, X_p$. We assume a linear model for each output

$$Y_k = \beta_{0k} + \sum_{j=1}^{p} X_j \beta_{jk} + \varepsilon_k \tag{3.34}$$

$$= f_k(X) + \varepsilon_k. \tag{3.35}$$

With N training cases we can write the model in matrix notation

$$\mathbf{Y} = \mathbf{XB} + \mathbf{E}. \tag{3.36}$$

Here $\mathbf{Y}$ is the $N \times K$ response matrix, with ik entry y_{ik}, $\mathbf{X}$ is the $N \times (p+1)$ input matrix, $\mathbf{B}$ is the $(p + 1) \times K$ matrix of parameters and $\mathbf{E}$ is the $N \times K$ matrix of errors. A straightforward generalization of the univariate loss function (3.2) is

$$\text{RSS}(\mathbf{B}) = \sum_{k=1}^{K} \sum_{i=1}^{N} (y_{ik} - f_k(x_i))^2 \tag{3.37}$$

$$= \text{tr}[(\mathbf{Y} - \mathbf{XB})^T (\mathbf{Y} - \mathbf{XB})]. \tag{3.38}$$

The least squares estimates have exactly the same form as before

$$\hat{\mathbf{B}} = (\mathbf{X}^T \mathbf{X})^{-1} \mathbf{X}^T \mathbf{Y}. \tag{3.39}$$

Hence the coefficients for the kth outcome are just the least squares estimates in the regression of $\mathbf{y}_k$ on $\mathbf{x}_0, \mathbf{x}_1, \ldots, \mathbf{x}_p$. Multiple outputs do not affect one another's least squares estimates.

If the errors $\varepsilon = (\varepsilon_1, \ldots, \varepsilon_K)$ in (3.34) are correlated, then it might seem appropriate to modify (3.37) in favor of a multivariate version. Specifically, suppose $\text{Cov}(\varepsilon) = \mathbf{\Sigma}$, then the multivariate weighted criterion

$$\text{RSS}(\mathbf{B}; \mathbf{\Sigma}) = \sum_{i=1}^{N} (y_i - f(x_i))^T \mathbf{\Sigma}^{-1} (y_i - f(x_i)) \tag{3.40}$$

arises naturally from multivariate Gaussian theory. Here $f(x)$ is the vector function $(f_1(x), \ldots, f_K(x))^T$, and y_i the vector of K responses for observation i. However, it can be shown that again the solution is given by (3.39); K separate regressions that ignore the correlations (Exercise 3.11). If the $\mathbf{\Sigma}_i$ vary among observations, then this is no longer the case, and the solution for $\mathbf{B}$ no longer decouples.

In Section 3.7 we pursue the multiple outcome problem, and consider situations where it does pay to combine the regressions.

3.3   Subset Selection

There are two reasons why we are often not satisfied with the least squares estimates (3.6).

- The first is *prediction accuracy*: the least squares estimates often have low bias but large variance. Prediction accuracy can sometimes be improved by shrinking or setting some coefficients to zero. By doing so we sacrifice a little bit of bias to reduce the variance of the predicted values, and hence may improve the overall prediction accuracy.

- The second reason is *interpretation*. With a large number of predictors, we often would like to determine a smaller subset that exhibit the strongest effects. In order to get the "big picture," we are willing to sacrifice some of the small details.

In this section we describe a number of approaches to variable subset selection with linear regression. In later sections we discuss shrinkage and hybrid approaches for controlling variance, as well as other dimension-reduction strategies. These all fall under the general heading *model selection*. Model selection is not restricted to linear models; Chapter 7 covers this topic in some detail.

With subset selection we retain only a subset of the variables, and eliminate the rest from the model. Least squares regression is used to estimate the coefficients of the inputs that are retained. There are a number of different strategies for choosing the subset.

3.3.1   Best-Subset Selection

Best subset regression finds for each $k \in \{0, 1, 2, \ldots, p\}$ the subset of size k that gives smallest residual sum of squares (3.2). An efficient algorithm—the *leaps and bounds* procedure (Furnival and Wilson, 1974)—makes this feasible for p as large as 30 or 40. Figure 3.5 shows all the subset models for the prostate cancer example. The lower boundary represents the models that are eligible for selection by the best-subsets approach. Note that the best subset of size 2, for example, need not include the variable that was in the best subset of size 1 (for this example all the subsets are nested). The best-subset curve (red lower boundary in Figure 3.5) is necessarily decreasing, so cannot be used to select the subset size k. The question of how to choose k involves the tradeoff between bias and variance, along with the more subjective desire for parsimony. There are a number of criteria that one may use; typically we choose the smallest model that minimizes an estimate of the expected prediction error.

Many of the other approaches that we discuss in this chapter are similar, in that they use the training data to produce a sequence of models varying in complexity and indexed by a single parameter. In the next section we use

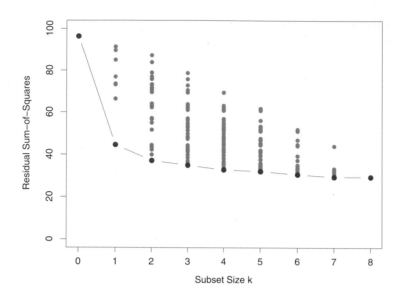

FIGURE 3.5. *All possible subset models for the prostate cancer example. At each subset size is shown the residual sum-of-squares for each model of that size.*

cross-validation to estimate prediction error and select k; the AIC criterion is a popular alternative. We defer more detailed discussion of these and other approaches to Chapter 7.

3.3.2 Forward- and Backward-Stepwise Selection

Rather than search through all possible subsets (which becomes infeasible for p much larger than 40), we can seek a good path through them. *Forward-stepwise selection* starts with the intercept, and then sequentially adds into the model the predictor that most improves the fit. With many candidate predictors, this might seem like a lot of computation; however, clever updating algorithms can exploit the QR decomposition for the current fit to rapidly establish the next candidate (Exercise 3.9). Like best-subset regression, forward stepwise produces a sequence of models indexed by k, the subset size, which must be determined.

Forward-stepwise selection is a *greedy algorithm*, producing a nested sequence of models. In this sense it might seem sub-optimal compared to best-subset selection. However, there are several reasons why it might be preferred:

- *Computational;* for large p we cannot compute the best subset sequence, but we can always compute the forward stepwise sequence (even when $p \gg N$).

- *Statistical;* a price is paid in variance for selecting the best subset of each size; forward stepwise is a more constrained search, and will have lower variance, but perhaps more bias.

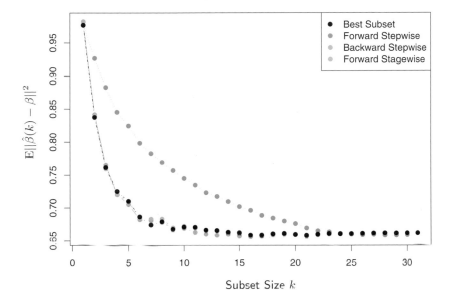

FIGURE 3.6. *Comparison of four subset-selection techniques on a simulated linear regression problem $Y = X^T \beta + \varepsilon$. There are $N = 300$ observations on $p = 31$ standard Gaussian variables, with pairwise correlations all equal to 0.85. For 10 of the variables, the coefficients are drawn at random from a $N(0, 0.4)$ distribution; the rest are zero. The noise $\varepsilon \sim N(0, 6.25)$, resulting in a signal-to-noise ratio of 0.64. Results are averaged over 50 simulations. Shown is the mean-squared error of the estimated coefficient $\hat{\beta}(k)$ at each step from the true β.*

Backward-stepwise selection starts with the full model, and sequentially deletes the predictor that has the least impact on the fit. The candidate for dropping is the variable with the smallest Z-score (Exercise 3.10). Backward selection can only be used when $N > p$, while forward stepwise can always be used.

Figure 3.6 shows the results of a small simulation study to compare best-subset regression with the simpler alternatives forward and backward selection. Their performance is very similar, as is often the case. Included in the figure is forward stagewise regression (next section), which takes longer to reach minimum error.

On the prostate cancer example, best-subset, forward and backward selection all gave exactly the same sequence of terms.

Some software packages implement hybrid stepwise-selection strategies that consider both forward and backward moves at each step, and select the "best" of the two. For example in the R package the `step` function uses the AIC criterion for weighing the choices, which takes proper account of the number of parameters fit; at each step an add or drop will be performed that minimizes the AIC score.

Other more traditional packages base the selection on F-statistics, adding "significant" terms, and dropping "non-significant" terms. These are out of fashion, since they do not take proper account of the multiple testing issues. It is also tempting after a model search to print out a summary of the chosen model, such as in Table 3.2; however, the standard errors are not valid, since they do not account for the search process. The bootstrap (Section 8.2) can be useful in such settings.

Finally, we note that often variables come in groups (such as the dummy variables that code a multi-level categorical predictor). Smart stepwise procedures (such as `step` in R) will add or drop whole groups at a time, taking proper account of their degrees-of-freedom.

3.3.3 Forward-Stagewise Regression

Forward-stagewise regression (FS) is even more constrained than forward-stepwise regression. It starts like forward-stepwise regression, with an intercept equal to $\bar{y}$, and centered predictors with coefficients initially all 0. At each step the algorithm identifies the variable most correlated with the current residual. It then computes the simple linear regression coefficient of the residual on this chosen variable, and then adds it to the current coefficient for that variable. This is continued till none of the variables have correlation with the residuals—i.e. the least-squares fit when $N > p$.

Unlike forward-stepwise regression, none of the other variables are adjusted when a term is added to the model. As a consequence, forward stagewise can take many more than p steps to reach the least squares fit, and historically has been dismissed as being inefficient. It turns out that this "slow fitting" can pay dividends in high-dimensional problems. We see in Section 3.8.1 that both forward stagewise and a variant which is slowed down even further are quite competitive, especially in very high-dimensional problems.

Forward-stagewise regression is included in Figure 3.6. In this example it takes over 1000 steps to get all the correlations below 10^{-4}. For subset size k, we plotted the error for the last step for which there where k nonzero coefficients. Although it catches up with the best fit, it takes longer to do so.

3.3.4 Prostate Cancer Data Example (Continued)

Table 3.3 shows the coefficients from a number of different selection and shrinkage methods. They are *best-subset selection* using an all-subsets search, *ridge regression*, the *lasso*, *principal components regression* and *partial least squares*. Each method has a complexity parameter, and this was chosen to minimize an estimate of prediction error based on tenfold cross-validation; full details are given in Section 7.10. Briefly, cross-validation works by dividing the training data randomly into ten equal parts. The learning method is fit—for a range of values of the complexity parameter—to nine-tenths of the data, and the prediction error is computed on the remaining one-tenth. This is done in turn for each one-tenth of the data, and the ten prediction error estimates are averaged. From this we obtain an estimated prediction error curve as a function of the complexity parameter.

Note that we have already divided these data into a training set of size 67 and a test set of size 30. Cross-validation is applied to the training set, since selecting the shrinkage parameter is part of the training process. The test set is there to judge the performance of the selected model.

The estimated prediction error curves are shown in Figure 3.7. Many of the curves are very flat over large ranges near their minimum. Included are estimated standard error bands for each estimated error rate, based on the ten error estimates computed by cross-validation. We have used the "one-standard-error" rule—we pick the most parsimonious model within one standard error of the minimum (Section 7.10, page 244). Such a rule acknowledges the fact that the tradeoff curve is estimated with error, and hence takes a conservative approach.

Best-subset selection chose to use the two predictors lcvol and lweight. The last two lines of the table give the average prediction error (and its estimated standard error) over the test set.

3.4 Shrinkage Methods

By retaining a subset of the predictors and discarding the rest, subset selection produces a model that is interpretable and has possibly lower prediction error than the full model. However, because it is a discrete process—variables are either retained or discarded—it often exhibits high variance, and so doesn't reduce the prediction error of the full model. Shrinkage methods are more continuous, and don't suffer as much from high variability.

3.4.1 Ridge Regression

Ridge regression shrinks the regression coefficients by imposing a penalty on their size. The ridge coefficients minimize a penalized residual sum of

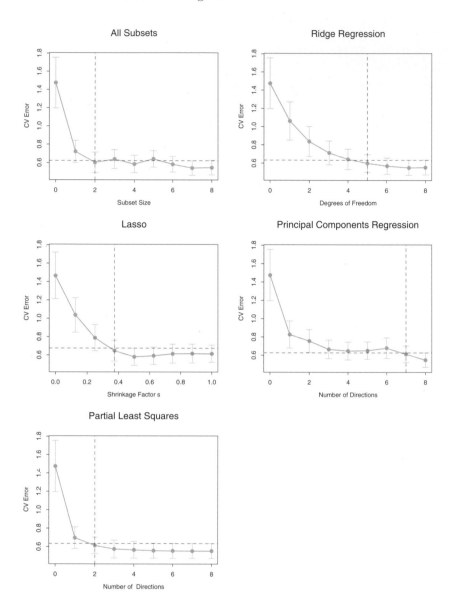

FIGURE 3.7. *Estimated prediction error curves and their standard errors for the various selection and shrinkage methods. Each curve is plotted as a function of the corresponding complexity parameter for that method. The horizontal axis has been chosen so that the model complexity increases as we move from left to right. The estimates of prediction error and their standard errors were obtained by tenfold cross-validation; full details are given in Section 7.10. The least complex model within one standard error of the best is chosen, indicated by the purple vertical broken lines.*

TABLE 3.3. *Estimated coefficients and test error results, for different subset and shrinkage methods applied to the prostate data. The blank entries correspond to variables omitted.*

Term	LS	Best Subset	Ridge	Lasso	PCR	PLS
Intercept	2.465	2.477	2.452	2.468	2.497	2.452
lcavol	0.680	0.740	0.420	0.533	0.543	0.419
lweight	0.263	0.316	0.238	0.169	0.289	0.344
age	−0.141		−0.046		−0.152	−0.026
lbph	0.210		0.162	0.002	0.214	0.220
svi	0.305		0.227	0.094	0.315	0.243
lcp	−0.288		0.000		−0.051	0.079
gleason	−0.021		0.040		0.232	0.011
pgg45	0.267		0.133		−0.056	0.084
Test Error	0.521	0.492	0.492	0.479	0.449	0.528
Std Error	0.179	0.143	0.165	0.164	0.105	0.152

squares,

$$\hat{\beta}^{\text{ridge}} = \underset{\beta}{\text{argmin}}\left\{\sum_{i=1}^{N}(y_i - \beta_0 - \sum_{j=1}^{p}x_{ij}\beta_j)^2 + \lambda\sum_{j=1}^{p}\beta_j^2\right\}. \quad (3.41)$$

Here $\lambda \geq 0$ is a complexity parameter that controls the amount of shrinkage: the larger the value of λ, the greater the amount of shrinkage. The coefficients are shrunk toward zero (and each other). The idea of penalizing by the sum-of-squares of the parameters is also used in neural networks, where it is known as *weight decay* (Chapter 11).

An equivalent way to write the ridge problem is

$$\hat{\beta}^{\text{ridge}} = \underset{\beta}{\text{argmin}}\sum_{i=1}^{N}\left(y_i - \beta_0 - \sum_{j=1}^{p}x_{ij}\beta_j\right)^2,$$

$$\text{subject to } \sum_{j=1}^{p}\beta_j^2 \leq t, \quad (3.42)$$

which makes explicit the size constraint on the parameters. There is a one-to-one correspondence between the parameters λ in (3.41) and t in (3.42). When there are many correlated variables in a linear regression model, their coefficients can become poorly determined and exhibit high variance. A wildly large positive coefficient on one variable can be canceled by a similarly large negative coefficient on its correlated cousin. By imposing a size constraint on the coefficients, as in (3.42), this problem is alleviated.

The ridge solutions are not equivariant under scaling of the inputs, and so one normally standardizes the inputs before solving (3.41). In addition, notice that the intercept β_0 has been left out of the penalty term. Penalization of the intercept would make the procedure depend on the origin chosen for Y; that is, adding a constant c to each of the targets y_i would not simply result in a shift of the predictions by the same amount c. It can be shown (Exercise 3.5) that the solution to (3.41) can be separated into two parts, after reparametrization using *centered* inputs: each x_{ij} gets replaced by $x_{ij} - \bar{x}_j$. We estimate β_0 by $\bar{y} = \frac{1}{N} \sum_1^N y_i$. The remaining coefficients get estimated by a ridge regression without intercept, using the centered x_{ij}. Henceforth we assume that this centering has been done, so that the input matrix $\mathbf{X}$ has p (rather than $p+1$) columns.

Writing the criterion in (3.41) in matrix form,

$$\text{RSS}(\lambda) = (\mathbf{y} - \mathbf{X}\beta)^T(\mathbf{y} - \mathbf{X}\beta) + \lambda\beta^T\beta, \qquad (3.43)$$

the ridge regression solutions are easily seen to be

$$\hat{\beta}^{\text{ridge}} = (\mathbf{X}^T\mathbf{X} + \lambda\mathbf{I})^{-1}\mathbf{X}^T\mathbf{y}, \qquad (3.44)$$

where $\mathbf{I}$ is the $p \times p$ identity matrix. Notice that with the choice of quadratic penalty $\beta^T\beta$, the ridge regression solution is again a linear function of $\mathbf{y}$. The solution adds a positive constant to the diagonal of $\mathbf{X}^T\mathbf{X}$ before inversion. This makes the problem nonsingular, even if $\mathbf{X}^T\mathbf{X}$ is not of full rank, and was the main motivation for ridge regression when it was first introduced in statistics (Hoerl and Kennard, 1970). Traditional descriptions of ridge regression start with definition (3.44). We choose to motivate it via (3.41) and (3.42), as these provide insight into how it works.

Figure 3.8 shows the ridge coefficient estimates for the prostate cancer example, plotted as functions of df(λ), the *effective degrees of freedom* implied by the penalty λ (defined in (3.50) on page 68). In the case of orthonormal inputs, the ridge estimates are just a scaled version of the least squares estimates, that is, $\hat{\beta}^{\text{ridge}} = \hat{\beta}/(1 + \lambda)$.

Ridge regression can also be derived as the mean or mode of a posterior distribution, with a suitably chosen prior distribution. In detail, suppose $y_i \sim N(\beta_0 + x_i^T\beta, \sigma^2)$, and the parameters β_j are each distributed as $N(0, \tau^2)$, independently of one another. Then the (negative) log-posterior density of β, with τ^2 and σ^2 assumed known, is equal to the expression in curly braces in (3.41), with $\lambda = \sigma^2/\tau^2$ (Exercise 3.6). Thus the ridge estimate is the mode of the posterior distribution; since the distribution is Gaussian, it is also the posterior mean.

The *singular value decomposition* (SVD) of the centered input matrix $\mathbf{X}$ gives us some additional insight into the nature of ridge regression. This decomposition is extremely useful in the analysis of many statistical methods. The SVD of the $N \times p$ matrix $\mathbf{X}$ has the form

$$\mathbf{X} = \mathbf{U}\mathbf{D}\mathbf{V}^T. \qquad (3.45)$$

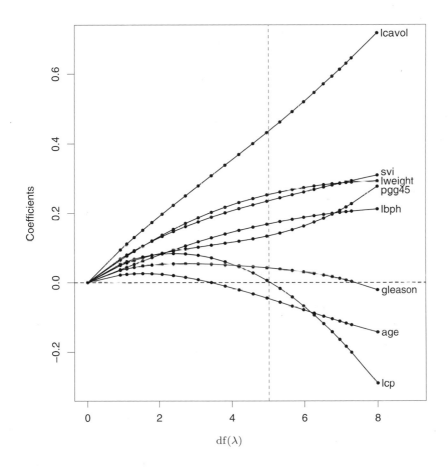

FIGURE 3.8. *Profiles of ridge coefficients for the prostate cancer example, as the tuning parameter λ is varied. Coefficients are plotted versus* $\mathrm{df}(\lambda)$*, the effective degrees of freedom. A vertical line is drawn at* $\mathrm{df} = 5.0$*, the value chosen by cross-validation.*

Here $\mathbf{U}$ and $\mathbf{V}$ are $N \times p$ and $p \times p$ orthogonal matrices, with the columns of $\mathbf{U}$ spanning the column space of $\mathbf{X}$, and the columns of $\mathbf{V}$ spanning the row space. $\mathbf{D}$ is a $p \times p$ diagonal matrix, with diagonal entries $d_1 \geq d_2 \geq \cdots \geq d_p \geq 0$ called the singular values of $\mathbf{X}$. If one or more values $d_j = 0$, $\mathbf{X}$ is singular.

Using the singular value decomposition we can write the least squares fitted vector as

$$
\begin{aligned}
\mathbf{X}\hat{\beta}^{\mathrm{ls}} &= \mathbf{X}(\mathbf{X}^T\mathbf{X})^{-1}\mathbf{X}^T\mathbf{y} \\
&= \mathbf{U}\mathbf{U}^T\mathbf{y},
\end{aligned}
\tag{3.46}
$$

after some simplification. Note that $\mathbf{U}^T\mathbf{y}$ are the coordinates of $\mathbf{y}$ with respect to the orthonormal basis $\mathbf{U}$. Note also the similarity with (3.33); $\mathbf{Q}$ and $\mathbf{U}$ are generally different orthogonal bases for the column space of $\mathbf{X}$ (Exercise 3.8).

Now the ridge solutions are

$$
\begin{aligned}
\mathbf{X}\hat{\beta}^{\mathrm{ridge}} &= \mathbf{X}(\mathbf{X}^T\mathbf{X} + \lambda\mathbf{I})^{-1}\mathbf{X}^T\mathbf{y} \\
&= \mathbf{U}\,\mathbf{D}(\mathbf{D}^2 + \lambda\mathbf{I})^{-1}\mathbf{D}\,\mathbf{U}^T\mathbf{y} \\
&= \sum_{j=1}^{p} \mathbf{u}_j \frac{d_j^2}{d_j^2 + \lambda} \mathbf{u}_j^T \mathbf{y},
\end{aligned}
\tag{3.47}
$$

where the $\mathbf{u}_j$ are the columns of $\mathbf{U}$. Note that since $\lambda \geq 0$, we have $d_j^2/(d_j^2 + \lambda) \leq 1$. Like linear regression, ridge regression computes the coordinates of $\mathbf{y}$ with respect to the orthonormal basis $\mathbf{U}$. It then shrinks these coordinates by the factors $d_j^2/(d_j^2 + \lambda)$. This means that a greater amount of shrinkage is applied to the coordinates of basis vectors with smaller d_j^2.

What does a small value of d_j^2 mean? The SVD of the centered matrix $\mathbf{X}$ is another way of expressing the *principal components* of the variables in $\mathbf{X}$. The sample covariance matrix is given by $\mathbf{S} = \mathbf{X}^T\mathbf{X}/N$, and from (3.45) we have

$$
\mathbf{X}^T\mathbf{X} = \mathbf{V}\mathbf{D}^2\mathbf{V}^T,
\tag{3.48}
$$

which is the *eigen decomposition* of $\mathbf{X}^T\mathbf{X}$ (and of $\mathbf{S}$, up to a factor N). The eigenvectors v_j (columns of $\mathbf{V}$) are also called the *principal components* (or Karhunen–Loeve) directions of $\mathbf{X}$. The first principal component direction v_1 has the property that $\mathbf{z}_1 = \mathbf{X}v_1$ has the largest sample variance amongst all normalized linear combinations of the columns of $\mathbf{X}$. This sample variance is easily seen to be

$$
\mathrm{Var}(\mathbf{z}_1) = \mathrm{Var}(\mathbf{X}v_1) = \frac{d_1^2}{N},
\tag{3.49}
$$

and in fact $\mathbf{z}_1 = \mathbf{X}v_1 = \mathbf{u}_1 d_1$. The derived variable $\mathbf{z}_1$ is called the first principal component of $\mathbf{X}$, and hence $\mathbf{u}_1$ is the normalized first principal

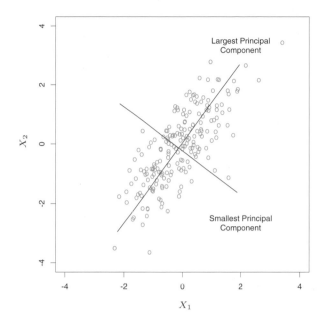

FIGURE 3.9. *Principal components of some input data points. The largest principal component is the direction that maximizes the variance of the projected data, and the smallest principal component minimizes that variance. Ridge regression projects* **y** *onto these components, and then shrinks the coefficients of the low–variance components more than the high variance components.*

component. Subsequent principal components $\mathbf{z}_j$ have maximum variance d_j^2/N, subject to being orthogonal to the earlier ones. Conversely the last principal component has *minimum* variance. Hence the small singular values d_j correspond to directions in the column space of $\mathbf{X}$ having small variance, and ridge regression shrinks these directions the most.

Figure 3.9 illustrates the principal components of some data points in two dimensions. If we consider fitting a linear surface over this domain (the Y-axis is sticking out of the page), the configuration of the data allow us to determine its gradient more accurately in the long direction than the short. Ridge regression protects against the potentially high variance of gradients estimated in the short directions. The implicit assumption is that the response will tend to vary most in the directions of high variance of the inputs. This is often a reasonable assumption, since predictors are often chosen for study because they vary with the response variable, but need not hold in general.

In Figure 3.7 we have plotted the estimated prediction error versus the quantity

$$
\begin{aligned}
\mathrm{df}(\lambda) &= \mathrm{tr}[\mathbf{X}(\mathbf{X}^T\mathbf{X} + \lambda\mathbf{I})^{-1}\mathbf{X}^T], \\
&= \mathrm{tr}(\mathbf{H}_\lambda) \\
&= \sum_{j=1}^{p} \frac{d_j^2}{d_j^2 + \lambda}.
\end{aligned}
\tag{3.50}
$$

This monotone decreasing function of λ is the *effective degrees of freedom* of the ridge regression fit. Usually in a linear-regression fit with p variables, the degrees-of-freedom of the fit is p, the number of free parameters. The idea is that although all p coefficients in a ridge fit will be non-zero, they are fit in a restricted fashion controlled by λ. Note that $\mathrm{df}(\lambda) = p$ when $\lambda = 0$ (no regularization) and $\mathrm{df}(\lambda) \to 0$ as $\lambda \to \infty$. Of course there is always an additional one degree of freedom for the intercept, which was removed *apriori*. This definition is motivated in more detail in Section 3.4.4 and Sections 7.4–7.6. In Figure 3.7 the minimum occurs at $\mathrm{df}(\lambda) = 5.0$. Table 3.3 shows that ridge regression reduces the test error of the full least squares estimates by a small amount.

3.4.2 The Lasso

The lasso is a shrinkage method like ridge, with subtle but important differences. The lasso estimate is defined by

$$
\hat{\beta}^{\text{lasso}} = \underset{\beta}{\mathrm{argmin}} \sum_{i-1}^{N}\left(y_i - \beta_0 - \sum_{j=1}^{p} x_{ij}\beta_j\right)^2
$$

$$
\text{subject to } \sum_{j=1}^{p} |\beta_j| \le t.
\tag{3.51}
$$

Just as in ridge regression, we can re-parametrize the constant β_0 by standardizing the predictors; the solution for $\hat{\beta}_0$ is $\bar{y}$, and thereafter we fit a model without an intercept (Exercise 3.5). In the signal processing literature, the lasso is also known as *basis pursuit* (Chen et al., 1998).

We can also write the lasso problem in the equivalent *Lagrangian form*

$$
\hat{\beta}^{\text{lasso}} = \underset{\beta}{\mathrm{argmin}}\left\{\frac{1}{2}\sum_{i=1}^{N}\left(y_i - \beta_0 - \sum_{j=1}^{p} x_{ij}\beta_j\right)^2 + \lambda\sum_{j=1}^{p} |\beta_j|\right\}.
\tag{3.52}
$$

Notice the similarity to the ridge regression problem (3.42) or (3.41): the L_2 ridge penalty $\sum_1^p \beta_j^2$ is replaced by the L_1 lasso penalty $\sum_1^p |\beta_j|$. This latter constraint makes the solutions nonlinear in the y_i, and there is no closed form expression as in ridge regression. Computing the lasso solution is a quadratic programming problem, although we see in Section 3.4.4 that efficient algorithms are available for computing the entire path of solutions as λ is varied, with the same computational cost as for ridge regression. Because of the nature of the constraint, making t sufficiently small will cause some of the coefficients to be exactly zero. Thus the lasso does a kind of continuous subset selection. If t is chosen larger than $t_0 = \sum_1^p |\hat{\beta}_j|$ (where $\hat{\beta}_j = \hat{\beta}_j^{ls}$, the least squares estimates), then the lasso estimates are the $\hat{\beta}_j$'s. On the other hand, for $t = t_0/2$ say, then the least squares coefficients are shrunk by about 50% on average. However, the nature of the shrinkage is not obvious, and we investigate it further in Section 3.4.4 below. Like the subset size in variable subset selection, or the penalty parameter in ridge regression, t should be adaptively chosen to minimize an estimate of expected prediction error.

In Figure 3.7, for ease of interpretation, we have plotted the lasso prediction error estimates versus the standardized parameter $s = t / \sum_1^p |\hat{\beta}_j|$. A value $\hat{s} \approx 0.36$ was chosen by 10-fold cross-validation; this caused four coefficients to be set to zero (fifth column of Table 3.3). The resulting model has the second lowest test error, slightly lower than the full least squares model, but the standard errors of the test error estimates (last line of Table 3.3) are fairly large.

Figure 3.10 shows the lasso coefficients as the standardized tuning parameter $s = t / \sum_1^p |\hat{\beta}_j|$ is varied. At $s = 1.0$ these are the least squares estimates; they decrease to 0 as $s \to 0$. This decrease is not always strictly monotonic, although it is in this example. A vertical line is drawn at $s = 0.36$, the value chosen by cross-validation.

3.4.3 Discussion: Subset Selection, Ridge Regression and the Lasso

In this section we discuss and compare the three approaches discussed so far for restricting the linear regression model: subset selection, ridge regression and the lasso.

In the case of an orthonormal input matrix $\mathbf{X}$ the three procedures have explicit solutions. Each method applies a simple transformation to the least squares estimate $\hat{\beta}_j$, as detailed in Table 3.4.

Ridge regression does a proportional shrinkage. Lasso translates each coefficient by a constant factor λ, truncating at zero. This is called "soft thresholding," and is used in the context of wavelet-based smoothing in Section 5.9. Best-subset selection drops all variables with coefficients smaller than the Mth largest; this is a form of "hard-thresholding."

Back to the nonorthogonal case; some pictures help understand their relationship. Figure 3.11 depicts the lasso (left) and ridge regression (right) when there are only two parameters. The residual sum of squares has elliptical contours, centered at the full least squares estimate. The constraint

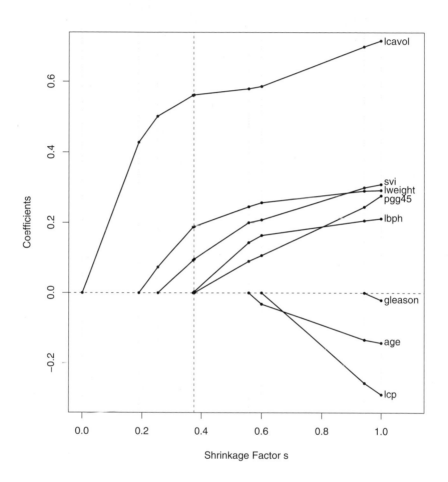

FIGURE 3.10. *Profiles of lasso coefficients, as the tuning parameter t is varied. Coefficients are plotted versus $s = t / \sum_1^p |\hat{\beta}_j|$. A vertical line is drawn at $s = 0.36$, the value chosen by cross-validation. Compare Figure 3.8 on page 65; the lasso profiles hit zero, while those for ridge do not. The profiles are piece-wise linear, and so are computed only at the points displayed; see Section 3.4.4 for details.*

TABLE 3.4. *Estimators of β_j in the case of orthonormal columns of* **X**. *M and λ are constants chosen by the corresponding techniques;* sign *denotes the sign of its argument (± 1), and x_+ denotes "positive part" of x. Below the table, estimators are shown by broken red lines. The 45° line in gray shows the unrestricted estimate for reference.*

Estimator	Formula
Best subset (size M)	$\hat{\beta}_j \cdot I(|\hat{\beta}_j| \geq |\hat{\beta}_{(M)}|)$
Ridge	$\hat{\beta}_j / (1 + \lambda)$
Lasso	$\mathrm{sign}(\hat{\beta}_j)(|\hat{\beta}_j| - \lambda)_+$

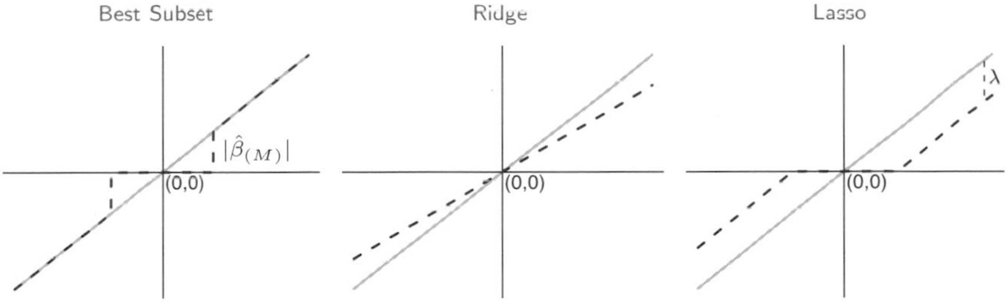

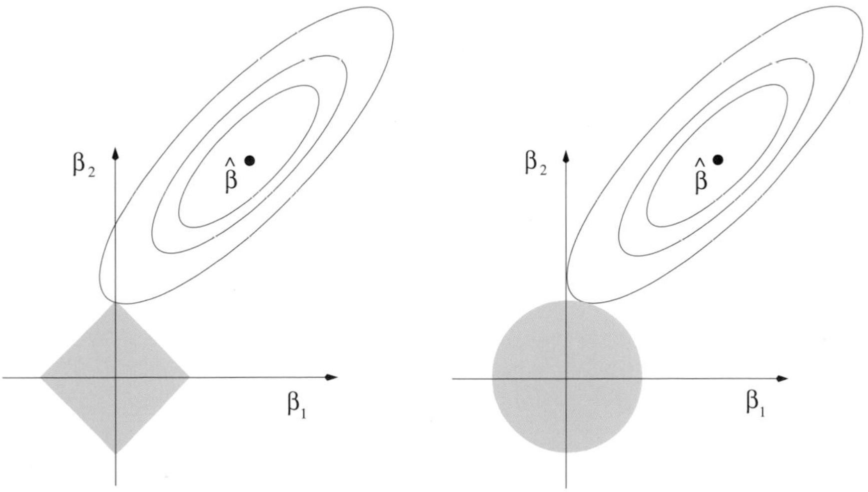

FIGURE 3.11. *Estimation picture for the lasso (left) and ridge regression (right). Shown are contours of the error and constraint functions. The solid blue areas are the constraint regions $|\beta_1| + |\beta_2| \leq t$ and $\beta_1^2 + \beta_2^2 \leq t^2$, respectively, while the red ellipses are the contours of the least squares error function.*

region for ridge regression is the disk $\beta_1^2 + \beta_2^2 \leq t$, while that for lasso is the diamond $|\beta_1| + |\beta_2| \leq t$. Both methods find the first point where the elliptical contours hit the constraint region. Unlike the disk, the diamond has corners; if the solution occurs at a corner, then it has one parameter β_j equal to zero. When $p > 2$, the diamond becomes a rhomboid, and has many corners, flat edges and faces; there are many more opportunities for the estimated parameters to be zero.

We can generalize ridge regression and the lasso, and view them as Bayes estimates. Consider the criterion

$$\tilde{\beta} = \underset{\beta}{\text{argmin}} \left\{ \sum_{i=1}^{N} (y_i - \beta_0 - \sum_{j=1}^{p} x_{ij}\beta_j)^2 + \lambda \sum_{j=1}^{p} |\beta_j|^q \right\} \qquad (3.53)$$

for $q \geq 0$. The contours of constant value of $\sum_j |\beta_j|^q$ are shown in Figure 3.12, for the case of two inputs.

Thinking of $|\beta_j|^q$ as the log-prior density for β_j, these are also the equicontours of the prior distribution of the parameters. The value $q = 0$ corresponds to variable subset selection, as the penalty simply counts the number of nonzero parameters; $q = 1$ corresponds to the lasso, while $q = 2$ to ridge regression. Notice that for $q \leq 1$, the prior is not uniform in direction, but concentrates more mass in the coordinate directions. The prior corresponding to the $q = 1$ case is an independent double exponential (or Laplace) distribution for each input, with density $(1/2\tau) \exp(-|\beta|/\tau)$ and $\tau = 1/\lambda$. The case $q = 1$ (lasso) is the smallest q such that the constraint region is convex; non-convex constraint regions make the optimization problem more difficult.

In this view, the lasso, ridge regression and best subset selection are Bayes estimates with different priors. Note, however, that they are derived as posterior modes, that is, maximizers of the posterior. It is more common to use the mean of the posterior as the Bayes estimate. Ridge regression is also the posterior mean, but the lasso and best subset selection are not.

Looking again at the criterion (3.53), we might try using other values of q besides 0, 1, or 2. Although one might consider estimating q from the data, our experience is that it is not worth the effort for the extra variance incurred. Values of $q \in (1, 2)$ suggest a compromise between the lasso and ridge regression. Although this is the case, with $q > 1$, $|\beta_j|^q$ is differentiable at 0, and so does not share the ability of lasso ($q = 1$) for

| $q = 4$ | $q = 2$ | $q = 1$ | $q = 0.5$ | $q = 0.1$ |

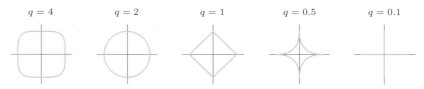

FIGURE 3.12. *Contours of constant value of $\sum_j |\beta_j|^q$ for given values of q.*

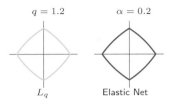

FIGURE 3.13. *Contours of constant value of $\sum_j |\beta_j|^q$ for $q = 1.2$ (left plot), and the elastic-net penalty $\sum_j (\alpha\beta_j^2 + (1-\alpha)|\beta_j|)$ for $\alpha = 0.2$ (right plot). Although visually very similar, the elastic-net has sharp (non-differentiable) corners, while the $q = 1.2$ penalty does not.*

setting coefficients exactly to zero. Partly for this reason as well as for computational tractability, Zou and Hastie (2005) introduced the *elastic-net* penalty

$$\lambda \sum_{j=1}^{p} \left(\alpha\beta_j^2 + (1 - \alpha)|\beta_j|\right), \tag{3.54}$$

a different compromise between ridge and lasso. Figure 3.13 compares the L_q penalty with $q = 1.2$ and the elastic-net penalty with $\alpha = 0.2$; it is hard to detect the difference by eye. The elastic-net selects variables like the lasso, and shrinks together the coefficients of correlated predictors like ridge. It also has considerable computational advantages over the L_q penalties. We discuss the elastic-net further in Section 18.4.

3.4.4 Least Angle Regression

Least angle regression (LAR) is a relative newcomer (Efron et al., 2004), and can be viewed as a kind of "democratic" version of forward stepwise regression (Section 3.3.2). As we will see, LAR is intimately connected with the lasso, and in fact provides an extremely efficient algorithm for computing the entire lasso path as in Figure 3.10.

Forward stepwise regression builds a model sequentially, adding one variable at a time. At each step, it identifies the best variable to include in the *active set*, and then updates the least squares fit to include all the active variables.

Least angle regression uses a similar strategy, but only enters "as much" of a predictor as it deserves. At the first step it identifies the variable most correlated with the response. Rather than fit this variable completely, LAR moves the coefficient of this variable continuously toward its least-squares value (causing its correlation with the evolving residual to decrease in absolute value). As soon as another variable "catches up" in terms of correlation with the residual, the process is paused. The second variable then joins the active set, and their coefficients are moved together in a way that keeps their correlations tied and decreasing. This process is continued until all the variables are in the model, and ends at the full least-squares fit. Algorithm 3.2 provides the details. The termination condition in step 5 requires some explanation. If $p > N - 1$, the LAR algorithm reaches a zero residual solution after $N - 1$ steps (the -1 is because we have centered the data).

Algorithm 3.2 *Least Angle Regression.*

---

1. Standardize the predictors to have mean zero and unit norm. Start with the residual $\mathbf{r} = \mathbf{y} - \bar{\mathbf{y}}$, $\beta_1, \beta_2, \ldots, \beta_p = 0$.

2. Find the predictor $\mathbf{x}_j$ most correlated with $\mathbf{r}$.

3. Move β_j from 0 towards its least-squares coefficient $\langle \mathbf{x}_j, \mathbf{r} \rangle$, until some other competitor $\mathbf{x}_k$ has as much correlation with the current residual as does $\mathbf{x}_j$.

4. Move β_j and β_k in the direction defined by their joint least squares coefficient of the current residual on $(\mathbf{x}_j, \mathbf{x}_k)$, until some other competitor $\mathbf{x}_l$ has as much correlation with the current residual.

5. Continue in this way until all p predictors have been entered. After $\min(N - 1, p)$ steps, we arrive at the full least-squares solution.

Suppose $\mathcal{A}_k$ is the active set of variables at the beginning of the kth step, and let $\beta_{\mathcal{A}_k}$ be the coefficient vector for these variables at this step; there will be $k - 1$ nonzero values, and the one just entered will be zero. If $\mathbf{r}_k = \mathbf{y} - \mathbf{X}_{\mathcal{A}_k} \beta_{\mathcal{A}_k}$ is the current residual, then the direction for this step is

$$\delta_k = (\mathbf{X}_{\mathcal{A}_k}^T \mathbf{X}_{\mathcal{A}_k})^{-1} \mathbf{X}_{\mathcal{A}_k}^T \mathbf{r}_k. \tag{3.55}$$

The coefficient profile then evolves as $\beta_{\mathcal{A}_k}(\alpha) = \beta_{\mathcal{A}_k} + \alpha \cdot \delta_k$. Exercise 3.23 verifies that the directions chosen in this fashion do what is claimed: keep the correlations tied and decreasing. If the fit vector at the beginning of this step is $\hat{\mathbf{f}}_k$, then it evolves as $\hat{\mathbf{f}}_k(\alpha) = \hat{\mathbf{f}}_k + \alpha \cdot \mathbf{u}_k$, where $\mathbf{u}_k = \mathbf{X}_{\mathcal{A}_k} \delta_k$ is the new fit direction. The name "least angle" arises from a geometrical interpretation of this process; $\mathbf{u}_k$ makes the smallest (and equal) angle with each of the predictors in $\mathcal{A}_k$ (Exercise 3.24). Figure 3.14 shows the absolute correlations decreasing and joining ranks with each step of the LAR algorithm, using simulated data.

By construction the coefficients in LAR change in a piecewise linear fashion. Figure 3.15 [left panel] shows the LAR coefficient profile evolving as a function of their L_1 arc length [2]. Note that we do not need to take small

---

[2]The L_1 arc-length of a differentiable curve $\beta(s)$ for $s \in [0, S]$ is given by $\mathrm{TV}(\beta, S) = \int_0^S \|\dot{\beta}(s)\|_1 ds$, where $\dot{\beta}(s) = \partial \beta(s)/\partial s$. For the piecewise-linear LAR coefficient profile, this amounts to summing the L_1 norms of the changes in coefficients from step to step.

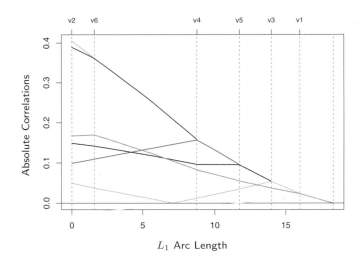

FIGURE 3.14. *Progression of the absolute correlations during each step of the LAR procedure, using a simulated data set with six predictors. The labels at the top of the plot indicate which variables enter the active set at each step. The step length are measured in units of L_1 arc length.*

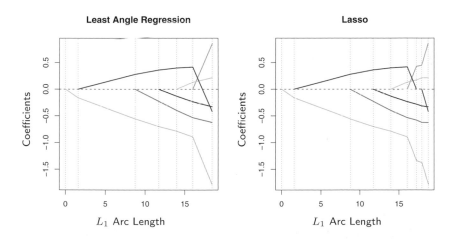

FIGURE 3.15. *Left panel shows the LAR coefficient profiles on the simulated data, as a function of the L_1 arc length. The right panel shows the Lasso profile. They are identical until the dark-blue coefficient crosses zero at an arc length of about 18.*

steps and recheck the correlations in step 3; using knowledge of the covariance of the predictors and the piecewise linearity of the algorithm, we can work out the exact step length at the beginning of each step (Exercise 3.25).

The right panel of Figure 3.15 shows the lasso coefficient profiles on the same data. They are almost identical to those in the left panel, and differ for the first time when the blue coefficient passes back through zero. For the prostate data, the LAR coefficient profile turns out to be identical to the lasso profile in Figure 3.10, which never crosses zero. These observations lead to a simple modification of the LAR algorithm that gives the entire lasso path, which is also piecewise-linear.

Algorithm 3.2a *Least Angle Regression: Lasso Modification.*

4a. If a non-zero coefficient hits zero, drop its variable from the active set of variables and recompute the current joint least squares direction.

The LAR(lasso) algorithm is extremely efficient, requiring the same order of computation as that of a single least squares fit using the p predictors. Least angle regression always takes p steps to get to the full least squares estimates. The lasso path can have more than p steps, although the two are often quite similar. Algorithm 3.2 with the lasso modification 3.2a is an efficient way of computing the solution to any lasso problem, especially when $p \gg N$. Osborne et al. (2000a) also discovered a piecewise-linear path for computing the lasso, which they called a *homotopy* algorithm.

We now give a heuristic argument for why these procedures are so similar. Although the LAR algorithm is stated in terms of correlations, if the input features are standardized, it is equivalent and easier to work with inner-products. Suppose $\mathcal{A}$ is the active set of variables at some stage in the algorithm, tied in their absolute inner-product with the current residuals $\mathbf{y} - \mathbf{X}\beta$. We can express this as

$$\mathbf{x}_j^T(\mathbf{y} - \mathbf{X}\beta) = \gamma \cdot s_j, \ \forall j \in \mathcal{A} \tag{3.56}$$

where $s_j \in \{-1, 1\}$ indicates the sign of the inner-product, and γ is the common value. Also $|\mathbf{x}_k^T(\mathbf{y} - \mathbf{X}\beta)| \leq \gamma \ \forall k \notin \mathcal{A}$. Now consider the lasso criterion (3.52), which we write in vector form

$$R(\beta) = \tfrac{1}{2}||\mathbf{y} - \mathbf{X}\beta||_2^2 + \lambda||\beta||_1. \tag{3.57}$$

Let $\mathcal{B}$ be the active set of variables in the solution for a given value of λ. For these variables $R(\beta)$ is differentiable, and the stationarity conditions give

$$\mathbf{x}_j^T(\mathbf{y} - \mathbf{X}\beta) = \lambda \cdot \text{sign}(\beta_j), \ \forall j \in \mathcal{B} \tag{3.58}$$

Comparing (3.58) with (3.56), we see that they are identical only if the sign of β_j matches the sign of the inner product. That is why the LAR algorithm and lasso start to differ when an active coefficient passes through zero; condition (3.58) is violated for that variable, and it is kicked out of the active set $\mathcal{B}$. Exercise 3.23 shows that these equations imply a piecewise-linear coefficient profile as λ decreases. The stationarity conditions for the non-active variables require that

$$|\mathbf{x}_k^T(\mathbf{y} - \mathbf{X}\beta)| \leq \lambda, \ \forall k \notin \mathcal{B}, \tag{3.59}$$

which again agrees with the LAR algorithm.

Figure 3.16 compares LAR and lasso to forward stepwise and stagewise regression. The setup is the same as in Figure 3.6 on page 59, except here $N = 100$ here rather than 300, so the problem is more difficult. We see that the more aggressive forward stepwise starts to overfit quite early (well before the 10 true variables can enter the model), and ultimately performs worse than the slower forward stagewise regression. The behavior of LAR and lasso is similar to that of forward stagewise regression. Incremental forward stagewise is similar to LAR and lasso, and is described in Section 3.8.1.

Degrees-of-Freedom Formula for LAR and Lasso

Suppose that we fit a linear model via the least angle regression procedure, stopping at some number of steps $k < p$, or equivalently using a lasso bound t that produces a constrained version of the full least squares fit. How many parameters, or "degrees of freedom" have we used?

Consider first a linear regression using a subset of k features. If this subset is prespecified in advance without reference to the training data, then the degrees of freedom used in the fitted model is defined to be k. Indeed, in classical statistics, the number of linearly independent parameters is what is meant by "degrees of freedom." Alternatively, suppose that we carry out a best subset selection to determine the "optimal" set of k predictors. Then the resulting model has k parameters, but in some sense we have used up more than k degrees of freedom.

We need a more general definition for the effective degrees of freedom of an adaptively fitted model. We define the degrees of freedom of the fitted vector $\hat{\mathbf{y}} = (\hat{y}_1, \hat{y}_2, \ldots, \hat{y}_N)$ as

$$\mathrm{df}(\hat{\mathbf{y}}) = \frac{1}{\sigma^2} \sum_{i=1}^{N} \mathrm{Cov}(\hat{y}_i, y_i). \tag{3.60}$$

Here $\mathrm{Cov}(\hat{y}_i, y_i)$ refers to the sampling covariance between the predicted value $\hat{y}_i$ and its corresponding outcome value y_i. This makes intuitive sense: the harder that we fit to the data, the larger this covariance and hence $\mathrm{df}(\hat{\mathbf{y}})$. Expression (3.60) is a useful notion of degrees of freedom, one that can be applied to any model prediction $\hat{\mathbf{y}}$. This includes models that are

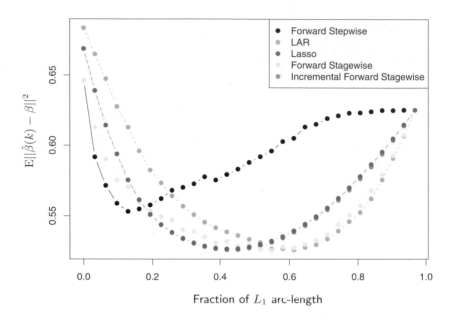

FIGURE 3.16. *Comparison of LAR and lasso with forward stepwise, forward stagewise (FS) and incremental forward stagewise (FS$_0$) regression. The setup is the same as in Figure 3.6, except $N = 100$ here rather than 300. Here the slower FS regression ultimately outperforms forward stepwise. LAR and lasso show similar behavior to FS and FS$_0$. Since the procedures take different numbers of steps (across simulation replicates and methods), we plot the MSE as a function of the fraction of total L_1 arc-length toward the least-squares fit.*

adaptively fitted to the training data. This definition is motivated and discussed further in Sections 7.4–7.6.

Now for a linear regression with k fixed predictors, it is easy to show that $\mathrm{df}(\hat{\mathbf{y}}) = k$. Likewise for ridge regression, this definition leads to the closed-form expression (3.50) on page 68: $\mathrm{df}(\hat{\mathbf{y}}) = \mathrm{tr}(\mathbf{S}_\lambda)$. In both these cases, (3.60) is simple to evaluate because the fit $\hat{\mathbf{y}} = \mathbf{H}_\lambda\mathbf{y}$ is linear in $\mathbf{y}$. If we think about definition (3.60) in the context of a best subset selection of size k, it seems clear that $\mathrm{df}(\hat{\mathbf{y}})$ will be larger than k, and this can be verified by estimating $\mathrm{Cov}(\hat{y}_i, y_i)/\sigma^2$ directly by simulation. However there is no closed form method for estimating $\mathrm{df}(\hat{\mathbf{y}})$ for best subset selection.

For LAR and lasso, something magical happens. These techniques are adaptive in a smoother way than best subset selection, and hence estimation of degrees of freedom is more tractable. Specifically it can be shown that after the kth step of the LAR procedure, the effective degrees of freedom of the fit vector is exactly k. Now for the lasso, the (modified) LAR procedure often takes more than p steps, since predictors can drop out. Hence the definition is a little different; for the lasso, at any stage $\mathrm{df}(\hat{\mathbf{y}})$ approximately equals the number of predictors in the model. While this approximation works reasonably well anywhere in the lasso path, for each k it works best at the *last* model in the sequence that contains k predictors. A detailed study of the degrees of freedom for the lasso may be found in Zou et al. (2007).

3.5 Methods Using Derived Input Directions

In many situations we have a large number of inputs, often very correlated. The methods in this section produce a small number of linear combinations Z_m, $m = 1, \ldots, M$ of the original inputs X_j, and the Z_m are then used in place of the X_j as inputs in the regression. The methods differ in how the linear combinations are constructed.

3.5.1 Principal Components Regression

In this approach the linear combinations Z_m used are the principal components as defined in Section 3.4.1 above.

Principal component regression forms the derived input columns $\mathbf{z}_m = \mathbf{X}v_m$, and then regresses $\mathbf{y}$ on $\mathbf{z}_1, \mathbf{z}_2, \ldots, \mathbf{z}_M$ for some $M \leq p$. Since the $\mathbf{z}_m$ are orthogonal, this regression is just a sum of univariate regressions:

$$\hat{\mathbf{y}}_{(M)}^{\mathrm{pcr}} = \bar{y}\mathbf{1} + \sum_{m=1}^{M} \hat{\theta}_m \mathbf{z}_m, \tag{3.61}$$

where $\hat{\theta}_m = \langle \mathbf{z}_m, \mathbf{y} \rangle / \langle \mathbf{z}_m, \mathbf{z}_m \rangle$. Since the $\mathbf{z}_m$ are each linear combinations of the original $\mathbf{x}_j$, we can express the solution (3.61) in terms of coefficients of the $\mathbf{x}_j$ (Exercise 3.13):

$$\hat{\beta}^{\mathrm{pcr}}(M) = \sum_{m=1}^{M} \hat{\theta}_m v_m. \tag{3.62}$$

As with ridge regression, principal components depend on the scaling of the inputs, so typically we first standardize them. Note that if $M = p$, we would just get back the usual least squares estimates, since the columns of $\mathbf{Z} = \mathbf{U}\mathbf{D}$ span the column space of $\mathbf{X}$. For $M < p$ we get a reduced regression. We see that principal components regression is very similar to ridge regression: both operate via the principal components of the input matrix. Ridge regression shrinks the coefficients of the principal components (Figure 3.17), shrinking more depending on the size of the corresponding eigenvalue; principal components regression discards the $p - M$ smallest eigenvalue components. Figure 3.17 illustrates this.

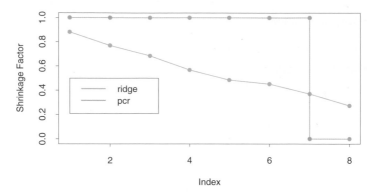

FIGURE 3.17. *Ridge regression shrinks the regression coefficients of the principal components, using shrinkage factors $d_j^2/(d_j^2 + \lambda)$ as in (3.47). Principal component regression truncates them. Shown are the shrinkage and truncation patterns corresponding to Figure 3.7, as a function of the principal component index.*

In Figure 3.7 we see that cross-validation suggests seven terms; the resulting model has the lowest test error in Table 3.3.

3.5.2 *Partial Least Squares*

This technique also constructs a set of linear combinations of the inputs for regression, but unlike principal components regression it uses $\mathbf{y}$ (in addition to $\mathbf{X}$) for this construction. Like principal component regression, partial least squares (PLS) is not scale invariant, so we assume that each $\mathbf{x}_j$ is standardized to have mean 0 and variance 1. PLS begins by computing $\hat{\varphi}_{1j} = \langle \mathbf{x}_j, \mathbf{y} \rangle$ for each j. From this we construct the derived input $\mathbf{z}_1 = \sum_j \hat{\varphi}_{1j} \mathbf{x}_j$, which is the first partial least squares direction. Hence in the construction of each $\mathbf{z}_m$, the inputs are weighted by the strength of their univariate effect on $\mathbf{y}$[3]. The outcome $\mathbf{y}$ is regressed on $\mathbf{z}_1$ giving coefficient $\hat{\theta}_1$, and then we orthogonalize $\mathbf{x}_1, \dots, \mathbf{x}_p$ with respect to $\mathbf{z}_1$. We continue this process, until $M \leq p$ directions have been obtained. In this manner, partial least squares produces a sequence of derived, orthogonal inputs or directions $\mathbf{z}_1, \mathbf{z}_2, \dots, \mathbf{z}_M$. As with principal-component regression, if we were to construct all $M = p$ directions, we would get back a solution equivalent to the usual least squares estimates; using $M < p$ directions produces a reduced regression. The procedure is described fully in Algorithm 3.3.

---

[3]Since the $\mathbf{x}_j$ are standardized, the first directions $\hat{\varphi}_{1j}$ are the univariate regression coefficients (up to an irrelevant constant); this is not the case for subsequent directions.

Algorithm 3.3 *Partial Least Squares.*

---

1. Standardize each $\mathbf{x}_j$ to have mean zero and variance one. Set $\hat{\mathbf{y}}^{(0)} = \bar{y}\mathbf{1}$, and $\mathbf{x}_j^{(0)} = \mathbf{x}_j$, $j = 1, \ldots, p$.

2. For $m = 1, 2, \ldots, p$

   (a) $\mathbf{z}_m = \sum_{j=1}^p \hat{\varphi}_{mj} \mathbf{x}_j^{(m-1)}$, where $\hat{\varphi}_{mj} = \langle \mathbf{x}_j^{(m-1)}, \mathbf{y} \rangle$.

   (b) $\hat{\theta}_m = \langle \mathbf{z}_m, \mathbf{y} \rangle / \langle \mathbf{z}_m, \mathbf{z}_m \rangle$.

   (c) $\hat{\mathbf{y}}^{(m)} = \hat{\mathbf{y}}^{(m-1)} + \hat{\theta}_m \mathbf{z}_m$.

   (d) Orthogonalize each $\mathbf{x}_j^{(m-1)}$ with respect to $\mathbf{z}_m$: $\mathbf{x}_j^{(m)} = \mathbf{x}_j^{(m-1)} - [\langle \mathbf{z}_m, \mathbf{x}_j^{(m-1)} \rangle / \langle \mathbf{z}_m, \mathbf{z}_m \rangle] \mathbf{z}_m$, $j = 1, 2, \ldots, p$.

3. Output the sequence of fitted vectors $\{\hat{\mathbf{y}}^{(m)}\}_1^p$. Since the $\{\mathbf{z}_\ell\}_1^m$ are linear in the original $\mathbf{x}_j$, so is $\hat{\mathbf{y}}^{(m)} = \mathbf{X}\hat{\beta}^{\text{pls}}(m)$. These linear coefficients can be recovered from the sequence of PLS transformations.

In the prostate cancer example, cross-validation chose $M = 2$ PLS directions in Figure 3.7. This produced the model given in the rightmost column of Table 3.3.

What optimization problem is partial least squares solving? Since it uses the response $\mathbf{y}$ to construct its directions, its solution path is a nonlinear function of $\mathbf{y}$. It can be shown (Exercise 3.15) that partial least squares seeks directions that have high variance *and* have high correlation with the response, in contrast to principal components regression which keys only on high variance (Stone and Brooks, 1990; Frank and Friedman, 1993). In particular, the mth principal component direction v_m solves:

$$\max_\alpha \text{Var}(\mathbf{X}\alpha) \tag{3.63}$$
$$\text{subject to } ||\alpha|| = 1, \ \alpha^T \mathbf{S} v_\ell = 0, \ \ell = 1, \ldots, m - 1,$$

where $\mathbf{S}$ is the sample covariance matrix of the $\mathbf{x}_j$. The conditions $\alpha^T \mathbf{S} v_\ell = 0$ ensures that $\mathbf{z}_m = \mathbf{X}\alpha$ is uncorrelated with all the previous linear combinations $\mathbf{z}_\ell = \mathbf{X} v_\ell$. The mth PLS direction $\hat{\varphi}_m$ solves:

$$\max_\alpha \text{Corr}^2(\mathbf{y}, \mathbf{X}\alpha) \text{Var}(\mathbf{X}\alpha) \tag{3.64}$$
$$\text{subject to } ||\alpha|| = 1, \ \alpha^T \mathbf{S}\hat{\varphi}_\ell = 0, \ \ell = 1, \ldots, m - 1.$$

Further analysis reveals that the variance aspect tends to dominate, and so partial least squares behaves much like ridge regression and principal components regression. We discuss this further in the next section.

If the input matrix $\mathbf{X}$ is orthogonal, then partial least squares finds the least squares estimates after $m = 1$ steps. Subsequent steps have no effect

since the $\hat{\varphi}_{mj}$ are zero for $m > 1$ (Exercise 3.14). It can also be shown that
the sequence of PLS coefficients for $m = 1, 2, \ldots, p$ represents the conjugate
gradient sequence for computing the least squares solutions (Exercise 3.18).

3.6   Discussion: A Comparison of the Selection and Shrinkage Methods

There are some simple settings where we can understand better the relationship between the different methods described above. Consider an example with two correlated inputs X_1 and X_2, with correlation ρ. We assume
that the true regression coefficients are $\beta_1 = 4$ and $\beta_2 = 2$. Figure 3.18
shows the coefficient profiles for the different methods, as their tuning parameters are varied. The top panel has $\rho = 0.5$, the bottom panel $\rho = -0.5$.
The tuning parameters for ridge and lasso vary over a continuous range,
while best subset, PLS and PCR take just two discrete steps to the least
squares solution. In the top panel, starting at the origin, ridge regression
shrinks the coefficients together until it finally converges to least squares.
PLS and PCR show similar behavior to ridge, although are discrete and
more extreme. Best subset overshoots the solution and then backtracks.
The behavior of the lasso is intermediate to the other methods. When the
correlation is negative (lower panel), again PLS and PCR roughly track
the ridge path, while all of the methods are more similar to one another.

It is interesting to compare the shrinkage behavior of these different
methods. Recall that ridge regression shrinks all directions, but shrinks
low-variance directions more. Principal components regression leaves M
high-variance directions alone, and discards the rest. Interestingly, it can
be shown that partial least squares also tends to shrink the low-variance
directions, but can actually inflate some of the higher variance directions.
This can make PLS a little unstable, and cause it to have slightly higher
prediction error compared to ridge regression. A full study is given in Frank
and Friedman (1993). These authors conclude that for minimizing prediction error, ridge regression is generally preferable to variable subset selection, principal components regression and partial least squares. However
the improvement over the latter two methods was only slight.

To summarize, PLS, PCR and ridge regression tend to behave similarly.
Ridge regression may be preferred because it shrinks smoothly, rather than
in discrete steps. Lasso falls somewhere between ridge regression and best
subset regression, and enjoys some of the properties of each.

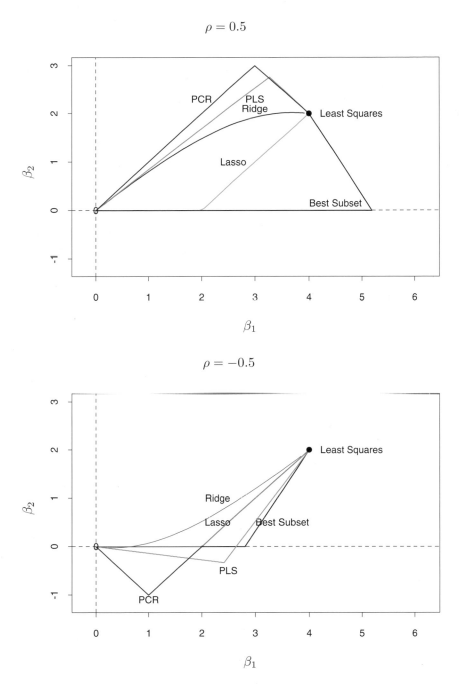

FIGURE 3.18. *Coefficient profiles from different methods for a simple problem: two inputs with correlation ± 0.5, and the true regression coefficients $\beta = (4, 2)$.*

3.7 Multiple Outcome Shrinkage and Selection

As noted in Section 3.2.4, the least squares estimates in a multiple-output linear model are simply the individual least squares estimates for each of the outputs.

To apply selection and shrinkage methods in the multiple output case, one could apply a univariate technique individually to each outcome or simultaneously to all outcomes. With ridge regression, for example, we could apply formula (3.44) to each of the K columns of the outcome matrix Y, using possibly different parameters λ, or apply it to all columns using the same value of λ. The former strategy would allow different amounts of regularization to be applied to different outcomes but require estimation of k separate regularization parameters $\lambda_1, \dots, \lambda_k$, while the latter would permit all k outputs to be used in estimating the sole regularization parameter λ.

Other more sophisticated shrinkage and selection strategies that exploit correlations in the different responses can be helpful in the multiple output case. Suppose for example that among the outputs we have

$$Y_k = f(X) + \varepsilon_k \tag{3.65}$$
$$Y_\ell = f(X) + \varepsilon_\ell; \tag{3.66}$$

i.e., (3.65) and (3.66) share the same structural part $f(X)$ in their models. It is clear in this case that we should pool our observations on Y_k and Y_l to estimate the common f.

Combining responses is at the heart of *canonical correlation analysis* (CCA), a data reduction technique developed for the multiple output case. Similar to PCA, CCA finds a sequence of uncorrelated linear combinations $\mathbf{X}v_m$, $m = 1, \dots, M$ of the $\mathbf{x}_j$, and a corresponding sequence of uncorrelated linear combinations $\mathbf{Y}u_m$ of the responses $\mathbf{y}_k$, such that the correlations

$$\mathrm{Corr}^2(\mathbf{Y}u_m, \mathbf{X}v_m) \tag{3.67}$$

are successively maximized. Note that at most $M = \min(K, p)$ directions can be found. The leading canonical response variates are those linear combinations (derived responses) best predicted by the $\mathbf{x}_j$; in contrast, the trailing canonical variates can be poorly predicted by the $\mathbf{x}_j$, and are candidates for being dropped. The CCA solution is computed using a generalized SVD of the sample cross-covariance matrix $\mathbf{Y}^T\mathbf{X}/N$ (assuming $\mathbf{Y}$ and $\mathbf{X}$ are centered; Exercise 3.20).

Reduced-rank regression (Izenman, 1975; van der Merwe and Zidek, 1980) formalizes this approach in terms of a regression model that explicitly pools information. Given an error covariance $\mathrm{Cov}(\varepsilon) = \mathbf{\Sigma}$, we solve the following restricted multivariate regression problem:

$$\hat{\mathbf{B}}^{\mathrm{rr}}(m) = \underset{\mathrm{rank}(\mathbf{B})=m}{\mathrm{argmin}} \sum_{i=1}^{N}(y_i - \mathbf{B}^T x_i)^T \boldsymbol{\Sigma}^{-1}(y_i - \mathbf{B}^T x_i). \tag{3.68}$$

With $\boldsymbol{\Sigma}$ replaced by the estimate $\mathbf{Y}^T\mathbf{Y}/N$, one can show (Exercise 3.21) that the solution is given by a CCA of $\mathbf{Y}$ and $\mathbf{X}$:

$$\hat{\mathbf{B}}^{\mathrm{rr}}(m) = \hat{\mathbf{B}}\mathbf{U}_m\mathbf{U}_m^-, \tag{3.69}$$

where $\mathbf{U}_m$ is the $K \times m$ sub-matrix of $\mathbf{U}$ consisting of the first m columns, and $\mathbf{U}$ is the $K \times M$ matrix of *left* canonical vectors $u_1, u_2, \ldots, u_M$. $\mathbf{U}_m^-$ is its generalized inverse. Writing the solution as

$$\hat{\mathbf{B}}^{\mathrm{rr}}(M) = (\mathbf{X}^T\mathbf{X})^{-1}\mathbf{X}^T(\mathbf{Y}\mathbf{U}_m)\mathbf{U}_m^-, \tag{3.70}$$

we see that reduced-rank regression performs a linear regression on the pooled response matrix $\mathbf{Y}\mathbf{U}_m$, and then maps the coefficients (and hence the fits as well) back to the original response space. The reduced-rank fits are given by

$$\begin{aligned}\hat{\mathbf{Y}}^{\mathrm{rr}}(m) &= \mathbf{X}(\mathbf{X}^T\mathbf{X})^{-1}\mathbf{X}^T\mathbf{Y}\mathbf{U}_m\mathbf{U}_m^- \\ &= \mathbf{H}\mathbf{Y}\mathbf{P}_m, \end{aligned} \tag{3.71}$$

where $\mathbf{H}$ is the usual linear regression projection operator, and $\mathbf{P}_m$ is the rank-m CCA response projection operator. Although a better estimate of $\boldsymbol{\Sigma}$ would be $(\mathbf{Y}-\mathbf{X}\hat{\mathbf{B}})^T(\mathbf{Y}-\mathbf{X}\hat{\mathbf{B}})/(N-pK)$, one can show that the solution remains the same (Exercise 3.22).

Reduced-rank regression borrows strength among responses by truncating the CCA. Breiman and Friedman (1997) explored with some success shrinkage of the canonical variates between $\mathbf{X}$ and $\mathbf{Y}$, a smooth version of *reduced rank* regression. Their proposal has the form (compare (3.69))

$$\hat{\mathbf{B}}^{\mathrm{c+w}} = \hat{\mathbf{B}}\mathbf{U}\boldsymbol{\Lambda}\mathbf{U}^{-1}, \tag{3.72}$$

where $\boldsymbol{\Lambda}$ is a diagonal shrinkage matrix (the "c+w" stands for "Curds and Whey," the name they gave to their procedure). Based on optimal prediction in the population setting, they show that $\boldsymbol{\Lambda}$ has diagonal entries

$$\lambda_m = \frac{c_m^2}{c_m^2 + \frac{p}{N}(1-c_m^2)}, \quad m = 1, \ldots, M, \tag{3.73}$$

where c_m is the mth canonical correlation coefficient. Note that as the ratio of the number of input variables to sample size p/N gets small, the shrinkage factors approach 1. Breiman and Friedman (1997) proposed modified versions of $\boldsymbol{\Lambda}$ based on training data and cross-validation, but the general form is the same. Here the fitted response has the form

$$\hat{\mathbf{Y}}^{\mathrm{c+w}} = \mathbf{H}\mathbf{Y}\mathbf{S}^{\mathrm{c+w}}, \tag{3.74}$$

where $\mathbf{S}^{\text{c+w}} = \mathbf{U}\boldsymbol{\Lambda}\mathbf{U}^{-1}$ is the response shrinkage operator.

Breiman and Friedman (1997) also suggested shrinking in both the Y space and X space. This leads to hybrid shrinkage models of the form

$$\hat{\mathbf{Y}}^{\text{ridge,c+w}} = \mathbf{A}_\lambda \mathbf{Y}\mathbf{S}^{\text{c+w}}, \tag{3.75}$$

where $\mathbf{A}_\lambda = \mathbf{X}(\mathbf{X}^T\mathbf{X} + \lambda\mathbf{I})^{-1}\mathbf{X}^T$ is the ridge regression shrinkage operator, as in (3.46) on page 66. Their paper and the discussions thereof contain many more details.

3.8 More on the Lasso and Related Path Algorithms

Since the publication of the LAR algorithm (Efron et al., 2004) there has been a lot of activity in developing algorithms for fitting regularization paths for a variety of different problems. In addition, L_1 regularization has taken on a life of its own, leading to the development of the field *compressed sensing* in the signal-processing literature. (Donoho, 2006a; Candes, 2006). In this section we discuss some related proposals and other path algorithms, starting off with a precursor to the LAR algorithm.

3.8.1 *Incremental Forward Stagewise Regression*

Here we present another LAR-like algorithm, this time focused on forward stagewise regression. Interestingly, efforts to understand a flexible nonlinear regression procedure (boosting) led to a new algorithm for linear models (LAR). In reading the first edition of this book and the forward stagewise

---

Algorithm 3.4 *Incremental Forward Stagewise Regression—FS_ϵ.*

---

1. Start with the residual $\mathbf{r}$ equal to $\mathbf{y}$ and $\beta_1, \beta_2, \ldots, \beta_p = 0$. All the predictors are standardized to have mean zero and unit norm.

2. Find the predictor $\mathbf{x}_j$ most correlated with $\mathbf{r}$

3. Update $\beta_j \leftarrow \beta_j + \delta_j$, where $\delta_j = \epsilon \cdot \text{sign}[\langle \mathbf{x}_j, \mathbf{r}\rangle]$ and $\epsilon > 0$ is a small step size, and set $\mathbf{r} \leftarrow \mathbf{r} - \delta_j\mathbf{x}_j$.

4. Repeat steps 2 and 3 many times, until the residuals are uncorrelated with all the predictors.

Algorithm 16.1 of Chapter 16[4], our colleague Brad Efron realized that with

---

[4]In the first edition, this was Algorithm 10.4 in Chapter 10.

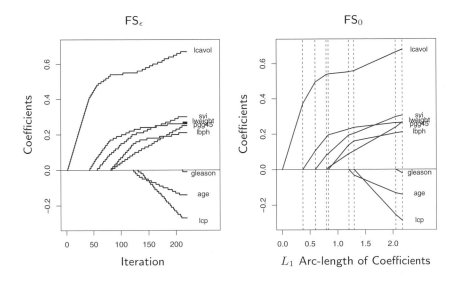

FIGURE 3.19. *Coefficient profiles for the prostate data. The left panel shows incremental forward stagewise regression with step size $\epsilon = 0.01$. The right panel shows the infinitesimal version FS_0 obtained letting $\epsilon \to 0$. This profile was fit by the modification 3.2b to the LAR Algorithm 3.2. In this example the FS_0 profiles are monotone, and hence identical to those of lasso and LAR.*

linear models, one could explicitly construct the piecewise-linear lasso paths of Figure 3.10. This led him to propose the LAR procedure of Section 3.4.4, as well as the incremental version of forward-stagewise regression presented here.

Consider the linear-regression version of the forward-stagewise boosting algorithm 16.1 proposed in Section 16.1 (page 608). It generates a coefficient profile by repeatedly updating (by a small amount ϵ) the coefficient of the variable most correlated with the current residuals. Algorithm 3.4 gives the details. Figure 3.19 (left panel) shows the progress of the algorithm on the prostate data with step size $\epsilon = 0.01$. If $\delta_j = \langle \mathbf{x}_j, \mathbf{r} \rangle$ (the least-squares coefficient of the residual on jth predictor), then this is exactly the usual forward stagewise procedure (FS) outlined in Section 3.3.3.

Here we are mainly interested in small values of ϵ. Letting $\epsilon \to 0$ gives the right panel of Figure 3.19, which in this case is identical to the lasso path in Figure 3.10. We call this limiting procedure *infinitesimal forward stagewise regression* or FS_0. This procedure plays an important role in non-linear, adaptive methods like boosting (Chapters 10 and 16) and is the version of incremental forward stagewise regression that is most amenable to theoretical analysis. Bühlmann and Hothorn (2007) refer to the same procedure as "L2boost", because of its connections to boosting.

88   3. Linear Methods for Regression

Efron originally thought that the LAR Algorithm 3.2 was an implementation of FS_0, allowing each tied predictor a chance to update their coefficients in a balanced way, while remaining tied in correlation. However, he then realized that the LAR least-squares fit amongst the tied predictors can result in coefficients moving in the *opposite* direction to their correlation, which cannot happen in Algorithm 3.4. The following modification of the LAR algorithm implements FS_0:

Algorithm 3.2b *Least Angle Regression: FS_0 Modification.*

---

4. Find the new direction by solving the constrained least squares problem

$$\min_b ||\mathbf{r} - \mathbf{X}_{\mathcal{A}} b||_2^2 \text{ subject to } b_j s_j \geq 0, \ j \in \mathcal{A},$$

where s_j is the sign of $\langle \mathbf{x}_j, \mathbf{r} \rangle$.

The modification amounts to a non-negative least squares fit, keeping the signs of the coefficients the same as those of the correlations. One can show that this achieves the optimal balancing of infinitesimal "update turns" for the variables tied for maximal correlation (Hastie et al., 2007). Like lasso, the entire FS_0 path can be computed very efficiently via the LAR algorithm.

As a consequence of these results, if the LAR profiles are monotone non-increasing or non-decreasing, as they are in Figure 3.19, then all three methods—LAR, lasso, and FS_0—give identical profiles. If the profiles are not monotone but do not cross the zero axis, then LAR and lasso are identical.

Since FS_0 is different from the lasso, it is natural to ask if it optimizes a criterion. The answer is more complex than for lasso; the FS_0 coefficient profile is the solution to a differential equation. While the lasso makes optimal progress in terms of reducing the residual sum-of-squares per unit increase in L_1-norm of the coefficient vector β, FS_0 is optimal per unit increase in L_1 arc-length traveled along the coefficient path. Hence its coefficient path is discouraged from changing directions too often.

FS_0 is more constrained than lasso, and in fact can be viewed as a monotone version of the lasso; see Figure 16.3 on page 614 for a dramatic example. FS_0 may be useful in $p \gg N$ situations, where its coefficient profiles are much smoother and hence have less variance than those of lasso. More details on FS_0 are given in Section 16.2.3 and Hastie et al. (2007). Figure 3.16 includes FS_0 where its performance is very similar to that of the lasso.

3.8.2 Piecewise-Linear Path Algorithms

The least angle regression procedure exploits the piecewise linear nature of the lasso solution paths. It has led to similar "path algorithms" for other regularized problems. Suppose we solve

$$\hat{\beta}(\lambda) = \text{argmin}_\beta \left[R(\beta) + \lambda J(\beta) \right], \tag{3.76}$$

with

$$R(\beta) = \sum_{i=1}^{N} L(y_i, \beta_0 + \sum_{j=1}^{p} x_{ij}\beta_j), \tag{3.77}$$

where both the loss function L and the penalty function J are convex. Then the following are sufficient conditions for the solution path $\hat{\beta}(\lambda)$ to be piecewise linear (Rosset and Zhu, 2007):

1. R is quadratic or piecewise-quadratic as a function of β, and

2. J is piecewise linear in β.

This also implies (in principle) that the solution path can be efficiently computed. Examples include squared- and absolute-error loss, "Huberized" losses, and the L_1, L_∞ penalties on β. Another example is the "hinge loss" function used in the support vector machine. There the loss is piecewise linear, and the penalty is quadratic. Interestingly, this leads to a piecewise-linear path algorithm in the *dual space*; more details are given in Section 12.3.5.

3.8.3 The Dantzig Selector

Candes and Tao (2007) proposed the following criterion:

$$\min_\beta ||\beta||_1 \text{ subject to } ||\mathbf{X}^T(\mathbf{y} - \mathbf{X}\beta)||_\infty \leq s. \tag{3.78}$$

They call the solution the *Dantzig selector* (DS). It can be written equivalently as

$$\min_\beta ||\mathbf{X}^T(\mathbf{y} - \mathbf{X}\beta)||_\infty \text{ subject to } ||\beta||_1 \leq t. \tag{3.79}$$

Here $|| \cdot ||_\infty$ denotes the L_∞ norm, the maximum absolute value of the components of the vector. In this form it resembles the lasso, replacing squared error loss by the maximum absolute value of its gradient. Note that as t gets large, both procedures yield the least squares solution if $N < p$. If $p \geq N$, they both yield the least squares solution with minimum L_1 norm. However for smaller values of t, the DS procedure produces a different path of solutions than the lasso.

Candes and Tao (2007) show that the solution to DS is a linear programming problem; hence the name Dantzig selector, in honor of the late

George Dantzig, the inventor of the simplex method for linear programming. They also prove a number of interesting mathematical properties for the method, related to its ability to recover an underlying sparse coefficient vector. These same properties also hold for the lasso, as shown later by Bickel et al. (2008).

Unfortunately the operating properties of the DS method are somewhat unsatisfactory. The method seems similar in spirit to the lasso, especially when we look at the lasso's stationary conditions (3.58). Like the LAR algorithm, the lasso maintains the same inner product (and correlation) with the current residual for all variables in the active set, and moves their coefficients to optimally decrease the residual sum of squares. In the process, this common correlation is decreased monotonically (Exercise 3.23), and at all times this correlation is larger than that for non-active variables. The Dantzig selector instead tries to minimize the maximum inner product of the current residual with all the predictors. Hence it can achieve a smaller maximum than the lasso, but in the process a curious phenomenon can occur. If the size of the active set is m, there will be m variables tied with maximum correlation. However, these need not coincide with the active set! Hence it can include a variable in the model that has smaller correlation with the current residual than some of the excluded variables (Efron et al., 2007). This seems unreasonable and may be responsible for its sometimes inferior prediction accuracy. Efron et al. (2007) also show that DS can yield extremely erratic coefficient paths as the regularization parameter s is varied.

3.8.4 The Grouped Lasso

In some problems, the predictors belong to pre-defined groups; for example genes that belong to the same biological pathway, or collections of indicator (dummy) variables for representing the levels of a categorical predictor. In this situation it may be desirable to shrink and select the members of a group together. The *grouped lasso* is one way to achieve this. Suppose that the p predictors are divided into L groups, with p_ℓ the number in group ℓ. For ease of notation, we use a matrix $\mathbf{X}_\ell$ to represent the predictors corresponding to the ℓth group, with corresponding coefficient vector β_ℓ. The grouped-lasso minimizes the convex criterion

$$\min_{\beta \in \mathbb{R}^p} \left(||\mathbf{y} - \beta_0 \mathbf{1} - \sum_{\ell=1}^{L} \mathbf{X}_\ell \beta_\ell ||_2^2 + \lambda \sum_{\ell=1}^{L} \sqrt{p_\ell} ||\beta_\ell||_2 \right), \qquad (3.80)$$

where the $\sqrt{p_\ell}$ terms accounts for the varying group sizes, and $|| \cdot ||_2$ is the Euclidean norm (not squared). Since the Euclidean norm of a vector β_ℓ is zero only if all of its components are zero, this procedure encourages sparsity at both the group and individual levels. That is, for some values of λ, an entire group of predictors may drop out of the model. This procedure was proposed by Bakin (1999) and Lin and Zhang (2006), and studied and generalized by Yuan and Lin (2007). Generalizations include more general L_2 norms $||\eta||_K = (\eta^T K \eta)^{1/2}$, as well as allowing overlapping groups of predictors (Zhao et al., 2008). There are also connections to methods for fitting sparse additive models (Lin and Zhang, 2006; Ravikumar et al., 2008).

3.8.5 Further Properties of the Lasso

A number of authors have studied the ability of the lasso and related procedures to recover the correct model, as N and p grow. Examples of this work include Knight and Fu (2000), Greenshtein and Ritov (2004), Tropp (2004), Donoho (2006b), Meinshausen (2007), Meinshausen and Bühlmann (2006), Tropp (2006), Zhao and Yu (2006), Wainwright (2006), and Bunea et al. (2007). For example Donoho (2006b) focuses on the $p > N$ case and considers the lasso solution as the bound t gets large. In the limit this gives the solution with minimum L_1 norm among all models with zero training error. He shows that under certain assumptions on the model matrix $\mathbf{X}$, if the true model is sparse, this solution identifies the correct predictors with high probability.

Many of the results in this area assume a condition on the model matrix of the form

$$\max_{j \in \mathcal{S}^c} ||\mathbf{x}_j^T \mathbf{X}_\mathcal{S} (\mathbf{X}_\mathcal{S}^T \mathbf{X}_\mathcal{S})^{-1}||_1 < (1 - \epsilon) \text{ for some } \epsilon \in (0, 1]. \qquad (3.81)$$

Here $\mathcal{S}$ indexes the subset of features with non-zero coefficients in the true underlying model, and $\mathbf{X}_\mathcal{S}$ are the columns of $\mathbf{X}$ corresponding to those features. Similarly $\mathcal{S}^c$ are the features with true coefficients equal to zero, and $\mathbf{X}_{\mathcal{S}^c}$ the corresponding columns. This says that the least squares coefficients for the columns of $\mathbf{X}_{\mathcal{S}^c}$ on $\mathbf{X}_\mathcal{S}$ are not too large, that is, the "good" variables $\mathcal{S}$ are not too highly correlated with the nuisance variables $\mathcal{S}^c$.

Regarding the coefficients themselves, the lasso shrinkage causes the estimates of the non-zero coefficients to be biased towards zero, and in general they are not consistent[5]. One approach for reducing this bias is to run the lasso to identify the set of non-zero coefficients, and then fit an unrestricted linear model to the selected set of features. This is not always feasible, if the selected set is large. Alternatively, one can use the lasso to select the set of non-zero predictors, and then apply the lasso again, but using only the selected predictors from the first step. This is known as the *relaxed lasso* (Meinshausen, 2007). The idea is to use cross-validation to estimate the initial penalty parameter for the lasso, and then again for a second penalty parameter applied to the selected set of predictors. Since

---

[5]Statistical consistency means as the sample size grows, the estimates converge to the true values.

the variables in the second step have less "competition" from noise variables, cross-validation will tend to pick a smaller value for λ, and hence their coefficients will be shrunken less than those in the initial estimate.

Alternatively, one can modify the lasso penalty function so that larger coefficients are shrunken less severely; the *smoothly clipped absolute deviation* (SCAD) penalty of Fan and Li (2005) replaces $\lambda|\beta|$ by $J_a(\beta, \lambda)$, where

$$\frac{dJ_a(\beta, \lambda)}{d\beta} = \lambda \cdot \text{sign}(\beta)\left[I(|\beta| \leq \lambda) + \frac{(a\lambda - |\beta|)_+}{(a-1)\lambda}I(|\beta| > \lambda)\right] \quad (3.82)$$

for some $a \geq 2$. The second term in square-braces reduces the amount of shrinkage in the lasso for larger values of β, with ultimately no shrinkage as $a \to \infty$. Figure 3.20 shows the SCAD penalty, along with the lasso and

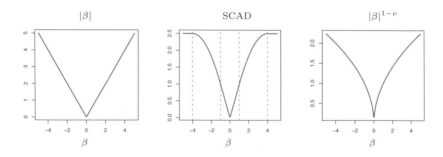

FIGURE 3.20. *The lasso and two alternative non-convex penalties designed to penalize large coefficients less. For SCAD we use $\lambda = 1$ and $a = 4$, and $\nu = \frac{1}{2}$ in the last panel.*

$|\beta|^{1-\nu}$. However this criterion is non-convex, which is a drawback since it makes the computation much more difficult. The *adaptive lasso* (Zou, 2006) uses a weighted penalty of the form $\sum_{j=1}^{p} w_j|\beta_j|$ where $w_j = 1/|\hat{\beta}_j|^\nu$, $\hat{\beta}_j$ is the ordinary least squares estimate and $\nu > 0$. This is a practical approximation to the $|\beta|^q$ penalties ($q = 1 - \nu$ here) discussed in Section 3.4.3. The adaptive lasso yields consistent estimates of the parameters while retaining the attractive convexity property of the lasso.

3.8.6 *Pathwise Coordinate Optimization*

An alternate approach to the LARS algorithm for computing the lasso solution is simple coordinate descent. This idea was proposed by Fu (1998) and Daubechies et al. (2004), and later studied and generalized by Friedman et al. (2007), Wu and Lange (2008) and others. The idea is to fix the penalty parameter λ in the Lagrangian form (3.52) and optimize successively over each parameter, holding the other parameters fixed at their current values.

Suppose the predictors are all standardized to have mean zero and unit norm. Denote by $\tilde{\beta}_k(\lambda)$ the current estimate for β_k at penalty parameter

λ. We can rearrange (3.52) to isolate β_j,

$$R(\tilde{\beta}(\lambda), \beta_j) = \frac{1}{2} \sum_{i=1}^{N} \left(y_i - \sum_{k \neq j} x_{ik} \tilde{\beta}_k(\lambda) - x_{ij} \beta_j \right)^2 + \lambda \sum_{k \neq j} |\tilde{\beta}_k(\lambda)| + \lambda |\beta_j|,$$

(3.83)

where we have suppressed the intercept and introduced a factor $\frac{1}{2}$ for convenience. This can be viewed as a univariate lasso problem with response variable the partial residual $y_i - \tilde{y}_i^{(j)} = y_i - \sum_{k \neq j} x_{ik} \tilde{\beta}_k(\lambda)$. This has an explicit solution, resulting in the update

$$\tilde{\beta}_j(\lambda) \leftarrow S\left(\sum_{i=1}^{N} x_{ij} (y_i - \tilde{y}_i^{(j)}), \lambda \right).$$

(3.84)

Here $S(t, \lambda) = \text{sign}(t)(|t| - \lambda)_+$ is the soft-thresholding operator in Table 3.4 on page 71. The first argument to $S(\cdot)$ is the simple least-squares coefficient of the partial residual on the standardized variable x_{ij}. Repeated iteration of (3.84)—cycling through each variable in turn until convergence—yields the lasso estimate $\hat{\beta}(\lambda)$.

We can also use this simple algorithm to efficiently compute the lasso solutions at a grid of values of λ. We start with the smallest value $\lambda_{\max}$ for which $\hat{\beta}(\lambda_{\max}) = 0$, decrease it a little and cycle through the variables until convergence. Then λ is decreased again and the process is repeated, using the previous solution as a "warm start" for the new value of λ. This can be faster than the LARS algorithm, especially in large problems. A key to its speed is the fact that the quantities in (3.84) can be updated quickly as j varies, and often the update is to leave $\tilde{\beta}_j = 0$. On the other hand, it delivers solutions over a grid of λ values, rather than the entire solution path. The same kind of algorithm can be applied to the elastic net, the grouped lasso and many other models in which the penalty is a sum of functions of the individual parameters (Friedman et al., 2010). It can also be applied, with some substantial modifications, to the fused lasso (Section 18.4.2); details are in Friedman et al. (2007).

3.9 Computational Considerations

Least squares fitting is usually done via the Cholesky decomposition of the matrix $\mathbf{X}^T \mathbf{X}$ or a QR decomposition of $\mathbf{X}$. With N observations and p features, the Cholesky decomposition requires $p^3 + Np^2/2$ operations, while the QR decomposition requires Np^2 operations. Depending on the relative size of N and p, the Cholesky can sometimes be faster; on the other hand, it can be less numerically stable (Lawson and Hansen, 1974). Computation of the lasso via the LAR algorithm has the same order of computation as a least squares fit.

Bibliographic Notes

Linear regression is discussed in many statistics books, for example, Seber (1984), Weisberg (1980) and Mardia et al. (1979). Ridge regression was introduced by Hoerl and Kennard (1970), while the lasso was proposed by Tibshirani (1996). Around the same time, lasso-type penalties were proposed in the *basis pursuit* method for signal processing (Chen et al., 1998). The least angle regression procedure was proposed in Efron et al. (2004); related to this is the earlier homotopy procedure of Osborne et al. (2000a) and Osborne et al. (2000b). Their algorithm also exploits the piecewise linearity used in the LAR/lasso algorithm, but lacks its transparency. The criterion for the forward stagewise criterion is discussed in Hastie et al. (2007). Park and Hastie (2007) develop a path algorithm similar to least angle regression for generalized regression models. Partial least squares was introduced by Wold (1975). Comparisons of shrinkage methods may be found in Copas (1983) and Frank and Friedman (1993).

Exercises

Ex. 3.1 Show that the F statistic (3.13) for dropping a single coefficient from a model is equal to the square of the corresponding z-score (3.12).

Ex. 3.2 Given data on two variables X and Y, consider fitting a cubic polynomial regression model $f(X) = \sum_{j=0}^{3} \beta_j X^j$. In addition to plotting the fitted curve, you would like a 95% confidence band about the curve. Consider the following two approaches:

1. At each point x_0, form a 95% confidence interval for the linear function $a^T \beta = \sum_{j=0}^{3} \beta_j x_0^j$.

2. Form a 95% confidence set for β as in (3.15), which in turn generates confidence intervals for $f(x_0)$.

How do these approaches differ? Which band is likely to be wider? Conduct a small simulation experiment to compare the two methods.

Ex. 3.3 Gauss–Markov theorem:

(a) Prove the Gauss–Markov theorem: the least squares estimate of a parameter $a^T \beta$ has variance no bigger than that of any other linear unbiased estimate of $a^T \beta$ (Section 3.2.2).

(b) The matrix inequality $\mathbf{B} \preceq \mathbf{A}$ holds if $\mathbf{A} - \mathbf{B}$ is positive semidefinite. Show that if $\hat{\mathbf{V}}$ is the variance-covariance matrix of the least squares estimate of β and $\tilde{\mathbf{V}}$ is the variance-covariance matrix of any other linear unbiased estimate, then $\hat{\mathbf{V}} \preceq \tilde{\mathbf{V}}$.

Ex. 3.4 Show how the vector of least squares coefficients can be obtained from a single pass of the Gram–Schmidt procedure (Algorithm 3.1). Represent your solution in terms of the QR decomposition of $\mathbf{X}$.

Ex. 3.5 Consider the ridge regression problem (3.41). Show that this problem is equivalent to the problem

$$\hat{\beta}^c = \underset{\beta^c}{\text{argmin}} \left\{ \sum_{i=1}^{N} [y_i - \beta_0^c - \sum_{j=1}^{p} (x_{ij} - \bar{x}_j)\beta_j^c]^2 + \lambda \sum_{j=1}^{p} \beta_j^{c2} \right\}. \quad (3.85)$$

Give the correspondence between β^c and the original β in (3.41). Characterize the solution to this modified criterion. Show that a similar result holds for the lasso.

Ex. 3.6 Show that the ridge regression estimate is the mean (and mode) of the posterior distribution, under a Gaussian prior $\beta \sim N(0, \tau \mathbf{I})$, and Gaussian sampling model $\mathbf{y} \sim N(\mathbf{X}\beta, \sigma^2 \mathbf{I})$. Find the relationship between the regularization parameter λ in the ridge formula, and the variances τ and σ^2.

Ex. 3.7 Assume $y_i \sim N(\beta_0 + x_i^T \beta, \sigma^2), i = 1, 2, \ldots, N$, and the parameters β_j, $j = 1, \ldots, p$ are each distributed as $N(0, \tau^2)$, independently of one another. Assuming σ^2 and τ^2 are known, and β_0 is not governed by a prior (or has a flat improper prior), show that the (minus) log-posterior density of β is proportional to $\sum_{i=1}^{N} (y_i - \beta_0 - \sum_j x_{ij}\beta_j)^2 + \lambda \sum_{j=1}^{p} \beta_j^2$ where $\lambda = \sigma^2 / \tau^2$.

Ex. 3.8 Consider the QR decomposition of the uncentered $N \times (p + 1)$ matrix $\mathbf{X}$ (whose first column is all ones), and the SVD of the $N \times p$ centered matrix $\tilde{\mathbf{X}}$. Show that $\mathbf{Q}_2$ and $\mathbf{U}$ span the same subspace, where $\mathbf{Q}_2$ is the sub-matrix of $\mathbf{Q}$ with the first column removed. Under what circumstances will they be the same, up to sign flips?

Ex. 3.9 *Forward stepwise regression.* Suppose we have the QR decomposition for the $N \times q$ matrix $\mathbf{X}_1$ in a multiple regression problem with response $\mathbf{y}$, and we have an additional $p - q$ predictors in the matrix $\mathbf{X}_2$. Denote the current residual by $\mathbf{r}$. We wish to establish which one of these additional variables will reduce the residual-sum-of squares the most when included with those in $\mathbf{X}_1$. Describe an efficient procedure for doing this.

Ex. 3.10 *Backward stepwise regression.* Suppose we have the multiple regression fit of $\mathbf{y}$ on $\mathbf{X}$, along with the standard errors and Z-scores as in Table 3.2. We wish to establish which variable, when dropped, will increase the residual sum-of-squares the least. How would you do this?

Ex. 3.11 Show that the solution to the multivariate linear regression problem (3.40) is given by (3.39). What happens if the covariance matrices $\mathbf{\Sigma}_i$ are different for each observation?

Ex. 3.12 Show that the ridge regression estimates can be obtained by ordinary least squares regression on an augmented data set. We augment the centered matrix $\mathbf{X}$ with p additional rows $\sqrt{\lambda}\mathbf{I}$, and augment $\mathbf{y}$ with p zeros. By introducing artificial data having response value zero, the fitting procedure is forced to shrink the coefficients toward zero. This is related to the idea of *hints* due to Abu-Mostafa (1995), where model constraints are implemented by adding artificial data examples that satisfy them.

Ex. 3.13 Derive the expression (3.62), and show that $\hat{\beta}^{\text{pcr}}(p) = \hat{\beta}^{\text{ls}}$.

Ex. 3.14 Show that in the orthogonal case, PLS stops after $m = 1$ steps, because subsequent $\hat{\varphi}_{mj}$ in step 2 in Algorithm 3.3 are zero.

Ex. 3.15 Verify expression (3.64), and hence show that the partial least squares directions are a compromise between the ordinary regression coefficient and the principal component directions.

Ex. 3.16 Derive the entries in Table 3.4, the explicit forms for estimators in the orthogonal case.

Ex. 3.17 Repeat the analysis of Table 3.3 on the spam data discussed in Chapter 1.

Ex. 3.18 Read about conjugate gradient algorithms (Murray et al., 1981, for example), and establish a connection between these algorithms and partial least squares.

Ex. 3.19 Show that $\|\hat{\beta}^{\text{ridge}}\|$ increases as its tuning parameter $\lambda \to 0$. Does the same property hold for the lasso and partial least squares estimates? For the latter, consider the "tuning parameter" to be the successive steps in the algorithm.

Ex. 3.20 Consider the canonical-correlation problem (3.67). Show that the leading pair of canonical variates u_1 and v_1 solve the problem

$$\max_{\substack{u^T(\mathbf{Y}^T\mathbf{Y})u=1 \\ v_T(\mathbf{X}^T\mathbf{X})v=1}} u^T(\mathbf{Y}^T\mathbf{X})v, \tag{3.86}$$

a generalized SVD problem. Show that the solution is given by $u_1 = (\mathbf{Y}^T\mathbf{Y})^{-\frac{1}{2}}u_1^*$, and $v_1 = (\mathbf{X}^T\mathbf{X})^{-\frac{1}{2}}v_1^*$, where u_1^* and v_1^* are the leading left and right singular vectors in

$$(\mathbf{Y}^T\mathbf{Y})^{-\frac{1}{2}}(\mathbf{Y}^T\mathbf{X})(\mathbf{X}^T\mathbf{X})^{-\frac{1}{2}} = \mathbf{U}^*\mathbf{D}^*\mathbf{V}^{*T}. \tag{3.87}$$

Show that the entire sequence u_m, v_m, $m = 1, \ldots, \min(K, p)$ is also given by (3.87).

Ex. 3.21 Show that the solution to the reduced-rank regression problem (3.68), with $\boldsymbol{\Sigma}$ estimated by $\mathbf{Y}^T\mathbf{Y}/N$, is given by (3.69). *Hint:* Transform

$\mathbf{Y}$ to $\mathbf{Y}^* = \mathbf{Y}\mathbf{\Sigma}^{-\frac{1}{2}}$, and solved in terms of the canonical vectors u_m^*. Show that $\mathbf{U}_m = \mathbf{\Sigma}^{-\frac{1}{2}}\mathbf{U}_m^*$, and a generalized inverse is $\mathbf{U}_m^- = \mathbf{U}_m^{*T}\mathbf{\Sigma}^{\frac{1}{2}}$.

Ex. 3.22 Show that the solution in Exercise 3.21 does not change if $\mathbf{\Sigma}$ is estimated by the more natural quantity $(\mathbf{Y} - \mathbf{X}\hat{\mathbf{B}})^T(\mathbf{Y} - \mathbf{X}\hat{\mathbf{B}})/(N - pK)$.

Ex. 3.23 Consider a regression problem with all variables and response having mean zero and standard deviation one. Suppose also that each variable has identical absolute correlation with the response:

$$\frac{1}{N}|\langle \mathbf{x}_j, \mathbf{y} \rangle| = \lambda, \ j = 1, \dots, p.$$

Let $\hat{\beta}$ be the least-squares coefficient of $\mathbf{y}$ on $\mathbf{X}$, and let $\mathbf{u}(\alpha) = \alpha\mathbf{X}\hat{\beta}$ for $\alpha \in [0, 1]$ be the vector that moves a fraction α toward the least squares fit $\mathbf{u}$. Let RSS be the residual sum-of-squares from the full least squares fit.

(a) Show that

$$\frac{1}{N}|\langle \mathbf{x}_j, \mathbf{y} - \mathbf{u}(\alpha) \rangle| = (1 - \alpha)\lambda, \ j = 1, \dots, p,$$

and hence the correlations of each $\mathbf{x}_j$ with the residuals remain equal in magnitude as we progress toward $\mathbf{u}$.

(b) Show that these correlations are all equal to

$$\lambda(\alpha) = \frac{(1 - \alpha)}{\sqrt{(1 - \alpha)^2 + \frac{\alpha(2-\alpha)}{N} \cdot RSS}} \cdot \lambda,$$

and hence they decrease monotonically to zero.

(c) Use these results to show that the LAR algorithm in Section 3.4.4 keeps the correlations tied and monotonically decreasing, as claimed in (3.55).

Ex. 3.24 *LAR directions.* Using the notation around equation (3.55) on page 74, show that the LAR direction makes an equal angle with each of the predictors in $\mathcal{A}_k$.

Ex. 3.25 *LAR look-ahead (Efron et al., 2004, Sec. 2).* Starting at the beginning of the kth step of the LAR algorithm, derive expressions to identify the next variable to enter the active set at step $k+1$, and the value of α at which this occurs (using the notation around equation (3.55) on page 74).

Ex. 3.26 Forward stepwise regression enters the variable at each step that most reduces the residual sum-of-squares. LAR adjusts variables that have the most (absolute) correlation with the current residuals. Show that these two entry criteria are not necessarily the same. [Hint: let $\mathbf{x}_{j.\mathcal{A}}$ be the jth variable, linearly adjusted for all the variables currently in the model. Show that the first criterion amounts to identifying the j for which $\mathrm{Cor}(\mathbf{x}_{j.\mathcal{A}}, \mathbf{r})$ is largest in magnitude.

Ex. 3.27 *Lasso and LAR*: Consider the lasso problem in Lagrange multiplier form: with $L(\beta) = \frac{1}{2}\sum_i (y_i - \sum_j x_{ij}\beta_j)^2$, we minimize

$$L(\beta) + \lambda \sum_j |\beta_j| \tag{3.88}$$

for fixed $\lambda > 0$.

(a) Setting $\beta_j = \beta_j^+ - \beta_j^-$ with $\beta_j^+, \beta_j^- \geq 0$, expression (3.88) becomes $L(\beta) + \lambda \sum_j (\beta_j^+ + \beta_j^-)$. Show that the Lagrange dual function is

$$L(\beta) + \lambda \sum_j (\beta_j^+ + \beta_j^-) - \sum_j \lambda_j^+ \beta_j^+ - \sum_j \lambda_j^- \beta_j^- \tag{3.89}$$

and the Karush–Kuhn–Tucker optimality conditions are

$$\begin{aligned}
\nabla L(\beta)_j + \lambda - \lambda_j^+ &= 0 \\
-\nabla L(\beta)_j + \lambda - \lambda_j^- &= 0 \\
\lambda_j^+ \beta_j^+ &= 0 \\
\lambda_j^- \beta_j^- &= 0,
\end{aligned}$$

along with the non-negativity constraints on the parameters and all the Lagrange multipliers.

(b) Show that $|\nabla L(\beta)_j| \leq \lambda\ \forall j$, and that the KKT conditions imply one of the following three scenarios:

$$\begin{aligned}
\lambda = 0 \quad &\Rightarrow \quad \nabla L(\beta)_j = 0\ \forall j \\
\beta_j^+ > 0,\ \lambda > 0 \quad &\Rightarrow \quad \lambda_j^+ = 0,\ \nabla L(\beta)_j = -\lambda < 0,\ \beta_j^- = 0 \\
\beta_j^- > 0,\ \lambda > 0 \quad &\Rightarrow \quad \lambda_j^- = 0,\ \nabla L(\beta)_j = \lambda > 0,\ \beta_j^+ = 0.
\end{aligned}$$

Hence show that for any "active" predictor having $\beta_j \neq 0$, we must have $\nabla L(\beta)_j = -\lambda$ if $\beta_j > 0$, and $\nabla L(\beta)_j = \lambda$ if $\beta_j < 0$. Assuming the predictors are standardized, relate λ to the correlation between the jth predictor and the current residuals.

(c) Suppose that the set of active predictors is unchanged for $\lambda_0 \geq \lambda \geq \lambda_1$. Show that there is a vector γ_0 such that

$$\hat{\beta}(\lambda) = \hat{\beta}(\lambda_0) - (\lambda - \lambda_0)\gamma_0 \tag{3.90}$$

Thus the lasso solution path is linear as λ ranges from λ_0 to λ_1 (Efron et al., 2004; Rosset and Zhu, 2007).

Ex. 3.28 Suppose for a given t in (3.51), the fitted lasso coefficient for variable X_j is $\hat{\beta}_j = a$. Suppose we augment our set of variables with an identical copy $X_j^* = X_j$. Characterize the effect of this exact collinearity by describing the set of solutions for $\hat{\beta}_j$ and $\hat{\beta}_j^*$, using the same value of t.

Ex. 3.29 Suppose we run a ridge regression with parameter λ on a single variable X, and get coefficient a. We now include an exact copy $X^* = X$, and refit our ridge regression. Show that both coefficients are identical, and derive their value. Show in general that if m copies of a variable X_j are included in a ridge regression, their coefficients are all the same.

Ex. 3.30 Consider the elastic-net optimization problem:

$$\min_{\beta} ||\mathbf{y} - \mathbf{X}\beta||^2 + \lambda\big[\alpha||\beta||_2^2 + (1 - \alpha)||\beta||_1\big]. \qquad (3.91)$$

Show how one can turn this into a lasso problem, using an augmented version of $\mathbf{X}$ and $\mathbf{y}$.

4
Linear Methods for Classification

4.1 Introduction

In this chapter we revisit the classification problem and focus on linear methods for classification. Since our predictor $G(x)$ takes values in a discrete set $\mathcal{G}$, we can always divide the input space into a collection of regions labeled according to the classification. We saw in Chapter 2 that the boundaries of these regions can be rough or smooth, depending on the prediction function. For an important class of procedures, these *decision boundaries* are linear; this is what we will mean by linear methods for classification.

There are several different ways in which linear decision boundaries can be found. In Chapter 2 we fit linear regression models to the class indicator variables, and classify to the largest fit. Suppose there are K classes, for convenience labeled $1, 2, \ldots, K$, and the fitted linear model for the kth indicator response variable is $\hat{f}_k(x) = \hat{\beta}_{k0} + \hat{\beta}_k^T x$. The decision boundary between class k and ℓ is that set of points for which $\hat{f}_k(x) = \hat{f}_\ell(x)$, that is, the set $\{x : (\hat{\beta}_{k0} - \hat{\beta}_{\ell0}) + (\hat{\beta}_k - \hat{\beta}_\ell)^T x = 0\}$, an affine set or hyperplane.[1] Since the same is true for any pair of classes, the input space is divided into regions of constant classification, with piecewise hyperplanar decision boundaries. This regression approach is a member of a class of methods that model *discriminant functions* $\delta_k(x)$ for each class, and then classify x to the class with the largest value for its discriminant function. Methods

---

[1]Strictly speaking, a hyperplane passes through the origin, while an affine set need not. We sometimes ignore the distinction and refer in general to hyperplanes.

T. Hastie et al., *The Elements of Statistical Learning, Second Edition,*
DOI: 10.1007/b94608_4,
© Springer Science+Business Media, LLC 2009

that model the posterior probabilities $\Pr(G = k|X = x)$ are also in this class. Clearly, if either the $\delta_k(x)$ or $\Pr(G = k|X = x)$ are linear in x, then the decision boundaries will be linear.

Actually, all we require is that some monotone transformation of δ_k or $\Pr(G = k|X = x)$ be linear for the decision boundaries to be linear. For example, if there are two classes, a popular model for the posterior probabilities is

$$
\Pr(G = 1|X = x) = \frac{\exp(\beta_0 + \beta^T x)}{1 + \exp(\beta_0 + \beta^T x)},
$$
$$
\Pr(G = 2|X = x) = \frac{1}{1 + \exp(\beta_0 + \beta^T x)}. \tag{4.1}
$$

Here the monotone transformation is the *logit* transformation: $\log[p/(1-p)]$, and in fact we see that

$$
\log \frac{\Pr(G = 1|X = x)}{\Pr(G = 2|X = x)} = \beta_0 + \beta^T x. \tag{4.2}
$$

The decision boundary is the set of points for which the *log-odds* are zero, and this is a hyperplane defined by $\{x|\beta_0 + \beta^T x = 0\}$. We discuss two very popular but different methods that result in linear log-odds or logits: linear discriminant analysis and linear logistic regression. Although they differ in their derivation, the essential difference between them is in the way the linear function is fit to the training data.

A more direct approach is to explicitly model the boundaries between the classes as linear. For a two-class problem in a p-dimensional input space, this amounts to modeling the decision boundary as a hyperplane—in other words, a normal vector and a cut-point. We will look at two methods that explicitly look for "separating hyperplanes." The first is the well-known *perceptron* model of Rosenblatt (1958), with an algorithm that finds a separating hyperplane in the training data, if one exists. The second method, due to Vapnik (1996), finds an *optimally separating hyperplane* if one exists, else finds a hyperplane that minimizes some measure of overlap in the training data. We treat the separable case here, and defer treatment of the nonseparable case to Chapter 12.

While this entire chapter is devoted to linear decision boundaries, there is considerable scope for generalization. For example, we can expand our variable set $X_1, \ldots, X_p$ by including their squares and cross-products $X_1^2, X_2^2, \ldots,$ $X_1 X_2, \ldots$, thereby adding $p(p+1)/2$ additional variables. Linear functions in the augmented space map down to quadratic functions in the original space—hence linear decision boundaries to quadratic decision boundaries. Figure 4.1 illustrates the idea. The data are the same: the left plot uses linear decision boundaries in the two-dimensional space shown, while the right plot uses linear decision boundaries in the augmented five-dimensional space described above. This approach can be used with any basis transfor-

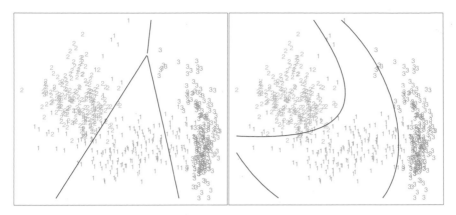

FIGURE 4.1. *The left plot shows some data from three classes, with linear decision boundaries found by linear discriminant analysis. The right plot shows quadratic decision boundaries. These were obtained by finding linear boundaries in the five-dimensional space* $X_1, X_2, X_1 X_2, X_1^2, X_2^2$. *Linear inequalities in this space are quadratic inequalities in the original space.*

mation $h(X)$ where $h : \mathbb{R}^p \mapsto \mathbb{R}^q$ with $q > p$, and will be explored in later chapters.

4.2 Linear Regression of an Indicator Matrix

Here each of the response categories are coded via an indicator variable. Thus if $\mathcal{G}$ has K classes, there will be K such indicators Y_k, $k = 1, \ldots, K$, with $Y_k = 1$ if $G = k$ else 0. These are collected together in a vector $Y = (Y_1, \ldots, Y_K)$, and the N training instances of these form an $N \times K$ *indicator response matrix* $\mathbf{Y}$. $\mathbf{Y}$ is a matrix of 0's and 1's, with each row having a single 1. We fit a linear regression model to each of the columns of $\mathbf{Y}$ simultaneously, and the fit is given by

$$\hat{\mathbf{Y}} = \mathbf{X}(\mathbf{X}^T\mathbf{X})^{-1}\mathbf{X}^T\mathbf{Y}. \tag{4.3}$$

Chapter 3 has more details on linear regression. Note that we have a coefficient vector for each response column $\mathbf{y}_k$, and hence a $(p+1) \times K$ coefficient matrix $\hat{\mathbf{B}} = (\mathbf{X}^T\mathbf{X})^{-1}\mathbf{X}^T\mathbf{Y}$. Here $\mathbf{X}$ is the model matrix with $p+1$ columns corresponding to the p inputs, and a leading column of 1's for the intercept.

A new observation with input x is classified as follows:

- compute the fitted output $\hat{f}(x)^T = (1, x^T)\hat{\mathbf{B}}$, a K vector;

- identify the largest component and classify accordingly:

$$\hat{G}(x) = \text{argmax}_{k \in \mathcal{G}} \hat{f}_k(x). \tag{4.4}$$

What is the rationale for this approach? One rather formal justification is to view the regression as an estimate of conditional expectation. For the random variable Y_k, $E(Y_k|X = x) = \Pr(G = k|X = x)$, so conditional expectation of each of the Y_k seems a sensible goal. The real issue is: how good an approximation to conditional expectation is the rather rigid linear regression model? Alternatively, are the $\hat{f}_k(x)$ reasonable estimates of the posterior probabilities $\Pr(G = k|X = x)$, and more importantly, does this matter?

It is quite straightforward to verify that $\sum_{k \in \mathcal{G}} \hat{f}_k(x) = 1$ for any x, as long as there is an intercept in the model (column of 1's in $\mathbf{X}$). However, the $\hat{f}_k(x)$ can be negative or greater than 1, and typically some are. This is a consequence of the rigid nature of linear regression, especially if we make predictions outside the hull of the training data. These violations in themselves do not guarantee that this approach will not work, and in fact on many problems it gives similar results to more standard linear methods for classification. If we allow linear regression onto basis expansions $h(X)$ of the inputs, this approach can lead to consistent estimates of the probabilities. As the size of the training set N grows bigger, we adaptively include more basis elements so that linear regression onto these basis functions approaches conditional expectation. We discuss such approaches in Chapter 5.

A more simplistic viewpoint is to construct *targets* t_k for each class, where t_k is the kth column of the $K \times K$ identity matrix. Our prediction problem is to try and reproduce the appropriate target for an observation. With the same coding as before, the response vector y_i (*i*th row of $\mathbf{Y}$) for observation i has the value $y_i = t_k$ if $g_i = k$. We might then fit the linear model by least squares:

$$\min_{\mathbf{B}} \sum_{i=1}^{N} ||y_i - [(1, x_i^T)\mathbf{B}]^T||^2. \tag{4.5}$$

The criterion is a sum-of-squared Euclidean distances of the fitted vectors from their targets. A new observation is classified by computing its fitted vector $\hat{f}(x)$ and classifying to the closest target:

$$\hat{G}(x) = \underset{k}{\operatorname{argmin}} ||\hat{f}(x) - t_k||^2. \tag{4.6}$$

This is exactly the same as the previous approach:

- The sum-of-squared-norm criterion is exactly the criterion for multiple response linear regression, just viewed slightly differently. Since a squared norm is itself a sum of squares, the components decouple and can be rearranged as a separate linear model for each element. Note that this is only possible because there is nothing in the model that binds the different responses together.

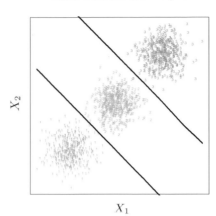

FIGURE 4.2. *The data come from three classes in* $\mathbb{R}^2$ *and are easily separated by linear decision boundaries. The right plot shows the boundaries found by linear discriminant analysis. The left plot shows the boundaries found by linear regression of the indicator response variables. The middle class is completely masked (never dominates).*

- The closest target classification rule (4.6) is easily seen to be exactly the same as the maximum fitted component criterion (4.4).

There is a serious problem with the regression approach when the number of classes $K \geq 3$, especially prevalent when K is large. Because of the rigid nature of the regression model, classes can be *masked* by others. Figure 4.2 illustrates an extreme situation when $K = 3$. The three classes are perfectly separated by linear decision boundaries, yet linear regression misses the middle class completely.

In Figure 4.3 we have projected the data onto the line joining the three centroids (there is no information in the orthogonal direction in this case), and we have included and coded the three response variables Y_1, Y_2 and Y_3. The three regression lines (left panel) are included, and we see that the line corresponding to the middle class is horizontal and its fitted values are never dominant! Thus, observations from class 2 are classified either as class 1 or class 3. The right panel uses quadratic regression rather than linear regression. For this simple example a quadratic rather than linear fit (for the middle class at least) would solve the problem. However, it can be seen that if there were four rather than three classes lined up like this, a quadratic would not come down fast enough, and a cubic would be needed as well. A loose but general rule is that if $K \geq 3$ classes are lined up, polynomial terms up to degree $K - 1$ might be needed to resolve them. Note also that these are polynomials along the derived direction passing through the centroids, which can have arbitrary orientation. So in

Degree = 1; Error = 0.33   Degree = 2; Error = 0.04

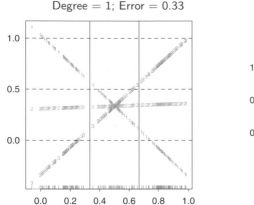 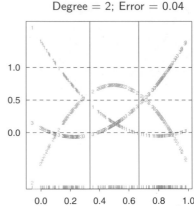

FIGURE 4.3. *The effects of masking on linear regression in $\mathbb{R}$ for a three-class problem. The* rug plot *at the base indicates the positions and class membership of each observation. The three curves in each panel are the fitted regressions to the three-class indicator variables; for example, for the blue class, y_{blue} is 1 for the blue observations, and 0 for the green and orange. The fits are linear and quadratic polynomials. Above each plot is the training error rate. The Bayes error rate is 0.025 for this problem, as is the LDA error rate.*

p-dimensional input space, one would need general polynomial terms and cross-products of total degree $K - 1$, $O(p^{K-1})$ terms in all, to resolve such worst-case scenarios.

The example is extreme, but for large K and small p such maskings naturally occur. As a more realistic illustration, Figure 4.4 is a projection of the training data for a vowel recognition problem onto an informative two-dimensional subspace. There are $K = 11$ classes in $p = 10$ dimensions. This is a difficult classification problem, and the best methods achieve around 40% errors on the test data. The main point here is summarized in Table 4.1; linear regression has an error rate of 67%, while a close relative, linear discriminant analysis, has an error rate of 56%. It seems that masking has hurt in this case. While all the other methods in this chapter are based on linear functions of x as well, they use them in such a way that avoids this masking problem.

4.3   Linear Discriminant Analysis

Decision theory for classification (Section 2.4) tells us that we need to know the class posteriors $\Pr(G|X)$ for optimal classification. Suppose $f_k(x)$ is the class-conditional density of X in class $G = k$, and let π_k be the prior probability of class k, with $\sum_{k=1}^{K} \pi_k = 1$. A simple application of Bayes

Linear Discriminant Analysis

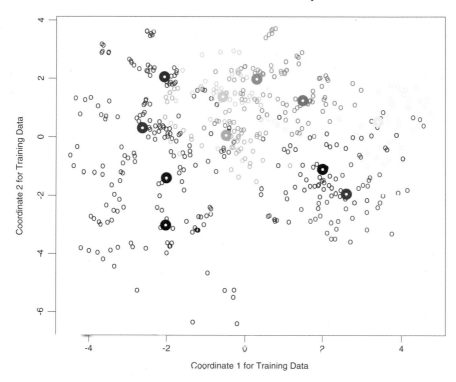

FIGURE 4.4. *A two-dimensional plot of the vowel training data. There are eleven classes with $X \in \mathbb{R}^{10}$, and this is the best view in terms of a LDA model (Section 4.3.3). The heavy circles are the projected mean vectors for each class. The class overlap is considerable.*

TABLE 4.1. *Training and test error rates using a variety of linear techniques on the vowel data. There are eleven classes in ten dimensions, of which three account for 90% of the variance (via a principal components analysis). We see that linear regression is hurt by masking, increasing the test and training error by over 10%.*

Technique	Error Rates	
	Training	Test
Linear regression	0.48	0.67
Linear discriminant analysis	0.32	0.56
Quadratic discriminant analysis	0.01	0.53
Logistic regression	0.22	0.51

theorem gives us

$$\Pr(G = k | X = x) = \frac{f_k(x)\pi_k}{\sum_{\ell=1}^{K} f_\ell(x)\pi_\ell}. \tag{4.7}$$

We see that in terms of ability to classify, having the $f_k(x)$ is almost equivalent to having the quantity $\Pr(G = k | X = x)$.

Many techniques are based on models for the class densities:

- linear and quadratic discriminant analysis use Gaussian densities;

- more flexible mixtures of Gaussians allow for nonlinear decision boundaries (Section 6.8);

- general nonparametric density estimates for each class density allow the most flexibility (Section 6.6.2);

- *Naive Bayes* models are a variant of the previous case, and assume that each of the class densities are products of marginal densities; that is, they assume that the inputs are conditionally independent in each class (Section 6.6.3).

Suppose that we model each class density as multivariate Gaussian

$$f_k(x) = \frac{1}{(2\pi)^{p/2}|\mathbf{\Sigma}_k|^{1/2}} e^{-\frac{1}{2}(x-\mu_k)^T \mathbf{\Sigma}_k^{-1}(x-\mu_k)}. \tag{4.8}$$

Linear discriminant analysis (LDA) arises in the special case when we assume that the classes have a common covariance matrix $\mathbf{\Sigma}_k = \mathbf{\Sigma} \,\forall k$. In comparing two classes k and ℓ, it is sufficient to look at the log-ratio, and we see that

$$
\begin{aligned}
\log \frac{\Pr(G = k | X = x)}{\Pr(G = \ell | X = x)} &= \log \frac{f_k(x)}{f_\ell(x)} + \log \frac{\pi_k}{\pi_\ell} \\
&= \log \frac{\pi_k}{\pi_\ell} - \frac{1}{2}(\mu_k + \mu_\ell)^T \mathbf{\Sigma}^{-1}(\mu_k - \mu_\ell) \\
&\quad + x^T \mathbf{\Sigma}^{-1}(\mu_k - \mu_\ell),
\end{aligned}
\tag{4.9}
$$

an equation linear in x. The equal covariance matrices cause the normalization factors to cancel, as well as the quadratic part in the exponents. This linear log-odds function implies that the decision boundary between classes k and ℓ—the set where $\Pr(G = k | X = x) = \Pr(G = \ell | X = x)$—is linear in x; in p dimensions a hyperplane. This is of course true for any pair of classes, so all the decision boundaries are linear. If we divide $\mathbb{R}^p$ into regions that are classified as class 1, class 2, etc., these regions will be separated by hyperplanes. Figure 4.5 (left panel) shows an idealized example with three classes and $p = 2$. Here the data do arise from three Gaussian distributions with a common covariance matrix. We have included in

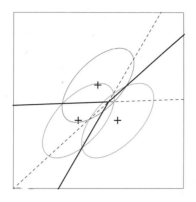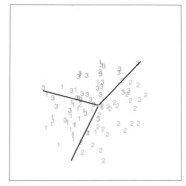

FIGURE 4.5. *The left panel shows three Gaussian distributions, with the same covariance and different means. Included are the contours of constant density enclosing 95% of the probability in each case. The Bayes decision boundaries between each pair of classes are shown (broken straight lines), and the Bayes decision boundaries separating all three classes are the thicker solid lines (a subset of the former). On the right we see a sample of 30 drawn from each Gaussian distribution, and the fitted LDA decision boundaries.*

the figure the contours corresponding to 95% highest probability density, as well as the class centroids. Notice that the decision boundaries are not the perpendicular bisectors of the line segments joining the centroids. This would be the case if the covariance $\mathbf{\Sigma}$ were spherical $\sigma^2\mathbf{I}$, and the class priors were equal. From (4.9) we see that the *linear discriminant functions*

$$\delta_k(x) = x^T \mathbf{\Sigma}^{-1}\mu_k - \frac{1}{2}\mu_k^T\mathbf{\Sigma}^{-1}\mu_k + \log \pi_k \qquad (4.10)$$

are an equivalent description of the decision rule, with $G(x) = \text{argmax}_k \delta_k(x)$.

In practice we do not know the parameters of the Gaussian distributions, and will need to estimate them using our training data:

- $\hat{\pi}_k = N_k/N$, where N_k is the number of class-k observations;

- $\hat{\mu}_k = \sum_{g_i=k} x_i/N_k$;

- $\hat{\mathbf{\Sigma}} = \sum_{k=1}^{K} \sum_{g_i=k}(x_i - \hat{\mu}_k)(x_i - \hat{\mu}_k)^T/(N - K)$.

Figure 4.5 (right panel) shows the estimated decision boundaries based on a sample of size 30 each from three Gaussian distributions. Figure 4.1 on page 103 is another example, but here the classes are not Gaussian.

With two classes there is a simple correspondence between linear discriminant analysis and classification by linear regression, as in (4.5). The LDA rule classifies to class 2 if

$$x^T\hat{\mathbf{\Sigma}}^{-1}(\hat{\mu}_2 - \hat{\mu}_1) > \frac{1}{2}(\hat{\mu}_2 + \hat{\mu}_1)^T\hat{\mathbf{\Sigma}}^{-1}(\hat{\mu}_2 - \hat{\mu}_1) - \log(N_2/N_1), \qquad (4.11)$$

and class 1 otherwise. Suppose we code the targets in the two classes as $+1$ and -1, respectively. It is easy to show that the coefficient vector from least squares is proportional to the LDA direction given in (4.11) (Exercise 4.2). [In fact, this correspondence occurs for any (distinct) coding of the targets; see Exercise 4.2]. However unless $N_1 = N_2$ the intercepts are different and hence the resulting decision rules are different.

Since this derivation of the LDA direction via least squares does not use a Gaussian assumption for the features, its applicability extends beyond the realm of Gaussian data. However the derivation of the particular intercept or cut-point given in (4.11) *does* require Gaussian data. Thus it makes sense to instead choose the cut-point that empirically minimizes training error for a given dataset. This is something we have found to work well in practice, but have not seen it mentioned in the literature.

With more than two classes, LDA is not the same as linear regression of the class indicator matrix, and it avoids the masking problems associated with that approach (Hastie et al., 1994). A correspondence between regression and LDA can be established through the notion of *optimal scoring*, discussed in Section 12.5.

Getting back to the general discriminant problem (4.8), if the Σ_k are not assumed to be equal, then the convenient cancellations in (4.9) do not occur; in particular the pieces quadratic in x remain. We then get *quadratic discriminant functions* (QDA),

$$\delta_k(x) = -\frac{1}{2}\log|\Sigma_k| - \frac{1}{2}(x - \mu_k)^T\Sigma_k^{-1}(x - \mu_k) + \log\pi_k. \qquad (4.12)$$

The decision boundary between each pair of classes k and ℓ is described by a quadratic equation $\{x : \delta_k(x) = \delta_\ell(x)\}$.

Figure 4.6 shows an example (from Figure 4.1 on page 103) where the three classes are Gaussian mixtures (Section 6.8) and the decision boundaries are approximated by quadratic equations in x. Here we illustrate two popular ways of fitting these quadratic boundaries. The right plot uses QDA as described here, while the left plot uses LDA in the enlarged five-dimensional quadratic polynomial space. The differences are generally small; QDA is the preferred approach, with the LDA method a convenient substitute [2].

The estimates for QDA are similar to those for LDA, except that separate covariance matrices must be estimated for each class. When p is large this can mean a dramatic increase in parameters. Since the decision boundaries are functions of the parameters of the densities, counting the number of parameters must be done with care. For LDA, it seems there are $(K - 1) \times (p + 1)$ parameters, since we only need the differences $\delta_k(x) - \delta_K(x)$

---

[2] For this figure and many similar figures in the book we compute the decision boundaries by an exhaustive contouring method. We compute the decision rule on a fine lattice of points, and then use contouring algorithms to compute the boundaries.

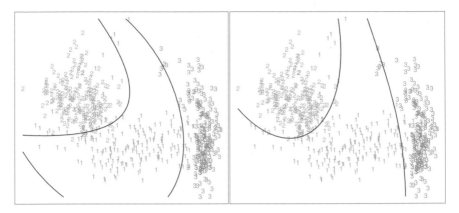

FIGURE 4.6. *Two methods for fitting quadratic boundaries. The left plot shows the quadratic decision boundaries for the data in Figure 4.1 (obtained using LDA in the five-dimensional space* $X_1, X_2, X_1X_2, X_1^2, X_2^2$*). The right plot shows the quadratic decision boundaries found by QDA. The differences are small, as is usually the case.*

between the discriminant functions where K is some pre-chosen class (here we have chosen the last), and each difference requires $p + 1$ parameters[3]. Likewise for QDA there will be $(K - 1) \times \{p(p + 3)/2 + 1\}$ parameters. Both LDA and QDA perform well on an amazingly large and diverse set of classification tasks. For example, in the STATLOG project (Michie et al., 1994) LDA was among the top three classifiers for 7 of the 22 datasets, QDA among the top three for four datasets, and one of the pair were in the top three for 10 datasets. Both techniques are widely used, and entire books are devoted to LDA. It seems that whatever exotic tools are the rage of the day, we should always have available these two simple tools. The question arises why LDA and QDA have such a good track record. The reason is not likely to be that the data are approximately Gaussian, and in addition for LDA that the covariances are approximately equal. More likely a reason is that the data can only support simple decision boundaries such as linear or quadratic, and the estimates provided via the Gaussian models are stable. This is a bias variance tradeoff—we can put up with the bias of a linear decision boundary because it can be estimated with much lower variance than more exotic alternatives. This argument is less believable for QDA, since it can have many parameters itself, although perhaps fewer than the non-parametric alternatives.

---

[3] Although we fit the covariance matrix $\hat{\mathbf{\Sigma}}$ to compute the LDA discriminant functions, a much reduced function of it is all that is required to estimate the $O(p)$ parameters needed to compute the decision boundaries.

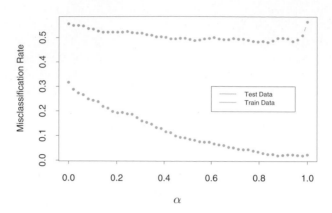

FIGURE 4.7. *Test and training errors for the vowel data, using regularized discriminant analysis with a series of values of $\alpha \in [0,1]$. The optimum for the test data occurs around $\alpha = 0.9$, close to quadratic discriminant analysis.*

4.3.1 Regularized Discriminant Analysis

Friedman (1989) proposed a compromise between LDA and QDA, which allows one to shrink the separate covariances of QDA toward a common covariance as in LDA. These methods are very similar in flavor to ridge regression. The regularized covariance matrices have the form

$$\hat{\boldsymbol{\Sigma}}_k(\alpha) = \alpha \hat{\boldsymbol{\Sigma}}_k + (1 - \alpha)\hat{\boldsymbol{\Sigma}}, \tag{4.13}$$

where $\hat{\boldsymbol{\Sigma}}$ is the pooled covariance matrix as used in LDA. Here $\alpha \in [0,1]$ allows a continuum of models between LDA and QDA, and needs to be specified. In practice α can be chosen based on the performance of the model on validation data, or by cross-validation.

Figure 4.7 shows the results of RDA applied to the vowel data. Both the training and test error improve with increasing α, although the test error increases sharply after $\alpha = 0.9$. The large discrepancy between the training and test error is partly due to the fact that there are many repeat measurements on a small number of individuals, different in the training and test set.

Similar modifications allow $\hat{\boldsymbol{\Sigma}}$ itself to be shrunk toward the scalar covariance,

$$\hat{\boldsymbol{\Sigma}}(\gamma) = \gamma \hat{\boldsymbol{\Sigma}} + (1 - \gamma)\hat{\sigma}^2 \mathbf{I} \tag{4.14}$$

for $\gamma \in [0,1]$. Replacing $\hat{\boldsymbol{\Sigma}}$ in (4.13) by $\hat{\boldsymbol{\Sigma}}(\gamma)$ leads to a more general family of covariances $\hat{\boldsymbol{\Sigma}}(\alpha, \gamma)$ indexed by a pair of parameters.

In Chapter 12, we discuss other regularized versions of LDA, which are more suitable when the data arise from digitized analog signals and images.

In these situations the features are high-dimensional and correlated, and the LDA coefficients can be regularized to be smooth or sparse in the original domain of the signal. This leads to better generalization and allows for easier interpretation of the coefficients. In Chapter 18 we also deal with very high-dimensional problems, where for example the features are gene-expression measurements in microarray studies. There the methods focus on the case $\gamma = 0$ in (4.14), and other severely regularized versions of LDA.

4.3.2 Computations for LDA

As a lead-in to the next topic, we briefly digress on the computations required for LDA and especially QDA. Their computations are simplified by diagonalizing $\hat{\boldsymbol{\Sigma}}$ or $\hat{\boldsymbol{\Sigma}}_k$. For the latter, suppose we compute the eigen-decomposition for each $\hat{\boldsymbol{\Sigma}}_k = \mathbf{U}_k \mathbf{D}_k \mathbf{U}_k^T$, where $\mathbf{U}_k$ is $p \times p$ orthonormal, and $\mathbf{D}_k$ a diagonal matrix of positive eigenvalues $d_{k\ell}$. Then the ingredients for $\delta_k(x)$ (4.12) are

- $(x - \hat{\mu}_k)^T \hat{\boldsymbol{\Sigma}}_k^{-1} (x - \hat{\mu}_k) = [\mathbf{U}_k^T (x - \hat{\mu}_k)]^T \mathbf{D}_k^{-1} [\mathbf{U}_k^T (x - \hat{\mu}_k)]$;

- $\log |\hat{\boldsymbol{\Sigma}}_k| = \sum_{\ell} \log d_{k\ell}$.

In light of the computational steps outlined above, the LDA classifier can be implemented by the following pair of steps:

- *Sphere* the data with respect to the common covariance estimate $\hat{\boldsymbol{\Sigma}}$: $X^* \leftarrow \mathbf{D}^{-\frac{1}{2}} \mathbf{U}^T X$, where $\hat{\boldsymbol{\Sigma}} = \mathbf{U} \mathbf{D} \mathbf{U}^T$. The common covariance estimate of X^* will now be the identity.

- Classify to the closest class centroid in the transformed space, modulo the effect of the class prior probabilities π_k.

4.3.3 Reduced-Rank Linear Discriminant Analysis

So far we have discussed LDA as a restricted Gaussian classifier. Part of its popularity is due to an additional restriction that allows us to view informative low-dimensional projections of the data.

The K centroids in p-dimensional input space lie in an affine subspace of dimension $\leq K - 1$, and if p is much larger than K, this will be a considerable drop in dimension. Moreover, in locating the closest centroid, we can ignore distances orthogonal to this subspace, since they will contribute equally to each class. Thus we might just as well project the X^* onto this centroid-spanning subspace H_{K-1}, and make distance comparisons there. Thus there is a fundamental dimension reduction in LDA, namely, that we need only consider the data in a subspace of dimension at most $K - 1$.

If $K = 3$, for instance, this could allow us to view the data in a two-dimensional plot, color-coding the classes. In doing so we would not have relinquished any of the information needed for LDA classification.

What if $K > 3$? We might then ask for a $L < K - 1$ dimensional subspace $H_L \subseteq H_{K-1}$ optimal for LDA in some sense. Fisher defined optimal to mean that the projected centroids were spread out as much as possible in terms of variance. This amounts to finding principal component subspaces of the centroids themselves (principal components are described briefly in Section 3.5.1, and in more detail in Section 14.5.1). Figure 4.4 shows such an optimal two-dimensional subspace for the vowel data. Here there are eleven classes, each a different vowel sound, in a ten-dimensional input space. The centroids require the full space in this case, since $K - 1 = p$, but we have shown an optimal two-dimensional subspace. The dimensions are ordered, so we can compute additional dimensions in sequence. Figure 4.8 shows four additional pairs of coordinates, also known as *canonical* or *discriminant* variables. In summary then, finding the sequences of optimal subspaces for LDA involves the following steps:

- compute the $K \times p$ matrix of class centroids $\mathbf{M}$ and the common covariance matrix $\mathbf{W}$ (for *within-class* covariance);

- compute $\mathbf{M}^* = \mathbf{M}\mathbf{W}^{-\frac{1}{2}}$ using the eigen-decomposition of $\mathbf{W}$;

- compute $\mathbf{B}^*$, the covariance matrix of $\mathbf{M}^*$ ($\mathbf{B}$ for *between-class* covariance), and its eigen-decomposition $\mathbf{B}^* = \mathbf{V}^* \mathbf{D}_B \mathbf{V}^{*T}$. The columns v_ℓ^* of $\mathbf{V}^*$ in sequence from first to last define the coordinates of the optimal subspaces.

Combining all these operations the ℓth *discriminant variable* is given by $Z_\ell = v_\ell^T X$ with $v_\ell = \mathbf{W}^{-\frac{1}{2}} v_\ell^*$.

Fisher arrived at this decomposition via a different route, without referring to Gaussian distributions at all. He posed the problem:

> *Find the linear combination $Z = a^T X$ such that the between-class variance is maximized relative to the within-class variance.*

Again, the between class variance is the variance of the class means of Z, and the within class variance is the pooled variance about the means. Figure 4.9 shows why this criterion makes sense. Although the direction joining the centroids separates the means as much as possible (i.e., maximizes the between-class variance), there is considerable overlap between the projected classes due to the nature of the covariances. By taking the covariance into account as well, a direction with minimum overlap can be found.

The between-class variance of Z is $a^T \mathbf{B} a$ and the within-class variance $a^T \mathbf{W} a$, where $\mathbf{W}$ is defined earlier, and $\mathbf{B}$ is the covariance matrix of the class centroid matrix $\mathbf{M}$. Note that $\mathbf{B} + \mathbf{W} = \mathbf{T}$, where $\mathbf{T}$ is the *total* covariance matrix of X, ignoring class information.

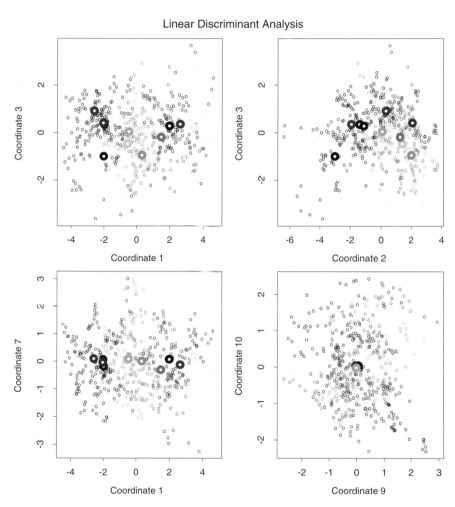

FIGURE 4.8. *Four projections onto pairs of canonical variates. Notice that as the rank of the canonical variates increases, the centroids become less spread out. In the lower right panel they appear to be superimposed, and the classes most confused.*

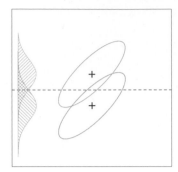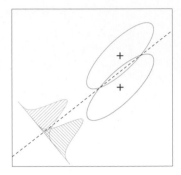

FIGURE 4.9. *Although the line joining the centroids defines the direction of greatest centroid spread, the projected data overlap because of the covariance (left panel). The discriminant direction minimizes this overlap for Gaussian data (right panel).*

Fisher's problem therefore amounts to maximizing the *Rayleigh quotient,*

$$\max_{a} \frac{a^T \mathbf{B} a}{a^T \mathbf{W} a}, \qquad (4.15)$$

or equivalently

$$\max_{a} a^T \mathbf{B} a \text{ subject to } a^T \mathbf{W} a = 1. \qquad (4.16)$$

This is a generalized eigenvalue problem, with a given by the largest eigenvalue of $\mathbf{W}^{-1}\mathbf{B}$. It is not hard to show (Exercise 4.1) that the optimal a_1 is identical to v_1 defined above. Similarly one can find the next direction a_2, orthogonal in $\mathbf{W}$ to a_1, such that $a_2^T \mathbf{B} a_2 / a_2^T \mathbf{W} a_2$ is maximized; the solution is $a_2 = v_2$, and so on. The a_ℓ are referred to as *discriminant coordinates,* not to be confused with discriminant functions. They are also referred to as *canonical variates,* since an alternative derivation of these results is through a canonical correlation analysis of the indicator response matrix $\mathbf{Y}$ on the predictor matrix $\mathbf{X}$. This line is pursued in Section 12.5.

To summarize the developments so far:

- Gaussian classification with common covariances leads to linear decision boundaries. Classification can be achieved by sphering the data with respect to $\mathbf{W}$, and classifying to the closest centroid (modulo $\log \pi_k$) in the sphered space.

- Since only the relative distances to the centroids count, one can confine the data to the subspace spanned by the centroids in the sphered space.

- This subspace can be further decomposed into successively optimal subspaces in term of centroid separation. This decomposition is identical to the decomposition due to Fisher.

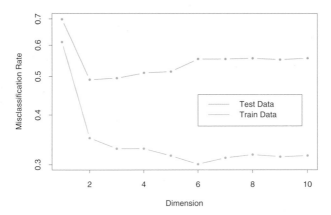

FIGURE 4.10. *Training and test error rates for the vowel data, as a function of the dimension of the discriminant subspace. In this case the best error rate is for dimension 2. Figure 4.11 shows the decision boundaries in this space.*

The reduced subspaces have been motivated as a data reduction (for viewing) tool. Can they also be used for classification, and what is the rationale? Clearly they can, as in our original derivation; we simply limit the distance-to-centroid calculations to the chosen subspace. One can show that this is a Gaussian classification rule with the additional restriction that the centroids of the Gaussians lie in a L-dimensional subspace of $\mathbb{R}^p$. Fitting such a model by maximum likelihood, and then constructing the posterior probabilities using Bayes' theorem amounts to the classification rule described above (Exercise 4.8).

Gaussian classification dictates the $\log \pi_k$ correction factor in the distance calculation. The reason for this correction can be seen in Figure 4.9. The misclassification rate is based on the area of overlap between the two densities. If the π_k are equal (implicit in that figure), then the optimal cut-point is midway between the projected means. If the π_k are not equal, moving the cut-point toward the *smaller* class will improve the error rate. As mentioned earlier for two classes, one can derive the linear rule using LDA (or any other method), and then choose the cut-point to minimize misclassification error over the training data.

As an example of the benefit of the reduced-rank restriction, we return to the vowel data. There are 11 classes and 10 variables, and hence 10 possible dimensions for the classifier. We can compute the training and test error in each of these hierarchical subspaces; Figure 4.10 shows the results. Figure 4.11 shows the decision boundaries for the classifier based on the two-dimensional LDA solution.

There is a close connection between Fisher's reduced rank discriminant analysis and regression of an indicator response matrix. It turns out that

Classification in Reduced Subspace

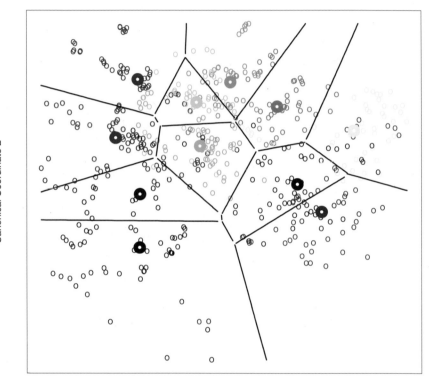

Canonical Coordinate 1

FIGURE 4.11. *Decision boundaries for the vowel training data, in the two-dimensional subspace spanned by the first two canonical variates. Note that in any higher-dimensional subspace, the decision boundaries are higher-dimensional affine planes, and could not be represented as lines.*

LDA amounts to the regression followed by an eigen-decomposition of $\hat{\mathbf{Y}}^T\mathbf{Y}$. In the case of two classes, there is a single discriminant variable that is identical up to a scalar multiplication to either of the columns of $\hat{\mathbf{Y}}$. These connections are developed in Chapter 12. A related fact is that if one transforms the original predictors $\mathbf{X}$ to $\hat{\mathbf{Y}}$, then LDA using $\hat{\mathbf{Y}}$ is identical to LDA in the original space (Exercise 4.3).

4.4 Logistic Regression

The logistic regression model arises from the desire to model the posterior probabilities of the K classes via linear functions in x, while at the same time ensuring that they sum to one and remain in $[0, 1]$. The model has the form

$$
\begin{aligned}
\log \frac{\Pr(G = 1|X = x)}{\Pr(G = K|X = x)} &= \beta_{10} + \beta_1^T x \\
\log \frac{\Pr(G = 2|X = x)}{\Pr(G = K|X = x)} &= \beta_{20} + \beta_2^T x \\
&\vdots \\
\log \frac{\Pr(G = K - 1|X = x)}{\Pr(G = K|X = x)} &= \beta_{(K-1)0} + \beta_{K-1}^T x.
\end{aligned}
\tag{4.17}
$$

The model is specified in terms of $K - 1$ log-odds or logit transformations (reflecting the constraint that the probabilities sum to one). Although the model uses the last class as the denominator in the odds-ratios, the choice of denominator is arbitrary in that the estimates are equivariant under this choice. A simple calculation shows that

$$
\begin{aligned}
\Pr(G = k|X = x) &= \frac{\exp(\beta_{k0} + \beta_k^T x)}{1 + \sum_{\ell=1}^{K-1} \exp(\beta_{\ell 0} + \beta_\ell^T x)}, \quad k = 1, \ldots, K - 1, \\
\Pr(G = K|X = x) &= \frac{1}{1 + \sum_{\ell=1}^{K-1} \exp(\beta_{\ell 0} + \beta_\ell^T x)},
\end{aligned}
\tag{4.18}
$$

and they clearly sum to one. To emphasize the dependence on the entire parameter set $\theta = \{\beta_{10}, \beta_1^T, \ldots, \beta_{(K-1)0}, \beta_{K-1}^T\}$, we denote the probabilities $\Pr(G = k|X = x) = p_k(x; \theta)$.

When $K = 2$, this model is especially simple, since there is only a single linear function. It is widely used in biostatistical applications where binary responses (two classes) occur quite frequently. For example, patients survive or die, have heart disease or not, or a condition is present or absent.

4.4.1 Fitting Logistic Regression Models

Logistic regression models are usually fit by maximum likelihood, using the conditional likelihood of G given X. Since $\Pr(G|X)$ completely specifies the conditional distribution, the *multinomial* distribution is appropriate. The log-likelihood for N observations is

$$\ell(\theta) = \sum_{i=1}^{N} \log p_{g_i}(x_i; \theta), \tag{4.19}$$

where $p_k(x_i; \theta) = \Pr(G = k|X = x_i; \theta)$.

We discuss in detail the two-class case, since the algorithms simplify considerably. It is convenient to code the two-class g_i via a 0/1 response y_i, where $y_i = 1$ when $g_i = 1$, and $y_i = 0$ when $g_i = 2$. Let $p_1(x; \theta) = p(x; \theta)$, and $p_2(x; \theta) = 1 - p(x; \theta)$. The log-likelihood can be written

$$
\begin{aligned}
\ell(\beta) &= \sum_{i=1}^{N} \Big\{ y_i \log p(x_i; \beta) + (1 - y_i) \log(1 - p(x_i; \beta)) \Big\} \\
&= \sum_{i=1}^{N} \Big\{ y_i \beta^T x_i - \log(1 + e^{\beta^T x_i}) \Big\}.
\end{aligned}
\tag{4.20}
$$

Here $\beta = \{\beta_{10}, \beta_1\}$, and we assume that the vector of inputs x_i includes the constant term 1 to accommodate the intercept.

To maximize the log-likelihood, we set its derivatives to zero. These *score* equations are

$$\frac{\partial \ell(\beta)}{\partial \beta} = \sum_{i=1}^{N} x_i(y_i - p(x_i; \beta)) = 0, \tag{4.21}$$

which are $p+1$ equations *nonlinear* in β. Notice that since the first component of x_i is 1, the first score equation specifies that $\sum_{i=1}^{N} y_i = \sum_{i=1}^{N} p(x_i; \beta)$; the *expected* number of class ones matches the observed number (and hence also class twos.)

To solve the score equations (4.21), we use the Newton–Raphson algorithm, which requires the second-derivative or Hessian matrix

$$\frac{\partial^2 \ell(\beta)}{\partial \beta \partial \beta^T} = -\sum_{i=1}^{N} x_i x_i^T p(x_i; \beta)(1 - p(x_i; \beta)). \tag{4.22}$$

Starting with β^{old}, a single Newton update is

$$\beta^{\text{new}} = \beta^{\text{old}} - \left(\frac{\partial^2 \ell(\beta)}{\partial \beta \partial \beta^T} \right)^{-1} \frac{\partial \ell(\beta)}{\partial \beta}, \tag{4.23}$$

where the derivatives are evaluated at β^{old}.

It is convenient to write the score and Hessian in matrix notation. Let $\mathbf{y}$ denote the vector of y_i values, $\mathbf{X}$ the $N \times (p+1)$ matrix of x_i values, $\mathbf{p}$ the vector of fitted probabilities with ith element $p(x_i; \beta^{\text{old}})$ and $\mathbf{W}$ a $N \times N$ diagonal matrix of weights with ith diagonal element $p(x_i; \beta^{\text{old}})(1 - p(x_i; \beta^{\text{old}}))$. Then we have

$$\frac{\partial \ell(\beta)}{\partial \beta} = \mathbf{X}^T(\mathbf{y} - \mathbf{p}) \tag{4.24}$$

$$\frac{\partial^2 \ell(\beta)}{\partial \beta \partial \beta^T} = -\mathbf{X}^T \mathbf{W} \mathbf{X} \tag{4.25}$$

The Newton step is thus

$$\begin{aligned}
\beta^{\text{new}} &= \beta^{\text{old}} + (\mathbf{X}^T \mathbf{W} \mathbf{X})^{-1} \mathbf{X}^T (\mathbf{y} - \mathbf{p}) \\
&= (\mathbf{X}^T \mathbf{W} \mathbf{X})^{-1} \mathbf{X}^T \mathbf{W} \left(\mathbf{X}\beta^{\text{old}} + \mathbf{W}^{-1}(\mathbf{y} - \mathbf{p}) \right) \\
&= (\mathbf{X}^T \mathbf{W} \mathbf{X})^{-1} \mathbf{X}^T \mathbf{W} \mathbf{z}. \tag{4.26}
\end{aligned}$$

In the second and third line we have re-expressed the Newton step as a weighted least squares step, with the response

$$\mathbf{z} = \mathbf{X}\beta^{\text{old}} + \mathbf{W}^{-1}(\mathbf{y} - \mathbf{p}), \tag{4.27}$$

sometimes known as the *adjusted response*. These equations get solved repeatedly, since at each iteration $\mathbf{p}$ changes, and hence so does $\mathbf{W}$ and $\mathbf{z}$. This algorithm is referred to as *iteratively reweighted least squares* or IRLS, since each iteration solves the weighted least squares problem:

$$\beta^{\text{new}} \leftarrow \arg \min_{\beta} (\mathbf{z} - \mathbf{X}\beta)^T \mathbf{W} (\mathbf{z} - \mathbf{X}\beta). \tag{4.28}$$

It seems that $\beta = 0$ is a good starting value for the iterative procedure, although convergence is never guaranteed. Typically the algorithm does converge, since the log-likelihood is concave, but overshooting can occur. In the rare cases that the log-likelihood decreases, step size halving will guarantee convergence.

For the multiclass case ($K \geq 3$) the Newton algorithm can also be expressed as an iteratively reweighted least squares algorithm, but with a *vector* of $K-1$ responses and a nondiagonal weight matrix per observation. The latter precludes any simplified algorithms, and in this case it is numerically more convenient to work with the expanded vector θ directly (Exercise 4.4). Alternatively coordinate-descent methods (Section 3.8.6) can be used to maximize the log-likelihood efficiently. The R package glmnet (Friedman et al., 2010) can fit very large logistic regression problems efficiently, both in N and p. Although designed to fit regularized models, options allow for unregularized fits.

Logistic regression models are used mostly as a data analysis and inference tool, where the goal is to understand the role of the input variables

TABLE 4.2. *Results from a logistic regression fit to the South African heart disease data.*

	Coefficient	Std. Error	Z Score
(Intercept)	−4.130	0.964	−4.285
sbp	0.006	0.006	1.023
tobacco	0.080	0.026	3.034
ldl	0.185	0.057	3.219
famhist	0.939	0.225	4.178
obesity	-0.035	0.029	−1.187
alcohol	0.001	0.004	0.136
age	0.043	0.010	4.184

in *explaining* the outcome. Typically many models are fit in a search for a parsimonious model involving a subset of the variables, possibly with some interactions terms. The following example illustrates some of the issues involved.

4.4.2 Example: South African Heart Disease

Here we present an analysis of binary data to illustrate the traditional statistical use of the logistic regression model. The data in Figure 4.12 are a subset of the Coronary Risk-Factor Study (CORIS) baseline survey, carried out in three rural areas of the Western Cape, South Africa (Rousseauw et al., 1983). The aim of the study was to establish the intensity of ischemic heart disease risk factors in that high-incidence region. The data represent white males between 15 and 64, and the response variable is the presence or absence of myocardial infarction (MI) at the time of the survey (the overall prevalence of MI was 5.1% in this region). There are 160 cases in our data set, and a sample of 302 controls. These data are described in more detail in Hastie and Tibshirani (1987).

We fit a logistic-regression model by maximum likelihood, giving the results shown in Table 4.2. This summary includes Z scores for each of the coefficients in the model (coefficients divided by their standard errors); a nonsignificant Z score suggests a coefficient can be dropped from the model. Each of these correspond formally to a test of the null hypothesis that the coefficient in question is zero, while all the others are not (also known as the Wald test). A Z score greater than approximately 2 in absolute value is significant at the 5% level.

There are some surprises in this table of coefficients, which must be interpreted with caution. Systolic blood pressure (sbp) is not significant! Nor is obesity, and its sign is negative. This confusion is a result of the correlation between the set of predictors. On their own, both sbp and obesity are significant, and with positive sign. However, in the presence of many

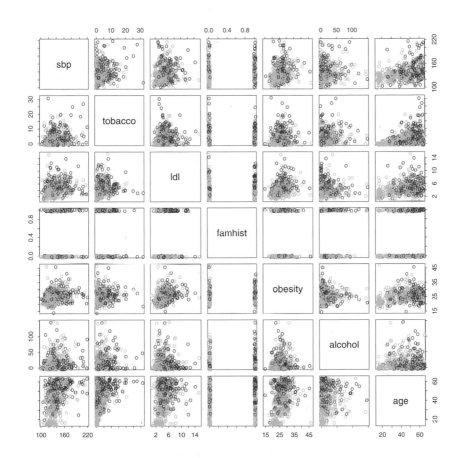

FIGURE 4.12. *A scatterplot matrix of the South African heart disease data. Each plot shows a pair of risk factors, and the cases and controls are color coded (red is a case). The variable* family history of heart disease (`famhist`) *is binary (yes or no).*

TABLE 4.3. *Results from stepwise logistic regression fit to South African heart disease data.*

	Coefficient	Std. Error	Z score
(Intercept)	−4.204	0.498	−8.45
tobacco	0.081	0.026	3.16
ldl	0.168	0.054	3.09
famhist	0.924	0.223	4.14
age	0.044	0.010	4.52

other correlated variables, they are no longer needed (and can even get a negative sign).

At this stage the analyst might do some model selection; find a subset of the variables that are sufficient for explaining their joint effect on the prevalence of chd. One way to proceed by is to drop the least significant coefficient, and refit the model. This is done repeatedly until no further terms can be dropped from the model. This gave the model shown in Table 4.3.

A better but more time-consuming strategy is to refit each of the models with one variable removed, and then perform an *analysis of deviance* to decide which variable to exclude. The residual deviance of a fitted model is minus twice its log-likelihood, and the deviance between two models is the difference of their individual residual deviances (in analogy to sums-of-squares). This strategy gave the same final model as above.

How does one interpret a coefficient of 0.081 (Std. Error = 0.026) for tobacco, for example? Tobacco is measured in total lifetime usage in kilograms, with a median of 1.0kg for the controls and 4.1kg for the cases. Thus an increase of 1kg in lifetime tobacco usage accounts for an increase in the odds of coronary heart disease of $\exp(0.081) = 1.084$ or 8.4%. Incorporating the standard error we get an approximate 95% confidence interval of $\exp(0.081 \pm 2 \times 0.026) = (1.03, 1.14)$.

We return to these data in Chapter 5, where we see that some of the variables have nonlinear effects, and when modeled appropriately, are not excluded from the model.

4.4.3 Quadratic Approximations and Inference

The maximum-likelihood parameter estimates $\hat{\beta}$ satisfy a self-consistency relationship: they are the coefficients of a weighted least squares fit, where the responses are

$$z_i = x_i^T \hat{\beta} + \frac{(y_i - \hat{p}_i)}{\hat{p}_i(1 - \hat{p}_i)}, \tag{4.29}$$

and the weights are $w_i = \hat{p}_i(1 - \hat{p}_i)$, both depending on $\hat{\beta}$ itself. Apart from providing a convenient algorithm, this connection with least squares has more to offer:

- The weighted residual sum-of-squares is the familiar Pearson chi-square statistic

$$\sum_{i=1}^{N} \frac{(y_i - \hat{p}_i)^2}{\hat{p}_i(1 - \hat{p}_i)}, \qquad (4.30)$$

  a quadratic approximation to the deviance.

- Asymptotic likelihood theory says that if the model is correct, then $\hat{\beta}$ is consistent (i.e., converges to the *true* β).

- A central limit theorem then shows that the distribution of $\hat{\beta}$ converges to $N(\beta, (\mathbf{X}^T \mathbf{W} \mathbf{X})^{-1})$. This and other asymptotics can be derived directly from the weighted least squares fit by mimicking normal theory inference.

- Model building can be costly for logistic regression models, because each model fitted requires iteration. Popular shortcuts are the *Rao score test* which tests for inclusion of a term, and the *Wald test* which can be used to test for exclusion of a term. Neither of these require iterative fitting, and are based on the maximum-likelihood fit of the current model. It turns out that both of these amount to adding or dropping a term from the weighted least squares fit, using the *same* weights. Such computations can be done efficiently, without recomputing the entire weighted least squares fit.

Software implementations can take advantage of these connections. For example, the generalized linear modeling software in R (which includes logistic regression as part of the binomial family of models) exploits them fully. GLM (generalized linear model) objects can be treated as linear model objects, and all the tools available for linear models can be applied automatically.

4.4.4 L_1 Regularized Logistic Regression

The L_1 penalty used in the lasso (Section 3.4.2) can be used for variable selection and shrinkage with any linear regression model. For logistic regression, we would maximize a penalized version of (4.20):

$$\max_{\beta_0, \beta} \left\{ \sum_{i=1}^{N} \left[y_i(\beta_0 + \beta^T x_i) - \log(1 + e^{\beta_0 + \beta^T x_i}) \right] - \lambda \sum_{j=1}^{p} |\beta_j| \right\}. \qquad (4.31)$$

As with the lasso, we typically do not penalize the intercept term, and standardize the predictors for the penalty to be meaningful. Criterion (4.31) is concave, and a solution can be found using nonlinear programming methods (Koh et al., 2007, for example). Alternatively, using the same quadratic approximations that were used in the Newton algorithm in Section 4.4.1, we can solve (4.31) by repeated application of a weighted lasso algorithm. Interestingly, the score equations [see (4.24)] for the variables with non-zero coefficients have the form

$$\mathbf{x}_j^T(\mathbf{y} - \mathbf{p}) = \lambda \cdot \text{sign}(\beta_j), \tag{4.32}$$

which generalizes (3.58) in Section 3.4.4; the active variables are tied in their *generalized* correlation with the residuals.

Path algorithms such as LAR for lasso are more difficult, because the coefficient profiles are piecewise smooth rather than linear. Nevertheless, progress can be made using quadratic approximations.

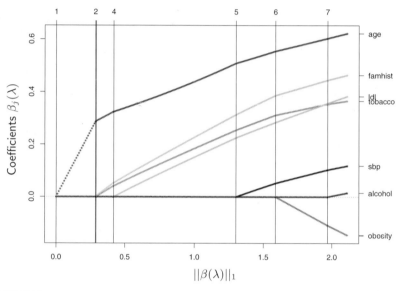

FIGURE 4.13. L_1 *regularized logistic regression coefficients for the South African heart disease data, plotted as a function of the L_1 norm. The variables were all standardized to have unit variance. The profiles are computed exactly at each of the plotted points.*

Figure 4.13 shows the L_1 regularization path for the South African heart disease data of Section 4.4.2. This was produced using the R package glmpath (Park and Hastie, 2007), which uses *predictor–corrector* methods of convex optimization to identify the exact values of λ at which the active set of non-zero coefficients changes (vertical lines in the figure). Here the profiles look almost linear; in other examples the curvature will be more visible.

Coordinate descent methods (Section 3.8.6) are very efficient for computing the coefficient profiles on a grid of values for λ. The R package glmnet

(Friedman et al., 2010) can fit coefficient paths for very large logistic regression problems efficiently (large in N or p). Their algorithms can exploit sparsity in the predictor matrix $\mathbf{X}$, which allows for even larger problems. See Section 18.4 for more details, and a discussion of L_1-regularized multinomial models.

4.4.5 Logistic Regression or LDA?

In Section 4.3 we find that the log-posterior odds between class k and K are linear functions of x (4.9):

$$
\begin{aligned}
\log \frac{\Pr(G=k|X=x)}{\Pr(G=K|X=x)} &= \log \frac{\pi_k}{\pi_K} - \frac{1}{2}(\mu_k + \mu_K)^T \mathbf{\Sigma}^{-1}(\mu_k - \mu_K) \\
&\quad + x^T \mathbf{\Sigma}^{-1}(\mu_k - \mu_K) \\
&= \alpha_{k0} + \alpha_k^T x.
\end{aligned}
\tag{4.33}
$$

This linearity is a consequence of the Gaussian assumption for the class densities, as well as the assumption of a common covariance matrix. The linear logistic model (4.17) by construction has linear logits:

$$
\log \frac{\Pr(G=k|X=x)}{\Pr(G=K|X=x)} = \beta_{k0} + \beta_k^T x.
\tag{4.34}
$$

It seems that the models are the same. Although they have exactly the same form, the difference lies in the way the linear coefficients are estimated. The logistic regression model is more general, in that it makes less assumptions. We can write the *joint density* of X and G as

$$
\Pr(X, G=k) = \Pr(X)\Pr(G=k|X),
\tag{4.35}
$$

where $\Pr(X)$ denotes the marginal density of the inputs X. For both LDA and logistic regression, the second term on the right has the logit-linear form

$$
\Pr(G=k|X=x) = \frac{e^{\beta_{k0}+\beta_k^T x}}{1 + \sum_{\ell=1}^{K-1} e^{\beta_{\ell 0}+\beta_\ell^T x}},
\tag{4.36}
$$

where we have again arbitrarily chosen the last class as the reference.

The logistic regression model leaves the marginal density of X as an arbitrary density function $\Pr(X)$, and fits the parameters of $\Pr(G|X)$ by maximizing the *conditional likelihood*—the multinomial likelihood with probabilities the $\Pr(G=k|X)$. Although $\Pr(X)$ is totally ignored, we can think of this marginal density as being estimated in a fully nonparametric and unrestricted fashion, using the empirical distribution function which places mass $1/N$ at each observation.

With LDA we fit the parameters by maximizing the full log-likelihood, based on the joint density

$$
\Pr(X, G=k) = \phi(X; \mu_k, \mathbf{\Sigma})\pi_k,
\tag{4.37}
$$

where ϕ is the Gaussian density function. Standard normal theory leads easily to the estimates $\hat{\mu}_k, \hat{\Sigma}$, and $\hat{\pi}_k$ given in Section 4.3. Since the linear parameters of the logistic form (4.33) are functions of the Gaussian parameters, we get their maximum-likelihood estimates by plugging in the corresponding estimates. However, unlike in the conditional case, the marginal density $\Pr(X)$ does play a role here. It is a mixture density

$$\Pr(X) = \sum_{k=1}^{K} \pi_k \phi(X; \mu_k, \Sigma), \tag{4.38}$$

which also involves the parameters.

What role can this additional component/restriction play? By relying on the additional model assumptions, we have more information about the parameters, and hence can estimate them more efficiently (lower variance). If in fact the true $f_k(x)$ are Gaussian, then in the worst case ignoring this marginal part of the likelihood constitutes a loss of efficiency of about 30% asymptotically in the error rate (Efron, 1975). Paraphrasing: with 30% more data, the conditional likelihood will do as well.

For example, observations far from the decision boundary (which are down-weighted by logistic regression) play a role in estimating the common covariance matrix. This is not all good news, because it also means that LDA is not robust to gross outliers.

From the mixture formulation, it is clear that even observations without class labels have information about the parameters. Often it is expensive to generate class labels, but unclassified observations come cheaply. By relying on strong model assumptions, such as here, we can use both types of information.

The marginal likelihood can be thought of as a regularizer, requiring in some sense that class densities be *visible* from this marginal view. For example, if the data in a two-class logistic regression model can be perfectly separated by a hyperplane, the maximum likelihood estimates of the parameters are undefined (i.e., infinite; see Exercise 4.5). The LDA coefficients for the same data will be well defined, since the marginal likelihood will not permit these degeneracies.

In practice these assumptions are never correct, and often some of the components of X are qualitative variables. It is generally felt that logistic regression is a safer, more robust bet than the LDA model, relying on fewer assumptions. It is our experience that the models give very similar results, even when LDA is used inappropriately, such as with qualitative predictors.

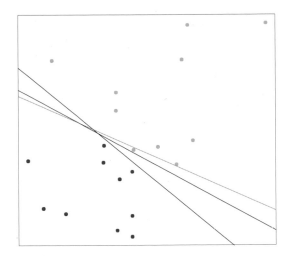

FIGURE 4.14. *A toy example with two classes separable by a hyperplane. The orange line is the least squares solution, which misclassifies one of the training points. Also shown are two blue separating hyperplanes found by the* perceptron learning algorithm *with different random starts.*

4.5 Separating Hyperplanes

We have seen that linear discriminant analysis and logistic regression both estimate linear decision boundaries in similar but slightly different ways. For the rest of this chapter we describe separating hyperplane classifiers. These procedures construct linear decision boundaries that explicitly try to separate the data into different classes as well as possible. They provide the basis for support vector classifiers, discussed in Chapter 12. The mathematical level of this section is somewhat higher than that of the previous sections.

Figure 4.14 shows 20 data points in two classes in $\mathbb{R}^2$. These data can be separated by a linear boundary. Included in the figure (blue lines) are two of the infinitely many possible *separating hyperplanes*. The orange line is the least squares solution to the problem, obtained by regressing the $-1/1$ response Y on X (with intercept); the line is given by

$$\{x : \hat{\beta}_0 + \hat{\beta}_1 x_1 + \hat{\beta}_2 x_2 = 0\}. \tag{4.39}$$

This least squares solution does not do a perfect job in separating the points, and makes one error. This is the same boundary found by LDA, in light of its equivalence with linear regression in the two-class case (Section 4.3 and Exercise 4.2).

Classifiers such as (4.39), that compute a linear combination of the input features and return the sign, were called *perceptrons* in the engineering liter-

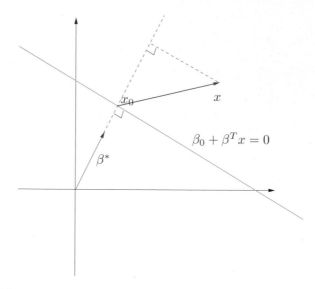

FIGURE 4.15. *The linear algebra of a hyperplane (affine set).*

ature in the late 1950s (Rosenblatt, 1958). Perceptrons set the foundations for the neural network models of the 1980s and 1990s.

Before we continue, let us digress slightly and review some vector algebra. Figure 4.15 depicts a hyperplane or *affine set* L defined by the equation $f(x) = \beta_0 + \beta^T x = 0$; since we are in $\mathbb{R}^2$ this is a line.

Here we list some properties:

1. For any two points x_1 and x_2 lying in L, $\beta^T(x_1 - x_2) = 0$, and hence $\beta^* = \beta/\|\beta\|$ is the vector normal to the surface of L.

2. For any point x_0 in L, $\beta^T x_0 = -\beta_0$.

3. The signed distance of any point x to L is given by

$$
\begin{aligned}
\beta^{*T}(x - x_0) &= \frac{1}{\|\beta\|}(\beta^T x + \beta_0) \\
&= \frac{1}{\|f'(x)\|}f(x). \quad (4.40)
\end{aligned}
$$

Hence $f(x)$ is proportional to the signed distance from x to the hyperplane defined by $f(x) = 0$.

4.5.1 Rosenblatt's Perceptron Learning Algorithm

The *perceptron learning algorithm* tries to find a separating hyperplane by minimizing the distance of misclassified points to the decision boundary. If a response $y_i = 1$ is misclassified, then $x_i^T \beta + \beta_0 < 0$, and the opposite for a misclassified response with $y_i = -1$. The goal is to minimize

$$D(\beta, \beta_0) = - \sum_{i \in \mathcal{M}} y_i (x_i^T \beta + \beta_0), \qquad (4.41)$$

where $\mathcal{M}$ indexes the set of misclassified points. The quantity is non-negative and proportional to the distance of the misclassified points to the decision boundary defined by $\beta^T x + \beta_0 = 0$. The gradient (assuming $\mathcal{M}$ is fixed) is given by

$$\partial \frac{D(\beta, \beta_0)}{\partial \beta} = - \sum_{i \in \mathcal{M}} y_i x_i, \qquad (4.42)$$

$$\partial \frac{D(\beta, \beta_0)}{\partial \beta_0} = - \sum_{i \in \mathcal{M}} y_i. \qquad (4.43)$$

The algorithm in fact uses *stochastic gradient descent* to minimize this piecewise linear criterion. This means that rather than computing the sum of the gradient contributions of each observation followed by a step in the negative gradient direction, a step is taken after each observation is visited. Hence the misclassified observations are visited in some sequence, and the parameters β are updated via

$$\begin{pmatrix} \beta \\ \beta_0 \end{pmatrix} \leftarrow \begin{pmatrix} \beta \\ \beta_0 \end{pmatrix} + \rho \begin{pmatrix} y_i x_i \\ y_i \end{pmatrix}. \qquad (4.44)$$

Here ρ is the learning rate, which in this case can be taken to be 1 without loss in generality. If the classes are linearly separable, it can be shown that the algorithm converges to a separating hyperplane in a finite number of steps (Exercise 4.6). Figure 4.14 shows two solutions to a toy problem, each started at a different random guess.

There are a number of problems with this algorithm, summarized in Ripley (1996):

- When the data are separable, there are many solutions, and which one is found depends on the starting values.

- The "finite" number of steps can be very large. The smaller the gap, the longer the time to find it.

- When the data are not separable, the algorithm will not converge, and cycles develop. The cycles can be long and therefore hard to detect.

The second problem can often be eliminated by seeking a hyperplane not in the original space, but in a much enlarged space obtained by creating many basis-function transformations of the original variables. This is analogous to driving the residuals in a polynomial regression problem down to zero by making the degree sufficiently large. Perfect separation cannot always be achieved: for example, if observations from two different classes share the same input. It may not be desirable either, since the resulting model is likely to be overfit and will not generalize well. We return to this point at the end of the next section.

A rather elegant solution to the first problem is to add additional constraints to the separating hyperplane.

4.5.2 Optimal Separating Hyperplanes

The *optimal separating hyperplane* separates the two classes and maximizes the distance to the closest point from either class (Vapnik, 1996). Not only does this provide a unique solution to the separating hyperplane problem, but by maximizing the margin between the two classes on the training data, this leads to better classification performance on test data.

We need to generalize criterion (4.41). Consider the optimization problem

$$\max_{\beta,\beta_0,||\beta||=1} M$$
$$\text{subject to } y_i(x_i^T\beta + \beta_0) \ge M, \; i = 1,\dots,N. \tag{4.45}$$

The set of conditions ensure that all the points are at least a signed distance M from the decision boundary defined by β and β_0, and we seek the largest such M and associated parameters. We can get rid of the $||\beta|| = 1$ constraint by replacing the conditions with

$$\frac{1}{||\beta||}y_i(x_i^T\beta + \beta_0) > M, \tag{4.46}$$

(which redefines β_0) or equivalently

$$y_i(x_i^T\beta + \beta_0) \ge M||\beta||. \tag{4.47}$$

Since for any β and β_0 satisfying these inequalities, any positively scaled multiple satisfies them too, we can arbitrarily set $||\beta|| = 1/M$. Thus (4.45) is equivalent to

$$\min_{\beta,\beta_0} \frac{1}{2}||\beta||^2$$
$$\text{subject to } y_i(x_i^T\beta + \beta_0) \ge 1, \; i = 1,\dots,N. \tag{4.48}$$

In light of (4.40), the constraints define an empty slab or margin around the linear decision boundary of thickness $1/||\beta||$. Hence we choose β and β_0 to maximize its thickness. This is a convex optimization problem (quadratic criterion with linear inequality constraints). The Lagrange (primal) function, to be minimized w.r.t. β and β_0, is

$$L_P = \frac{1}{2}||\beta||^2 - \sum_{i=1}^{N} \alpha_i [y_i(x_i^T \beta + \beta_0) - 1]. \tag{4.49}$$

Setting the derivatives to zero, we obtain:

$$\beta = \sum_{i=1}^{N} \alpha_i y_i x_i, \tag{4.50}$$

$$0 = \sum_{i=1}^{N} \alpha_i y_i, \tag{4.51}$$

and substituting these in (4.49) we obtain the so-called Wolfe dual

$$L_D = \sum_{i=1}^{N} \alpha_i - \frac{1}{2}\sum_{i=1}^{N}\sum_{k=1}^{N} \alpha_i \alpha_k y_i y_k x_i^T x_k$$

$$\text{subject to } \alpha_i \geq 0 \text{ and } \sum_{i=1}^{N} \alpha_i y_i = 0. \tag{4.52}$$

The solution is obtained by maximizing L_D in the positive orthant, a simpler convex optimization problem, for which standard software can be used. In addition the solution must satisfy the Karush–Kuhn–Tucker conditions, which include (4.50), (4.51), (4.52) and

$$\alpha_i [y_i(x_i^T \beta + \beta_0) - 1] = 0 \ \forall i. \tag{4.53}$$

From these we can see that

- if $\alpha_i > 0$, then $y_i(x_i^T \beta + \beta_0) = 1$, or in other words, x_i is on the boundary of the slab;

- if $y_i(x_i^T \beta + \beta_0) > 1$, x_i is not on the boundary of the slab, and $\alpha_i = 0$.

From (4.50) we see that the solution vector β is defined in terms of a linear combination of the *support points* x_i—those points defined to be on the boundary of the slab via $\alpha_i > 0$. Figure 4.16 shows the optimal separating hyperplane for our toy example; there are three support points. Likewise, β_0 is obtained by solving (4.53) for any of the support points.

The optimal separating hyperplane produces a function $\hat{f}(x) = x^T \hat{\beta} + \hat{\beta}_0$ for classifying new observations:

$$\hat{G}(x) = \text{sign}\hat{f}(x). \tag{4.54}$$

Although none of the training observations fall in the margin (by construction), this will not necessarily be the case for test observations. The

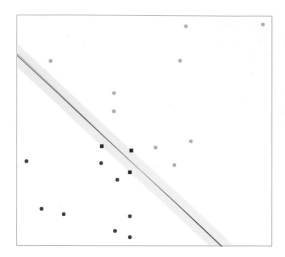

FIGURE 4.16. *The same data as in Figure 4.14. The shaded region delineates the maximum margin separating the two classes. There are three support points indicated, which lie on the boundary of the margin, and the optimal separating hyperplane (blue line) bisects the slab. Included in the figure is the boundary found using logistic regression (red line), which is very close to the optimal separating hyperplane (see Section 12.3.3).*

intuition is that a large margin on the training data will lead to good separation on the test data.

The description of the solution in terms of support points seems to suggest that the optimal hyperplane focuses more on the points that count, and is more robust to model misspecification. The LDA solution, on the other hand, depends on all of the data, even points far away from the decision boundary. Note, however, that the identification of these support points required the use of all the data. Of course, if the classes are really Gaussian, then LDA is optimal, and separating hyperplanes will pay a price for focusing on the (noisier) data at the boundaries of the classes.

Included in Figure 4.16 is the logistic regression solution to this problem, fit by maximum likelihood. Both solutions are similar in this case. When a separating hyperplane exists, logistic regression will always find it, since the log-likelihood can be driven to 0 in this case (Exercise 4.5). The logistic regression solution shares some other qualitative features with the separating hyperplane solution. The coefficient vector is defined by a weighted least squares fit of a zero-mean linearized response on the input features, and the weights are larger for points near the decision boundary than for those further away.

When the data are not separable, there will be no feasible solution to this problem, and an alternative formulation is needed. Again one can enlarge the space using basis transformations, but this can lead to artificial separation through over-fitting. In Chapter 12 we discuss a more attractive alternative known as the *support vector machine*, which allows for overlap, but minimizes a measure of the extent of this overlap.

Bibliographic Notes

Good general texts on classification include Duda et al. (2000), Hand (1981), McLachlan (1992) and Ripley (1996). Mardia et al. (1979) have a concise discussion of linear discriminant analysis. Michie et al. (1994) compare a large number of popular classifiers on benchmark datasets. Linear separating hyperplanes are discussed in Vapnik (1996). Our account of the perceptron learning algorithm follows Ripley (1996).

Exercises

Ex. 4.1 Show how to solve the generalized eigenvalue problem $\max a^T \mathbf{B} a$ subject to $a^T \mathbf{W} a = 1$ by transforming to a standard eigenvalue problem.

Ex. 4.2 Suppose we have features $x \in \mathbb{R}^p$, a two-class response, with class sizes N_1, N_2, and the target coded as $-N/N_1, N/N_2$.

(a) Show that the LDA rule classifies to class 2 if

$$x^T \hat{\boldsymbol{\Sigma}}^{-1}(\hat{\mu}_2 - \hat{\mu}_1) > \frac{1}{2}(\hat{\mu}_2 + \hat{\mu}_1)^T \hat{\boldsymbol{\Sigma}}^{-1}(\hat{\mu}_2 - \hat{\mu}_1) - \log(N_2/N_1),$$

and class 1 otherwise.

(b) Consider minimization of the least squares criterion

$$\sum_{i=1}^{N}(y_i - \beta_0 - x_i^T \beta)^2. \tag{4.55}$$

Show that the solution $\hat{\beta}$ satisfies

$$\left[(N-2)\hat{\boldsymbol{\Sigma}} + N\hat{\boldsymbol{\Sigma}}_B\right]\beta = N(\hat{\mu}_2 - \hat{\mu}_1) \tag{4.56}$$

(after simplification), where $\hat{\boldsymbol{\Sigma}}_B = \frac{N_1 N_2}{N^2}(\hat{\mu}_2 - \hat{\mu}_1)(\hat{\mu}_2 - \hat{\mu}_1)^T$.

(c) Hence show that $\hat{\boldsymbol{\Sigma}}_B \beta$ is in the direction $(\hat{\mu}_2 - \hat{\mu}_1)$ and thus

$$\hat{\beta} \propto \hat{\boldsymbol{\Sigma}}^{-1}(\hat{\mu}_2 - \hat{\mu}_1). \tag{4.57}$$

Therefore the least-squares regression coefficient is identical to the LDA coefficient, up to a scalar multiple.

(d) Show that this result holds for any (distinct) coding of the two classes.

(e) Find the solution $\hat{\beta}_0$ (up to the same scalar multiple as in (c), and hence the predicted value $\hat{f}(x) = \hat{\beta}_0 + x^T \hat{\beta}$. Consider the following rule: classify to class 2 if $\hat{f}(x) > 0$ and class 1 otherwise. Show this is not the same as the LDA rule unless the classes have equal numbers of observations.

(Fisher, 1936; Ripley, 1996)

Ex. 4.3 Suppose we transform the original predictors $\mathbf{X}$ to $\hat{\mathbf{Y}}$ via linear regression. In detail, let $\hat{\mathbf{Y}} = \mathbf{X}(\mathbf{X}^T\mathbf{X})^{-1}\mathbf{X}^T\mathbf{Y} = \mathbf{X}\hat{\mathbf{B}}$, where $\mathbf{Y}$ is the indicator response matrix. Similarly for any input $x \in \mathbb{R}^p$, we get a transformed vector $\hat{y} = \hat{\mathbf{B}}^T x \in \mathbb{R}^K$. Show that LDA using $\hat{\mathbf{Y}}$ is identical to LDA in the original space.

Ex. 4.4 Consider the multilogit model with K classes (4.17). Let β be the $(p+1)(K-1)$-vector consisting of all the coefficients. Define a suitably enlarged version of the input vector x to accommodate this vectorized coefficient matrix. Derive the Newton-Raphson algorithm for maximizing the multinomial log-likelihood, and describe how you would implement this algorithm.

Ex. 4.5 Consider a two-class logistic regression problem with $x \in \mathbb{R}$. Characterize the maximum-likelihood estimates of the slope and intercept parameter if the sample x_i for the two classes are separated by a point $x_0 \in \mathbb{R}$. Generalize this result to (a) $x \in \mathbb{R}^p$ (see Figure 4.16), and (b) more than two classes.

Ex. 4.6 Suppose we have N points x_i in $\mathbb{R}^p$ in general position, with class labels $y_i \in \{-1, 1\}$. Prove that the perceptron learning algorithm converges to a separating hyperplane in a finite number of steps:

(a) Denote a hyperplane by $f(x) = \beta_1^T x + \beta_0 = 0$, or in more compact notation $\beta^T x^* = 0$, where $x^* = (x, 1)$ and $\beta = (\beta_1, \beta_0)$. Let $z_i = x_i^*/||x_i^*||$. Show that separability implies the existence of a β_{sep} such that $y_i \beta_{\text{sep}}^T z_i \geq 1 \; \forall i$

(b) Given a current β_{old}, the perceptron algorithm identifies a point z_i that is misclassified, and produces the update $\beta_{\text{new}} \leftarrow \beta_{\text{old}} + y_i z_i$. Show that $||\beta_{\text{new}} - \beta_{\text{sep}}||^2 \leq ||\beta_{\text{old}} - \beta_{\text{sep}}||^2 - 1$, and hence that the algorithm converges to a separating hyperplane in no more than $||\beta_{\text{start}} - \beta_{\text{sep}}||^2$ steps (Ripley, 1996).

Ex. 4.7 Consider the criterion

$$D^*(\beta, \beta_0) = -\sum_{i=1}^{N} y_i (x_i^T \beta + \beta_0), \qquad (4.58)$$

a generalization of (4.41) where we sum over all the observations. Consider minimizing D^* subject to $||\beta|| = 1$. Describe this criterion in words. Does it solve the optimal separating hyperplane problem?

Ex. 4.8 Consider the multivariate Gaussian model $X|G = k \sim N(\mu_k, \Sigma)$, with the additional restriction that $\text{rank}\{\mu_k\}_1^K = L < \max(K - 1, p)$. Derive the constrained MLEs for the μ_k and Σ. Show that the Bayes classification rule is equivalent to classifying in the reduced subspace computed by LDA (Hastie and Tibshirani, 1996b).

Ex. 4.9 Write a computer program to perform a quadratic discriminant analysis by fitting a separate Gaussian model per class. Try it out on the vowel data, and compute the misclassification error for the test data. The data can be found in the book website `www-stat.stanford.edu/ElemStatLearn`.

5
Basis Expansions and Regularization

5.1 Introduction

We have already made use of models linear in the input features, both for
regression and classification. Linear regression, linear discriminant analysis,
logistic regression and separating hyperplanes all rely on a linear model.
It is extremely unlikely that the true function $f(X)$ is actually linear in
X. In regression problems, $f(X) = \mathrm{E}(Y|X)$ will typically be nonlinear and
nonadditive in X, and representing $f(X)$ by a linear model is usually a convenient, and sometimes a necessary, approximation. Convenient because a
linear model is easy to interpret, and is the first-order Taylor approximation to $f(X)$. Sometimes necessary, because with N small and/or p large,
a linear model might be all we are able to fit to the data without overfitting. Likewise in classification, a linear, Bayes-optimal decision boundary
implies that some monotone transformation of $\Pr(Y = 1|X)$ is linear in X.
This is inevitably an approximation.

In this chapter and the next we discuss popular methods for moving
beyond linearity. The core idea in this chapter is to augment/replace the
vector of inputs X with additional variables, which are transformations of
X, and then use linear models in this new space of derived input features.

Denote by $h_m(X) : \mathbb{R}^p \mapsto \mathbb{R}$ the mth transformation of X, $m =
1, \ldots, M$. We then model

$$f(X) = \sum_{m=1}^{M} \beta_m h_m(X), \tag{5.1}$$

T. Hastie et al., *The Elements of Statistical Learning, Second Edition*,
DOI: 10.1007/b94608_5,
© Springer Science+Business Media, LLC 2009

a *linear basis expansion* in X. The beauty of this approach is that once the basis functions h_m have been determined, the models are linear in these new variables, and the fitting proceeds as before.

Some simple and widely used examples of the h_m are the following:

- $h_m(X) = X_m$, $m = 1, \ldots, p$ recovers the original linear model.

- $h_m(X) = X_j^2$ or $h_m(X) = X_j X_k$ allows us to augment the inputs with polynomial terms to achieve higher-order Taylor expansions. Note, however, that the number of variables grows exponentially in the degree of the polynomial. A full quadratic model in p variables requires $O(p^2)$ square and cross-product terms, or more generally $O(p^d)$ for a degree-d polynomial.

- $h_m(X) = \log(X_j)$, $\sqrt{X_j}, \ldots$ permits other nonlinear transformations of single inputs. More generally one can use similar functions involving several inputs, such as $h_m(X) = ||X||$.

- $h_m(X) = I(L_m \leq X_k < U_m)$, an indicator for a region of X_k. By breaking the range of X_k up into M_k such nonoverlapping regions results in a model with a piecewise constant contribution for X_k.

Sometimes the problem at hand will call for particular basis functions h_m, such as logarithms or power functions. More often, however, we use the basis expansions as a device to achieve more flexible representations for $f(X)$. Polynomials are an example of the latter, although they are limited by their global nature—tweaking the coefficients to achieve a functional form in one region can cause the function to flap about madly in remote regions. In this chapter we consider more useful families of *piecewise-polynomials* and *splines* that allow for local polynomial representations. We also discuss the *wavelet* bases, especially useful for modeling signals and images. These methods produce a *dictionary* $\mathcal{D}$ consisting of typically a very large number $|\mathcal{D}|$ of basis functions, far more than we can afford to fit to our data. Along with the dictionary we require a method for controlling the complexity of our model, using basis functions from the dictionary. There are three common approaches:

- Restriction methods, where we decide before-hand to limit the class of functions. Additivity is an example, where we assume that our model has the form

$$f(X) = \sum_{j=1}^{p} f_j(X_j)$$

$$= \sum_{j=1}^{p} \sum_{m=1}^{M_j} \beta_{jm} h_{jm}(X_j). \tag{5.2}$$

The size of the model is limited by the number of basis functions M_j used for each component function f_j.

- Selection methods, which adaptively scan the dictionary and include only those basis functions h_m that contribute significantly to the fit of the model. Here the variable selection techniques discussed in Chapter 3 are useful. The stagewise greedy approaches such as CART, MARS and boosting fall into this category as well.

- Regularization methods where we use the entire dictionary but restrict the coefficients. Ridge regression is a simple example of a regularization approach, while the lasso is both a regularization and selection method. Here we discuss these and more sophisticated methods for regularization.

5.2   Piecewise Polynomials and Splines

We assume until Section 5.7 that X is one-dimensional. A piecewise polynomial function $f(X)$ is obtained by dividing the domain of X into contiguous intervals, and representing f by a separate polynomial in each interval. Figure 5.1 shows two simple piecewise polynomials. The first is piecewise constant, with three basis functions:

$$h_1(X) = I(X < \xi_1), \quad h_2(X) = I(\xi_1 \le X < \xi_2), \quad h_3(X) = I(\xi_2 \le X).$$

Since these are positive over disjoint regions, the least squares estimate of the model $f(X) = \sum_{m=1}^{3} \beta_m h_m(X)$ amounts to $\hat{\beta}_m = \bar{Y}_m$, the mean of Y in the mth region.

The top right panel shows a piecewise linear fit. Three additional basis functions are needed: $h_{m+3} = h_m(X)X$, $m = 1, \ldots, 3$. Except in special cases, we would typically prefer the third panel, which is also piecewise linear, but restricted to be continuous at the two knots. These continuity restrictions lead to linear constraints on the parameters; for example, $f(\xi_1^-) = f(\xi_1^+)$ implies that $\beta_1 + \xi_1 \beta_4 = \beta_2 + \xi_1 \beta_5$. In this case, since there are two restrictions, we expect to *get back* two parameters, leaving four free parameters.

A more direct way to proceed in this case is to use a basis that incorporates the constraints:

$$h_1(X) = 1, \quad h_2(X) = X, \quad h_3(X) = (X - \xi_1)_+, \quad h_4(X) = (X - \xi_2)_+,$$

where t_+ denotes the positive part. The function h_3 is shown in the lower right panel of Figure 5.1. We often prefer smoother functions, and these can be achieved by increasing the order of the local polynomial. Figure 5.2 shows a series of piecewise-cubic polynomials fit to the same data, with

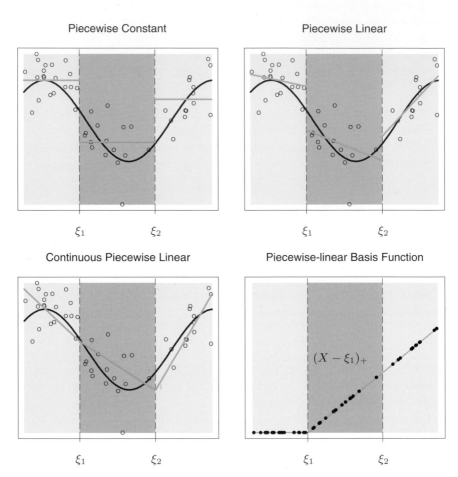

FIGURE 5.1. *The top left panel shows a piecewise constant function fit to some artificial data. The broken vertical lines indicate the positions of the two knots ξ_1 and ξ_2. The blue curve represents the true function, from which the data were generated with Gaussian noise. The remaining two panels show piecewise linear functions fit to the same data—the top right unrestricted, and the lower left restricted to be continuous at the knots. The lower right panel shows a piecewise-linear basis function, $h_3(X) = (X - \xi_1)_+$, continuous at ξ_1. The black points indicate the sample evaluations $h_3(x_i)$, $i = 1, \ldots, N$.*

Piecewise Cubic Polynomials

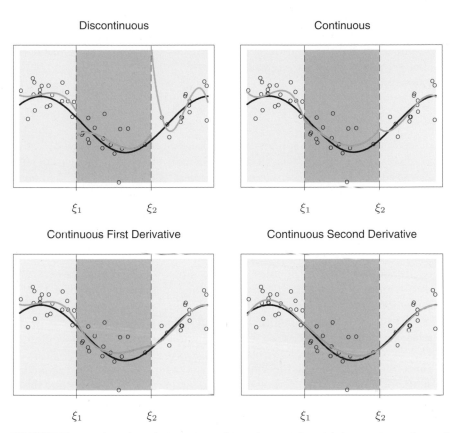

FIGURE 5.2. *A series of piecewise-cubic polynomials, with increasing orders of continuity.*

increasing orders of continuity at the knots. The function in the lower right panel is continuous, and has continuous first and second derivatives at the knots. It is known as a *cubic spline*. Enforcing one more order of continuity would lead to a global cubic polynomial. It is not hard to show (Exercise 5.1) that the following basis represents a cubic spline with knots at ξ_1 and ξ_2:

$$
\begin{aligned}
h_1(X) &= 1, & h_3(X) &= X^2, & h_5(X) &= (X - \xi_1)^3_+, \\
h_2(X) &= X, & h_4(X) &= X^3, & h_6(X) &= (X - \xi_2)^3_+.
\end{aligned}
\tag{5.3}
$$

There are six basis functions corresponding to a six-dimensional linear space of functions. A quick check confirms the parameter count: (3 regions)×(4 parameters per region) −(2 knots)×(3 constraints per knot)= 6.

More generally, an order-M spline with knots ξ_j, $j = 1, \ldots, K$ is a piecewise-polynomial of order M, and has continuous derivatives up to order $M - 2$. A cubic spline has $M = 4$. In fact the piecewise-constant function in Figure 5.1 is an order-1 spline, while the continuous piecewise linear function is an order-2 spline. Likewise the general form for the truncated-power basis set would be

$$
\begin{aligned}
h_j(X) &= X^{j-1}, \; j = 1, \ldots, M, \\
h_{M+\ell}(X) &= (X - \xi_\ell)_+^{M-1}, \; \ell = 1, \ldots, K.
\end{aligned}
$$

It is claimed that cubic splines are the lowest-order spline for which the knot-discontinuity is not visible to the human eye. There is seldom any good reason to go beyond cubic-splines, unless one is interested in smooth derivatives. In practice the most widely used orders are $M = 1, 2$ and 4.

These fixed-knot splines are also known as *regression splines*. One needs to select the order of the spline, the number of knots and their placement. One simple approach is to parameterize a family of splines by the number of basis functions or degrees of freedom, and have the observations x_i determine the positions of the knots. For example, the expression bs(x,df=7) in R generates a basis matrix of cubic-spline functions evaluated at the N observations in x, with the $7 - 3 = 4$[1] interior knots at the appropriate percentiles of x (20, 40, 60 and 80th.) One can be more explicit, however; bs(x, degree=1, knots = c(0.2, 0.4, 0.6)) generates a basis for linear splines, with three interior knots, and returns an $N \times 4$ matrix.

Since the space of spline functions of a particular order and knot sequence is a vector space, there are many equivalent bases for representing them (just as there are for ordinary polynomials.) While the truncated power basis is conceptually simple, it is not too attractive numerically: powers of large numbers can lead to severe rounding problems. The *B-spline* basis, described in the Appendix to this chapter, allows for efficient computations even when the number of knots K is large.

5.2.1 *Natural Cubic Splines*

We know that the behavior of polynomials fit to data tends to be erratic near the boundaries, and extrapolation can be dangerous. These problems are exacerbated with splines. The polynomials fit beyond the boundary knots behave even more wildly than the corresponding global polynomials in that region. This can be conveniently summarized in terms of the pointwise variance of spline functions fit by least squares (see the example in the next section for details on these variance calculations). Figure 5.3 compares

---

[1] A cubic spline with four knots is eight-dimensional. The bs() function omits by default the constant term in the basis, since terms like this are typically included with other terms in the model.

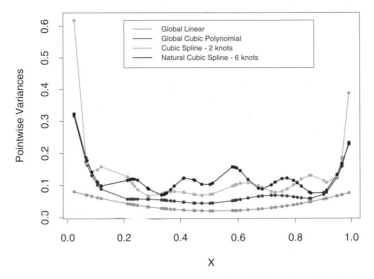

FIGURE 5.3. *Pointwise variance curves for four different models, with X con-sisting of 50 points drawn at random from $U[0,1]$, and an assumed error model with constant variance. The linear and cubic polynomial fits have two and four degrees of freedom, respectively, while the cubic spline and natural cubic spline each have six degrees of freedom. The cubic spline has two knots at 0.33 and 0.66, while the natural spline has boundary knots at 0.1 and 0.9, and four interior knots uniformly spaced between them.*

the pointwise variances for a variety of different models. The explosion of the variance near the boundaries is clear, and inevitably is worst for cubic splines.

A *natural cubic spline* adds additional constraints, namely that the function is linear beyond the boundary knots. This frees up four degrees of freedom (two constraints each in both boundary regions), which can be spent more profitably by sprinkling more knots in the interior region. This tradeoff is illustrated in terms of variance in Figure 5.3. There will be a price paid in bias near the boundaries, but assuming the function is lin-ear near the boundaries (where we have less information anyway) is often considered reasonable.

A natural cubic spline with K knots is represented by K basis functions. One can start from a basis for cubic splines, and derive the reduced ba-sis by imposing the boundary constraints. For example, starting from the truncated power series basis described in Section 5.2, we arrive at (Exer-cise 5.4):

$$N_1(X) = 1, \quad N_2(X) = X, \quad N_{k+2}(X) = d_k(X) - d_{K-1}(X), \quad (5.4)$$

where

$$d_k(X) = \frac{(X - \xi_k)_+^3 - (X - \xi_K)_+^3}{\xi_K - \xi_k}. \tag{5.5}$$

Each of these basis functions can be seen to have zero second and third derivative for $X \geq \xi_K$.

5.2.2 Example: South African Heart Disease (Continued)

In Section 4.4.2 we fit linear logistic regression models to the South African heart disease data. Here we explore nonlinearities in the functions using natural splines. The functional form of the model is

$$\text{logit}[\Pr(\text{chd}|X)] = \theta_0 + h_1(X_1)^T\theta_1 + h_2(X_2)^T\theta_2 + \cdots + h_p(X_p)^T\theta_p, \tag{5.6}$$

where each of the θ_j are vectors of coefficients multiplying their associated vector of natural spline basis functions h_j.

We use four natural spline bases for each term in the model. For example, with X_1 representing sbp, $h_1(X_1)$ is a basis consisting of four basis functions. This actually implies three rather than two interior knots (chosen at uniform quantiles of sbp), plus two boundary knots at the extremes of the data, since we exclude the constant term from each of the h_j.

Since famhist is a two-level factor, it is coded by a simple binary or dummy variable, and is associated with a single coefficient in the fit of the model.

More compactly we can combine all p vectors of basis functions (and the constant term) into one big vector $h(X)$, and then the model is simply $h(X)^T\theta$, with total number of parameters $\text{df} = 1 + \sum_{j=1}^{p} \text{df}_j$, the sum of the parameters in each component term. Each basis function is evaluated at each of the N samples, resulting in a $N \times \text{df}$ basis matrix $\mathbf{H}$. At this point the model is like any other linear logistic model, and the algorithms described in Section 4.4.1 apply.

We carried out a backward stepwise deletion process, dropping terms from this model while preserving the group structure of each term, rather than dropping one coefficient at a time. The AIC statistic (Section 7.5) was used to drop terms, and all the terms remaining in the final model would cause AIC to increase if deleted from the model (see Table 5.1). Figure 5.4 shows a plot of the final model selected by the stepwise regression. The functions displayed are $\hat{f}_j(X_j) = h_j(X_j)^T\hat{\theta}_j$ for each variable X_j. The covariance matrix $\text{Cov}(\hat{\theta}) = \Sigma$ is estimated by $\hat{\Sigma} = (\mathbf{H}^T\mathbf{W}\mathbf{H})^{-1}$, where $\mathbf{W}$ is the diagonal weight matrix from the logistic regression. Hence $v_j(X_j) = \text{Var}[\hat{f}_j(X_j)] = h_j(X_j)^T\hat{\Sigma}_{jj}h_j(X_j)$ is the pointwise variance function of $\hat{f}_j$, where $\text{Cov}(\hat{\theta}_j) = \hat{\Sigma}_{jj}$ is the appropriate sub-matrix of $\hat{\Sigma}$. The shaded region in each panel is defined by $\hat{f}_j(X_j) \pm 2\sqrt{v_j(X_j)}$.

The AIC statistic is slightly more generous than the likelihood-ratio test (deviance test). Both sbp and obesity are included in this model, while

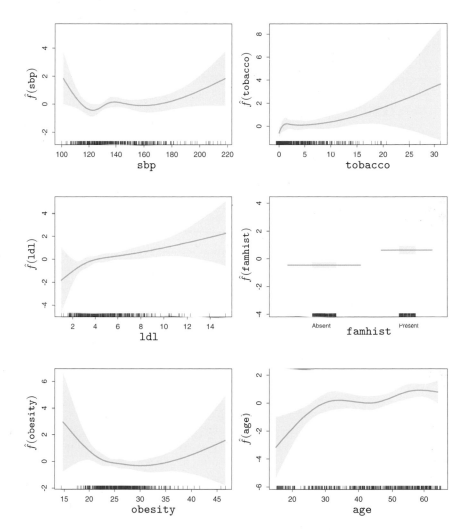

FIGURE 5.4. *Fitted natural-spline functions for each of the terms in the final model selected by the stepwise procedure. Included are pointwise standard-error bands. The* rug plot *at the base of each figure indicates the location of each of the sample values for that variable (jittered to break ties).*

TABLE 5.1. *Final logistic regression model, after stepwise deletion of natural splines terms. The column labeled "LRT" is the likelihood-ratio test statistic when that term is deleted from the model, and is the change in deviance from the full model (labeled "none").*

Terms	Df	Deviance	AIC	LRT	P-value
none		458.09	502.09		
sbp	4	467.16	503.16	9.076	0.059
tobacco	4	470.48	506.48	12.387	0.015
ldl	4	472.39	508.39	14.307	0.006
famhist	1	479.44	521.44	21.356	0.000
obesity	4	466.24	502.24	8.147	0.086
age	4	481.86	517.86	23.768	0.000

they were not in the linear model. The figure explains why, since their contributions are inherently nonlinear. These effects at first may come as a surprise, but an explanation lies in the nature of the retrospective data. These measurements were made sometime after the patients suffered a heart attack, and in many cases they had already benefited from a healthier diet and lifestyle, hence the apparent *increase* in risk at low values for obesity and sbp. Table 5.1 shows a summary of the selected model.

5.2.3 Example: Phoneme Recognition

In this example we use splines to reduce flexibility rather than increase it; the application comes under the general heading of *functional* modeling. In the top panel of Figure 5.5 are displayed a sample of 15 log-periodograms for each of the two phonemes "aa" and "ao" measured at 256 frequencies. The goal is to use such data to classify a spoken phoneme. These two phonemes were chosen because they are difficult to separate.

The input feature is a vector x of length 256, which we can think of as a vector of evaluations of a function $X(f)$ over a grid of frequencies f. In reality there is a continuous analog signal which is a function of frequency, and we have a sampled version of it.

The gray lines in the lower panel of Figure 5.5 show the coefficients of a linear logistic regression model fit by maximum likelihood to a training sample of 1000 drawn from the total of 695 "aa"s and 1022 "ao"s. The coefficients are also plotted as a function of frequency, and in fact we can think of the model in terms of its continuous counterpart

$$\log \frac{\Pr(\text{aa}|X)}{\Pr(\text{ao}|X)} = \int X(f)\beta(f)df, \qquad (5.7)$$

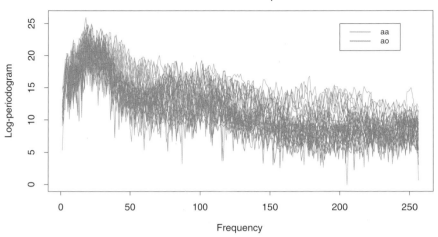

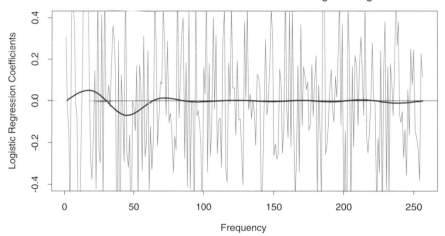

FIGURE 5.5. *The top panel displays the log-periodogram as a function of frequency for* 15 *examples each of the phonemes "aa" and "ao" sampled from a total of* 695 *"aa"s and* 1022 *"ao"s. Each log-periodogram is measured at* 256 *uniformly spaced frequencies. The lower panel shows the coefficients (as a function of frequency) of a logistic regression fit to the data by maximum likelihood, using the* 256 *log-periodogram values as inputs. The coefficients are restricted to be smooth in the red curve, and are unrestricted in the jagged gray curve.*

which we approximate by

$$\sum_{j=1}^{256} X(f_j)\beta(f_j) = \sum_{j=1}^{256} x_j \beta_j. \qquad (5.8)$$

The coefficients compute a contrast functional, and will have appreciable values in regions of frequency where the log-periodograms differ between the two classes.

The gray curves are very rough. Since the input signals have fairly strong positive autocorrelation, this results in negative autocorrelation in the coefficients. In addition the sample size effectively provides only four observations per coefficient.

Applications such as this permit a natural regularization. We force the coefficients to vary smoothly as a function of frequency. The red curve in the lower panel of Figure 5.5 shows such a smooth coefficient curve fit to these data. We see that the lower frequencies offer the most discriminatory power. Not only does the smoothing allow easier interpretation of the contrast, it also produces a more accurate classifier:

	Raw	Regularized
Training error	0.080	0.185
Test error	0.255	0.158

The smooth red curve was obtained through a very simple use of natural cubic splines. We can represent the coefficient function as an expansion of splines $\beta(f) = \sum_{m=1}^{M} h_m(f)\theta_m$. In practice this means that $\beta = \mathbf{H}\theta$ where, $\mathbf{H}$ is a $p \times M$ basis matrix of natural cubic splines, defined on the set of frequencies. Here we used $M = 12$ basis functions, with knots uniformly placed over the integers $1, 2, \ldots, 256$ representing the frequencies. Since $x^T\beta = x^T\mathbf{H}\theta$, we can simply replace the input features x by their *filtered* versions $x^* = \mathbf{H}^T x$, and fit θ by linear logistic regression on the x^*. The red curve is thus $\hat{\beta}(f) = h(f)^T\hat{\theta}$.

5.3 Filtering and Feature Extraction

In the previous example, we constructed a $p \times M$ basis matrix $\mathbf{H}$, and then transformed our features x into new features $x^* = \mathbf{H}^T x$. These filtered versions of the features were then used as inputs into a learning procedure: in the previous example, this was linear logistic regression.

Preprocessing of high-dimensional features is a very general and powerful method for improving the performance of a learning algorithm. The preprocessing need not be linear as it was above, but can be a general (nonlinear) function of the form $x^* = g(x)$. The derived features x^* can then be used as inputs into any (linear or nonlinear) learning procedure.

For example, for signal or image recognition a popular approach is to first transform the raw features via a wavelet transform $x^* = \mathbf{H}^T x$ (Section 5.9) and then use the features x^* as inputs into a neural network (Chapter 11). Wavelets are effective in capturing discrete jumps or edges, and the neural network is a powerful tool for constructing nonlinear functions of these features for predicting the target variable. By using domain knowledge to construct appropriate features, one can often improve upon a learning method that has only the raw features x at its disposal.

5.4   Smoothing Splines

Here we discuss a spline basis method that avoids the knot selection problem completely by using a maximal set of knots. The complexity of the fit is controlled by regularization. Consider the following problem: among all functions $f(x)$ with two continuous derivatives, find one that minimizes the penalized residual sum of squares

$$\text{RSS}(f, \lambda) = \sum_{i=1}^{N} \{y_i - f(x_i)\}^2 + \lambda \int \{f''(t)\}^2 dt, \qquad (5.9)$$

where λ is a fixed *smoothing parameter*. The first term measures closeness to the data, while the second term penalizes curvature in the function, and λ establishes a tradeoff between the two. Two special cases are:

$\lambda = 0$: f can be any function that interpolates the data.

$\lambda = \infty$: the simple least squares line fit, since no second derivative can be tolerated.

These vary from very rough to very smooth, and the hope is that $\lambda \in (0, \infty)$ indexes an interesting class of functions in between.

The criterion (5.9) is defined on an infinite-dimensional function space—in fact, a Sobolev space of functions for which the second term is defined. Remarkably, it can be shown that (5.9) has an explicit, finite-dimensional, unique minimizer which is a natural cubic spline with knots at the unique values of the x_i, $i = 1, \ldots, N$ (Exercise 5.7). At face value it seems that the family is still over-parametrized, since there are as many as N knots, which implies N degrees of freedom. However, the penalty term translates to a penalty on the spline coefficients, which are shrunk some of the way toward the linear fit.

Since the solution is a natural spline, we can write it as

$$f(x) = \sum_{j=1}^{N} N_j(x)\theta_j, \qquad (5.10)$$

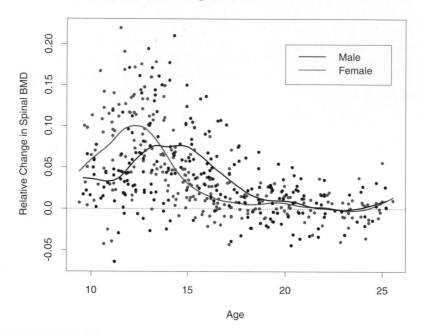

FIGURE 5.6. *The response is the relative change in bone mineral density measured at the spine in adolescents, as a function of age. A separate smoothing spline was fit to the males and females, with $\lambda \approx 0.00022$. This choice corresponds to about* 12 *degrees of freedom.*

where the $N_j(x)$ are an N-dimensional set of basis functions for representing this family of natural splines (Section 5.2.1 and Exercise 5.4). The criterion thus reduces to

$$\text{RSS}(\theta, \lambda) = (\mathbf{y} - \mathbf{N}\theta)^T(\mathbf{y} - \mathbf{N}\theta) + \lambda\theta^T\mathbf{\Omega}_N\theta, \tag{5.11}$$

where $\{\mathbf{N}\}_{ij} = N_j(x_i)$ and $\{\mathbf{\Omega}_N\}_{jk} = \int N_j''(t)N_k''(t)dt$. The solution is easily seen to be

$$\hat{\theta} = (\mathbf{N}^T\mathbf{N} + \lambda\mathbf{\Omega}_N)^{-1}\mathbf{N}^T\mathbf{y}, \tag{5.12}$$

a generalized ridge regression. The fitted smoothing spline is given by

$$\hat{f}(x) \;=\; \sum_{j=1}^{N} N_j(x)\hat{\theta}_j. \tag{5.13}$$

Efficient computational techniques for smoothing splines are discussed in the Appendix to this chapter.

Figure 5.6 shows a smoothing spline fit to some data on bone mineral density (BMD) in adolescents. The response is relative change in spinal BMD over two consecutive visits, typically about one year apart. The data are color coded by gender, and two separate curves were fit. This simple summary reinforces the evidence in the data that the growth spurt for females precedes that for males by about two years. In both cases the smoothing parameter λ was approximately 0.00022; this choice is discussed in the next section.

5.4.1 Degrees of Freedom and Smoother Matrices

We have not yet indicated how λ is chosen for the smoothing spline. Later in this chapter we describe automatic methods using techniques such as cross-validation. In this section we discuss intuitive ways of prespecifying the amount of smoothing.

A smoothing spline with prechosen λ is an example of a *linear smoother* (as in linear operator). This is because the estimated parameters in (5.12) are a linear combination of the y_i. Denote by $\hat{\mathbf{f}}$ the N-vector of fitted values $\hat{f}(x_i)$ at the training predictors x_i. Then

$$\begin{aligned} \hat{\mathbf{f}} &= \mathbf{N}(\mathbf{N}^T\mathbf{N} + \lambda\mathbf{\Omega}_N)^{-1}\mathbf{N}^T\mathbf{y} \\ &= \mathbf{S}_\lambda\mathbf{y}. \end{aligned} \tag{5.14}$$

Again the fit is linear in $\mathbf{y}$, and the finite linear operator $\mathbf{S}_\lambda$ is known as the *smoother matrix*. One consequence of this linearity is that the recipe for producing $\hat{\mathbf{f}}$ from $\mathbf{y}$ does not depend on $\mathbf{y}$ itself; $\mathbf{S}_\lambda$ depends only on the x_i and λ.

Linear operators are familiar in more traditional least squares fitting as well. Suppose $\mathbf{B}_\xi$ is a $N \times M$ matrix of M cubic-spline basis functions evaluated at the N training points x_i, with knot sequence ξ, and $M \ll N$. Then the vector of fitted spline values is given by

$$\begin{aligned} \hat{\mathbf{f}} &= \mathbf{B}_\xi(\mathbf{B}_\xi^T\mathbf{B}_\xi)^{-1}\mathbf{B}_\xi^T\mathbf{y} \\ &= \mathbf{H}_\xi\mathbf{y}. \end{aligned} \tag{5.15}$$

Here the linear operator $\mathbf{H}_\xi$ is a projection operator, also known as the *hat* matrix in statistics. There are some important similarities and differences between $\mathbf{H}_\xi$ and $\mathbf{S}_\lambda$:

- Both are symmetric, positive semidefinite matrices.

- $\mathbf{H}_\xi\mathbf{H}_\xi = \mathbf{H}_\xi$ (idempotent), while $\mathbf{S}_\lambda\mathbf{S}_\lambda \preceq \mathbf{S}_\lambda$, meaning that the right-hand side exceeds the left-hand side by a positive semidefinite matrix. This is a consequence of the *shrinking* nature of $\mathbf{S}_\lambda$, which we discuss further below.

- $\mathbf{H}_\xi$ has rank M, while $\mathbf{S}_\lambda$ has rank N.

The expression $M = \text{trace}(\mathbf{H}_\xi)$ gives the dimension of the projection space, which is also the number of basis functions, and hence the number of parameters involved in the fit. By analogy we define the *effective degrees of freedom* of a smoothing spline to be

$$\text{df}_\lambda = \text{trace}(\mathbf{S}_\lambda), \tag{5.16}$$

the sum of the diagonal elements of $\mathbf{S}_\lambda$. This very useful definition allows us a more intuitive way to parameterize the smoothing spline, and indeed many other smoothers as well, in a consistent fashion. For example, in Figure 5.6 we specified $\text{df}_\lambda = 12$ for each of the curves, and the corresponding $\lambda \approx 0.00022$ was derived numerically by solving $\text{trace}(\mathbf{S}_\lambda) = 12$. There are many arguments supporting this definition of degrees of freedom, and we cover some of them here.

Since $\mathbf{S}_\lambda$ is symmetric (and positive semidefinite), it has a real eigen-decomposition. Before we proceed, it is convenient to rewrite $\mathbf{S}_\lambda$ in the *Reinsch* form

$$\mathbf{S}_\lambda = (\mathbf{I} + \lambda\mathbf{K})^{-1}, \tag{5.17}$$

where $\mathbf{K}$ does not depend on λ (Exercise 5.9). Since $\hat{\mathbf{f}} = \mathbf{S}_\lambda\mathbf{y}$ solves

$$\min_{\mathbf{f}}(\mathbf{y} - \mathbf{f})^T(\mathbf{y} - \mathbf{f}) + \lambda\mathbf{f}^T\mathbf{K}\mathbf{f}, \tag{5.18}$$

$\mathbf{K}$ is known as the *penalty matrix*, and indeed a quadratic form in $\mathbf{K}$ has a representation in terms of a weighted sum of squared (divided) second differences. The eigen-decomposition of $\mathbf{S}_\lambda$ is

$$\mathbf{S}_\lambda = \sum_{k=1}^N \rho_k(\lambda)\mathbf{u}_k\mathbf{u}_k^T \tag{5.19}$$

with

$$\rho_k(\lambda) = \frac{1}{1 + \lambda d_k}, \tag{5.20}$$

and d_k the corresponding eigenvalue of $\mathbf{K}$. Figure 5.7 (top) shows the results of applying a cubic smoothing spline to some air pollution data (128 observations). Two fits are given: a *smoother* fit corresponding to a larger penalty λ and a *rougher* fit for a smaller penalty. The lower panels represent the eigenvalues (lower left) and some eigenvectors (lower right) of the corresponding smoother matrices. Some of the highlights of the eigenrepresentation are the following:

- The eigenvectors are not affected by changes in λ, and hence the whole family of smoothing splines (for a particular sequence $\mathbf{x}$) indexed by λ have the same eigenvectors.

- $\mathbf{S}_\lambda\mathbf{y} = \sum_{k=1}^N \mathbf{u}_k\rho_k(\lambda)\langle\mathbf{u}_k, \mathbf{y}\rangle$, and hence the smoothing spline operates by decomposing $\mathbf{y}$ w.r.t. the (complete) basis $\{\mathbf{u}_k\}$, and differentially shrinking the contributions using $\rho_k(\lambda)$. This is to be contrasted with a basis-regression method, where the components are

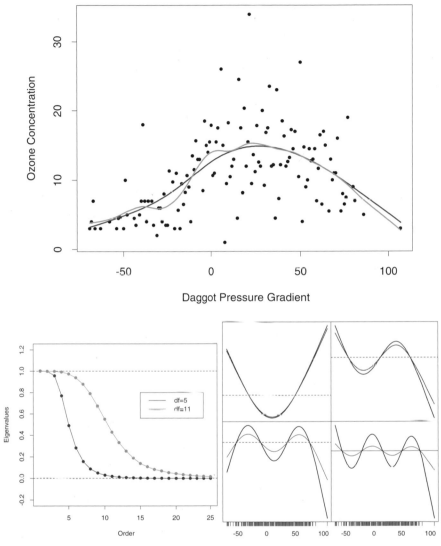

FIGURE 5.7. *(Top:) Smoothing spline fit of ozone concentration versus Daggot pressure gradient. The two fits correspond to different values of the smoothing parameter, chosen to achieve five and eleven effective degrees of freedom, defined by $df_\lambda = trace(\mathbf{S}_\lambda)$. (Lower left:) First 25 eigenvalues for the two smoothing-spline matrices. The first two are exactly 1, and all are ≥ 0. (Lower right:) Third to sixth eigenvectors of the spline smoother matrices. In each case, $\mathbf{u}_k$ is plotted against $\mathbf{x}$, and as such is viewed as a function of x. The rug at the base of the plots indicate the occurrence of data points. The damped functions represent the smoothed versions of these functions (using the 5 df smoother).*

either left alone, or shrunk to zero—that is, a projection matrix such as $\mathbf{H}_\xi$ above has M eigenvalues equal to 1, and the rest are 0. For this reason smoothing splines are referred to as *shrinking* smoothers, while regression splines are *projection* smoothers (see Figure 3.17 on page 80).

- The sequence of $\mathbf{u}_k$, ordered by decreasing $\rho_k(\lambda)$, appear to increase in complexity. Indeed, they have the zero-crossing behavior of polynomials of increasing degree. Since $\mathbf{S}_\lambda \mathbf{u}_k = \rho_k(\lambda)\mathbf{u}_k$, we see how each of the eigenvectors themselves are shrunk by the smoothing spline: the higher the complexity, the more they are shrunk. If the domain of X is periodic, then the $\mathbf{u}_k$ are sines and cosines at different frequencies.

- The first two eigenvalues are *always* one, and they correspond to the two-dimensional eigenspace of functions linear in x (Exercise 5.11), which are never shrunk.

- The eigenvalues $\rho_k(\lambda) = 1/(1 + \lambda d_k)$ are an inverse function of the eigenvalues d_k of the penalty matrix $\mathbf{K}$, moderated by λ; λ controls the rate at which the $\rho_k(\lambda)$ decrease to zero. $d_1 = d_2 = 0$ and again linear functions are not penalized.

- One can reparametrize the smoothing spline using the basis vectors $\mathbf{u}_k$ (the *Demmler–Reinsch* basis). In this case the smoothing spline solves

$$\min_{\boldsymbol{\theta}} \|\mathbf{y} - \mathbf{U}\boldsymbol{\theta}\|^2 + \lambda \boldsymbol{\theta}^T \mathbf{D} \boldsymbol{\theta}, \tag{5.21}$$

where $\mathbf{U}$ has columns $\mathbf{u}_k$ and $\mathbf{D}$ is a diagonal matrix with elements d_k.

- $\mathrm{df}_\lambda = \mathrm{trace}(\mathbf{S}_\lambda) = \sum_{k=1}^{N} \rho_k(\lambda)$. For projection smoothers, all the eigenvalues are 1, each one corresponding to a dimension of the projection subspace.

Figure 5.8 depicts a smoothing spline matrix, with the rows ordered with x. The banded nature of this representation suggests that a smoothing spline is a local fitting method, much like the locally weighted regression procedures in Chapter 6. The right panel shows in detail selected rows of $\mathbf{S}$, which we call the *equivalent kernels*. As $\lambda \to 0$, $\mathrm{df}_\lambda \to N$, and $\mathbf{S}_\lambda \to \mathbf{I}$, the N-dimensional identity matrix. As $\lambda \to \infty$, $\mathrm{df}_\lambda \to 2$, and $\mathbf{S}_\lambda \to \mathbf{H}$, the hat matrix for linear regression on $\mathbf{x}$.

5.5 Automatic Selection of the Smoothing Parameters

The smoothing parameters for regression splines encompass the degree of the splines, and the number and placement of the knots. For smoothing

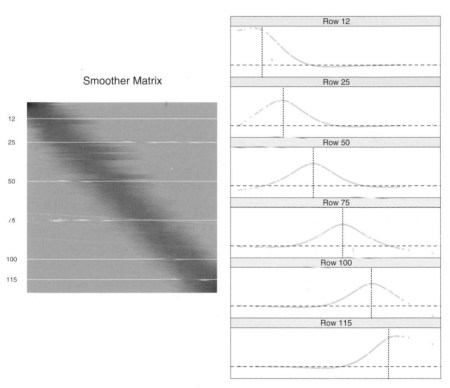

FIGURE 5.8. *The smoother matrix for a smoothing spline is nearly banded, indicating an equivalent kernel with local support. The left panel represents the elements of* **S** *as an image. The right panel shows the equivalent kernel or weighting function in detail for the indicated rows.*

splines, we have only the penalty parameter λ to select, since the knots are at all the unique training X's, and cubic degree is almost always used in practice.

Selecting the placement and number of knots for regression splines can be a combinatorially complex task, unless some simplifications are enforced. The MARS procedure in Chapter 9 uses a greedy algorithm with some additional approximations to achieve a practical compromise. We will not discuss this further here.

5.5.1 Fixing the Degrees of Freedom

Since $\mathrm{df}_\lambda = \mathrm{trace}(\mathbf{S}_\lambda)$ is monotone in λ for smoothing splines, we can invert the relationship and specify λ by fixing df. In practice this can be achieved by simple numerical methods. So, for example, in R one can use `smooth.spline(x,y,df=6)` to specify the amount of smoothing. This encourages a more traditional mode of model selection, where we might try a couple of different values of df, and select one based on approximate F-tests, residual plots and other more subjective criteria. Using df in this way provides a uniform approach to compare many different smoothing methods. It is particularly useful in *generalized additive models* (Chapter 9), where several smoothing methods can be simultaneously used in one model.

5.5.2 The Bias–Variance Tradeoff

Figure 5.9 shows the effect of the choice of df_λ when using a smoothing spline on a simple example:

$$Y = f(X) + \varepsilon,$$
$$f(X) = \frac{\sin(12(X + 0.2))}{X + 0.2}, \tag{5.22}$$

with $X \sim U[0,1]$ and $\varepsilon \sim N(0,1)$. Our training sample consists of $N = 100$ pairs x_i, y_i drawn independently from this model.

The fitted splines for three different values of df_λ are shown. The yellow shaded region in the figure represents the pointwise standard error of $\hat{f}_\lambda$, that is, we have shaded the region between $\hat{f}_\lambda(x) \pm 2 \cdot \mathrm{se}(\hat{f}_\lambda(x))$. Since $\hat{\mathbf{f}} = \mathbf{S}_\lambda \mathbf{y}$,

$$
\begin{aligned}
\mathrm{Cov}(\hat{\mathbf{f}}) &= \mathbf{S}_\lambda \mathrm{Cov}(\mathbf{y}) \mathbf{S}_\lambda^T \\
&= \mathbf{S}_\lambda \mathbf{S}_\lambda^T.
\end{aligned}
\tag{5.23}
$$

The diagonal contains the pointwise variances at the training x_i. The bias is given by

$$
\begin{aligned}
\mathrm{Bias}(\hat{\mathbf{f}}) &= \mathbf{f} - \mathrm{E}(\hat{\mathbf{f}}) \\
&= \mathbf{f} - \mathbf{S}_\lambda \mathbf{f},
\end{aligned}
\tag{5.24}
$$

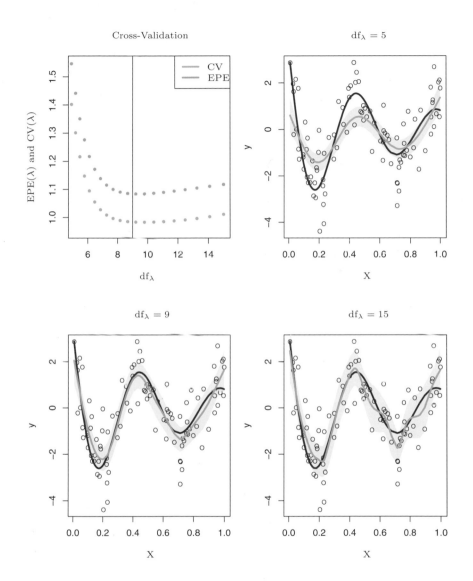

FIGURE 5.9. *The top left panel shows the* $\text{EPE}(\lambda)$ *and* $\text{CV}(\lambda)$ *curves for a realization from a nonlinear additive error model (5.22). The remaining panels show the data, the true functions (in purple), and the fitted curves (in green) with yellow shaded* $\pm 2\times$ *standard error bands, for three different values of* df_λ.

where $\mathbf{f}$ is the (unknown) vector of evaluations of the true f at the training X's. The expectations and variances are with respect to repeated draws of samples of size $N = 100$ from the model (5.22). In a similar fashion $\text{Var}(\hat{f}_\lambda(x_0))$ and $\text{Bias}(\hat{f}_\lambda(x_0))$ can be computed at any point x_0 (Exercise 5.10). The three fits displayed in the figure give a visual demonstration of the bias-variance tradeoff associated with selecting the smoothing parameter.

$\text{df}_\lambda = 5$: The spline under fits, and clearly *trims down the hills and fills in the valleys*. This leads to a bias that is most dramatic in regions of high curvature. The standard error band is very narrow, so we estimate a badly biased version of the true function with great reliability!

$\text{df}_\lambda = 9$: Here the fitted function is close to the true function, although a slight amount of bias seems evident. The variance has not increased appreciably.

$\text{df}_\lambda = 15$: The fitted function is somewhat wiggly, but close to the true function. The wiggliness also accounts for the increased width of the standard error bands—the curve is starting to follow some individual points too closely.

Note that in these figures we are seeing a single realization of data and hence fitted spline $\hat{f}$ in each case, while the bias involves an expectation $\text{E}(\hat{f})$. We leave it as an exercise (5.10) to compute similar figures where the bias is shown as well. The middle curve seems "just right," in that it has achieved a good compromise between bias and variance.

The integrated squared prediction error (EPE) combines both bias and variance in a single summary:

$$
\begin{aligned}
\text{EPE}(\hat{f}_\lambda) &= \text{E}(Y - \hat{f}_\lambda(X))^2 \\
&= \text{Var}(Y) + \text{E}\left[\text{Bias}^2(\hat{f}_\lambda(X)) + \text{Var}(\hat{f}_\lambda(X))\right] \\
&= \sigma^2 + \text{MSE}(\hat{f}_\lambda).
\end{aligned}
\tag{5.25}
$$

Note that this is averaged both over the training sample (giving rise to $\hat{f}_\lambda$), and the values of the (independently chosen) prediction points (X, Y). EPE is a natural quantity of interest, and does create a tradeoff between bias and variance. The blue points in the top left panel of Figure 5.9 suggest that $\text{df}_\lambda = 9$ is spot on!

Since we don't know the true function, we do not have access to EPE, and need an estimate. This topic is discussed in some detail in Chapter 7, and techniques such as K-fold cross-validation, GCV and C_p are all in common use. In Figure 5.9 we include the N-fold (leave-one-out) cross-validation curve:

$$\text{CV}(\hat{f}_\lambda) \;=\; \frac{1}{N} \sum_{i=1}^{N} (y_i - \hat{f}_\lambda^{(-i)}(x_i))^2 \tag{5.26}$$

$$= \; \frac{1}{N} \sum_{i=1}^{N} \left(\frac{y_i - \hat{f}_\lambda(x_i)}{1 - S_\lambda(i,i)} \right)^2, \tag{5.27}$$

which can (remarkably) be computed for each value of λ from the original fitted values and the diagonal elements $S_\lambda(i,i)$ of $\mathbf{S}_\lambda$ (Exercise 5.13).

The EPE and CV curves have a similar shape, but the entire CV curve is above the EPE curve. For some realizations this is reversed, and overall the CV curve is approximately unbiased as an estimate of the EPE curve.

5.6 Nonparametric Logistic Regression

The smoothing spline problem (5.9) in Section 5.4 is posed in a regression setting. It is typically straightforward to transfer this technology to other domains. Here we consider logistic regression with a single quantitative input X. The model is

$$\log \frac{\Pr(Y = 1 | X = x)}{\Pr(Y = 0 | X = x)} = f(x), \tag{5.28}$$

which implies

$$\Pr(Y = 1 | X = x) = \frac{e^{f(x)}}{1 + e^{f(x)}}. \tag{5.29}$$

Fitting $f(x)$ in a smooth fashion leads to a smooth estimate of the conditional probability $\Pr(Y = 1 | x)$, which can be used for classification or risk scoring.

We construct the penalized log-likelihood criterion

$$\ell(f; \lambda) \;=\; \sum_{i=1}^{N} [y_i \log p(x_i) + (1 - y_i) \log(1 - p(x_i))] - \frac{1}{2} \lambda \int \{f''(t)\}^2 dt$$

$$= \; \sum_{i=1}^{N} \left[y_i f(x_i) - \log(1 + e^{f(x_i)}) \right] - \frac{1}{2} \lambda \int \{f''(t)\}^2 dt, \tag{5.30}$$

where we have abbreviated $p(x) = \Pr(Y = 1 | x)$. The first term in this expression is the log-likelihood based on the binomial distribution (c.f. Chapter 4, page 120). Arguments similar to those used in Section 5.4 show that the optimal f is a finite-dimensional natural spline with knots at the unique values of x. This means that we can represent $f(x) = \sum_{j=1}^{N} N_j(x)\theta_j$. We compute the first and second derivatives

$$\frac{\partial \ell(\theta)}{\partial \theta} = \mathbf{N}^T(\mathbf{y} - \mathbf{p}) - \lambda \mathbf{\Omega} \theta, \tag{5.31}$$

$$\frac{\partial^2 \ell(\theta)}{\partial \theta \partial \theta^T} = -\mathbf{N}^T \mathbf{W} \mathbf{N} - \lambda \mathbf{\Omega}, \tag{5.32}$$

where $\mathbf{p}$ is the N-vector with elements $p(x_i)$, and $\mathbf{W}$ is a diagonal matrix of weights $p(x_i)(1 - p(x_i))$. The first derivative (5.31) is nonlinear in θ, so we need to use an iterative algorithm as in Section 4.4.1. Using Newton–Raphson as in (4.23) and (4.26) for linear logistic regression, the update equation can be written

$$
\begin{aligned}
\theta^{\text{new}} &= (\mathbf{N}^T \mathbf{W} \mathbf{N} + \lambda \mathbf{\Omega})^{-1} \mathbf{N}^T \mathbf{W} \left(\mathbf{N} \theta^{\text{old}} + \mathbf{W}^{-1}(\mathbf{y} - \mathbf{p}) \right) \\
&= (\mathbf{N}^T \mathbf{W} \mathbf{N} + \lambda \mathbf{\Omega})^{-1} \mathbf{N}^T \mathbf{W} \mathbf{z}.
\end{aligned}
\tag{5.33}
$$

We can also express this update in terms of the fitted values

$$
\begin{aligned}
\mathbf{f}^{\text{new}} &= \mathbf{N}(\mathbf{N}^T \mathbf{W} \mathbf{N} + \lambda \mathbf{\Omega})^{-1} \mathbf{N}^T \mathbf{W} \left(\mathbf{f}^{\text{old}} + \mathbf{W}^{-1}(\mathbf{y} - \mathbf{p}) \right) \\
&= \mathbf{S}_{\lambda, w} \mathbf{z}.
\end{aligned}
\tag{5.34}
$$

Referring back to (5.12) and (5.14), we see that the update fits a weighted smoothing spline to the working response $\mathbf{z}$ (Exercise 5.12).

The form of (5.34) is suggestive. It is tempting to replace $\mathbf{S}_{\lambda, w}$ by any nonparametric (weighted) regression operator, and obtain general families of nonparametric logistic regression models. Although here x is one-dimensional, this procedure generalizes naturally to higher-dimensional x. These extensions are at the heart of *generalized additive models*, which we pursue in Chapter 9.

5.7 Multidimensional Splines

So far we have focused on one-dimensional spline models. Each of the approaches have multidimensional analogs. Suppose $X \in \mathbb{R}^2$, and we have a basis of functions $h_{1k}(X_1)$, $k = 1, \ldots, M_1$ for representing functions of coordinate X_1, and likewise a set of M_2 functions $h_{2k}(X_2)$ for coordinate X_2. Then the $M_1 \times M_2$ dimensional *tensor product basis* defined by

$$g_{jk}(X) = h_{1j}(X_1)h_{2k}(X_2), \ j = 1, \ldots, M_1, \ k = 1, \ldots, M_2 \tag{5.35}$$

can be used for representing a two-dimensional function:

$$g(X) = \sum_{j=1}^{M_1} \sum_{k=1}^{M_2} \theta_{jk} g_{jk}(X). \tag{5.36}$$

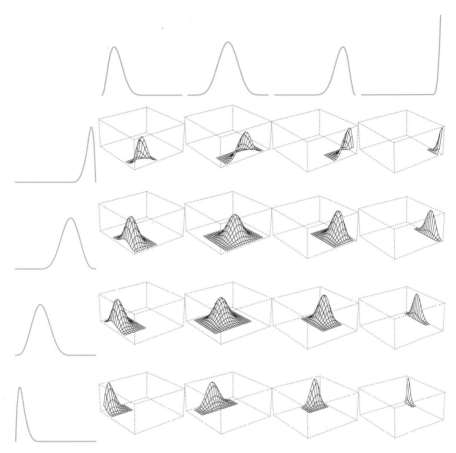

FIGURE 5.10. *A tensor product basis of B-splines, showing some selected pairs. Each two-dimensional function is the tensor product of the corresponding one dimensional marginals.*

Figure 5.10 illustrates a tensor product basis using B-splines. The coefficients can be fit by least squares, as before. This can be generalized to d dimensions, but note that the dimension of the basis grows exponentially fast—yet another manifestation of the curse of dimensionality. The MARS procedure discussed in Chapter 9 is a greedy forward algorithm for including only those tensor products that are deemed necessary by least squares.

Figure 5.11 illustrates the difference between additive and tensor product (natural) splines on the simulated classification example from Chapter 2. A logistic regression model $\text{logit}[\Pr(T|x)] = h(x)^T\theta$ is fit to the binary response, and the estimated decision boundary is the contour $h(x)^T\hat{\theta} = 0$. The tensor product basis can achieve more flexibility at the decision boundary, but introduces some spurious structure along the way.

Additive Natural Cubic Splines - 4 df each

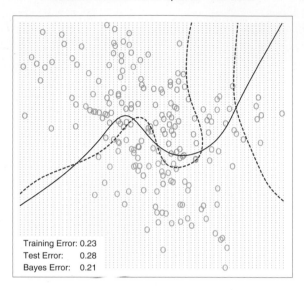

Training Error: 0.23
Test Error: 0.28
Bayes Error: 0.21

Natural Cubic Splines - Tensor Product - 4 df each

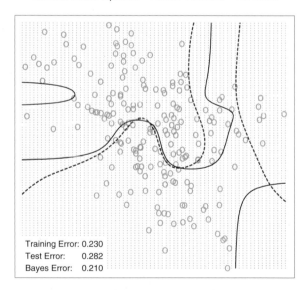

Training Error: 0.230
Test Error: 0.282
Bayes Error: 0.210

FIGURE 5.11. *The simulation example of Figure 2.1. The upper panel shows the decision boundary of an additive logistic regression model, using natural splines in each of the two coordinates (total df = $1 + (4 - 1) + (4 - 1) = 7$). The lower panel shows the results of using a tensor product of natural spline bases in each coordinate (total df = $4 \times 4 = 16$). The broken purple boundary is the Bayes decision boundary for this problem.*

One-dimensional smoothing splines (via regularization) generalize to higher dimensions as well. Suppose we have pairs y_i, x_i with $x_i \in \mathbb{R}^d$, and we
seek a d-dimensional regression function $f(x)$. The idea is to set up the
problem

$$\min_f \sum_{i=1}^{N} \{y_i - f(x_i)\}^2 + \lambda J[f], \qquad (5.37)$$

where J is an appropriate penalty functional for stabilizing a function f in
$\mathbb{R}^d$. For example, a natural generalization of the one-dimensional roughness
penalty (5.9) for functions on $\mathbb{R}^2$ is

$$J[f] = \int \int_{\mathbb{R}^2} \left[\left(\frac{\partial^2 f(x)}{\partial x_1^2} \right)^2 + 2 \left(\frac{\partial^2 f(x)}{\partial x_1 \partial x_2} \right)^2 + \left(\frac{\partial^2 f(x)}{\partial x_2^2} \right)^2 \right] dx_1 dx_2. \qquad (5.38)$$

Optimizing (5.37) with this penalty leads to a smooth two-dimensional
surface, known as a thin-plate spline. It shares many properties with the
one-dimensional cubic smoothing spline:

- as $\lambda \to 0$, the solution approaches an interpolating function [the one
  with smallest penalty (5.38)];

- as $\lambda \to \infty$, the solution approaches the least squares plane;

- for intermediate values of λ, the solution can be represented as a
  linear expansion of basis functions, whose coefficients are obtained
  by a form of generalized ridge regression.

The solution has the form

$$f(x) = \beta_0 + \beta^T x + \sum_{j=1}^{N} \alpha_j h_j(x), \qquad (5.39)$$

where $h_j(x) = ||x - x_j||^2 \log ||x - x_j||$. These h_j are examples of *radial
basis functions*, which are discussed in more detail in the next section. The
coefficients are found by plugging (5.39) into (5.37), which reduces to a
finite-dimensional penalized least squares problem. For the penalty to be
finite, the coefficients α_j have to satisfy a set of linear constraints; see
Exercise 5.14.

Thin-plate splines are defined more generally for arbitrary dimension d,
for which an appropriately more general J is used.

There are a number of hybrid approaches that are popular in practice,
both for computational and conceptual simplicity. Unlike one-dimensional
smoothing splines, the computational complexity for thin-plate splines is
$O(N^3)$, since there is not in general any sparse structure that can be exploited. However, as with univariate smoothing splines, we can get away
with substantially less than the N knots prescribed by the solution (5.39).

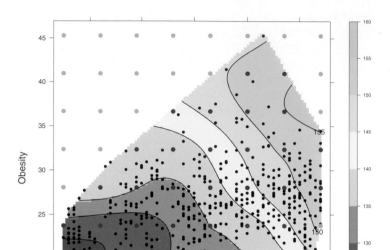

FIGURE 5.12. *A thin-plate spline fit to the heart disease data, displayed as a contour plot. The response is* systolic blood pressure, *modeled as a function of* age *and* obesity. *The data points are indicated, as well as the lattice of points used as knots. Care should be taken to use knots from the lattice inside the convex hull of the data (red), and ignore those outside (green).*

In practice, it is usually sufficient to work with a lattice of knots covering the domain. The penalty is computed for the reduced expansion just as before. Using K knots reduces the computations to $O(NK^2 + K^3)$. Figure 5.12 shows the result of fitting a thin-plate spline to some heart disease risk factors, representing the surface as a contour plot. Indicated are the location of the input features, as well as the knots used in the fit. Note that λ was specified via $\mathrm{df}_\lambda = \mathrm{trace}(S_\lambda) = 15$.

More generally one can represent $f \in \mathbb{R}^d$ as an expansion in any arbitrarily large collection of basis functions, and control the complexity by applying a regularizer such as (5.38). For example, we could construct a basis by forming the tensor products of all pairs of univariate smoothing-spline basis functions as in (5.35), using, for example, the univariate B-splines recommended in Section 5.9.2 as ingredients. This leads to an exponential growth in basis functions as the dimension increases, and typically we have to reduce the number of functions per coordinate accordingly.

The additive spline models discussed in Chapter 9 are a restricted class of multidimensional splines. They can be represented in this general formulation as well; that is, there exists a penalty $J[f]$ that guarantees that the solution has the form $f(X) = \alpha + f_1(X_1) + \cdots + f_d(X_d)$ and that each of the functions f_j are univariate splines. In this case the penalty is somewhat degenerate, and it is more natural to *assume* that f is additive, and then simply impose an additional penalty on each of the component functions:

$$
\begin{aligned}
J[f] &= J(f_1 + f_2 + \cdots + f_d) \\
&= \sum_{j=1}^{d} \int f_j''(t_j)^2 dt_j.
\end{aligned}
\tag{5.40}
$$

These are naturally extended to ANOVA spline decompositions,

$$
f(X) = \alpha + \sum_j f_j(X_j) + \sum_{j<k} f_{jk}(X_j, X_k) + \cdots ,
\tag{5.41}
$$

where each of the components are splines of the required dimension. There are many choices to be made:

- The maximum order of interaction—we have shown up to order 2 above.

- Which terms to include—not all main effects and interactions are necessarily needed.

- What representation to use—some choices are:

  - regression splines with a relatively small number of basis functions per coordinate, and their tensor products for interactions;

  - a complete basis as in smoothing splines, and include appropriate regularizers for each term in the expansion.

In many cases when the number of potential dimensions (features) is large, automatic methods are more desirable. The MARS and MART procedures (Chapters 9 and 10, respectively), both fall into this category.

5.8 Regularization and Reproducing Kernel Hilbert Spaces

In this section we cast splines into the larger context of regularization methods and reproducing kernel Hilbert spaces. This section is quite technical and can be skipped by the disinterested or intimidated reader.

A general class of regularization problems has the form

$$\min_{f \in \mathcal{H}} \left[\sum_{i=1}^{N} L(y_i, f(x_i)) + \lambda J(f) \right] \qquad (5.42)$$

where $L(y, f(x))$ is a loss function, $J(f)$ is a penalty functional, and $\mathcal{H}$ is a space of functions on which $J(f)$ is defined. Girosi et al. (1995) describe quite general penalty functionals of the form

$$J(f) = \int_{\mathbb{R}^d} \frac{|\tilde{f}(s)|^2}{\tilde{G}(s)} ds, \qquad (5.43)$$

where $\tilde{f}$ denotes the Fourier transform of f, and $\tilde{G}$ is some positive function that falls off to zero as $||s|| \to \infty$. The idea is that $1/\tilde{G}$ increases the penalty for high-frequency components of f. Under some additional assumptions they show that the solutions have the form

$$f(X) = \sum_{k=1}^{K} \alpha_k \phi_k(X) + \sum_{i=1}^{N} \theta_i G(X - x_i), \qquad (5.44)$$

where the ϕ_k span the null space of the penalty functional J, and G is the inverse Fourier transform of $\tilde{G}$. Smoothing splines and thin-plate splines fall into this framework. The remarkable feature of this solution is that while the criterion (5.42) is defined over an infinite-dimensional space, the solution is finite-dimensional. In the next sections we look at some specific examples.

5.8.1 Spaces of Functions Generated by Kernels

An important subclass of problems of the form (5.42) are generated by a positive definite kernel $K(x, y)$, and the corresponding space of functions $\mathcal{H}_K$ is called a *reproducing kernel Hilbert space* (RKHS). The penalty functional J is defined in terms of the kernel as well. We give a brief and simplified introduction to this class of models, adapted from Wahba (1990) and Girosi et al. (1995), and nicely summarized in Evgeniou et al. (2000).

Let $x, y \in \mathbb{R}^p$. We consider the space of functions generated by the linear span of $\{K(\cdot, y), \ y \in \mathbb{R}^p)\}$; i.e arbitrary linear combinations of the form $f(x) = \sum_m \alpha_m K(x, y_m)$, where each kernel term is viewed as a function of the first argument, and indexed by the second. Suppose that K has an eigen-expansion

$$K(x, y) = \sum_{i=1}^{\infty} \gamma_i \phi_i(x) \phi_i(y) \qquad (5.45)$$

with $\gamma_i \geq 0$, $\sum_{i=1}^{\infty} \gamma_i^2 < \infty$. Elements of $\mathcal{H}_K$ have an expansion in terms of these eigen-functions,

$$f(x) = \sum_{i=1}^{\infty} c_i \phi_i(x), \qquad (5.46)$$

with the constraint that

$$||f||^2_{\mathcal{H}_K} \stackrel{\text{def}}{=} \sum_{i=1}^{\infty} c_i^2/\gamma_i < \infty, \tag{5.47}$$

where $||f||_{\mathcal{H}_K}$ is the norm induced by K. The penalty functional in (5.42) for the space $\mathcal{H}_K$ is defined to be the squared norm $J(f) = ||f||^2_{\mathcal{H}_K}$. The quantity $J(f)$ can be interpreted as a generalized ridge penalty, where functions with large eigenvalues in the expansion (5.45) get penalized less, and vice versa.

Rewriting (5.42) we have

$$\min_{f \in \mathcal{H}_K} \left[\sum_{i=1}^{N} L(y_i, f(x_i)) + \lambda ||f||^2_{\mathcal{H}_K} \right] \tag{5.48}$$

or equivalently

$$\min_{\{c_j\}_1^{\infty}} \left[\sum_{i=1}^{N} L(y_i, \sum_{j=1}^{\infty} c_j \phi_j(x_i)) + \lambda \sum_{j=1}^{\infty} c_j^2/\gamma_j \right]. \tag{5.49}$$

It can be shown (Wahba, 1990, see also Exercise 5.15) that the solution to (5.48) is finite-dimensional, and has the form

$$f(x) = \sum_{i=1}^{N} \alpha_i K(x, x_i). \tag{5.50}$$

The basis function $h_i(x) = K(x, x_i)$ (as a function of the first argument) is known as the *representer of evaluation* at x_i in $\mathcal{H}_K$, since for $f \in \mathcal{H}_K$, it is easily seen that $\langle K(\cdot, x_i), f \rangle_{\mathcal{H}_K} = f(x_i)$. Similarly $\langle K(\cdot, x_i), K(\cdot, x_j) \rangle_{\mathcal{H}_K} = K(x_i, x_j)$ (the *reproducing* property of $\mathcal{H}_K$), and hence

$$J(f) = \sum_{i=1}^{N} \sum_{j=1}^{N} K(x_i, x_j) \alpha_i \alpha_j \tag{5.51}$$

for $f(x) = \sum_{i=1}^{N} \alpha_i K(x, x_i)$.

In light of (5.50) and (5.51), (5.48) reduces to a finite-dimensional criterion

$$\min_{\boldsymbol{\alpha}} L(\mathbf{y}, \mathbf{K}\boldsymbol{\alpha}) + \lambda \boldsymbol{\alpha}^T \mathbf{K}\boldsymbol{\alpha}. \tag{5.52}$$

We are using a vector notation, in which $\mathbf{K}$ is the $N \times N$ matrix with ijth entry $K(x_i, x_j)$ and so on. Simple numerical algorithms can be used to optimize (5.52). This phenomenon, whereby the infinite-dimensional problem (5.48) or (5.49) reduces to a finite dimensional optimization problem, has been dubbed the *kernel property* in the literature on support-vector machines (see Chapter 12).

There is a Bayesian interpretation of this class of models, in which f is interpreted as a realization of a zero-mean stationary Gaussian process, with prior covariance function K. The eigen-decomposition produces a series of orthogonal eigen-functions $\phi_j(x)$ with associated variances γ_j. The typical scenario is that "smooth" functions ϕ_j have large prior variance, while "rough" ϕ_j have small prior variances. The penalty in (5.48) is the contribution of the prior to the joint likelihood, and penalizes more those components with smaller prior variance (compare with (5.43)).

For simplicity we have dealt with the case here where all members of $\mathcal{H}$ are penalized, as in (5.48). More generally, there may be some components in $\mathcal{H}$ that we wish to leave alone, such as the linear functions for cubic smoothing splines in Section 5.4. The multidimensional thin-plate splines of Section 5.7 and tensor product splines fall into this category as well. In these cases there is a more convenient representation $\mathcal{H} = \mathcal{H}_0 \oplus \mathcal{H}_1$, with the *null space* $\mathcal{H}_0$ consisting of, for example, low degree polynomials in x that do not get penalized. The penalty becomes $J(f) = \|P_1 f\|$, where P_1 is the orthogonal projection of f onto $\mathcal{H}_1$. The solution has the form $f(x) = \sum_{j=1}^{M} \beta_j h_j(x) + \sum_{i=1}^{N} \alpha_i K(x, x_i)$, where the first term represents an expansion in $\mathcal{H}_0$. From a Bayesian perspective, the coefficients of components in $\mathcal{H}_0$ have improper priors, with infinite variance.

5.8.2 Examples of RKHS

The machinery above is driven by the choice of the kernel K and the loss function L. We consider first regression using squared-error loss. In this case (5.48) specializes to penalized least squares, and the solution can be characterized in two equivalent ways corresponding to (5.49) or (5.52):

$$\min_{\{c_j\}_1^\infty} \sum_{i=1}^{N} \left(y_i - \sum_{j=1}^{\infty} c_j \phi_j(x_i) \right)^2 + \lambda \sum_{j=1}^{\infty} \frac{c_j^2}{\gamma_j} \qquad (5.53)$$

an infinite-dimensional, generalized ridge regression problem, or

$$\min_{\boldsymbol{\alpha}} (\mathbf{y} - \mathbf{K}\boldsymbol{\alpha})^T (\mathbf{y} - \mathbf{K}\boldsymbol{\alpha}) + \lambda \boldsymbol{\alpha}^T \mathbf{K} \boldsymbol{\alpha}. \qquad (5.54)$$

The solution for $\boldsymbol{\alpha}$ is obtained simply as

$$\hat{\boldsymbol{\alpha}} = (\mathbf{K} + \lambda \mathbf{I})^{-1} \mathbf{y}, \qquad (5.55)$$

and

$$\hat{f}(x) = \sum_{j=1}^{N} \hat{\alpha}_j K(x, x_j). \qquad (5.56)$$

The vector of N fitted values is given by

$$
\begin{aligned}
\hat{\mathbf{f}} &= \mathbf{K}\hat{\alpha} \\
&= \mathbf{K}(\mathbf{K} + \lambda\mathbf{I})^{-1}\mathbf{y} \qquad\qquad (5.57) \\
&= (\mathbf{I} + \lambda\mathbf{K}^{-1})^{-1}\mathbf{y}. \qquad\qquad (5.58)
\end{aligned}
$$

The estimate (5.57) also arises as the *kriging* estimate of a Gaussian random field in spatial statistics (Cressie, 1993). Compare also (5.58) with the smoothing spline fit (5.17) on page 154.

Penalized Polynomial Regression

The kernel $K(x, y) = (\langle x, y \rangle + 1)^d$ (Vapnik, 1996), for $x, y \in \mathbb{R}^p$, has $M = \binom{p+d}{d}$ eigen-functions that span the space of polynomials in $\mathbb{R}^p$ of total degree d. For example, with $p = 2$ and $d = 2$, $M = 6$ and

$$
\begin{aligned}
K(x, y) &= 1 + 2x_1y_1 + 2x_2y_2 + x_1^2y_1^2 + x_2^2y_2^2 + 2x_1x_2y_1y_2 \quad (5.59) \\
&= \sum_{m=1}^{M} h_m(x)h_m(y) \qquad\qquad\qquad\qquad\qquad (5.60)
\end{aligned}
$$

with

$$
h(x)^T = (1, \sqrt{2}x_1, \sqrt{2}x_2, x_1^2, x_2^2, \sqrt{2}x_1x_2). \qquad (5.61)
$$

One can represent h in terms of the M orthogonal eigen-functions and eigenvalues of K,

$$
h(x) = \mathbf{V}\mathbf{D}_\gamma^{\frac{1}{2}}\phi(x), \qquad (5.62)
$$

where $\mathbf{D}_\gamma = \mathrm{diag}(\gamma_1, \gamma_2, \ldots, \gamma_M)$, and $\mathbf{V}$ is $M \times M$ and orthogonal.

Suppose we wish to solve the penalized polynomial regression problem

$$
\min_{\{\beta_m\}_1^M} \sum_{i=1}^{N} \left(y_i - \sum_{m=1}^{M} \beta_m h_m(x_i) \right)^2 + \lambda \sum_{m=1}^{M} \beta_m^2. \qquad (5.63)
$$

Substituting (5.62) into (5.63), we get an expression of the form (5.53) to optimize (Exercise 5.16).

The number of basis functions $M = \binom{p+d}{d}$ can be very large, often much larger than N. Equation (5.55) tells us that if we use the kernel representation for the solution function, we have only to evaluate the kernel N^2 times, and can compute the solution in $O(N^3)$ operations.

This simplicity is not without implications. Each of the polynomials h_m in (5.61) inherits a scaling factor from the particular form of K, which has a bearing on the impact of the penalty in (5.63). We elaborate on this in the next section.

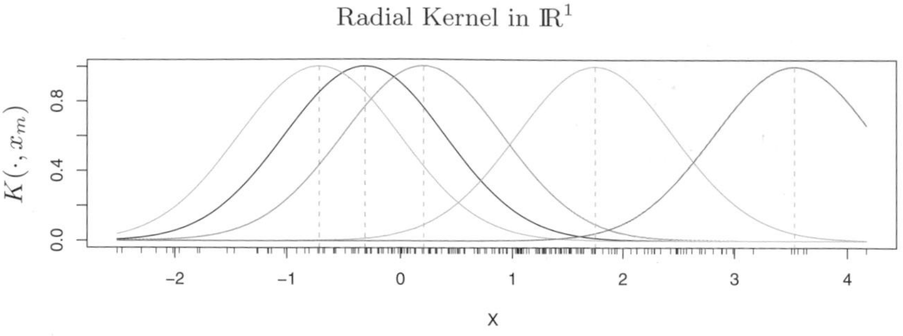

FIGURE 5.13. *Radial kernels $k_k(x)$ for the mixture data, with scale parameter $\nu = 1$. The kernels are centered at five points x_m chosen at random from the 200.*

Gaussian Radial Basis Functions

In the preceding example, the kernel is chosen because it represents an expansion of polynomials and can conveniently compute high-dimensional inner products. In this example the kernel is chosen because of its functional form in the representation (5.50).

The Gaussian kernel $K(x,y) = e^{-\nu||x-y||^2}$ along with squared-error loss, for example, leads to a regression model that is an expansion in Gaussian radial basis functions,

$$k_m(x) = e^{-\nu||x-x_m||^2}, \ m = 1, \ldots, N, \tag{5.64}$$

each one centered at one of the training feature vectors x_m. The coefficients are estimated using (5.54).

Figure 5.13 illustrates radial kernels in $\mathbb{R}^1$ using the first coordinate of the mixture example from Chapter 2. We show five of the 200 kernel basis functions $k_m(x) = K(x, x_m)$.

Figure 5.14 illustrates the implicit feature space for the radial kernel with $x \in \mathbb{R}^1$. We computed the 200×200 kernel matrix $\mathbf{K}$, and its eigendecomposition $\mathbf{\Phi D_\gamma \Phi}^T$. We can think of the columns of $\mathbf{\Phi}$ and the corresponding eigenvalues in $\mathbf{D}_\gamma$ as empirical estimates of the eigen expansion (5.45)[2]. Although the eigenvectors are discrete, we can represent them as functions on $\mathbb{R}^1$ (Exercise 5.17). Figure 5.15 shows the largest 50 eigenvalues of $\mathbf{K}$. The leading eigenfunctions are smooth, and they are successively more wiggly as the order increases. This brings to life the penalty in (5.49), where we see the coefficients of higher-order functions get penalized more than lower-order ones. The right panel in Figure 5.14 shows the correspond-

---

[2]The ℓth column of $\mathbf{\Phi}$ is an estimate of ϕ_ℓ, evaluated at each of the N observations. Alternatively, the ith row of $\mathbf{\Phi}$ is the estimated vector of basis functions $\phi(x_i)$, evaluated at the point x_i. Although in principle, there can be infinitely many elements in ϕ, our estimate has at most N elements.

Orthonormal Basis **Φ**   Feature Space **H**

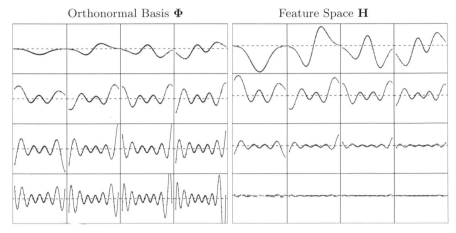

FIGURE 5.14. *(Left panel) The first 16 normalized eigenvectors of* **K**, *the* 200×200 *kernel matrix for the first coordinate of the mixture data. These are viewed as estimates* $\hat{\phi}_\ell$ *of the eigenfunctions in (5.45), and are represented as functions in* $\mathbb{R}^1$ *with the observed values superimposed in color. They are arranged in rows, starting at the top left. (Right panel) Rescaled versions* $h_\ell = \sqrt{\hat{\gamma}_\ell}\hat{\phi}_\ell$ *of the functions in the left panel, for which the kernel computes the "inner product."*

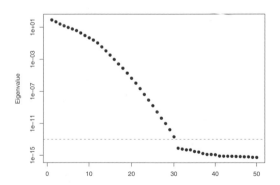

FIGURE 5.15. *The largest 50 eigenvalues of* **K**; *all those beyond the 30th are effectively zero.*

ing *feature space* representation of the eigenfunctions

$$h_\ell(x) = \sqrt{\hat{\gamma}_\ell}\hat{\phi}_\ell(x), \ \ell = 1, \dots, N. \tag{5.65}$$

Note that $\langle h(x_i), h(x_{i'}) \rangle = K(x_i, x_{i'})$. The scaling by the eigenvalues quickly shrinks most of the functions down to zero, leaving an effective dimension of about 12 in this case. The corresponding optimization problem is a standard ridge regression, as in (5.63). So although in principle the implicit feature space is infinite dimensional, the effective dimension is dramatically lower because of the relative amounts of shrinkage applied to each basis function. The kernel scale parameter ν plays a role here as well; larger ν implies more local k_m functions, and increases the effective dimension of the feature space. See Hastie and Zhu (2006) for more details.

It is also known (Girosi et al., 1995) that a thin-plate spline (Section 5.7) is an expansion in radial basis functions, generated by the kernel

$$K(x, y) = \|x - y\|^2 \log(\|x - y\|). \tag{5.66}$$

Radial basis functions are discussed in more detail in Section 6.7.

Support Vector Classifiers

The support vector machines of Chapter 12 for a two-class classification problem have the form $f(x) = \alpha_0 + \sum_{i=1}^{N} \alpha_i K(x, x_i)$, where the parameters are chosen to minimize

$$\min_{\alpha_0, \boldsymbol{\alpha}} \left\{ \sum_{i=1}^{N} [1 - y_i f(x_i)]_+ + \frac{\lambda}{2}\boldsymbol{\alpha}^T \mathbf{K}\boldsymbol{\alpha} \right\}, \tag{5.67}$$

where $y_i \in \{-1, 1\}$, and $[z]_+$ denotes the positive part of z. This can be viewed as a quadratic optimization problem with linear constraints, and requires a quadratic programming algorithm for its solution. The name *support vector* arises from the fact that typically many of the $\hat{\alpha}_i = 0$ [due to the piecewise-zero nature of the loss function in (5.67)], and so $\hat{f}$ is an expansion in a subset of the $K(\cdot, x_i)$. See Section 12.3.3 for more details.

5.9 Wavelet Smoothing

We have seen two different modes of operation with dictionaries of basis functions. With regression splines, we select a subset of the bases, using either subject-matter knowledge, or else automatically. The more adaptive procedures such as MARS (Chapter 9) can capture both smooth and nonsmooth behavior. With smoothing splines, we use a complete basis, but then shrink the coefficients toward smoothness.

Haar Wavelets   Symmlet-8 Wavelets

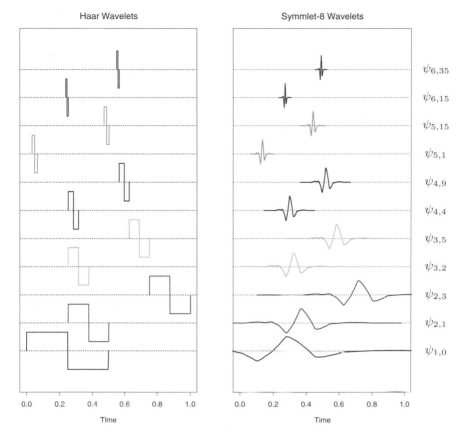

FIGURE 5.16. *Some selected wavelets at different translations and dilations for the Haar and symmlet families. The functions have been scaled to suit the display.*

Wavelets typically use a complete orthonormal basis to represent functions, but then shrink and select the coefficients toward a *sparse* representation. Just as a smooth function can be represented by a few spline basis functions, a mostly flat function with a few isolated bumps can be represented with a few (bumpy) basis functions. Wavelets bases are very popular in signal processing and compression, since they are able to represent both smooth and/or locally bumpy functions in an efficient way—a phenomenon dubbed *time and frequency localization*. In contrast, the traditional Fourier basis allows only frequency localization.

Before we give details, let's look at the Haar wavelets in the left panel of Figure 5.16 to get an intuitive idea of how wavelet smoothing works. The vertical axis indicates the scale (frequency) of the wavelets, from low scale at the bottom to high scale at the top. At each scale the wavelets are "packed in" side-by-side to completely fill the time axis: we have only shown a selected subset. Wavelet smoothing fits the coefficients for this basis by least squares, and then thresholds (discards, filters) the smaller coefficients. Since there are many basis functions at each scale, it can use bases where it needs them and discard the ones it does not need, to achieve time and frequency localization. The Haar wavelets are simple to understand, but not smooth enough for most purposes. The *symmlet* wavelets in the right panel of Figure 5.16 have the same orthonormal properties, but are smoother.

Figure 5.17 displays an NMR (nuclear magnetic resonance) signal, which appears to be composed of smooth components and isolated spikes, plus some noise. The wavelet transform, using a symmlet basis, is shown in the lower left panel. The wavelet coefficients are arranged in rows, from lowest scale at the bottom, to highest scale at the top. The length of each line segment indicates the size of the coefficient. The bottom right panel shows the wavelet coefficients after they have been thresholded. The threshold procedure, given below in equation (5.69), is the same soft-thresholding rule that arises in the lasso procedure for linear regression (Section 3.4.2). Notice that many of the smaller coefficients have been set to zero. The green curve in the top panel shows the back-transform of the thresholded coefficients: this is the smoothed version of the original signal. In the next section we give the details of this process, including the construction of wavelets and the thresholding rule.

5.9.1 Wavelet Bases and the Wavelet Transform

In this section we give details on the construction and filtering of wavelets. Wavelet bases are generated by translations and dilations of a single scaling function $\phi(x)$ (also known as the *father*). The red curves in Figure 5.18 are the *Haar* and *symmlet-8* scaling functions. The Haar basis is particularly easy to understand, especially for anyone with experience in analysis of variance or trees, since it produces a piecewise-constant representation. Thus if $\phi(x) = I(x \in [0,1])$, then $\phi_{0,k}(x) = \phi(x-k)$, k an integer, generates an orthonormal basis for functions with jumps at the integers. Call this *reference* space V_0. The dilations $\phi_{1,k}(x) = \sqrt{2}\phi(2x-k)$ form an orthonormal basis for a space $V_1 \supset V_0$ of functions piecewise constant on intervals of length $\frac{1}{2}$. In fact, more generally we have $\cdots \supset V_1 \supset V_0 \supset V_{-1} \supset \cdots$ where each V_j is spanned by $\phi_{j,k} = 2^{j/2}\phi(2^j x - k)$.

Now to the definition of wavelets. In analysis of variance, we often represent a pair of means μ_1 and μ_2 by their grand mean $\mu = \frac{1}{2}(\mu_1 + \mu_2)$, and then a contrast $\alpha = \frac{1}{2}(\mu_1 - \mu_2)$. A simplification occurs if the contrast α is very small, because then we can set it to zero. In a similar manner we might represent a function in V_{j+1} by a component in V_j plus the component in the orthogonal complement W_j of V_j to V_{j+1}, written as $V_{j+1} = V_j \oplus W_j$. The component in W_j represents *detail*, and we might wish to set some elements of this component to zero. It is easy to see that the functions $\psi(x-k)$

NMR Signal

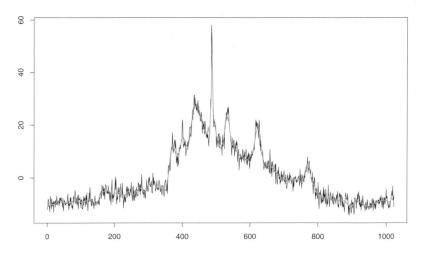

Wavelet Transform - Original Signal   Wavelet Transform - WaveShrunk Signal

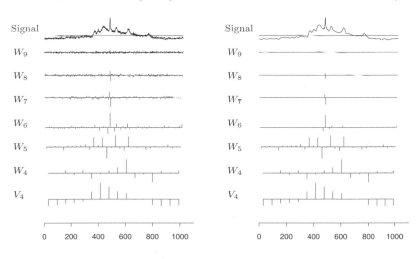

FIGURE 5.17. *The top panel shows an NMR signal, with the wavelet-shrunk version superimposed in green. The lower left panel represents the wavelet transform of the original signal, down to V_4, using the* symmlet-8 *basis. Each coefficient is represented by the height (positive or negative) of the vertical bar. The lower right panel represents the wavelet coefficients after being shrunken using the* waveshrink *function in S-PLUS, which implements the* SureShrink *method of wavelet adaptation of Donoho and Johnstone.*

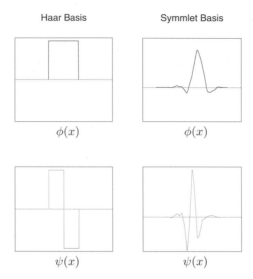

Haar Basis   Symmlet Basis

$\phi(x)$ $\phi(x)$

$\psi(x)$ $\psi(x)$

FIGURE 5.18. *The* Haar *and* symmlet *father (scaling) wavelet* $\phi(x)$ *and mother wavelet* $\psi(x)$.

generated by the *mother wavelet* $\psi(x) = \phi(2x) - \phi(2x-1)$ form an orthonormal basis for W_0 for the Haar family. Likewise $\psi_{j,k} = 2^{j/2}\psi(2^j x - k)$ form a basis for W_j.

Now $V_{j+1} = V_j \oplus W_j = V_{j-1} \oplus W_{j-1} \oplus W_j$, so besides representing a function by its level-j detail and level-j rough components, the latter can be broken down to level-$(j-1)$ detail and rough, and so on. Finally we get a representation of the form $V_J = V_0 \oplus W_0 \oplus W_1 \cdots \oplus W_{J-1}$. Figure 5.16 on page 175 shows particular wavelets $\psi_{j,k}(x)$.

Notice that since these spaces are orthogonal, all the basis functions are orthonormal. In fact, if the domain is discrete with $N = 2^J$ (time) points, this is as far as we can go. There are 2^j basis elements at level j, and adding up, we have a total of $2^J - 1$ elements in the W_j, and one in V_0. This structured orthonormal basis allows for a *multiresolution analysis*, which we illustrate in the next section.

While helpful for understanding the construction above, the Haar basis is often too coarse for practical purposes. Fortunately, many clever wavelet bases have been invented. Figures 5.16 and 5.18 include the *Daubechies symmlet-8* basis. This basis has smoother elements than the corresponding Haar basis, but there is a tradeoff:

- Each wavelet has a support covering 15 consecutive time intervals, rather than one for the Haar basis. More generally, the symmlet-p family has a support of $2p - 1$ consecutive intervals. The wider the support, the more time the wavelet has to die to zero, and so it can achieve this more smoothly. Note that the effective support seems to be much narrower.

- The symmlet-p wavelet $\psi(x)$ has p vanishing moments; that is,

$$\int \psi(x)x^j dx = 0, \; j = 0, \ldots, p-1.$$

One implication is that any order-p polynomial over the $N = 2^J$ times points is reproduced exactly in V_0 (Exercise 5.18). In this sense V_0 is equivalent to the null space of the smoothing-spline penalty. The Haar wavelets have one vanishing moment, and V_0 can reproduce any constant function.

The symmlet-p scaling functions are one of many families of wavelet generators. The operations are similar to those for the Haar basis:

- If V_0 is spanned by $\phi(x-k)$, then $V_1 \supset V_0$ is spanned by $\phi_{1,k}(x) = \sqrt{2}\phi(2x-k)$ and $\phi(x) = \sum_{k \in \mathcal{Z}} h(k)\phi_{1,k}(x)$, for some filter coefficients $h(k)$.

- W_0 is spanned by $\psi(x) = \sum_{k \in \mathcal{Z}} g(k)\phi_{1,k}(x)$, with filter coefficients $g(k) = (-1)^{1-k}h(1-k)$.

5.9.2 Adaptive Wavelet Filtering

Wavelets are particularly useful when the data are measured on a uniform lattice, such as a discretized signal, image, or a time series. We will focus on the one-dimensional case, and having $N = 2^J$ lattice-points is convenient. Suppose $\mathbf{y}$ is the response vector, and $\mathbf{W}$ is the $N \times N$ orthonormal wavelet basis matrix evaluated at the N uniformly spaced observations. Then $\mathbf{y}^* = \mathbf{W}^T\mathbf{y}$ is called the *wavelet transform* of $\mathbf{y}$ (and is the full least squares regression coefficient). A popular method for adaptive wavelet fitting is known as *SURE shrinkage* (Stein Unbiased Risk Estimation, Donoho and Johnstone (1994)). We start with the criterion

$$\min_{\boldsymbol{\theta}} ||\mathbf{y} - \mathbf{W}\boldsymbol{\theta}||_2^2 + 2\lambda||\boldsymbol{\theta}||_1, \tag{5.68}$$

which is the same as the lasso criterion in Chapter 3. Because $\mathbf{W}$ is orthonormal, this leads to the simple solution:

$$\hat{\theta}_j = \text{sign}(y_j^*)(|y_j^*| - \lambda)_+. \tag{5.69}$$

The least squares coefficients are translated toward zero, and truncated at zero. The fitted function (vector) is then given by the *inverse wavelet transform* $\hat{\mathbf{f}} = \mathbf{W}\hat{\boldsymbol{\theta}}$.

A simple choice for λ is $\lambda = \sigma\sqrt{2\log N}$, where σ is an estimate of the standard deviation of the noise. We can give some motivation for this choice. Since $\mathbf{W}$ is an orthonormal transformation, if the elements of $\mathbf{y}$ are white noise (independent Gaussian variates with mean 0 and variance σ^2), then so are $\mathbf{y}^*$. Furthermore if random variables $Z_1, Z_2, \ldots, Z_N$ are white noise, the expected maximum of $|Z_j|, j = 1, \ldots, N$ is approximately $\sigma\sqrt{2\log N}$. Hence all coefficients below $\sigma\sqrt{2\log N}$ are likely to be noise and are set to zero.

The space $\mathbf{W}$ could be any basis of orthonormal functions: polynomials, natural splines or cosinusoids. What makes wavelets special is the particular form of basis functions used, which allows for a representation *localized in time and in frequency*.

Let's look again at the NMR signal of Figure 5.17. The wavelet transform was computed using a *symmlet*-8 basis. Notice that the coefficients do not descend all the way to V_0, but stop at V_4 which has 16 basis functions. As we ascend to each level of detail, the coefficients get smaller, except in locations where spiky behavior is present. The wavelet coefficients represent characteristics of the signal localized in time (the basis functions at each level are translations of each other) and localized in frequency. Each dilation increases the detail by a factor of two, and in this sense corresponds to doubling the frequency in a traditional Fourier representation. In fact, a more mathematical understanding of wavelets reveals that the wavelets at a particular scale have a Fourier transform that is restricted to a limited range or octave of frequencies.

The shrinking/truncation in the right panel was achieved using the SURE approach described in the introduction to this section. The orthonormal $N \times N$ basis matrix $\mathbf{W}$ has columns which are the wavelet basis functions evaluated at the N time points. In particular, in this case there will be 16 columns corresponding to the $\phi_{4,k}(x)$, and the remainder devoted to the $\psi_{j,k}(x)$, $j = 4, \ldots, 11$. In practice λ depends on the noise variance, and has to be estimated from the data (such as the variance of the coefficients at the highest level).

Notice the similarity between the SURE criterion (5.68) on page 179, and the smoothing spline criterion (5.21) on page 156:

- Both are hierarchically structured from coarse to fine detail, although wavelets are also localized in time within each resolution level.

- The splines build in a bias toward smooth functions by imposing differential shrinking constants d_k. Early versions of SURE shrinkage treated all scales equally. The S+wavelets function waveshrink() has many options, some of which allow for differential shrinkage.

- The spline L_2 penalty cause pure shrinkage, while the SURE L_1 penalty does shrinkage and selection.

More generally smoothing splines achieve compression of the original signal by imposing smoothness, while wavelets impose sparsity. Figure 5.19 compares a wavelet fit (using SURE shrinkage) to a smoothing spline fit (using cross-validation) on two examples different in nature. For the NMR data in the upper panel, the smoothing spline introduces detail everywhere in order to capture the detail in the isolated spikes; the wavelet fit nicely localizes the spikes. In the lower panel, the true function is smooth, and the noise is relatively high. The wavelet fit has let in some additional and unnecessary wiggles—a price it pays in variance for the additional adaptivity.

The wavelet transform is not performed by matrix multiplication as in $\mathbf{y}^* = \mathbf{W}^T\mathbf{y}$. In fact, using clever pyramidal schemes $\mathbf{y}^*$ can be obtained in $O(N)$ computations, which is even faster than the $N\log(N)$ of the fast Fourier transform (FFT). While the general construction is beyond the scope of this book, it is easy to see for the Haar basis (Exercise 5.19). Likewise, the inverse wavelet transform $\mathbf{W}\hat{\boldsymbol{\theta}}$ is also $O(N)$.

This has been a very brief glimpse of this vast and growing field. There is a very large mathematical and computational base built on wavelets. Modern image compression is often performed using two-dimensional wavelet representations.

Bibliographic Notes

Splines and B-splines are discussed in detail in de Boor (1978). Green and Silverman (1994) and Wahba (1990) give a thorough treatment of smoothing splines and thin-plate splines; the latter also covers reproducing kernel Hilbert spaces. See also Girosi et al. (1995) and Evgeniou et al. (2000) for connections between many nonparametric regression techniques using RKHS approaches. Modeling functional data, as in Section 5.2.3, is covered in detail in Ramsay and Silverman (1997).

Daubechies (1992) is a classic and mathematical treatment of wavelets. Other useful sources are Chui (1992) and Wickerhauser (1994). Donoho and Johnstone (1994) developed the SURE shrinkage and selection technology from a statistical estimation framework; see also Vidakovic (1999). Bruce and Gao (1996) is a useful applied introduction, which also describes the wavelet software in S-PLUS.

Exercises

Ex. 5.1 Show that the truncated power basis functions in (5.3) represent a basis for a cubic spline with the two knots as indicated.

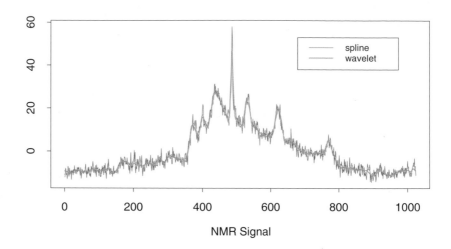

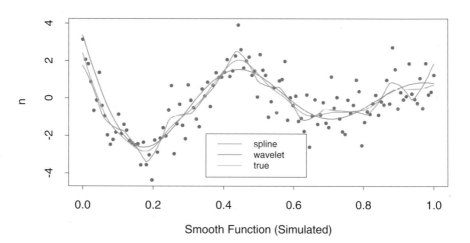

FIGURE 5.19. *Wavelet smoothing compared with smoothing splines on two examples. Each panel compares the SURE-shrunk wavelet fit to the cross-validated smoothing spline fit.*

Ex. 5.2 Suppose that $B_{i,M}(x)$ is an order-M B-spline defined in the Appendix on page 186 through the sequence (5.77)–(5.78).

(a) Show by induction that $B_{i,M}(x) = 0$ for $x \notin [\tau_i, \tau_{i+M}]$. This shows, for example, that the support of cubic B-splines is at most 5 knots.

(b) Show by induction that $B_{i,M}(x) > 0$ for $x \in (\tau_i, \tau_{i+M})$. The B-splines are positive in the interior of their support.

(c) Show by induction that $\sum_{i=1}^{K+M} B_{i,M}(x) = 1 \, \forall x \in [\xi_0, \xi_{K+1}]$.

(d) Show that $B_{i,M}$ is a piecewise polynomial of order M (degree $M-1$) on $[\xi_0, \xi_{K+1}]$, with breaks only at the knots $\xi_1, \ldots, \xi_K$.

(e) Show that an order-M B-spline basis function is the density function of a convolution of M uniform random variables.

Ex. 5.3 Write a program to reproduce Figure 5.3 on page 145.

Ex. 5.4 Consider the truncated power series representation for cubic splines with K interior knots. Let

$$f(X) = \sum_{j=0}^{3} \beta_j X^j + \sum_{k=1}^{K} \theta_k (X - \xi_k)_+^3. \tag{5.70}$$

Prove that the natural boundary conditions for natural cubic splines (Section 5.2.1) imply the following linear constraints on the coefficients:

$$\begin{array}{ll} \beta_2 = 0, & \sum_{k=1}^{K} \theta_k = 0, \\ \beta_3 = 0, & \sum_{k=1}^{K} \xi_k \theta_k = 0. \end{array} \tag{5.71}$$

Hence derive the basis (5.4) and (5.5).

Ex. 5.5 Write a program to classify the phoneme data using a quadratic discriminant analysis (Section 4.3). Since there are many correlated features, you should filter them using a smooth basis of natural cubic splines (Section 5.2.3). Decide beforehand on a series of five different choices for the number and position of the knots, and use tenfold cross-validation to make the final selection. The phoneme data are available from the book website www-stat.stanford.edu/ElemStatLearn.

Ex. 5.6 Suppose you wish to fit a periodic function, with a known period T. Describe how you could modify the truncated power series basis to achieve this goal.

Ex. 5.7 *Derivation of smoothing splines* (Green and Silverman, 1994). Suppose that $N \geq 2$, and that g is the natural cubic spline interpolant to the pairs $\{x_i, z_i\}_1^N$, with $a < x_1 < \cdots < x_N < b$. This is a natural spline with a knot at every x_i; being an N-dimensional space of functions, we can determine the coefficients such that it interpolates the sequence z_i exactly. Let $\tilde{g}$ be any other differentiable function on $[a, b]$ that interpolates the N pairs.

(a) Let $h(x) = \tilde{g}(x) - g(x)$. Use integration by parts and the fact that g is a natural cubic spline to show that

$$\int_a^b g''(x)h''(x)dx = -\sum_{j=1}^{N-1} g'''(x_j^+)\{h(x_{j+1}) - h(x_j)\} \quad (5.72)$$

$$= 0.$$

(b) Hence show that

$$\int_a^b \tilde{g}''(t)^2 dt \geq \int_a^b g''(t)^2 dt,$$

and that equality can only hold if h is identically zero in $[a, b]$.

(c) Consider the penalized least squares problem

$$\min_f \left[\sum_{i=1}^N (y_i - f(x_i))^2 + \lambda \int_a^b f''(t)^2 dt \right].$$

Use (b) to argue that the minimizer must be a cubic spline with knots at each of the x_i.

Ex. 5.8 In the appendix to this chapter we show how the smoothing spline computations could be more efficiently carried out using a $(N + 4)$ dimensional basis of B-splines. Describe a slightly simpler scheme using a $(N+2)$ dimensional B-spline basis defined on the $N - 2$ interior knots.

Ex. 5.9 Derive the Reinsch form $\mathbf{S}_\lambda = (\mathbf{I} + \lambda \mathbf{K})^{-1}$ for the smoothing spline.

Ex. 5.10 Derive an expression for $\text{Var}(\hat{f}_\lambda(x_0))$ and $\text{bias}(\hat{f}_\lambda(x_0))$. Using the example (5.22), create a version of Figure 5.9 where the mean and several (pointwise) quantiles of $\hat{f}_\lambda(x)$ are shown.

Ex. 5.11 Prove that for a smoothing spline the null space of $\mathbf{K}$ is spanned by functions linear in X.

Ex. 5.12 Characterize the solution to the following problem,

$$\min_f \text{RSS}(f, \lambda) = \sum_{i=1}^N w_i \{y_i - f(x_i)\}^2 + \lambda \int \{f''(t)\}^2 dt, \quad (5.73)$$

where the $w_i \geq 0$ are observation weights.

Characterize the solution to the smoothing spline problem (5.9) when the training data have ties in X.

Ex. 5.13 You have fitted a smoothing spline $\hat{f}_\lambda$ to a sample of N pairs (x_i, y_i). Suppose you augment your original sample with the pair $x_0, \hat{f}_\lambda(x_0)$, and refit; describe the result. Use this to derive the N-fold cross-validation formula (5.26).

Ex. 5.14 Derive the constraints on the α_j in the thin-plate spline expansion (5.39) to guarantee that the penalty $J(f)$ is finite. How else could one ensure that the penalty was finite?

Ex. 5.15 This exercise derives some of the results quoted in Section 5.8.1. Suppose $K(x, y)$ satisfying the conditions (5.45) and let $f(x) \in \mathcal{H}_K$. Show that

(a) $\langle K(\cdot, x_i), f \rangle_{\mathcal{H}_K} = f(x_i)$.

(b) $\langle K(\cdot, x_i), K(\cdot, x_j) \rangle_{\mathcal{H}_K} = K(x_i, x_j)$.

(c) If $g(x) = \sum_{i=1}^{N} \alpha_i K(x, x_i)$, then

$$J(g) = \sum_{i=1}^{N} \sum_{j=1}^{N} K(x_i, x_j) \alpha_i \alpha_j.$$

Suppose that $\tilde{g}(x) = g(x) + \rho(x)$, with $\rho(x) \in \mathcal{H}_K$, and orthogonal in $\mathcal{H}_K$ to each of $K(x, x_i)$, $i = 1, \ldots, N$. Show that

(d)

$$\sum_{i=1}^{N} L(y_i, \tilde{g}(x_i)) + \lambda J(\tilde{g}) \geq \sum_{i=1}^{N} L(y_i, g(x_i)) + \lambda J(g) \qquad (5.74)$$

with equality iff $\rho(x) = 0$.

Ex. 5.16 Consider the ridge regression problem (5.53), and assume $M \geq N$. Assume you have a kernel K that computes the inner product $K(x, y) = \sum_{m=1}^{M} h_m(x) h_m(y)$.

(a) Derive (5.62) on page 171 in the text. How would you compute the matrices $\mathbf{V}$ and $\mathbf{D}_\gamma$, given K? Hence show that (5.63) is equivalent to (5.53).

(b) Show that

$$\begin{aligned} \hat{\mathbf{f}} &= \mathbf{H}\hat{\beta} \\ &= \mathbf{K}(\mathbf{K} + \lambda\mathbf{I})^{-1}\mathbf{y}, \end{aligned} \qquad (5.75)$$

where $\mathbf{H}$ is the $N \times M$ matrix of evaluations $h_m(x_i)$, and $\mathbf{K} = \mathbf{H}\mathbf{H}^T$ the $N \times N$ matrix of inner-products $h(x_i)^T h(x_j)$.

(c) Show that

$$\hat{f}(x) \;=\; h(x)^T \hat{\boldsymbol{\beta}}$$
$$=\; \sum_{i=1}^{N} K(x, x_i)\hat{\alpha}_i \tag{5.76}$$

and $\hat{\boldsymbol{\alpha}} = (\mathbf{K} + \lambda \mathbf{I})^{-1}\mathbf{y}$.

(d) How would you modify your solution if $M < N$?

Ex. 5.17 Show how to convert the discrete eigen-decomposition of $\mathbf{K}$ in Section 5.8.2 to estimates of the eigenfunctions of K.

Ex. 5.18 The wavelet function $\psi(x)$ of the symmlet-p wavelet basis has vanishing moments up to order p. Show that this implies that polynomials of order p are represented exactly in V_0, defined on page 176.

Ex. 5.19 Show that the Haar wavelet transform of a signal of length $N = 2^J$ can be computed in $O(N)$ computations.

Appendix: Computations for Splines

In this Appendix, we describe the B-spline basis for representing polynomial splines. We also discuss their use in the computations of smoothing splines.

B-splines

Before we can get started, we need to augment the knot sequence defined in Section 5.2. Let $\xi_0 < \xi_1$ and $\xi_K < \xi_{K+1}$ be two *boundary* knots, which typically define the domain over which we wish to evaluate our spline. We now define the augmented knot sequence τ such that

- $\tau_1 \le \tau_2 \le \cdots \le \tau_M \le \xi_0$;

- $\tau_{j+M} = \xi_j$, $j = 1, \cdots, K$;

- $\xi_{K+1} \le \tau_{K+M+1} \le \tau_{K+M+2} \le \cdots \le \tau_{K+2M}$.

The actual values of these additional knots beyond the boundary are arbitrary, and it is customary to make them all the same and equal to ξ_0 and ξ_{K+1}, respectively.

   Denote by $B_{i,m}(x)$ the ith B-spline basis function of order m for the knot-sequence τ, $m \le M$. They are defined recursively in terms of divided differences as follows:

$$B_{i,1}(x) = \begin{cases} 1 & \text{if } \tau_i \le x < \tau_{i+1} \\ 0 & \text{otherwise} \end{cases} \tag{5.77}$$

for $i = 1, \ldots, K + 2M - 1$. These are also known as Haar basis functions.

$$B_{i,m}(x) = \frac{x - \tau_i}{\tau_{i+m-1} - \tau_i} B_{i,m-1}(x) + \frac{\tau_{i+m} - x}{\tau_{i+m} - \tau_{i+1}} B_{i+1,m-1}(x)$$

for $i = 1, \ldots, K + 2M - m$.

$$\tag{5.78}$$

Thus with $M = 4$, $B_{i,4}$, $i = 1, \cdots, K + 4$ are the $K + 4$ cubic B-spline basis functions for the knot sequence ξ. This recursion can be continued and will generate the B-spline basis for any order spline. Figure 5.20 shows the sequence of B-splines up to order four with knots at the points $0.0, 0.1, \ldots, 1.0$. Since we have created some duplicate knots, some care has to be taken to avoid division by zero. If we adopt the convention that $B_{i,1} = 0$ if $\tau_i = \tau_{i+1}$, then by induction $B_{i,m} = 0$ if $\tau_i = \tau_{i+1} = \ldots = \tau_{i+m}$. Note also that in the construction above, only the subset $B_{i,m}$, $i = M - m + 1, \ldots, M + K$ are required for the B-spline basis of order $m < M$ with knots ξ.

To fully understand the properties of these functions, and to show that they do indeed span the space of cubic splines for the knot sequence, requires additional mathematical machinery, including the properties of divided differences. Exercise 5.2 explores these issues.

The scope of B-splines is in fact bigger than advertised here, and has to do with knot duplication. If we duplicate an interior knot in the construction of the τ sequence above, and then generate the B-spline sequence as before, the resulting basis spans the space of piecewise polynomials with one less continuous derivative at the duplicated knot. In general, if in addition to the repeated boundary knots, we include the interior knot ξ_j $1 \le r_j \le M$ times, then the lowest-order derivative to be discontinuous at $x = \xi_j$ will be order $M - r_j$. Thus for cubic splines with no repeats, $r_j = 1$, $j = 1, \ldots, K$, and at each interior knot the third derivatives $(4 - 1)$ are discontinuous. Repeating the jth knot three times leads to a discontinuous 1st derivative; repeating it four times leads to a discontinuous zeroth derivative, i.e., the function is discontinuous at $x = \xi_j$. This is exactly what happens at the boundary knots; we repeat the knots M times, so the spline becomes discontinuous at the boundary knots (i.e., undefined beyond the boundary).

The local support of B-splines has important computational implications, especially when the number of knots K is large. Least squares computations with N observations and $K + M$ variables (basis functions) take $O(N(K + M)^2 + (K + M)^3)$ flops (floating point operations.) If K is some appreciable fraction of N, this leads to $O(N^3)$ algorithms which becomes

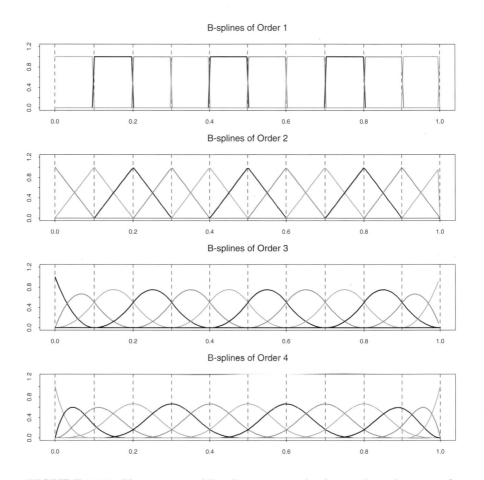

FIGURE 5.20. *The sequence of B-splines up to order four with ten knots evenly spaced from 0 to 1. The B-splines have* local support; *they are nonzero on an interval spanned by $M + 1$ knots.*

unacceptable for large N. If the N observations are sorted, the $N \times (K+M)$ regression matrix consisting of the $K + M$ B-spline basis functions evaluated at the N points has many zeros, which can be exploited to reduce the computational complexity back to $O(N)$. We take this up further in the next section.

Computations for Smoothing Splines

Although natural splines (Section 5.2.1) provide a basis for smoothing splines, it is computationally more convenient to operate in the larger space of unconstrained B-splines. We write $f(x) = \sum_1^{N+4} \gamma_j B_j(x)$, where γ_j are coefficients and the B_j are the cubic B-spline basis functions. The solution looks the same as before,

$$\hat{\gamma} = (\mathbf{B}^T \mathbf{B} + \lambda \mathbf{\Omega}_B)^{-1} \mathbf{B}^T \mathbf{y}, \tag{5.79}$$

except now the $N \times N$ matrix $\mathbf{N}$ is replaced by the $N \times (N+4)$ matrix $\mathbf{B}$, and similarly the $(N+4) \times (N+4)$ penalty matrix $\mathbf{\Omega}_B$ replaces the $N \times N$ dimensional $\mathbf{\Omega}_N$. Although at face value it seems that there are no boundary derivative constraints, it turns out that the penalty term automatically imposes them by giving effectively infinite weight to any non zero derivative beyond the boundary. In practice, $\hat{\gamma}$ is restricted to a linear subspace for which the penalty is always finite.

Since the columns of $\mathbf{B}$ are the evaluated B-splines, in order from left to right and evaluated at the *sorted* values of X, and the cubic B-splines have local support, $\mathbf{B}$ is lower 4-banded. Consequently the matrix $\mathbf{M} = (\mathbf{B}^T \mathbf{B} + \lambda \mathbf{\Omega})$ is 4-banded and hence its Cholesky decomposition $\mathbf{M} = \mathbf{L}\mathbf{L}^\mathbf{T}$ can be computed easily. One then solves $\mathbf{L}\mathbf{L}^T \gamma = \mathbf{B}^T \mathbf{y}$ by back-substitution to give γ and hence the solution $\hat{f}$ in $O(N)$ operations.

In practice, when N is large, it is unnecessary to use all N interior knots, and any reasonable *thinning* strategy will save in computations and have negligible effect on the fit. For example, the `smooth.spline` function in S-PLUS uses an approximately logarithmic strategy: if $N < 50$ all knots are included, but even at $N = 5,000$ only 204 knots are used.

6
Kernel Smoothing Methods

In this chapter we describe a class of regression techniques that achieve flexibility in estimating the regression function $f(X)$ over the domain $\mathbb{R}^p$ by fitting a different but simple model separately at each query point x_0. This is done by using only those observations close to the target point x_0 to fit the simple model, and in such a way that the resulting estimated function $\hat{f}(X)$ is *smooth* in $\mathbb{R}^p$. This localization is achieved via a weighting function or *kernel* $K_\lambda(x_0, x_i)$, which assigns a weight to x_i based on its distance from x_0. The kernels K_λ are typically indexed by a parameter λ that dictates the width of the neighborhood. These *memory-based* methods require in principle little or no training; all the work gets done at evaluation time. The only parameter that needs to be determined from the training data is λ. The model, however, is the entire training data set.

We also discuss more general classes of kernel-based techniques , which tie in with structured methods in other chapters, and are useful for density estimation and classification.

The techniques in this chapter should not be confused with those associated with the more recent usage of the phrase "kernel methods". In this chapter kernels are mostly used as a device for localization. We discuss kernel methods in Sections 5.8, 14.5.4, 18.5 and Chapter 12; in those contexts the kernel computes an inner product in a high-dimensional (implicit) feature space, and is used for regularized nonlinear modeling. We make some connections to the methodology in this chapter at the end of Section 6.7.

T. Hastie et al., *The Elements of Statistical Learning, Second Edition*,
DOI: 10.1007/b94608_6,
© Springer Science+Business Media, LLC 2009

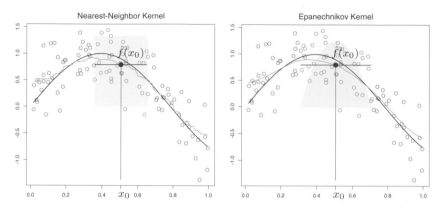

FIGURE 6.1. *In each panel* 100 *pairs* x_i, y_i *are generated at random from the blue curve with Gaussian errors:* $Y = \sin(4X) + \varepsilon$, $X \sim U[0,1]$, $\varepsilon \sim N(0, 1/3)$. *In the left panel the green curve is the result of a 30-nearest-neighbor running-mean smoother. The red point is the fitted constant* $\hat{f}(x_0)$, *and the red circles indicate those observations contributing to the fit at* x_0. *The solid yellow region indicates the weights assigned to observations. In the right panel, the green curve is the kernel-weighted average, using an Epanechnikov kernel with (half) window width* $\lambda = 0.2$.

6.1   One-Dimensional Kernel Smoothers

In Chapter 2, we motivated the k–nearest-neighbor average

$$\hat{f}(x) = \text{Ave}(y_i | x_i \in N_k(x)) \tag{6.1}$$

as an estimate of the regression function $\text{E}(Y | X = x)$. Here $N_k(x)$ is the set of k points nearest to x in squared distance, and Ave denotes the average (mean). The idea is to relax the definition of conditional expectation, as illustrated in the left panel of Figure 6.1, and compute an average in a neighborhood of the target point. In this case we have used the 30-nearest neighborhood—the fit at x_0 is the average of the 30 pairs whose x_i values are closest to x_0. The green curve is traced out as we apply this definition at different values x_0. The green curve is bumpy, since $\hat{f}(x)$ is discontinuous in x. As we move x_0 from left to right, the k-nearest neighborhood remains constant, until a point x_i to the right of x_0 becomes closer than the furthest point $x_{i'}$ in the neighborhood to the left of x_0, at which time x_i replaces $x_{i'}$. The average in (6.1) changes in a discrete way, leading to a discontinuous $\hat{f}(x)$.

This discontinuity is ugly and unnecessary. Rather than give all the points in the neighborhood equal weight, we can assign weights that die off smoothly with distance from the target point. The right panel shows an example of this, using the so-called Nadaraya–Watson kernel-weighted average

$$\hat{f}(x_0) = \frac{\sum_{i=1}^{N} K_\lambda(x_0, x_i) y_i}{\sum_{i=1}^{N} K_\lambda(x_0, x_i)}, \tag{6.2}$$

with the *Epanechnikov* quadratic kernel

$$K_\lambda(x_0, x) = D\left(\frac{|x - x_0|}{\lambda}\right), \tag{6.3}$$

with

$$D(t) = \begin{cases} \frac{3}{4}(1 - t^2) & \text{if } |t| \le 1; \\ 0 & \text{otherwise.} \end{cases} \tag{6.4}$$

The fitted function is now continuous, and quite smooth in the right panel of Figure 6.1. As we move the target from left to right, points enter the neighborhood initially with weight zero, and then their contribution slowly increases (see Exercise 6.1).

In the right panel we used a metric window size $\lambda = 0.2$ for the kernel fit, which does not change as we move the target point x_0, while the size of the 30-nearest-neighbor smoothing window adapts to the local density of the x_i. One can, however, also use such adaptive neighborhoods with kernels, but we need to use a more general notation. Let $h_\lambda(x_0)$ be a width function (indexed by λ) that determines the width of the neighborhood at x_0. Then more generally we have

$$K_\lambda(x_0, x) = D\left(\frac{|x - x_0|}{h_\lambda(x_0)}\right). \tag{6.5}$$

In (6.3), $h_\lambda(x_0) = \lambda$ is constant. For k-nearest neighborhoods, the neighborhood size k replaces λ, and we have $h_k(x_0) - |x_0 - x_{[k]}|$ where $x_{[k]}$ is the kth closest x_i to x_0.

There are a number of details that one has to attend to in practice:

• The smoothing parameter λ, which determines the width of the local neighborhood, has to be determined. Large λ implies lower variance (averages over more observations) but higher bias (we essentially assume the true function is constant within the window).

• Metric window widths (constant $h_\lambda(x)$) tend to keep the bias of the estimate constant, but the variance is inversely proportional to the local density. Nearest-neighbor window widths exhibit the opposite behavior; the variance stays constant and the absolute bias varies inversely with local density.

• Issues arise with nearest-neighbors when there are ties in the x_i. With most smoothing techniques one can simply reduce the data set by averaging the y_i at tied values of X, and supplementing these new observations at the unique values of x_i with an additional weight w_i (which multiples the kernel weight).

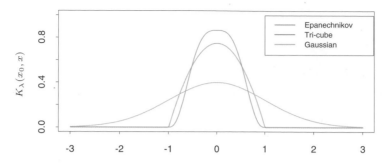

FIGURE 6.2. *A comparison of three popular kernels for local smoothing. Each has been calibrated to integrate to 1. The tri-cube kernel is compact and has two continuous derivatives at the boundary of its support, while the Epanechnikov kernel has none. The Gaussian kernel is continuously differentiable, but has infinite support.*

- This leaves a more general problem to deal with: observation weights w_i. Operationally we simply multiply them by the kernel weights before computing the weighted average. With nearest neighborhoods, it is now natural to insist on neighborhoods with a total weight content k (relative to $\sum w_i$). In the event of overflow (the last observation needed in a neighborhood has a weight w_j which causes the sum of weights to exceed the budget k), then fractional parts can be used.

- Boundary issues arise. The metric neighborhoods tend to contain less points on the boundaries, while the nearest-neighborhoods get wider.

- The Epanechnikov kernel has compact support (needed when used with nearest-neighbor window size). Another popular compact kernel is based on the tri-cube function

$$D(t) = \begin{cases} (1 - |t|^3)^3 & \text{if } |t| \leq 1; \\ 0 & \text{otherwise} \end{cases} \qquad (6.6)$$

This is flatter on the top (like the nearest-neighbor box) and is differentiable at the boundary of its support. The Gaussian density function $D(t) = \phi(t)$ is a popular noncompact kernel, with the standard-deviation playing the role of the window size. Figure 6.2 compares the three.

6.1.1 Local Linear Regression

We have progressed from the raw moving average to a smoothly varying locally weighted average by using kernel weighting. The smooth kernel fit still has problems, however, as exhibited in Figure 6.3 (left panel). Locally-weighted averages can be badly biased on the boundaries of the domain,

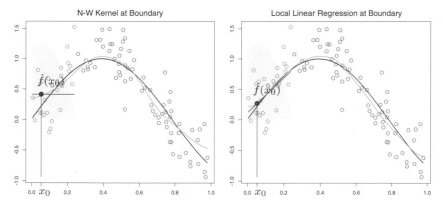

FIGURE 6.3. *The locally weighted average has bias problems at or near the boundaries of the domain. The true function is approximately linear here, but most of the observations in the neighborhood have a higher mean than the target point, so despite weighting, their mean will be biased upwards. By fitting a locally weighted linear regression (right panel), this bias is removed to first order.*

because of the asymmetry of the kernel in that region. By fitting straight lines rather than constants locally, we can remove this bias exactly to first order; see Figure 6.3 (right panel). Actually, this bias can be present in the interior of the domain as well, if the X values are not equally spaced (for the same reasons, but usually less severe). Again locally weighted linear regression will make a first-order correction.

Locally weighted regression solves a separate weighted least squares problem at each target point x_0:

$$\min_{\alpha(x_0),\beta(x_0)} \sum_{i=1}^{N} K_\lambda(x_0, x_i)\left[y_i - \alpha(x_0) - \beta(x_0)x_i\right]^2. \tag{6.7}$$

The estimate is then $\hat{f}(x_0) = \hat{\alpha}(x_0) + \hat{\beta}(x_0)x_0$. Notice that although we fit an entire linear model to the data in the region, we only use it to evaluate the fit at the single point x_0.

Define the vector-valued function $b(x)^T = (1, x)$. Let $\mathbf{B}$ be the $N \times 2$ regression matrix with ith row $b(x_i)^T$, and $\mathbf{W}(x_0)$ the $N \times N$ diagonal matrix with ith diagonal element $K_\lambda(x_0, x_i)$. Then

$$\hat{f}(x_0) = b(x_0)^T(\mathbf{B}^T\mathbf{W}(x_0)\mathbf{B})^{-1}\mathbf{B}^T\mathbf{W}(x_0)\mathbf{y} \tag{6.8}$$

$$= \sum_{i=1}^{N} l_i(x_0)y_i. \tag{6.9}$$

Equation (6.8) gives an explicit expression for the local linear regression estimate, and (6.9) highlights the fact that the estimate is *linear* in the

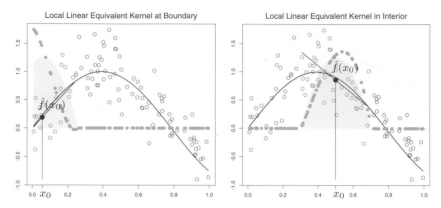

FIGURE 6.4. *The green points show the equivalent kernel $l_i(x_0)$ for local regression. These are the weights in $\hat{f}(x_0) = \sum_{i=1}^{N} l_i(x_0)y_i$, plotted against their corresponding x_i. For display purposes, these have been rescaled, since in fact they sum to 1. Since the yellow shaded region is the (rescaled) equivalent kernel for the Nadaraya–Watson local average, we see how local regression automatically modifies the weighting kernel to correct for biases due to asymmetry in the smoothing window.*

y_i (the $l_i(x_0)$ do not involve $\mathbf{y}$). These weights $l_i(x_0)$ combine the weighting kernel $K_\lambda(x_0, \cdot)$ and the least squares operations, and are sometimes referred to as the *equivalent kernel*. Figure 6.4 illustrates the effect of local linear regression on the equivalent kernel. Historically, the bias in the Nadaraya–Watson and other local average kernel methods were corrected by modifying the kernel. These modifications were based on theoretical asymptotic mean-square-error considerations, and besides being tedious to implement, are only approximate for finite sample sizes. Local linear regression *automatically* modifies the kernel to correct the bias *exactly* to first order, a phenomenon dubbed as *automatic kernel carpentry*. Consider the following expansion for $\mathrm{E}\hat{f}(x_0)$, using the linearity of local regression and a series expansion of the true function f around x_0,

$$
\begin{aligned}
\mathrm{E}\hat{f}(x_0) &= \sum_{i=1}^{N} l_i(x_0)f(x_i) \\
&= f(x_0)\sum_{i=1}^{N} l_i(x_0) + f'(x_0)\sum_{i=1}^{N}(x_i - x_0)l_i(x_0) \\
&\quad + \frac{f''(x_0)}{2}\sum_{i=1}^{N}(x_i - x_0)^2 l_i(x_0) + R,
\end{aligned}
\tag{6.10}
$$

where the remainder term R involves third- and higher-order derivatives of f, and is typically small under suitable smoothness assumptions. It can be

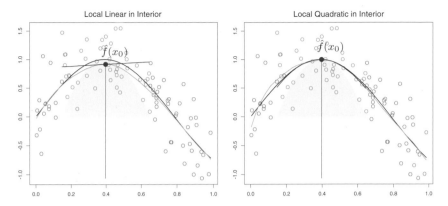

FIGURE 6.5. *Local linear fits exhibit bias in regions of curvature of the true function. Local quadratic fits tend to eliminate this bias.*

shown (Exercise 6.2) that for local linear regression, $\sum_{i=1}^{N} l_i(x_0) = 1$ and $\sum_{i=1}^{N} (x_i - x_0) l_i(x_0) = 0$. Hence the middle term equals $f(x_0)$, and since the bias is $\mathrm{E}\hat{f}(x_0) - f(x_0)$, we see that it depends only on quadratic and higher–order terms in the expansion of f.

6.1.2 Local Polynomial Regression

Why stop at local linear fits? We can fit local polynomial fits of any degree d,

$$\min_{\alpha(x_0),\beta_j(x_0),\, j=1,...,d} \sum_{i=1}^{N} K_\lambda(x_0, x_i) \left[y_i - \alpha(x_0) - \sum_{j=1}^{d} \beta_j(x_0) x_i^j \right]^2 \quad (6.11)$$

with solution $\hat{f}(x_0) = \hat{\alpha}(x_0) + \sum_{j=1}^{d} \hat{\beta}_j(x_0) x_0^j$. In fact, an expansion such as (6.10) will tell us that the bias will only have components of degree $d+1$ and higher (Exercise 6.2). Figure 6.5 illustrates local quadratic regression. Local linear fits tend to be biased in regions of curvature of the true function, a phenomenon referred to as *trimming the hills* and *filling the valleys*. Local quadratic regression is generally able to correct this bias.

There is of course a price to be paid for this bias reduction, and that is increased variance. The fit in the right panel of Figure 6.5 is slightly more wiggly, especially in the tails. Assuming the model $y_i = f(x_i) + \varepsilon_i$, with ε_i independent and identically distributed with mean zero and variance σ^2, $\mathrm{Var}(\hat{f}(x_0)) = \sigma^2 ||l(x_0)||^2$, where $l(x_0)$ is the vector of equivalent kernel weights at x_0. It can be shown (Exercise 6.3) that $||l(x_0)||$ increases with d, and so there is a bias–variance tradeoff in selecting the polynomial degree. Figure 6.6 illustrates these variance curves for degree zero, one and two

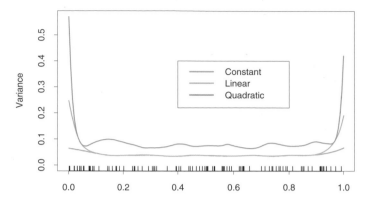

FIGURE 6.6. *The variances functions* $||l(x)||^2$ *for local constant, linear and quadratic regression, for a metric bandwidth ($\lambda = 0.2$) tri-cube kernel.*

local polynomials. To summarize some collected wisdom on this issue:

- Local linear fits can help bias dramatically at the boundaries at a modest cost in variance. Local quadratic fits do little at the boundaries for bias, but increase the variance a lot.

- Local quadratic fits tend to be most helpful in reducing bias due to curvature in the interior of the domain.

- Asymptotic analysis suggest that local polynomials of odd degree dominate those of even degree. This is largely due to the fact that asymptotically the MSE is dominated by boundary effects.

While it may be helpful to tinker, and move from local linear fits at the boundary to local quadratic fits in the interior, we do not recommend such strategies. Usually the application will dictate the degree of the fit. For example, if we are interested in extrapolation, then the boundary is of more interest, and local linear fits are probably more reliable.

6.2 Selecting the Width of the Kernel

In each of the kernels K_λ, λ is a parameter that controls its width:

- For the Epanechnikov or tri-cube kernel with metric width, λ is the radius of the support region.

- For the Gaussian kernel, λ is the standard deviation.

- λ is the number k of nearest neighbors in k-nearest neighborhoods, often expressed as a fraction or *span* k/N of the total training sample.

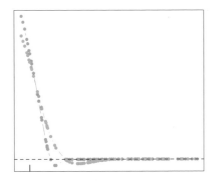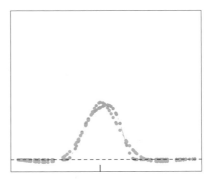

FIGURE 6.7. *Equivalent kernels for a local linear regression smoother (tri-cube kernel; orange) and a smoothing spline (blue), with matching degrees of freedom. The vertical spikes indicates the target points.*

There is a natural bias–variance tradeoff as we change the width of the averaging window, which is most explicit for local averages:

- If the window is narrow, $\hat{f}(x_0)$ is an average of a small number of y_i close to x_0, and its variance will be relatively large—close to that of an individual y_i. The bias will tend to be small, again because each of the $E(y_i) = f(x_i)$ should be close to $f(x_0)$.

- If the window is wide, the variance of $\hat{f}(x_0)$ will be small relative to the variance of any y_i, because of the effects of averaging. The bias will be higher, because we are now using observations x_i further from x_0, and there is no guarantee that $f(x_i)$ will be close to $f(x_0)$.

Similar arguments apply to local regression estimates, say local linear: as the width goes to zero, the estimates approach a piecewise-linear function that interpolates the training data[1]; as the width gets infinitely large, the fit approaches the global linear least-squares fit to the data.

The discussion in Chapter 5 on selecting the regularization parameter for smoothing splines applies here, and will not be repeated. Local regression smoothers are linear estimators; the smoother matrix in $\hat{\mathbf{f}} = \mathbf{S}_\lambda \mathbf{y}$ is built up from the equivalent kernels (6.8), and has ijth entry $\{\mathbf{S}_\lambda\}_{ij} = l_i(x_j)$. Leave-one-out cross-validation is particularly simple (Exercise 6.7), as is generalized cross-validation, C_p (Exercise 6.10), and k-fold cross-validation. The effective degrees of freedom is again defined as trace($\mathbf{S}_\lambda$), and can be used to calibrate the amount of smoothing. Figure 6.7 compares the equivalent kernels for a smoothing spline and local linear regression. The local regression smoother has a span of 40%, which results in df = trace($\mathbf{S}_\lambda$) = 5.86. The smoothing spline was calibrated to have the same df, and their equivalent kernels are qualitatively quite similar.

---

[1] With uniformly spaced x_i; with irregularly spaced x_i, the behavior can deteriorate.

6.3 Local Regression in $\mathbb{R}^p$

Kernel smoothing and local regression generalize very naturally to two or more dimensions. The Nadaraya–Watson kernel smoother fits a constant locally with weights supplied by a p-dimensional kernel. Local linear regression will fit a hyperplane locally in X, by weighted least squares, with weights supplied by a p-dimensional kernel. It is simple to implement and is generally preferred to the local constant fit for its superior performance on the boundaries.

Let $b(X)$ be a vector of polynomial terms in X of maximum degree d. For example, with $d = 1$ and $p = 2$ we get $b(X) = (1, X_1, X_2)$; with $d = 2$ we get $b(X) = (1, X_1, X_2, X_1^2, X_2^2, X_1 X_2)$; and trivially with $d = 0$ we get $b(X) = 1$. At each $x_0 \in \mathbb{R}^p$ solve

$$\min_{\beta(x_0)} \sum_{i=1}^{N} K_\lambda(x_0, x_i)(y_i - b(x_i)^T \beta(x_0))^2 \tag{6.12}$$

to produce the fit $\hat{f}(x_0) = b(x_0)^T \hat{\beta}(x_0)$. Typically the kernel will be a radial function, such as the radial Epanechnikov or tri-cube kernel

$$K_\lambda(x_0, x) = D\left(\frac{||x - x_0||}{\lambda}\right), \tag{6.13}$$

where $||\cdot||$ is the Euclidean norm. Since the Euclidean norm depends on the units in each coordinate, it makes most sense to standardize each predictor, for example, to unit standard deviation, prior to smoothing.

While boundary effects are a problem in one-dimensional smoothing, they are a much bigger problem in two or higher dimensions, since the fraction of points on the boundary is larger. In fact, one of the manifestations of the curse of dimensionality is that the fraction of points close to the boundary increases to one as the dimension grows. Directly modifying the kernel to accommodate two-dimensional boundaries becomes very messy, especially for irregular boundaries. Local polynomial regression seamlessly performs boundary correction to the desired order in any dimensions. Figure 6.8 illustrates local linear regression on some measurements from an astronomical study with an unusual predictor design (star-shaped). Here the boundary is extremely irregular, and the fitted surface must also interpolate over regions of increasing data sparsity as we approach the boundary.

Local regression becomes less useful in dimensions much higher than two or three. We have discussed in some detail the problems of dimensionality, for example, in Chapter 2. It is impossible to simultaneously maintain localness ($\Rightarrow$ low bias) and a sizable sample in the neighborhood ($\Rightarrow$ low variance) as the dimension increases, without the total sample size increasing exponentially in p. Visualization of $\hat{f}(X)$ also becomes difficult in higher dimensions, and this is often one of the primary goals of smoothing.

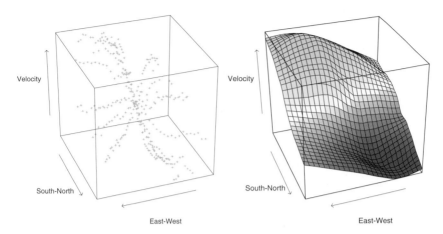

FIGURE 6.8. *The left panel shows three-dimensional data, where the response is the velocity measurements on a galaxy, and the two predictors record positions on the celestial sphere. The unusual "star"-shaped design indicates the way the measurements were made, and results in an extremely irregular boundary. The right panel shows the results of local linear regression smoothing in $\mathbb{R}^2$, using a nearest-neighbor window with 15% of the data.*

Although the scatter-cloud and wire-frame pictures in Figure 6.8 look attractive, it is quite difficult to interpret the results except at a gross level. From a data analysis perspective, conditional plots are far more useful.

Figure 6.9 shows an analysis of some environmental data with three predictors. The *trellis* display here shows ozone as a function of radiation, conditioned on the other two variables, temperature and wind speed. However, conditioning on the value of a variable really implies local to that value (as in local regression). Above each of the panels in Figure 6.9 is an indication of the range of values present in that panel for each of the conditioning values. In the panel itself the data subsets are displayed (response versus remaining variable), and a one-dimensional local linear regression is fit to the data. Although this is not quite the same as looking at slices of a fitted three-dimensional surface, it is probably more useful in terms of understanding the joint behavior of the data.

6.4 Structured Local Regression Models in $\mathbb{R}^p$

When the dimension to sample-size ratio is unfavorable, local regression does not help us much, unless we are willing to make some structural assumptions about the model. Much of this book is about structured regression and classification models. Here we focus on some approaches directly related to kernel methods.

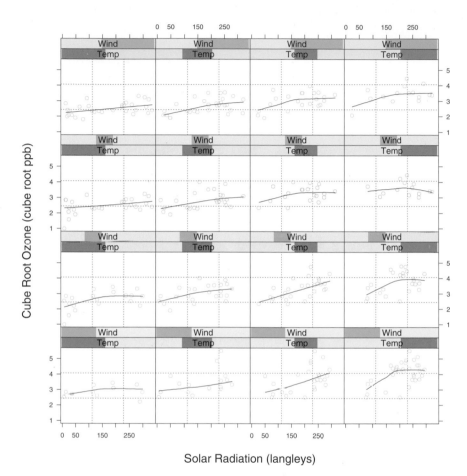

FIGURE 6.9. *Three-dimensional smoothing example. The response is (cube-root of) ozone concentration, and the three predictors are temperature, wind speed and radiation. The* trellis *display shows ozone as a function of radiation, conditioned on intervals of temperature and wind speed (indicated by darker green or orange shaded bars). Each panel contains about 40% of the range of each of the conditioned variables. The curve in each panel is a univariate local linear regression, fit to the data in the panel.*

6.4.1 Structured Kernels

One line of approach is to modify the kernel. The default spherical kernel (6.13) gives equal weight to each coordinate, and so a natural default strategy is to standardize each variable to unit standard deviation. A more general approach is to use a positive semidefinite matrix $\mathbf{A}$ to weigh the different coordinates:

$$K_{\lambda,A}(x_0, x) = D\left(\frac{(x - x_0)^T \mathbf{A}(x - x_0)}{\lambda}\right). \qquad (6.14)$$

Entire coordinates or directions can be downgraded or omitted by imposing appropriate restrictions on $\mathbf{A}$. For example, if $\mathbf{A}$ is diagonal, then we can increase or decrease the influence of individual predictors X_j by increasing or decreasing A_{jj}. Often the predictors are many and highly correlated, such as those arising from digitized analog signals or images. The covariance function of the predictors can be used to tailor a metric $\mathbf{A}$ that focuses less, say, on high-frequency contrasts (Exercise 6.4). Proposals have been made for learning the parameters for multidimensional kernels. For example, the projection-pursuit regression model discussed in Chapter 11 is of this flavor, where low-rank versions of $\mathbf{A}$ imply ridge functions for $\hat{f}(X)$. More general models for $\mathbf{A}$ are cumbersome, and we favor instead the structured forms for the regression function discussed next.

6.4.2 Structured Regression Functions

We are trying to fit a regression function $E(Y|X) = f(X_1, X_2, \ldots, X_p)$ in $\mathbb{R}^p$, in which every level of interaction is potentially present. It is natural to consider analysis-of-variance (ANOVA) decompositions of the form

$$f(X_1, X_2, \ldots, X_p) = \alpha + \sum_j g_j(X_j) + \sum_{k < \ell} g_{k\ell}(X_k, X_\ell) + \cdots \qquad (6.15)$$

and then introduce structure by eliminating some of the higher-order terms. Additive models assume only main effect terms: $f(X) = \alpha + \sum_{j=1}^p g_j(X_j)$; second-order models will have terms with interactions of order at most two, and so on. In Chapter 9, we describe iterative *backfitting* algorithms for fitting such low-order interaction models. In the additive model, for example, if all but the kth term is assumed known, then we can estimate g_k by local regression of $Y - \sum_{j \neq k} g_j(X_j)$ on X_k. This is done for each function in turn, repeatedly, until convergence. The important detail is that at any stage, one-dimensional local regression is all that is needed. The same ideas can be used to fit low-dimensional ANOVA decompositions.

An important special case of these structured models are the class of *varying coefficient models*. Suppose, for example, that we divide the p predictors in X into a set $(X_1, X_2, \ldots, X_q)$ with $q < p$, and the remainder of

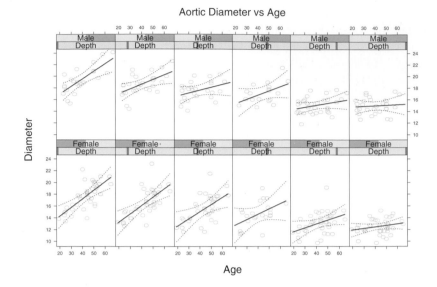

FIGURE 6.10. *In each panel the* `aorta diameter` *is modeled as a linear function of* `age`. *The coefficients of this model vary with* `gender` *and* `depth` *down the* `aorta` *(left is near the top, right is low down). There is a clear trend in the coefficients of the linear model.*

the variables we collect in the vector Z. We then assume the conditionally linear model

$$f(X) = \alpha(Z) + \beta_1(Z)X_1 + \cdots + \beta_q(Z)X_q. \tag{6.16}$$

For given Z, this is a linear model, but each of the coefficients can vary with Z. It is natural to fit such a model by locally weighted least squares:

$$\min_{\alpha(z_0),\beta(z_0)} \sum_{i=1}^{N} K_\lambda(z_0, z_i) \left(y_i - \alpha(z_0) - x_{1i}\beta_1(z_0) - \cdots - x_{qi}\beta_q(z_0)\right)^2. \tag{6.17}$$

Figure 6.10 illustrates the idea on measurements of the human aorta. A longstanding claim has been that the aorta thickens with `age`. Here we model the `diameter` of the aorta as a linear function of `age`, but allow the coefficients to vary with `gender` and `depth` down the aorta. We used a local regression model separately for males and females. While the aorta clearly does thicken with age at the higher regions of the aorta, the relationship fades with distance down the aorta. Figure 6.11 shows the intercept and slope as a function of depth.

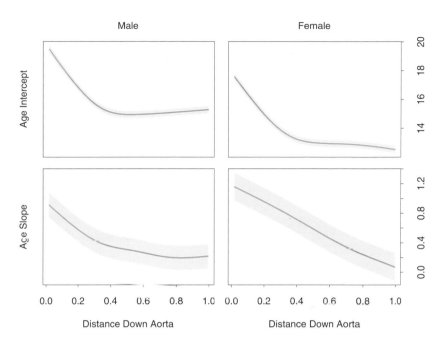

FIGURE 6.11. *The intercept and slope of* age *as a function of* distance *down the aorta, separately for males and females. The yellow bands indicate one standard error.*

6.5   Local Likelihood and Other Models

The concept of local regression and varying coefficient models is extremely broad: any parametric model can be made local if the fitting method accommodates observation weights. Here are some examples:

- Associated with each observation y_i is a parameter $\theta_i = \theta(x_i) = x_i^T \beta$ linear in the covariate(s) x_i, and inference for β is based on the log-likelihood $l(\beta) = \sum_{i=1}^{N} l(y_i, x_i^T \beta)$. We can model $\theta(X)$ more flexibly by using the likelihood local to x_0 for inference of $\theta(x_0) = x_0^T \beta(x_0)$:

$$l(\beta(x_0)) = \sum_{i=1}^{N} K_\lambda(x_0, x_i) l(y_i, x_i^T \beta(x_0)).$$

  Many likelihood models, in particular the family of generalized linear models including logistic and log-linear models, involve the covariates in a linear fashion. Local likelihood allows a relaxation from a globally linear model to one that is locally linear.

- As above, except different variables are associated with θ from those used for defining the local likelihood:

$$l(\theta(z_0)) = \sum_{i=1}^{N} K_\lambda(z_0, z_i) l(y_i, \eta(x_i, \theta(z_0))).$$

For example, $\eta(x, \theta) = x^T \theta$ could be a linear model in x. This will fit a varying coefficient model $\theta(z)$ by maximizing the local likelihood.

- Autoregressive time series models of order k have the form $y_t = \beta_0 + \beta_1 y_{t-1} + \beta_2 y_{t-2} + \cdots + \beta_k y_{t-k} + \varepsilon_t$. Denoting the *lag set* by $z_t = (y_{t-1}, y_{t-2}, \ldots, y_{t-k})$, the model looks like a standard linear model $y_t = z_t^T \beta + \varepsilon_t$, and is typically fit by least squares. Fitting by local least squares with a kernel $K(z_0, z_t)$ allows the model to vary according to the short-term history of the series. This is to be distinguished from the more traditional dynamic linear models that vary by windowing time.

As an illustration of local likelihood, we consider the local version of the multiclass linear logistic regression model (4.36) of Chapter 4. The data consist of features x_i and an associated categorical response $g_i \in \{1, 2, \ldots, J\}$, and the linear model has the form

$$\Pr(G = j | X = x) = \frac{e^{\beta_{j0} + \beta_j^T x}}{1 + \sum_{k=1}^{J-1} e^{\beta_{k0} + \beta_k^T x}}. \tag{6.18}$$

The local log-likelihood for this J class model can be written

$$\sum_{i=1}^{N} K_\lambda(x_0, x_i) \Bigg\{ \beta_{g_i 0}(x_0) + \beta_{g_i}(x_0)^T (x_i - x_0)$$

$$- \log \left[1 + \sum_{k=1}^{J-1} \exp \left(\beta_{k0}(x_0) + \beta_k(x_0)^T (x_i - x_0) \right) \right] \Bigg\}. \tag{6.19}$$

Notice that

- we have used g_i as a subscript in the first line to pick out the appropriate numerator;

- $\beta_{J0} = 0$ and $\beta_J = 0$ by the definition of the model;

- we have centered the local regressions at x_0, so that the fitted posterior probabilities at x_0 are simply

$$\hat{\Pr}(G = j | X = x_0) = \frac{e^{\hat{\beta}_{j0}(x_0)}}{1 + \sum_{k=1}^{J-1} e^{\hat{\beta}_{k0}(x_0)}}. \tag{6.20}$$

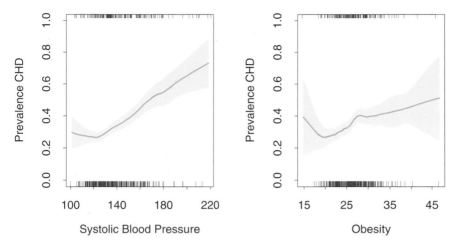

FIGURE 6.12. *Each plot shows the binary response CHD (coronary heart disease) as a function of a risk factor for the South African heart disease data. For each plot we have computed the fitted prevalence of CHD using a local linear logistic regression model. The unexpected increase in the prevalence of CHD at the lower ends of the ranges is because these are retrospective data, and some of the subjects had already undergone treatment to reduce their blood pressure and weight. The shaded region in the plot indicates an estimated pointwise standard error band.*

This model can be used for flexible multiclass classification in moderately low dimensions, although successes have been reported with the high-dimensional ZIP-code classification problem. Generalized additive models (Chapter 9) using kernel smoothing methods are closely related, and avoid dimensionality problems by assuming an additive structure for the regression function.

As a simple illustration we fit a two-class local linear logistic model to the heart disease data of Chapter 4. Figure 6.12 shows the univariate local logistic models fit to two of the risk factors (separately). This is a useful screening device for detecting nonlinearities, when the data themselves have little visual information to offer. In this case an unexpected anomaly is uncovered in the data, which may have gone unnoticed with traditional methods.

Since CHD is a binary indicator, we could estimate the conditional prevalence $\Pr(G = j|x_0)$ by simply smoothing this binary response directly without resorting to a likelihood formulation. This amounts to fitting a locally constant logistic regression model (Exercise 6.5). In order to enjoy the bias-correction of local-linear smoothing, it is more natural to operate on the unrestricted logit scale.

Typically with logistic regression, we compute parameter estimates as well as their standard errors. This can be done locally as well, and so

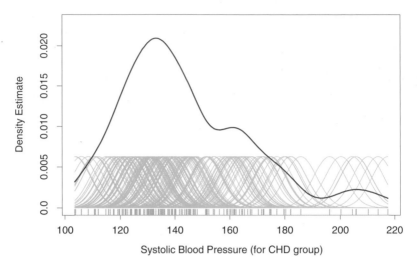

FIGURE 6.13. *A kernel density estimate for systolic blood pressure (for the CHD group). The density estimate at each point is the average contribution from each of the kernels at that point. We have scaled the kernels down by a factor of 10 to make the graph readable.*

we can produce, as shown in the plot, estimated pointwise standard-error bands about our fitted prevalence.

6.6 Kernel Density Estimation and Classification

Kernel density estimation is an unsupervised learning procedure, which historically precedes kernel regression. It also leads naturally to a simple family of procedures for nonparametric classification.

6.6.1 *Kernel Density Estimation*

Suppose we have a random sample $x_1, \ldots, x_N$ drawn from a probability density $f_X(x)$, and we wish to estimate f_X at a point x_0. For simplicity we assume for now that $X \in \mathbb{R}$. Arguing as before, a natural local estimate has the form

$$\hat{f}_X(x_0) = \frac{\#x_i \in \mathcal{N}(x_0)}{N\lambda}, \qquad (6.21)$$

where $\mathcal{N}(x_0)$ is a small metric neighborhood around x_0 of width λ. This estimate is bumpy, and the smooth *Parzen* estimate is preferred

$$\hat{f}_X(x_0) = \frac{1}{N\lambda} \sum_{i=1}^{N} K_\lambda(x_0, x_i), \qquad (6.22)$$

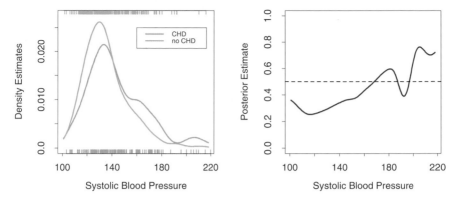

FIGURE 6.14. *The left panel shows the two separate density estimates for systolic blood pressure in the CHD versus no-CHD groups, using a Gaussian kernel density estimate in each. The right panel shows the estimated posterior probabilities for CHD, using (6.25).*

because it counts observations close to x_0 with weights that decrease with distance from x_0. In this case a popular choice for K_λ is the Gaussian kernel $K_\lambda(x_0, x) = \phi(|x - x_0|/\lambda)$. Figure 6.13 shows a Gaussian kernel density fit to the sample values for `systolic blood pressure` for the `CHD` group. Letting ϕ_λ denote the Gaussian density with mean zero and standard-deviation λ, then (6.22) has the form

$$
\begin{aligned}
\hat{f}_X(x) &= \frac{1}{N} \sum_{i=1}^{N} \phi_\lambda(x - x_i) \\
&= (\hat{F} \star \phi_\lambda)(x),
\end{aligned}
\tag{6.23}
$$

the convolution of the sample empirical distribution $\hat{F}$ with ϕ_λ. The distribution $\hat{F}(x)$ puts mass $1/N$ at each of the observed x_i, and is jumpy; in $\hat{f}_X(x)$ we have smoothed $\hat{F}$ by adding independent Gaussian noise to each observation x_i.

The Parzen density estimate is the equivalent of the local average, and improvements have been proposed along the lines of local regression [on the log scale for densities; see Loader (1999)]. We will not pursue these here. In $\mathbb{R}^p$ the natural generalization of the Gaussian density estimate amounts to using the Gaussian product kernel in (6.23),

$$
\hat{f}_X(x_0) = \frac{1}{N(2\lambda^2\pi)^{\frac{p}{2}}} \sum_{i=1}^{N} e^{-\frac{1}{2}(||x_i - x_0||/\lambda)^2}.
\tag{6.24}
$$

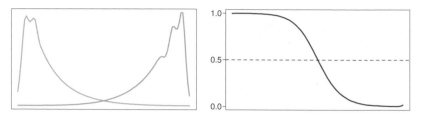

FIGURE 6.15. *The population class densities may have interesting structure (left) that disappears when the posterior probabilities are formed (right).*

6.6.2 Kernel Density Classification

One can use nonparametric density estimates for classification in a straightforward fashion using Bayes' theorem. Suppose for a J class problem we fit nonparametric density estimates $\hat{f}_j(X)$, $j = 1, \ldots, J$ separately in each of the classes, and we also have estimates of the class priors $\hat{\pi}_j$ (usually the sample proportions). Then

$$\hat{\Pr}(G = j | X = x_0) = \frac{\hat{\pi}_j \hat{f}_j(x_0)}{\sum_{k=1}^{J} \hat{\pi}_k \hat{f}_k(x_0)}. \qquad (6.25)$$

Figure 6.14 uses this method to estimate the prevalence of CHD for the heart risk factor study, and should be compared with the left panel of Figure 6.12. The main difference occurs in the region of high SBP in the right panel of Figure 6.14. In this region the data are sparse for both classes, and since the Gaussian kernel density estimates use metric kernels, the density estimates are low and of poor quality (high variance) in these regions. The local logistic regression method (6.20) uses the tri-cube kernel with k-NN bandwidth; this effectively widens the kernel in this region, and makes use of the local linear assumption to smooth out the estimate (on the logit scale).

If classification is the ultimate goal, then learning the separate class densities well may be unnecessary, and can in fact be misleading. Figure 6.15 shows an example where the densities are both multimodal, but the posterior ratio is quite smooth. In learning the separate densities from data, one might decide to settle for a rougher, high-variance fit to capture these features, which are irrelevant for the purposes of estimating the posterior probabilities. In fact, if classification is the ultimate goal, then we need only to estimate the posterior well near the decision boundary (for two classes, this is the set $\{x | \Pr(G = 1 | X = x) = \frac{1}{2}\}$).

6.6.3 The Naive Bayes Classifier

This is a technique that has remained popular over the years, despite its name (also known as "Idiot's Bayes"!) It is especially appropriate when the dimension p of the feature space is high, making density estimation unattractive. The naive Bayes model assumes that given a class $G = j$, the features X_k are independent:

$$f_j(X) = \prod_{k=1}^{p} f_{jk}(X_k). \qquad (6.26)$$

While this assumption is generally not true, it does simplify the estimation dramatically:

- The individual class-conditional marginal densities f_{jk} can each be estimated separately using one-dimensional kernel density estimates. This is in fact a generalization of the original naive Bayes procedures, which used univariate Gaussians to represent these marginals.

- If a component X_j of X is discrete, then an appropriate histogram estimate can be used. This provides a seamless way of mixing variable types in a feature vector.

Despite these rather optimistic assumptions, naive Bayes classifiers often outperform far more sophisticated alternatives. The reasons are related to Figure 6.15: although the individual class density estimates may be biased, this bias might not hurt the posterior probabilities as much, especially near the decision regions. In fact, the problem may be able to withstand considerable bias for the savings in variance such a "naive" assumption earns.

Starting from (6.26) we can derive the logit-transform (using class J as the base):

$$
\begin{aligned}
\log \frac{\Pr(G = \ell | X)}{\Pr(G = J | X)} &= \log \frac{\pi_\ell f_\ell(X)}{\pi_J f_J(X)} \\
&= \log \frac{\pi_\ell \prod_{k=1}^{p} f_{\ell k}(X_k)}{\pi_J \prod_{k=1}^{p} f_{Jk}(X_k)} \\
&= \log \frac{\pi_\ell}{\pi_J} + \sum_{k=1}^{p} \log \frac{f_{\ell k}(X_k)}{f_{Jk}(X_k)} \\
&= \alpha_\ell + \sum_{k=1}^{p} g_{\ell k}(X_k).
\end{aligned}
\qquad (6.27)
$$

This has the form of a *generalized additive model*, which is described in more detail in Chapter 9. The models are fit in quite different ways though; their differences are explored in Exercise 6.9. The relationship between naive Bayes and generalized additive models is analogous to that between linear discriminant analysis and logistic regression (Section 4.4.5).

6.7 Radial Basis Functions and Kernels

In Chapter 5, functions are represented as expansions in basis functions: $f(x) = \sum_{j=1}^{M} \beta_j h_j(x)$. The art of flexible modeling using basis expansions consists of picking an appropriate family of basis functions, and then controlling the complexity of the representation by selection, regularization, or both. Some of the families of basis functions have elements that are defined locally; for example, B-splines are defined locally in $\mathbb{R}$. If more flexibility is desired in a particular region, then that region needs to be represented by more basis functions (which in the case of B-splines translates to more knots). Tensor products of $\mathbb{R}$-local basis functions deliver basis functions local in $\mathbb{R}^p$. Not all basis functions are local—for example, the truncated power bases for splines, or the sigmoidal basis functions $\sigma(\alpha_0 + \alpha x)$ used in neural-networks (see Chapter 11). The composed function $f(x)$ can nevertheless show local behavior, because of the particular signs and values of the coefficients causing cancellations of global effects. For example, the truncated power basis has an equivalent B-spline basis for the same space of functions; the cancellation is exact in this case.

Kernel methods achieve flexibility by fitting simple models in a region local to the target point x_0. Localization is achieved via a weighting kernel K_λ, and individual observations receive weights $K_\lambda(x_0, x_i)$.

Radial basis functions combine these ideas, by treating the kernel functions $K_\lambda(\xi, x)$ as basis functions. This leads to the model

$$
\begin{aligned}
f(x) &= \sum_{j=1}^{M} K_{\lambda_j}(\xi_j, x)\beta_j \\
&= \sum_{j=1}^{M} D\left(\frac{\|x - \xi_j\|}{\lambda_j}\right)\beta_j,
\end{aligned}
\tag{6.28}
$$

where each basis element is indexed by a location or *prototype* parameter ξ_j and a scale parameter λ_j. A popular choice for D is the standard Gaussian density function. There are several approaches to learning the parameters $\{\lambda_j, \xi_j, \beta_j\}$, $j = 1, \ldots, M$. For simplicity we will focus on least squares methods for regression, and use the Gaussian kernel.

- Optimize the sum-of-squares with respect to all the parameters:

$$
\min_{\{\lambda_j, \xi_j, \beta_j\}_1^M} \sum_{i=1}^{N} \left(y_i - \beta_0 - \sum_{j=1}^{M} \beta_j \exp\left\{-\frac{(x_i - \xi_j)^T(x_i - \xi_j)}{\lambda_j^2}\right\}\right)^2.
\tag{6.29}
$$

This model is commonly referred to as an RBF network, an alternative to the sigmoidal neural network discussed in Chapter 11; the ξ_j and λ_j playing the role of the weights. This criterion is nonconvex

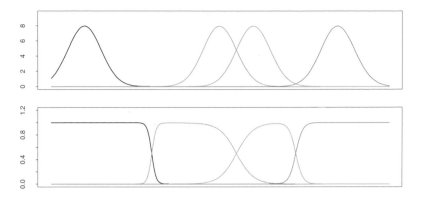

FIGURE 6.16. *Gaussian radial basis functions in* $\mathbb{R}$ *with fixed width can leave holes (top panel). Renormalized Gaussian radial basis functions avoid this problem, and produce basis functions similar in some respects to B-splines.*

with multiple local minima, and the algorithms for optimization are similar to those used for neural networks.

- Estimate the $\{\lambda_j, \xi_j\}$ separately from the β_j. Given the former, the estimation of the latter is a simple least squares problem. Often the kernel parameters λ_j and ξ_j are chosen in an unsupervised way using the X distribution alone. One of the methods is to fit a Gaussian mixture density model to the training x_i, which provides both the centers ξ_j and the scales λ_j. Other even more adhoc approaches use clustering methods to locate the prototypes ξ_j, and treat $\lambda_j = \lambda$ as a hyper-parameter. The obvious drawback of these approaches is that the conditional distribution $\Pr(Y|X)$ and in particular $E(Y|X)$ is having no say in where the action is concentrated. On the positive side, they are much simpler to implement.

While it would seem attractive to reduce the parameter set and assume a constant value for $\lambda_j = \lambda$, this can have an undesirable side effect of creating *holes*—regions of $\mathbb{R}^p$ where none of the kernels has appreciable support, as illustrated in Figure 6.16 (upper panel). *Renormalized* radial basis functions,

$$h_j(x) = \frac{D(||x - \xi_j||/\lambda)}{\sum_{k=1}^{M} D(||x - \xi_k||/\lambda)}, \tag{6.30}$$

avoid this problem (lower panel).

The Nadaraya–Watson kernel regression estimator (6.2) in $\mathbb{R}^p$ can be viewed as an expansion in renormalized radial basis functions,

$$\begin{aligned}
\hat{f}(x_0) &= \sum_{i=1}^{N} y_i \frac{K_\lambda(x_0, x_i)}{\sum_{i=1}^{N} K_\lambda(x_0, x_i)} \\
&= \sum_{i=1}^{N} y_i h_i(x_0)
\end{aligned} \tag{6.31}$$

with a basis function h_i located at every observation and coefficients y_i; that is, $\xi_i = x_i$, $\hat{\beta}_i = y_i$, $i = 1, \ldots, N$.

Note the similarity between the expansion (6.31) and the solution (5.50) on page 169 to the regularization problem induced by the kernel K. Radial basis functions form the bridge between the modern "kernel methods" and local fitting technology.

6.8 Mixture Models for Density Estimation and Classification

The mixture model is a useful tool for density estimation, and can be viewed as a kind of kernel method. The Gaussian mixture model has the form

$$f(x) = \sum_{m=1}^{M} \alpha_m \phi(x; \mu_m, \boldsymbol{\Sigma}_m) \tag{6.32}$$

with mixing proportions α_m, $\sum_m \alpha_m = 1$, and each Gaussian density has a mean μ_m and covariance matrix $\boldsymbol{\Sigma}_m$. In general, mixture models can use any component densities in place of the Gaussian in (6.32): the Gaussian mixture model is by far the most popular.

The parameters are usually fit by maximum likelihood, using the EM algorithm as described in Chapter 8. Some special cases arise:

- If the covariance matrices are constrained to be scalar: $\boldsymbol{\Sigma}_m = \sigma_m \mathbf{I}$, then (6.32) has the form of a radial basis expansion.

- If in addition $\sigma_m = \sigma > 0$ is fixed, and $M \uparrow N$, then the maximum likelihood estimate for (6.32) approaches the kernel density estimate (6.22) where $\hat{\alpha}_m = 1/N$ and $\hat{\mu}_m = x_m$.

Using Bayes' theorem, separate mixture densities in each class lead to flexible models for $\Pr(G|X)$; this is taken up in some detail in Chapter 12.

Figure 6.17 shows an application of mixtures to the heart disease risk-factor study. In the top row are histograms of Age for the no CHD and CHD groups separately, and then combined on the right. Using the combined data, we fit a two-component mixture of the form (6.32) with the (scalars) Σ_1 and Σ_2 not constrained to be equal. Fitting was done via the EM algorithm (Chapter 8): note that the procedure does not use knowledge of the CHD labels. The resulting estimates were

$$\begin{aligned} \hat{\mu}_1 &= 36.4, & \hat{\Sigma}_1 &= 157.7, & \hat{\alpha}_1 &= 0.7, \\ \hat{\mu}_2 &= 58.0, & \hat{\Sigma}_2 &= 15.6, & \hat{\alpha}_2 &= 0.3. \end{aligned}$$

The component densities $\phi(\hat{\mu}_1, \hat{\Sigma}_1)$ and $\phi(\hat{\mu}_2, \hat{\Sigma}_2)$ are shown in the lower-left and middle panels. The lower-right panel shows these component densities (orange and blue) along with the estimated mixture density (green).

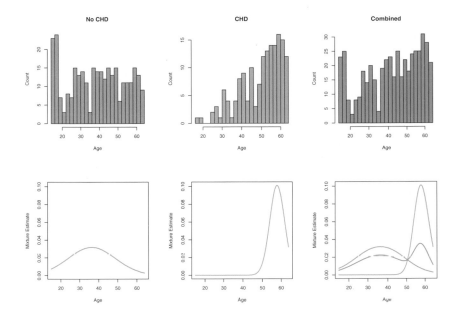

FIGURE 6.17. *Application of mixtures to the heart disease risk-factor study. (Top row:) Histograms of* Age *for the* no CHD *and* CHD *groups separately, and combined. (Bottom row:) estimated component densities from a Gaussian mixture model, (bottom left, bottom middle); (bottom right:) Estimated component densities (blue and orange) along with the estimated mixture density (green). The orange density has a very large standard deviation, and approximates a uniform density.*

The mixture model also provides an estimate of the probability that observation i belongs to component m,

$$\hat{r}_{im} = \frac{\hat{\alpha}_m \phi(x_i; \hat{\mu}_m, \hat{\Sigma}_m)}{\sum_{k=1}^{M} \hat{\alpha}_k \phi(x_i; \hat{\mu}_k, \hat{\Sigma}_k)}, \qquad (6.33)$$

where x_i is Age in our example. Suppose we threshold each value $\hat{r}_{i2}$ and hence define $\hat{\delta}_i = I(\hat{r}_{i2} > 0.5)$. Then we can compare the classification of each observation by CHD and the mixture model:

		Mixture model	
		$\hat{\delta}=0$	$\hat{\delta}=1$
CHD	No	232	70
	Yes	76	84

Although the mixture model did not use the CHD labels, it has done a fair job in discovering the two CHD subpopulations. Linear logistic regression, using the CHD as a response, achieves the same error rate (32%) when fit to these data using maximum-likelihood (Section 4.4).

6.9 Computational Considerations

Kernel and local regression and density estimation are *memory-based* methods: the model is the entire training data set, and the fitting is done at evaluation or prediction time. For many real-time applications, this can make this class of methods infeasible.

The computational cost to fit at a single observation x_0 is $O(N)$ flops, except in oversimplified cases (such as square kernels). By comparison, an expansion in M basis functions costs $O(M)$ for one evaluation, and typically $M \sim O(\log N)$. Basis function methods have an initial cost of at least $O(NM^2 + M^3)$.

The smoothing parameter(s) λ for kernel methods are typically determined off-line, for example using cross-validation, at a cost of $O(N^2)$ flops.

Popular implementations of local regression, such as the `loess` function in S-PLUS and R and the `locfit` procedure (Loader, 1999), use triangulation schemes to reduce the computations. They compute the fit exactly at M carefully chosen locations ($O(NM)$), and then use blending techniques to interpolate the fit elsewhere ($O(M)$ per evaluation).

Bibliographic Notes

There is a vast literature on kernel methods which we will not attempt to summarize. Rather we will point to a few good references that themselves have extensive bibliographies. Loader (1999) gives excellent coverage of local regression and likelihood, and also describes state-of-the-art software for fitting these models. Fan and Gijbels (1996) cover these models from a more theoretical aspect. Hastie and Tibshirani (1990) discuss local regression in the context of additive modeling. Silverman (1986) gives a good overview of density estimation, as does Scott (1992).

Exercises

Ex. 6.1 Show that the Nadaraya–Watson kernel smooth with fixed metric bandwidth λ and a Gaussian kernel is differentiable. What can be said for the Epanechnikov kernel? What can be said for the Epanechnikov kernel with adaptive nearest-neighbor bandwidth $\lambda(x_0)$?

Ex. 6.2 Show that $\sum_{i=1}^{N}(x_i - x_0)l_i(x_0) = 0$ for local linear regression. Define $b_j(x_0) = \sum_{i=1}^{N}(x_i - x_0)^j l_i(x_0)$. Show that $b_0(x_0) = 1$ for local polynomial regression of any degree (including local constants). Show that $b_j(x_0) = 0$ for all $j \in \{1, 2, \ldots, k\}$ for local polynomial regression of degree k. What are the implications of this on the bias?

Ex. 6.3 Show that $||l(x)||$ (Section 6.1.2) increases with the degree of the local polynomial.

Ex. 6.4 Suppose that the p predictors X arise from sampling relatively smooth analog curves at p uniformly spaced abscissa values. Denote by $\text{Cov}(X|Y) = \boldsymbol{\Sigma}$ the conditional covariance matrix of the predictors, and assume this does not change much with Y. Discuss the nature of *Mahalanobis* choice $\mathbf{A} = \boldsymbol{\Sigma}^{-1}$ for the metric in (6.14). How does this compare with $\mathbf{A} = \mathbf{I}$? How might you construct a kernel $\mathbf{A}$ that (a) downweights high-frequency components in the distance metric; (b) ignores them completely?

Ex. 6.5 Show that fitting a locally constant multinomial logit model of the form (6.19) amounts to smoothing the binary response indicators for each class separately using a Nadaraya–Watson kernel smoother with kernel weights $K_\lambda(x_0, x_i)$.

Ex. 6.6 Suppose that all you have is software for fitting local regression, but you can specify exactly which monomials are included in the fit. How could you use this software to fit a varying-coefficient model in some of the variables?

Ex. 6.7 Derive an expression for the leave-one-out cross-validated residual sum-of-squares for local polynomial regression.

Ex. 6.8 Suppose that for continuous response Y and predictor X, we model the joint density of X, Y using a multivariate Gaussian kernel estimator. Note that the kernel in this case would be the product kernel $\phi_\lambda(X)\phi_\lambda(Y)$. Show that the conditional mean $E(Y|X)$ derived from this estimate is a Nadaraya–Watson estimator. Extend this result to classification by providing a suitable kernel for the estimation of the joint distribution of a continuous X and discrete Y.

Ex. 6.9 Explore the differences between the naive Bayes model (6.27) and a generalized additive logistic regression model, in terms of (a) model assumptions and (b) estimation. If all the variables X_k are discrete, what can you say about the corresponding GAM?

Ex. 6.10 Suppose we have N samples generated from the model $y_i = f(x_i) + \varepsilon_i$, with ε_i independent and identically distributed with mean zero and variance σ^2, the x_i assumed fixed (non random). We estimate f using a linear smoother (local regression, smoothing spline, etc.) with smoothing parameter λ. Thus the vector of fitted values is given by $\hat{\mathbf{f}} = \mathbf{S}_\lambda \mathbf{y}$. Consider the *in-sample* prediction error

$$\text{PE}(\lambda) = \text{E}\frac{1}{N}\sum_{i=1}^{N}(y_i^* - \hat{f}_\lambda(x_i))^2 \tag{6.34}$$

for predicting new responses at the N input values. Show that the average squared residual on the training data, $\text{ASR}(\lambda)$, is a biased estimate (optimistic) for $\text{PE}(\lambda)$, while

$$C_\lambda = \text{ASR}(\lambda) + \frac{2\sigma^2}{N}\text{trace}(\mathbf{S}_\lambda) \qquad (6.35)$$

is unbiased.

Ex. 6.11 Show that for the Gaussian mixture model (6.32) the likelihood is maximized at $+\infty$, and describe how.

Ex. 6.12 Write a computer program to perform a local linear discriminant analysis. At each query point x_0, the training data receive weights $K_\lambda(x_0, x_i)$ from a weighting kernel, and the ingredients for the linear decision boundaries (see Section 4.3) are computed by weighted averages. Try out your program on the zipcode data, and show the training and test errors for a series of five pre-chosen values of λ. The zipcode data are available from the book website www-stat.stanford.edu/ElemStatLearn.

7
Model Assessment and Selection

7.1 Introduction

The *generalization* performance of a learning method relates to its prediction capability on independent test data. Assessment of this performance
is extremely important in practice, since it guides the choice of learning
method or model, and gives us a measure of the quality of the ultimately
chosen model.

In this chapter we describe and illustrate the key methods for performance assessment, and show how they are used to select models. We begin
the chapter with a discussion of the interplay between bias, variance and
model complexity.

7.2 Bias, Variance and Model Complexity

Figure 7.1 illustrates the important issue in assessing the ability of a learning method to generalize. Consider first the case of a quantitative or interval
scale response. We have a target variable Y, a vector of inputs X, and a
prediction model $\hat{f}(X)$ that has been estimated from a training set $\mathcal{T}$.
The loss function for measuring errors between Y and $\hat{f}(X)$ is denoted by
$L(Y, \hat{f}(X))$. Typical choices are

$$
L(Y, \hat{f}(X)) = \begin{cases} (Y - \hat{f}(X))^2 & \text{squared error} \\ |Y - \hat{f}(X)| & \text{absolute error}. \end{cases} \tag{7.1}
$$

T. Hastie et al., *The Elements of Statistical Learning, Second Edition*,
DOI: 10.1007/b94608_7,
© Springer Science+Business Media, LLC 2009

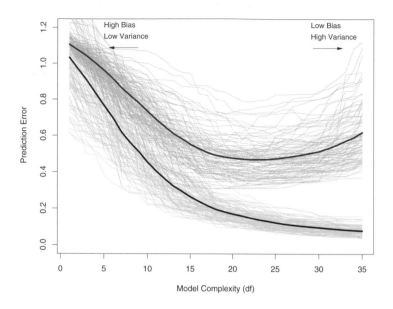

FIGURE 7.1. *Behavior of test sample and training sample error as the model complexity is varied. The light blue curves show the training error* $\overline{\mathrm{err}}$, *while the light red curves show the conditional test error* $\mathrm{Err}_{\mathcal{T}}$ *for* 100 *training sets of size* 50 *each, as the model complexity is increased. The solid curves show the expected test error* Err *and the expected training error* $\mathrm{E}[\overline{\mathrm{err}}]$.

Test error, also referred to as *generalization error*, is the prediction error over an independent test sample

$$\mathrm{Err}_{\mathcal{T}} = \mathrm{E}[L(Y, \hat{f}(X))|\mathcal{T}] \qquad (7.2)$$

where both X and Y are drawn randomly from their joint distribution (population). Here the training set $\mathcal{T}$ is fixed, and test error refers to the error for this specific training set. A related quantity is the expected prediction error (or expected test error)

$$\mathrm{Err} = \mathrm{E}[L(Y, \hat{f}(X))] = \mathrm{E}[\mathrm{Err}_{\mathcal{T}}]. \qquad (7.3)$$

Note that this expectation averages over everything that is random, including the randomness in the training set that produced $\hat{f}$.

Figure 7.1 shows the prediction error (light red curves) $\mathrm{Err}_{\mathcal{T}}$ for 100 simulated training sets each of size 50. The lasso (Section 3.4.2) was used to produce the sequence of fits. The solid red curve is the average, and hence an estimate of Err.

Estimation of $\mathrm{Err}_{\mathcal{T}}$ will be our goal, although we will see that Err is more amenable to statistical analysis, and most methods effectively estimate the expected error. It does not seem possible to estimate conditional error effectively, given only the information in the same training set. Some discussion of this point is given in Section 7.12.

Training error is the average loss over the training sample

$$\overline{\text{err}} = \frac{1}{N} \sum_{i=1}^{N} L(y_i, \hat{f}(x_i)). \tag{7.4}$$

We would like to know the expected test error of our estimated model $\hat{f}$. As the model becomes more and more complex, it uses the training data more and is able to adapt to more complicated underlying structures. Hence there is a decrease in bias but an increase in variance. There is some intermediate model complexity that gives minimum expected test error.

Unfortunately training error is not a good estimate of the test error, as seen in Figure 7.1. Training error consistently decreases with model complexity, typically dropping to zero if we increase the model complexity enough. However, a model with zero training error is overfit to the training data and will typically generalize poorly.

The story is similar for a qualitative or categorical response G taking one of K values in a set $\mathcal{G}$, labeled for convenience as $1, 2, \ldots, K$. Typically we model the probabilities $p_k(X) = \Pr(G = k|X)$ (or some monotone transformations $f_k(X)$), and then $\hat{G}(X) = \arg\max_k \hat{p}_k(X)$. In some cases, such as 1-nearest neighbor classification (Chapters 2 and 13) we produce $\hat{G}(X)$ directly. Typical loss functions are

$$L(G, \hat{G}(X)) = I(G \neq \hat{G}(X)) \quad \text{(0–1 loss)}, \tag{7.5}$$

$$L(G, \hat{p}(X)) = -2 \sum_{k=1}^{K} I(G = k) \log \hat{p}_k(X)$$

$$= -2 \log \hat{p}_G(X) \quad (-2 \times \text{log-likelihood}). \tag{7.6}$$

The quantity $-2 \times$ the log-likelihood is sometimes referred to as the *deviance*.

Again, test error here is $\text{Err}_{\mathcal{T}} = \text{E}[L(G, \hat{G}(X))|\mathcal{T}]$, the population misclassification error of the classifier trained on $\mathcal{T}$, and Err is the expected misclassification error.

Training error is the sample analogue, for example,

$$\overline{\text{err}} = -\frac{2}{N} \sum_{i=1}^{N} \log \hat{p}_{g_i}(x_i), \tag{7.7}$$

the sample log-likelihood for the model.

The log-likelihood can be used as a loss-function for general response densities, such as the Poisson, gamma, exponential, log-normal and others. If $\Pr_{\theta(X)}(Y)$ is the density of Y, indexed by a parameter $\theta(X)$ that depends on the predictor X, then

$$L(Y, \theta(X)) = -2 \cdot \log \Pr_{\theta(X)}(Y). \tag{7.8}$$

The "-2" in the definition makes the log-likelihood loss for the Gaussian distribution match squared-error loss.

For ease of exposition, for the remainder of this chapter we will use Y and $f(X)$ to represent all of the above situations, since we focus mainly on the quantitative response (squared-error loss) setting. For the other situations, the appropriate translations are obvious.

In this chapter we describe a number of methods for estimating the expected test error for a model. Typically our model will have a tuning parameter or parameters α and so we can write our predictions as $\hat{f}_\alpha(x)$. The tuning parameter varies the complexity of our model, and we wish to find the value of α that minimizes error, that is, produces the minimum of the average test error curve in Figure 7.1. Having said this, for brevity we will often suppress the dependence of $\hat{f}(x)$ on α.

It is important to note that there are in fact two separate goals that we might have in mind:

Model selection: estimating the performance of different models in order to choose the best one.

Model assessment: having chosen a final model, estimating its prediction error (generalization error) on new data.

If we are in a data-rich situation, the best approach for both problems is to randomly divide the dataset into three parts: a training set, a validation set, and a test set. The training set is used to fit the models; the validation set is used to estimate prediction error for model selection; the test set is used for assessment of the generalization error of the final chosen model. Ideally, the test set should be kept in a "vault," and be brought out only at the end of the data analysis. Suppose instead that we use the test-set repeatedly, choosing the model with smallest test-set error. Then the test set error of the final chosen model will underestimate the true test error, sometimes substantially.

It is difficult to give a general rule on how to choose the number of observations in each of the three parts, as this depends on the signal-to-noise ratio in the data and the training sample size. A typical split might be 50% for training, and 25% each for validation and testing:

The methods in this chapter are designed for situations where there is insufficient data to split it into three parts. Again it is too difficult to give a general rule on how much training data is enough; among other things, this depends on the signal-to-noise ratio of the underlying function, and the complexity of the models being fit to the data.

The methods of this chapter approximate the validation step either analytically (AIC, BIC, MDL, SRM) or by efficient sample re-use (cross-validation and the bootstrap). Besides their use in model selection, we also examine to what extent each method provides a reliable estimate of test error of the final chosen model.

Before jumping into these topics, we first explore in more detail the nature of test error and the bias–variance tradeoff.

7.3   The Bias–Variance Decomposition

As in Chapter 2, if we assume that $Y = f(X) + \varepsilon$ where $E(\varepsilon) = 0$ and $\text{Var}(\varepsilon) = \sigma_\varepsilon^2$, we can derive an expression for the expected prediction error of a regression fit $\hat{f}(X)$ at an input point $X = x_0$, using squared-error loss:

$$
\begin{aligned}
\text{Err}(x_0) &= E[(Y - \hat{f}(x_0))^2 | X = x_0] \\
&= \sigma_\varepsilon^2 + [E\hat{f}(x_0) - f(x_0)]^2 + E[\hat{f}(x_0) - E\hat{f}(x_0)]^2 \\
&= \sigma_\varepsilon^2 + \text{Bias}^2(\hat{f}(x_0)) + \text{Var}(\hat{f}(x_0)) \\
&= \text{Irreducible Error} + \text{Bias}^2 + \text{Variance}.
\end{aligned}
\tag{7.9}
$$

The first term is the variance of the target around its true mean $f(x_0)$, and cannot be avoided no matter how well we estimate $f(x_0)$, unless $\sigma_\varepsilon^2 = 0$. The second term is the squared bias, the amount by which the average of our estimate differs from the true mean; the last term is the variance; the expected squared deviation of $\hat{f}(x_0)$ around its mean. Typically the more complex we make the model $\hat{f}$, the lower the (squared) bias but the higher the variance.

For the k-nearest-neighbor regression fit, these expressions have the simple form

$$
\begin{aligned}
\text{Err}(x_0) &= E[(Y - \hat{f}_k(x_0))^2 | X = x_0] \\
&= \sigma_\varepsilon^2 + \left[f(x_0) - \frac{1}{k} \sum_{\ell=1}^{k} f(x_{(\ell)}) \right]^2 + \frac{\sigma_\varepsilon^2}{k}.
\end{aligned}
\tag{7.10}
$$

Here we assume for simplicity that training inputs x_i are fixed, and the randomness arises from the y_i. The number of neighbors k is inversely related to the model complexity. For small k, the estimate $\hat{f}_k(x)$ can potentially adapt itself better to the underlying $f(x)$. As we increase k, the bias—the squared difference between $f(x_0)$ and the average of $f(x)$ at the k-nearest neighbors—will typically increase, while the variance decreases.

For a linear model fit $\hat{f}_p(x) = x^T\hat{\beta}$, where the parameter vector β with p components is fit by least squares, we have

$$
\text{Err}(x_0) = E[(Y - \hat{f}_p(x_0))^2 | X = x_0]
$$

$$= \sigma_\varepsilon^2 + [f(x_0) - \mathrm{E}\hat{f}_p(x_0)]^2 + ||\mathbf{h}(x_0)||^2 \sigma_\varepsilon^2. \qquad (7.11)$$

Here $\mathbf{h}(x_0) = \mathbf{X}(\mathbf{X}^T\mathbf{X})^{-1}x_0$, the N-vector of linear weights that produce the fit $\hat{f}_p(x_0) = x_0{}^T(\mathbf{X}^T\mathbf{X})^{-1}\mathbf{X}^T\mathbf{y}$, and hence $\mathrm{Var}[\hat{f}_p(x_0)] = ||\mathbf{h}(x_0)||^2\sigma_\varepsilon^2$. While this variance changes with x_0, its average (with x_0 taken to be each of the sample values x_i) is $(p/N)\sigma_\varepsilon^2$, and hence

$$\frac{1}{N}\sum_{i=1}^N \mathrm{Err}(x_i) = \sigma_\varepsilon^2 + \frac{1}{N}\sum_{i=1}^N [f(x_i) - \mathrm{E}\hat{f}(x_i)]^2 + \frac{p}{N}\sigma_\varepsilon^2, \qquad (7.12)$$

the *in-sample* error. Here model complexity is directly related to the number of parameters p.

The test error $\mathrm{Err}(x_0)$ for a ridge regression fit $\hat{f}_\alpha(x_0)$ is identical in form to (7.11), except the linear weights in the variance term are different: $\mathbf{h}(x_0) = \mathbf{X}(\mathbf{X}^T\mathbf{X} + \alpha\mathbf{I})^{-1}x_0$. The bias term will also be different.

For a linear model family such as ridge regression, we can break down the bias more finely. Let β_* denote the parameters of the best-fitting linear approximation to f:

$$\beta_* = \arg\min_\beta \mathrm{E}\left(f(X) - X^T\beta\right)^2. \qquad (7.13)$$

Here the expectation is taken with respect to the distribution of the input variables X. Then we can write the average squared bias as

$$
\begin{aligned}
\mathrm{E}_{x_0}\left[f(x_0) - \mathrm{E}\hat{f}_\alpha(x_0)\right]^2 &= \mathrm{E}_{x_0}\left[f(x_0) - x_0^T\beta_*\right]^2 + \mathrm{E}_{x_0}\left[x_0^T\beta_* - \mathrm{E}x_0^T\hat{\beta}_\alpha\right]^2 \\
&= \mathrm{Ave}[\text{Model Bias}]^2 + \mathrm{Ave}[\text{Estimation Bias}]^2
\end{aligned}
$$
$$(7.14)$$

The first term on the right-hand side is the average squared *model bias*, the error between the best-fitting linear approximation and the true function. The second term is the average squared *estimation bias*, the error between the average estimate $\mathrm{E}(x_0^T\hat{\beta})$ and the best-fitting linear approximation.

For linear models fit by ordinary least squares, the estimation bias is zero. For restricted fits, such as ridge regression, it is positive, and we trade it off with the benefits of a reduced variance. The model bias can only be reduced by enlarging the class of linear models to a richer collection of models, by including interactions and transformations of the variables in the model.

Figure 7.2 shows the bias–variance tradeoff schematically. In the case of linear models, the model space is the set of all linear predictions from p inputs and the black dot labeled "closest fit" is $x^T\beta_*$. The blue-shaded region indicates the error σ_ε with which we see the truth in the training sample.

Also shown is the variance of the least squares fit, indicated by the large yellow circle centered at the black dot labeled "closest fit in population,'

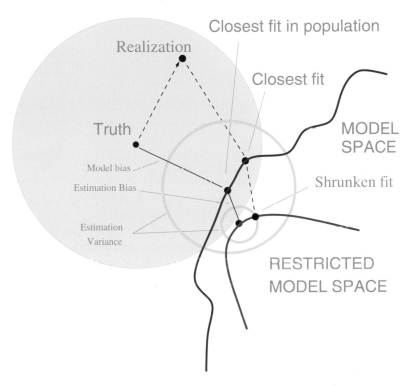

FIGURE 7.2. *Schematic of the behavior of bias and variance. The model space is the set of all possible predictions from the model, with the "closest fit" labeled with a black dot. The model bias from the truth is shown, along with the variance, indicated by the large yellow circle centered at the black dot labeled "closest fit in population." A shrunken or regularized fit is also shown, having additional estimation bias, but smaller prediction error due to its decreased variance.*

Now if we were to fit a model with fewer predictors, or regularize the coefficients by shrinking them toward zero (say), we would get the "shrunken fit" shown in the figure. This fit has an additional estimation bias, due to the fact that it is not the closest fit in the model space. On the other hand, it has smaller variance. If the decrease in variance exceeds the increase in (squared) bias, then this is worthwhile.

7.3.1 Example: Bias–Variance Tradeoff

Figure 7.3 shows the bias–variance tradeoff for two simulated examples. There are 80 observations and 20 predictors, uniformly distributed in the hypercube $[0, 1]^{20}$. The situations are as follows:

Left panels: Y is 0 if $X_1 \leq 1/2$ and 1 if $X_1 > 1/2$, and we apply k-nearest neighbors.

Right panels: Y is 1 if $\sum_{j=1}^{10} X_j$ is greater than 5 and 0 otherwise, and we use best subset linear regression of size p.

The top row is regression with squared error loss; the bottom row is classification with 0–1 loss. The figures show the prediction error (red), squared bias (green) and variance (blue), all computed for a large test sample.

In the regression problems, bias and variance add to produce the prediction error curves, with minima at about $k = 5$ for k-nearest neighbors, and $p \geq 10$ for the linear model. For classification loss (bottom figures), some interesting phenomena can be seen. The bias and variance curves are the same as in the top figures, and prediction error now refers to misclassification rate. We see that prediction error is no longer the sum of squared bias and variance. For the k-nearest neighbor classifier, prediction error decreases or stays the same as the number of neighbors is increased to 20, despite the fact that the squared bias is rising. For the linear model classifier the minimum occurs for $p \geq 10$ as in regression, but the improvement over the $p = 1$ model is more dramatic. We see that bias and variance seem to interact in determining prediction error.

Why does this happen? There is a simple explanation for the first phenomenon. Suppose at a given input point, the true probability of class 1 is 0.9 while the expected value of our estimate is 0.6. Then the squared bias—$(0.6 - 0.9)^2$—is considerable, but the prediction error is zero since we make the correct decision. In other words, estimation errors that leave us on the right side of the decision boundary don't hurt. Exercise 7.2 demonstrates this phenomenon analytically, and also shows the interaction effect between bias and variance.

The overall point is that the bias–variance tradeoff behaves differently for 0–1 loss than it does for squared error loss. This in turn means that the best choices of tuning parameters may differ substantially in the two

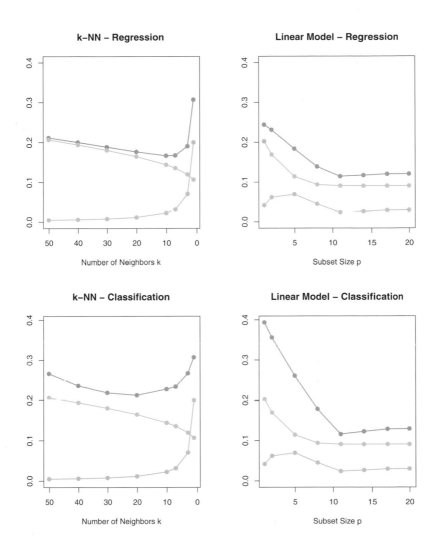

FIGURE 7.3. *Expected prediction error (orange), squared bias (green) and variance (blue) for a simulated example. The top row is regression with squared error loss; the bottom row is classification with 0–1 loss. The models are k-nearest neighbors (left) and best subset regression of size p (right). The variance and bias curves are the same in regression and classification, but the prediction error curve is different.*

settings. One should base the choice of tuning parameter on an estimate of prediction error, as described in the following sections.

7.4 Optimism of the Training Error Rate

Discussions of error rate estimation can be confusing, because we have to make clear which quantities are fixed and which are random[1]. Before we continue, we need a few definitions, elaborating on the material of Section 7.2. Given a training set $\mathcal{T} = \{(x_1, y_1), (x_2, y_2), \ldots (x_N, y_N)\}$ the generalization error of a model $\hat{f}$ is

$$\text{Err}_{\mathcal{T}} = \text{E}_{X^0, Y^0}[L(Y^0, \hat{f}(X^0))|\mathcal{T}]; \tag{7.15}$$

Note that the training set $\mathcal{T}$ is fixed in expression (7.15). The point (X^0, Y^0) is a new test data point, drawn from F, the joint distribution of the data. Averaging over training sets $\mathcal{T}$ yields the expected error

$$\text{Err} = \text{E}_{\mathcal{T}} \text{E}_{X^0, Y^0}[L(Y^0, \hat{f}(X^0))|\mathcal{T}], \tag{7.16}$$

which is more amenable to statistical analysis. As mentioned earlier, it turns out that most methods effectively estimate the expected error rather than $\text{E}_{\mathcal{T}}$; see Section 7.12 for more on this point.

Now typically, the training error

$$\overline{\text{err}} = \frac{1}{N} \sum_{i=1}^{N} L(y_i, \hat{f}(x_i)) \tag{7.17}$$

will be less than the true error $\text{Err}_{\mathcal{T}}$, because the same data is being used to fit the method and assess its error (see Exercise 2.9). A fitting method typically adapts to the training data, and hence the apparent or training error $\overline{\text{err}}$ will be an overly optimistic estimate of the generalization error $\text{Err}_{\mathcal{T}}$.

Part of the discrepancy is due to where the evaluation points occur. The quantity $\text{Err}_{\mathcal{T}}$ can be thought of as *extra-sample* error, since the test input vectors don't need to coincide with the training input vectors. The nature of the optimism in $\overline{\text{err}}$ is easiest to understand when we focus instead on the *in-sample* error

$$\text{Err}_{\text{in}} = \frac{1}{N} \sum_{i=1}^{N} \text{E}_{Y^0}[L(Y_i^0, \hat{f}(x_i))|\mathcal{T}] \tag{7.18}$$

The Y^0 notation indicates that we observe N new response values at each of the training points x_i, $i = 1, 2, \ldots, N$. We define the *optimism* as

---

[1]Indeed, in the first edition of our book, this section wasn't sufficiently clear.

the difference between Err_{in} and the training error $\overline{\text{err}}$:

$$\text{op} \equiv \text{Err}_{\text{in}} - \overline{\text{err}}. \qquad (7.19)$$

This is typically positive since $\overline{\text{err}}$ is usually biased downward as an estimate of prediction error. Finally, the average optimism is the expectation of the optimism over training sets

$$\omega \equiv \text{E}_{\mathbf{y}}(\text{op}). \qquad (7.20)$$

Here the predictors in the training set are fixed, and the expectation is over the training set outcome values; hence we have used the notation $\text{E}_{\mathbf{y}}$ instead of $\text{E}_{\mathcal{T}}$. We can usually estimate only the expected error ω rather than op, in the same way that we can estimate the expected error Err rather than the conditional error $\text{Err}_{\mathcal{T}}$.

For squared error, 0–1, and other loss functions, one can show quite generally that

$$\omega = \frac{2}{N} \sum_{i=1}^{N} \text{Cov}(\hat{y}_i, y_i), \qquad (7.21)$$

where Cov indicates covariance. Thus the amount by which $\overline{\text{err}}$ underestimates the true error depends on how strongly y_i affects its own prediction. The harder we fit the data, the greater $\text{Cov}(\hat{y}_i, y_i)$ will be, thereby increasing the optimism. Exercise 7.4 proves this result for squared error loss where $\hat{y}_i$ is the fitted value from the regression. For 0–1 loss, $\hat{y}_i \in \{0, 1\}$ is the classification at x_i, and for entropy loss, $\hat{y}_i \in [0, 1]$ is the fitted probability of class 1 at x_i.

In summary, we have the important relation

$$\text{E}_{\mathbf{y}}(\text{Err}_{\text{in}}) = \text{E}_{\mathbf{y}}(\overline{\text{err}}) + \frac{2}{N} \sum_{i=1}^{N} \text{Cov}(\hat{y}_i, y_i). \qquad (7.22)$$

This expression simplifies if $\hat{y}_i$ is obtained by a linear fit with d inputs or basis functions. For example,

$$\sum_{i=1}^{N} \text{Cov}(\hat{y}_i, y_i) = d\sigma_{\varepsilon}^2 \qquad (7.23)$$

for the additive error model $Y = f(X) + \varepsilon$, and so

$$\text{E}_{\mathbf{y}}(\text{Err}_{\text{in}}) = \text{E}_{\mathbf{y}}(\overline{\text{err}}) + 2 \cdot \frac{d}{N} \sigma_{\varepsilon}^2. \qquad (7.24)$$

Expression (7.23) is the basis for the definition of the *effective number of parameters* discussed in Section 7.6 The optimism increases linearly with the number d of inputs or basis functions we use, but decreases as the training sample size increases. Versions of (7.24) hold approximately for other error models, such as binary data and entropy loss.

An obvious way to estimate prediction error is to estimate the optimism and then add it to the training error $\overline{\text{err}}$. The methods described in the next section—C_p, AIC, BIC and others—work in this way, for a special class of estimates that are linear in their parameters.

In contrast, cross-validation and bootstrap methods, described later in the chapter, are direct estimates of the extra-sample error Err. These general tools can be used with any loss function, and with nonlinear, adaptive fitting techniques.

In-sample error is not usually of direct interest since future values of the features are not likely to coincide with their training set values. But for comparison between models, in-sample error is convenient and often leads to effective model selection. The reason is that the relative (rather than absolute) size of the error is what matters.

7.5   Estimates of In-Sample Prediction Error

The general form of the in-sample estimates is

$$\widehat{\text{Err}}_{\text{in}} = \overline{\text{err}} + \hat{\omega}, \tag{7.25}$$

where $\hat{\omega}$ is an estimate of the average optimism.

Using expression (7.24), applicable when d parameters are fit under squared error loss, leads to a version of the so-called C_p statistic,

$$C_p = \overline{\text{err}} + 2 \cdot \frac{d}{N} \hat{\sigma}_\varepsilon^2. \tag{7.26}$$

Here $\hat{\sigma}_\varepsilon^2$ is an estimate of the noise variance, obtained from the mean-squared error of a low-bias model. Using this criterion we adjust the training error by a factor proportional to the number of basis functions used.

The *Akaike information criterion* is a similar but more generally applicable estimate of Err_{in} when a log-likelihood loss function is used. It relies on a relationship similar to (7.24) that holds asymptotically as $N \to \infty$:

$$-2 \cdot \text{E}[\log \text{Pr}_{\hat{\theta}}(Y)] \approx -\frac{2}{N} \cdot \text{E}[\text{loglik}] + 2 \cdot \frac{d}{N}. \tag{7.27}$$

Here $\text{Pr}_\theta(Y)$ is a family of densities for Y (containing the "true" density), $\hat{\theta}$ is the maximum-likelihood estimate of θ, and "loglik" is the maximized log-likelihood:

$$\text{loglik} = \sum_{i=1}^{N} \log \text{Pr}_{\hat{\theta}}(y_i). \tag{7.28}$$

For example, for the logistic regression model, using the binomial log-likelihood, we have

$$\text{AIC} = -\frac{2}{N} \cdot \text{loglik} + 2 \cdot \frac{d}{N}. \tag{7.29}$$

For the Gaussian model (with variance $\sigma_\varepsilon^2 = \hat{\sigma}_\varepsilon^{\,2}$ assumed known), the AIC statistic is equivalent to C_p, and so we refer to them collectively as AIC.

To use AIC for model selection, we simply choose the model giving smallest AIC over the set of models considered. For nonlinear and other complex models, we need to replace d by some measure of model complexity. We discuss this in Section 7.6.

Given a set of models $f_\alpha(x)$ indexed by a tuning parameter α, denote by $\overline{\text{err}}(\alpha)$ and $d(\alpha)$ the training error and number of parameters for each model. Then for this set of models we define

$$\text{AIC}(\alpha) = \overline{\text{err}}(\alpha) + 2 \cdot \frac{d(\alpha)}{N} \hat{\sigma}_\varepsilon^{\,2}. \tag{7.30}$$

The function $\text{AIC}(\alpha)$ provides an estimate of the test error curve, and we find the tuning parameter $\hat{\alpha}$ that minimizes it. Our final chosen model is $f_{\hat{\alpha}}(x)$. Note that if the basis functions are chosen adaptively, (7.23) no longer holds. For example, if we have a total of p inputs, and we choose the best-fitting linear model with $d < p$ inputs, the optimism will exceed $(2d/N)\sigma_\varepsilon^2$. Put another way, by choosing the best-fitting model with d inputs, the *effective number of parameters* fit is more than d.

Figure 7.4 shows AIC in action for the phoneme recognition example of Section 5.2.3 on page 148. The input vector is the log-periodogram of the spoken vowel, quantized to 256 uniformly spaced frequencies. A linear logistic regression model is used to predict the phoneme class, with coefficient function $\beta(f) = \sum_{m=1}^{M} h_m(f)\theta_m$, an expansion in M spline basis functions. For any given M, a basis of natural cubic splines is used for the h_m, with knots chosen uniformly over the range of frequencies (so $d(\alpha) = d(M) = M$). Using AIC to select the number of basis functions will approximately minimize $\text{Err}(M)$ for both entropy and 0–1 loss.

The simple formula

$$(2/N) \sum_{i=1}^{N} \text{Cov}(\hat{y}_i, y_i) = (2d/N)\sigma_\varepsilon^2$$

holds exactly for linear models with additive errors and squared error loss, and approximately for linear models and log-likelihoods. In particular, the formula does not hold in general for 0–1 loss (Efron, 1986), although many authors nevertheless use it in that context (right panel of Figure 7.4).

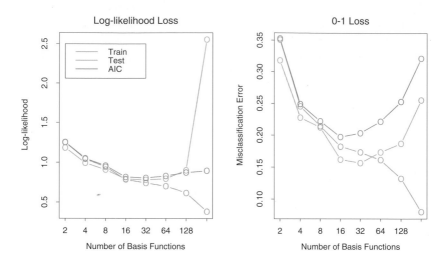

FIGURE 7.4. *AIC used for model selection for the phoneme recognition example of Section 5.2.3. The logistic regression coefficient function $\beta(f) = \sum_{m=1}^{M} h_m(f)\theta_m$ is modeled as an expansion in M spline basis functions. In the left panel we see the AIC statistic used to estimate $\mathrm{Err_{in}}$ using log-likelihood loss. Included is an estimate of Err based on an independent test sample. It does well except for the extremely over-parametrized case ($M = 256$ parameters for $N = 1000$ observations). In the right panel the same is done for 0–1 loss. Although the AIC formula does not strictly apply here, it does a reasonable job in this case.*

7.6   The Effective Number of Parameters

The concept of "number of parameters" can be generalized, especially to models where regularization is used in the fitting. Suppose we stack the outcomes $y_1, y_2, \ldots, y_N$ into a vector $\mathbf{y}$, and similarly for the predictions $\hat{\mathbf{y}}$. Then a linear fitting method is one for which we can write

$$\hat{\mathbf{y}} = \mathbf{S}\mathbf{y}, \tag{7.31}$$

where $\mathbf{S}$ is an $N \times N$ matrix depending on the input vectors x_i but not on the y_i. Linear fitting methods include linear regression on the original features or on a derived basis set, and smoothing methods that use quadratic shrinkage, such as ridge regression and cubic smoothing splines. Then the *effective number of parameters* is defined as

$$\mathrm{df}(\mathbf{S}) = \mathrm{trace}(\mathbf{S}), \tag{7.32}$$

the sum of the diagonal elements of $\mathbf{S}$ (also known as the *effective degrees-of-freedom*). Note that if $\mathbf{S}$ is an orthogonal-projection matrix onto a basis set spanned by M features, then trace$(\mathbf{S}) = M$. It turns out that trace$(\mathbf{S})$ is exactly the correct quantity to replace d as the number of parameters in the C_p statistic (7.26). If $\mathbf{y}$ arises from an additive-error model $Y = f(X) + \varepsilon$ with Var$(\varepsilon) = \sigma_\varepsilon^2$, then one can show that $\sum_{i=1}^{N} \mathrm{Cov}(\hat{y}_i, y_i) = \mathrm{trace}(\mathbf{S})\sigma_\varepsilon^2$, which motivates the more general definition

$$\mathrm{df}(\hat{\mathbf{y}}) = \frac{\sum_{i=1}^{N} \mathrm{Cov}(\hat{y}_i, y_i)}{\sigma_\varepsilon^2} \tag{7.33}$$

(Exercises 7.4 and 7.5). Section 5.4.1 on page 153 gives some more intuition for the definition df $= \mathrm{trace}(\mathbf{S})$ in the context of smoothing splines.

For models like neural networks, in which we minimize an error function $R(w)$ with weight decay penalty (regularization) $\alpha \sum_m w_m^2$, the effective number of parameters has the form

$$\mathrm{df}(\alpha) = \sum_{m=1}^{M} \frac{\theta_m}{\theta_m + \alpha}, \tag{7.34}$$

where the θ_m are the eigenvalues of the Hessian matrix $\partial^2 R(w) / \partial w \partial w^T$. Expression (7.34) follows from (7.32) if we make a quadratic approximation to the error function at the solution (Bishop, 1995).

7.7   The Bayesian Approach and BIC

The Bayesian information criterion (BIC), like AIC, is applicable in settings where the fitting is carried out by maximization of a log-likelihood. The generic form of BIC is

$$\mathrm{BIC} = -2 \cdot \mathrm{loglik} + (\log N) \cdot d. \tag{7.35}$$

The BIC statistic (times $1/2$) is also known as the Schwarz criterion (Schwarz, 1978).

Under the Gaussian model, assuming the variance σ_ε^2 is known, $-2\cdot$loglik equals (up to a constant) $\sum_i (y_i - \hat{f}(x_i))^2 / \sigma_\varepsilon^2$, which is $N \cdot \overline{\mathrm{err}}/\sigma_\varepsilon^2$ for squared error loss. Hence we can write

$$\mathrm{BIC} = \frac{N}{\sigma_\varepsilon^2}\left[\overline{\mathrm{err}} + (\log N) \cdot \frac{d}{N}\sigma_\varepsilon^2\right]. \tag{7.36}$$

Therefore BIC is proportional to AIC (C_p), with the factor 2 replaced by $\log N$. Assuming $N > e^2 \approx 7.4$, BIC tends to penalize complex models more heavily, giving preference to simpler models in selection. As with AIC, σ_ε^2 is typically estimated by the mean squared error of a low-bias model. For classification problems, use of the multinomial log-likelihood leads to a similar relationship with the AIC, using cross-entropy as the error measure.

Note however that the misclassification error measure does not arise in the BIC context, since it does not correspond to the log-likelihood of the data under any probability model.

Despite its similarity with AIC, BIC is motivated in quite a different way. It arises in the Bayesian approach to model selection, which we now describe.

Suppose we have a set of candidate models $\mathcal{M}_m, m = 1, \ldots, M$ and corresponding model parameters θ_m, and we wish to choose a best model from among them. Assuming we have a prior distribution $\Pr(\theta_m|\mathcal{M}_m)$ for the parameters of each model $\mathcal{M}_m$, the posterior probability of a given model is

$$\Pr(\mathcal{M}_m|\mathbf{Z}) \quad \propto \quad \Pr(\mathcal{M}_m) \cdot \Pr(\mathbf{Z}|\mathcal{M}_m) \tag{7.37}$$

$$\propto \quad \Pr(\mathcal{M}_m) \cdot \int \Pr(\mathbf{Z}|\theta_m, \mathcal{M}_m)\Pr(\theta_m|\mathcal{M}_m)d\theta_m,$$

where $\mathbf{Z}$ represents the training data $\{x_i, y_i\}_1^N$. To compare two models $\mathcal{M}_m$ and $\mathcal{M}_\ell$, we form the posterior odds

$$\frac{\Pr(\mathcal{M}_m|\mathbf{Z})}{\Pr(\mathcal{M}_\ell|\mathbf{Z})} = \frac{\Pr(\mathcal{M}_m)}{\Pr(\mathcal{M}_\ell)} \cdot \frac{\Pr(\mathbf{Z}|\mathcal{M}_m)}{\Pr(\mathbf{Z}|\mathcal{M}_\ell)}. \tag{7.38}$$

If the odds are greater than one we choose model m, otherwise we choose model ℓ. The rightmost quantity

$$\text{BF}(\mathbf{Z}) = \frac{\Pr(\mathbf{Z}|\mathcal{M}_m)}{\Pr(\mathbf{Z}|\mathcal{M}_\ell)} \tag{7.39}$$

is called the *Bayes factor*, the contribution of the data toward the posterior odds.

Typically we assume that the prior over models is uniform, so that $\Pr(\mathcal{M}_m)$ is constant. We need some way of approximating $\Pr(\mathbf{Z}|\mathcal{M}_m)$. A so-called Laplace approximation to the integral followed by some other simplifications (Ripley, 1996, page 64) to (7.37) gives

$$\log \Pr(\mathbf{Z}|\mathcal{M}_m) = \log \Pr(\mathbf{Z}|\hat{\theta}_m, \mathcal{M}_m) - \frac{d_m}{2} \cdot \log N + O(1). \tag{7.40}$$

Here $\hat{\theta}_m$ is a maximum likelihood estimate and d_m is the number of free parameters in model $\mathcal{M}_m$. If we define our loss function to be

$$-2 \log \Pr(\mathbf{Z}|\hat{\theta}_m, \mathcal{M}_m),$$

this is equivalent to the BIC criterion of equation (7.35).

Therefore, choosing the model with minimum BIC is equivalent to choosing the model with largest (approximate) posterior probability. But this framework gives us more. If we compute the BIC criterion for a set of M, models, giving BIC_m, $m = 1, 2, \ldots, M$, then we can estimate the posterior probability of each model $\mathcal{M}_m$ as

$$\frac{e^{-\frac{1}{2} \cdot \text{BIC}_m}}{\sum_{\ell=1}^{M} e^{-\frac{1}{2} \cdot \text{BIC}_\ell}}. \tag{7.41}$$

Thus we can estimate not only the best model, but also assess the relative merits of the models considered.

For model selection purposes, there is no clear choice between AIC and BIC. BIC is asymptotically consistent as a selection criterion. What this means is that given a family of models, including the true model, the probability that BIC will select the correct model approaches one as the sample size $N \to \infty$. This is not the case for AIC, which tends to choose models which are too complex as $N \to \infty$. On the other hand, for finite samples, BIC often chooses models that are too simple, because of its heavy penalty on complexity.

7.8   Minimum Description Length

The minimum description length (MDL) approach gives a selection criterion formally identical to the BIC approach, but is motivated from an optimal coding viewpoint. We first review the theory of coding for data compression, and then apply it to model selection.

We think of our datum z as a message that we want to encode and send to someone else (the "receiver"). We think of our model as a way of encoding the datum, and will choose the most parsimonious model, that is the shortest code, for the transmission.

Suppose first that the possible messages we might want to transmit are $z_1, z_2, \ldots, z_m$. Our code uses a finite alphabet of length A: for example, we might use a binary code $\{0, 1\}$ of length $A = 2$. Here is an example with four possible messages and a binary coding:

Message	z_1	z_2	z_3	z_4	
Code	0	10	110	111	(7.42)

This code is known as an instantaneous prefix code: no code is the prefix of any other, and the receiver (who knows all of the possible codes), knows exactly when the message has been completely sent. We restrict our discussion to such instantaneous prefix codes.

One could use the coding in (7.42) or we could permute the codes, for example use codes $110, 10, 111, 0$ for z_1, z_2, z_3, z_4. How do we decide which to use? It depends on how often we will be sending each of the messages. If, for example, we will be sending z_1 most often, it makes sense to use the shortest code 0 for z_1. Using this kind of strategy—shorter codes for more frequent messages—the average message length will be shorter.

In general, if messages are sent with probabilities $\Pr(z_i), i = 1, 2, \ldots, 4$, a famous theorem due to Shannon says we should use code lengths $l_i = -\log_2 \Pr(z_i)$ and the average message length satisfies

$$\text{E(length)} \geq -\sum \Pr(z_i) \log_2 (\Pr(z_i)). \qquad (7.43)$$

The right-hand side above is also called the entropy of the distribution $\Pr(z_i)$. The inequality is an equality when the probabilities satisfy $p_i = A^{-l_i}$. In our example, if $\Pr(z_i) = 1/2, 1/4, 1/8, 1/8$, respectively, then the coding shown in (7.42) is optimal and achieves the entropy lower bound.

In general the lower bound cannot be achieved, but procedures like the Huffman coding scheme can get close to the bound. Note that with an infinite set of messages, the entropy is replaced by $-\int \Pr(z) \log_2 \Pr(z) dz$.

From this result we glean the following:

> *To transmit a random variable z having probability density function $\Pr(z)$, we require about $-\log_2 \Pr(z)$ bits of information.*

We henceforth change notation from $\log_2 \Pr(z)$ to $\log \Pr(z) = \log_e \Pr(z)$; this is for convenience, and just introduces an unimportant multiplicative constant.

Now we apply this result to the problem of model selection. We have a model M with parameters θ, and data $\mathbf{Z} = (\mathbf{X}, \mathbf{y})$ consisting of both inputs and outputs. Let the (conditional) probability of the outputs under the model be $\Pr(\mathbf{y}|\theta, M, \mathbf{X})$, assume the receiver knows all of the inputs, and we wish to transmit the outputs. Then the message length required to transmit the outputs is

$$\text{length} = -\log \Pr(\mathbf{y}|\theta, M, \mathbf{X}) - \log \Pr(\theta|M), \qquad (7.44)$$

the log-probability of the target values given the inputs. The second term is the average code length for transmitting the model parameters θ, while the first term is the average code length for transmitting the discrepancy between the model and actual target values. For example suppose we have a single target y with $y \sim N(\theta, \sigma^2)$, parameter $\theta \sim N(0,1)$ and no input (for simplicity). Then the message length is

$$\text{length} = \text{constant} + \log \sigma + \frac{(y - \theta)^2}{2\sigma^2} + \frac{\theta^2}{2}. \qquad (7.45)$$

Note that the smaller σ is, the shorter on average is the message length, since y is more concentrated around θ.

The MDL principle says that we should choose the model that minimizes (7.44). We recognize (7.44) as the (negative) log-posterior distribution, and hence minimizing description length is equivalent to maximizing posterior probability. Hence the BIC criterion, derived as approximation to log-posterior probability, can also be viewed as a device for (approximate) model choice by minimum description length.

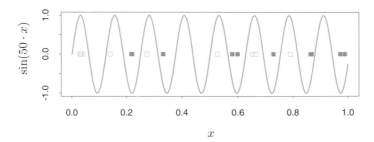

FIGURE 7.5. *The solid curve is the function* $\sin(50x)$ *for* $x \in [0, 1]$. *The green (solid) and blue (hollow) points illustrate how the associated indicator function* $I(\sin(\alpha x) > 0)$ *can shatter (separate) an arbitrarily large number of points by choosing an appropriately high frequency* α.

Note that we have ignored the precision with which a random variable z is coded. With a finite code length we cannot code a continuous variable exactly. However, if we code z within a tolerance δz, the message length needed is the log of the probability in the interval $[z, z+\delta z]$ which is well approximated by $\delta z \Pr(z)$ if δz is small. Since $\log \delta z \Pr(z) = \log \delta z + \log \Pr(z)$, this means we can just ignore the constant $\log \delta z$ and use $\log \Pr(z)$ as our measure of message length, as we did above.

The preceding view of MDL for model selection says that we should choose the model with highest posterior probability. However, many Bayesians would instead do inference by sampling from the posterior distribution.

7.9   Vapnik–Chervonenkis Dimension

A difficulty in using estimates of in-sample error is the need to specify the number of parameters (or the complexity) d used in the fit. Although the effective number of parameters introduced in Section 7.6 is useful for some nonlinear models, it is not fully general. The Vapnik–Chervonenkis (VC) theory provides such a general measure of complexity, and gives associated bounds on the optimism. Here we give a brief review of this theory.

Suppose we have a class of functions $\{f(x, \alpha)\}$ indexed by a parameter vector α, with $x \in \mathbb{R}^p$. Assume for now that f is an indicator function, that is, takes the values 0 or 1. If $\alpha = (\alpha_0, \alpha_1)$ and f is the linear indicator function $I(\alpha_0 + \alpha_1^T x > 0)$, then it seems reasonable to say that the complexity of the class f is the number of parameters $p + 1$. But what about $f(x, \alpha) = I(\sin \alpha \cdot x)$ where α is any real number and $x \in \mathbb{R}$? The function $\sin(50 \cdot x)$ is shown in Figure 7.5. This is a very wiggly function that gets even rougher as the frequency α increases, but it has only one parameter: despite this, it doesn't seem reasonable to conclude that it has less complexity than the linear indicator function $I(\alpha_0 + \alpha_1 x)$ in $p = 1$ dimension.

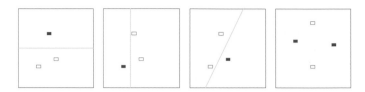

FIGURE 7.6. *The first three panels show that the class of lines in the plane can shatter three points. The last panel shows that this class cannot shatter four points, as no line will put the hollow points on one side and the solid points on the other. Hence the VC dimension of the class of straight lines in the plane is three. Note that a class of nonlinear curves could shatter four points, and hence has VC dimension greater than three.*

The Vapnik–Chervonenkis dimension is a way of measuring the complexity of a class of functions by assessing how wiggly its members can be.

> *The* VC *dimension of the class* $\{f(x, \alpha)\}$ *is defined to be the largest number of points (in some configuration) that can be shattered by members of* $\{f(x, \alpha)\}$.

A set of points is said to be shattered by a class of functions if, no matter how we assign a binary label to each point, a member of the class can perfectly separate them.

Figure 7.6 shows that the VC dimension of linear indicator functions in the plane is 3 but not 4, since no four points can be shattered by a set of lines. In general, a linear indicator function in p dimensions has VC dimension $p + 1$, which is also the number of free parameters. On the other hand, it can be shown that the family $\sin(\alpha x)$ has infinite VC dimension, as Figure 7.5 suggests. By appropriate choice of α, any set of points can be shattered by this class (Exercise 7.8).

So far we have discussed the VC dimension only of indicator functions, but this can be extended to real-valued functions. The VC dimension of a class of real-valued functions $\{g(x, \alpha)\}$ is defined to be the VC dimension of the indicator class $\{I(g(x, \alpha) - \beta > 0)\}$, where β takes values over the range of g.

One can use the VC dimension in constructing an estimate of (extra-sample) prediction error; different types of results are available. Using the concept of VC dimension, one can prove results about the optimism of the training error when using a class of functions. An example of such a result is the following. If we fit N training points using a class of functions $\{f(x, \alpha)\}$ having VC dimension h, then with probability at least $1 - \eta$ over training sets:

$$\mathrm{Err}_\mathcal{T} \;\leq\; \overline{\mathrm{err}} + \frac{\epsilon}{2}\left(1 + \sqrt{1 + \frac{4 \cdot \overline{\mathrm{err}}}{\epsilon}}\right) \quad \text{(binary classification)}$$

$$\mathrm{Err}_\mathcal{T} \;\leq\; \frac{\overline{\mathrm{err}}}{(1 - c\sqrt{\epsilon})_+} \quad \text{(regression)} \tag{7.46}$$

$$\text{where } \epsilon = a_1 \frac{h[\log\left(a_2 N/h\right) + 1] - \log\left(\eta/4\right)}{N},$$

$$\text{and } 0 < a_1 \leq 4,\; 0 < a_2 \leq 2$$

These bounds hold simultaneously for all members $f(x, \alpha)$, and are taken from Cherkassky and Mulier (2007, pages 116–118). They recommend the value $c = 1$. For regression they suggest $a_1 = a_2 = 1$, and for classification they make no recommendation, with $a_1 = 4$ and $a_2 = 2$ corresponding to worst-case scenarios. They also give an alternative *practical* bound for regression

$$\mathrm{Err}_\mathcal{T} \leq \overline{\mathrm{err}} \left(1 - \sqrt{\rho - \rho\log\rho + \frac{\log N}{2N}}\right)_+^{-1} \tag{7.47}$$

with $\rho = \frac{h}{N}$, which is free of tuning constants. The bounds suggest that the optimism increases with h and decreases with N in qualitative agreement with the AIC correction d/N given in (7.24). However, the results in (7.46) are stronger: rather than giving the expected optimism for each fixed function $f(x, \alpha)$, they give probabilistic upper bounds for all functions $f(x, \alpha)$, and hence allow for searching over the class.

Vapnik's *structural risk minimization* (SRM) approach fits a nested sequence of models of increasing VC dimensions $h_1 < h_2 < \cdots$, and then chooses the model with the smallest value of the upper bound.

We note that upper bounds like the ones in (7.46) are often very loose, but that doesn't rule them out as good criteria for model selection, where the relative (not absolute) size of the test error is important. The main drawback of this approach is the difficulty in calculating the VC dimension of a class of functions. Often only a crude upper bound for VC dimension is obtainable, and this may not be adequate. An example in which the structural risk minimization program can be successfully carried out is the support vector classifier, discussed in Section 12.2.

7.9.1 Example (Continued)

Figure 7.7 shows the results when AIC, BIC and SRM are used to select the model size for the examples of Figure 7.3. For the examples labeled KNN, the model index α refers to neighborhood size, while for those labeled REG α refers to subset size. Using each selection method (e.g., AIC) we estimated the best model $\hat{\alpha}$ and found its true prediction error $\mathrm{Err}_\mathcal{T}(\hat{\alpha})$ on a test set. For the same training set we computed the prediction error of the best

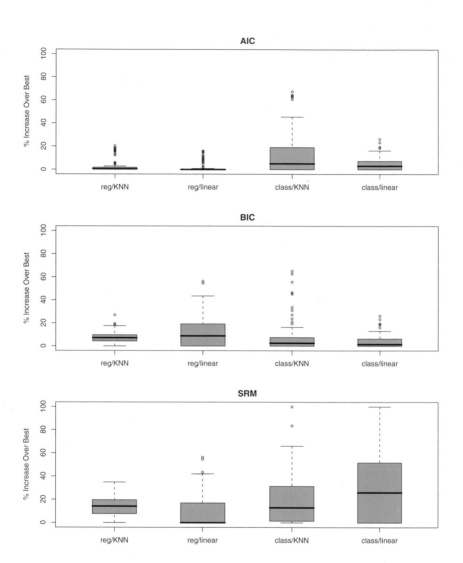

FIGURE 7.7. *Boxplots show the distribution of the relative error* $100 \times [\mathrm{Err}_{\mathcal{T}}(\hat{\alpha}) - \min_\alpha \mathrm{Err}_{\mathcal{T}}(\alpha)]/[\max_\alpha \mathrm{Err}_{\mathcal{T}}(\alpha) - \min_\alpha \mathrm{Err}_{\mathcal{T}}(\alpha)]$ *over the four scenarios of Figure 7.3. This is the error in using the chosen model relative to the best model. There are* 100 *training sets each of size* 80 *represented in each boxplot, with the errors computed on test sets of size* 10,000.

and worst possible model choices: $\min_\alpha \mathrm{Err}_{\mathcal{T}}(\alpha)$ and $\max_\alpha \mathrm{Err}_{\mathcal{T}}(\alpha)$. The boxplots show the distribution of the quantity

$$100 \times \frac{\mathrm{Err}_{\mathcal{T}}(\hat{\alpha}) - \min_\alpha \mathrm{Err}_{\mathcal{T}}(\alpha)}{\max_\alpha \mathrm{Err}_{\mathcal{T}}(\alpha) - \min_\alpha \mathrm{Err}_{\mathcal{T}}(\alpha)},$$

which represents the error in using the chosen model relative to the best model. For linear regression the model complexity was measured by the number of features; as mentioned in Section 7.5, this underestimates the df, since it does not *charge* for the search for the best model of that size. This was also used for the VC dimension of the linear classifier. For k-nearest neighbors, we used the quantity N/k. Under an additive-error regression model, this can be justified as the exact effective degrees of freedom (Exercise 7.6); we do not know if it corresponds to the VC dimension. We used $a_1 = a_2 = 1$ for the constants in (7.46); the results for SRM changed with different constants, and this choice gave the most favorable results. We repeated the SRM selection using the alternative practical bound (7.47), and got almost identical results. For misclassification error we used $\hat{\sigma}_\varepsilon^2 = [N/(N - d)] \cdot \overline{\mathrm{err}}(\alpha)$ for the least restrictive model ($k = 5$ for KNN, since $k = 1$ results in zero training error). The AIC criterion seems to work well in all four scenarios, despite the lack of theoretical support with 0–1 loss. BIC does nearly as well, while the performance of SRM is mixed.

7.10   Cross-Validation

Probably the simplest and most widely used method for estimating prediction error is cross-validation. This method directly estimates the expected extra-sample error $\mathrm{Err} = \mathrm{E}[L(Y, \hat{f}(X))]$, the average generalization error when the method $\hat{f}(X)$ is applied to an independent test sample from the joint distribution of X and Y. As mentioned earlier, we might hope that cross-validation estimates the conditional error, with the training set $\mathcal{T}$ held fixed. But as we will see in Section 7.12, cross-validation typically estimates well only the expected prediction error.

7.10.1   K-Fold Cross-Validation

Ideally, if we had enough data, we would set aside a validation set and use it to assess the performance of our prediction model. Since data are often scarce, this is usually not possible. To finesse the problem, K-fold cross-validation uses part of the available data to fit the model, and a different part to test it. We split the data into K roughly equal-sized parts; for example, when $K = 5$, the scenario looks like this:

1	2	3	4	5
Train	Train	Validation	Train	Train

For the kth part (third above), we fit the model to the other $K-1$ parts of the data, and calculate the prediction error of the fitted model when predicting the kth part of the data. We do this for $k = 1, 2, \ldots, K$ and combine the K estimates of prediction error.

Here are more details. Let $\kappa : \{1, \ldots, N\} \mapsto \{1, \ldots, K\}$ be an indexing function that indicates the partition to which observation i is allocated by the randomization. Denote by $\hat{f}^{-k}(x)$ the fitted function, computed with the kth part of the data removed. Then the cross-validation estimate of prediction error is

$$\mathrm{CV}(\hat{f}) = \frac{1}{N} \sum_{i=1}^{N} L(y_i, \hat{f}^{-\kappa(i)}(x_i)). \tag{7.48}$$

Typical choices of K are 5 or 10 (see below). The case $K = N$ is known as *leave-one-out* cross-validation. In this case $\kappa(i) = i$, and for the ith observation the fit is computed using all the data except the ith.

Given a set of models $f(x, \alpha)$ indexed by a tuning parameter α, denote by $\hat{f}^{-k}(x, \alpha)$ the αth model fit with the kth part of the data removed. Then for this set of models we define

$$\mathrm{CV}(\hat{f}, \alpha) = \frac{1}{N} \sum_{i=1}^{N} L(y_i, \hat{f}^{-\kappa(i)}(x_i, \alpha)). \tag{7.49}$$

The function $\mathrm{CV}(\hat{f}, \alpha)$ provides an estimate of the test error curve, and we find the tuning parameter $\hat{\alpha}$ that minimizes it. Our final chosen model is $f(x, \hat{\alpha})$, which we then fit to all the data.

It is interesting to wonder about what quantity K-fold cross-validation estimates. With $K = 5$ or 10, we might guess that it estimates the expected error Err, since the training sets in each fold are quite different from the original training set. On the other hand, if $K = N$ we might guess that cross-validation estimates the conditional error $\mathrm{Err}_{\mathcal{T}}$. It turns out that cross-validation only estimates effectively the average error Err, as discussed in Section 7.12.

What value should we choose for K? With $K = N$, the cross-validation estimator is approximately unbiased for the true (expected) prediction error, but can have high variance because the N "training sets" are so similar to one another. The computational burden is also considerable, requiring N applications of the learning method. In certain special problems, this computation can be done quickly—see Exercises 7.3 and 5.13.

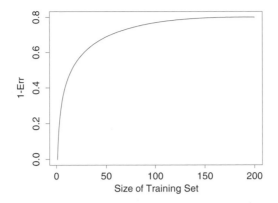

FIGURE 7.8. *Hypothetical learning curve for a classifier on a given task: a plot of* 1 − Err *versus the size of the training set* N. *With a dataset of* 200 *observations, 5-fold cross-validation would use training sets of size* 160, *which would behave much like the full set. However, with a dataset of* 50 *observations fivefold cross-validation would use training sets of size* 40, *and this would result in a considerable overestimate of prediction error.*

On the other hand, with $K = 5$ say, cross-validation has lower variance. But bias could be a problem, depending on how the performance of the learning method varies with the size of the training set. Figure 7.8 shows a hypothetical "learning curve" for a classifier on a given task, a plot of $1 - $ Err versus the size of the training set N. The performance of the classifier improves as the training set size increases to 100 observations; increasing the number further to 200 brings only a small benefit. If our training set had 200 observations, fivefold cross-validation would estimate the performance of our classifier over training sets of size 160, which from Figure 7.8 is virtually the same as the performance for training set size 200. Thus cross-validation would not suffer from much bias. However if the training set had 50 observations, fivefold cross-validation would estimate the performance of our classifier over training sets of size 40, and from the figure that would be an underestimate of $1 - $ Err. Hence as an estimate of Err, cross-validation would be biased upward.

To summarize, if the learning curve has a considerable slope at the given training set size, five- or tenfold cross-validation will overestimate the true prediction error. Whether this bias is a drawback in practice depends on the objective. On the other hand, leave-one-out cross-validation has low bias but can have high variance. Overall, five- or tenfold cross-validation are recommended as a good compromise: see Breiman and Spector (1992) and Kohavi (1995).

Figure 7.9 shows the prediction error and tenfold cross-validation curve estimated from a single training set, from the scenario in the bottom right panel of Figure 7.3. This is a two-class classification problem, using a lin-

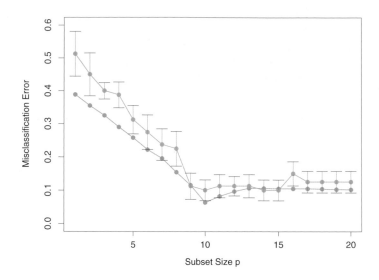

FIGURE 7.9. *Prediction error (orange) and tenfold cross-validation curve (blue) estimated from a single training set, from the scenario in the bottom right panel of Figure 7.3.*

ear model with best subsets regression of subset size p. Standard error bars are shown, which are the standard errors of the individual misclassification error rates for each of the ten parts. Both curves have minima at $p = 10$, although the CV curve is rather flat beyond 10. Often a "one-standard error" rule is used with cross-validation, in which we choose the most parsimonious model whose error is no more than one standard error above the error of the best model. Here it looks like a model with about $p = 9$ predictors would be chosen, while the true model uses $p = 10$.

Generalized cross-validation provides a convenient approximation to leave-one out cross-validation, for linear fitting under squared-error loss. As defined in Section 7.6, a linear fitting method is one for which we can write

$$\hat{\mathbf{y}} = \mathbf{S}\mathbf{y}. \tag{7.50}$$

Now for many linear fitting methods,

$$\frac{1}{N}\sum_{i=1}^{N}[y_i - \hat{f}^{-i}(x_i)]^2 = \frac{1}{N}\sum_{i=1}^{N}\left[\frac{y_i - \hat{f}(x_i)}{1 - S_{ii}}\right]^2, \tag{7.51}$$

where S_{ii} is the ith diagonal element of $\mathbf{S}$ (see Exercise 7.3). The GCV approximation is

$$\text{GCV}(\hat{f}) = \frac{1}{N}\sum_{i=1}^{N}\left[\frac{y_i - \hat{f}(x_i)}{1 - \text{trace}(\mathbf{S})/N}\right]^2. \tag{7.52}$$

The quantity trace($\mathbf{S}$) is the effective number of parameters, as defined in Section 7.6.

GCV can have a computational advantage in some settings, where the trace of $\mathbf{S}$ can be computed more easily than the individual elements S_{ii}. In smoothing problems, GCV can also alleviate the tendency of cross-validation to undersmooth. The similarity between GCV and AIC can be seen from the approximation $1/(1-x)^2 \approx 1 + 2x$ (Exercise 7.7).

7.10.2   The Wrong and Right Way to Do Cross-validation

Consider a classification problem with a large number of predictors, as may arise, for example, in genomic or proteomic applications. A typical strategy for analysis might be as follows:

1. Screen the predictors: find a subset of "good" predictors that show fairly strong (univariate) correlation with the class labels

2. Using just this subset of predictors, build a multivariate classifier.

3. Use cross-validation to estimate the unknown tuning parameters and to estimate the prediction error of the final model.

Is this a correct application of cross-validation? Consider a scenario with $N = 50$ samples in two equal-sized classes, and $p = 5000$ quantitative predictors (standard Gaussian) that are independent of the class labels. The true (test) error rate of any classifier is 50%. We carried out the above recipe, choosing in step (1) the 100 predictors having highest correlation with the class labels, and then using a 1-nearest neighbor classifier, based on just these 100 predictors, in step (2). Over 50 simulations from this setting, the average CV error rate was 3%. This is far lower than the true error rate of 50%.

What has happened? The problem is that the predictors have an unfair advantage, as they were chosen in step (1) on the basis of *all of the samples*. Leaving samples out *after* the variables have been selected does not correctly mimic the application of the classifier to a completely independent test set, since these predictors "have already seen" the left out samples.

Figure 7.10 (top panel) illustrates the problem. We selected the 100 predictors having largest correlation with the class labels over all 50 samples. Then we chose a random set of 10 samples, as we would do in five-fold cross-validation, and computed the correlations of the pre-selected 100 predictors with the class labels over just these 10 samples (top panel). We see that the correlations average about 0.28, rather than 0, as one might expect.

Here is the correct way to carry out cross-validation in this example:

1. Divide the samples into K cross-validation folds (groups) at random.

2. For each fold $k = 1, 2, \ldots, K$

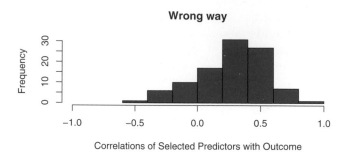

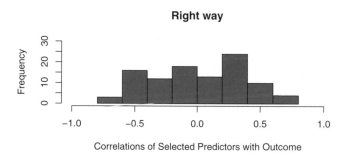

FIGURE 7.10. *Cross-validation the wrong and right way: histograms shows the correlation of class labels, in 10 randomly chosen samples, with the 100 predictors chosen using the incorrect (upper red) and correct (lower green) versions of cross-validation.*

(a) Find a subset of "good" predictors that show fairly strong (univariate) correlation with the class labels, using all of the samples except those in fold k.

(b) Using just this subset of predictors, build a multivariate classifier, using all of the samples except those in fold k.

(c) Use the classifier to predict the class labels for the samples in fold k.

The error estimates from step 2(c) are then accumulated over all K folds, to produce the cross-validation estimate of prediction error. The lower panel of Figure 7.10 shows the correlations of class labels with the 100 predictors chosen in step 2(a) of the correct procedure, over the samples in a typical fold k. We see that they average about zero, as they should.

In general, with a multistep modeling procedure, cross-validation must be applied to the entire sequence of modeling steps. In particular, samples must be "left out" before any selection or filtering steps are applied. There is one qualification: initial *unsupervised* screening steps can be done before samples are left out. For example, we could select the 1000 predictors with highest variance across all 50 samples, before starting cross-validation. Since this filtering does not involve the class labels, it does not give the predictors an unfair advantage.

While this point may seem obvious to the reader, we have seen this blunder committed many times in published papers in top rank journals. With the large numbers of predictors that are so common in genomic and other areas, the potential consequences of this error have also increased dramatically; see Ambroise and McLachlan (2002) for a detailed discussion of this issue.

7.10.3 Does Cross-Validation Really Work?

We once again examine the behavior of cross-validation in a high-dimensional classification problem. Consider a scenario with $N = 20$ samples in two equal-sized classes, and $p = 500$ quantitative predictors that are independent of the class labels. Once again, the true error rate of any classifier is 50%. Consider a simple univariate classifier: a single split that minimizes the misclassification error (a "stump"). Stumps are trees with a single split, and are used in boosting methods (Chapter 10). A simple argument suggests that cross-validation will not work properly in this setting[2]:

> Fitting to the entire training set, we will find a predictor that splits the data very well. If we do 5-fold cross-validation, this same predictor should split any 4/5ths and 1/5th of the data well too, and hence its cross-validation error will be small (much less than 50%.) Thus CV does not give an accurate estimate of error.

To investigate whether this argument is correct, Figure 7.11 shows the result of a simulation from this setting. There are 500 predictors and 20 samples, in each of two equal-sized classes, with all predictors having a standard Gaussian distribution. The panel in the top left shows the number of training errors for each of the 500 stumps fit to the training data. We have marked in color the six predictors yielding the fewest errors. In the top right panel, the training errors are shown for stumps fit to a random 4/5ths partition of the data (16 samples), and tested on the remaining 1/5th (four samples). The colored points indicate the same predictors marked in the top left panel. We see that the stump for the blue predictor (whose stump was the best in the top left panel), makes two out of four test errors (50%), and is no better than random.

What has happened? The preceding argument has ignored the fact that in cross-validation, the model must be completely retrained for each fold

---

[2]This argument was made to us by a scientist at a proteomics lab meeting, and led to material in this section.

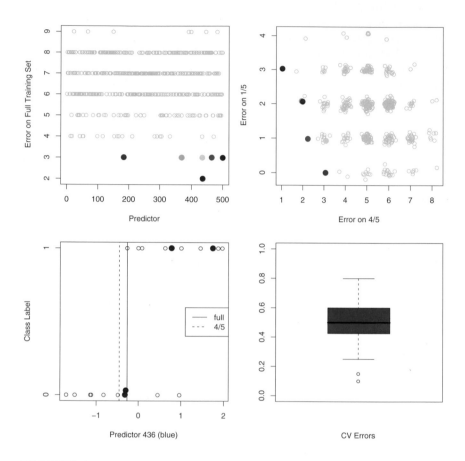

FIGURE 7.11. *Simulation study to investigate the performance of cross validation in a high-dimensional problem where the predictors are independent of the class labels. The top-left panel shows the number of errors made by individual stump classifiers on the full training set (20 observations). The top right panel shows the errors made by individual stumps trained on a random split of the dataset into 4/5ths (16 observations) and tested on the remaining 1/5th (4 observations). The best performers are depicted by colored dots in each panel. The bottom left panel shows the effect of re-estimating the split point in each fold: the colored points correspond to the four samples in the 1/5th validation set. The split point derived from the full dataset classifies all four samples correctly, but when the split point is re-estimated on the 4/5ths data (as it should be), it commits two errors on the four validation samples. In the bottom right we see the overall result of five-fold cross-validation applied to 50 simulated datasets. The average error rate is about 50%, as it should be.*

of the process. In the present example, this means that the best predictor
and corresponding split point are found from 4/5ths of the data. The effect
of predictor choice is seen in the top right panel. Since the class labels are
independent of the predictors, the performance of a stump on the 4/5ths
training data contains no information about its performance in the remaining 1/5th. The effect of the choice of split point is shown in the bottom left
panel. Here we see the data for predictor 436, corresponding to the blue
dot in the top left plot. The colored points indicate the 1/5th data, while
the remaining points belong to the 4/5ths. The optimal split points for this
predictor based on both the full training set and 4/5ths data are indicated.
The split based on the full data makes no errors on the 1/5ths data. But
cross-validation must base its split on the 4/5ths data, and this incurs two
errors out of four samples.

The results of applying five-fold cross-validation to each of 50 simulated
datasets is shown in the bottom right panel. As we would hope, the average
cross-validation error is around 50%, which is the true expected prediction
error for this classifier. Hence cross-validation has behaved as it should.
On the other hand, there is considerable variability in the error, underscor-
ing the importance of reporting the estimated standard error of the CV
estimate. See Exercise 7.10 for another variation of this problem.

7.11   Bootstrap Methods

The bootstrap is a general tool for assessing statistical accuracy. First we
describe the bootstrap in general, and then show how it can be used to
estimate extra-sample prediction error. As with cross-validation, the bootstrap seeks to estimate the conditional error $\text{Err}_{\mathcal{T}}$, but typically estimates
well only the expected prediction error Err.

Suppose we have a model fit to a set of training data. We denote the
training set by $\mathbf{Z} = (z_1, z_2, \ldots, z_N)$ where $z_i = (x_i, y_i)$. The basic idea is
to randomly draw datasets with replacement from the training data, each
sample the same size as the original training set. This is done B times
($B = 100$ say), producing B bootstrap datasets, as shown in Figure 7.12.
Then we refit the model to each of the bootstrap datasets, and examine
the behavior of the fits over the B replications.

In the figure, $S(\mathbf{Z})$ is any quantity computed from the data $\mathbf{Z}$, for example, the prediction at some input point. From the bootstrap sampling
we can estimate any aspect of the distribution of $S(\mathbf{Z})$, for example, its
variance,

$$\widehat{\text{Var}}[S(\mathbf{Z})] = \frac{1}{B-1} \sum_{b=1}^{B} (S(\mathbf{Z}^{*b}) - \bar{S}^*)^2, \tag{7.53}$$

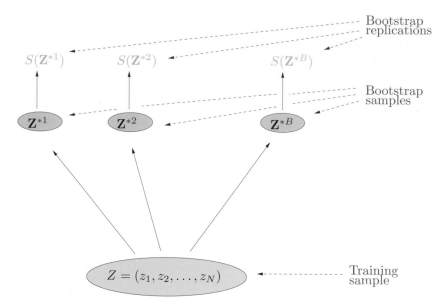

FIGURE 7.12. *Schematic of the bootstrap process. We wish to assess the statistical accuracy of a quantity $S(\mathbf{Z})$ computed from our dataset. B training sets $\mathbf{Z}^{*b}$, $b = 1, \ldots, B$ each of size N are drawn with replacement from the original dataset. The quantity of interest $S(\mathbf{Z})$ is computed from each bootstrap training set, and the values $S(\mathbf{Z}^{*1}), \ldots, S(\mathbf{Z}^{*B})$ are used to assess the statistical accuracy of $S(\mathbf{Z})$.*

where $\bar{S}^{*} = \sum_{b} S(\mathbf{Z}^{*b})/B$. Note that $\widehat{\mathrm{Var}}[S(\mathbf{Z})]$ can be thought of as a Monte-Carlo estimate of the variance of $S(\mathbf{Z})$ under sampling from the empirical distribution function $\hat{F}$ for the data $(z_1, z_2, \ldots, z_N)$.

How can we apply the bootstrap to estimate prediction error? One approach would be to fit the model in question on a set of bootstrap samples, and then keep track of how well it predicts the original training set. If $\hat{f}^{*b}(x_i)$ is the predicted value at x_i, from the model fitted to the bth bootstrap dataset, our estimate is

$$\widehat{\mathrm{Err}}_{\mathrm{boot}} = \frac{1}{B}\frac{1}{N}\sum_{b=1}^{B}\sum_{i=1}^{N} L(y_i, \hat{f}^{*b}(x_i)). \tag{7.54}$$

However, it is easy to see that $\widehat{\mathrm{Err}}_{\mathrm{boot}}$ does not provide a good estimate in general. The reason is that the bootstrap datasets are acting as the training samples, while the original training set is acting as the test sample, and these two samples have observations in common. This overlap can make overfit predictions look unrealistically good, and is the reason that cross-validation explicitly uses non-overlapping data for the training and test samples. Consider for example a 1-nearest neighbor classifier applied to a two-class classification problem with the same number of observations in each class, in which the predictors and class labels are in fact independent. Then the true error rate is 0.5. But the contributions to the bootstrap estimate $\widehat{\mathrm{Err}}_{\mathrm{boot}}$ will be zero unless the observation i does not appear in the bootstrap sample b. In this latter case it will have the correct expectation 0.5. Now

$$
\begin{aligned}
\Pr\{\text{observation } i \in \text{bootstrap sample } b\} \;&=\; 1 - \left(1 - \frac{1}{N}\right)^N \\
&\approx\; 1 - e^{-1} \\
&=\; 0.632. \qquad (7.55)
\end{aligned}
$$

Hence the expectation of $\widehat{\mathrm{Err}}_{\mathrm{boot}}$ is about $0.5 \times 0.368 = 0.184$, far below the correct error rate 0.5.

By mimicking cross-validation, a better bootstrap estimate can be obtained. For each observation, we only keep track of predictions from bootstrap samples not containing that observation. The leave-one-out bootstrap estimate of prediction error is defined by

$$
\widehat{\mathrm{Err}}^{(1)} = \frac{1}{N} \sum_{i=1}^{N} \frac{1}{|C^{-i}|} \sum_{b \in C^{-i}} L(y_i, \hat{f}^{*b}(x_i)). \qquad (7.56)
$$

Here C^{-i} is the set of indices of the bootstrap samples b that do *not* contain observation i, and $|C^{-i}|$ is the number of such samples. In computing $\widehat{\mathrm{Err}}^{(1)}$, we either have to choose B large enough to ensure that all of the $|C^{-i}|$ are greater than zero, or we can just leave out the terms in (7.56) corresponding to $|C^{-i}|$'s that are zero.

The leave-one out bootstrap solves the overfitting problem suffered by $\widehat{\mathrm{Err}}_{\mathrm{boot}}$, but has the training-set-size bias mentioned in the discussion of cross-validation. The average number of distinct observations in each bootstrap sample is about $0.632 \cdot N$, so its bias will roughly behave like that of twofold cross-validation. Thus if the learning curve has considerable slope at sample size $N/2$, the leave-one out bootstrap will be biased upward as an estimate of the true error.

The ".632 estimator" is designed to alleviate this bias. It is defined by

$$
\widehat{\mathrm{Err}}^{(.632)} = .368 \cdot \overline{\mathrm{err}} + .632 \cdot \widehat{\mathrm{Err}}^{(1)}. \qquad (7.57)
$$

The derivation of the .632 estimator is complex; intuitively it pulls the leave-one out bootstrap estimate down toward the training error rate, and hence reduces its upward bias. The use of the constant .632 relates to (7.55).

The .632 estimator works well in "light fitting" situations, but can break down in overfit ones. Here is an example due to Breiman et al. (1984). Suppose we have two equal-size classes, with the targets independent of the class labels, and we apply a one-nearest neighbor rule. Then $\overline{\mathrm{err}} = 0$,

$\widehat{\text{Err}}^{(1)} = 0.5$ and so $\widehat{\text{Err}}^{(.632)} = .632 \times 0.5 = .316$. However, the true error rate is 0.5.

One can improve the .632 estimator by taking into account the amount of overfitting. First we define γ to be the *no-information error rate*: this is the error rate of our prediction rule if the inputs and class labels were independent. An estimate of γ is obtained by evaluating the prediction rule on all possible combinations of targets y_i and predictors $x_{i'}$

$$\hat{\gamma} = \frac{1}{N^2} \sum_{i=1}^{N} \sum_{i'=1}^{N} L(y_i, \hat{f}(x_{i'})). \tag{7.58}$$

For example, consider the dichotomous classification problem: let $\hat{p}_1$ be the observed proportion of responses y_i equaling 1, and let $\hat{q}_1$ be the observed proportion of predictions $\hat{f}(x_{i'})$ equaling 1. Then

$$\hat{\gamma} = \hat{p}_1(1 - \hat{q}_1) + (1 - \hat{p}_1)\hat{q}_1. \tag{7.59}$$

With a rule like 1-nearest neighbors for which $\hat{q}_1 = \hat{p}_1$ the value of $\hat{\gamma}$ is $2\hat{p}_1(1-\hat{p}_1)$. The multi-category generalization of (7.59) is $\hat{\gamma} = \sum_\ell \hat{p}_\ell(1-\hat{q}_\ell)$.

Using this, the *relative overfitting rate* is defined to be

$$\hat{R} = \frac{\widehat{\text{Err}}^{(1)} - \overline{\text{err}}}{\hat{\gamma} - \overline{\text{err}}}, \tag{7.60}$$

a quantity that ranges from 0 if there is no overfitting ($\widehat{\text{Err}}^{(1)} = \overline{\text{err}}$) to 1 if the overfitting equals the no-information value $\hat{\gamma} - \overline{\text{err}}$. Finally, we define the ".632+" estimator by

$$\widehat{\text{Err}}^{(.632+)} = (1 - \hat{w}) \cdot \overline{\text{err}} + \hat{w} \cdot \widehat{\text{Err}}^{(1)} \tag{7.61}$$

$$\text{with } \hat{w} = \frac{.632}{1 - .368\hat{R}}.$$

The weight w ranges from .632 if $\hat{R} = 0$ to 1 if $\hat{R} = 1$, so $\widehat{\text{Err}}^{(.632+)}$ ranges from $\widehat{\text{Err}}^{(.632)}$ to $\widehat{\text{Err}}^{(1)}$. Again, the derivation of (7.61) is complicated: roughly speaking, it produces a compromise between the leave-one-out bootstrap and the training error rate that depends on the amount of overfitting. For the 1-nearest-neighbor problem with class labels independent of the inputs, $\hat{w} = \hat{R} = 1$, so $\widehat{\text{Err}}^{(.632+)} = \widehat{\text{Err}}^{(1)}$, which has the correct expectation of 0.5. In other problems with less overfitting, $\widehat{\text{Err}}^{(.632+)}$ will lie somewhere between $\overline{\text{err}}$ and $\widehat{\text{Err}}^{(1)}$.

7.11.1 Example (Continued)

Figure 7.13 shows the results of tenfold cross-validation and the .632+ bootstrap estimate in the same four problems of Figures 7.7. As in that figure,

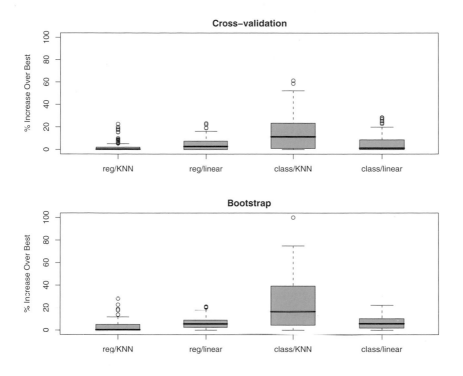

FIGURE 7.13. *Boxplots show the distribution of the relative error* $100 \cdot [\mathrm{Err}_{\hat{\alpha}} - \min_{\alpha} \mathrm{Err}(\alpha)]/[\max_{\alpha} \mathrm{Err}(\alpha) - \min_{\alpha} \mathrm{Err}(\alpha)]$ *over the four scenarios of Figure 7.3. This is the error in using the chosen model relative to the best model. There are* 100 *training sets represented in each boxplot.*

Figure 7.13 shows boxplots of $100 \cdot [\mathrm{Err}_{\hat{\alpha}} - \min_{\alpha} \mathrm{Err}(\alpha)]/[\max_{\alpha} \mathrm{Err}(\alpha) - \min_{\alpha} \mathrm{Err}(\alpha)]$, the error in using the chosen model relative to the best model. There are 100 different training sets represented in each boxplot. Both measures perform well overall, perhaps the same or slightly worse than the AIC in Figure 7.7.

Our conclusion is that for these particular problems and fitting methods, minimization of either AIC, cross-validation or bootstrap yields a model fairly close to the best available. Note that for the purpose of model selection, any of the measures could be biased and it wouldn't affect things, as long as the bias did not change the relative performance of the methods. For example, the addition of a constant to any of the measures would not change the resulting chosen model. However, for many adaptive, nonlinear techniques (like trees), estimation of the effective number of parameters is very difficult. This makes methods like AIC impractical and leaves us with cross-validation or bootstrap as the methods of choice.

A different question is: how well does each method estimate test error? On the average the AIC criterion overestimated prediction error of its chosen model by 38%, 37%, 51%, and 30%, respectively, over the four scenarios, with BIC performing similarly. In contrast, cross-validation overestimated the error by 1%, 4%, 0%, and 4%, with the bootstrap doing about the same. Hence the extra work involved in computing a cross-validation or bootstrap measure is worthwhile, if an accurate estimate of test error is required. With other fitting methods like trees, cross-validation and bootstrap can underestimate the true error by 10%, because the search for best tree is strongly affected by the validation set. In these situations only a separate test set will provide an unbiased estimate of test error.

7.12 Conditional or Expected Test Error?

Figures 7.14 and 7.15 examine the question of whether cross-validation does a good job in estimating $\text{Err}_{\mathcal{T}}$, the error conditional on a given training set $\mathcal{T}$ (expression (7.15) on page 228), as opposed to the expected test error. For each of 100 training sets generated from the "reg/linear" setting in the top-right panel of Figure 7.3, Figure 7.14 shows the conditional error curves $\text{Err}_{\mathcal{T}}$ as a function of subset size (top left). The next two panels show 10-fold and N-fold cross-validation, the latter also known as leave-one-out (LOO). The thick red curve in each plot is the expected error Err, while the thick black curves are the expected cross-validation curves. The lower right panel shows how well cross-validation approximates the conditional and expected error.

One might have expected N-fold CV to approximate $\text{Err}_{\mathcal{T}}$ well, since it *almost* uses the full training sample to fit a new test point. 10-fold CV, on the other hand, might be expected to estimate Err well, since it averages over somewhat different training sets. From the figure it appears 10-fold does a better job than N-fold in estimating $\text{Err}_{\mathcal{T}}$, and estimates Err even better. Indeed, the similarity of the two black curves with the red curve suggests both CV curves are approximately unbiased for Err, with 10-fold having less variance. Similar trends were reported by Efron (1983).

Figure 7.15 shows scatterplots of both 10-fold and N-fold cross-validation error estimates versus the true conditional error for the 100 simulations. Although the scatterplots do not indicate much correlation, the lower right panel shows that for the most part the correlations are negative, a curious phenomenon that has been observed before. This negative correlation explains why neither form of CV estimates $\text{Err}_{\mathcal{T}}$ well. The broken lines in each plot are drawn at $\text{Err}(p)$, the expected error for the best subset of size p. We see again that both forms of CV are approximately unbiased for expected error, but the variation in test error for different training sets is quite substantial.

Among the four experimental conditions in 7.3, this "reg/linear" scenario showed the highest correlation between actual and predicted test error. This

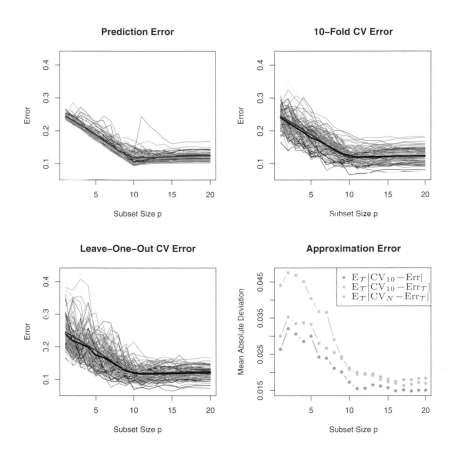

FIGURE 7.14. *Conditional prediction-error* $\mathrm{Err}_{\mathcal{T}}$, *10-fold cross-validation, and leave-one-out cross-validation curves for a* 100 *simulations from the top-right panel in Figure 7.3. The thick red curve is the expected prediction error* Err, *while the thick black curves are the expected CV curves* $\mathrm{E}_{\mathcal{T}}\mathrm{CV}_{10}$ *and* $\mathrm{E}_{\mathcal{T}}\mathrm{CV}_{N}$. *The lower-right panel shows the mean absolute deviation of the CV curves from the conditional error,* $\mathrm{E}_{\mathcal{T}}|\mathrm{CV}_{K} - \mathrm{Err}_{\mathcal{T}}|$ *for* $K = 10$ *(blue) and* $K = N$ *(green), as well as from the expected error* $\mathrm{E}_{\mathcal{T}}|\mathrm{CV}_{10} - \mathrm{Err}|$ *(orange).*

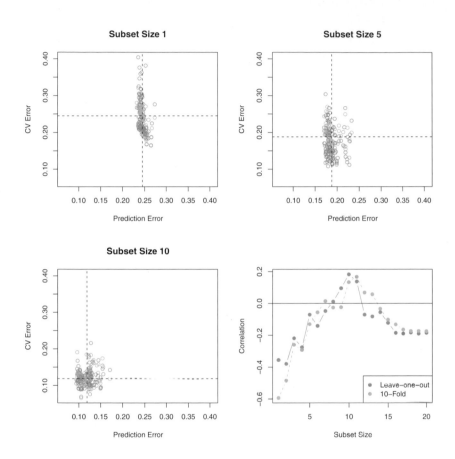

FIGURE 7.15. *Plots of the CV estimates of error versus the true conditional error for each of the* 100 *training sets, for the simulation setup in the top right panel Figure 7.3. Both* 10-*fold and leave-one-out CV are depicted in different colors. The first three panels correspond to different subset sizes p, and vertical and horizontal lines are drawn at* Err(p). *Although there appears to be little correlation in these plots, we see in the lower right panel that for the most part the correlation is* negative.

phenomenon also occurs for bootstrap estimates of error, and we would guess, for any other estimate of conditional prediction error.

We conclude that estimation of test error for a particular training set is not easy in general, given just the data from that same training set. Instead, cross-validation and related methods may provide reasonable estimates of the *expected* error Err.

Bibliographic Notes

Key references for cross-validation are Stone (1974), Stone (1977) and Allen (1974). The AIC was proposed by Akaike (1973), while the BIC was introduced by Schwarz (1978). Madigan and Raftery (1994) give an overview of Bayesian model selection. The MDL criterion is due to Rissanen (1983). Cover and Thomas (1991) contains a good description of coding theory and complexity. VC dimension is described in Vapnik (1996). Stone (1977) showed that the AIC and leave-one out cross-validation are asymptotically equivalent. Generalized cross-validation is described by Golub et al. (1979) and Wahba (1980); a further discussion of the topic may be found in the monograph by Wahba (1990). See also Hastie and Tibshirani (1990), Chapter 3. The bootstrap is due to Efron (1979); see Efron and Tibshirani (1993) for an overview. Efron (1983) proposes a number of bootstrap estimates of prediction error, including the optimism and .632 estimates. Efron (1986) compares CV, GCV and bootstrap estimates of error rates. The use of cross-validation and the bootstrap for model selection is studied by Breiman and Spector (1992), Breiman (1992), Shao (1996), Zhang (1993) and Kohavi (1995). The .632+ estimator was proposed by Efron and Tibshirani (1997).

Cherkassky and Ma (2003) published a study on the performance of SRM for model selection in regression, in response to our study of section 7.9.1. They complained that we had been unfair to SRM because had not applied it properly. Our response can be found in the same issue of the journal (Hastie et al. (2003)).

Exercises

Ex. 7.1 Derive the estimate of in-sample error (7.24).

Ex. 7.2 For 0–1 loss with $Y \in \{0, 1\}$ and $\Pr(Y = 1|x_0) = f(x_0)$, show that

$$
\begin{aligned}
\text{Err}(x_0) &= \Pr(Y \neq \hat{G}(x_0)|X = x_0) \\
&= \text{Err}_B(x_0) + |2f(x_0) - 1|\Pr(\hat{G}(x_0) \neq G(x_0)|X = x_0),
\end{aligned}
\tag{7.62}
$$

where $\hat{G}(x) = I(\hat{f}(x) > \frac{1}{2})$, $G(x) = I(f(x) > \frac{1}{2})$ is the Bayes classifier, and $\text{Err}_B(x_0) = \Pr(Y \neq G(x_0)|X = x_0)$, the irreducible *Bayes error* at x_0. Using the approximation $\hat{f}(x_0) \sim N(\text{E}\hat{f}(x_0), \text{Var}(\hat{f}(x_0))$, show that

$$\Pr(\hat{G}(x_0) \neq G(x_0)|X = x_0) \approx \Phi\left(\frac{\text{sign}(\frac{1}{2} - f(x_0))(\text{E}\hat{f}(x_0) - \frac{1}{2})}{\sqrt{\text{Var}(\hat{f}(x_0))}}\right). \quad (7.63)$$

In the above,

$$\Phi(t) = \frac{1}{\sqrt{2\pi}} \int_{-\infty}^{t} \exp(-t^2/2)dt,$$

the cumulative Gaussian distribution function. This is an increasing function, with value 0 at $t = -\infty$ and value 1 at $t = +\infty$.

We can think of $\text{sign}(\frac{1}{2} - f(x_0))(\text{E}\hat{f}(x_0) - \frac{1}{2})$ as a kind of *boundary-bias* term, as it depends on the true $f(x_0)$ only through which side of the boundary $(\frac{1}{2})$ that it lies. Notice also that the bias and variance combine in a multiplicative rather than additive fashion. If $\text{E}\hat{f}(x_0)$ is on the same side of $\frac{1}{2}$ as $f(x_0)$, then the bias is negative, and decreasing the variance will decrease the misclassification error. On the other hand, if $\text{E}\hat{f}(x_0)$ is on the opposite side of $\frac{1}{2}$ to $f(x_0)$, then the bias is positive and it pays to increase the variance! Such an increase will improve the chance that $\hat{f}(x_0)$ falls on the correct side of $\frac{1}{2}$ (Friedman, 1997).

Ex. 7.3 Let $\hat{\mathbf{f}} = \mathbf{S}\mathbf{y}$ be a linear smoothing of $\mathbf{y}$.

(a) If S_{ii} is the ith diagonal element of $\mathbf{S}$, show that for $\mathbf{S}$ arising from least squares projections and cubic smoothing splines, the cross-validated residual can be written as

$$y_i - \hat{f}^{-i}(x_i) = \frac{y_i - \hat{f}(x_i)}{1 - S_{ii}}. \quad (7.64)$$

(b) Use this result to show that $|y_i - \hat{f}^{-i}(x_i)| \geq |y_i - \hat{f}(x_i)|$.

(c) Find general conditions on any smoother $\mathbf{S}$ to make result (7.64) hold.

Ex. 7.4 Consider the in-sample prediction error (7.18) and the training error $\overline{\text{err}}$ in the case of squared-error loss:

$$\text{Err}_{\text{in}} = \frac{1}{N}\sum_{i=1}^{N} \text{E}_{Y^0}(Y_i^0 - \hat{f}(x_i))^2$$

$$\overline{\text{err}} = \frac{1}{N}\sum_{i=1}^{N}(y_i - \hat{f}(x_i))^2.$$

Add and subtract $f(x_i)$ and $\mathrm{E}\hat{f}(x_i)$ in each expression and expand. Hence establish that the average optimism in the training error is

$$\frac{2}{N} \sum_{i=1}^{N} \mathrm{Cov}(\hat{y}_i, y_i),$$

as given in (7.21).

Ex. 7.5 For a linear smoother $\hat{\mathbf{y}} = \mathbf{S}\mathbf{y}$, show that

$$\sum_{i=1}^{N} \mathrm{Cov}(\hat{y}_i, y_i) = \mathrm{trace}(\mathbf{S})\sigma_\varepsilon^2, \qquad (7.65)$$

which justifies its use as the effective number of parameters.

Ex. 7.6 Show that for an additive-error model, the effective degrees-of-freedom for the k-nearest-neighbors regression fit is N/k.

Ex. 7.7 Use the approximation $1/(1-x)^2 \approx 1+2x$ to expose the relationship between $C_p/$AIC (7.26) and GCV (7.52), the main difference being the model used to estimate the noise variance σ_ε^2.

Ex. 7.8 Show that the set of functions $\{I(\sin(\alpha x) > 0)\}$ can shatter the following points on the line:

$$z^1 = 10^{-1}, \ldots, z^\ell = 10^{-\ell}, \qquad (7.66)$$

for any ℓ. Hence the VC dimension of the class $\{I(\sin(\alpha x) > 0)\}$ is infinite.

Ex. 7.9 For the prostate data of Chapter 3, carry out a best-subset linear regression analysis, as in Table 3.3 (third column from left). Compute the AIC, BIC, five- and tenfold cross-validation, and bootstrap .632 estimates of prediction error. Discuss the results.

Ex. 7.10 Referring to the example in Section 7.10.3, suppose instead that all of the p predictors are binary, and hence there is no need to estimate split points. The predictors are independent of the class labels as before. Then if p is very large, we can probably find a predictor that splits the entire training data perfectly, and hence would split the validation data (one-fifth of data) perfectly as well. This predictor would therefore have zero cross-validation error. Does this mean that cross-validation does not provide a good estimate of test error in this situation? [This question was suggested by Li Ma.]

8
Model Inference and Averaging

8.1 Introduction

For most of this book, the fitting (learning) of models has been achieved by minimizing a sum of squares for regression, or by minimizing cross-entropy for classification. In fact, both of these minimizations are instances of the maximum likelihood approach to fitting.

In this chapter we provide a general exposition of the maximum likelihood approach, as well as the Bayesian method for inference. The bootstrap, introduced in Chapter 7, is discussed in this context, and its relation to maximum likelihood and Bayes is described. Finally, we present some related techniques for model averaging and improvement, including committee methods, bagging, stacking and bumping.

8.2 The Bootstrap and Maximum Likelihood Methods

8.2.1 A Smoothing Example

The bootstrap method provides a direct computational way of assessing uncertainty, by sampling from the training data. Here we illustrate the bootstrap in a simple one-dimensional smoothing problem, and show its connection to maximum likelihood.

T. Hastie et al., *The Elements of Statistical Learning, Second Edition,*
DOI: 10.1007/b94608_8,
© Springer Science+Business Media, LLC 2009

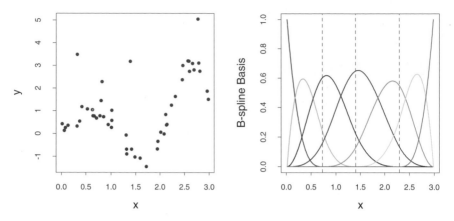

FIGURE 8.1. *(Left panel): Data for smoothing example. (Right panel:) Set of seven B-spline basis functions. The broken vertical lines indicate the placement of the three knots.*

Denote the training data by $\mathbf{Z} = \{z_1, z_2, \ldots, z_N\}$, with $z_i = (x_i, y_i)$, $i = 1, 2, \ldots, N$. Here x_i is a one-dimensional input, and y_i the outcome, either continuous or categorical. As an example, consider the $N = 50$ data points shown in the left panel of Figure 8.1.

Suppose we decide to fit a cubic spline to the data, with three knots placed at the quartiles of the X values. This is a seven-dimensional linear space of functions, and can be represented, for example, by a linear expansion of B-spline basis functions (see Section 5.9.2):

$$\mu(x) = \sum_{j=1}^{7} \beta_j h_j(x). \tag{8.1}$$

Here the $h_j(x)$, $j = 1, 2, \ldots, 7$ are the seven functions shown in the right panel of Figure 8.1. We can think of $\mu(x)$ as representing the conditional mean $E(Y|X = x)$.

Let $\mathbf{H}$ be the $N \times 7$ matrix with ijth element $h_j(x_i)$. The usual estimate of β, obtained by minimizing the squared error over the training set, is given by

$$\hat{\beta} = (\mathbf{H}^T \mathbf{H})^{-1} \mathbf{H}^T \mathbf{y}. \tag{8.2}$$

The corresponding fit $\hat{\mu}(x) = \sum_{j=1}^{7} \hat{\beta}_j h_j(x)$ is shown in the top left panel of Figure 8.2.

The estimated covariance matrix of $\hat{\beta}$ is

$$\widehat{\text{Var}}(\hat{\beta}) = (\mathbf{H}^T \mathbf{H})^{-1} \hat{\sigma}^2, \tag{8.3}$$

where we have estimated the noise variance by $\hat{\sigma}^2 = \sum_{i=1}^{N} (y_i - \hat{\mu}(x_i))^2 / N$. Letting $h(x)^T = (h_1(x), h_2(x), \ldots, h_7(x))$, the standard error of a predic-

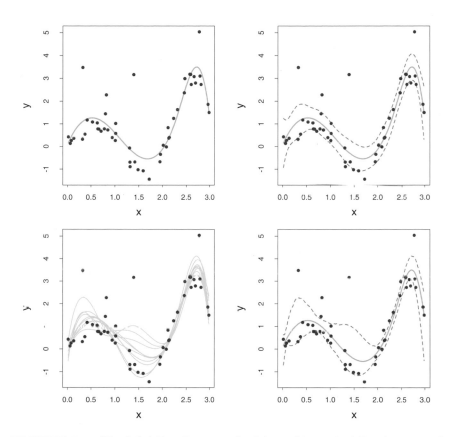

FIGURE 8.2. *(Top left:) B-spline smooth of data. (Top right:) B-spline smooth plus and minus 1.96× standard error bands. (Bottom left:) Ten bootstrap replicates of the B-spline smooth. (Bottom right:) B-spline smooth with 95% standard error bands computed from the bootstrap distribution.*

tion $\hat{\mu}(x) = h(x)^T \hat{\beta}$ is

$$\widehat{\text{se}}[\hat{\mu}(x)] = [h(x)^T (\mathbf{H}^T \mathbf{H})^{-1} h(x)]^{\frac{1}{2}} \hat{\sigma}. \tag{8.4}$$

In the top right panel of Figure 8.2 we have plotted $\hat{\mu}(x) \pm 1.96 \cdot \widehat{\text{se}}[\hat{\mu}(x)]$. Since 1.96 is the 97.5% point of the standard normal distribution, these represent approximate $100 - 2 \times 2.5\% = 95\%$ pointwise confidence bands for $\mu(x)$.

Here is how we could apply the bootstrap in this example. We draw B datasets each of size $N = 50$ with replacement from our training data, the sampling unit being the pair $z_i = (x_i, y_i)$. To each bootstrap dataset $\mathbf{Z}^*$ we fit a cubic spline $\hat{\mu}^*(x)$; the fits from ten such samples are shown in the bottom left panel of Figure 8.2. Using $B = 200$ bootstrap samples, we can form a 95% pointwise confidence band from the percentiles at each x: we find the $2.5\% \times 200 =$ fifth largest and smallest values at each x. These are plotted in the bottom right panel of Figure 8.2. The bands look similar to those in the top right, being a little wider at the endpoints.

There is actually a close connection between the least squares estimates (8.2) and (8.3), the bootstrap, and maximum likelihood. Suppose we further assume that the model errors are Gaussian,

$$\begin{aligned} Y &= \mu(X) + \varepsilon; \quad \varepsilon \sim N(0, \sigma^2), \\ \mu(x) &= \sum_{j=1}^{7} \beta_j h_j(x). \end{aligned} \tag{8.5}$$

The bootstrap method described above, in which we sample with replacement from the training data, is called the *nonparametric bootstrap*. This really means that the method is "model-free," since it uses the raw data, not a specific parametric model, to generate new datasets. Consider a variation of the bootstrap, called the *parametric bootstrap*, in which we simulate new responses by adding Gaussian noise to the predicted values:

$$y_i^* = \hat{\mu}(x_i) + \varepsilon_i^*; \quad \varepsilon_i^* \sim N(0, \hat{\sigma}^2); \quad i = 1, 2, \dots, N. \tag{8.6}$$

This process is repeated B times, where $B = 200$ say. The resulting bootstrap datasets have the form $(x_1, y_1^*), \dots, (x_N, y_N^*)$ and we recompute the B-spline smooth on each. The confidence bands from this method will exactly equal the least squares bands in the top right panel, as the number of bootstrap samples goes to infinity. A function estimated from a bootstrap sample $\mathbf{y}^*$ is given by $\hat{\mu}^*(x) = h(x)^T (\mathbf{H}^T \mathbf{H})^{-1} \mathbf{H}^T \mathbf{y}^*$, and has distribution

$$\hat{\mu}^*(x) \sim N(\hat{\mu}(x), h(x)^T (\mathbf{H}^T \mathbf{H})^{-1} h(x) \hat{\sigma}^2). \tag{8.7}$$

Notice that the mean of this distribution is the least squares estimate, and the standard deviation is the same as the approximate formula (8.4).

8.2.2   Maximum Likelihood Inference

It turns out that the parametric bootstrap agrees with least squares in the previous example because the model (8.5) has additive Gaussian errors. In general, the parametric bootstrap agrees not with least squares but with maximum likelihood, which we now review.

We begin by specifying a probability density or probability mass function for our observations

$$z_i \sim g_\theta(z). \tag{8.8}$$

In this expression θ represents one or more unknown parameters that govern the distribution of Z. This is called a *parametric model* for Z. As an example, if Z has a normal distribution with mean μ and variance σ^2, then

$$\theta = (\mu, \sigma^2), \tag{8.9}$$

and

$$g_\theta(z) = \frac{1}{\sqrt{2\pi}\sigma} e^{-\frac{1}{2}(z-\mu)^2/\sigma^2}. \tag{8.10}$$

Maximum likelihood is based on the *likelihood function*, given by

$$L(\theta; \mathbf{Z}) = \prod_{i=1}^{N} g_\theta(z_i), \tag{8.11}$$

the probability of the observed data under the model g_θ. The likelihood is defined only up to a positive multiplier, which we have taken to be one. We think of $L(\theta; \mathbf{Z})$ as a function of θ, with our data $\mathbf{Z}$ fixed.

Denote the logarithm of $L(\theta; \mathbf{Z})$ by

$$\begin{aligned} \ell(\theta; \mathbf{Z}) &= \sum_{i=1}^{N} \ell(\theta; z_i) \\ &= \sum_{i=1}^{N} \log g_\theta(z_i), \end{aligned} \tag{8.12}$$

which we will sometimes abbreviate as $\ell(\theta)$. This expression is called the log-likelihood, and each value $\ell(\theta; z_i) = \log g_\theta(z_i)$ is called a log-likelihood component. The method of maximum likelihood chooses the value $\theta = \hat{\theta}$ to maximize $\ell(\theta; \mathbf{Z})$.

The likelihood function can be used to assess the precision of $\hat{\theta}$. We need a few more definitions. The *score function* is defined by

$$\dot{\ell}(\theta; \mathbf{Z}) = \sum_{i=1}^{N} \dot{\ell}(\theta; z_i), \tag{8.13}$$

where $\dot{\ell}(\theta; z_i) = \partial\ell(\theta; z_i)/\partial\theta$. Assuming that the likelihood takes its maximum in the interior of the parameter space, $\dot{\ell}(\hat{\theta}; \mathbf{Z}) = 0$. The *information matrix* is

$$\mathbf{I}(\theta) = -\sum_{i=1}^{N} \frac{\partial^2\ell(\theta; z_i)}{\partial\theta\partial\theta^T}. \tag{8.14}$$

When $\mathbf{I}(\theta)$ is evaluated at $\theta = \hat{\theta}$, it is often called the *observed information*. The *Fisher information* (or expected information) is

$$\mathbf{i}(\theta) = \mathrm{E}_\theta[\mathbf{I}(\theta)]. \tag{8.15}$$

Finally, let θ_0 denote the true value of θ.

A standard result says that the sampling distribution of the maximum likelihood estimator has a limiting normal distribution

$$\hat{\theta} \to N(\theta_0, \mathbf{i}(\theta_0)^{-1}), \tag{8.16}$$

as $N \to \infty$. Here we are independently sampling from $g_{\theta_0}(z)$. This suggests that the sampling distribution of $\hat{\theta}$ may be approximated by

$$N(\hat{\theta}, \mathbf{i}(\hat{\theta})^{-1}) \text{ or } N(\hat{\theta}, \mathbf{I}(\hat{\theta})^{-1}), \tag{8.17}$$

where $\hat{\theta}$ represents the maximum likelihood estimate from the observed data.

The corresponding estimates for the standard errors of $\hat{\theta}_j$ are obtained from

$$\sqrt{\widehat{\mathbf{i}(\hat{\theta})}_{jj}^{-1}} \quad \text{and} \quad \sqrt{\mathbf{I}(\hat{\theta})_{jj}^{-1}}. \tag{8.18}$$

Confidence points for θ_j can be constructed from either approximation in (8.17). Such a confidence point has the form

$$\hat{\theta}_j - z^{(1-\alpha)} \cdot \sqrt{\mathbf{i}(\hat{\theta})_{jj}^{-1}} \quad \text{or} \quad \hat{\theta}_j - z^{(1-\alpha)} \cdot \sqrt{\mathbf{I}(\hat{\theta})_{jj}^{-1}},$$

respectively, where $z^{(1-\alpha)}$ is the $1 - \alpha$ percentile of the standard normal distribution. More accurate confidence intervals can be derived from the likelihood function, by using the chi-squared approximation

$$2[\ell(\hat{\theta}) - \ell(\theta_0)] \sim \chi_p^2, \tag{8.19}$$

where p is the number of components in θ. The resulting $1 - 2\alpha$ confidence interval is the set of all θ_0 such that $2[\ell(\hat{\theta}) - \ell(\theta_0)] \leq \chi_p^{2(1-2\alpha)}$, where $\chi_p^{2(1-2\alpha)}$ is the $1 - 2\alpha$ percentile of the chi-squared distribution with p degrees of freedom.

Let's return to our smoothing example to see what maximum likelihood yields. The parameters are $\theta = (\beta, \sigma^2)$. The log-likelihood is

$$\ell(\theta) = -\frac{N}{2} \log \sigma^2 2\pi - \frac{1}{2\sigma^2} \sum_{i=1}^{N} (y_i - h(x_i)^T \beta)^2. \qquad (8.20)$$

The maximum likelihood estimate is obtained by setting $\partial \ell/\partial \beta = 0$ and $\partial \ell/\partial \sigma^2 = 0$, giving

$$\hat{\beta} = (\mathbf{H}^T \mathbf{H})^{-1} \mathbf{H}^T \mathbf{y},$$
$$\hat{\sigma}^2 = \frac{1}{N} \sum (y_i - \hat{\mu}(x_i))^2, \qquad (8.21)$$

which are the same as the usual estimates given in (8.2) and below (8.3).

The information matrix for $\theta = (\beta, \sigma^2)$ is block-diagonal, and the block corresponding to β is

$$\mathbf{I}(\beta) = (\mathbf{H}^T \mathbf{H})/\sigma^2, \qquad (8.22)$$

so that the estimated variance $(\mathbf{H}^T \mathbf{H})^{-1} \hat{\sigma}^2$ agrees with the least squares estimate (8.3).

8.2.3 *Bootstrap versus Maximum Likelihood*

In essence the bootstrap is a computer implementation of nonparametric or parametric maximum likelihood. The advantage of the bootstrap over the maximum likelihood formula is that it allows us to compute maximum likelihood estimates of standard errors and other quantities in settings where no formulas are available.

In our example, suppose that we adaptively choose by cross-validation the number and position of the knots that define the B-splines, rather than fix them in advance. Denote by λ the collection of knots and their positions. Then the standard errors and confidence bands should account for the adaptive choice of λ, but there is no way to do this analytically. With the bootstrap, we compute the B-spline smooth with an adaptive choice of knots for each bootstrap sample. The percentiles of the resulting curves capture the variability from both the noise in the targets as well as that from $\hat{\lambda}$. In this particular example the confidence bands (not shown) don't look much different than the fixed λ bands. But in other problems, where more adaptation is used, this can be an important effect to capture.

8.3 Bayesian Methods

In the Bayesian approach to inference, we specify a sampling model $\Pr(\mathbf{Z}|\theta)$ (density or probability mass function) for our data given the parameters, and a prior distribution for the parameters $\Pr(\theta)$ reflecting our knowledge about θ before we see the data. We then compute the posterior distribution

$$\Pr(\theta|\mathbf{Z}) = \frac{\Pr(\mathbf{Z}|\theta) \cdot \Pr(\theta)}{\int \Pr(\mathbf{Z}|\theta) \cdot \Pr(\theta)d\theta}, \tag{8.23}$$

which represents our updated knowledge about θ after we see the data. To understand this posterior distribution, one might draw samples from it or summarize by computing its mean or mode. The Bayesian approach differs from the standard ("frequentist") method for inference in its use of a prior distribution to express the uncertainty present before seeing the data, and to allow the uncertainty remaining after seeing the data to be expressed in the form of a posterior distribution.

The posterior distribution also provides the basis for predicting the values of a future observation z^{new}, via the *predictive distribution*:

$$\Pr(z^{\text{new}}|\mathbf{Z}) = \int \Pr(z^{\text{new}}|\theta) \cdot \Pr(\theta|\mathbf{Z})d\theta. \tag{8.24}$$

In contrast, the maximum likelihood approach would use $\Pr(z^{\text{new}}|\hat{\theta})$, the data density evaluated at the maximum likelihood estimate, to predict future data. Unlike the predictive distribution (8.24), this does not account for the uncertainty in estimating θ.

Let's walk through the Bayesian approach in our smoothing example. We start with the parametric model given by equation (8.5), and assume for the moment that σ^2 is known. We assume that the observed feature values $x_1, x_2, \ldots, x_N$ are fixed, so that the randomness in the data comes solely from y varying around its mean $\mu(x)$.

The second ingredient we need is a prior distribution. Distributions on functions are fairly complex entities: one approach is to use a Gaussian process prior in which we specify the prior covariance between any two function values $\mu(x)$ and $\mu(x')$ (Wahba, 1990; Neal, 1996).

Here we take a simpler route: by considering a finite B-spline basis for $\mu(x)$, we can instead provide a prior for the coefficients β, and this implicitly defines a prior for $\mu(x)$. We choose a Gaussian prior centered at zero

$$\beta \sim N(0, \tau\mathbf{\Sigma}) \tag{8.25}$$

with the choices of the prior correlation matrix $\mathbf{\Sigma}$ and variance τ to be discussed below. The implicit process prior for $\mu(x)$ is hence Gaussian, with covariance kernel

$$\begin{aligned} K(x, x') &= \text{cov}[\mu(x), \mu(x')] \\ &= \tau \cdot h(x)^T \mathbf{\Sigma} h(x'). \end{aligned} \tag{8.26}$$

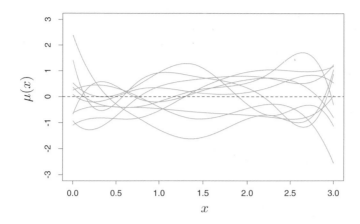

FIGURE 8.3. *Smoothing example: Ten draws from the Gaussian prior distribution for the function $\mu(x)$.*

The posterior distribution for β is also Gaussian, with mean and covariance

$$
\mathrm{E}(\beta|\mathbf{Z}) = \left(\mathbf{H}^T\mathbf{H} + \frac{\sigma^2}{\tau}\boldsymbol{\Sigma}^{-1}\right)^{-1}\mathbf{H}^T\mathbf{y},
$$

$$
\mathrm{cov}(\beta|\mathbf{Z}) = \left(\mathbf{H}^T\mathbf{H} + \frac{\sigma^2}{\tau}\boldsymbol{\Sigma}^{-1}\right)^{-1}\sigma^2,
$$

(8.27)

with the corresponding posterior values for $\mu(x)$,

$$
\mathrm{E}(\mu(x)|\mathbf{Z}) = h(x)^T\left(\mathbf{H}^T\mathbf{H} + \frac{\sigma^2}{\tau}\boldsymbol{\Sigma}^{-1}\right)^{-1}\mathbf{H}^T\mathbf{y},
$$

$$
\mathrm{cov}[\mu(x),\mu(x')|\mathbf{Z}] = h(x)^T\left(\mathbf{H}^T\mathbf{H} + \frac{\sigma^2}{\tau}\boldsymbol{\Sigma}^{-1}\right)^{-1}h(x')\sigma^2.
$$

(8.28)

How do we choose the prior correlation matrix $\boldsymbol{\Sigma}$? In some settings the prior can be chosen from subject matter knowledge about the parameters. Here we are willing to say the function $\mu(x)$ should be smooth, and have guaranteed this by expressing μ in a smooth low-dimensional basis of B-splines. Hence we can take the prior correlation matrix to be the identity $\boldsymbol{\Sigma} = \mathbf{I}$. When the number of basis functions is large, this might not be sufficient, and additional smoothness can be enforced by imposing restrictions on $\boldsymbol{\Sigma}$; this is exactly the case with smoothing splines (Section 5.8.1).

Figure 8.3 shows ten draws from the corresponding prior for $\mu(x)$. To generate posterior values of the function $\mu(x)$, we generate values β' from its posterior (8.27), giving corresponding posterior value $\mu'(x) = \sum_1^7 \beta'_j h_j(x)$. Ten such posterior curves are shown in Figure 8.4. Two different values were used for the prior variance τ, 1 and 1000. Notice how similar the right panel looks to the bootstrap distribution in the bottom left panel

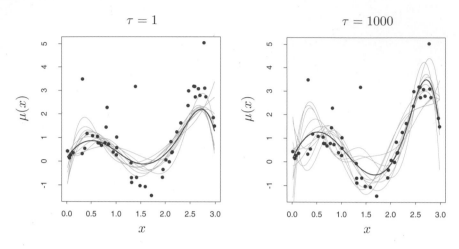

FIGURE 8.4. *Smoothing example: Ten draws from the posterior distribution for the function $\mu(x)$, for two different values of the prior variance τ. The purple curves are the posterior means.*

of Figure 8.2 on page 263. This similarity is no accident. As $\tau \to \infty$, the posterior distribution (8.27) and the bootstrap distribution (8.7) coincide. On the other hand, for $\tau = 1$, the posterior curves $\mu(x)$ in the left panel of Figure 8.4 are smoother than the bootstrap curves, because we have imposed more prior weight on smoothness.

The distribution (8.25) with $\tau \to \infty$ is called a *noninformative prior* for θ. In Gaussian models, maximum likelihood and parametric bootstrap analyses tend to agree with Bayesian analyses that use a noninformative prior for the free parameters. These tend to agree, because with a constant prior, the posterior distribution is proportional to the likelihood. This correspondence also extends to the nonparametric case, where the nonparametric bootstrap approximates a noninformative Bayes analysis; Section 8.4 has the details.

We have, however, done some things that are not proper from a Bayesian point of view. We have used a noninformative (constant) prior for σ^2 and replaced it with the maximum likelihood estimate $\hat{\sigma}^2$ in the posterior. A more standard Bayesian analysis would also put a prior on σ (typically $g(\sigma) \propto 1/\sigma$), calculate a joint posterior for $\mu(x)$ and σ, and then integrate out σ, rather than just extract the maximum of the posterior distribution ("MAP" estimate).

8.4 Relationship Between the Bootstrap and Bayesian Inference

Consider first a very simple example, in which we observe a single observation z from a normal distribution

$$z \sim N(\theta, 1). \tag{8.29}$$

To carry out a Bayesian analysis for θ, we need to specify a prior. The most convenient and common choice would be $\theta \sim N(0, \tau)$ giving posterior distribution

$$\theta | z \sim N \left(\frac{z}{1 + 1/\tau}, \frac{1}{1 + 1/\tau} \right). \tag{8.30}$$

Now the larger we take τ, the more concentrated the posterior becomes around the maximum likelihood estimate $\hat{\theta} = z$. In the limit as $\tau \to \infty$ we obtain a noninformative (constant) prior, and the posterior distribution is

$$\theta | z \sim N(z, 1). \tag{8.31}$$

This is the same as a parametric bootstrap distribution in which we generate bootstrap values z^* from the maximum likelihood estimate of the sampling density $N(z, 1)$.

There are three ingredients that make this correspondence work:

1. The choice of noninformative prior for θ.

2. The dependence of the log-likelihood $\ell(\theta; \mathbf{Z})$ on the data $\mathbf{Z}$ only through the maximum likelihood estimate $\hat{\theta}$. Hence we can write the log-likelihood as $\ell(\theta; \hat{\theta})$.

3. The symmetry of the log-likelihood in θ and $\hat{\theta}$, that is, $\ell(\theta; \hat{\theta}) = \ell(\hat{\theta}; \theta) + \text{constant}$.

Properties (2) and (3) essentially only hold for the Gaussian distribution. However, they also hold approximately for the multinomial distribution, leading to a correspondence between the nonparametric bootstrap and Bayes inference, which we outline next.

Assume that we have a discrete sample space with L categories. Let w_j be the probability that a sample point falls in category j, and $\hat{w}_j$ the observed proportion in category j. Let $w = (w_1, w_2, \ldots, w_L)$, $\hat{w} = (\hat{w}_1, \hat{w}_2, \ldots, \hat{w}_L)$. Denote our estimator by $S(\hat{w})$; take as a prior distribution for w a symmetric Dirichlet distribution with parameter a:

$$w \sim \text{Di}_L(a1), \tag{8.32}$$

that is, the prior probability mass function is proportional to $\prod_{\ell=1}^{L} w_{\ell}^{a-1}$. Then the posterior density of w is

$$w \sim \mathrm{Di}_L(a1 + N\hat{w}), \tag{8.33}$$

where N is the sample size. Letting $a \to 0$ to obtain a noninformative prior gives

$$w \sim \mathrm{Di}_L(N\hat{w}). \tag{8.34}$$

Now the bootstrap distribution, obtained by sampling with replacement from the data, can be expressed as sampling the category proportions from a multinomial distribution. Specifically,

$$N\hat{w}^* \sim \mathrm{Mult}(N, \hat{w}), \tag{8.35}$$

where $\mathrm{Mult}(N, \hat{w})$ denotes a multinomial distribution, having probability mass function $\binom{N}{N\hat{w}_1^*, \ldots, N\hat{w}_L^*} \prod \hat{w}_\ell^{N\hat{w}_\ell^*}$. This distribution is similar to the posterior distribution above, having the same support, same mean, and nearly the same covariance matrix. Hence the bootstrap distribution of $S(\hat{w}^*)$ will closely approximate the posterior distribution of $S(w)$.

In this sense, the bootstrap distribution represents an (approximate) nonparametric, noninformative posterior distribution for our parameter. But this bootstrap distribution is obtained painlessly—without having to formally specify a prior and without having to sample from the posterior distribution. Hence we might think of the bootstrap distribution as a "poor man's" Bayes posterior. By perturbing the data, the bootstrap approximates the Bayesian effect of perturbing the parameters, and is typically much simpler to carry out.

8.5 The EM Algorithm

The EM algorithm is a popular tool for simplifying difficult maximum likelihood problems. We first describe it in the context of a simple mixture model.

8.5.1 *Two-Component Mixture Model*

In this section we describe a simple mixture model for density estimation, and the associated EM algorithm for carrying out maximum likelihood estimation. This has a natural connection to Gibbs sampling methods for Bayesian inference. Mixture models are discussed and demonstrated in several other parts of the book, in particular Sections 6.8, 12.7 and 13.2.3.

The left panel of Figure 8.5 shows a histogram of the 20 fictitious data points in Table 8.1.

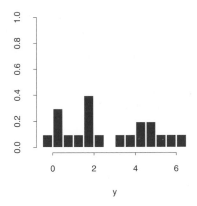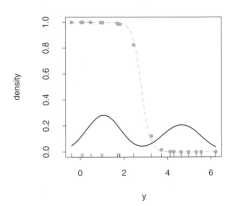

FIGURE 8.5. *Mixture example. (Left panel:) Histogram of data. (Right panel:) Maximum likelihood fit of Gaussian densities (solid red) and responsibility (dotted green) of the left component density for observation y, as a function of y.*

TABLE 8.1. *Twenty fictitious data points used in the two-component mixture example in Figure 8.5.*

-0.39	0.12	0.94	1.67	1.76	2.44	3.72	4.28	4.92	5.53
0.06	0.48	1.01	1.68	1.80	3.25	4.12	4.60	5.28	6.22

We would like to model the density of the data points, and due to the apparent bi-modality, a Gaussian distribution would not be appropriate. There seems to be two separate underlying regimes, so instead we model Y as a mixture of two normal distributions:

$$
\begin{aligned}
Y_1 &\sim N(\mu_1, \sigma_1^2), \\
Y_2 &\sim N(\mu_2, \sigma_2^2), \\
Y &= (1 - \Delta) \cdot Y_1 + \Delta \cdot Y_2,
\end{aligned}
\tag{8.36}
$$

where $\Delta \in \{0, 1\}$ with $\Pr(\Delta = 1) = \pi$. This *generative* representation is explicit: generate a $\Delta \in \{0, 1\}$ with probability π, and then depending on the outcome, deliver either Y_1 or Y_2. Let $\phi_\theta(x)$ denote the normal density with parameters $\theta = (\mu, \sigma^2)$. Then the density of Y is

$$
g_Y(y) = (1 - \pi)\phi_{\theta_1}(y) + \pi\phi_{\theta_2}(y).
\tag{8.37}
$$

Now suppose we wish to fit this model to the data in Figure 8.5 by maximum likelihood. The parameters are

$$
\theta = (\pi, \theta_1, \theta_2) = (\pi, \mu_1, \sigma_1^2, \mu_2, \sigma_2^2).
\tag{8.38}
$$

The log-likelihood based on the N training cases is

$$
\ell(\theta; \mathbf{Z}) = \sum_{i=1}^{N} \log[(1 - \pi)\phi_{\theta_1}(y_i) + \pi\phi_{\theta_2}(y_i)].
\tag{8.39}
$$

Direct maximization of $\ell(\theta; \mathbf{Z})$ is quite difficult numerically, because of the sum of terms inside the logarithm. There is, however, a simpler approach. We consider unobserved latent variables Δ_i taking values 0 or 1 as in (8.36): if $\Delta_i = 1$ then Y_i comes from model 2, otherwise it comes from model 1. Suppose we knew the values of the Δ_i's. Then the log-likelihood would be

$$\ell_0(\theta; \mathbf{Z}, \mathbf{\Delta}) = \sum_{i=1}^{N} [(1 - \Delta_i) \log \phi_{\theta_1}(y_i) + \Delta_i \log \phi_{\theta_2}(y_i)]$$

$$+ \sum_{i=1}^{N} [(1 - \Delta_i) \log(1 - \pi) + \Delta_i \log \pi], \quad (8.40)$$

and the maximum likelihood estimates of μ_1 and σ_1^2 would be the sample mean and variance for those data with $\Delta_i = 0$, and similarly those for μ_2 and σ_2^2 would be the sample mean and variance of the data with $\Delta_i = 1$. The estimate of π would be the proportion of $\Delta_i = 1$.

Since the values of the Δ_i's are actually unknown, we proceed in an iterative fashion, substituting for each Δ_i in (8.40) its expected value

$$\gamma_i(\theta) = \mathrm{E}(\Delta_i | \theta, \mathbf{Z}) = \mathrm{Pr}(\Delta_i = 1 | \theta, \mathbf{Z}), \quad (8.41)$$

also called the *responsibility* of model 2 for observation i. We use a procedure called the EM algorithm, given in Algorithm 8.1 for the special case of Gaussian mixtures. In the *expectation* step, we do a soft assignment of each observation to each model: the current estimates of the parameters are used to assign responsibilities according to the relative density of the training points under each model. In the *maximization* step, these responsibilities are used in weighted maximum-likelihood fits to update the estimates of the parameters.

A good way to construct initial guesses for $\hat{\mu}_1$ and $\hat{\mu}_2$ is simply to choose two of the y_i at random. Both $\hat{\sigma}_1^2$ and $\hat{\sigma}_2^2$ can be set equal to the overall sample variance $\sum_{i=1}^{N} (y_i - \bar{y})^2 / N$. The mixing proportion $\hat{\pi}$ can be started at the value 0.5.

Note that the actual maximizer of the likelihood occurs when we put a spike of infinite height at any one data point, that is, $\hat{\mu}_1 = y_i$ for some i and $\hat{\sigma}_1^2 = 0$. This gives infinite likelihood, but is not a useful solution. Hence we are actually looking for a good local maximum of the likelihood, one for which $\hat{\sigma}_1^2, \hat{\sigma}_2^2 > 0$. To further complicate matters, there can be more than one local maximum having $\hat{\sigma}_1^2, \hat{\sigma}_2^2 > 0$. In our example, we ran the EM algorithm with a number of different initial guesses for the parameters, all having $\hat{\sigma}_k^2 > 0.5$, and chose the run that gave us the highest maximized likelihood. Figure 8.6 shows the progress of the EM algorithm in maximizing the log-likelihood. Table 8.2 shows $\hat{\pi} = \sum_i \hat{\gamma}_i / N$, the maximum likelihood estimate of the proportion of observations in class 2, at selected iterations of the EM procedure.

Algorithm 8.1 *EM Algorithm for Two-component Gaussian Mixture.*

1. Take initial guesses for the parameters $\hat{\mu}_1, \hat{\sigma}_1^2, \hat{\mu}_2, \hat{\sigma}_2^2, \hat{\pi}$ (see text).

2. *Expectation Step*: compute the responsibilities

$$\hat{\gamma}_i = \frac{\hat{\pi}\phi_{\hat{\theta}_2}(y_i)}{(1-\hat{\pi})\phi_{\hat{\theta}_1}(y_i) + \hat{\pi}\phi_{\hat{\theta}_2}(y_i)}, \quad i = 1, 2, \ldots, N. \quad (8.42)$$

3. *Maximization Step*: compute the weighted means and variances:

$$\hat{\mu}_1 = \frac{\sum_{i=1}^N (1-\hat{\gamma}_i)y_i}{\sum_{i=1}^N (1-\hat{\gamma}_i)}, \qquad \hat{\sigma}_1^2 = \frac{\sum_{i=1}^N (1-\hat{\gamma}_i)(y_i-\hat{\mu}_1)^2}{\sum_{i=1}^N (1-\hat{\gamma}_i)},$$

$$\hat{\mu}_2 = \frac{\sum_{i=1}^N \hat{\gamma}_i y_i}{\sum_{i=1}^N \hat{\gamma}_i}, \qquad \hat{\sigma}_2^2 = \frac{\sum_{i=1}^N \hat{\gamma}_i(y_i-\hat{\mu}_2)^2}{\sum_{i=1}^N \hat{\gamma}_i},$$

and the mixing probability $\hat{\pi} = \sum_{i=1}^N \hat{\gamma}_i/N$.

4. Iterate steps 2 and 3 until convergence.

TABLE 8.2. *Selected iterations of the EM algorithm for mixture example.*

Iteration	$\hat{\pi}$
1	0.485
5	0.493
10	0.523
15	0.544
20	0.546

The final maximum likelihood estimates are

$$\hat{\mu}_1 = 4.62, \qquad\qquad \hat{\sigma}_1^2 = 0.87,$$
$$\hat{\mu}_2 = 1.06, \qquad\qquad \hat{\sigma}_2^2 = 0.77,$$
$$\hat{\pi} = 0.546.$$

The right panel of Figure 8.5 shows the estimated Gaussian mixture density from this procedure (solid red curve), along with the responsibilities (dotted green curve). Note that mixtures are also useful for supervised learning; in Section 6.7 we show how the Gaussian mixture model leads to a version of radial basis functions.

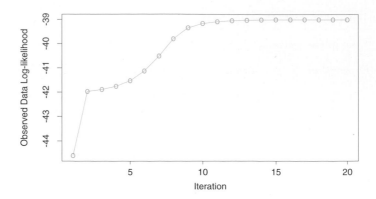

FIGURE 8.6. *EM algorithm: observed data log-likelihood as a function of the iteration number.*

8.5.2   The EM Algorithm in General

The above procedure is an example of the EM (or Baum–Welch) algorithm for maximizing likelihoods in certain classes of problems. These problems are ones for which maximization of the likelihood is difficult, but made easier by enlarging the sample with latent (unobserved) data. This is called *data augmentation*. Here the latent data are the model memberships Δ_i. In other problems, the latent data are actual data that should have been observed but are missing.

Algorithm 8.2 gives the general formulation of the EM algorithm. Our observed data is $\mathbf{Z}$, having log-likelihood $\ell(\theta; \mathbf{Z})$ depending on parameters θ. The latent or missing data is $\mathbf{Z}^m$, so that the complete data is $\mathbf{T} = (\mathbf{Z}, \mathbf{Z}^m)$ with log-likelihood $\ell_0(\theta; \mathbf{T})$, ℓ_0 based on the complete density. In the mixture problem $(\mathbf{Z}, \mathbf{Z}^m) = (\mathbf{y}, \Delta)$, and $\ell_0(\theta; \mathbf{T})$ is given in (8.40).

In our mixture example, $\mathrm{E}(\ell_0(\theta'; \mathbf{T})|\mathbf{Z}, \hat{\theta}^{(j)})$ is simply (8.40) with the Δ_i replaced by the responsibilities $\hat{\gamma}_i(\hat{\theta})$, and the maximizers in step 3 are just weighted means and variances.

We now give an explanation of why the EM algorithm works in general. Since

$$\Pr(\mathbf{Z}^m|\mathbf{Z}, \theta') = \frac{\Pr(\mathbf{Z}^m, \mathbf{Z}|\theta')}{\Pr(\mathbf{Z}|\theta')}, \tag{8.44}$$

we can write

$$\Pr(\mathbf{Z}|\theta') = \frac{\Pr(\mathbf{T}|\theta')}{\Pr(\mathbf{Z}^m|\mathbf{Z}, \theta')}. \tag{8.45}$$

In terms of log-likelihoods, we have $\ell(\theta'; \mathbf{Z}) = \ell_0(\theta'; \mathbf{T}) - \ell_1(\theta'; \mathbf{Z}^m|\mathbf{Z})$, where ℓ_1 is based on the conditional density $\Pr(\mathbf{Z}^m|\mathbf{Z}, \theta')$. Taking conditional expectations with respect to the distribution of $\mathbf{T}|\mathbf{Z}$ governed by parameter θ gives

$$\ell(\theta'; \mathbf{Z}) \;=\; \mathrm{E}[\ell_0(\theta'; \mathbf{T})|\mathbf{Z}, \theta] - \mathrm{E}[\ell_1(\theta'; \mathbf{Z}^m|\mathbf{Z})|\mathbf{Z}, \theta]$$

Algorithm 8.2 *The EM Algorithm.*

---

1. Start with initial guesses for the parameters $\hat{\theta}^{(0)}$.

2. *Expectation Step*: at the jth step, compute

$$Q(\theta', \hat{\theta}^{(j)}) = \mathrm{E}(\ell_0(\theta'; \mathbf{T}) | \mathbf{Z}, \hat{\theta}^{(j)}) \tag{8.43}$$

   as a function of the dummy argument θ'.

3. *Maximization Step*: determine the new estimate $\hat{\theta}^{(j+1)}$ as the maximizer of $Q(\theta', \hat{\theta}^{(j)})$ over θ'.

4. Iterate steps 2 and 3 until convergence.

$$\equiv \quad Q(\theta', \theta) - R(\theta', \theta). \tag{8.46}$$

In the M step, the EM algorithm maximizes $Q(\theta', \theta)$ over θ', rather than the actual objective function $\ell(\theta'; \mathbf{Z})$. Why does it succeed in maximizing $\ell(\theta'; \mathbf{Z})$? Note that $R(\theta^*, \theta)$ is the expectation of a log-likelihood of a density (indexed by θ^*), with respect to the same density indexed by θ, and hence (by Jensen's inequality) is maximized as a function of θ^*, when $\theta^* = \theta$ (see Exercise 8.1). So if θ' maximizes $Q(\theta', \theta)$, we see that

$$\begin{aligned} \ell(\theta'; \mathbf{Z}) - \ell(\theta; \mathbf{Z}) &= [Q(\theta', \theta) - Q(\theta, \theta)] - [R(\theta', \theta) - R(\theta, \theta)] \\ &\geq 0. \end{aligned} \tag{8.47}$$

Hence the EM iteration never decreases the log-likelihood.

This argument also makes it clear that a full maximization in the M step is not necessary: we need only to find a value $\hat{\theta}^{(j+1)}$ so that $Q(\theta', \hat{\theta}^{(j)})$ increases as a function of the first argument, that is, $Q(\hat{\theta}^{(j+1)}, \hat{\theta}^{(j)}) > Q(\hat{\theta}^{(j)}, \hat{\theta}^{(j)})$. Such procedures are called *GEM (generalized EM)* algorithms. The EM algorithm can also be viewed as a minorization procedure: see Exercise 8.7.

8.5.3 EM as a Maximization–Maximization Procedure

Here is a different view of the EM procedure, as a joint maximization algorithm. Consider the function

$$F(\theta', \tilde{P}) = \mathrm{E}_{\tilde{P}}[\ell_0(\theta'; \mathbf{T})] - \mathrm{E}_{\tilde{P}}[\log \tilde{P}(\mathbf{Z}^m)]. \tag{8.48}$$

Here $\tilde{P}(\mathbf{Z}^m)$ is any distribution over the latent data $\mathbf{Z}^m$. In the mixture example, $\tilde{P}(\mathbf{Z}^m)$ comprises the set of probabilities $\gamma_i = \mathrm{Pr}(\Delta_i = 1 | \theta, \mathbf{Z})$. Note that F evaluated at $\tilde{P}(\mathbf{Z}^m) = \mathrm{Pr}(\mathbf{Z}^m | \mathbf{Z}, \theta')$, is the log-likelihood of

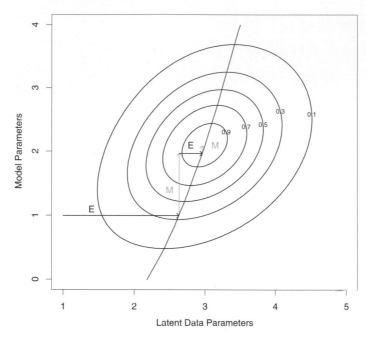

FIGURE 8.7. *Maximization–maximization view of the EM algorithm. Shown are the contours of the (augmented) observed data log-likelihood $F(\theta', \tilde{P})$. The E step is equivalent to maximizing the log-likelihood over the parameters of the latent data distribution. The M step maximizes it over the parameters of the log-likelihood. The red curve corresponds to the observed data log-likelihood, a profile obtained by maximizing $F(\theta', \tilde{P})$ for each value of θ'.*

the observed data, from $(8.46)^1$. The function F expands the domain of the log-likelihood, to facilitate its maximization.

The EM algorithm can be viewed as a joint maximization method for F over θ' and $\tilde{P}(\mathbf{Z}^m)$, by fixing one argument and maximizing over the other. The maximizer over $\tilde{P}(\mathbf{Z}^m)$ for fixed θ' can be shown to be

$$\tilde{P}(\mathbf{Z}^m) = \Pr(\mathbf{Z}^m|\mathbf{Z}, \theta') \tag{8.49}$$

(Exercise 8.2). This is the distribution computed by the E step, for example, (8.42) in the mixture example. In the M step, we maximize $F(\theta', \tilde{P})$ over θ' with $\tilde{P}$ fixed: this is the same as maximizing the first term $\mathrm{E}_{\tilde{P}}[\ell_0(\theta'; \mathbf{T})|\mathbf{Z}, \theta]$ since the second term does not involve θ'.

Finally, since $F(\theta', \tilde{P})$ and the observed data log-likelihood agree when $\tilde{P}(\mathbf{Z}^m) = \Pr(\mathbf{Z}^m|\mathbf{Z}, \theta')$, maximization of the former accomplishes maximization of the latter. Figure 8.7 shows a schematic view of this process. This view of the EM algorithm leads to alternative maximization proce-

---

[1] (8.46) holds for all θ, including $\theta = \theta'$.

Algorithm 8.3 *Gibbs Sampler.*

---

1. Take some initial values $U_k^{(0)}, k = 1, 2, \ldots, K$.

2. Repeat for $t = 1, 2, \ldots, . :$

   For $k = 1, 2, \ldots, K$ generate $U_k^{(t)}$ from
   $$\Pr(U_k^{(t)}|U_1^{(t)}, \ldots, U_{k-1}^{(t)}, U_{k+1}^{(t-1)}, \ldots, U_K^{(t-1)}).$$

3. Continue step 2 until the joint distribution of $(U_1^{(t)}, U_2^{(t)}, \ldots, U_K^{(t)})$ *does not change.*

---

dures. For example, one does not need to maximize with respect to all of the latent data parameters at once, but could instead maximize over one of them at a time, alternating with the M step.

8.6   MCMC for Sampling from the Posterior

Having defined a Bayesian model, one would like to draw samples from the resulting posterior distribution, in order to make inferences about the parameters. Except for simple models, this is often a difficult computational problem. In this section we discuss the *Markov chain Monte Carlo* (MCMC) approach to posterior sampling. We will see that Gibbs sampling, an MCMC procedure, is closely related to the EM algorithm: the main difference is that it samples from the conditional distributions rather than maximizing over them.

Consider first the following abstract problem. We have random variables $U_1, U_2, \ldots, U_K$ and we wish to draw a sample from their joint distribution. Suppose this is difficult to do, but it is easy to simulate from the conditional distributions $\Pr(U_j|U_1, U_2, \ldots, U_{j-1}, U_{j+1}, \ldots, U_K)$, $j = 1, 2, \ldots, K$. The Gibbs sampling procedure alternatively simulates from each of these distributions and when the process stabilizes, provides a sample from the desired joint distribution. The procedure is defined in Algorithm 8.3.

Under regularity conditions it can be shown that this procedure eventually stabilizes, and the resulting random variables are indeed a sample from the joint distribution of $U_1, U_2, \ldots, U_K$. This occurs despite the fact that the samples $(U_1^{(t)}, U_2^{(t)}, \ldots, U_K^{(t)})$ are clearly not independent for different t. More formally, Gibbs sampling produces a Markov chain whose stationary distribution is the true joint distribution, and hence the term "Markov chain Monte Carlo." It is not surprising that the true joint distribution is stationary under this process, as the successive steps leave the marginal distributions of the U_k's unchanged.

Note that we don't need to know the explicit form of the conditional densities, but just need to be able to sample from them. After the procedure reaches stationarity, the marginal density of any subset of the variables can be approximated by a density estimate applied to the sample values. However if the explicit form of the conditional density $\Pr(U_k, |U_\ell, \ell \neq k)$ is available, a better estimate of say the marginal density of U_k can be obtained from (Exercise 8.3):

$$\widehat{\Pr}_{U_k}(u) = \frac{1}{(M - m + 1)} \sum_{t=m}^{M} \Pr(u|U_\ell^{(t)}, \ell \neq k). \tag{8.50}$$

Here we have averaged over the last $M - m + 1$ members of the sequence, to allow for an initial "burn-in" period before stationarity is reached.

Now getting back to Bayesian inference, our goal is to draw a sample from the joint posterior of the parameters given the data $\mathbf{Z}$. Gibbs sampling will be helpful if it is easy to sample from the conditional distribution of each parameter given the other parameters and $\mathbf{Z}$. An example—the Gaussian mixture problem—is detailed next.

There is a close connection between Gibbs sampling from a posterior and the EM algorithm in exponential family models. The key is to consider the latent data $\mathbf{Z}^m$ from the EM procedure to be another parameter for the Gibbs sampler. To make this explicit for the Gaussian mixture problem, we take our parameters to be $(\theta, \mathbf{Z}^m)$. For simplicity we fix the variances σ_1^2, σ_2^2 and mixing proportion π at their maximum likelihood values so that the only unknown parameters in θ are the means μ_1 and μ_2. The Gibbs sampler for the mixture problem is given in Algorithm 8.4. We see that steps 2(a) and 2(b) are the same as the E and M steps of the EM procedure, except that we sample rather than maximize. In step 2(a), rather than compute the maximum likelihood responsibilities $\gamma_i = \mathrm{E}(\Delta_i|\theta, \mathbf{Z})$, the Gibbs sampling procedure simulates the latent data Δ_i from the distributions $\Pr(\Delta_i|\theta, \mathbf{Z})$. In step 2(b), rather than compute the maximizers of the posterior $\Pr(\mu_1, \mu_2, \boldsymbol{\Delta}|\mathbf{Z})$ we simulate from the conditional distribution $\Pr(\mu_1, \mu_2|\boldsymbol{\Delta}, \mathbf{Z})$.

Figure 8.8 shows 200 iterations of Gibbs sampling, with the mean parameters μ_1 (lower) and μ_2 (upper) shown in the left panel, and the proportion of class 2 observations $\sum_i \Delta_i/N$ on the right. Horizontal broken lines have been drawn at the maximum likelihood estimate values $\hat{\mu}_1, \hat{\mu}_2$ and $\sum_i \hat{\gamma}_i/N$ in each case. The values seem to stabilize quite quickly, and are distributed evenly around the maximum likelihood values.

The above mixture model was simplified, in order to make the clear connection between Gibbs sampling and the EM algorithm. More realistically, one would put a prior distribution on the variances σ_1^2, σ_2^2 and mixing proportion π, and include separate Gibbs sampling steps in which we sample from their posterior distributions, conditional on the other parameters. One can also incorporate proper (informative) priors for the mean param-

---

Algorithm 8.4 *Gibbs sampling for mixtures.*

---

1. Take some initial values $\theta^{(0)} = (\mu_1^{(0)}, \mu_2^{(0)})$.

2. Repeat for $t = 1, 2, \ldots,$.

   (a) For $i = 1, 2, \ldots, N$ generate $\Delta_i^{(t)} \in \{0, 1\}$ with $\Pr(\Delta_i^{(t)} = 1) = \hat{\gamma}_i(\theta^{(t)})$, from equation (8.42).

   (b) Set

   $$\hat{\mu}_1 = \frac{\sum_{i=1}^{N}(1 - \Delta_i^{(t)}) \cdot y_i}{\sum_{i=1}^{N}(1 - \Delta_i^{(t)})},$$

   $$\hat{\mu}_2 = \frac{\sum_{i=1}^{N} \Delta_i^{(t)} \cdot y_i}{\sum_{i=1}^{N} \Delta_i^{(t)}},$$

   and generate $\mu_1^{(t)} \sim N(\hat{\mu}_1, \hat{\sigma}_1^2)$ and $\mu_2^{(t)} \sim N(\hat{\mu}_2, \hat{\sigma}_2^2)$.

3. Continue step 2 until the joint distribution of $(\boldsymbol{\Delta}^{(t)}, \mu_1^{(t)}, \mu_2^{(t)})$ doesn't change

---

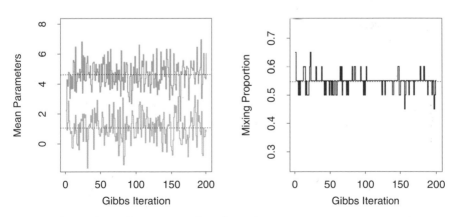

FIGURE 8.8. *Mixture example. (Left panel:) 200 values of the two mean parameters from Gibbs sampling; horizontal lines are drawn at the maximum likelihood estimates $\hat{\mu}_1$, $\hat{\mu}_2$. (Right panel:) Proportion of values with $\Delta_i = 1$, for each of the 200 Gibbs sampling iterations; a horizontal line is drawn at $\sum_i \hat{\gamma}_i/N$.*

eters. These priors must not be improper as this will lead to a degenerate posterior, with all the mixing weight on one component.

Gibbs sampling is just one of a number of recently developed procedures for sampling from posterior distributions. It uses conditional sampling of each parameter given the rest, and is useful when the structure of the problem makes this sampling easy to carry out. Other methods do not require such structure, for example the *Metropolis–Hastings* algorithm. These and other computational Bayesian methods have been applied to sophisticated learning algorithms such as Gaussian process models and neural networks. Details may be found in the references given in the Bibliographic Notes at the end of this chapter.

8.7 Bagging

Earlier we introduced the bootstrap as a way of assessing the accuracy of a parameter estimate or a prediction. Here we show how to use the bootstrap to improve the estimate or prediction itself. In Section 8.4 we investigated the relationship between the bootstrap and Bayes approaches, and found that the bootstrap mean is approximately a posterior average. Bagging further exploits this connection.

Consider first the regression problem. Suppose we fit a model to our training data $\mathbf{Z} = \{(x_1, y_1), (x_2, y_2), \ldots, (x_N, y_N)\}$, obtaining the prediction $\hat{f}(x)$ at input x. Bootstrap aggregation or *bagging* averages this prediction over a collection of bootstrap samples, thereby reducing its variance. For each bootstrap sample $\mathbf{Z}^{*b}$, $b = 1, 2, \ldots, B$, we fit our model, giving prediction $\hat{f}^{*b}(x)$. The bagging estimate is defined by

$$\hat{f}_{\text{bag}}(x) = \frac{1}{B} \sum_{b=1}^{B} \hat{f}^{*b}(x). \tag{8.51}$$

Denote by $\hat{\mathcal{P}}$ the empirical distribution putting equal probability $1/N$ on each of the data points (x_i, y_i). In fact the "true" bagging estimate is defined by $E_{\hat{\mathcal{P}}} \hat{f}^*(x)$, where $\mathbf{Z}^* = \{(x_1^*, y_1^*), (x_2^*, y_2^*), \ldots, (x_N^*, y_N^*)\}$ and each $(x_i^*, y_i^*) \sim \hat{\mathcal{P}}$. Expression (8.51) is a Monte Carlo estimate of the true bagging estimate, approaching it as $B \to \infty$.

The bagged estimate (8.51) will differ from the original estimate $\hat{f}(x)$ only when the latter is a nonlinear or adaptive function of the data. For example, to bag the B-spline smooth of Section 8.2.1, we average the curves in the bottom left panel of Figure 8.2 at each value of x. The B-spline smoother is linear in the data if we fix the inputs; hence if we sample using the parametric bootstrap in equation (8.6), then $\hat{f}_{\text{bag}}(x) \to \hat{f}(x)$ as $B \to \infty$ (Exercise 8.4). Hence bagging just reproduces the original smooth in the top left panel of Figure 8.2. The same is approximately true if we were to bag using the nonparametric bootstrap.

A more interesting example is a regression tree, where $\hat{f}(x)$ denotes the tree's prediction at input vector x (regression trees are described in Chapter 9). Each bootstrap tree will typically involve different features than the original, and might have a different number of terminal nodes. The bagged estimate is the average prediction at x from these B trees.

Now suppose our tree produces a classifier $\hat{G}(x)$ for a K-class response. Here it is useful to consider an underlying indicator-vector function $\hat{f}(x)$, with value a single one and $K-1$ zeroes, such that $\hat{G}(x) = \arg\max_k \hat{f}(x)$. Then the bagged estimate $\hat{f}_{\text{bag}}(x)$ (8.51) is a K-vector $[p_1(x), p_2(x), \ldots, p_K(x)]$, with $p_k(x)$ equal to the proportion of trees predicting class k at x. The bagged classifier selects the class with the most "votes" from the B trees, $\hat{G}_{\text{bag}}(x) = \arg\max_k \hat{f}_{\text{bag}}(x)$.

Often we require the class-probability estimates at x, rather than the classifications themselves. It is tempting to treat the voting proportions $p_k(x)$ as estimates of these probabilities. A simple two-class example shows that they fail in this regard. Suppose the true probability of class 1 at x is 0.75, and each of the bagged classifiers accurately predict a 1. Then $p_1(x) = 1$, which is incorrect. For many classifiers $\hat{G}(x)$, however, there is already an underlying function $\hat{f}(x)$ that estimates the class probabilities at x (for trees, the class proportions in the terminal node). An alternative bagging strategy is to average these instead, rather than the vote indicator vectors. Not only does this produce improved estimates of the class probabilities, but it also tends to produce bagged classifiers with lower variance, especially for small B (see Figure 8.10 in the next example).

8.7.1 Example: Trees with Simulated Data

We generated a sample of size $N = 30$, with two classes and $p = 5$ features, each having a standard Gaussian distribution with pairwise correlation 0.95. The response Y was generated according to $\Pr(Y = 1|x_1 \leq 0.5) = 0.2$, $\Pr(Y = 1|x_1 > 0.5) = 0.8$. The Bayes error is 0.2. A test sample of size 2000 was also generated from the same population. We fit classification trees to the training sample and to each of 200 bootstrap samples (classification trees are described in Chapter 9). No pruning was used. Figure 8.9 shows the original tree and eleven bootstrap trees. Notice how the trees are all different, with different splitting features and cutpoints. The test error for the original tree and the bagged tree is shown in Figure 8.10. In this example the trees have high variance due to the correlation in the predictors. Bagging succeeds in smoothing out this variance and hence reducing the test error.

Bagging can dramatically reduce the variance of unstable procedures like trees, leading to improved prediction. A simple argument shows why

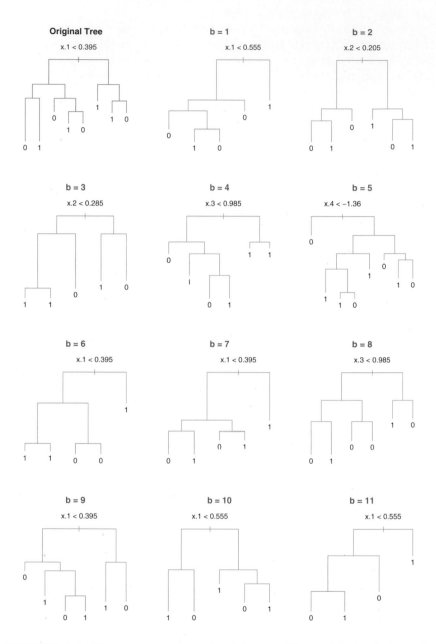

FIGURE 8.9. *Bagging trees on simulated dataset. The top left panel shows the original tree. Eleven trees grown on bootstrap samples are shown. For each tree, the top split is annotated.*

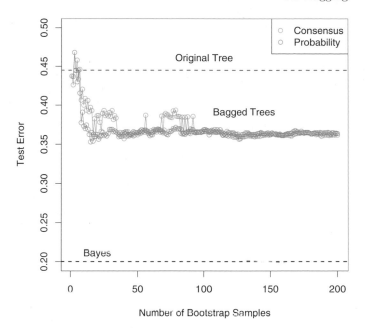

FIGURE 8.10. *Error curves for the bagging example of Figure 8.9. Shown is the test error of the original tree and bagged trees as a function of the number of bootstrap samples. The orange points correspond to the consensus vote, while the green points average the probabilities.*

bagging helps under squared-error loss, in short because averaging reduces variance and leaves bias unchanged.

Assume our training observations (x_i, y_i), $i = 1, \ldots, N$ are independently drawn from a distribution $\mathcal{P}$, and consider the ideal aggregate estimator $f_{\mathrm{ag}}(x) = \mathrm{E}_{\mathcal{P}} \hat{f}^*(x)$. Here x is fixed and the bootstrap dataset $\mathbf{Z}^*$ consists of observations x_i^*, y_i^*, $i = 1, 2, \ldots, N$ sampled from $\mathcal{P}$. Note that $f_{\mathrm{ag}}(x)$ is a bagging estimate, drawing bootstrap samples from the actual population $\mathcal{P}$ rather than the data. It is not an estimate that we can use in practice, but is convenient for analysis. We can write

$$
\begin{aligned}
\mathrm{E}_{\mathcal{P}}[Y - \hat{f}^*(x)]^2 &= \mathrm{E}_{\mathcal{P}}[Y - f_{\mathrm{ag}}(x) + f_{\mathrm{ag}}(x) - \hat{f}^*(x)]^2 \\
&= \mathrm{E}_{\mathcal{P}}[Y - f_{\mathrm{ag}}(x)]^2 + \mathrm{E}_{\mathcal{P}}[\hat{f}^*(x) - f_{\mathrm{ag}}(x)]^2 \\
&\geq \mathrm{E}_{\mathcal{P}}[Y - f_{\mathrm{ag}}(x)]^2.
\end{aligned}
\tag{8.52}
$$

The extra error on the right-hand side comes from the variance of $\hat{f}^*(x)$ around its mean $f_{\mathrm{ag}}(x)$. Therefore true population aggregation never increases mean squared error. This suggests that bagging—drawing samples from the training data— will often decrease mean-squared error.

The above argument does not hold for classification under 0-1 loss, because of the nonadditivity of bias and variance. In that setting, bagging a good classifier can make it better, but bagging a bad classifier can make it worse. Here is a simple example, using a randomized rule. Suppose $Y = 1$ for all x, and the classifier $\hat{G}(x)$ predicts $Y = 1$ (for all x) with probability 0.4 and predicts $Y = 0$ (for all x) with probability 0.6. Then the misclassification error of $\hat{G}(x)$ is 0.6 but that of the bagged classifier is 1.0.

For classification we can understand the bagging effect in terms of a consensus of independent *weak learners* (Dieterich, 2000a). Let the Bayes optimal decision at x be $G(x) = 1$ in a two-class example. Suppose each of the weak learners G_b^* have an error-rate $e_b = e < 0.5$, and let $S_1(x) = \sum_{b=1}^{B} I(G_b^*(x) = 1)$ be the consensus vote for class 1. Since the weak learners are assumed to be independent, $S_1(x) \sim \text{Bin}(B, 1 - e)$, and $\Pr(S_1 > B/2) \to 1$ as B gets large. This concept has been popularized outside of statistics as the "Wisdom of Crowds" (Surowiecki, 2004) — the collective knowledge of a diverse and independent body of people typically exceeds the knowledge of any single individual, and can be harnessed by voting. Of course, the main caveat here is "independent," and bagged trees are not. Figure 8.11 illustrates the power of a consensus vote in a simulated example, where only 30% of the voters have some knowledge.

In Chapter 15 we see how random forests improve on bagging by reducing the correlation between the sampled trees.

Note that when we bag a model, any simple structure in the model is lost. As an example, a bagged tree is no longer a tree. For interpretation of the model this is clearly a drawback. More stable procedures like nearest neighbors are typically not affected much by bagging. Unfortunately, the unstable models most helped by bagging are unstable because of the emphasis on interpretability, and this is lost in the bagging process.

Figure 8.12 shows an example where bagging doesn't help. The 100 data points shown have two features and two classes, separated by the gray linear boundary $x_1 + x_2 = 1$. We choose as our classifier $\hat{G}(x)$ a single axis-oriented split, choosing the split along either x_1 or x_2 that produces the largest decrease in training misclassification error.

The decision boundary obtained from bagging the 0-1 decision rule over $B = 50$ bootstrap samples is shown by the blue curve in the left panel. It does a poor job of capturing the true boundary. The single split rule, derived from the training data, splits near 0 (the middle of the range of x_1 or x_2), and hence has little contribution away from the center. Averaging the probabilities rather than the classifications does not help here. Bagging estimates the expected class probabilities from the single split rule, that is, averaged over many replications. Note that the expected class probabilities computed by bagging cannot be realized on any single replication, in the same way that a woman cannot have 2.4 children. In this sense, bagging increases somewhat the space of models of the individual base classifier. However, it doesn't help in this and many other examples where a greater enlargement of the model class is needed. "Boosting" is a way of doing this

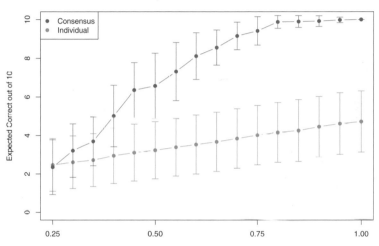

FIGURE 8.11. *Simulated academy awards voting.* 50 *members vote in 10 categories, each with 4 nominations. For any category, only* 15 *voters have some knowledge, represented by their probability of selecting the "correct" candidate in that category (so P = 0.25 means they have no knowledge). For each category, the* 15 *experts are chosen at random from the 50. Results show the expected correct (based on 50 simulations) for the consensus, as well as for the individuals. The error bars indicate one standard deviation. We see, for example, that if the 15 informed for a category have a 50% chance of selecting the correct candidate, the consensus doubles the expected performance of an individual.*

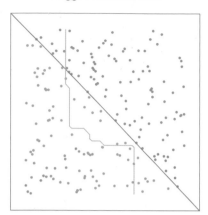
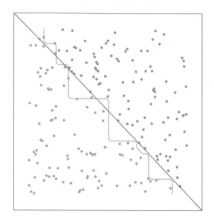

FIGURE 8.12. *Data with two features and two classes, separated by a linear boundary. (Left panel:) Decision boundary estimated from bagging the decision rule from a single split, axis-oriented classifier. (Right panel:) Decision boundary from boosting the decision rule of the same classifier. The test error rates are 0.166, and 0.065, respectively. Boosting is described in Chapter 10.*

and is described in Chapter 10. The decision boundary in the right panel is the result of the boosting procedure, and it roughly captures the diagonal boundary.

8.8 Model Averaging and Stacking

In Section 8.4 we viewed bootstrap values of an estimator as approximate posterior values of a corresponding parameter, from a kind of nonparametric Bayesian analysis. Viewed in this way, the bagged estimate (8.51) is an approximate posterior Bayesian mean. In contrast, the training sample estimate $\hat{f}(x)$ corresponds to the mode of the posterior. Since the posterior mean (not mode) minimizes squared-error loss, it is not surprising that bagging can often reduce mean squared-error.

Here we discuss Bayesian model averaging more generally. We have a set of candidate models $\mathcal{M}_m$, $m = 1, \ldots, M$ for our training set $\mathbf{Z}$. These models may be of the same type with different parameter values (e.g., subsets in linear regression), or different models for the same task (e.g., neural networks and regression trees).

Suppose ζ is some quantity of interest, for example, a prediction $f(x)$ at some fixed feature value x. The posterior distribution of ζ is

$$\Pr(\zeta|\mathbf{Z}) = \sum_{m=1}^{M} \Pr(\zeta|\mathcal{M}_m, \mathbf{Z})\Pr(\mathcal{M}_m|\mathbf{Z}), \qquad (8.53)$$

with posterior mean

$$E(\zeta|\mathbf{Z}) = \sum_{m=1}^{M} E(\zeta|\mathcal{M}_m, \mathbf{Z})\Pr(\mathcal{M}_m|\mathbf{Z}). \qquad (8.54)$$

This Bayesian prediction is a weighted average of the individual predictions, with weights proportional to the posterior probability of each model.

This formulation leads to a number of different model-averaging strategies. *Committee methods* take a simple unweighted average of the predictions from each model, essentially giving equal probability to each model. More ambitiously, the development in Section 7.7 shows the BIC criterion can be used to estimate posterior model probabilities. This is applicable in cases where the different models arise from the same parametric model, with different parameter values. The BIC gives weight to each model depending on how well it fits and how many parameters it uses. One can also carry out the Bayesian recipe in full. If each model $\mathcal{M}_m$ has parameters θ_m, we write

$$\begin{aligned} \Pr(\mathcal{M}_m|\mathbf{Z}) \quad &\propto \quad \Pr(\mathcal{M}_m) \cdot \Pr(\mathbf{Z}|\mathcal{M}_m) \\ &\propto \quad \Pr(\mathcal{M}_m) \cdot \int \Pr(\mathbf{Z}|\theta_m, \mathcal{M}_m)\Pr(\theta_m|\mathcal{M}_m)d\theta_m. \end{aligned} \qquad (8.55)$$

In principle one can specify priors $\Pr(\theta_m|\mathcal{M}_m)$ and numerically compute the posterior probabilities from (8.55), to be used as model-averaging weights. However, we have seen no real evidence that this is worth all of the effort, relative to the much simpler BIC approximation.

How can we approach model averaging from a frequentist viewpoint? Given predictions $\hat{f}_1(x), \hat{f}_2(x), \ldots, \hat{f}_M(x)$, under squared-error loss, we can seek the weights $w = (w_1, w_2, \ldots, w_M)$ such that

$$\hat{w} = \operatorname*{argmin}_{w} E_{\mathcal{P}}\left[Y - \sum_{m=1}^{M} w_m \hat{f}_m(x)\right]^2. \qquad (8.56)$$

Here the input value x is fixed and the N observations in the dataset $\mathbf{Z}$ (and the target Y) are distributed according to $\mathcal{P}$. The solution is the population linear regression of Y on $\hat{F}(x)^T \equiv [\hat{f}_1(x), \hat{f}_2(x), \ldots, \hat{f}_M(x)]$:

$$\hat{w} = E_{\mathcal{P}}[\hat{F}(x)\hat{F}(x)^T]^{-1}E_{\mathcal{P}}[\hat{F}(x)Y]. \qquad (8.57)$$

Now the full regression has smaller error than any single model

$$E_{\mathcal{P}}\left[Y - \sum_{m=1}^{M} \hat{w}_m \hat{f}_m(x)\right]^2 \leq E_{\mathcal{P}}\left[Y - \hat{f}_m(x)\right]^2 \forall m \qquad (8.58)$$

so combining models never makes things worse, at the population level.

Of course the population linear regression (8.57) is not available, and it is natural to replace it with the linear regression over the training set. But there are simple examples where this does not work well. For example, if $\hat{f}_m(x)$, $m = 1, 2, \ldots, M$ represent the prediction from the best subset of inputs of size m among M total inputs, then linear regression would put all of the weight on the largest model, that is, $\hat{w}_M = 1$, $\hat{w}_m = 0$, $m < M$. The problem is that we have not put each of the models on the same footing by taking into account their complexity (the number of inputs m in this example).

Stacked generalization, or *stacking*, is a way of doing this. Let $\hat{f}_m^{-i}(x)$ be the prediction at x, using model m, applied to the dataset with the ith training observation removed. The stacking estimate of the weights is obtained from the least squares linear regression of y_i on $\hat{f}_m^{-i}(x_i)$, $m = 1, 2, \ldots, M$. In detail the stacking weights are given by

$$\hat{w}^{\text{st}} = \underset{w}{\text{argmin}} \sum_{i=1}^{N} \left[y_i - \sum_{m=1}^{M} w_m \hat{f}_m^{-i}(x_i) \right]^2 . \tag{8.59}$$

The final prediction is $\sum_m \hat{w}_m^{\text{st}} \hat{f}_m(x)$. By using the cross-validated predictions $\hat{f}_m^{-i}(x)$, stacking avoids giving unfairly high weight to models with higher complexity. Better results can be obtained by restricting the weights to be nonnegative, and to sum to 1. This seems like a reasonable restriction if we interpret the weights as posterior model probabilities as in equation (8.54), and it leads to a tractable quadratic programming problem.

There is a close connection between stacking and model selection via leave-one-out cross-validation (Section 7.10). If we restrict the minimization in (8.59) to weight vectors w that have one unit weight and the rest zero, this leads to a model choice $\hat{m}$ with smallest leave-one-out cross-validation error. Rather than choose a single model, stacking combines them with estimated optimal weights. This will often lead to better prediction, but less interpretability than the choice of only one of the M models.

The stacking idea is actually more general than described above. One can use any learning method, not just linear regression, to combine the models as in (8.59); the weights could also depend on the input location x. In this way, learning methods are "stacked" on top of one another, to improve prediction performance.

8.9   Stochastic Search: Bumping

The final method described in this chapter does not involve averaging or combining models, but rather is a technique for finding a better single model. *Bumping* uses bootstrap sampling to move randomly through model space. For problems where fitting method finds many local minima, bumping can help the method to avoid getting stuck in poor solutions.

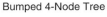

Regular 4-Node Tree   Bumped 4-Node Tree

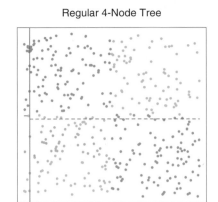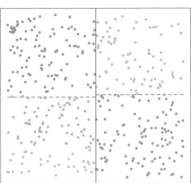

FIGURE 8.13. *Data with two features and two classes (blue and orange), displaying a pure interaction. The left panel shows the partition found by three splits of a standard, greedy, tree-growing algorithm. The vertical grey line near the left edge is the first split, and the broken lines are the two subsequent splits. The algorithm has no idea where to make a good initial split, and makes a poor choice. The right panel shows the near-optimal splits found by bumping the tree-growing algorithm* 20 *times.*

As in bagging, we draw bootstrap samples and fit a model to each. But rather than average the predictions, we choose the model estimated from a bootstrap sample that best fits the training data. In detail, we draw bootstrap samples $\mathbf{Z}^{*1}, \dots, \mathbf{Z}^{*B}$ and fit our model to each, giving predictions $\hat{f}^{*b}(x)$, $b = 1, 2, \dots, B$ at input point x. We then choose the model that produces the smallest prediction error, averaged over the *original training set*. For squared error, for example, we choose the model obtained from bootstrap sample $\hat{b}$, where

$$\hat{b} = \arg\min_b \sum_{i=1}^N [y_i - \hat{f}^{*b}(x_i)]^2. \tag{8.60}$$

The corresponding model predictions are $\hat{f}^{*\hat{b}}(x)$. By convention we also include the original training sample in the set of bootstrap samples, so that the method is free to pick the original model if it has the lowest training error.

By perturbing the data, bumping tries to move the fitting procedure around to good areas of model space. For example, if a few data points are causing the procedure to find a poor solution, any bootstrap sample that omits those data points should procedure a better solution.

For another example, consider the classification data in Figure 8.13, the notorious *exclusive or (XOR)* problem. There are two classes (blue and orange) and two input features, with the features exhibiting a pure interaction. By splitting the data at $x_1 = 0$ and then splitting each resulting strata at $x_2 = 0$, (or vice versa) a tree-based classifier could achieve perfect discrimination. However, the greedy, short-sighted CART algorithm (Section 9.2) tries to find the best split on either feature, and then splits the resulting strata. Because of the balanced nature of the data, all initial splits on x_1 or x_2 appear to be useless, and the procedure essentially generates a random split at the top level. The actual split found for these data is shown in the left panel of Figure 8.13. By bootstrap sampling from the data, bumping breaks the balance in the classes, and with a reasonable number of bootstrap samples (here 20), it will by chance produce at least one tree with initial split near either $x_1 = 0$ or $x_2 = 0$. Using just 20 bootstrap samples, bumping found the near optimal splits shown in the right panel of Figure 8.13. This shortcoming of the greedy tree-growing algorithm is exacerbated if we add a number of noise features that are independent of the class label. Then the tree-growing algorithm cannot distinguish x_1 or x_2 from the others, and gets seriously lost.

Since bumping compares different models on the training data, one must ensure that the models have roughly the same complexity. In the case of trees, this would mean growing trees with the same number of terminal nodes on each bootstrap sample. Bumping can also help in problems where it is difficult to optimize the fitting criterion, perhaps because of a lack of smoothness. The trick is to optimize a different, more convenient criterion over the bootstrap samples, and then choose the model producing the best results for the desired criterion on the training sample.

Bibliographic Notes

There are many books on classical statistical inference: Cox and Hinkley (1974) and Silvey (1975) give nontechnical accounts. The bootstrap is due to Efron (1979) and is described more fully in Efron and Tibshirani (1993) and Hall (1992). A good modern book on Bayesian inference is Gelman et al. (1995). A lucid account of the application of Bayesian methods to neural networks is given in Neal (1996). The statistical application of Gibbs sampling is due to Geman and Geman (1984), and Gelfand and Smith (1990), with related work by Tanner and Wong (1987). Markov chain Monte Carlo methods, including Gibbs sampling and the Metropolis–Hastings algorithm, are discussed in Spiegelhalter et al. (1996). The EM algorithm is due to Dempster et al. (1977); as the discussants in that paper make clear, there was much related, earlier work. The view of EM as a joint maximization scheme for a penalized complete-data log-likelihood was elucidated by Neal and Hinton (1998); they credit Csiszar and Tusnády (1984) and Hathaway (1986) as having noticed this connection earlier. Bagging was proposed by Breiman (1996a). Stacking is due to Wolpert (1992);

Breiman (1996b) contains an accessible discussion for statisticians. Leblanc and Tibshirani (1996) describe variations on stacking based on the bootstrap. Model averaging in the Bayesian framework has been recently advocated by Madigan and Raftery (1994). Bumping was proposed by Tibshirani and Knight (1999).

Exercises

Ex. 8.1 Let $r(y)$ and $q(y)$ be probability density functions. Jensen's inequality states that for a random variable X and a convex function $\phi(x)$, $E[\phi(X)] \geq \phi[E(X)]$. Use Jensen's inequality to show that

$$E_q \log[r(Y)/q(Y)] \tag{8.61}$$

is maximized as a function of $r(y)$ when $r(y) = q(y)$. Hence show that $R(\theta, \theta) \geq R(\theta', \theta)$ as stated below equation (8.46).

Ex. 8.2 Consider the maximization of the log-likelihood (8.48), over distributions $\tilde{P}(\mathbf{Z}^m)$ such that $\tilde{P}(\mathbf{Z}^m) \geq 0$ and $\sum_{\mathbf{Z}^m} \tilde{P}(\mathbf{Z}^m) = 1$. Use Lagrange multipliers to show that the solution is the conditional distribution $\tilde{P}(\mathbf{Z}^m) = \Pr(\mathbf{Z}^m | \mathbf{Z}, \theta')$, as in (8.49).

Ex. 8.3 Justify the estimate (8.50), using the relationship

$$\Pr(A) = \int \Pr(A|B) d(\Pr(B)).$$

Ex. 8.4 Consider the bagging method of Section 8.7. Let our estimate $\hat{f}(x)$ be the B-spline smoother $\hat{\mu}(x)$ of Section 8.2.1. Consider the parametric bootstrap of equation (8.6), applied to this estimator. Show that if we bag $\hat{f}(x)$, using the parametric bootstrap to generate the bootstrap samples, the bagging estimate $\hat{f}_{\text{bag}}(x)$ converges to the original estimate $\hat{f}(x)$ as $B \to \infty$.

Ex. 8.5 Suggest generalizations of each of the loss functions in Figure 10.4 to more than two classes, and design an appropriate plot to compare them.

Ex. 8.6 Consider the bone mineral density data of Figure 5.6.

(a) Fit a cubic smooth spline to the relative change in spinal BMD, as a function of age. Use cross-validation to estimate the optimal amount of smoothing. Construct pointwise 90% confidence bands for the underlying function.

(b) Compute the posterior mean and covariance for the true function via (8.28), and compare the posterior bands to those obtained in (a).

(c) Compute 100 bootstrap replicates of the fitted curves, as in the bottom left panel of Figure 8.2. Compare the results to those obtained in (a) and (b).

Ex. 8.7 *EM as a minorization algorithm*(Hunter and Lange, 2004; Wu and Lange, 2007). A function $g(x, y)$ to said to *minorize* a function $f(x)$ if

$$g(x, y) \leq f(x), \quad g(x, x) = f(x) \tag{8.62}$$

for all x, y in the domain. This is useful for maximizing $f(x)$ since it is easy to show that $f(x)$ is non-decreasing under the update

$$x^{s+1} = \operatorname{argmax}_x g(x, x^s) \tag{8.63}$$

There are analogous definitions for *majorization*, for minimizing a function $f(x)$. The resulting algorithms are known as *MM* algorithms, for "Minorize-Maximize" or "Majorize-Minimize."

Show that the EM algorithm (Section 8.5.2) is an example of an MM algorithm, using $Q(\theta', \theta) + \log \Pr(\mathbf{Z}|\theta) - Q(\theta, \theta)$ to minorize the observed data log-likelihood $\ell(\theta'; \mathbf{Z})$. (Note that only the first term involves the relevant parameter θ').

9
Additive Models, Trees, and Related Methods

In this chapter we begin our discussion of some specific methods for supervised learning. These techniques each assume a (different) structured form for the unknown regression function, and by doing so they finesse the curse of dimensionality. Of course, they pay the possible price of misspecifying the model, and so in each case there is a tradeoff that has to be made. They take off where Chapters 3–6 left off. We describe five related techniques: generalized additive models, trees, multivariate adaptive regression splines, the patient rule induction method, and hierarchical mixtures of experts.

9.1 Generalized Additive Models

Regression models play an important role in many data analyses, providing prediction and classification rules, and data analytic tools for understanding the importance of different inputs.

Although attractively simple, the traditional linear model often fails in these situations: in real life, effects are often not linear. In earlier chapters we described techniques that used predefined basis functions to achieve nonlinearities. This section describes more automatic flexible statistical methods that may be used to identify and characterize nonlinear regression effects. These methods are called "generalized additive models."

In the regression setting, a generalized additive model has the form

$$E(Y|X_1, X_2, \ldots, X_p) = \alpha + f_1(X_1) + f_2(X_2) + \cdots + f_p(X_p). \qquad (9.1)$$

T. Hastie et al., *The Elements of Statistical Learning, Second Edition,*
DOI: 10.1007/b94608_9,
© Springer Science+Business Media, LLC 2009

As usual $X_1, X_2, \ldots, X_p$ represent predictors and Y is the outcome; the f_j's are unspecified smooth ("nonparametric") functions. If we were to model each function using an expansion of basis functions (as in Chapter 5), the resulting model could then be fit by simple least squares. Our approach here is different: we fit each function using a scatterplot smoother (e.g., a cubic smoothing spline or kernel smoother), and provide an algorithm for simultaneously estimating all p functions (Section 9.1.1).

For two-class classification, recall the logistic regression model for binary data discussed in Section 4.4. We relate the mean of the binary response $\mu(X) = \Pr(Y = 1|X)$ to the predictors via a linear regression model and the *logit* link function:

$$\log\left(\frac{\mu(X)}{1 - \mu(X)}\right) = \alpha + \beta_1 X_1 + \cdots + \beta_p X_p. \tag{9.2}$$

The *additive* logistic regression model replaces each linear term by a more general functional form

$$\log\left(\frac{\mu(X)}{1 - \mu(X)}\right) = \alpha + f_1(X_1) + \cdots + f_p(X_p), \tag{9.3}$$

where again each f_j is an unspecified smooth function. While the nonparametric form for the functions f_j makes the model more flexible, the additivity is retained and allows us to interpret the model in much the same way as before. The additive logistic regression model is an example of a generalized additive model. In general, the conditional mean $\mu(X)$ of a response Y is related to an additive function of the predictors via a *link* function g:

$$g[\mu(X)] = \alpha + f_1(X_1) + \cdots + f_p(X_p). \tag{9.4}$$

Examples of classical link functions are the following:

- $g(\mu) = \mu$ is the identity link, used for linear and additive models for Gaussian response data.

- $g(\mu) = \text{logit}(\mu)$ as above, or $g(\mu) = \text{probit}(\mu)$, the *probit* link function, for modeling binomial probabilities. The probit function is the inverse Gaussian cumulative distribution function: $\text{probit}(\mu) = \Phi^{-1}(\mu)$.

- $g(\mu) = \log(\mu)$ for log-linear or log-additive models for Poisson count data.

All three of these arise from exponential family sampling models, which in addition include the gamma and negative-binomial distributions. These families generate the well-known class of generalized linear models, which are all extended in the same way to generalized additive models.

The functions f_j are estimated in a flexible manner, using an algorithm whose basic building block is a scatterplot smoother. The estimated function $\hat{f}_j$ can then reveal possible nonlinearities in the effect of X_j. Not all of the functions f_j need to be nonlinear. We can easily mix in linear and other parametric forms with the nonlinear terms, a necessity when some of the inputs are qualitative variables (factors). The nonlinear terms are not restricted to main effects either; we can have nonlinear components in two or more variables, or separate curves in X_j for each level of the factor X_k. Thus each of the following would qualify:

- $g(\mu) = X^T \beta + \alpha_k + f(Z)$—a *semiparametric* model, where X is a vector of predictors to be modeled linearly, α_k the effect for the kth level of a qualitative input V, and the effect of predictor Z is modeled nonparametrically.

- $g(\mu) = f(X) + g_k(Z)$—again k indexes the levels of a qualitative input V, and thus creates an interaction term $g(V, Z) = g_k(Z)$ for the effect of V and Z.

- $g(\mu) = f(X) + g(Z, W)$ where g is a nonparametric function in two features.

Additive models can replace linear models in a wide variety of settings, for example an additive decomposition of time series,

$$Y_t = S_t + T_t + \varepsilon_t, \tag{9.5}$$

where S_t is a seasonal component, T_t is a trend and ε is an error term.

9.1.1 *Fitting Additive Models*

In this section we describe a modular algorithm for fitting additive models and their generalizations. The building block is the scatterplot smoother for fitting nonlinear effects in a flexible way. For concreteness we use as our scatterplot smoother the cubic smoothing spline described in Chapter 5.

The additive model has the form

$$Y = \alpha + \sum_{j=1}^{p} f_j(X_j) + \varepsilon, \tag{9.6}$$

where the error term ε has mean zero. Given observations x_i, y_i, a criterion like the penalized sum of squares (5.9) of Section 5.4 can be specified for this problem,

$$\text{PRSS}(\alpha, f_1, f_2, \ldots, f_p) = \sum_{i=1}^{N} \left(y_i - \alpha - \sum_{j=1}^{p} f_j(x_{ij}) \right)^2 + \sum_{j=1}^{p} \lambda_j \int f_j''(t_j)^2 dt_j, \tag{9.7}$$

where the $\lambda_j \geq 0$ are tuning parameters. It can be shown that the minimizer of (9.7) is an additive cubic spline model; each of the functions f_j is a

---

Algorithm 9.1 *The Backfitting Algorithm for Additive Models.*

1. Initialize: $\hat{\alpha} = \frac{1}{N} \sum_1^N y_i$, $\hat{f}_j \equiv 0, \forall i, j$.

2. Cycle: $j = 1, 2, \ldots, p, \ldots, 1, 2, \ldots, p, \ldots,$

$$\hat{f}_j \leftarrow \mathcal{S}_j \left[\{y_i - \hat{\alpha} - \sum_{k \neq j} \hat{f}_k(x_{ik})\}_1^N \right],$$

$$\hat{f}_j \leftarrow \hat{f}_j - \frac{1}{N} \sum_{i=1}^N \hat{f}_j(x_{ij}).$$

until the functions $\hat{f}_j$ change less than a prespecified threshold.

---

cubic spline in the component X_j, with knots at each of the unique values of x_{ij}, $i = 1, \ldots, N$. However, without further restrictions on the model, the solution is not unique. The constant α is not identifiable, since we can add or subtract any constants to each of the functions f_j, and adjust α accordingly. The standard convention is to assume that $\sum_1^N f_j(x_{ij}) = 0 \ \forall j$—the functions average zero over the data. It is easily seen that $\hat{\alpha} = \text{ave}(y_i)$ in this case. If in addition to this restriction, the matrix of input values (having ijth entry x_{ij}) has full column rank, then (9.7) is a strictly convex criterion and the minimizer is unique. If the matrix is singular, then the *linear part* of the components f_j cannot be uniquely determined (while the nonlinear parts can!)(Buja et al., 1989).

Furthermore, a simple iterative procedure exists for finding the solution. We set $\hat{\alpha} = \text{ave}(y_i)$, and it never changes. We apply a cubic smoothing spline $\mathcal{S}_j$ to the targets $\{y_i - \hat{\alpha} - \sum_{k \neq j} \hat{f}_k(x_{ik})\}_1^N$, as a function of x_{ij}, to obtain a new estimate $\hat{f}_j$. This is done for each predictor in turn, using the current estimates of the other functions $\hat{f}_k$ when computing $y_i - \hat{\alpha} - \sum_{k \neq j} \hat{f}_k(x_{ik})$. The process is continued until the estimates $\hat{f}_j$ stabilize. This procedure, given in detail in Algorithm 9.1, is known as "backfitting" and the resulting fit is analogous to a multiple regression for linear models.

In principle, the second step in (2) of Algorithm 9.1 is not needed, since the smoothing spline fit to a mean-zero response has mean zero (Exercise 9.1). In practice, machine rounding can cause slippage, and the adjustment is advised.

This same algorithm can accommodate other fitting methods in exactly the same way, by specifying appropriate smoothing operators $\mathcal{S}_j$:

- other univariate regression smoothers such as local polynomial regression and kernel methods;

- linear regression operators yielding polynomial fits, piecewise constant fits, parametric spline fits, series and Fourier fits;

- more complicated operators such as surface smoothers for second or higher-order interactions or periodic smoothers for seasonal effects.

If we consider the operation of smoother $\mathcal{S}_j$ only at the training points, it can be represented by an $N \times N$ operator matrix $\mathbf{S}_j$ (see Section 5.4.1). Then the degrees of freedom for the jth term are (approximately) computed as $\mathrm{df}_j = \mathrm{trace}[\mathbf{S}_j] - 1$, by analogy with degrees of freedom for smoothers discussed in Chapters 5 and 6.

For a large class of linear smoothers $\mathbf{S}_j$, backfitting is equivalent to a Gauss–Seidel algorithm for solving a certain linear system of equations. Details are given in Exercise 9.2.

For the logistic regression model and other generalized additive models, the appropriate criterion is a penalized log-likelihood. To maximize it, the backfitting procedure is used in conjunction with a likelihood maximizer. The usual Newton–Raphson routine for maximizing log-likelihoods in generalized linear models can be recast as an IRLS (iteratively reweighted least squares) algorithm. This involves repeatedly fitting a weighted linear regression of a working response variable on the covariates; each regression yields a new value of the parameter estimates, which in turn give new working responses and weights, and the process is iterated (see Section 4.4.1). In the generalized additive model, the weighted linear regression is simply replaced by a weighted backfitting algorithm. We describe the algorithm in more detail for logistic regression below, and more generally in Chapter 6 of Hastie and Tibshirani (1990).

9.1.2 Example: Additive Logistic Regression

Probably the most widely used model in medical research is the logistic model for binary data. In this model the outcome Y can be coded as 0 or 1, with 1 indicating an event (like death or relapse of a disease) and 0 indicating no event. We wish to model $\Pr(Y = 1|X)$, the probability of an event given values of the prognostic factors $X^T = (X_1, \ldots, X_p)$. The goal is usually to understand the roles of the prognostic factors, rather than to classify new individuals. Logistic models are also used in applications where one is interested in estimating the class probabilities, for use in risk screening. Apart from medical applications, credit risk screening is a popular application.

The generalized additive logistic model has the form

$$\log \frac{\Pr(Y = 1|X)}{\Pr(Y = 0|X)} = \alpha + f_1(X_1) + \cdots + f_p(X_p). \tag{9.8}$$

The functions $f_1, f_2, \ldots, f_p$ are estimated by a backfitting algorithm within a Newton–Raphson procedure, shown in Algorithm 9.2.

Algorithm 9.2 *Local Scoring Algorithm for the Additive Logistic Regression Model.*

---

1. Compute starting values: $\hat{\alpha} = \log[\bar{y}/(1 - \bar{y})]$, where $\bar{y} = \text{ave}(y_i)$, the sample proportion of ones, and set $\hat{f}_j \equiv 0 \; \forall j$.

2. Define $\hat{\eta}_i = \hat{\alpha} + \sum_j \hat{f}_j(x_{ij})$ and $\hat{p}_i = 1/[1 + \exp(-\hat{\eta}_i)]$.

   Iterate:

   (a) Construct the working target variable
   $$z_i = \hat{\eta}_i + \frac{(y_i - \hat{p}_i)}{\hat{p}_i(1 - \hat{p}_i)}.$$

   (b) Construct weights $w_i = \hat{p}_i(1 - \hat{p}_i)$

   (c) Fit an additive model to the targets z_i with weights w_i, using a weighted backfitting algorithm. This gives new estimates $\hat{\alpha}, \hat{f}_j, \; \forall j$

3. Continue step 2. until the change in the functions falls below a pre-specified threshold.

The additive model fitting in step (2) of Algorithm 9.2 requires a weighted scatterplot smoother. Most smoothing procedures can accept observation weights (Exercise 5.12); see Chapter 3 of Hastie and Tibshirani (1990) for further details.

The additive logistic regression model can be generalized further to handle more than two classes, using the multilogit formulation as outlined in Section 4.4. While the formulation is a straightforward extension of (9.8), the algorithms for fitting such models are more complex. See Yee and Wild (1996) for details, and the VGAM software currently available from:

http://www.stat.auckland.ac.nz/~yee.

Example: Predicting Email Spam

We apply a generalized additive model to the spam data introduced in Chapter 1. The data consists of information from 4601 email messages, in a study to screen email for "spam" (i.e., junk email). The data is publicly available at ftp.ics.uci.edu, and was donated by George Forman from Hewlett-Packard laboratories, Palo Alto, California.

The response variable is binary, with values email or spam, and there are 57 predictors as described below:

- 48 quantitative predictors—the percentage of words in the email that match a given word. Examples include business, address, internet,

TABLE 9.1. *Test data confusion matrix for the additive logistic regression model fit to the spam training data. The overall test error rate is 5.5%.*

	Predicted Class	
True Class	email (0)	spam (1)
email (0)	58.3%	2.5%
spam (1)	3.0%	36.3%

free, and george. The idea was that these could be customized for individual users.

- 6 quantitative predictors—the percentage of characters in the email that match a given character. The characters are ch;, ch(, ch[, ch!, ch$, and ch#.

- The average length of uninterrupted sequences of capital letters: CAPAVE.

- The length of the longest uninterrupted sequence of capital letters: CAPMAX.

- The sum of the length of uninterrupted sequences of capital letters: CAPTOT.

We coded spam as 1 and email as zero. A test set of size 1536 was randomly chosen, leaving 3065 observations in the training set. A generalized additive model was fit, using a cubic smoothing spline with a nominal four degrees of freedom for each predictor. What this means is that for each predictor X_j, the smoothing-spline parameter λ_j was chosen so that $\text{trace}[\mathbf{S}_j(\lambda_j)] - 1 = 4$, where $\mathbf{S}_j(\lambda)$ is the smoothing spline operator matrix constructed using the observed values x_{ij}, $i = 1, \ldots, N$. This is a convenient way of specifying the amount of smoothing in such a complex model.

Most of the spam predictors have a very long-tailed distribution. Before fitting the GAM model, we log-transformed each variable (actually $\log(x + 0.1)$), but the plots in Figure 9.1 are shown as a function of the original variables.

The test error rates are shown in Table 9.1; the overall error rate is 5.3%. By comparison, a linear logistic regression has a test error rate of 7.6%. Table 9.2 shows the predictors that are highly significant in the additive model.

For ease of interpretation, in Table 9.2 the contribution for each variable is decomposed into a linear component and the remaining nonlinear component. The top block of predictors are positively correlated with spam, while the bottom block is negatively correlated. The linear component is a weighted least squares linear fit of the fitted curve on the predictor, while the nonlinear part is the residual. The linear component of an estimated

TABLE 9.2. *Significant predictors from the additive model fit to the spam training data. The coefficients represent the linear part of $\hat{f}_j$, along with their standard errors and Z-score. The nonlinear P-value is for a test of nonlinearity of $\hat{f}_j$.*

Name	Num.	df	Coefficient	Std. Error	Z Score	Nonlinear P-value
Positive effects						
our	5	3.9	0.566	0.114	4.970	0.052
over	6	3.9	0.244	0.195	1.249	0.004
remove	7	4.0	0.949	0.183	5.201	0.093
internet	8	4.0	0.524	0.176	2.974	0.028
free	16	3.9	0.507	0.127	4.010	0.065
business	17	3.8	0.779	0.186	4.179	0.194
hpl	26	3.8	0.045	0.250	0.181	0.002
ch!	52	4.0	0.674	0.128	5.283	0.164
ch$	53	3.9	1.419	0.280	5.062	0.354
CAPMAX	56	3.8	0.247	0.228	1.080	0.000
CAPTOT	57	4.0	0.755	0.165	4.566	0.063
Negative effects						
hp	25	3.9	−1.404	0.224	−6.262	0.140
george	27	3.7	−5.003	0.744	−6.722	0.045
1999	37	3.8	−0.672	0.191	−3.512	0.011
re	45	3.9	−0.620	0.133	−4.649	0.597
edu	46	4.0	−1.183	0.209	−5.647	0.000

function is summarized by the coefficient, standard error and Z-score; the latter is the coefficient divided by its standard error, and is considered significant if it exceeds the appropriate quantile of a standard normal distribution. The column labeled *nonlinear P-value* is a test of nonlinearity of the estimated function. Note, however, that the effect of each predictor is fully adjusted for the entire effects of the other predictors, not just for their linear parts. The predictors shown in the table were judged significant by at least one of the tests (linear or nonlinear) at the $p = 0.01$ level (two-sided).

Figure 9.1 shows the estimated functions for the significant predictors appearing in Table 9.2. Many of the nonlinear effects appear to account for a strong discontinuity at zero. For example, the probability of spam drops significantly as the frequency of george increases from zero, but then does not change much after that. This suggests that one might replace each of the frequency predictors by an indicator variable for a zero count, and resort to a linear logistic model. This gave a test error rate of 7.4%; including the linear effects of the frequencies as well dropped the test error to 6.6%. It appears that the nonlinearities in the additive model have an additional predictive power.

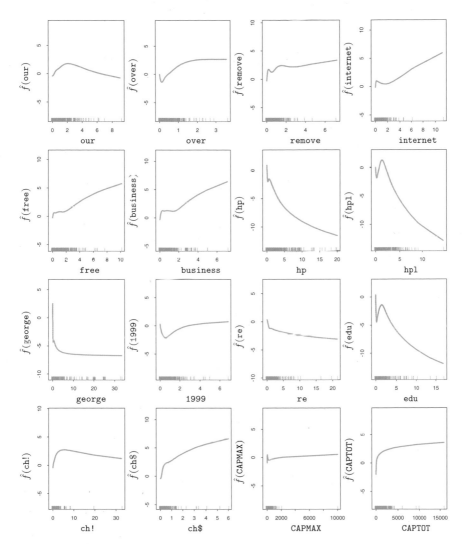

FIGURE 9.1. *Spam analysis: estimated functions for significant predictors. The* rug plot *along the bottom of each frame indicates the observed values of the corresponding predictor. For many of the predictors the nonlinearity picks up the discontinuity at zero.*

It is more serious to classify a genuine `email` message as `spam`, since then a good email would be filtered out and would not reach the user. We can alter the balance between the class error rates by changing the losses (see Section 2.4). If we assign a loss L_{01} for predicting a true class 0 as class 1, and L_{10} for predicting a true class 1 as class 0, then the estimated Bayes rule predicts class 1 if its probability is greater than $L_{01}/(L_{01} + L_{10})$. For example, if we take $L_{01} = 10, L_{10} = 1$ then the (true) class 0 and class 1 error rates change to 0.8% and 8.7%.

More ambitiously, we can encourage the model to fit better data in the class 0 by using weights L_{01} for the class 0 observations and L_{10} for the class 1 observations. As above, we then use the estimated Bayes rule to predict. This gave error rates of 1.2% and 8.0% in (true) class 0 and class 1, respectively. We discuss below the issue of unequal losses further, in the context of tree-based models.

After fitting an additive model, one should check whether the inclusion of some interactions can significantly improve the fit. This can be done "manually," by inserting products of some or all of the significant inputs, or automatically via the MARS procedure (Section 9.4).

This example uses the additive model in an automatic fashion. As a data analysis tool, additive models are often used in a more interactive fashion, adding and dropping terms to determine their effect. By calibrating the amount of smoothing in terms of df_j, one can move seamlessly between linear models ($df_j = 1$) and partially linear models, where some terms are modeled more flexibly. See Hastie and Tibshirani (1990) for more details.

9.1.3 Summary

Additive models provide a useful extension of linear models, making them more flexible while still retaining much of their interpretability. The familiar tools for modeling and inference in linear models are also available for additive models, seen for example in Table 9.2. The backfitting procedure for fitting these models is simple and modular, allowing one to choose a fitting method appropriate for each input variable. As a result they have become widely used in the statistical community.

However additive models can have limitations for large data-mining applications. The backfitting algorithm fits all predictors, which is not feasible or desirable when a large number are available. The BRUTO procedure (Hastie and Tibshirani, 1990, Chapter 9) combines backfitting with selection of inputs, but is not designed for large data-mining problems. There has also been recent work using lasso-type penalties to estimate sparse additive models, for example the COSSO procedure of Lin and Zhang (2006) and the SpAM proposal of Ravikumar et al. (2008). For large problems a forward stagewise approach such as boosting (Chapter 10) is more effective, and also allows for interactions to be included in the model.

9.2 Tree-Based Methods

9.2.1 Background

Tree-based methods partition the feature space into a set of rectangles, and then fit a simple model (like a constant) in each one. They are conceptually simple yet powerful. We first describe a popular method for tree-based regression and classification called CART, and later contrast it with C4.5, a major competitor.

Let's consider a regression problem with continuous response Y and inputs X_1 and X_2, each taking values in the unit interval. The top left panel of Figure 9.2 shows a partition of the feature space by lines that are parallel to the coordinate axes. In each partition element we can model Y with a different constant. However, there is a problem: although each partitioning line has a simple description like $X_1 = c$, some of the resulting regions are complicated to describe.

To simplify matters, we restrict attention to recursive binary partitions like that in the top right panel of Figure 9.2. We first split the space into two regions, and model the response by the mean of Y in each region. We choose the variable and split-point to achieve the best fit. Then one or both of these regions are split into two more regions, and this process is continued, until some stopping rule is applied. For example, in the top right panel of Figure 9.2, we first split at $X_1 = t_1$. Then the region $X_1 \leq t_1$ is split at $X_2 = t_2$ and the region $X_1 > t_1$ is split at $X_1 = t_3$. Finally, the region $X_1 > t_3$ is split at $X_2 = t_4$. The result of this process is a partition into the five regions $R_1, R_2, \ldots, R_5$ shown in the figure. The corresponding regression model predicts Y with a constant c_m in region R_m, that is,

$$\hat{f}(X) = \sum_{m=1}^{5} c_m I\{(X_1, X_2) \in R_m\}. \tag{9.9}$$

This same model can be represented by the binary tree in the bottom left panel of Figure 9.2. The full dataset sits at the top of the tree. Observations satisfying the condition at each junction are assigned to the left branch, and the others to the right branch. The terminal nodes or leaves of the tree correspond to the regions $R_1, R_2, \ldots, R_5$. The bottom right panel of Figure 9.2 is a perspective plot of the regression surface from this model. For illustration, we chose the node means $c_1 = -5, c_2 = -7, c_3 = 0, c_4 = 2, c_5 = 4$ to make this plot.

A key advantage of the recursive binary tree is its interpretability. The feature space partition is fully described by a single tree. With more than two inputs, partitions like that in the top right panel of Figure 9.2 are difficult to draw, but the binary tree representation works in the same way. This representation is also popular among medical scientists, perhaps because it mimics the way that a doctor thinks. The tree stratifies the

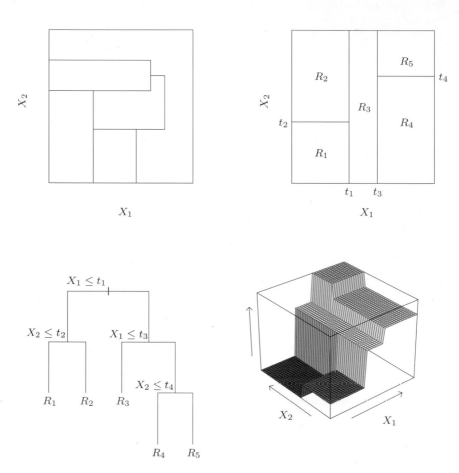

FIGURE 9.2. *Partitions and CART. Top right panel shows a partition of a two-dimensional feature space by recursive binary splitting, as used in CART, applied to some fake data. Top left panel shows a general partition that cannot be obtained from recursive binary splitting. Bottom left panel shows the tree corresponding to the partition in the top right panel, and a perspective plot of the prediction surface appears in the bottom right panel.*

population into strata of high and low outcome, on the basis of patient characteristics.

9.2.2 Regression Trees

We now turn to the question of how to grow a regression tree. Our data consists of p inputs and a response, for each of N observations: that is, (x_i, y_i) for $i = 1, 2, \ldots, N$, with $x_i = (x_{i1}, x_{i2}, \ldots, x_{ip})$. The algorithm needs to automatically decide on the splitting variables and split points, and also what topology (shape) the tree should have. Suppose first that we have a partition into M regions $R_1, R_2, \ldots, R_M$, and we model the response as a constant c_m in each region:

$$f(x) = \sum_{m=1}^{M} c_m I(x \in R_m). \tag{9.10}$$

If we adopt as our criterion minimization of the sum of squares $\sum(y_i - f(x_i))^2$, it is easy to see that the best $\hat{c}_m$ is just the average of y_i in region R_m:

$$\hat{c}_m = \text{ave}(y_i | x_i \in R_m). \tag{9.11}$$

Now finding the best binary partition in terms of minimum sum of squares is generally computationally infeasible. Hence we proceed with a greedy algorithm. Starting with all of the data, consider a splitting variable j and split point s, and define the pair of half-planes

$$R_1(j, s) = \{X | X_j \leq s\} \quad \text{and} \quad R_2(j, s) = \{X | X_j > s\}. \tag{9.12}$$

Then we seek the splitting variable j and split point s that solve

$$\min_{j, s} \left[\min_{c_1} \sum_{x_i \in R_1(j,s)} (y_i - c_1)^2 + \min_{c_2} \sum_{x_i \in R_2(j,s)} (y_i - c_2)^2 \right]. \tag{9.13}$$

For any choice j and s, the inner minimization is solved by

$$\hat{c}_1 = \text{ave}(y_i | x_i \in R_1(j, s)) \quad \text{and} \quad \hat{c}_2 = \text{ave}(y_i | x_i \in R_2(j, s)). \tag{9.14}$$

For each splitting variable, the determination of the split point s can be done very quickly and hence by scanning through all of the inputs, determination of the best pair (j, s) is feasible.

Having found the best split, we partition the data into the two resulting regions and repeat the splitting process on each of the two regions. Then this process is repeated on all of the resulting regions.

How large should we grow the tree? Clearly a very large tree might overfit the data, while a small tree might not capture the important structure.

Tree size is a tuning parameter governing the model's complexity, and the optimal tree size should be adaptively chosen from the data. One approach would be to split tree nodes only if the decrease in sum-of-squares due to the split exceeds some threshold. This strategy is too short-sighted, however, since a seemingly worthless split might lead to a very good split below it.

The preferred strategy is to grow a large tree T_0, stopping the splitting process only when some minimum node size (say 5) is reached. Then this large tree is pruned using *cost-complexity pruning*, which we now describe.

We define a subtree $T \subset T_0$ to be any tree that can be obtained by pruning T_0, that is, collapsing any number of its internal (non-terminal) nodes. We index terminal nodes by m, with node m representing region R_m. Let $|T|$ denote the number of terminal nodes in T. Letting

$$N_m = \#\{x_i \in R_m\},$$

$$\hat{c}_m = \frac{1}{N_m} \sum_{x_i \in R_m} y_i,$$

$$Q_m(T) = \frac{1}{N_m} \sum_{x_i \in R_m} (y_i - \hat{c}_m)^2,$$

(9.15)

we define the cost complexity criterion

$$C_\alpha(T) = \sum_{m=1}^{|T|} N_m Q_m(T) + \alpha |T|.$$

(9.16)

The idea is to find, for each α, the subtree $T_\alpha \subseteq T_0$ to minimize $C_\alpha(T)$. The tuning parameter $\alpha \geq 0$ governs the tradeoff between tree size and its goodness of fit to the data. Large values of α result in smaller trees T_α, and conversely for smaller values of α. As the notation suggests, with $\alpha = 0$ the solution is the full tree T_0. We discuss how to adaptively choose α below.

For each α one can show that there is a unique smallest subtree T_α that minimizes $C_\alpha(T)$. To find T_α we use *weakest link pruning*: we successively collapse the internal node that produces the smallest per-node increase in $\sum_m N_m Q_m(T)$, and continue until we produce the single-node (root) tree. This gives a (finite) sequence of subtrees, and one can show this sequence must contain T_α. See Breiman et al. (1984) or Ripley (1996) for details. Estimation of α is achieved by five- or tenfold cross-validation: we choose the value $\hat{\alpha}$ to minimize the cross-validated sum of squares. Our final tree is $T_{\hat{\alpha}}$.

9.2.3 Classification Trees

If the target is a classification outcome taking values $1, 2, \ldots, K$, the only changes needed in the tree algorithm pertain to the criteria for splitting nodes and pruning the tree. For regression we used the squared-error node

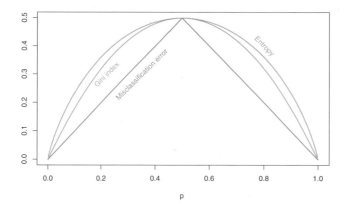

FIGURE 9.3. *Node impurity measures for two-class classification, as a function of the proportion p in class 2. Cross-entropy has been scaled to pass through* $(0.5, 0.5)$.

impurity measure $Q_m(T)$ defined in (9.15), but this is not suitable for classification. In a node m, representing a region R_m with N_m observations, let

$$\hat{p}_{mk} = \frac{1}{N_m} \sum_{x_i \in R_m} I(y_i = k),$$

the proportion of class k observations in node m. We classify the observations in node m to class $k(m) = \arg\max_k \hat{p}_{mk}$, the majority class in node m. Different measures $Q_m(T)$ of node impurity include the following:

Misclassification error: $\quad \frac{1}{N_m} \sum_{i \in R_m} I(y_i \neq k(m)) = 1 - \hat{p}_{mk(m)}.$

Gini index: $\quad \sum_{k \neq k'} \hat{p}_{mk}\hat{p}_{mk'} = \sum_{k=1}^{K} \hat{p}_{mk}(1 - \hat{p}_{mk}).$

Cross-entropy or deviance: $\quad -\sum_{k=1}^{K} \hat{p}_{mk} \log \hat{p}_{mk}.$

$$(9.17)$$

For two classes, if p is the proportion in the second class, these three measures are $1 - \max(p, 1 - p)$, $2p(1 - p)$ and $-p \log p - (1 - p) \log (1 - p)$, respectively. They are shown in Figure 9.3. All three are similar, but cross-entropy and the Gini index are differentiable, and hence more amenable to numerical optimization. Comparing (9.13) and (9.15), we see that we need to weight the node impurity measures by the number N_{m_L} and N_{m_R} of observations in the two child nodes created by splitting node m.

In addition, cross-entropy and the Gini index are more sensitive to changes in the node probabilities than the misclassification rate. For example, in a two-class problem with 400 observations in each class (denote this by $(400, 400)$), suppose one split created nodes $(300, 100)$ and $(100, 300)$, while the other created nodes $(200, 400)$ and $(200, 0)$. Both splits produce a misclassification rate of 0.25, but the second split produces a pure node and is probably preferable. Both the Gini index and cross-entropy are lower for the second split. For this reason, either the Gini index or cross-entropy should be used when growing the tree. To guide cost-complexity pruning, any of the three measures can be used, but typically it is the misclassification rate.

The Gini index can be interpreted in two interesting ways. Rather than classify observations to the majority class in the node, we could classify them to class k with probability $\hat{p}_{mk}$. Then the expected training error rate of this rule in the node is $\sum_{k \neq k'} \hat{p}_{mk}\hat{p}_{mk'}$—the Gini index. Similarly, if we code each observation as 1 for class k and zero otherwise, the variance over the node of this 0-1 response is $\hat{p}_{mk}(1 - \hat{p}_{mk})$. Summing over classes k again gives the Gini index.

9.2.4   Other Issues

Categorical Predictors

When splitting a predictor having q possible unordered values, there are $2^{q-1} - 1$ possible partitions of the q values into two groups, and the computations become prohibitive for large q. However, with a $0 - 1$ outcome, this computation simplifies. We order the predictor classes according to the proportion falling in outcome class 1. Then we split this predictor as if it were an ordered predictor. One can show this gives the optimal split, in terms of cross-entropy or Gini index, among all possible $2^{q-1} - 1$ splits. This result also holds for a quantitative outcome and square error loss—the categories are ordered by increasing mean of the outcome. Although intuitive, the proofs of these assertions are not trivial. The proof for binary outcomes is given in Breiman et al. (1984) and Ripley (1996); the proof for quantitative outcomes can be found in Fisher (1958). For multicategory outcomes, no such simplifications are possible, although various approximations have been proposed (Loh and Vanichsetakul, 1988).

The partitioning algorithm tends to favor categorical predictors with many levels q; the number of partitions grows exponentially in q, and the more choices we have, the more likely we can find a good one for the data at hand. This can lead to severe overfitting if q is large, and such variables should be avoided.

The Loss Matrix

In classification problems, the consequences of misclassifying observations are more serious in some classes than others. For example, it is probably worse to predict that a person will not have a heart attack when he/she actually will, than vice versa. To account for this, we define a $K \times K$ loss matrix $\mathbf{L}$, with $L_{kk'}$ being the loss incurred for classifying a class k observation as class k'. Typically no loss is incurred for correct classifications, that is, $L_{kk} = 0 \; \forall k$. To incorporate the losses into the modeling process, we could modify the Gini index to $\sum_{k \neq k'} L_{kk'} \hat{p}_{mk} \hat{p}_{mk'}$; this would be the expected loss incurred by the randomized rule. This works for the multi-class case, but in the two-class case has no effect, since the coefficient of $\hat{p}_{mk} \hat{p}_{mk'}$ is $L_{kk'} + L_{k'k}$. For two classes a better approach is to weight the observations in class k by $L_{kk'}$. This can be used in the multiclass case only if, as a function of k, $L_{kk'}$ doesn't depend on k'. Observation weighting can be used with the deviance as well. The effect of observation weighting is to alter the prior probability on the classes. In a terminal node, the empirical Bayes rule implies that we classify to class $k(m) = \arg \min_k \sum_{\ell} L_{\ell k} \hat{p}_{m\ell}$.

Missing Predictor Values

Suppose our data has some missing predictor values in some or all of the variables. We might discard any observation with some missing values, but this could lead to serious depletion of the training set. Alternatively we might try to fill in (impute) the missing values, with say the mean of that predictor over the nonmissing observations. For tree-based models, there are two better approaches. The first is applicable to categorical predictors: we simply make a new category for "missing." From this we might discover that observations with missing values for some measurement behave differently than those with nonmissing values. The second more general approach is the construction of surrogate variables. When considering a predictor for a split, we use only the observations for which that predictor is not missing. Having chosen the best (primary) predictor and split point, we form a list of surrogate predictors and split points. The first surrogate is the predictor and corresponding split point that best mimics the split of the training data achieved by the primary split. The second surrogate is the predictor and corresponding split point that does second best, and so on. When sending observations down the tree either in the training phase or during prediction, we use the surrogate splits in order, if the primary splitting predictor is missing. Surrogate splits exploit correlations between predictors to try and alleviate the effect of missing data. The higher the correlation between the missing predictor and the other predictors, the smaller the loss of information due to the missing value. The general problem of missing data is discussed in Section 9.6.

Why Binary Splits?

Rather than splitting each node into just two groups at each stage (as above), we might consider multiway splits into more than two groups. While this can sometimes be useful, it is not a good general strategy. The problem is that multiway splits fragment the data too quickly, leaving insufficient data at the next level down. Hence we would want to use such splits only when needed. Since multiway splits can be achieved by a series of binary splits, the latter are preferred.

Other Tree-Building Procedures

The discussion above focuses on the CART (classification and regression tree) implementation of trees. The other popular methodology is ID3 and its later versions, C4.5 and C5.0 (Quinlan, 1993). Early versions of the program were limited to categorical predictors, and used a top-down rule with no pruning. With more recent developments, C5.0 has become quite similar to CART. The most significant feature unique to C5.0 is a scheme for deriving rule sets. After a tree is grown, the splitting rules that define the terminal nodes can sometimes be simplified: that is, one or more condition can be dropped without changing the subset of observations that fall in the node. We end up with a simplified set of rules defining each terminal node; these no longer follow a tree structure, but their simplicity might make them more attractive to the user.

Linear Combination Splits

Rather than restricting splits to be of the form $X_j \leq s$, one can allow splits along linear combinations of the form $\sum a_j X_j \leq s$. The weights a_j and split point s are optimized to minimize the relevant criterion (such as the Gini index). While this can improve the predictive power of the tree, it can hurt interpretability. Computationally, the discreteness of the split point search precludes the use of a smooth optimization for the weights. A better way to incorporate linear combination splits is in the hierarchical mixtures of experts (HME) model, the topic of Section 9.5.

Instability of Trees

One major problem with trees is their high variance. Often a small change in the data can result in a very different series of splits, making interpretation somewhat precarious. The major reason for this instability is the hierarchical nature of the process: the effect of an error in the top split is propagated down to all of the splits below it. One can alleviate this to some degree by trying to use a more stable split criterion, but the inherent instability is not removed. It is the price to be paid for estimating a simple, tree-based structure from the data. *Bagging* (Section 8.7) averages many trees to reduce this variance.

Lack of Smoothness

Another limitation of trees is the lack of smoothness of the prediction surface, as can be seen in the bottom right panel of Figure 9.2. In classification with 0/1 loss, this doesn't hurt much, since bias in estimation of the class probabilities has a limited effect. However, this can degrade performance in the regression setting, where we would normally expect the underlying function to be smooth. The MARS procedure, described in Section 9.4,

TABLE 9.3. *Spam data: confusion rates for the 17-node tree (chosen by cross–validation) on the test data. Overall error rate is 9.3%.*

	Predicted	
True	email	spam
email	57.3%	4.0%
spam	5.3%	33.4%

can be viewed as a modification of CART designed to alleviate this lack of smoothness.

Difficulty in Capturing Additive Structure

Another problem with trees is their difficulty in modeling additive structure. In regression, suppose, for example, that $Y = c_1 I(X_1 < t_1) + c_2 I(X_2 < t_2) + \varepsilon$ where ε is zero-mean noise. Then a binary tree might make its first split on X_1 near t_1. At the next level down it would have to split both nodes on X_2 at t_2 in order to capture the additive structure. This might happen with sufficient data, but the model is given no special encouragement to find such structure. If there were ten rather than two additive effects, it would take many fortuitous splits to recreate the structure, and the data analyst would be hard pressed to recognize it in the estimated tree. The "blame" here can again be attributed to the binary tree structure, which has both advantages and drawbacks. Again the MARS method (Section 9.4) gives up this tree structure in order to capture additive structure.

9.2.5 Spam Example (Continued)

We applied the classification tree methodology to the **spam** example introduced earlier. We used the deviance measure to grow the tree and misclassification rate to prune it. Figure 9.4 shows the 10-fold cross-validation error rate as a function of the size of the pruned tree, along with ± 2 standard errors of the mean, from the ten replications. The test error curve is shown in orange. Note that the cross-validation error rates are indexed by a sequence of values of α and *not* tree size; for trees grown in different folds, a value of α might imply different sizes. The sizes shown at the base of the plot refer to $|T_\alpha|$, the sizes of the pruned *original* tree.

The error flattens out at around 17 terminal nodes, giving the pruned tree in Figure 9.5. Of the 13 distinct features chosen by the tree, 11 overlap with the 16 significant features in the additive model (Table 9.2). The overall error rate shown in Table 9.3 is about 50% higher than for the additive model in Table 9.1.

Consider the rightmost branches of the tree. We branch to the right with a **spam** warning if more than 5.5% of the characters are the $ sign.

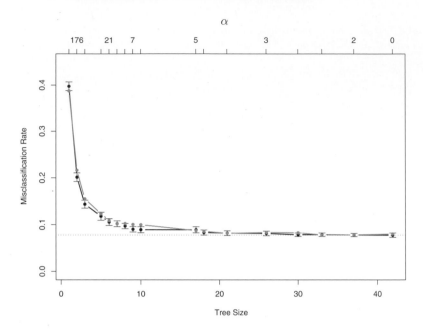

FIGURE 9.4. *Results for* spam *example. The blue curve is the* 10-*fold cross-validation estimate of misclassification rate as a function of tree size, with standard error bars. The minimum occurs at a tree size with about* 17 *terminal nodes (using the "one-standard-error" rule). The orange curve is the test error, which tracks the CV error quite closely. The cross-validation is indexed by values of* α, *shown above. The tree sizes shown below refer to* $|T_\alpha|$, *the size of the original tree indexed by* α.

However, if in addition the phrase hp occurs frequently, then this is likely to be company business and we classify as email. All of the 22 cases in the test set satisfying these criteria were correctly classified. If the second condition is not met, and in addition the average length of repeated capital letters CAPAVE is larger than 2.9, then we classify as spam. Of the 227 test cases, only seven were misclassified.

In medical classification problems, the terms *sensitivity* and *specificity* are used to characterize a rule. They are defined as follows:

Sensitivity: probability of predicting disease given true state is disease.

Specificity: probability of predicting non-disease given true state is non-disease.

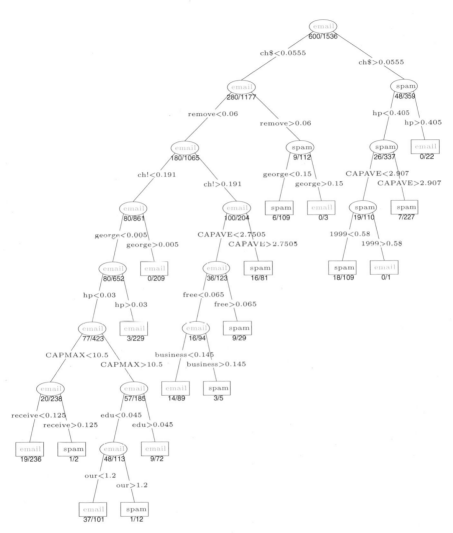

FIGURE 9.5. *The pruned tree for the* spam *example. The split variables are shown in blue on the branches, and the classification is shown in every node. The numbers under the terminal nodes indicate misclassification rates on the test data.*

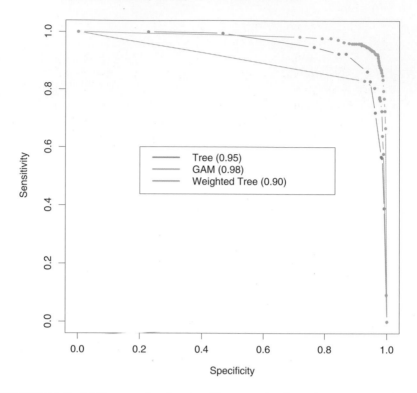

FIGURE 9.6. *ROC curves for the classification rules fit to the* spam *data. Curves that are closer to the northeast corner represent better classifiers. In this case the GAM classifier dominates the trees. The weighted tree achieves better sensitivity for higher specificity than the unweighted tree. The numbers in the legend represent the area under the curve.*

If we think of spam and email as the presence and absence of disease, respectively, then from Table 9.3 we have

$$
\begin{aligned}
Sensitivity &= 100 \times \frac{33.4}{33.4 + 5.3} = 86.3\%, \\
Specificity &= 100 \times \frac{57.3}{57.3 + 4.0} = 93.4\%.
\end{aligned}
$$

In this analysis we have used equal losses. As before let $L_{kk'}$ be the loss associated with predicting a class k object as class k'. By varying the relative sizes of the losses L_{01} and L_{10}, we increase the sensitivity and decrease the specificity of the rule, or vice versa. In this example, we want to avoid marking good email as spam, and thus we want the specificity to be very high. We can achieve this by setting $L_{01} > 1$ say, with $L_{10} = 1$. The Bayes' rule in each terminal node classifies to class 1 (spam) if the proportion of spam is $\geq L_{01}/(L_{10} + L_{01})$, and class zero otherwise. The receiver operating characteristic curve (ROC) is a commonly used summary for assessing the tradeoff between sensitivity and specificity. It is a plot of the sensitivity versus specificity as we vary the parameters of a classification rule. Varying the loss L_{01} between 0.1 and 10, and applying Bayes' rule to the 17-node tree selected in Figure 9.4, produced the ROC curve shown in Figure 9.6. The standard error of each curve near 0.9 is approximately $\sqrt{0.9(1-0.9)/1536} = 0.008$, and hence the standard error of the difference is about 0.01. We see that in order to achieve a specificity of close to 100%, the sensitivity has to drop to about 50%. The area under the curve is a commonly used quantitative summary; extending the curve linearly in each direction so that it is defined over $[0, 100]$, the area is approximately 0.95. For comparison, we have included the ROC curve for the GAM model fit to these data in Section 9.2; it gives a better classification rule for any loss, with an area of 0.98.

Rather than just modifying the Bayes rule in the nodes, it is better to take full account of the unequal losses in growing the tree, as was done in Section 9.2. With just two classes 0 and 1, losses may be incorporated into the tree-growing process by using weight $L_{k,1-k}$ for an observation in class k. Here we chose $L_{01} = 5, L_{10} = 1$ and fit the same size tree as before ($|T_\alpha| = 17$). This tree has higher sensitivity at high values of the specificity than the original tree, but does more poorly at the other extreme. Its top few splits are the same as the original tree, and then it departs from it. For this application the tree grown using $L_{01} = 5$ is clearly better than the original tree.

The area under the ROC curve, used above, is sometimes called the *c-statistic*. Interestingly, it can be shown that the area under the ROC curve is equivalent to the Mann-Whitney U statistic (or Wilcoxon rank-sum test), for the median difference between the prediction scores in the two groups (Hanley and McNeil, 1982). For evaluating the contribution of an additional predictor when added to a standard model, the *c*-statistic may not be an informative measure. The new predictor can be very significant in terms of the change in model deviance, but show only a small increase in the *c*-statistic. For example, removal of the highly significant term `george` from the model of Table 9.2 results in a decrease in the *c*-statistic of less than 0.01. Instead, it is useful to examine how the additional predictor changes the classification on an individual sample basis. A good discussion of this point appears in Cook (2007).

9.3 PRIM: Bump Hunting

Tree-based methods (for regression) partition the feature space into box-shaped regions, to try to make the response averages in each box as different as possible. The splitting rules defining the boxes are related to each through a binary tree, facilitating their interpretation.

The patient rule induction method (PRIM) also finds boxes in the feature space, but seeks boxes in which the response average is high. Hence it looks for maxima in the target function, an exercise known as *bump hunting*. (If minima rather than maxima are desired, one simply works with the negative response values.)

PRIM also differs from tree-based partitioning methods in that the box definitions are not described by a binary tree. This makes interpretation of the collection of rules more difficult; however, by removing the binary tree constraint, the individual rules are often simpler.

The main box construction method in PRIM works from the top down, starting with a box containing all of the data. The box is compressed along one face by a small amount, and the observations then falling outside the box are *peeled* off. The face chosen for compression is the one resulting in the largest box mean, after the compression is performed. Then the process is repeated, stopping when the current box contains some minimum number of data points.

This process is illustrated in Figure 9.7. There are 200 data points uniformly distributed over the unit square. The color-coded plot indicates the response Y taking the value 1 (red) when $0.5 < X_1 < 0.8$ and $0.4 < X_2 < 0.6$. and zero (blue) otherwise. The panels shows the successive boxes found by the top-down peeling procedure, peeling off a proportion $\alpha = 0.1$ of the remaining data points at each stage.

Figure 9.8 shows the mean of the response values in the box, as the box is compressed.

After the top-down sequence is computed, PRIM reverses the process, expanding along any edge, if such an expansion increases the box mean. This is called *pasting*. Since the top-down procedure is greedy at each step, such an expansion is often possible.

The result of these steps is a sequence of boxes, with different numbers of observation in each box. Cross-validation, combined with the judgment of the data analyst, is used to choose the optimal box size.

Denote by B_1 the indices of the observations in the box found in step 1. The PRIM procedure then removes the observations in B_1 from the training set, and the two-step process—top down peeling, followed by bottom-up pasting—is repeated on the remaining dataset. This entire process is repeated several times, producing a sequence of boxes $B_1, B_2, \ldots, B_k$. Each box is defined by a set of rules involving a subset of predictors like

$$(a_1 \leq X_1 \leq b_1) \text{ and } (b_1 \leq X_3 \leq b_2).$$

A summary of the PRIM procedure is given Algorithm 9.3.

PRIM can handle a categorical predictor by considering all partitions of the predictor, as in CART. Missing values are also handled in a manner similar to CART. PRIM is designed for regression (quantitative response

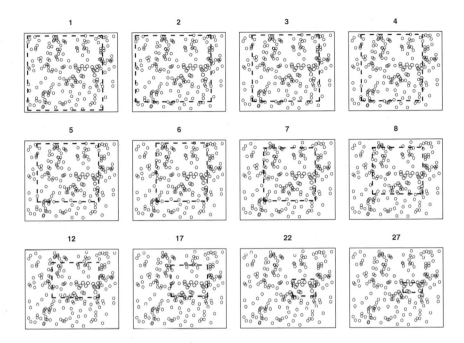

FIGURE 9.7. *Illustration of PRIM algorithm. There are two classes, indicated by the blue (class 0) and red (class 1) points. The procedure starts with a rectangle (broken black lines) surrounding all of the data, and then peels away points along one edge by a prespecified amount in order to maximize the mean of the points remaining in the box. Starting at the top left panel, the sequence of peelings is shown, until a pure red region is isolated in the bottom right panel. The iteration number is indicated at the top of each panel.*

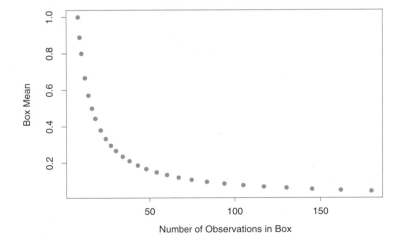

FIGURE 9.8. *Box mean as a function of number of observations in the box.*

Algorithm 9.3 *Patient Rule Induction Method.*

---

1. Start with all of the training data, and a maximal box containing all of the data.

2. Consider shrinking the box by compressing one face, so as to peel off the proportion α of observations having either the highest values of a predictor X_j, or the lowest. Choose the peeling that produces the highest response mean in the remaining box. (Typically $\alpha = 0.05$ or 0.10.)

3. Repeat step 2 until some minimal number of observations (say 10) remain in the box.

4. Expand the box along any face, as long as the resulting box mean increases.

5. Steps 1–4 give a sequence of boxes, with different numbers of observations in each box. Use cross-validation to choose a member of the sequence. Call the box B_1.

6. Remove the data in box B_1 from the dataset and repeat steps 2–5 to obtain a second box, and continue to get as many boxes as desired.

---

variable); a two-class outcome can be handled simply by coding it as 0 and 1. There is no simple way to deal with $k > 2$ classes simultaneously: one approach is to run PRIM separately for each class versus a baseline class.

An advantage of PRIM over CART is its patience. Because of its binary splits, CART fragments the data quite quickly. Assuming splits of equal size, with N observations it can only make $\log_2(N) - 1$ splits before running out of data. If PRIM peels off a proportion α of training points at each stage, it can perform approximately $-\log(N)/\log(1 - \alpha)$ peeling steps before running out of data. For example, if $N = 128$ and $\alpha = 0.10$, then $\log_2(N) - 1 = 6$ while $-\log(N)/\log(1 - \alpha) \approx 46$. Taking into account that there must be an integer number of observations at each stage, PRIM in fact can peel only 29 times. In any case, the ability of PRIM to be more patient should help the top-down greedy algorithm find a better solution.

9.3.1 Spam Example (Continued)

We applied PRIM to the spam data, with the response coded as 1 for spam and 0 for email.

The first two boxes found by PRIM are summarized below:

Rule 1	Global Mean	Box Mean	Box Support
Training	0.3931	0.9607	0.1413
Test	0.3958	1.0000	0.1536

$$
\text{Rule 1} \begin{cases}
\texttt{ch!} & > & 0.029 \\
\texttt{CAPAVE} & > & 2.331 \\
\texttt{your} & > & 0.705 \\
\texttt{1999} & < & 0.040 \\
\texttt{CAPTOT} & > & 79.50 \\
\texttt{edu} & < & 0.070 \\
\texttt{re} & < & 0.535 \\
\texttt{ch;} & < & 0.030
\end{cases}
$$

Rule 2	Remain Mean	Box Mean	Box Support
Training	0.2998	0.9560	0.1043
Test	0.2862	0.9264	0.1061

$$
\text{Rule 2} \begin{cases}
\texttt{remove} & > & 0.010 \\
\texttt{george} & < & 0.110
\end{cases}
$$

The box support is the proportion of observations falling in the box. The first box is purely spam, and contains about 15% of the test data. The second box contains 10.6% of the test observations, 92.6% of which are spam. Together the two boxes contain 26% of the data and are about 97% spam. The next few boxes (not shown) are quite small, containing only about 3% of the data.

The predictors are listed in order of importance. Interestingly the top splitting variables in the CART tree (Figure 9.5) do not appear in PRIM's first box.

9.4 MARS: Multivariate Adaptive Regression Splines

MARS is an adaptive procedure for regression, and is well suited for high-dimensional problems (i.e., a large number of inputs). It can be viewed as a generalization of stepwise linear regression or a modification of the CART method to improve the latter's performance in the regression setting. We introduce MARS from the first point of view, and later make the connection to CART.

MARS uses expansions in piecewise linear basis functions of the form $(x - t)_+$ and $(t - x)_+$. The "+" means positive part, so

$$
(x-t)_+ = \begin{cases} x - t, & \text{if } x > t, \\ 0, & \text{otherwise,} \end{cases} \quad \text{and} \quad (t-x)_+ = \begin{cases} t - x, & \text{if } x < t, \\ 0, & \text{otherwise.} \end{cases}
$$

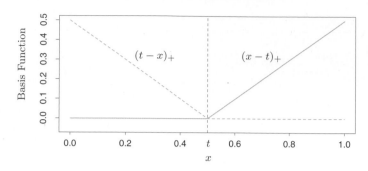

FIGURE 9.9. *The basis functions* $(x - t)_+$ *(solid orange) and* $(t - x)_+$ *(broken blue) used by MARS.*

As an example, the functions $(x - 0.5)_+$ and $(0.5 - x)_+$ are shown in Figure 9.9.

Each function is piecewise linear, with a *knot* at the value t. In the terminology of Chapter 5, these are linear splines. We call the two functions a *reflected pair* in the discussion below. The idea is to form reflected pairs for each input X_j with knots at each observed value x_{ij} of that input. Therefore, the collection of basis functions is

$$\mathcal{C} = \{(X_j - t)_+, \ (t - X_j)_+\}_{\substack{t \in \{x_{1j}, x_{2j}, \ldots, x_{Nj}\} \\ j = 1, 2, \ldots, p.}} \tag{9.18}$$

If all of the input values are distinct, there are $2Np$ basis functions altogether. Note that although each basis function depends only on a single X_j, for example, $h(X) = (X_j - t)_+$, it is considered as a function over the entire input space $\mathbb{R}^p$.

The model-building strategy is like a forward stepwise linear regression, but instead of using the original inputs, we are allowed to use functions from the set $\mathcal{C}$ and their products. Thus the model has the form

$$f(X) = \beta_0 + \sum_{m=1}^{M} \beta_m h_m(X), \tag{9.19}$$

where each $h_m(X)$ is a function in $\mathcal{C}$, or a product of two or more such functions.

Given a choice for the h_m, the coefficients β_m are estimated by minimizing the residual sum-of-squares, that is, by standard linear regression. The real art, however, is in the construction of the functions $h_m(x)$. We start with only the constant function $h_0(X) = 1$ in our model, and all functions in the set $\mathcal{C}$ are candidate functions. This is depicted in Figure 9.10.

At each stage we consider as a new basis function pair all products of a function h_m in the model set $\mathcal{M}$ with one of the reflected pairs in $\mathcal{C}$. We add to the model $\mathcal{M}$ the term of the form

$$\hat{\beta}_{M+1} h_\ell(X) \cdot (X_j - t)_+ + \hat{\beta}_{M+2} h_\ell(X) \cdot (t - X_j)_+, \ h_\ell \in \mathcal{M},$$

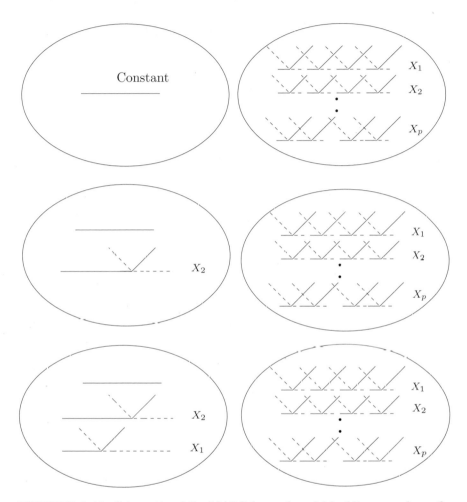

FIGURE 9.10. *Schematic of the MARS forward model-building procedure. On the left are the basis functions currently in the model: initially, this is the constant function $h(X) = 1$. On the right are all candidate basis functions to be considered in building the model. These are pairs of piecewise linear basis functions as in Figure 9.9, with knots t at all unique observed values x_{ij} of each predictor X_j. At each stage we consider all products of a candidate pair with a basis function in the model. The product that decreases the residual error the most is added into the current model. Above we illustrate the first three steps of the procedure, with the selected functions shown in red.*

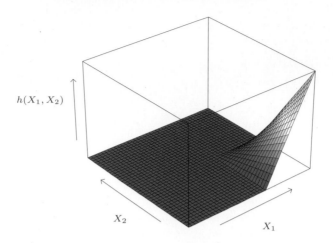

FIGURE 9.11. *The function* $h(X_1, X_2) = (X_1 - x_{51})_+ \cdot (x_{72} - X_2)_+$, *resulting from multiplication of two piecewise linear MARS basis functions.*

that produces the largest decrease in training error. Here $\hat{\beta}_{M+1}$ and $\hat{\beta}_{M+2}$ are coefficients estimated by least squares, along with all the other $M + 1$ coefficients in the model. Then the winning products are added to the model and the process is continued until the model set $\mathcal{M}$ contains some preset maximum number of terms.

For example, at the first stage we consider adding to the model a function of the form $\beta_1(X_j - t)_+ + \beta_2(t - X_j)_+$; $t \in \{x_{ij}\}$, since multiplication by the constant function just produces the function itself. Suppose the best choice is $\hat{\beta}_1(X_2 - x_{72})_+ + \hat{\beta}_2(x_{72} - X_2)_+$. Then this pair of basis functions is added to the set $\mathcal{M}$, and at the next stage we consider including a pair of products the form

$$h_m(X) \cdot (X_j - t)_+ \quad \text{and} \quad h_m(X) \cdot (t - X_j)_+, \; t \in \{x_{ij}\},$$

where for h_m we have the choices

$$
\begin{aligned}
h_0(X) &= 1, \\
h_1(X) &= (X_2 - x_{72})_+, \text{ or} \\
h_2(X) &= (x_{72} - X_2)_+.
\end{aligned}
$$

The third choice produces functions such as $(X_1 - x_{51})_+ \cdot (x_{72} - X_2)_+$, depicted in Figure 9.11.

At the end of this process we have a large model of the form (9.19). This model typically overfits the data, and so a backward deletion procedure is applied. The term whose removal causes the smallest increase in residual squared error is deleted from the model at each stage, producing an estimated best model $\hat{f}_\lambda$ of each size (number of terms) λ. One could use cross-validation to estimate the optimal value of λ, but for computational savings the MARS procedure instead uses generalized cross-validation. This criterion is defined as

$$\text{GCV}(\lambda) = \frac{\sum_{i=1}^{N}(y_i - \hat{f}_\lambda(x_i))^2}{(1 - M(\lambda)/N)^2}. \qquad (9.20)$$

The value $M(\lambda)$ is the effective number of parameters in the model: this accounts both for the number of terms in the models, plus the number of parameters used in selecting the optimal positions of the knots. Some mathematical and simulation results suggest that one should pay a price of three parameters for selecting a knot in a piecewise linear regression.

Thus if there are r linearly independent basis functions in the model, and K knots were selected in the forward process, the formula is $M(\lambda) = r + cK$, where $c = 3$. (When the model is restricted to be additive—details below— a penalty of $c = 2$ is used). Using this, we choose the model along the backward sequence that minimizes $\text{GCV}(\lambda)$.

Why these piecewise linear basis functions, and why this particular model strategy? A key property of the functions of Figure 9.9 is their ability to operate locally; they are zero over part of their range. When they are multiplied together, as in Figure 9.11, the result is nonzero only over the small part of the feature space where both component functions are nonzero. As a result, the regression surface is built up parsimoniously, using nonzero components locally—only where they are needed. This is important, since one should "spend" parameters carefully in high dimensions, as they can run out quickly. The use of other basis functions such as polynomials, would produce a nonzero product everywhere, and would not work as well.

The second important advantage of the piecewise linear basis function concerns computation. Consider the product of a function in $\mathcal{M}$ with each of the N reflected pairs for an input X_j. This appears to require the fitting of N single-input linear regression models, each of which uses $O(N)$ operations, making a total of $O(N^2)$ operations. However, we can exploit the simple form of the piecewise linear function. We first fit the reflected pair with rightmost knot. As the knot is moved successively one position at a time to the left, the basis functions differ by zero over the left part of the domain, and by a constant over the right part. Hence after each such move we can update the fit in $O(1)$ operations. This allows us to try every knot in only $O(N)$ operations.

The forward modeling strategy in MARS is hierarchical, in the sense that multiway products are built up from products involving terms already in the model. For example, a four-way product can only be added to the model if one of its three-way components is already in the model. The philosophy here is that a high-order interaction will likely only exist if some of its lower-order "footprints" exist as well. This need not be true, but is a reasonable working assumption and avoids the search over an exponentially growing space of alternatives.

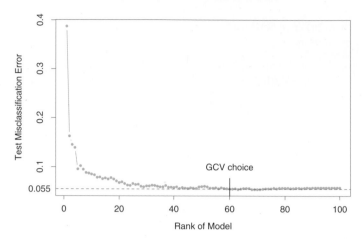

FIGURE 9.12. *Spam data: test error misclassification rate for the MARS procedure, as a function of the rank (number of independent basis functions) in the model.*

There is one restriction put on the formation of model terms: each input can appear at most once in a product. This prevents the formation of higher-order powers of an input, which increase or decrease too sharply near the boundaries of the feature space. Such powers can be approximated in a more stable way with piecewise linear functions.

A useful option in the MARS procedure is to set an upper limit on the order of interaction. For example, one can set a limit of two, allowing pairwise products of piecewise linear functions, but not three- or higher-way products. This can aid in the interpretation of the final model. An upper limit of one results in an additive model.

9.4.1 Spam Example (Continued)

We applied MARS to the "spam" data analyzed earlier in this chapter. To enhance interpretability, we restricted MARS to second-degree interactions. Although the target is a two-class variable, we used the squared-error loss function nonetheless (see Section 9.4.3). Figure 9.12 shows the test error misclassification rate as a function of the rank (number of independent basis functions) in the model. The error rate levels off at about 5.5%, which is slightly higher than that of the generalized additive model (5.3%) discussed earlier. GCV chose a model size of 60, which is roughly the smallest model giving optimal performance. The leading interactions found by MARS involved inputs (ch$, remove), (ch$, free) and (hp, CAPTOT). However, these interactions give no improvement in performance over the generalized additive model.

9.4.2 Example (Simulated Data)

Here we examine the performance of MARS in three contrasting scenarios. There are $N = 100$ observations, and the predictors $X_1, X_2, \ldots, X_p$ and errors ε have independent standard normal distributions.

Scenario 1: The data generation model is

$$Y = (X_1 - 1)_+ + (X_1 - 1)_+ \cdot (X_2 - .8)_+ + 0.12 \cdot \varepsilon. \qquad (9.21)$$

The noise standard deviation 0.12 was chosen so that the signal-to-noise ratio was about 5. We call this the tensor-product scenario; the product term gives a surface that looks like that of Figure 9.11.

Scenario 2: This is the same as scenario 1, but with $p = 20$ total predictors; that is, there are 18 inputs that are independent of the response.

Scenario 3: This has the structure of a neural network:

$$
\begin{aligned}
\ell_1 &= X_1 + X_2 + X_3 + X_4 + X_5, \\
\ell_2 &= X_6 - X_7 + X_8 - X_9 + X_{10}, \\
\sigma(t) &= 1/(1 + e^{-t}), \\
Y &= \sigma(\ell_1) + \sigma(\ell_2) + 0.12 \cdot \varepsilon.
\end{aligned}
\qquad (9.22)
$$

Scenarios 1 and 2 are ideally suited for MARS, while scenario 3 contains high-order interactions and may be difficult for MARS to approximate. We ran five simulations from each model, and recorded the results.

In scenario 1, MARS typically uncovered the correct model almost perfectly. In scenario 2, it found the correct structure but also found a few extraneous terms involving other predictors.

Let $\mu(x)$ be the true mean of Y, and let

$$
\begin{aligned}
\text{MSE}_0 &= \text{ave}_{x \in \text{Test}} (\bar{y} - \mu(x))^2, \\
\text{MSE} &= \text{ave}_{x \in \text{Test}} (\hat{f}(x) - \mu(x))^2.
\end{aligned}
\qquad (9.23)
$$

These represent the mean-square error of the constant model and the fitted MARS model, estimated by averaging at the 1000 test values of x. Table 9.4 shows the proportional decrease in model error or R^2 for each scenario:

$$R^2 = \frac{\text{MSE}_0 - \text{MSE}}{\text{MSE}_0}. \qquad (9.24)$$

The values shown are means and standard error over the five simulations. The performance of MARS is degraded only slightly by the inclusion of the useless inputs in scenario 2; it performs substantially worse in scenario 3.

TABLE 9.4. *Proportional decrease in model error (R^2) when MARS is applied to three different scenarios.*

Scenario	Mean (S.E.)
1: Tensor product $p = 2$	0.97 (0.01)
2: Tensor product $p = 20$	0.96 (0.01)
3: Neural network	0.79 (0.01)

9.4.3 Other Issues

MARS for Classification

The MARS method and algorithm can be extended to handle classification problems. Several strategies have been suggested.

For two classes, one can code the output as 0/1 and treat the problem as a regression; we did this for the spam example. For more than two classes, one can use the indicator response approach described in Section 4.2. One codes the K response classes via 0/1 indicator variables, and then performs a multi-response MARS regression. For the latter we use a common set of basis functions for all response variables. Classification is made to the class with the largest predicted response value. There are, however, potential masking problems with this approach, as described in Section 4.2. A generally superior approach is the "optimal scoring" method discussed in Section 12.5.

Stone et al. (1997) developed a hybrid of MARS called PolyMARS specifically designed to handle classification problems. It uses the multiple logistic framework described in Section 4.4. It grows the model in a forward stagewise fashion like MARS, but at each stage uses a quadratic approximation to the multinomial log-likelihood to search for the next basis-function pair. Once found, the enlarged model is fit by maximum likelihood, and the process is repeated.

Relationship of MARS to CART

Although they might seem quite different, the MARS and CART strategies actually have strong similarities. Suppose we take the MARS procedure and make the following changes:

- Replace the piecewise linear basis functions by step functions $I(x-t > 0)$ and $I(x - t \le 0)$.

- When a model term is involved in a multiplication by a candidate term, it gets replaced by the interaction, and hence is not available for further interactions.

With these changes, the MARS forward procedure is the same as the CART tree-growing algorithm. Multiplying a step function by a pair of reflected step functions is equivalent to splitting a node at the step. The second restriction implies that a node may not be split more than once, and leads to the attractive binary-tree representation of the CART model. On the other hand, it is this restriction that makes it difficult for CART to model additive structures. MARS forgoes the tree structure and gains the ability to capture additive effects.

Mixed Inputs

Mars can handle "mixed" predictors—quantitative and qualitative—in a natural way, much like CART does. MARS considers all possible binary partitions of the categories for a qualitative predictor into two groups. Each such partition generates a pair of piecewise constant basis functions—indicator functions for the two sets of categories. This basis pair is now treated as any other, and is used in forming tensor products with other basis functions already in the model.

9.5 Hierarchical Mixtures of Experts

The hierarchical mixtures of experts (HME) procedure can be viewed as a variant of tree-based methods. The main difference is that the tree splits are not hard decisions but rather soft probabilistic ones. At each node an observation goes left or right with probabilities depending on its input values. This has some computational advantages since the resulting parameter optimization problem is smooth, unlike the discrete split point search in the tree-based approach. The soft splits might also help in prediction accuracy and provide a useful alternative description of the data.

There are other differences between HMEs and the CART implementation of trees. In an HME, a linear (or logistic regression) model is fit in each terminal node, instead of a constant as in CART. The splits can be multiway, not just binary, and the splits are probabilistic functions of a linear combination of inputs, rather than a single input as in the standard use of CART. However, the relative merits of these choices are not clear, and most were discussed at the end of Section 9.2.

A simple two-level HME model in shown in Figure 9.13. It can be thought of as a tree with soft splits at each non-terminal node. However, the inventors of this methodology use a different terminology. The terminal nodes are called *experts*, and the non-terminal nodes are called *gating networks*. The idea is that each expert provides an opinion (prediction) about the response, and these are combined together by the gating networks. As we will see, the model is formally a mixture model, and the two-level model in the figure can be extend to multiple levels, hence the name *hierarchical mixtures of experts*.

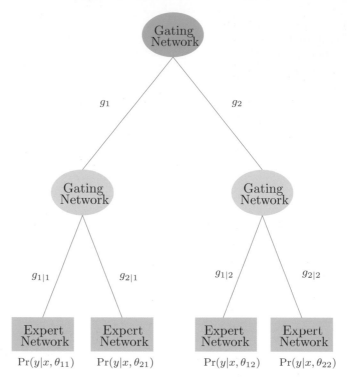

FIGURE 9.13. *A two-level hierarchical mixture of experts (HME) model.*

Consider the regression or classification problem, as described earlier in the chapter. The data is $(x_i, y_i), i = 1, 2, \ldots, N$, with y_i either a continuous or binary-valued response, and x_i a vector-valued input. For ease of notation we assume that the first element of x_i is one, to account for intercepts.

Here is how an HME is defined. The top gating network has the output

$$g_j(x, \gamma_j) = \frac{e^{\gamma_j^T x}}{\sum_{k=1}^{K} e^{\gamma_k^T x}}, \ j = 1, 2, \ldots, K, \tag{9.25}$$

where each γ_j is a vector of unknown parameters. This represents a soft K-way split ($K = 2$ in Figure 9.13.) Each $g_j(x, \gamma_j)$ is the probability of assigning an observation with feature vector x to the jth branch. Notice that with $K = 2$ groups, if we take the coefficient of one of the elements of x to be $+\infty$, then we get a logistic curve with infinite slope. In this case, the gating probabilities are either 0 or 1, corresponding to a hard split on that input.

At the second level, the gating networks have a similar form:

$$g_{\ell|j}(x, \gamma_{j\ell}) = \frac{e^{\gamma_{j\ell}^T x}}{\sum_{k=1}^{K} e^{\gamma_{jk}^T x}}, \ \ell = 1, 2, \ldots, K. \tag{9.26}$$

This is the probability of assignment to the ℓth branch, given assignment to the jth branch at the level above.

At each expert (terminal node), we have a model for the response variable of the form

$$Y \sim \Pr(y|x, \theta_{j\ell}). \tag{9.27}$$

This differs according to the problem.

Regression: The Gaussian linear regression model is used, with $\theta_{j\ell} = (\beta_{j\ell}, \sigma_{j\ell}^2)$:

$$Y = \beta_{j\ell}^T x + \varepsilon \text{ and } \varepsilon \sim N(0, \sigma_{j\ell}^2). \tag{9.28}$$

Classification: The linear logistic regression model is used:

$$\Pr(Y = 1|x, \theta_{j\ell}) = \frac{1}{1 + e^{-\theta_{j\ell}^T x}}. \tag{9.29}$$

Denoting the collection of all parameters by $\Psi = \{\gamma_j, \gamma_{j\ell}, \theta_{j\ell}\}$, the total probability that $Y - y$ is

$$\Pr(y|x, \Psi) = \sum_{j=1}^{K} g_j(x, \gamma_j) \sum_{\ell=1}^{K} g_{\ell|j}(x, \gamma_{j\ell}) \Pr(y|x, \theta_{j\ell}). \tag{9.30}$$

This is a mixture model, with the mixture probabilities determined by the gating network models.

To estimate the parameters, we maximize the log-likelihood of the data, $\sum_i \log \Pr(y_i|x_i, \Psi)$, over the parameters in Ψ. The most convenient method for doing this is the EM algorithm, which we describe for mixtures in Section 8.5. We define latent variables Δ_j, all of which are zero except for a single one. We interpret these as the branching decisions made by the top level gating network. Similarly we define latent variables $\Delta_{\ell|j}$ to describe the gating decisions at the second level.

In the E-step, the EM algorithm computes the expectations of the Δ_j and $\Delta_{\ell|j}$ given the current values of the parameters. These expectations are then used as observation weights in the M-step of the procedure, to estimate the parameters in the expert networks. The parameters in the internal nodes are estimated by a version of multiple logistic regression. The expectations of the Δ_j or $\Delta_{\ell|j}$ are probability profiles, and these are used as the response vectors for these logistic regressions.

The hierarchical mixtures of experts approach is a promising competitor to CART trees. By using *soft splits* rather than hard decision rules it can capture situations where the transition from low to high response is gradual. The log-likelihood is a smooth function of the unknown weights and hence is amenable to numerical optimization. The model is similar to CART with linear combination splits, but the latter is more difficult to optimize. On the other hand, to our knowledge there are no methods for finding a good tree topology for the HME model, as there are in CART. Typically one uses a fixed tree of some depth, possibly the output of the CART procedure. The emphasis in the research on HMEs has been on prediction rather than interpretation of the final model. A close cousin of the HME is the *latent class model* (Lin et al., 2000), which typically has only one layer; here the nodes or latent classes are interpreted as groups of subjects that show similar response behavior.

9.6   Missing Data

It is quite common to have observations with missing values for one or more input features. The usual approach is to impute (fill-in) the missing values in some way.

However, the first issue in dealing with the problem is determining whether the missing data mechanism has distorted the observed data. Roughly speaking, data are missing at random if the mechanism resulting in its omission is independent of its (unobserved) value. A more precise definition is given in Little and Rubin (2002). Suppose $\mathbf{y}$ is the response vector and $\mathbf{X}$ is the $N \times p$ matrix of inputs (some of which are missing). Denote by $\mathbf{X}_{\text{obs}}$ the observed entries in $\mathbf{X}$ and let $\mathbf{Z} = (\mathbf{y}, \mathbf{X})$, $\mathbf{Z}_{\text{obs}} = (\mathbf{y}, \mathbf{X}_{\text{obs}})$. Finally, if $\mathbf{R}$ is an indicator matrix with ijth entry 1 if x_{ij} is missing and zero otherwise, then the data is said to be *missing at random* (MAR) if the distribution of $\mathbf{R}$ depends on the data $\mathbf{Z}$ only through $\mathbf{Z}_{\text{obs}}$:

$$\Pr(\mathbf{R}|\mathbf{Z}, \theta) = \Pr(\mathbf{R}|\mathbf{Z}_{\text{obs}}, \theta). \tag{9.31}$$

Here θ are any parameters in the distribution of $\mathbf{R}$. Data are said to be *missing completely at random* (MCAR) if the distribution of $\mathbf{R}$ doesn't depend on the observed or missing data:

$$\Pr(\mathbf{R}|\mathbf{Z}, \theta) = \Pr(\mathbf{R}|\theta). \tag{9.32}$$

MCAR is a stronger assumption than MAR: most imputation methods rely on MCAR for their validity.

For example, if a patient's measurement was not taken because the doctor felt he was too sick, that observation would not be MAR or MCAR. In this case the missing data mechanism causes our observed training data to give a distorted picture of the true population, and data imputation is dangerous in this instance. Often the determination of whether features are MCAR must be made from information about the data collection process. For categorical features, one way to diagnose this problem is to code "missing" as an additional class. Then we fit our model to the training data and see if class "missing" is predictive of the response.

Assuming the features are missing completely at random, there are a number of ways of proceeding:

1. Discard observations with any missing values.

2. Rely on the learning algorithm to deal with missing values in its training phase.

3. Impute all missing values before training.

Approach (1) can be used if the relative amount of missing data is small, but otherwise should be avoided. Regarding (2), CART is one learning algorithm that deals effectively with missing values, through *surrogate splits* (Section 9.2.4). MARS and PRIM use similar approaches. In generalized additive modeling, all observations missing for a given input feature are omitted when the partial residuals are smoothed against that feature in the backfitting algorithm, and their fitted values are set to zero. Since the fitted curves have mean zero (when the model includes an intercept), this amounts to assigning the average fitted value to the missing observations.

For most learning methods, the imputation approach (3) is necessary. The simplest tactic is to impute the missing value with the mean or median of the nonmissing values for that feature. (Note that the above procedure for generalized additive models is analogous to this.)

If the features have at least some moderate degree of dependence, one can do better by estimating a predictive model for each feature given the other features and then imputing each missing value by its prediction from the model. In choosing the learning method for imputation of the features, one must remember that this choice is distinct from the method used for predicting $\mathbf{y}$ from $\mathbf{X}$. Thus a flexible, adaptive method will often be preferred, even for the eventual purpose of carrying out a linear regression of $\mathbf{y}$ on $\mathbf{X}$. In addition, if there are many missing feature values in the training set, the learning method must itself be able to deal with missing feature values. CART therefore is an ideal choice for this imputation "engine."

After imputation, missing values are typically treated as if they were actually observed. This ignores the uncertainty due to the imputation, which will itself introduce additional uncertainty into estimates and predictions from the response model. One can measure this additional uncertainty by doing multiple imputations and hence creating many different training sets. The predictive model for $\mathbf{y}$ can be fit to each training set, and the variation across training sets can be assessed. If CART was used for the imputation engine, the multiple imputations could be done by sampling from the values in the corresponding terminal nodes.

9.7 Computational Considerations

With N observations and p predictors, additive model fitting requires some number mp of applications of a one-dimensional smoother or regression method. The required number of cycles m of the backfitting algorithm is usually less than 20 and often less than 10, and depends on the amount of correlation in the inputs. With cubic smoothing splines, for example, $N \log N$ operations are needed for an initial sort and N operations for the spline fit. Hence the total operations for an additive model fit is $pN \log N + mpN$.

Trees require $pN \log N$ operations for an initial sort for each predictor, and typically another $pN \log N$ operations for the split computations. If the splits occurred near the edges of the predictor ranges, this number could increase to $N^2 p$.

MARS requires $Nm^2 + pmN$ operations to add a basis function to a model with m terms already present, from a pool of p predictors. Hence to build an M-term model requires $NM^3 + pM^2 N$ computations, which can be quite prohibitive if M is a reasonable fraction of N.

Each of the components of an HME are typically inexpensive to fit at each M-step: Np^2 for the regressions, and $Np^2 K^2$ for a K-class logistic regression. The EM algorithm, however, can take a long time to converge, and so sizable HME models are considered costly to fit.

Bibliographic Notes

The most comprehensive source for generalized additive models is the text of that name by Hastie and Tibshirani (1990). Different applications of this work in medical problems are discussed in Hastie et al. (1989) and Hastie and Herman (1990), and the software implementation in Splus is described in Chambers and Hastie (1991). Green and Silverman (1994) discuss penalization and spline models in a variety of settings. Efron and Tibshirani (1991) give an exposition of modern developments in statistics (including generalized additive models), for a nonmathematical audience. Classification and regression trees date back at least as far as Morgan and Sonquist (1963). We have followed the modern approaches of Breiman et al. (1984) and Quinlan (1993). The PRIM method is due to Friedman and Fisher (1999), while MARS is introduced in Friedman (1991), with an additive precursor in Friedman and Silverman (1989). Hierarchical mixtures of experts were proposed in Jordan and Jacobs (1994); see also Jacobs et al. (1991).

Exercises

Ex. 9.1 Show that a smoothing spline fit of y_i to x_i preserves the *linear part* of the fit. In other words, if $y_i = \hat{y}_i + r_i$, where $\hat{y}_i$ represents the linear regression fits, and $\mathbf{S}$ is the smoothing matrix, then $\mathbf{Sy} = \hat{\mathbf{y}} + \mathbf{Sr}$. Show that the same is true for local linear regression (Section 6.1.1). Hence argue that the adjustment step in the second line of (2) in Algorithm 9.1 is unnecessary.

Ex. 9.2 Let $\mathbf{A}$ be a known $k \times k$ matrix, $\mathbf{b}$ be a known k-vector, and $\mathbf{z}$ be an unknown k-vector. A Gauss–Seidel algorithm for solving the linear system of equations $\mathbf{Az} = \mathbf{b}$ works by successively solving for element z_j in the jth equation, fixing all other z_j's at their current guesses. This process is repeated for $j = 1, 2, \ldots, k, 1, 2, \ldots, k, \ldots$, until convergence (Golub and Van Loan, 1983).

(a) Consider an additive model with N observations and p terms, with the jth term to be fit by a linear smoother $\mathbf{S}_j$. Consider the following system of equations:

$$
\begin{pmatrix}
\mathbf{I} & \mathbf{S}_1 & \mathbf{S}_1 & \cdots & \mathbf{S}_1 \\
\mathbf{S}_2 & \mathbf{I} & \mathbf{S}_2 & \cdots & \mathbf{S}_2 \\
\vdots & \vdots & \vdots & \ddots & \vdots \\
\mathbf{S}_p & \mathbf{S}_p & \mathbf{S}_p & \cdots & \mathbf{I}
\end{pmatrix}
\begin{pmatrix}
\mathbf{f}_1 \\ \mathbf{f}_2 \\ \vdots \\ \mathbf{f}_p
\end{pmatrix}
=
\begin{pmatrix}
\mathbf{S}_1\mathbf{y} \\ \mathbf{S}_2\mathbf{y} \\ \vdots \\ \mathbf{S}_p\mathbf{y}
\end{pmatrix}.
\tag{9.33}
$$

Here each $\mathbf{f}_j$ is an N-vector of evaluations of the jth function at the data points, and $\mathbf{y}$ is an N-vector of the response values. Show that backfitting is a blockwise Gauss–Seidel algorithm for solving this system of equations.

(b) Let $\mathbf{S}_1$ and $\mathbf{S}_2$ be symmetric smoothing operators (matrices) with eigenvalues in $[0, 1)$. Consider a backfitting algorithm with response vector $\mathbf{y}$ and smoothers $\mathbf{S}_1, \mathbf{S}_2$. Show that with any starting values, the algorithm converges and give a formula for the final iterates.

Ex. 9.3 *Backfitting equations.* Consider a backfitting procedure with orthogonal projections, and let $\mathbf{D}$ be the overall regression matrix whose columns span $V = \mathcal{L}_{\mathrm{col}}(\mathbf{S}_1) \oplus \mathcal{L}_{\mathrm{col}}(\mathbf{S}_2) \oplus \cdots \oplus \mathcal{L}_{\mathrm{col}}(\mathbf{S}_p)$, where $\mathcal{L}_{\mathrm{col}}(\mathbf{S})$ denotes the column space of a matrix S. Show that the estimating equations

$$
\begin{pmatrix}
\mathbf{I} & \mathbf{S}_1 & \mathbf{S}_1 & \cdots & \mathbf{S}_1 \\
\mathbf{S}_2 & \mathbf{I} & \mathbf{S}_2 & \cdots & \mathbf{S}_2 \\
\vdots & \vdots & \vdots & \ddots & \vdots \\
\mathbf{S}_p & \mathbf{S}_p & \mathbf{S}_p & \cdots & \mathbf{I}
\end{pmatrix}
\begin{pmatrix}
\mathbf{f}_1 \\ \mathbf{f}_2 \\ \vdots \\ \mathbf{f}_p
\end{pmatrix}
=
\begin{pmatrix}
\mathbf{S}_1\mathbf{y} \\ \mathbf{S}_2\mathbf{y} \\ \vdots \\ \mathbf{S}_p\mathbf{y}
\end{pmatrix}
$$

are equivalent to the least squares normal equations $\mathbf{D}^T\mathbf{D}\beta = \mathbf{D}^T\mathbf{y}$ where β is the vector of coefficients.

Ex. 9.4 Suppose the same smoother $\mathbf{S}$ is used to estimate both terms in a two-term additive model (i.e., both variables are identical). Assume that $\mathbf{S}$ is symmetric with eigenvalues in $[0, 1)$. Show that the backfitting residual converges to $(\mathbf{I} + \mathbf{S})^{-1}(\mathbf{I} - \mathbf{S})\mathbf{y}$, and that the residual sum of squares converges upward. Can the residual sum of squares converge upward in less structured situations? How does this fit compare to the fit with a single term fit by $\mathbf{S}$? [*Hint*: Use the eigen-decomposition of $\mathbf{S}$ to help with this comparison.]

Ex. 9.5 *Degrees of freedom of a tree.* Given data y_i with mean $f(x_i)$ and variance σ^2, and a fitting operation $\mathbf{y} \to \hat{\mathbf{y}}$, let's define the degrees of freedom of a fit by $\sum_i \text{cov}(y_i, \hat{y}_i)/\sigma^2$.

Consider a fit $\hat{\mathbf{y}}$ estimated by a regression tree, fit to a set of predictors $X_1, X_2, \ldots, X_p$.

(a) In terms of the number of terminal nodes m, give a rough formula for the degrees of freedom of the fit.

(b) Generate 100 observations with predictors $X_1, X_2, \ldots, X_{10}$ as independent standard Gaussian variates and fix these values.

(c) Generate response values also as standard Gaussian ($\sigma^2 = 1$), independent of the predictors. Fit regression trees to the data of fixed size 1,5 and 10 terminal nodes and hence estimate the degrees of freedom of each fit. [Do ten simulations of the response and average the results, to get a good estimate of degrees of freedom.]

(d) Compare your estimates of degrees of freedom in (a) and (c) and discuss.

(e) If the regression tree fit were a linear operation, we could write $\hat{\mathbf{y}} = \mathbf{S}\mathbf{y}$ for some matrix $\mathbf{S}$. Then the degrees of freedom would be $\text{tr}(\mathbf{S})$. Suggest a way to compute an approximate $\mathbf{S}$ matrix for a regression tree, compute it and compare the resulting degrees of freedom to those in (a) and (c).

Ex. 9.6 Consider the ozone data of Figure 6.9.

(a) Fit an additive model to the cube root of ozone concentration. as a function of temperature, wind speed, and radiation. Compare your results to those obtained via the trellis display in Figure 6.9.

(b) Fit trees, MARS, and PRIM to the same data, and compare the results to those found in (a) and in Figure 6.9.

10
Boosting and Additive Trees

10.1 Boosting Methods

Boosting is one of the most powerful learning ideas introduced in the last twenty years. It was originally designed for classification problems, but as will be seen in this chapter, it can profitably be extended to regression as well. The motivation for boosting was a procedure that combines the outputs of many "weak" classifiers to produce a powerful "committee." From this perspective boosting bears a resemblance to bagging and other committee-based approaches (Section 8.8). However we shall see that the connection is at best superficial and that boosting is fundamentally different.

We begin by describing the most popular boosting algorithm due to Freund and Schapire (1997) called "AdaBoost.M1." Consider a two-class problem, with the output variable coded as $Y \in \{-1, 1\}$. Given a vector of predictor variables X, a classifier $G(X)$ produces a prediction taking one of the two values $\{-1, 1\}$. The error rate on the training sample is

$$\overline{\mathrm{err}} = \frac{1}{N} \sum_{i=1}^{N} I(y_i \neq G(x_i)),$$

and the expected error rate on future predictions is $\mathrm{E}_{XY} I(Y \neq G(X))$.

A weak classifier is one whose error rate is only slightly better than random guessing. The purpose of boosting is to sequentially apply the weak classification algorithm to repeatedly modified versions of the data, thereby producing a sequence of weak classifiers $G_m(x), m = 1, 2, \ldots, M$.

T. Hastie et al., *The Elements of Statistical Learning, Second Edition*,
DOI: 10.1007/b94608_10,
© Springer Science+Business Media, LLC 2009

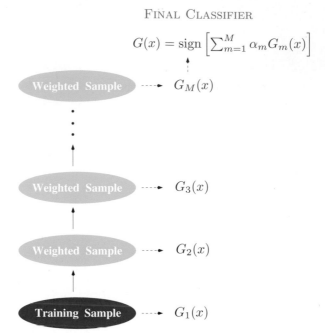

FINAL CLASSIFIER

$$G(x) = \text{sign}\left[\sum_{m=1}^{M} \alpha_m G_m(x)\right]$$

Weighted Sample ----→ $G_M(x)$

Weighted Sample ----→ $G_3(x)$

Weighted Sample ----→ $G_2(x)$

Training Sample ----→ $G_1(x)$

FIGURE 10.1. *Schematic of AdaBoost. Classifiers are trained on weighted versions of the dataset, and then combined to produce a final prediction.*

The predictions from all of them are then combined through a weighted majority vote to produce the final prediction:

$$G(x) = \text{sign}\left(\sum_{m=1}^{M} \alpha_m G_m(x)\right). \tag{10.1}$$

Here $\alpha_1, \alpha_2, \ldots, \alpha_M$ are computed by the boosting algorithm, and weight the contribution of each respective $G_m(x)$. Their effect is to give higher influence to the more accurate classifiers in the sequence. Figure 10.1 shows a schematic of the AdaBoost procedure.

The data modifications at each boosting step consist of applying weights $w_1, w_2, \ldots, w_N$ to each of the training observations (x_i, y_i), $i = 1, 2, \ldots, N$. Initially all of the weights are set to $w_i = 1/N$, so that the first step simply trains the classifier on the data in the usual manner. For each successive iteration $m = 2, 3, \ldots, M$ the observation weights are individually modified and the classification algorithm is reapplied to the weighted observations. At step m, those observations that were misclassified by the classifier $G_{m-1}(x)$ induced at the previous step have their weights increased, whereas the weights are decreased for those that were classified correctly. Thus as iterations proceed, observations that are difficult to classify correctly receive ever-increasing influence. Each successive classifier is thereby forced

---

Algorithm 10.1 *AdaBoost.M1.*

---

1. Initialize the observation weights $w_i = 1/N$, $i = 1, 2, \ldots, N$.

2. For $m = 1$ to M:

   (a) Fit a classifier $G_m(x)$ to the training data using weights w_i.

   (b) Compute
   $$\text{err}_m = \frac{\sum_{i=1}^{N} w_i I(y_i \neq G_m(x_i))}{\sum_{i=1}^{N} w_i}.$$

   (c) Compute $\alpha_m = \log((1 - \text{err}_m)/\text{err}_m)$.

   (d) Set $w_i \leftarrow w_i \cdot \exp[\alpha_m \cdot I(y_i \neq G_m(x_i))]$, $i = 1, 2, \ldots, N$.

3. Output $G(x) = \text{sign}\left[\sum_{m=1}^{M} \alpha_m G_m(x)\right]$.

---

to concentrate on those training observations that are missed by previous ones in the sequence.

Algorithm 10.1 shows the details of the AdaBoost.M1 algorithm. The current classifier $G_m(x)$ is induced on the weighted observations at line 2a. The resulting weighted error rate is computed at line 2b. Line 2c calculates the weight α_m given to $G_m(x)$ in producing the final classifier $G(x)$ (line 3). The individual weights of each of the observations are updated for the next iteration at line 2d. Observations misclassified by $G_m(x)$ have their weights scaled by a factor $\exp(\alpha_m)$, increasing their relative influence for inducing the next classifier $G_{m+1}(x)$ in the sequence.

The AdaBoost.M1 algorithm is known as "Discrete AdaBoost" in Friedman et al. (2000), because the base classifier $G_m(x)$ returns a discrete class label. If the base classifier instead returns a real-valued prediction (e.g., a probability mapped to the interval $[-1, 1]$), AdaBoost can be modified appropriately (see "Real AdaBoost" in Friedman et al. (2000)).

The power of AdaBoost to dramatically increase the performance of even a very weak classifier is illustrated in Figure 10.2. The features $X_1, \ldots, X_{10}$ are standard independent Gaussian, and the deterministic target Y is defined by

$$Y = \begin{cases} 1 & \text{if } \sum_{j=1}^{10} X_j^2 > \chi_{10}^2(0.5), \\ -1 & \text{otherwise.} \end{cases} \tag{10.2}$$

Here $\chi_{10}^2(0.5) = 9.34$ is the median of a chi-squared random variable with 10 degrees of freedom (sum of squares of 10 standard Gaussians). There are 2000 training cases, with approximately 1000 cases in each class, and 10,000 test observations. Here the weak classifier is just a "stump": a two terminal-node classification tree. Applying this classifier alone to the training data set yields a very poor test set error rate of 45.8%, compared to 50% for

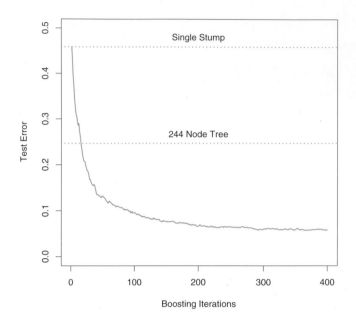

FIGURE 10.2. *Simulated data (10.2): test error rate for boosting with stumps, as a function of the number of iterations. Also shown are the test error rate for a single stump, and a 244-node classification tree.*

random guessing. However, as boosting iterations proceed the error rate steadily decreases, reaching 5.8% after 400 iterations. Thus, boosting this simple very weak classifier reduces its prediction error rate by almost a factor of four. It also outperforms a single large classification tree (error rate 24.7%). Since its introduction, much has been written to explain the success of AdaBoost in producing accurate classifiers. Most of this work has centered on using classification trees as the "base learner" $G(x)$, where improvements are often most dramatic. In fact, Breiman (NIPS Workshop, 1996) referred to AdaBoost with trees as the "best off-the-shelf classifier in the world" (see also Breiman (1998)). This is especially the case for data-mining applications, as discussed more fully in Section 10.7 later in this chapter.

10.1.1 Outline of This Chapter

Here is an outline of the developments in this chapter:

- We show that AdaBoost fits an additive model in a base learner, optimizing a novel exponential loss function. This loss function is very similar to the (negative) binomial log-likelihood (Sections 10.2–10.4).

- The population minimizer of the exponential loss function is shown to be the log-odds of the class probabilities (Section 10.5).

- We describe loss functions for regression and classification that are more robust than squared error or exponential loss (Section 10.6).

- It is argued that decision trees are an ideal base learner for data mining applications of boosting (Sections 10.7 and 10.9).

- We develop a class of gradient boosted models (GBMs), for boosting trees with any loss function (Section 10.10).

- The importance of "slow learning" is emphasized, and implemented by shrinkage of each new term that enters the model (Section 10.12), as well as randomization (Section 10.12.2).

- Tools for interpretation of the fitted model are described (Section 10.13).

10.2 Boosting Fits an Additive Model

The success of boosting is really not very mysterious. The key lies in expression (10.1). Boosting is a way of fitting an additive expansion in a set of elementary "basis" functions. Here the basis functions are the individual classifiers $G_m(x) \in \{-1, 1\}$. More generally, basis function expansions take the form

$$f(x) = \sum_{m=1}^{M} \beta_m b(x; \gamma_m), \tag{10.3}$$

where $\beta_m, m = 1, 2, \ldots, M$ are the expansion coefficients, and $b(x; \gamma) \in \mathbb{R}$ are usually simple functions of the multivariate argument x, characterized by a set of parameters γ. We discuss basis expansions in some detail in Chapter 5.

Additive expansions like this are at the heart of many of the learning techniques covered in this book:

- In single-hidden-layer neural networks (Chapter 11), $b(x; \gamma) = \sigma(\gamma_0 + \gamma_1^T x)$, where $\sigma(t) = 1/(1+e^{-t})$ is the sigmoid function, and γ parameterizes a linear combination of the input variables.

- In signal processing, wavelets (Section 5.9.1) are a popular choice with γ parameterizing the location and scale shifts of a "mother" wavelet.

- Multivariate adaptive regression splines (Section 9.4) uses truncated-power spline basis functions where γ parameterizes the variables and values for the knots.

Algorithm 10.2 *Forward Stagewise Additive Modeling.*

---

1. Initialize $f_0(x) = 0$.

2. For $m = 1$ to M:

   (a) Compute

   $$(\beta_m, \gamma_m) = \arg\min_{\beta, \gamma} \sum_{i=1}^{N} L(y_i, f_{m-1}(x_i) + \beta b(x_i; \gamma)).$$

   (b) Set $f_m(x) = f_{m-1}(x) + \beta_m b(x; \gamma_m)$.

- For trees, γ parameterizes the split variables and split points at the internal nodes, and the predictions at the terminal nodes.

Typically these models are fit by minimizing a loss function averaged over the training data, such as the squared-error or a likelihood-based loss function,

$$\min_{\{\beta_m, \gamma_m\}_1^M} \sum_{i=1}^{N} L\left(y_i, \sum_{m=1}^{M} \beta_m b(x_i; \gamma_m)\right). \tag{10.4}$$

For many loss functions $L(y, f(x))$ and/or basis functions $b(x; \gamma)$, this requires computationally intensive numerical optimization techniques. However, a simple alternative often can be found when it is feasible to rapidly solve the subproblem of fitting just a single basis function,

$$\min_{\beta, \gamma} \sum_{i=1}^{N} L\left(y_i, \beta b(x_i; \gamma)\right). \tag{10.5}$$

10.3   Forward Stagewise Additive Modeling

Forward stagewise modeling approximates the solution to (10.4) by sequentially adding new basis functions to the expansion without adjusting the parameters and coefficients of those that have already been added. This is outlined in Algorithm 10.2. At each iteration m, one solves for the optimal basis function $b(x; \gamma_m)$ and corresponding coefficient β_m to add to the current expansion $f_{m-1}(x)$. This produces $f_m(x)$, and the process is repeated. Previously added terms are not modified.

For squared-error loss

$$L(y, f(x)) = (y - f(x))^2, \tag{10.6}$$

one has

$$
\begin{aligned}
L(y_i, f_{m-1}(x_i) + \beta b(x_i; \gamma)) &= (y_i - f_{m-1}(x_i) - \beta b(x_i; \gamma))^2 \\
&= (r_{im} - \beta b(x_i; \gamma))^2, \quad (10.7)
\end{aligned}
$$

where $r_{im} = y_i - f_{m-1}(x_i)$ is simply the residual of the current model on the ith observation. Thus, for squared-error loss, the term $\beta_m b(x; \gamma_m)$ that best fits the current residuals is added to the expansion at each step. This idea is the basis for "least squares" regression boosting discussed in Section 10.10.2. However, as we show near the end of the next section, squared-error loss is generally not a good choice for classification; hence the need to consider other loss criteria.

10.4 Exponential Loss and AdaBoost

We now show that AdaBoost.M1 (Algorithm 10.1) is equivalent to forward stagewise additive modeling (Algorithm 10.2) using the loss function

$$
L(y, f(x)) = \exp(-y f(x)). \quad (10.8)
$$

The appropriateness of this criterion is addressed in the next section.

For AdaBoost the basis functions are the individual classifiers $G_m(x) \in \{-1, 1\}$. Using the exponential loss function, one must solve

$$
(\beta_m, G_m) = \arg\min_{\beta, G} \sum_{i=1}^{N} \exp[-y_i(f_{m-1}(x_i) + \beta\, G(x_i))]
$$

for the classifier G_m and corresponding coefficient β_m to be added at each step. This can be expressed as

$$
(\beta_m, G_m) = \arg\min_{\beta, G} \sum_{i=1}^{N} w_i^{(m)} \exp(-\beta\, y_i\, G(x_i)) \quad (10.9)
$$

with $w_i^{(m)} = \exp(-y_i\, f_{m-1}(x_i))$. Since each $w_i^{(m)}$ depends neither on β nor $G(x)$, it can be regarded as a weight that is applied to each observation. This weight depends on $f_{m-1}(x_i)$, and so the individual weight values change with each iteration m.

The solution to (10.9) can be obtained in two steps. First, for any value of $\beta > 0$, the solution to (10.9) for $G_m(x)$ is

$$
G_m = \arg\min_{G} \sum_{i=1}^{N} w_i^{(m)} I(y_i \neq G(x_i)), \quad (10.10)
$$

which is the classifier that minimizes the weighted error rate in predicting y. This can be easily seen by expressing the criterion in (10.9) as

$$e^{-\beta} \cdot \sum_{y_i = G(x_i)} w_i^{(m)} + e^{\beta} \cdot \sum_{y_i \neq G(x_i)} w_i^{(m)},$$

which in turn can be written as

$$\left(e^{\beta} - e^{-\beta}\right) \cdot \sum_{i=1}^{N} w_i^{(m)} I(y_i \neq G(x_i)) + e^{-\beta} \cdot \sum_{i=1}^{N} w_i^{(m)}. \tag{10.11}$$

Plugging this G_m into (10.9) and solving for β one obtains

$$\beta_m = \frac{1}{2} \log \frac{1 - \text{err}_m}{\text{err}_m}, \tag{10.12}$$

where err_m is the minimized weighted error rate

$$\text{err}_m = \frac{\sum_{i=1}^{N} w_i^{(m)} I(y_i \neq G_m(x_i))}{\sum_{i=1}^{N} w_i^{(m)}}. \tag{10.13}$$

The approximation is then updated

$$f_m(x) = f_{m-1}(x) + \beta_m G_m(x),$$

which causes the weights for the next iteration to be

$$w_i^{(m+1)} = w_i^{(m)} \cdot e^{-\beta_m y_i G_m(x_i)}. \tag{10.14}$$

Using the fact that $-y_i G_m(x_i) = 2 \cdot I(y_i \neq G_m(x_i)) - 1$, (10.14) becomes

$$w_i^{(m+1)} = w_i^{(m)} \cdot e^{\alpha_m I(y_i \neq G_m(x_i))} \cdot e^{-\beta_m}, \tag{10.15}$$

where $\alpha_m = 2\beta_m$ is the quantity defined at line 2(c) of AdaBoost.M1 (Algorithm 10.1). The factor $e^{-\beta_m}$ in (10.15) multiplies all weights by the same value, so it has no effect. Thus (10.15) is equivalent to line 2(d) of Algorithm 10.1.

One can view line 2(a) of the Adaboost.M1 algorithm as a method for approximately solving the minimization in (10.11) and hence (10.10). Hence we conclude that AdaBoost.M1 minimizes the exponential loss criterion (10.8) via a forward-stagewise additive modeling approach.

Figure 10.3 shows the training-set misclassification error rate and average exponential loss for the simulated data problem (10.2) of Figure 10.2. The training-set misclassification error decreases to zero at around 250 iterations (and remains there), but the exponential loss keeps decreasing. Notice also in Figure 10.2 that the test-set misclassification error continues to improve after iteration 250. Clearly Adaboost is not optimizing training-set misclassification error; the exponential loss is more sensitive to changes in the estimated class probabilities.

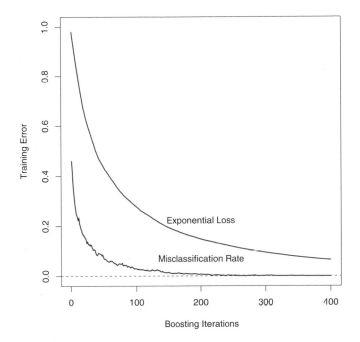

FIGURE 10.3. *Simulated data, boosting with stumps: misclassification error rate on the training set, and average exponential loss:* $(1/N)\sum_{i=1}^{N}\exp(-y_i f(x_i))$. *After about* 250 *iterations, the misclassification error is zero, while the exponential loss continues to decrease.*

10.5 Why Exponential Loss?

The AdaBoost.M1 algorithm was originally motivated from a very different perspective than presented in the previous section. Its equivalence to forward stagewise additive modeling based on exponential loss was only discovered five years after its inception. By studying the properties of the exponential loss criterion, one can gain insight into the procedure and discover ways it might be improved.

The principal attraction of exponential loss in the context of additive modeling is computational; it leads to the simple modular reweighting AdaBoost algorithm. However, it is of interest to inquire about its statistical properties. What does it estimate and how well is it being estimated? The first question is answered by seeking its population minimizer.

It is easy to show (Friedman et al., 2000) that

$$f^*(x) = \arg\min_{f(x)} \mathrm{E}_{Y|x}(e^{-Yf(x)}) = \frac{1}{2}\log\frac{\Pr(Y=1|x)}{\Pr(Y=-1|x)}, \qquad (10.16)$$

or equivalently

$$\Pr(Y = 1|x) = \frac{1}{1 + e^{-2f^*(x)}}.$$

Thus, the additive expansion produced by AdaBoost is estimating one-half the log-odds of $P(Y = 1|x)$. This justifies using its sign as the classification rule in (10.1).

Another loss criterion with the same population minimizer is the binomial negative log-likelihood or *deviance* (also known as cross-entropy), interpreting f as the logit transform. Let

$$p(x) = \Pr(Y = 1\,|\,x) = \frac{e^{f(x)}}{e^{-f(x)} + e^{f(x)}} = \frac{1}{1 + e^{-2f(x)}} \tag{10.17}$$

and define $Y' = (Y + 1)/2 \in \{0, 1\}$. Then the binomial log-likelihood loss function is

$$l(Y, p(x)) = Y' \log p(x) + (1 - Y') \log(1 - p(x)),$$

or equivalently the deviance is

$$-l(Y, f(x)) = \log\left(1 + e^{-2Y f(x)}\right). \tag{10.18}$$

Since the population maximizer of log-likelihood is at the true probabilities $p(x) = \Pr(Y = 1\,|\,x)$, we see from (10.17) that the population minimizers of the deviance $E_{Y|x}[-l(Y, f(x))]$ and $E_{Y|x}[e^{-Yf(x)}]$ are the same. Thus, using either criterion leads to the same solution at the population level. Note that e^{-Yf} itself is not a proper log-likelihood, since it is not the logarithm of any probability mass function for a binary random variable $Y \in \{-1, 1\}$.

10.6 Loss Functions and Robustness

In this section we examine the different loss functions for classification and regression more closely, and characterize them in terms of their robustness to extreme data.

Robust Loss Functions for Classification

Although both the exponential (10.8) and binomial deviance (10.18) yield the same solution when applied to the population joint distribution, the same is not true for finite data sets. Both criteria are monotone decreasing functions of the "margin" $yf(x)$. In classification (with a $-1/1$ response) the margin plays a role analogous to the residuals $y - f(x)$ in regression. The classification rule $G(x) = \text{sign}[f(x)]$ implies that observations with positive margin $y_i f(x_i) > 0$ are classified correctly whereas those with negative margin $y_i f(x_i) < 0$ are misclassified. The decision boundary is defined by

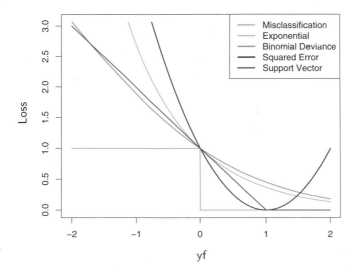

FIGURE 10.4. *Loss functions for two-class classification. The response is* $y = \pm 1$; *the prediction is* f, *with class prediction* $\text{sign}(f)$. *The losses are misclassification:* $I(\text{sign}(f) \neq y)$; *exponential:* $\exp(-yf)$; *binomial deviance:* $\log(1 + \exp(-2yf))$; *squared error:* $(y - f)^2$; *and support vector:* $(1 - yf)_+$ *(see Section 12.3). Each function has been scaled so that it passes through the point* $(0, 1)$.

$f(x) = 0$. The goal of the classification algorithm is to produce positive margins as frequently as possible. Any loss criterion used for classification should penalize negative margins more heavily than positive ones since positive margin observations are already correctly classified.

Figure 10.4 shows both the exponential (10.8) and binomial deviance criteria as a function of the margin $yf(x)$. Also shown is misclassification loss $L(y, f(x)) = I(yf(x) < 0)$, which gives unit penalty for negative margin values, and no penalty at all for positive ones. Both the exponential and deviance loss can be viewed as monotone continuous approximations to misclassification loss. They continuously penalize increasingly negative margin values more heavily than they reward increasingly positive ones. The difference between them is in degree. The penalty associated with binomial deviance increases linearly for large increasingly negative margin, whereas the exponential criterion increases the influence of such observations exponentially.

At any point in the training process the exponential criterion concentrates much more influence on observations with large negative margins. Binomial deviance concentrates relatively less influence on such observations, more evenly spreading the influence among all of the data. It is therefore far more robust in noisy settings where the Bayes error rate is not close to zero, and especially in situations where there is misspecification of the class labels in the training data. The performance of AdaBoost has been empirically observed to dramatically degrade in such situations.

Also shown in the figure is squared-error loss. The minimizer of the corresponding risk on the population is

$$f^*(x) = \arg\min_{f(x)} \mathrm{E}_{Y|x}(Y - f(x))^2 = \mathrm{E}(Y \,|\, x) = 2{\cdot}\Pr(Y = 1 \,|\, x) - 1. \quad (10.19)$$

As before the classification rule is $G(x) = \mathrm{sign}[f(x)]$. Squared-error loss is not a good surrogate for misclassification error. As seen in Figure 10.4, it is not a monotone decreasing function of increasing margin $yf(x)$. For margin values $y_i f(x_i) > 1$ it increases quadratically, thereby placing increasing influence (error) on observations that are correctly classified with increasing certainty, thereby reducing the relative influence of those incorrectly classified $y_i f(x_i) < 0$. Thus, if class assignment is the goal, a monotone decreasing criterion serves as a better surrogate loss function. Figure 12.4 on page 426 in Chapter 12 includes a modification of quadratic loss, the "Huberized" square hinge loss (Rosset et al., 2004b), which enjoys the favorable properties of the binomial deviance, quadratic loss and the SVM hinge loss. It has the same population minimizer as the quadratic (10.19), is zero for $yf(x) > 1$, and becomes linear for $yf(x) < -1$. Since quadratic functions are easier to compute with than exponentials, our experience suggests this to be a useful alternative to the binomial deviance.

With K-class classification, the response Y takes values in the unordered set $\mathcal{G} = \{\mathcal{G}_1, \ldots, \mathcal{G}_k\}$ (see Sections 2.4 and 4.4). We now seek a classifier $G(x)$ taking values in $\mathcal{G}$. It is sufficient to know the class conditional probabilities $p_k(x) = \Pr(Y = \mathcal{G}_k|x), k = 1, 2, \ldots, K$, for then the Bayes classifier is

$$G(x) = \mathcal{G}_k \text{ where } k = \arg\max_\ell p_\ell(x). \quad (10.20)$$

In principal, though, we need not learn the $p_k(x)$, but simply which one is largest. However, in data mining applications the interest is often more in the class probabilities $p_\ell(x)$, $\ell = 1, \ldots, K$ themselves, rather than in performing a class assignment. As in Section 4.4, the logistic model generalizes naturally to K classes,

$$p_k(x) = \frac{e^{f_k(x)}}{\sum_{l=1}^{K} e^{f_l(x)}}, \quad (10.21)$$

which ensures that $0 \le p_k(x) \le 1$ and that they sum to one. Note that here we have K different functions, one per class. There is a redundancy in the functions $f_k(x)$, since adding an arbitrary $h(x)$ to each leaves the model unchanged. Traditionally one of them is set to zero: for example,

$f_K(x) = 0$, as in (4.17). Here we prefer to retain the symmetry, and impose the constraint $\sum_{k=1}^{K} f_k(x) = 0$. The binomial deviance extends naturally to the K-class *multinomial deviance* loss function:

$$
\begin{aligned}
L(y, p(x)) &= -\sum_{k=1}^{K} I(y = \mathcal{G}_k) \log p_k(x) \\
&= -\sum_{k=1}^{K} I(y = \mathcal{G}_k) f_k(x) + \log \left(\sum_{\ell=1}^{K} e^{f_\ell(x)} \right). \quad (10.22)
\end{aligned}
$$

As in the two-class case, the criterion (10.22) penalizes incorrect predictions only linearly in their degree of incorrectness.

Zhu et al. (2005) generalize the exponential loss for K-class problems. See Exercise 10.5 for details.

Robust Loss Functions for Regression

In the regression setting, analogous to the relationship between exponential loss and binomial log-likelihood is the relationship between squared-error loss $L(y, f(x)) = (y - f(x))^2$ and absolute loss $L(y, f(x)) = |y - f(x)|$. The population solutions are $f(x) = E(Y|x)$ for squared-error loss, and $f(x) = \text{median}(Y|x)$ for absolute loss; for symmetric error distributions these are the same. However, on finite samples squared-error loss places much more emphasis on observations with large absolute residuals $|y_i - f(x_i)|$ during the fitting process. It is thus far less robust, and its performance severely degrades for long-tailed error distributions and especially for grossly mis-measured y-values ("outliers"). Other more robust criteria, such as absolute loss, perform much better in these situations. In the statistical robustness literature, a variety of regression loss criteria have been proposed that provide strong resistance (if not absolute immunity) to gross outliers while being nearly as efficient as least squares for Gaussian errors. They are often better than either for error distributions with moderately heavy tails. One such criterion is the Huber loss criterion used for M-regression (Huber, 1964)

$$
L(y, f(x)) = \begin{cases} [y - f(x)]^2 & \text{for } |y - f(x)| \le \delta, \\ 2\delta |y - f(x)| - \delta^2 & \text{otherwise.} \end{cases} \quad (10.23)
$$

Figure 10.5 compares these three loss functions.

These considerations suggest that when robustness is a concern, as is especially the case in data mining applications (see Section 10.7), squared-error loss for regression and exponential loss for classification are not the best criteria from a statistical perspective. However, they both lead to the elegant modular boosting algorithms in the context of forward stagewise additive modeling. For squared-error loss one simply fits the base learner to the residuals from the current model $y_i - f_{m-1}(x_i)$ at each step. For

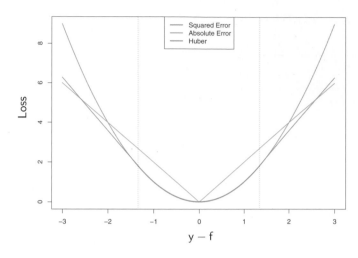

FIGURE 10.5. *A comparison of three loss functions for regression, plotted as a function of the margin $y-f$. The Huber loss function combines the good properties of squared-error loss near zero and absolute error loss when $|y - f|$ is large.*

exponential loss one performs a weighted fit of the base learner to the output values y_i, with weights $w_i = \exp(-y_i f_{m-1}(x_i))$. Using other more robust criteria directly in their place does not give rise to such simple feasible boosting algorithms. However, in Section 10.10.2 we show how one can derive simple elegant boosting algorithms based on any differentiable loss criterion, thereby producing highly robust boosting procedures for data mining.

10.7 "Off-the-Shelf" Procedures for Data Mining

Predictive learning is an important aspect of data mining. As can be seen from this book, a wide variety of methods have been developed for predictive learning from data. For each particular method there are situations for which it is particularly well suited, and others where it performs badly compared to the best that can be done with that data. We have attempted to characterize appropriate situations in our discussions of each of the respective methods. However, it is seldom known in advance which procedure will perform best or even well for any given problem. Table 10.1 summarizes some of the characteristics of a number of learning methods.

Industrial and commercial data mining applications tend to be especially challenging in terms of the requirements placed on learning procedures. Data sets are often very large in terms of number of observations and number of variables measured on each of them. Thus, computational con-

TABLE 10.1. *Some characteristics of different learning methods. Key:* ▲ = *good,* ◆ =*fair, and* ▼ =*poor.*

Characteristic	Neural Nets	SVM	Trees	MARS	k-NN, Kernels
Natural handling of data of "mixed" type	▼	▼	▲	▲	▼
Handling of missing values	▼	▼	▲	▲	▲
Robustness to outliers in input space	▼	▼	▲	▼	▲
Insensitive to monotone transformations of inputs	▼	▼	▲	▼	▼
Computational scalability (large N)	▼	▼	▲	▲	▼
Ability to deal with irrelevant inputs	▼	▼	▲	▲	▼
Ability to extract linear combinations of features	▲	▲	▼	▼	◆
Interpretability	▼	▼	◆	▲	▼
Predictive power	▲	▲	▼	◆	▲

siderations play an important role. Also, the data are usually *messy*: the inputs tend to be mixtures of quantitative, binary, and categorical variables, the latter often with many levels. There are generally many missing values, complete observations being rare. Distributions of numeric predictor and response variables are often long-tailed and highly skewed. This is the case for the spam data (Section 9.1.2); when fitting a generalized additive model, we first log-transformed each of the predictors in order to get a reasonable fit. In addition they usually contain a substantial fraction of gross mis-measurements (outliers). The predictor variables are generally measured on very different scales.

In data mining applications, usually only a small fraction of the large number of predictor variables that have been included in the analysis are actually relevant to prediction. Also, unlike many applications such as pattern recognition, there is seldom reliable domain knowledge to help create especially relevant features and/or filter out the irrelevant ones, the inclusion of which dramatically degrades the performance of many methods.

In addition, data mining applications generally require interpretable models. It is not enough to simply produce predictions. It is also desirable to have information providing qualitative understanding of the relationship between joint values of the input variables and the resulting predicted response value. Thus, *black box* methods such as neural networks, which can be quite useful in purely predictive settings such as pattern recognition, are far less useful for data mining.

These requirements of speed, interpretability and the messy nature of the data sharply limit the usefulness of most learning procedures as off-the-shelf methods for data mining. An "off-the-shelf" method is one that can be directly applied to the data without requiring a great deal of time-consuming data preprocessing or careful tuning of the learning procedure.

Of all the well-known learning methods, decision trees come closest to meeting the requirements for serving as an off-the-shelf procedure for data mining. They are relatively fast to construct and they produce interpretable models (if the trees are small). As discussed in Section 9.2, they naturally incorporate mixtures of numeric and categorical predictor variables and missing values. They are invariant under (strictly monotone) transformations of the individual predictors. As a result, scaling and/or more general transformations are not an issue, and they are immune to the effects of predictor outliers. They perform internal feature selection as an integral part of the procedure. They are thereby resistant, if not completely immune, to the inclusion of many irrelevant predictor variables. These properties of decision trees are largely the reason that they have emerged as the most popular learning method for data mining.

Trees have one aspect that prevents them from being the ideal tool for predictive learning, namely inaccuracy. They seldom provide predictive accuracy comparable to the best that can be achieved with the data at hand. As seen in Section 10.1, boosting decision trees improves their accuracy, often dramatically. At the same time it maintains most of their desirable properties for data mining. Some advantages of trees that are sacrificed by boosting are speed, interpretability, and, for AdaBoost, robustness against overlapping class distributions and especially mislabeling of the training data. A gradient boosted model (GBM) is a generalization of tree boosting that attempts to mitigate these problems, so as to produce an accurate and effective off-the-shelf procedure for data mining.

10.8   Example: Spam Data

Before we go into the details of gradient boosting, we demonstrate its abilities on a two-class classification problem. The spam data are introduced in Chapter 1, and used as an example for many of the procedures in Chapter 9 (Sections 9.1.2, 9.2.5, 9.3.1 and 9.4.1).

Applying gradient boosting to these data resulted in a test error rate of 4.5%, using the same test set as was used in Section 9.1.2. By comparison, an additive logistic regression achieved 5.5%, a CART tree fully grown and pruned by cross-validation 8.7%, and MARS 5.5%. The standard error of these estimates is around 0.6%, although gradient boosting is significantly better than all of them using the McNemar test (Exercise 10.6).

In Section 10.13 below we develop a relative importance measure for each predictor, as well as a partial dependence plot describing a predictor's contribution to the fitted model. We now illustrate these for the spam data.

Figure 10.6 displays the relative importance spectrum for all 57 predictor variables. Clearly some predictors are more important than others in separating spam from email. The frequencies of the character strings !, $, hp, and remove are estimated to be the four most relevant predictor variables. At the other end of the spectrum, the character strings 857, 415, table, and 3d have virtually no relevance.

The quantity being modeled here is the log-odds of spam versus email

$$f(x) = \log \frac{\Pr(\texttt{spam}|x)}{\Pr(\texttt{email}|x)} \tag{10.24}$$

(see Section 10.13 below). Figure 10.7 shows the partial dependence of the log-odds on selected important predictors, two positively associated with spam (! and remove), and two negatively associated (edu and hp). These particular dependencies are seen to be essentially monotonic. There is a general agreement with the corresponding functions found by the additive logistic regression model; see Figure 9.1 on page 303.

Running a gradient boosted model on these data with $J = 2$ terminal-node trees produces a purely additive (main effects) model for the log-odds, with a corresponding error rate of 4.7%, as compared to 4.5% for the full gradient boosted model (with $J = 5$ terminal-node trees). Although not significant, this slightly higher error rate suggests that there may be interactions among some of the important predictor variables. This can be diagnosed through two-variable partial dependence plots. Figure 10.8 shows one of the several such plots displaying strong interaction effects.

One sees that for very low frequencies of hp, the log-odds of spam are greatly increased. For high frequencies of hp, the log-odds of spam tend to be much lower and roughly constant as a function of !. As the frequency of hp decreases, the functional relationship with ! strengthens.

10.9 Boosting Trees

Regression and classification trees are discussed in detail in Section 9.2. They partition the space of all joint predictor variable values into disjoint regions $R_j, j = 1, 2, \ldots, J$, as represented by the terminal nodes of the tree. A constant γ_j is assigned to each such region and the predictive rule is

$$x \in R_j \Rightarrow f(x) = \gamma_j.$$

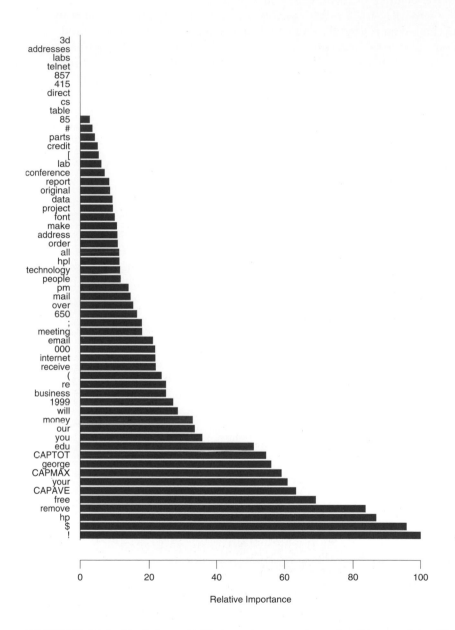

FIGURE 10.6. *Predictor variable importance spectrum for the* spam *data. The variable names are written on the vertical axis.*

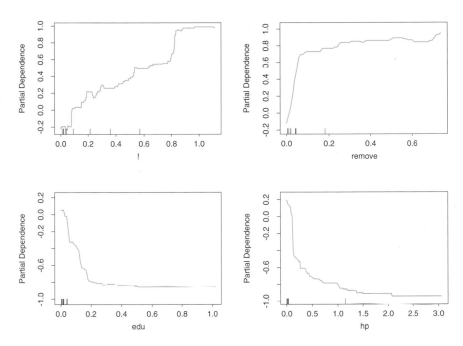

FIGURE 10.7. *Partial dependence of log-odds of* `spam` *on four important predictors. The red ticks at the base of the plots are deciles of the input variable.*

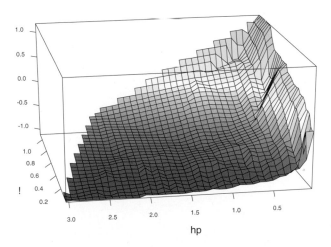

FIGURE 10.8. *Partial dependence of the log-odds of* `spam` *vs.* `email` *as a function of joint frequencies of* `hp` *and the character* !.

Thus a tree can be formally expressed as

$$T(x; \Theta) = \sum_{j=1}^{J} \gamma_j I(x \in R_j),$$ (10.25)

with parameters $\Theta = \{R_j, \gamma_j\}_1^J$. J is usually treated as a meta-parameter. The parameters are found by minimizing the empirical risk

$$\hat{\Theta} = \arg\min_{\Theta} \sum_{j=1}^{J} \sum_{x_i \in R_j} L(y_i, \gamma_j).$$ (10.26)

This is a formidable combinatorial optimization problem, and we usually settle for approximate suboptimal solutions. It is useful to divide the optimization problem into two parts:

Finding γ_j given R_j: Given the R_j, estimating the γ_j is typically trivial, and often $\hat{\gamma}_j = \bar{y}_j$, the mean of the y_i falling in region R_j. For misclassification loss, $\hat{\gamma}_j$ is the modal class of the observations falling in region R_j.

Finding R_j: This is the difficult part, for which approximate solutions are found. Note also that finding the R_j entails estimating the γ_j as well. A typical strategy is to use a greedy, top-down recursive partitioning algorithm to find the R_j. In addition, it is sometimes necessary to approximate (10.26) by a smoother and more convenient criterion for optimizing the R_j:

$$\tilde{\Theta} = \arg\min_{\Theta} \sum_{i=1}^{N} \tilde{L}(y_i, T(x_i, \Theta)).$$ (10.27)

Then given the $\hat{R}_j = \tilde{R}_j$, the γ_j can be estimated more precisely using the original criterion.

In Section 9.2 we described such a strategy for classification trees. The Gini index replaced misclassification loss in the growing of the tree (identifying the R_j).

The boosted tree model is a sum of such trees,

$$f_M(x) = \sum_{m=1}^{M} T(x; \Theta_m),$$ (10.28)

induced in a forward stagewise manner (Algorithm 10.2). At each step in the forward stagewise procedure one must solve

$$\hat{\Theta}_m = \arg\min_{\Theta_m} \sum_{i=1}^{N} L(y_i, f_{m-1}(x_i) + T(x_i; \Theta_m))$$ (10.29)

for the region set and constants $\Theta_m = \{R_{jm}, \gamma_{jm}\}_1^{J_m}$ of the next tree, given the current model $f_{m-1}(x)$.

Given the regions R_{jm}, finding the optimal constants γ_{jm} in each region is typically straightforward:

$$\hat{\gamma}_{jm} = \arg\min_{\gamma_{jm}} \sum_{x_i \in R_{jm}} L\left(y_i, f_{m-1}(x_i) + \gamma_{jm}\right). \qquad (10.30)$$

Finding the regions is difficult, and even more difficult than for a single tree. For a few special cases, the problem simplifies.

For squared-error loss, the solution to (10.29) is no harder than for a single tree. It is simply the regression tree that best predicts the current residuals $y_i - f_{m-1}(x_i)$, and $\hat{\gamma}_{jm}$ is the mean of these residuals in each corresponding region.

For two-class classification and exponential loss, this stagewise approach gives rise to the AdaBoost method for boosting classification trees (Algorithm 10.1). In particular, if the trees $T(x; \Theta_m)$ are restricted to be scaled classification trees, then we showed in Section 10.4 that the solution to (10.29) is the tree that minimizes the weighted error rate $\sum_{i=1}^N w_i^{(m)} I(y_i \neq T(x_i; \Theta_m))$ with weights $w_i^{(m)} = e^{-y_i f_{m-1}(x_i)}$. By a scaled classification tree, we mean $\beta_m T(x; \Theta_m)$, with the restriction that $\gamma_{jm} \in \{-1, 1\}$).

Without this restriction, (10.29) still simplifies for exponential loss to a weighted exponential criterion for the new tree:

$$\hat{\Theta}_m = \arg\min_{\Theta_m} \sum_{i=1}^N w_i^{(m)} \exp[-y_i T(x_i; \Theta_m)]. \qquad (10.31)$$

It is straightforward to implement a greedy recursive-partitioning algorithm using this weighted exponential loss as a splitting criterion. Given the R_{jm}, one can show (Exercise 10.7) that the solution to (10.30) is the weighted log-odds in each corresponding region

$$\hat{\gamma}_{jm} = \frac{1}{2} \log \frac{\sum_{x_i \in R_{jm}} w_i^{(m)} I(y_i = 1)}{\sum_{x_i \in R_{jm}} w_i^{(m)} I(y_i = -1)}. \qquad (10.32)$$

This requires a specialized tree-growing algorithm; in practice, we prefer the approximation presented below that uses a weighted least squares regression tree.

Using loss criteria such as the absolute error or the Huber loss (10.23) in place of squared-error loss for regression, and the deviance (10.22) in place of exponential loss for classification, will serve to robustify boosting trees. Unfortunately, unlike their nonrobust counterparts, these robust criteria do not give rise to simple fast boosting algorithms.

For more general loss criteria the solution to (10.30), given the R_{jm}, is typically straightforward since it is a simple "location" estimate. For absolute loss it is just the median of the residuals in each respective region. For the other criteria fast iterative algorithms exist for solving (10.30), and usually their faster "single-step" approximations are adequate. The problem is tree induction. Simple fast algorithms do not exist for solving (10.29) for these more general loss criteria, and approximations like (10.27) become essential.

10.10 Numerical Optimization via Gradient Boosting

Fast approximate algorithms for solving (10.29) with any differentiable loss criterion can be derived by analogy to numerical optimization. The loss in using $f(x)$ to predict y on the training data is

$$L(f) = \sum_{i=1}^{N} L(y_i, f(x_i)). \tag{10.33}$$

The goal is to minimize $L(f)$ with respect to f, where here $f(x)$ is constrained to be a sum of trees (10.28). Ignoring this constraint, minimizing (10.33) can be viewed as a numerical optimization

$$\hat{\mathbf{f}} = \arg\min_{\mathbf{f}} L(\mathbf{f}), \tag{10.34}$$

where the "parameters" $\mathbf{f} \in \mathbb{R}^N$ are the values of the approximating function $f(x_i)$ at each of the N data points x_i:

$$\mathbf{f} = \{f(x_1), f(x_2), \ldots, f(x_N)\}^T.$$

Numerical optimization procedures solve (10.34) as a sum of component vectors

$$\mathbf{f}_M = \sum_{m=0}^{M} \mathbf{h}_m, \quad \mathbf{h}_m \in \mathbb{R}^N,$$

where $\mathbf{f}_0 = \mathbf{h}_0$ is an initial guess, and each successive $\mathbf{f}_m$ is induced based on the current parameter vector $\mathbf{f}_{m-1}$, which is the sum of the previously induced updates. Numerical optimization methods differ in their prescriptions for computing each increment vector $\mathbf{h}_m$ ("step").

10.10.1 Steepest Descent

Steepest descent chooses $\mathbf{h}_m = -\rho_m \mathbf{g}_m$ where ρ_m is a scalar and $\mathbf{g}_m \in \mathbb{R}^N$ is the gradient of $L(\mathbf{f})$ evaluated at $\mathbf{f} = \mathbf{f}_{m-1}$. The components of the gradient $\mathbf{g}_m$ are

$$g_{im} = \left[\frac{\partial L(y_i, f(x_i))}{\partial f(x_i)} \right]_{f(x_i)=f_{m-1}(x_i)} \tag{10.35}$$

The *step length* ρ_m is the solution to

$$\rho_m = \arg\min_{\rho} L(\mathbf{f}_{m-1} - \rho\mathbf{g}_m). \qquad (10.36)$$

The current solution is then updated

$$\mathbf{f}_m = \mathbf{f}_{m-1} - \rho_m\mathbf{g}_m$$

and the process repeated at the next iteration. Steepest descent can be viewed as a very greedy strategy, since $-\mathbf{g}_m$ is the local direction in $\mathrm{I\!R}^N$ for which $L(\mathbf{f})$ is most rapidly decreasing at $\mathbf{f} = \mathbf{f}_{m-1}$.

10.10.2 *Gradient Boosting*

Forward stagewise boosting (Algorithm 10.2) is also a very greedy strategy. At each step the solution tree is the one that maximally reduces (10.29), given the current model f_{m-1} and its fits $f_{m-1}(x_i)$. Thus, the tree predictions $T(x_i; \Theta_m)$ are analogous to the components of the negative gradient (10.35). The principal difference between them is that the tree components $\mathbf{t}_m = \{T(x_1; \Theta_m), \ldots, T(x_N; \Theta_m)\}^T$ are not independent. They are constrained to be the predictions of a J_m-terminal node decision tree, whereas the negative gradient is the unconstrained maximal descent direction.

The solution to (10.30) in the stagewise approach is analogous to the line search (10.36) in steepest descent. The difference is that (10.30) performs a separate line search for those components of $\mathbf{t}_m$ that correspond to each separate terminal region $\{T(x_i; \Theta_m)\}_{x_i \in R_{jm}}$.

If minimizing loss on the training data (10.33) were the only goal, steepest descent would be the preferred strategy. The gradient (10.35) is trivial to calculate for any differentiable loss function $L(y, f(x))$, whereas solving (10.29) is difficult for the robust criteria discussed in Section 10.6. Unfortunately the gradient (10.35) is defined only at the training data points x_i, whereas the ultimate goal is to generalize $f_M(x)$ to new data not represented in the training set.

A possible resolution to this dilemma is to induce a tree $T(x; \Theta_m)$ at the mth iteration whose predictions $\mathbf{t}_m$ are as close as possible to the negative gradient. Using squared error to measure closeness, this leads us to

$$\tilde{\Theta}_m = \arg\min_{\Theta} \sum_{i=1}^{N} (-g_{im} - T(x_i; \Theta))^2. \qquad (10.37)$$

That is, one fits the tree T to the negative gradient values (10.35) by least squares. As noted in Section 10.9 fast algorithms exist for least squares decision tree induction. Although the solution regions $\tilde{R}_{jm}$ to (10.37) will not be identical to the regions R_{jm} that solve (10.29), it is generally similar enough to serve the same purpose. In any case, the forward stagewise

TABLE 10.2. *Gradients for commonly used loss functions.*

Setting	Loss Function	$-\partial L(y_i, f(x_i))/\partial f(x_i)$
Regression	$\frac{1}{2}[y_i - f(x_i)]^2$	$y_i - f(x_i)$
Regression	$\|y_i - f(x_i)\|$	$\text{sign}[y_i - f(x_i)]$
Regression	Huber	$y_i - f(x_i)$ for $\|y_i - f(x_i)\| \leq \delta_m$
		$\delta_m \text{sign}[y_i - f(x_i)]$ for $\|y_i - f(x_i)\| > \delta_m$
		where $\delta_m = \alpha$th-quantile$\{\|y_i - f(x_i)\|\}$
Classification	Deviance	kth component: $I(y_i = \mathcal{G}_k) - p_k(x_i)$

boosting procedure, and top-down decision tree induction, are themselves approximation procedures. After constructing the tree (10.37), the corresponding constants in each region are given by (10.30).

Table 10.2 summarizes the gradients for commonly used loss functions. For squared error loss, the negative gradient is just the ordinary residual $-g_{im} = y_i - f_{m-1}(x_i)$, so that (10.37) on its own is equivalent to standard least-squares boosting. With absolute error loss, the negative gradient is the *sign* of the residual, so at each iteration (10.37) fits the tree to the sign of the current residuals by least squares. For Huber M-regression, the negative gradient is a compromise between these two (see the table).

For classification the loss function is the multinomial deviance (10.22), and K least squares trees are constructed at each iteration. Each tree T_{km} is fit to its respective negative gradient vector $\mathbf{g}_{km}$,

$$
\begin{aligned}
-g_{ikm} &= \left[\frac{\partial L\left(y_i, f_1(x_i), \ldots, f_K(x_i)\right)}{\partial f_k(x_i)}\right]_{\mathbf{f}(x_i)=\mathbf{f}_{m-1}(x_i)} \\
&= I(y_i = \mathcal{G}_k) - p_k(x_i), \qquad (10.38)
\end{aligned}
$$

with $p_k(x)$ given by (10.21). Although K separate trees are built at each iteration, they are related through (10.21). For binary classification ($K = 2$), only one tree is needed (exercise 10.10).

10.10.3 *Implementations of Gradient Boosting*

Algorithm 10.3 presents the generic gradient tree-boosting algorithm for regression. Specific algorithms are obtained by inserting different loss criteria $L(y, f(x))$. The first line of the algorithm initializes to the optimal constant model, which is just a single terminal node tree. The components of the negative gradient computed at line 2(a) are referred to as generalized or *pseudo* residuals, r. Gradients for commonly used loss functions are summarized in Table 10.2.

Algorithm 10.3 *Gradient Tree-Boosting Algorithm.*

---

1. Initialize $f_0(x) = \arg\min_\gamma \sum_{i=1}^N L(y_i, \gamma)$.

2. For $m = 1$ to M:

   (a) For $i = 1, 2, \ldots, N$ compute

   $$r_{im} = -\left[\frac{\partial L(y_i, f(x_i))}{\partial f(x_i)}\right]_{f=f_{m-1}}.$$

   (b) Fit a regression tree to the targets r_{im} giving terminal regions R_{jm}, $j = 1, 2, \ldots, J_m$.

   (c) For $j = 1, 2, \ldots, J_m$ compute

   $$\gamma_{jm} = \arg\min_\gamma \sum_{x_i \in R_{jm}} L\left(y_i, f_{m-1}(x_i) + \gamma\right).$$

   (d) Update $f_m(x) = f_{m-1}(x) + \sum_{j=1}^{J_m} \gamma_{jm} I(x \in R_{jm})$.

3. Output $\hat{f}(x) = f_M(x)$.

The algorithm for classification is similar. Lines 2(a)–(d) are repeated K times at each iteration m, once for each class using (10.38). The result at line 3 is K different (coupled) tree expansions $f_{kM}(x), k = 1, 2, \ldots, K$. These produce probabilities via (10.21) or do classification as in (10.20). Details are given in Exercise 10.9. Two basic tuning parameters are the number of iterations M and the sizes of each of the constituent trees J_m, $m = 1, 2, \ldots, M$.

The original implementation of this algorithm was called MART for "multiple additive regression trees," and was referred to in the first edition of this book. Many of the figures in this chapter were produced by MART. Gradient boosting as described here is implemented in the R **gbm** package (Ridgeway, 1999, "Gradient Boosted Models"), and is freely available. The **gbm** package is used in Section 10.14.2, and extensively in Chapters 16 and 15. Another R implementation of boosting is **mboost** (Hothorn and Bühlmann, 2006). A commercial implementation of gradient boosting/MART called *TreeNet*® is available from Salford Systems, Inc.

10.11 Right-Sized Trees for Boosting

Historically, boosting was considered to be a technique for combining models, here trees. As such, the tree building algorithm was regarded as a

primitive that produced models to be combined by the boosting procedure. In this scenario, the optimal size of each tree is estimated separately
in the usual manner when it is built (Section 9.2). A very large (oversized)
tree is first induced, and then a bottom-up procedure is employed to prune
it to the estimated optimal number of terminal nodes. This approach assumes implicitly that each tree is the last one in the expansion (10.28).
Except perhaps for the very last tree, this is clearly a very poor assumption. The result is that trees tend to be much too large, especially during
the early iterations. This substantially degrades performance and increases
computation.

The simplest strategy for avoiding this problem is to restrict all trees
to be the same size, $J_m = J \ \forall m$. At each iteration a J-terminal node
regression tree is induced. Thus J becomes a meta-parameter of the entire
boosting procedure, to be adjusted to maximize estimated performance for
the data at hand.

One can get an idea of useful values for J by considering the properties
of the "target" function

$$\eta = \arg \min_f \mathrm{E}_{XY} L(Y, f(X)). \qquad (10.39)$$

Here the expected value is over the population joint distribution of (X, Y).
The target function $\eta(x)$ is the one with minimum prediction risk on future
data. This is the function we are trying to approximate.

One relevant property of $\eta(X)$ is the degree to which the coordinate variables $X^T = (X_1, X_2, \ldots, X_p)$ interact with one another. This is captured
by its ANOVA (analysis of variance) expansion

$$\eta(X) = \sum_j \eta_j(X_j) + \sum_{jk} \eta_{jk}(X_j, X_k) + \sum_{jkl} \eta_{jkl}(X_j, X_k, X_l) + \cdots. \quad (10.40)$$

The first sum in (10.40) is over functions of only a single predictor variable
X_j. The particular functions $\eta_j(X_j)$ are those that jointly best approximate
$\eta(X)$ under the loss criterion being used. Each such $\eta_j(X_j)$ is called the
"main effect" of X_j. The second sum is over those two-variable functions
that when added to the main effects best fit $\eta(X)$. These are called the
second-order interactions of each respective variable pair (X_j, X_k). The
third sum represents third-order interactions, and so on. For many problems
encountered in practice, low-order interaction effects tend to dominate.
When this is the case, models that produce strong higher-order interaction
effects, such as large decision trees, suffer in accuracy.

The interaction level of tree-based approximations is limited by the tree
size J. Namely, no interaction effects of level greater than $J - 1$ are possible. Since boosted models are additive in the trees (10.28), this limit
extends to them as well. Setting $J = 2$ (single split "decision stump")
produces boosted models with only main effects; no interactions are permitted. With $J = 3$, two-variable interaction effects are also allowed, and

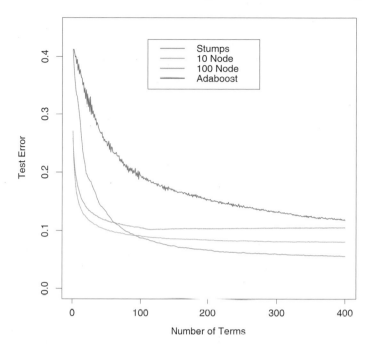

FIGURE 10.9. *Boosting with different sized trees, applied to the example (10.2) used in Figure 10.2. Since the generative model is additive, stumps perform the best. The boosting algorithm used the binomial deviance loss in Algorithm 10.3; shown for comparison is the AdaBoost Algorithm 10.1.*

so on. This suggests that the value chosen for J should reflect the level of dominant interactions of $\eta(x)$. This is of course generally unknown, but in most situations it will tend to be low. Figure 10.9 illustrates the effect of interaction order (choice of J) on the simulation example (10.2). The generative function is additive (sum of quadratic monomials), so boosting models with $J > 2$ incurs unnecessary variance and hence the higher test error. Figure 10.10 compares the coordinate functions found by boosted stumps with the true functions.

Although in many applications $J = 2$ will be insufficient, it is unlikely that $J > 10$ will be required. Experience so far indicates that $4 \le J \le 8$ works well in the context of boosting, with results being fairly insensitive to particular choices in this range. One can fine-tune the value for J by trying several different values and choosing the one that produces the lowest risk on a validation sample. However, this seldom provides significant improvement over using $J \simeq 6$.

Coordinate Functions for Additive Logistic Trees

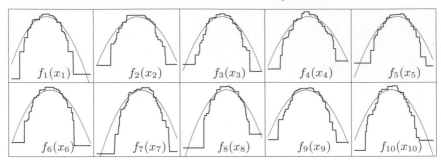

FIGURE 10.10. *Coordinate functions estimated by boosting stumps for the simulated example used in Figure 10.9. The true quadratic functions are shown for comparison.*

10.12   Regularization

Besides the size of the constituent trees, J, the other meta-parameter of gradient boosting is the number of boosting iterations M. Each iteration usually reduces the training risk $L(f_M)$, so that for M large enough this risk can be made arbitrarily small. However, fitting the training data too well can lead to overfitting, which degrades the risk on future predictions. Thus, there is an optimal number M^* minimizing future risk that is application dependent. A convenient way to estimate M^* is to monitor prediction risk as a function of M on a validation sample. The value of M that minimizes this risk is taken to be an estimate of M^*. This is analogous to the early stopping strategy often used with neural networks (Section 11.4).

10.12.1   Shrinkage

Controlling the value of M is not the only possible regularization strategy. As with ridge regression and neural networks, shrinkage techniques can be employed as well (see Sections 3.4.1 and 11.5). The simplest implementation of shrinkage in the context of boosting is to scale the contribution of each tree by a factor $0 < \nu < 1$ when it is added to the current approximation. That is, line 2(d) of Algorithm 10.3 is replaced by

$$f_m(x) = f_{m-1}(x) + \nu \cdot \sum_{j=1}^{J} \gamma_{jm} I(x \in R_{jm}). \qquad (10.41)$$

The parameter ν can be regarded as controlling the learning rate of the boosting procedure. Smaller values of ν (more shrinkage) result in larger training risk for the same number of iterations M. Thus, both ν and M control prediction risk on the training data. However, these parameters do not operate independently. Smaller values of ν lead to larger values of M for the same training risk, so that there is a tradeoff between them.

Empirically it has been found (Friedman, 2001) that smaller values of ν favor better test error, and require correspondingly larger values of M. In fact, the best strategy appears to be to set ν to be very small ($\nu < 0.1$) and then choose M by early stopping. This yields dramatic improvements (over no shrinkage $\nu = 1$) for regression and for probability estimation. The corresponding improvements in misclassification risk via (10.20) are less, but still substantial. The price paid for these improvements is computational: smaller values of ν give rise to larger values of M, and computation is proportional to the latter. However, as seen below, many iterations are generally computationally feasible even on very large data sets. This is partly due to the fact that small trees are induced at each step with no pruning.

Figure 10.11 shows test error curves for the simulated example (10.2) of Figure 10.2. A gradient boosted model (MART) was trained using binomial deviance, using either stumps or six terminal-node trees, and with or without shrinkage. The benefits of shrinkage are evident, especially when the binomial deviance is tracked. With shrinkage, each test error curve reaches a lower value, and stays there for many iterations.

Section 16.2.1 draws a connection between forward stagewise shrinkage in boosting and the use of an L_1 penalty for regularizing model parameters (the "lasso"). We argue that L_1 penalties may be superior to the L_2 penalties used by methods such as the support vector machine.

10.12.2 Subsampling

We saw in Section 8.7 that bootstrap averaging (bagging) improves the performance of a noisy classifier through averaging. Chapter 15 discusses in some detail the variance-reduction mechanism of this sampling followed by averaging. We can exploit the same device in gradient boosting, both to improve performance and computational efficiency.

With *stochastic gradient boosting* (Friedman, 1999), at each iteration we sample a fraction η of the training observations (without replacement), and grow the next tree using that subsample. The rest of the algorithm is identical. A typical value for η can be $\frac{1}{2}$, although for large N, η can be substantially smaller than $\frac{1}{2}$.

Not only does the sampling reduce the computing time by the same fraction η, but in many cases it actually produces a more accurate model.

Figure 10.12 illustrates the effect of subsampling using the simulated example (10.2), both as a classification and as a regression example. We see in both cases that sampling along with shrinkage slightly outperformed the rest. It appears here that subsampling without shrinkage does poorly.

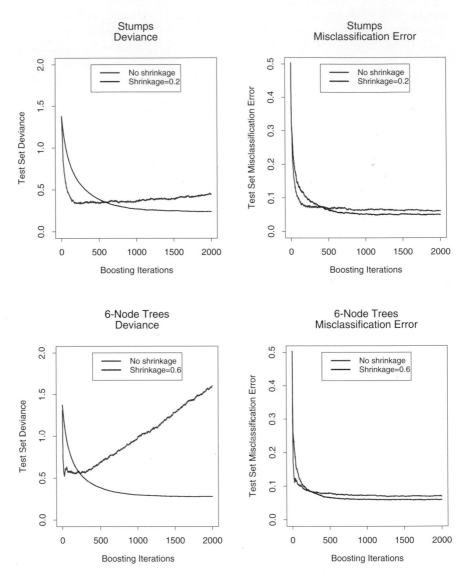

FIGURE 10.11. *Test error curves for simulated example (10.2) of Figure 10.9, using gradient boosting (MART). The models were trained using binomial deviance, either stumps or six terminal-node trees, and with or without shrinkage. The left panels report test deviance, while the right panels show misclassification error. The beneficial effect of shrinkage can be seen in all cases, especially for deviance in the left panels.*

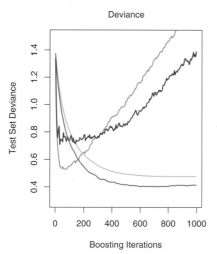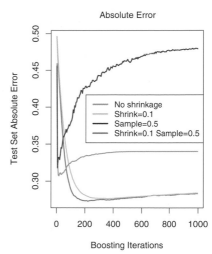

FIGURE 10.12. *Test-error curves for the simulated example (10.2), showing the effect of stochasticity. For the curves labeled "Sample= 0.5", a different 50% subsample of the training data was used each time a tree was grown. In the left panel the models were fit by* **gbm** *using a binomial deviance loss function; in the right-hand panel using square-error loss.*

The downside is that we now have four parameters to set: J, M, ν and η. Typically some early explorations determine suitable values for J, ν and η, leaving M as the primary parameter.

10.13 Interpretation

Single decision trees are highly interpretable. The entire model can be completely represented by a simple two-dimensional graphic (binary tree) that is easily visualized. Linear combinations of trees (10.28) lose this important feature, and must therefore be interpreted in a different way.

10.13.1 Relative Importance of Predictor Variables

In data mining applications the input predictor variables are seldom equally relevant. Often only a few of them have substantial influence on the response; the vast majority are irrelevant and could just as well have not been included. It is often useful to learn the relative importance or contribution of each input variable in predicting the response.

For a single decision tree T, Breiman et al. (1984) proposed

$$\mathcal{I}_\ell^2(T) = \sum_{t=1}^{J-1} \hat{\imath}_t^2 \, I(v(t) = \ell) \qquad (10.42)$$

as a measure of relevance for each predictor variable X_ℓ. The sum is over the $J-1$ internal nodes of the tree. At each such node t, one of the input variables $X_{v(t)}$ is used to partition the region associated with that node into two subregions; within each a separate constant is fit to the response values. The particular variable chosen is the one that gives maximal estimated improvement $\hat{\imath}_t^2$ in squared error risk over that for a constant fit over the entire region. The squared relative importance of variable X_ℓ is the sum of such squared improvements over all internal nodes for which it was chosen as the splitting variable.

This importance measure is easily generalized to additive tree expansions (10.28); it is simply averaged over the trees

$$\mathcal{I}_\ell^2 = \frac{1}{M} \sum_{m=1}^{M} \mathcal{I}_\ell^2(T_m). \qquad (10.43)$$

Due to the stabilizing effect of averaging, this measure turns out to be more reliable than is its counterpart (10.42) for a single tree. Also, because of shrinkage (Section 10.12.1) the masking of important variables by others with which they are highly correlated is much less of a problem. Note that (10.42) and (10.43) refer to *squared* relevance; the actual relevances are their respective square roots. Since these measures are relative, it is customary to assign the largest a value of 100 and then scale the others accordingly. Figure 10.6 shows the relevant importance of the 57 inputs in predicting spam versus email.

For K-class classification, K separate models $f_k(x), k = 1, 2, \ldots, K$ are induced, each consisting of a sum of trees

$$f_k(x) = \sum_{m=1}^{M} T_{km}(x). \qquad (10.44)$$

In this case (10.43) generalizes to

$$\mathcal{I}_{\ell k}^2 = \frac{1}{M} \sum_{m=1}^{M} \mathcal{I}_\ell^2(T_{km}). \qquad (10.45)$$

Here $\mathcal{I}_{\ell k}$ is the relevance of X_ℓ in separating the class k observations from the other classes. The overall relevance of X_ℓ is obtained by averaging over all of the classes

$$\mathcal{I}_\ell^2 = \frac{1}{K} \sum_{k=1}^{K} \mathcal{I}_{\ell k}^2. \qquad (10.46)$$

Figures 10.23 and 10.24 illustrate the use of these averaged and separate relative importances.

10.13.2 Partial Dependence Plots

After the most relevant variables have been identified, the next step is to attempt to understand the nature of the dependence of the approximation $f(X)$ on their joint values. Graphical renderings of the $f(X)$ as a function of its arguments provides a comprehensive summary of its dependence on the joint values of the input variables.

Unfortunately, such visualization is limited to low-dimensional views. We can easily display functions of one or two arguments, either continuous or discrete (or mixed), in a variety of different ways; this book is filled with such displays. Functions of slightly higher dimensions can be plotted by conditioning on particular sets of values of all but one or two of the arguments, producing a *trellis* of plots (Becker et al., 1996).[1]

For more than two or three variables, viewing functions of the corresponding higher-dimensional arguments is more difficult. A useful alternative can sometimes be to view a collection of plots, each one of which shows the partial dependence of the approximation $f(X)$ on a selected small subset of the input variables. Although such a collection can seldom provide a comprehensive depiction of the approximation, it can often produce helpful clues, especially when $f(x)$ is dominated by low-order interactions (10.40).

Consider the subvector $X_{\mathcal{S}}$ of $\ell < p$ of the input predictor variables $X^T = (X_1, X_2, \ldots, X_p)$, indexed by $\mathcal{S} \subset \{1, 2, \ldots, p\}$. Let $\mathcal{C}$ be the complement set, with $\mathcal{S} \cup \mathcal{C} = \{1, 2, \ldots, p\}$. A general function $f(X)$ will in principle depend on all of the input variables: $f(X) = f(X_{\mathcal{S}}, X_{\mathcal{C}})$. One way to define the average or *partial* dependence of $f(X)$ on $X_{\mathcal{S}}$ is

$$f_{\mathcal{S}}(X_{\mathcal{S}}) = \mathrm{E}_{X_{\mathcal{C}}} f(X_{\mathcal{S}}, X_{\mathcal{C}}). \qquad (10.47)$$

This is a marginal average of f, and can serve as a useful description of the effect of the chosen subset on $f(X)$ when, for example, the variables in $X_{\mathcal{S}}$ do not have strong interactions with those in $X_{\mathcal{C}}$.

Partial dependence functions can be used to interpret the results of any "black box" learning method. They can be estimated by

$$\bar{f}_{\mathcal{S}}(X_{\mathcal{S}}) = \frac{1}{N} \sum_{i=1}^{N} f(X_{\mathcal{S}}, x_{i\mathcal{C}}), \qquad (10.48)$$

where $\{x_{1\mathcal{C}}, x_{2\mathcal{C}}, \ldots, x_{N\mathcal{C}}\}$ are the values of $X_{\mathcal{C}}$ occurring in the training data. This requires a pass over the data for each set of joint values of $X_{\mathcal{S}}$ for which $\bar{f}_{\mathcal{S}}(X_{\mathcal{S}})$ is to be evaluated. This can be computationally intensive,

[1] lattice in R.

even for moderately sized data sets. Fortunately with decision trees, $\bar{f}_\mathcal{S}(X_\mathcal{S})$ (10.48) can be rapidly computed from the tree itself without reference to the data (Exercise 10.11).

It is important to note that partial dependence functions defined in (10.47) represent the effect of $X_\mathcal{S}$ on $f(X)$ after accounting for the (average) effects of the other variables $X_\mathcal{C}$ on $f(X)$. They are *not* the effect of $X_\mathcal{S}$ on $f(X)$ *ignoring* the effects of $X_\mathcal{C}$. The latter is given by the conditional expectation

$$\tilde{f}_\mathcal{S}(X_\mathcal{S}) = \mathrm{E}(f(X_\mathcal{S}, X_\mathcal{C})|X_\mathcal{S}), \tag{10.49}$$

and is the best least squares approximation to $f(X)$ by a function of $X_\mathcal{S}$ alone. The quantities $\tilde{f}_\mathcal{S}(X_\mathcal{S})$ and $\bar{f}_\mathcal{S}(X_\mathcal{S})$ will be the same only in the unlikely event that $X_\mathcal{S}$ and $X_\mathcal{C}$ are independent. For example, if the effect of the chosen variable subset happens to be purely additive,

$$f(X) = h_1(X_\mathcal{S}) + h_2(X_\mathcal{C}). \tag{10.50}$$

Then (10.47) produces the $h_1(X_\mathcal{S})$ up to an additive constant. If the effect is purely multiplicative,

$$f(X) = h_1(X_\mathcal{S}) \cdot h_2(X_\mathcal{C}), \tag{10.51}$$

then (10.47) produces $h_1(X_\mathcal{S})$ up to a multiplicative constant factor. On the other hand, (10.49) will not produce $h_1(X_\mathcal{S})$ in either case. In fact, (10.49) can produce strong effects on variable subsets for which $f(X)$ has no dependence at all.

Viewing plots of the partial dependence of the boosted-tree approximation (10.28) on selected variables subsets can help to provide a qualitative description of its properties. Illustrations are shown in Sections 10.8 and 10.14. Owing to the limitations of computer graphics, and human perception, the size of the subsets $X_\mathcal{S}$ must be small ($l \approx 1, 2, 3$). There are of course a large number of such subsets, but only those chosen from among the usually much smaller set of highly relevant predictors are likely to be informative. Also, those subsets whose effect on $f(X)$ is approximately additive (10.50) or multiplicative (10.51) will be most revealing.

For K-class classification, there are K separate models (10.44), one for each class. Each one is related to the respective probabilities (10.21) through

$$f_k(X) = \log p_k(X) - \frac{1}{K} \sum_{l=1}^{K} \log p_l(X). \tag{10.52}$$

Thus each $f_k(X)$ is a monotone increasing function of its respective probability on a logarithmic scale. Partial dependence plots of each respective $f_k(X)$ (10.44) on its most relevant predictors (10.45) can help reveal how the log-odds of realizing that class depend on the respective input variables.

10.14   Illustrations

In this section we illustrate gradient boosting on a number of larger datasets, using different loss functions as appropriate.

10.14.1   *California Housing*

This data set (Pace and Barry, 1997) is available from the Carnegie-Mellon *StatLib* repository[2]. It consists of aggregated data from each of 20,460 neighborhoods (1990 census block groups) in California. The response variable Y is the median house value in each neighborhood measured in units of \$100,000. The predictor variables are demographics such as median income MedInc, housing density as reflected by the number of houses House, and the average occupancy in each house AveOccup. Also included as predictors are the location of each neighborhood (longitude and latitude), and several quantities reflecting the properties of the houses in the neighborhood: average number of rooms AveRooms and bedrooms AveBedrms. There are thus a total of eight predictors, all numeric.

We fit a gradient boosting model using the MART procedure, with $J = 6$ terminal nodes, a learning rate (10.41) of $\nu = 0.1$, and the Huber loss criterion for predicting the numeric response. We randomly divided the dataset into a training set (80%) and a test set (20%).

Figure 10.13 shows the average absolute error

$$\text{AAE} - \text{E}\,|y - \hat{f}_M(x)| \tag{10.53}$$

as a function for number of iterations M on both the training data and test data. The test error is seen to decrease monotonically with increasing M, more rapidly during the early stages and then leveling off to being nearly constant as iterations increase. Thus, the choice of a particular value of M is not critical, as long as it is not too small. This tends to be the case in many applications. The shrinkage strategy (10.41) tends to eliminate the problem of overfitting, especially for larger data sets.

The value of AAE after 800 iterations is 0.31. This can be compared to that of the optimal constant predictor median$\{y_i\}$ which is 0.89. In terms of more familiar quantities, the squared multiple correlation coefficient of this model is $R^2 = 0.84$. Pace and Barry (1997) use a sophisticated spatial autoregression procedure, where prediction for each neighborhood is based on median house values in nearby neighborhoods, using the other predictors as covariates. Experimenting with transformations they achieved $R^2 = 0.85$, predicting $\log Y$. Using $\log Y$ as the response the corresponding value for gradient boosting was $R^2 = 0.86$.

---

[2]http://lib.stat.cmu.edu.

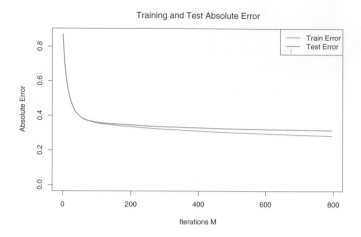

FIGURE 10.13. *Average-absolute error as a function of number of iterations for the California housing data.*

Figure 10.14 displays the relative variable importances for each of the eight predictor variables. Not surprisingly, median income in the neighborhood is the most relevant predictor. Longitude, latitude, and average occupancy all have roughly half the relevance of income, whereas the others are somewhat less influential.

Figure 10.15 shows single-variable partial dependence plots on the most relevant nonlocation predictors. Note that the plots are not strictly smooth. This is a consequence of using tree-based models. Decision trees produce discontinuous piecewise constant models (10.25). This carries over to sums of trees (10.28), with of course many more pieces. Unlike most of the methods discussed in this book, there is no smoothness constraint imposed on the result. Arbitrarily sharp discontinuities can be modeled. The fact that these curves generally exhibit a smooth trend is because that is what is estimated to best predict the response for this problem. This is often the case.

The hash marks at the base of each plot delineate the deciles of the data distribution of the corresponding variables. Note that here the data density is lower near the edges, especially for larger values. This causes the curves to be somewhat less well determined in those regions. The vertical scales of the plots are the same, and give a visual comparison of the relative importance of the different variables.

The partial dependence of median house value on median income is monotonic increasing, being nearly linear over the main body of data. House value is generally monotonic decreasing with increasing average occupancy, except perhaps for average occupancy rates less than one. Median house

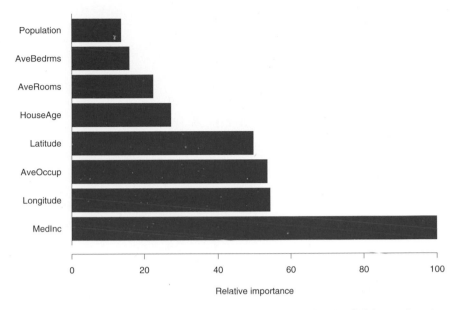

FIGURE 10.14. *Relative importance of the predictors for the California housing data.*

value has a nonmonotonic partial dependence on average number of rooms. It has a minimum at approximately three rooms and is increasing both for smaller and larger values.

Median house value is seen to have a very weak partial dependence on house age that is inconsistent with its importance ranking (Figure 10.14). This suggests that this weak main effect may be masking stronger interaction effects with other variables. Figure 10.16 shows the two-variable partial dependence of housing value on joint values of median age and average occupancy. An interaction between these two variables is apparent. For values of average occupancy greater than two, house value is nearly independent of median age, whereas for values less than two there is a strong dependence on age.

Figure 10.17 shows the two-variable partial dependence of the fitted model on joint values of longitude and latitude, displayed as a shaded contour plot. There is clearly a very strong dependence of median house value on the neighborhood location in California. Note that Figure 10.17 is *not* a plot of house value versus location *ignoring* the effects of the other predictors (10.49). Like all partial dependence plots, it represents the effect of location after accounting for the effects of the other neighborhood and house attributes (10.47). It can be viewed as representing an extra premium one pays for location. This premium is seen to be relatively large near the Pacific coast especially in the Bay Area and Los Angeles–San Diego re-

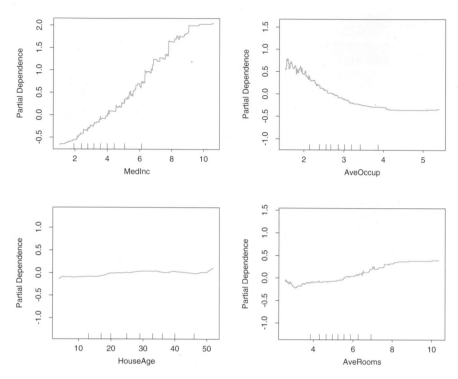

FIGURE 10.15. *Partial dependence of housing value on the nonlocation variables for the California housing data. The red ticks at the base of the plot are deciles of the input variables.*

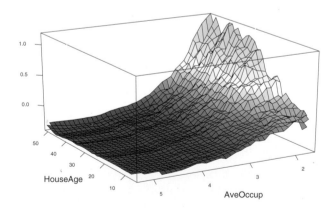

FIGURE 10.16. *Partial dependence of house value on median age and average occupancy. There appears to be a strong interaction effect between these two variables.*

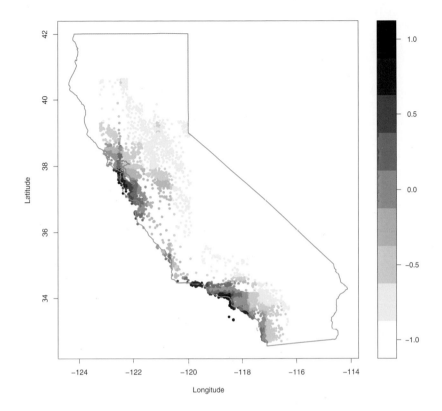

FIGURE 10.17. *Partial dependence of median house value on location in California. One unit is $100,000, at 1990 prices, and the values plotted are relative to the overall median of $180,000.*

gions. In the northern, central valley, and southeastern desert regions of California, location costs considerably less.

10.14.2 New Zealand Fish

Plant and animal ecologists use regression models to predict species presence, abundance and richness as a function of environmental variables. Although for many years simple linear and parametric models were popular, recent literature shows increasing interest in more sophisticated models such as generalized additive models (Section 9.1, GAM), multivariate adaptive regression splines (Section 9.4, MARS) and boosted regression trees (Leathwick et al., 2005; Leathwick et al., 2006). Here we model the presence and abundance of the *Black Oreo Dory*, a marine fish found in the oceanic waters around New Zealand.[3]

Figure 10.18 shows the locations of 17,000 trawls (deep-water net fishing, with a maximum depth of 2km), and the red points indicate those 2353 trawls for which the Black Oreo was present, one of over a hundred species regularly recorded. The catch size in kg for each species was recorded for each trawl. Along with the species catch, a number of environmental measurements are available for each trawl. These include the average depth of the trawl (`AvgDepth`), and the temperature and salinity of the water. Since the latter two are strongly correlated with depth, Leathwick et al. (2006) derived instead `TempResid` and `SalResid`, the residuals obtained when these two measures are adjusted for depth (via separate non-parametric regressions). `SSTGrad` is a measure of the gradient of the sea surface temperature, and `Chla` is a broad indicator of ecosytem productivity via satellite-image measurements. `SusPartMatter` provides a measure of suspended particulate matter, particularly in coastal waters, and is also satellite derived.

The goal of this analysis is to estimate the probability of finding Black Oreo in a trawl, as well as the expected catch size, standardized to take into account the effects of variation in trawl speed and distance, as well as the mesh size of the trawl net. The authors used logistic regression for estimating the probability. For the catch size, it might seem natural to assume a Poisson distribution and model the log of the mean count, but this is often not appropriate because of the excessive number of zeros. Although specialized approaches have been developed, such as the *zero-inflated* Poisson (Lambert, 1992), they chose a simpler approach. If Y is the (non-negative) catch size,

$$E(Y|X) - E(Y|Y > 0, X) \cdot \Pr(Y > 0|X). \qquad (10.54)$$

The second term is estimated by the logistic regression, and the first term can be estimated using only the 2353 trawls with a positive catch.

For the logistic regression the authors used a gradient boosted model (GBM)[4] with binomial deviance loss function, depth-10 trees, and a shrinkage factor $\nu = 0.025$. For the positive-catch regression, they modeled $\log(Y)$ using a GBM with squared-error loss (also depth-10 trees, but $\nu = 0.01$), and un-logged the predictions. In both cases they used 10-fold cross-validation for selecting the number of terms, as well as the shrinkage factor.

---

[3]The models, data, and maps shown here were kindly provided by Dr John Leathwick of the National Institute of Water and Atmospheric Research in New Zealand, and Dr Jane Elith, School of Botany, University of Melbourne. The collection of the research trawl data took place from 1979–2005, and was funded by the New Zealand Ministry of Fisheries.

[4]Version 1.5-7 of package **gbm** in R, ver. 2.2.0.

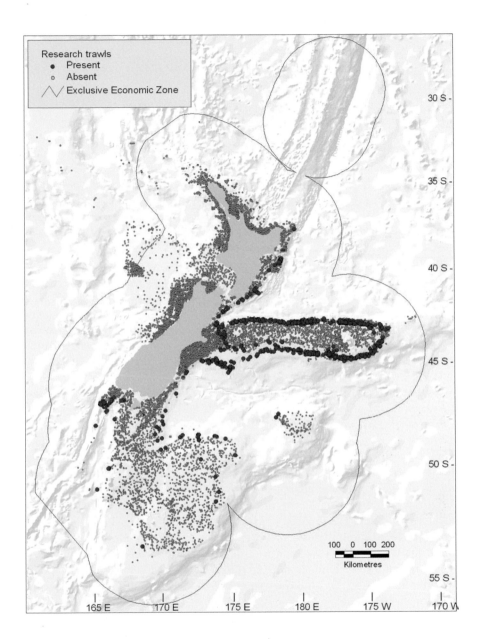

FIGURE 10.18. *Map of New Zealand and its surrounding exclusive economic zone, showing the locations of 17,000 trawls (small blue dots) taken between 1979 and 2005. The red points indicate trawls for which the species* Black Oreo Dory *were present.*

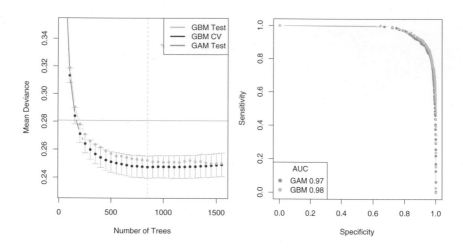

FIGURE 10.19. *The left panel shows the mean deviance as a function of the number of trees for the GBM logistic regression model fit to the presence/absence data. Shown are 10-fold cross-validation on the training data (and $1 \times$ s.e. bars), and test deviance on the test data. Also shown for comparison is the test deviance using a GAM model with 8 df for each term. The right panel shows ROC curves on the test data for the chosen GBM model (vertical line in left plot) and the GAM model.*

Figure 10.19 (left panel) shows the mean binomial deviance for the sequence of GBM models, both for 10-fold CV and test data. There is a modest improvement over the performance of a GAM model, fit using smoothing splines with 8 degrees-of-freedom (df) per term. The right panel shows the ROC curves (see Section 9.2.5) for both models, which measures predictive performance. From this point of view, the performance looks very similar, with GBM perhaps having a slight edge as summarized by the AUC (area under the curve). At the point of equal sensitivity/specificity, GBM achieves 91%, and GAM 90%.

Figure 10.20 summarizes the contributions of the variables in the logistic GBM fit. We see that there is a well-defined depth range over which Black Oreo are caught, with much more frequent capture in colder waters. We do not give details of the quantitative catch model; the important variables were much the same.

All the predictors used in these models are available on a fine geographical grid; in fact they were derived from environmental atlases, satellite images and the like—see Leathwick et al. (2006) for details. This also means that predictions can be made on this grid, and imported into GIS mapping systems. Figure 10.21 shows prediction maps for both presence and catch size, with both standardized to a common set of trawl conditions; since the predictors vary in a continuous fashion with geographical location, so do the predictions.

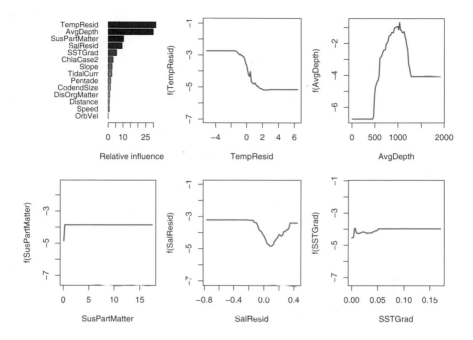

FIGURE 10.20. *The top-left panel shows the relative influence computed from the GBM logistic regression model. The remaining panels show the partial dependence plots for the leading five variables, all plotted on the same scale for comparison.*

Because of their ability to model interactions and automatically select variables, as well as robustness to outliers and missing data, GBM models are rapidly gaining popularity in this data-rich and enthusiastic community.

10.14.3 Demographics Data

In this section we illustrate gradient boosting on a multiclass classification problem, using MART. The data come from 9243 questionnaires filled out by shopping mall customers in the San Francisco Bay Area (Impact Resources, Inc., Columbus, OH). Among the questions are 14 concerning demographics. For this illustration the goal is to predict occupation using the other 13 variables as predictors, and hence identify demographic variables that discriminate between different occupational categories. We randomly divided the data into a training set (80%) and test set (20%), and used $J = 6$ node trees with a learning rate $\nu = 0.1$.

Figure 10.22 shows the $K = 9$ occupation class values along with their corresponding error rates. The overall error rate is 42.5%, which can be compared to the null rate of 69% obtained by predicting the most numerous

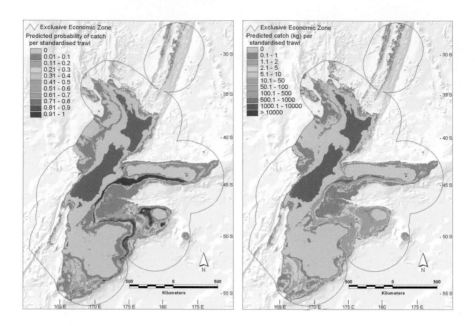

FIGURE 10.21. *Geological prediction maps of the presence probability (left map) and catch size (right map) obtained from the gradient boosted models.*

class `Prof/Man` (Professional/Managerial). The four best predicted classes are seen to be `Retired`, `Student`, `Prof/Man`, and `Homemaker`.

Figure 10.23 shows the relative predictor variable importances as averaged over all classes (10.46). Figure 10.24 displays the individual relative importance distributions (10.45) for each of the four best predicted classes. One sees that the most relevant predictors are generally different for each respective class. An exception is `age` which is among the three most relevant for predicting `Retired`, `Student`, and `Prof/Man`.

Figure 10.25 shows the partial dependence of the log-odds (10.52) on `age` for these three classes. The abscissa values are ordered codes for respective equally spaced age intervals. One sees that after accounting for the contributions of the other variables, the odds of being retired are higher for older people, whereas the opposite is the case for being a student. The odds of being professional/managerial are highest for middle-aged people. These results are of course not surprising. They illustrate that inspecting partial dependences separately for each class can lead to sensible results.

Bibliographic Notes

Schapire (1990) developed the first simple boosting procedure in the PAC learning framework (Valiant, 1984; Kearns and Vazirani, 1994). Schapire

Overall Error Rate = 0.425

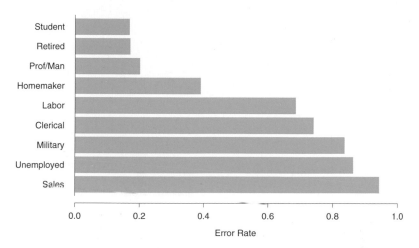

FIGURE 10.22. *Error rate for each occupation in the demographics data.*

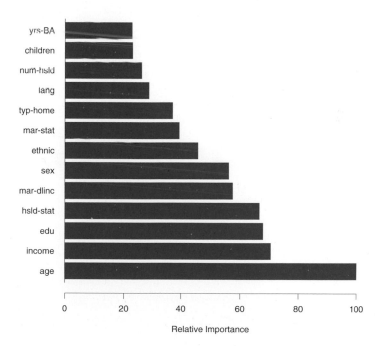

FIGURE 10.23. *Relative importance of the predictors as averaged over all classes for the demographics data.*

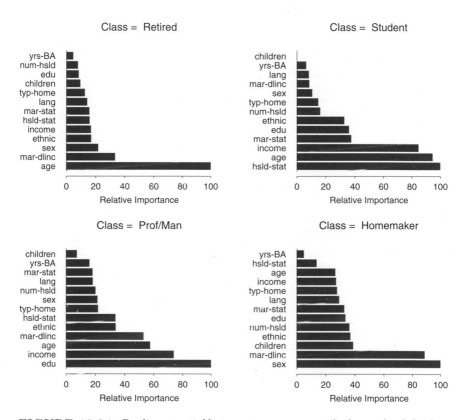

FIGURE 10.24. *Predictor variable importances separately for each of the four classes with lowest error rate for the demographics data.*

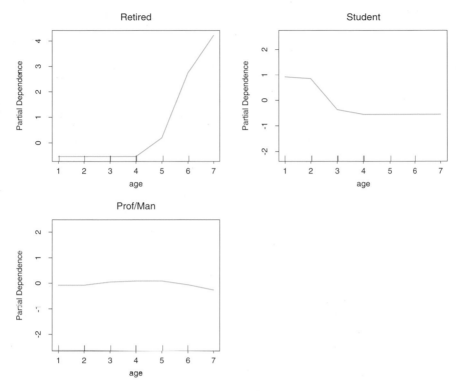

FIGURE 10.25. *Partial dependence of the odds of three different occupations on age, for the demographics data.*

showed that a *weak learner* could always improve its performance by training two additional classifiers on filtered versions of the input data stream. A weak learner is an algorithm for producing a two-class classifier with performance guaranteed (with high probability) to be significantly better than a coin-flip. After learning an initial classifier G_1 on the first N training points,

- G_2 is learned on a new sample of N points, half of which are misclassified by G_1;

- G_3 is learned on N points for which G_1 and G_2 disagree;

- the boosted classifier is $G_B = majority\ vote(G_1, G_2, G_3)$.

Schapire's "Strength of Weak Learnability" theorem proves that G_B has improved performance over G_1.

Freund (1995) proposed a "boost by majority" variation which combined many weak learners simultaneously and improved the performance of the simple boosting algorithm of Schapire. The theory supporting both of these algorithms requires the weak learner to produce a classifier with a fixed error rate. This led to the more adaptive and realistic AdaBoost (Freund and Schapire, 1996a) and its offspring, where this assumption was dropped.

Freund and Schapire (1996a) and Schapire and Singer (1999) provide some theory to support their algorithms, in the form of upper bounds on generalization error. This theory has evolved in the computational learning community, initially based on the concepts of PAC learning. Other theories attempting to explain boosting come from game theory (Freund and Schapire, 1996b; Breiman, 1999; Breiman, 1998), and VC theory (Schapire et al., 1998). The bounds and the theory associated with the AdaBoost algorithms are interesting, but tend to be too loose to be of practical importance. In practice, boosting achieves results far more impressive than the bounds would imply. Schapire (2002) and Meir and Rätsch (2003) give useful overviews more recent than the first edition of this book.

Friedman et al. (2000) and Friedman (2001) form the basis for our exposition in this chapter. Friedman et al. (2000) analyze AdaBoost statistically, derive the exponential criterion, and show that it estimates the log-odds of the class probability. They propose additive tree models, the right-sized trees and ANOVA representation of Section 10.11, and the multiclass logit formulation. Friedman (2001) developed gradient boosting and shrinkage for classification and regression, while Friedman (1999) explored stochastic variants of boosting. Mason et al. (2000) also embraced a gradient approach to boosting. As the published discussions of Friedman et al. (2000) shows, there is some controversy about how and why boosting works.

Since the publication of the first edition of this book, these debates have continued, and spread into the statistical community with a series of papers on consistency of boosting (Jiang, 2004; Lugosi and Vayatis, 2004; Zhang and Yu, 2005; Bartlett and Traskin, 2007). Mease and Wyner (2008), through a series of simulation examples, challenge some of our interpretations of boosting; our response (Friedman et al., 2008a) puts most of these objections to rest. A recent survey by Bühlmann and Hothorn (2007) supports our approach to boosting.

Exercises

Ex. 10.1 Derive expression (10.12) for the update parameter in AdaBoost.

Ex. 10.2 Prove result (10.16), that is, the minimizer of the population version of the AdaBoost criterion, is one-half of the log odds.

Ex. 10.3 Show that the marginal average (10.47) recovers additive and multiplicative functions (10.50) and (10.51), while the conditional expectation (10.49) does not.

Ex. 10.4

(a) Write a program implementing AdaBoost with trees.

(b) Redo the computations for the example of Figure 10.2. Plot the training error as well as test error, and discuss its behavior.

(c) Investigate the number of iterations needed to make the test error finally start to rise.

(d) Change the setup of this example as follows: define two classes, with the features in Class 1 being $X_1, X_2, \ldots, X_{10}$, standard independent Gaussian variates. In Class 2, the features $X_1, X_2, \ldots, X_{10}$ are also standard independent Gaussian, but conditioned on the event $\sum_j X_j^2 > 12$. Now the classes have significant overlap in feature space. Repeat the AdaBoost experiments as in Figure 10.2 and discuss the results.

Ex. 10.5 *Multiclass exponential loss* (Zhu et al., 2005). For a K-class classification problem, consider the coding $Y = (Y_1, \ldots, Y_K)^T$ with

$$Y_k = \begin{cases} 1, & \text{if } G = \mathcal{G}_k \\ -\frac{1}{K-1}, & \text{otherwise.} \end{cases} \tag{10.55}$$

Let $f = (f_1, \ldots, f_K)^T$ with $\sum_{k=1}^K f_k = 0$, and define

$$L(Y, f) = \exp\left(-\frac{1}{K} Y^T f\right). \tag{10.56}$$

(a) Using Lagrange multipliers, derive the population minimizer f^* of $L(Y, f)$, subject to the zero-sum constraint, and relate these to the class probabilities.

(b) Show that a multiclass boosting using this loss function leads to a reweighting algorithm similar to Adaboost, as in Section 10.4.

Ex. 10.6 *McNemar test* (Agresti, 1996). We report the test error rates on the spam data to be 5.5% for a generalized additive model (GAM), and 4.5% for gradient boosting (GBM), with a test sample of size 1536.

(a) Show that the standard error of these estimates is about 0.6%.

Since the same test data are used for both methods, the error rates are correlated, and we cannot perform a two-sample t-test. We can compare the methods directly on each test observation, leading to the summary

		GBM
GAM	Correct	Error
Correct	1434	18
Error	33	51

The McNemar test focuses on the discordant errors, 33 vs. 18.

(b) Conduct a test to show that GAM makes significantly more errors than gradient boosting, with a two-sided p-value of 0.036.

Ex. 10.7 Derive expression (10.32).

Ex. 10.8 Consider a K-class problem where the targets y_{ik} are coded as 1 if observation i is in class k and zero otherwise. Suppose we have a current model $f_k(x)$, $k = 1, \ldots, K$, with $\sum_{k=1}^{K} f_k(x) = 0$ (see (10.21) in Section 10.6). We wish to update the model for observations in a region R in predictor space, by adding constants $f_k(x) + \gamma_k$, with $\gamma_K = 0$.

(a) Write down the multinomial log-likelihood for this problem, and its first and second derivatives.

(b) Using only the diagonal of the Hessian matrix in (1), and starting from $\gamma_k = 0 \ \forall k$, show that a one-step approximate Newton update for γ_k is

$$\gamma_k^1 = \frac{\sum_{x_i \in R} (y_{ik} - p_{ik})}{\sum_{x_i \in R} p_{ik}(1 - p_{ik})}, \quad k = 1, \ldots, K - 1, \qquad (10.57)$$

where $p_{ik} = \exp(f_k(x_i))/\exp(\sum_{\ell=1}^{K} f_\ell(x_i))$.

(c) We prefer our update to sum to zero, as the current model does. Using symmetry arguments, show that

$$\hat{\gamma}_k = \frac{K - 1}{K}(\gamma_k^1 - \frac{1}{K}\sum_{\ell=1}^{K} \gamma_\ell^1), \quad k = 1, \ldots, K \qquad (10.58)$$

is an appropriate update, where γ_k^1 is defined as in (10.57) for all $k = 1, \ldots, K$.

Ex. 10.9 Consider a K-class problem where the targets y_{ik} are coded as 1 if observation i is in class k and zero otherwise. Using the multinomial deviance loss function (10.22) and the symmetric logistic transform, use the arguments leading to the gradient boosting Algorithm 10.3 to derive Algorithm 10.4. *Hint*: See exercise 10.8 for step 2(b)iii.

Ex. 10.10 Show that for $K = 2$ class classification, only one tree needs to be grown at each gradient-boosting iteration.

Ex. 10.11 Show how to compute the partial dependence function $f_S(X_S)$ in (10.47) efficiently.

Ex. 10.12 Referring to (10.49), let $S = \{1\}$ and $C = \{2\}$, with $f(X_1, X_2) = X_1$. Assume X_1 and X_2 are bivariate Gaussian, each with mean zero, variance one, and $E(X_1 X_2) = \rho$. Show that $E(f(X_1, X_2)|X_2) = \rho X_2$, even though f is not a function of X_2.

Algorithm 10.4 *Gradient Boosting for K-class Classification.*

---

1. Initialize $f_{k0}(x) = 0$, $k = 1, 2, \ldots, K$.

2. For $m=1$ to M:

   (a) Set

   $$p_k(x) = \frac{e^{f_k(x)}}{\sum_{\ell=1}^{K} e^{f_\ell(x)}}, \; k = 1, 2, \ldots, K.$$

   (b) For $k = 1$ to K:

       i. Compute $r_{ikm} = y_{ik} - p_k(x_i)$, $i = 1, 2, \ldots, N$.

       ii. Fit a regression tree to the targets r_{ikm}, $i = 1, 2, \ldots, N$, giving terminal regions R_{jkm}, $j = 1, 2, \ldots, J_m$.

       iii. Compute

   $$\gamma_{jkm} = \frac{K-1}{K} \frac{\sum_{x_i \in R_{jkm}} r_{ikm}}{\sum_{x_i \in R_{jkm}} |r_{ikm}|(1 - |r_{ikm}|)}, \; j = 1, 2, \ldots, J_m.$$

       iv. Update $f_{km}(x) = f_{k,m-1}(x) + \sum_{j=1}^{J_m} \gamma_{jkm} I(x \in R_{jkm})$.

3. Output $\hat{f}_k(x) = f_{kM}(x)$, $k = 1, 2, \ldots, K$.

11
Neural Networks

11.1 Introduction

In this chapter we describe a class of learning methods that was developed separately in different fields—statistics and artificial intelligence—based on essentially identical models. The central idea is to extract linear combinations of the inputs as derived features, and then model the target as a nonlinear function of these features. The result is a powerful learning method, with widespread applications in many fields. We first discuss the projection pursuit model, which evolved in the domain of semiparametric statistics and smoothing. The rest of the chapter is devoted to neural network models.

11.2 Projection Pursuit Regression

As in our generic supervised learning problem, assume we have an input vector X with p components, and a target Y. Let ω_m, $m = 1, 2, \ldots, M$, be unit p-vectors of unknown parameters. The projection pursuit regression (PPR) model has the form

$$f(X) = \sum_{m=1}^{M} g_m(\omega_m^T X). \tag{11.1}$$

This is an additive model, but in the derived features $V_m = \omega_m^T X$ rather than the inputs themselves. The functions g_m are unspecified and are esti-

T. Hastie et al., *The Elements of Statistical Learning, Second Edition,*
DOI: 10.1007/b94608_11,
© Springer Science+Business Media, LLC 2009

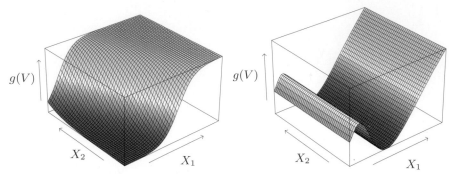

FIGURE 11.1. *Perspective plots of two ridge functions.*
(Left:) $g(V) = 1/[1 + \exp(-5(V - 0.5))]$, *where* $V = (X_1 + X_2)/\sqrt{2}$.
(Right:) $g(V) = (V + 0.1)\sin(1/(V/3 + 0.1))$, *where* $V = X_1$.

mated along with the directions ω_m using some flexible smoothing method
(see below).

The function $g_m(\omega_m^T X)$ is called a *ridge function* in $\mathbb{R}^p$. It varies only
in the direction defined by the vector ω_m. The scalar variable $V_m = \omega_m^T X$
is the projection of X onto the unit vector ω_m, and we seek ω_m so that
the model fits well, hence the name "projection pursuit." Figure 11.1 shows
some examples of ridge functions. In the example on the left $\omega = (1/\sqrt{2})(1, 1)^T$,
so that the function only varies in the direction $X_1 + X_2$. In the example
on the right, $\omega = (1, 0)$.

The PPR model (11.1) is very general, since the operation of forming
nonlinear functions of linear combinations generates a surprisingly large
class of models. For example, the product $X_1 \cdot X_2$ can be written as $[(X_1 + X_2)^2 - (X_1 - X_2)^2]/4$, and higher-order products can be represented similarly.

In fact, if M is taken arbitrarily large, for appropriate choice of g_m the
PPR model can approximate any continuous function in $\mathbb{R}^p$ arbitrarily
well. Such a class of models is called a *universal approximator*. However
this generality comes at a price. Interpretation of the fitted model is usually
difficult, because each input enters into the model in a complex and multi-
faceted way. As a result, the PPR model is most useful for prediction, and
not very useful for producing an understandable model for the data. The
$M = 1$ model, known as the *single index model* in econometrics, is an
exception. It is slightly more general than the linear regression model, and
offers a similar interpretation.

How do we fit a PPR model, given training data (x_i, y_i), $i = 1, 2, \ldots, N$?
We seek the approximate minimizers of the error function

$$\sum_{i=1}^{N} \left[y_i - \sum_{m=1}^{M} g_m(\omega_m^T x_i) \right]^2 \qquad (11.2)$$

over functions g_m and direction vectors ω_m, $m = 1, 2, \ldots, M$. As in other smoothing problems, we need either explicitly or implicitly to impose complexity constraints on the g_m, to avoid overfit solutions.

Consider just one term ($M = 1$, and drop the subscript). Given the direction vector ω, we form the derived variables $v_i = \omega^T x_i$. Then we have a one-dimensional smoothing problem, and we can apply any scatterplot smoother, such as a smoothing spline, to obtain an estimate of g.

On the other hand, given g, we want to minimize (11.2) over ω. A Gauss–Newton search is convenient for this task. This is a quasi-Newton method, in which the part of the Hessian involving the second derivative of g is discarded. It can be simply derived as follows. Let ω_{old} be the current estimate for ω. We write

$$g(\omega^T x_i) \approx g(\omega_{\text{old}}^T x_i) + g'(\omega_{\text{old}}^T x_i)(\omega - \omega_{\text{old}})^T x_i \qquad (11.3)$$

to give

$$\sum_{i=1}^{N} \left[y_i - g(\omega^T x_i) \right]^2 \approx \sum_{i=1}^{N} g'(\omega_{\text{old}}^T x_i)^2 \left[\left(\omega_{\text{old}}^T x_i + \frac{y_i - g(\omega_{\text{old}}^T x_i)}{g'(\omega_{\text{old}}^T x_i)} \right) - \omega^T x_i \right]^2 .$$
$$(11.4)$$

To minimize the right-hand side, we carry out a least squares regression with target $\omega_{\text{old}}^T x_i + (y_i - g(\omega_{\text{old}}^T x_i))/g'(\omega_{\text{old}}^T x_i)$ on the input x_i, with weights $g'(\omega_{\text{old}}^T x_i)^2$ and no intercept (bias) term. This produces the updated coefficient vector ω_{new}.

These two steps, estimation of g and ω, are iterated until convergence. With more than one term in the PPR model, the model is built in a forward stage-wise manner, adding a pair (ω_m, g_m) at each stage.

There are a number of implementation details.

- Although any smoothing method can in principle be used, it is convenient if the method provides derivatives. Local regression and smoothing splines are convenient.

- After each step the g_m's from previous steps can be readjusted using the backfitting procedure described in Chapter 9. While this may lead ultimately to fewer terms, it is not clear whether it improves prediction performance.

- Usually the ω_m are not readjusted (partly to avoid excessive computation), although in principle they could be as well.

- The number of terms M is usually estimated as part of the forward stage-wise strategy. The model building stops when the next term does not appreciably improve the fit of the model. Cross-validation can also be used to determine M.

There are many other applications, such as density estimation (Friedman et al., 1984; Friedman, 1987), where the projection pursuit idea can be used. In particular, see the discussion of ICA in Section 14.7 and its relationship with exploratory projection pursuit. However the projection pursuit regression model has not been widely used in the field of statistics, perhaps because at the time of its introduction (1981), its computational demands exceeded the capabilities of most readily available computers. But it does represent an important intellectual advance, one that has blossomed in its reincarnation in the field of neural networks, the topic of the rest of this chapter.

11.3 Neural Networks

The term *neural network* has evolved to encompass a large class of models and learning methods. Here we describe the most widely used "vanilla" neural net, sometimes called the single hidden layer back-propagation network, or single layer perceptron. There has been a great deal of *hype* surrounding neural networks, making them seem magical and mysterious. As we make clear in this section, they are just nonlinear statistical models, much like the projection pursuit regression model discussed above.

A neural network is a two-stage regression or classification model, typically represented by a *network diagram* as in Figure 11.2. This network applies both to regression or classification. For regression, typically $K = 1$ and there is only one output unit Y_1 at the top. However, these networks can handle multiple quantitative responses in a seamless fashion, so we will deal with the general case.

For K-class classification, there are K units at the top, with the kth unit modeling the probability of class k. There are K target measurements Y_k, $k = 1, \ldots, K$, each being coded as a $0 - 1$ variable for the kth class.

Derived features Z_m are created from linear combinations of the inputs, and then the target Y_k is modeled as a function of linear combinations of the Z_m,

$$
\begin{aligned}
Z_m &= \sigma(\alpha_{0m} + \alpha_m^T X), \; m = 1, \ldots, M, \\
T_k &= \beta_{0k} + \beta_k^T Z, \; k = 1, \ldots, K, \\
f_k(X) &= g_k(T), \; k = 1, \ldots, K,
\end{aligned}
\tag{11.5}
$$

where $Z = (Z_1, Z_2, \ldots, Z_M)$, and $T = (T_1, T_2, \ldots, T_K)$.

The activation function $\sigma(v)$ is usually chosen to be the *sigmoid* $\sigma(v) = 1/(1 + e^{-v})$; see Figure 11.3 for a plot of $1/(1 + e^{-v})$. Sometimes Gaussian radial basis functions (Chapter 6) are used for the $\sigma(v)$, producing what is known as a *radial basis function network*.

Neural network diagrams like Figure 11.2 are sometimes drawn with an additional *bias* unit feeding into every unit in the hidden and output layers.

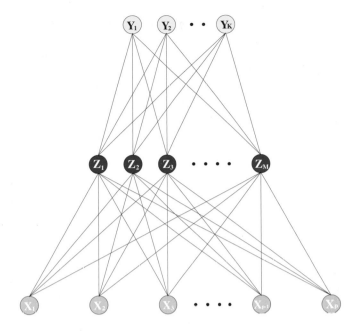

FIGURE 11.2. *Schematic of a single hidden layer, feed-forward neural network.*

Thinking of the constant "1" as an additional input feature, this bias unit captures the intercepts α_{0m} and β_{0k} in model (11.5)

The output function $g_k(T)$ allows a final transformation of the vector of outputs T. For regression we typically choose the identity function $g_k(T) = T_k$. Early work in K-class classification also used the identity function, but this was later abandoned in favor of the *softmax* function

$$g_k(T) = \frac{e^{T_k}}{\sum_{\ell=1}^{K} e^{T_\ell}}. \tag{11.6}$$

This is of course exactly the transformation used in the multilogit model (Section 4.4), and produces positive estimates that sum to one. In Section 4.2 we discuss other problems with linear activation functions, in particular potentially severe masking effects.

The units in the middle of the network, computing the derived features Z_m, are called *hidden units* because the values Z_m are not directly observed. In general there can be more than one hidden layer, as illustrated in the example at the end of this chapter. We can think of the Z_m as a basis expansion of the original inputs X; the neural network is then a standard linear model, or linear multilogit model, using these transformations as inputs. There is, however, an important enhancement over the basis-expansion techniques discussed in Chapter 5; here the parameters of the basis functions are learned from the data.

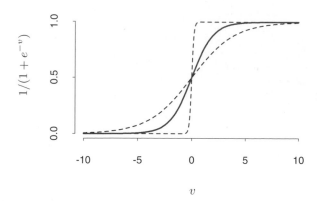

FIGURE 11.3. *Plot of the sigmoid function* $\sigma(v) = 1/(1+\exp(-v))$ *(red curve), commonly used in the hidden layer of a neural network. Included are* $\sigma(sv)$ *for* $s = \frac{1}{2}$ *(blue curve) and* $s = 10$ *(purple curve). The scale parameter* s *controls the activation rate, and we can see that large* s *amounts to a hard activation at* $v = 0$. *Note that* $\sigma(s(v - v_0))$ *shifts the activation threshold from 0 to* v_0.

Notice that if σ is the identity function, then the entire model collapses to a linear model in the inputs. Hence a neural network can be thought of as a nonlinear generalization of the linear model, both for regression and classification. By introducing the nonlinear transformation σ, it greatly enlarges the class of linear models. In Figure 11.3 we see that the rate of activation of the sigmoid depends on the norm of α_m, and if $\|\alpha_m\|$ is very small, the unit will indeed be operating in the *linear part* of its activation function.

Notice also that the neural network model with one hidden layer has exactly the same form as the projection pursuit model described above. The difference is that the PPR model uses nonparametric functions $g_m(v)$, while the neural network uses a far simpler function based on $\sigma(v)$, with three free parameters in its argument. In detail, viewing the neural network model as a PPR model, we identify

$$
\begin{aligned}
g_m(\omega_m^T X) &= \beta_m \sigma(\alpha_{0m} + \alpha_m^T X) \\
&= \beta_m \sigma(\alpha_{0m} + \|\alpha_m\|(\omega_m^T X)), \quad (11.7)
\end{aligned}
$$

where $\omega_m = \alpha_m/\|\alpha_m\|$ is the mth unit-vector. Since $\sigma_{\beta,\alpha_0,s}(v) = \beta\sigma(\alpha_0 + sv)$ has lower complexity than a more general nonparametric $g(v)$, it is not surprising that a neural network might use 20 or 100 such functions, while the PPR model typically uses fewer terms ($M = 5$ or 10, for example).

Finally, we note that the name "neural networks" derives from the fact that they were first developed as models for the human brain. Each unit represents a neuron, and the connections (links in Figure 11.2) represent synapses. In early models, the neurons fired when the total signal passed to that unit exceeded a certain threshold. In the model above, this corresponds to use of a step function for $\sigma(Z)$ and $g_m(T)$. Later the neural network was recognized as a useful tool for nonlinear statistical modeling, and for this purpose the step function is not smooth enough for optimization. Hence the step function was replaced by a smoother threshold function, the sigmoid in Figure 11.3.

11.4 Fitting Neural Networks

The neural network model has unknown parameters, often called *weights*, and we seek values for them that make the model fit the training data well. We denote the complete set of weights by θ, which consists of

$$\begin{aligned} \{\alpha_{0m}, \alpha_m;\ m = 1, 2, \ldots, M\} &\quad M(p+1) \text{ weights}, \\ \{\beta_{0k}, \beta_k;\ k = 1, 2, \ldots, K\} &\quad K(M+1) \text{ weights}. \end{aligned} \tag{11.8}$$

For regression, we use sum-of-squared errors as our measure of fit (error function)

$$R(\theta) = \sum_{k=1}^{K} \sum_{i=1}^{N} (y_{ik} - f_k(x_i))^2. \tag{11.9}$$

For classification we use either squared error or cross-entropy (deviance):

$$R(\theta) = -\sum_{i=1}^{N} \sum_{k=1}^{K} y_{ik} \log f_k(x_i), \tag{11.10}$$

and the corresponding classifier is $G(x) = \text{argmax}_k f_k(x)$. With the softmax activation function and the cross-entropy error function, the neural network model is exactly a linear logistic regression model in the hidden units, and all the parameters are estimated by maximum likelihood.

Typically we don't want the global minimizer of $R(\theta)$, as this is likely to be an overfit solution. Instead some regularization is needed: this is achieved directly through a penalty term, or indirectly by early stopping. Details are given in the next section.

The generic approach to minimizing $R(\theta)$ is by gradient descent, called *back-propagation* in this setting. Because of the compositional form of the model, the gradient can be easily derived using the chain rule for differentiation. This can be computed by a forward and backward sweep over the network, keeping track only of quantities local to each unit.

Here is back-propagation in detail for squared error loss. Let $z_{mi} = \sigma(\alpha_{0m} + \alpha_m^T x_i)$, from (11.5) and let $z_i = (z_{1i}, z_{2i}, \ldots, z_{Mi})$. Then we have

$$
\begin{aligned}
R(\theta) &\equiv \sum_{i=1}^{N} R_i \\
&= \sum_{i=1}^{N} \sum_{k=1}^{K} (y_{ik} - f_k(x_i))^2,
\end{aligned}
\tag{11.11}
$$

with derivatives

$$
\begin{aligned}
\frac{\partial R_i}{\partial \beta_{km}} &= -2(y_{ik} - f_k(x_i)) g_k'(\beta_k^T z_i) z_{mi}, \\
\frac{\partial R_i}{\partial \alpha_{m\ell}} &= -\sum_{k=1}^{K} 2(y_{ik} - f_k(x_i)) g_k'(\beta_k^T z_i) \beta_{km} \sigma'(\alpha_m^T x_i) x_{i\ell}.
\end{aligned}
\tag{11.12}
$$

Given these derivatives, a gradient descent update at the $(r+1)$st iteration has the form

$$
\begin{aligned}
\beta_{km}^{(r+1)} &= \beta_{km}^{(r)} - \gamma_r \sum_{i=1}^{N} \frac{\partial R_i}{\partial \beta_{km}^{(r)}}, \\
\alpha_{m\ell}^{(r+1)} &= \alpha_{m\ell}^{(r)} - \gamma_r \sum_{i=1}^{N} \frac{\partial R_i}{\partial \alpha_{m\ell}^{(r)}},
\end{aligned}
\tag{11.13}
$$

where γ_r is the *learning rate*, discussed below.

Now write (11.12) as

$$
\begin{aligned}
\frac{\partial R_i}{\partial \beta_{km}} &= \delta_{ki} z_{mi}, \\
\frac{\partial R_i}{\partial \alpha_{m\ell}} &= s_{mi} x_{i\ell}.
\end{aligned}
\tag{11.14}
$$

The quantities δ_{ki} and s_{mi} are "errors" from the current model at the output and hidden layer units, respectively. From their definitions, these errors satisfy

$$
s_{mi} = \sigma'(\alpha_m^T x_i) \sum_{k=1}^{K} \beta_{km} \delta_{ki},
\tag{11.15}
$$

known as the *back-propagation equations*. Using this, the updates in (11.13) can be implemented with a two-pass algorithm. In the *forward pass*, the current weights are fixed and the predicted values $\hat{f}_k(x_i)$ are computed from formula (11.5). In the *backward pass*, the errors δ_{ki} are computed, and then back-propagated via (11.15) to give the errors s_{mi}. Both sets of errors are then used to compute the gradients for the updates in (11.13), via (11.14).

This two-pass procedure is what is known as back-propagation. It has also been called the *delta rule* (Widrow and Hoff, 1960). The computational components for cross-entropy have the same form as those for the sum of squares error function, and are derived in Exercise 11.3.

The advantages of back-propagation are its simple, local nature. In the back propagation algorithm, each hidden unit passes and receives information only to and from units that share a connection. Hence it can be implemented efficiently on a parallel architecture computer.

The updates in (11.13) are a kind of *batch learning*, with the parameter updates being a sum over all of the training cases. Learning can also be carried out online—processing each observation one at a time, updating the gradient after each training case, and cycling through the training cases many times. In this case, the sums in equations (11.13) are replaced by a single summand. A *training epoch* refers to one sweep through the entire training set. Online training allows the network to handle very large training sets, and also to update the weights as new observations come in.

The learning rate γ_r for batch learning is usually taken to be a constant, and can also be optimized by a line search that minimizes the error function at each update. With online learning γ_r should decrease to zero as the iteration $r \to \infty$. This learning is a form of *stochastic approximation* (Robbins and Munro, 1951); results in this field ensure convergence if $\gamma_r \to 0$, $\sum_r \gamma_r = \infty$, and $\sum_r \gamma_r^2 < \infty$ (satisfied, for example, by $\gamma_r = 1/r$).

Back-propagation can be very slow, and for that reason is usually not the method of choice. Second-order techniques such as Newton's method are not attractive here, because the second derivative matrix of R (the Hessian) can be very large. Better approaches to fitting include conjugate gradients and variable metric methods. These avoid explicit computation of the second derivative matrix while still providing faster convergence.

11.5 Some Issues in Training Neural Networks

There is quite an art in training neural networks. The model is generally overparametrized, and the optimization problem is nonconvex and unstable unless certain guidelines are followed. In this section we summarize some of the important issues.

11.5.1 Starting Values

Note that if the weights are near zero, then the operative part of the sigmoid (Figure 11.3) is roughly linear, and hence the neural network collapses into an approximately linear model (Exercise 11.2). Usually starting values for weights are chosen to be random values near zero. Hence the model starts out nearly linear, and becomes nonlinear as the weights increase. Individual units localize to directions and introduce nonlinearities where needed. Use of exact zero weights leads to zero derivatives and perfect symmetry, and the algorithm never moves. Starting instead with large weights often leads to poor solutions.

11.5.2 Overfitting

Often neural networks have too many weights and will overfit the data at the global minimum of R. In early developments of neural networks, either by design or by accident, an early stopping rule was used to avoid overfitting. Here we train the model only for a while, and stop well before we approach the global minimum. Since the weights start at a highly regularized (linear) solution, this has the effect of shrinking the final model toward a linear model. A validation dataset is useful for determining when to stop, since we expect the validation error to start increasing.

A more explicit method for regularization is *weight decay*, which is analogous to ridge regression used for linear models (Section 3.4.1). We add a penalty to the error function $R(\theta) + \lambda J(\theta)$, where

$$J(\theta) = \sum_{k,m} \beta_{km}^2 + \sum_{m,\ell} \alpha_{m\ell}^2 \qquad (11.16)$$

and $\lambda \geq 0$ is a tuning parameter. Larger values of λ will tend to shrink the weights toward zero: typically cross-validation is used to estimate λ. The effect of the penalty is to simply add terms $2\beta_{km}$ and $2\alpha_{m\ell}$ to the respective gradient expressions (11.13). Other forms for the penalty have been proposed, for example,

$$J(\theta) = \sum_{k,m} \frac{\beta_{km}^2}{1 + \beta_{km}^2} + \sum_{m,\ell} \frac{\alpha_{m\ell}^2}{1 + \alpha_{m\ell}^2}, \qquad (11.17)$$

known as the *weight elimination* penalty. This has the effect of shrinking smaller weights more than (11.16) does.

Figure 11.4 shows the result of training a neural network with ten hidden units, without weight decay (upper panel) and with weight decay (lower panel), to the mixture example of Chapter 2. Weight decay has clearly improved the prediction. Figure 11.5 shows heat maps of the estimated weights from the training (grayscale versions of these are called *Hinton diagrams.*) We see that weight decay has dampened the weights in both layers: the resulting weights are spread fairly evenly over the ten hidden units.

11.5.3 Scaling of the Inputs

Since the scaling of the inputs determines the effective scaling of the weights in the bottom layer, it can have a large effect on the quality of the final

Neural Network - 10 Units, No Weight Decay

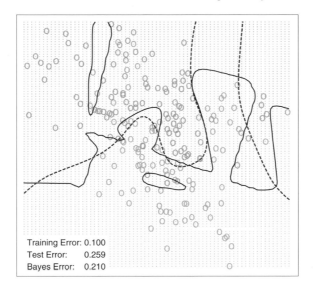

Neural Network - 10 Units, Weight Decay=0.02

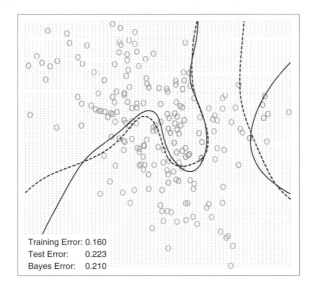

FIGURE 11.4. *A neural network on the mixture example of Chapter 2. The upper panel uses no weight decay, and overfits the training data. The lower panel uses weight decay, and achieves close to the Bayes error rate (broken purple boundary). Both use the softmax activation function and cross-entropy error.*

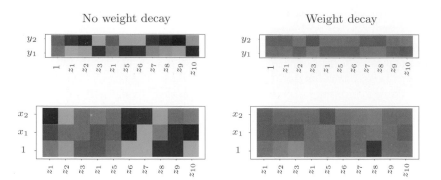

FIGURE 11.5. *Heat maps of the estimated weights from the training of neural networks from Figure 11.4. The display ranges from bright green (negative) to bright red (positive).*

solution. At the outset it is best to standardize all inputs to have mean zero and standard deviation one. This ensures all inputs are treated equally in the regularization process, and allows one to choose a meaningful range for the random starting weights. With standardized inputs, it is typical to take random uniform weights over the range $[-0.7, +0.7]$.

11.5.4 *Number of Hidden Units and Layers*

Generally speaking it is better to have too many hidden units than too few. With too few hidden units, the model might not have enough flexibility to capture the nonlinearities in the data; with too many hidden units, the extra weights can be shrunk toward zero if appropriate regularization is used. Typically the number of hidden units is somewhere in the range of 5 to 100, with the number increasing with the number of inputs and number of training cases. It is most common to put down a reasonably large number of units and train them with regularization. Some researchers use cross-validation to estimate the optimal number, but this seems unnecessary if cross-validation is used to estimate the regularization parameter. Choice of the number of hidden layers is guided by background knowledge and experimentation. Each layer extracts features of the input for regression or classification. Use of multiple hidden layers allows construction of hierarchical features at different levels of resolution. An example of the effective use of multiple layers is given in Section 11.6.

11.5.5 *Multiple Minima*

The error function $R(\theta)$ is nonconvex, possessing many local minima. As a result, the final solution obtained is quite dependent on the choice of starting weights. One must at least try a number of random starting configurations, and choose the solution giving lowest (penalized) error. Probably a better approach is to use the average predictions over the collection of networks as the final prediction (Ripley, 1996). This is preferable to averaging the weights, since the nonlinearity of the model implies that this averaged solution could be quite poor. Another approach is via *bagging*, which averages the predictions of networks training from randomly perturbed versions of the training data. This is described in Section 8.7.

11.6   Example: Simulated Data

We generated data from two additive error models $Y = f(X) + \varepsilon$:

$$\text{Sum of sigmoids: } Y = \sigma(a_1^T X) + \sigma(a_2^T X) + \varepsilon_1;$$

$$\text{Radial: } Y = \prod_{m=1}^{10} \phi(X_m) + \varepsilon_2.$$

Here $X^T = (X_1, X_2, \ldots, X_p)$, each X_j being a standard Gaussian variate, with $p = 2$ in the first model, and $p = 10$ in the second.

For the sigmoid model, $a_1 = (3, 3)$, $a_2 = (3, -3)$; for the radial model, $\phi(t) = (1/2\pi)^{1/2} \exp(-t^2/2)$. Both ε_1 and ε_2 are Gaussian errors, with variance chosen so that the signal-to noise ratio

$$\frac{\text{Var}(\text{E}(Y|X))}{\text{Var}(Y - \text{E}(Y|X))} - \frac{\text{Var}(f(X))}{\text{Var}(\varepsilon)} \qquad (11.18)$$

is 4 in both models. We took a training sample of size 100 and a test sample of size 10,000. We fit neural networks with weight decay and various numbers of hidden units, and recorded the average test error $\text{E}_{\text{Test}}(Y - \hat{f}(X))^2$ for each of 10 random starting weights. Only one training set was generated, but the results are typical for an "average" training set. The test errors are shown in Figure 11.6. Note that the zero hidden unit model refers to linear least squares regression. The neural network is perfectly suited to the sum of sigmoids model, and the two-unit model does perform the best, achieving an error close to the Bayes rate. (Recall that the Bayes rate for regression with squared error is the error variance; in the figures, we report test error relative to the Bayes error). Notice, however, that with more hidden units, overfitting quickly creeps in, and with some starting weights the model does worse than the linear model (zero hidden unit) model. Even with two hidden units, two of the ten starting weight configurations produced results no better than the linear model, confirming the importance of multiple starting values.

A radial function is in a sense the most difficult for the neural net, as it is spherically symmetric and with no preferred directions. We see in the right

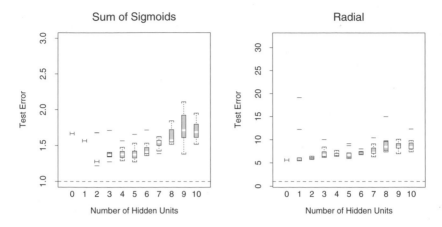

FIGURE 11.6. *Boxplots of test error, for simulated data example, relative to the Bayes error (broken horizontal line). True function is a sum of two sigmoids on the left, and a radial function is on the right. The test error is displayed for 10 different starting weights, for a single hidden layer neural network with the number of units as indicated.*

panel of Figure 11.6 that it does poorly in this case, with the test error staying well above the Bayes error (note the different vertical scale from the left panel). In fact, since a constant fit (such as the sample average) achieves a relative error of 5 (when the SNR is 4), we see that the neural networks perform increasingly worse than the mean.

In this example we used a fixed weight decay parameter of 0.0005, representing a mild amount of regularization. The results in the left panel of Figure 11.6 suggest that more regularization is needed with greater numbers of hidden units.

In Figure 11.7 we repeated the experiment for the sum of sigmoids model, with no weight decay in the left panel, and stronger weight decay ($\lambda = 0.1$) in the right panel. With no weight decay, overfitting becomes even more severe for larger numbers of hidden units. The weight decay value $\lambda = 0.1$ produces good results for all numbers of hidden units, and there does not appear to be overfitting as the number of units increase. Finally, Figure 11.8 shows the test error for a ten hidden unit network, varying the weight decay parameter over a wide range. The value 0.1 is approximately optimal.

In summary, there are two free parameters to select: the weight decay λ and number of hidden units M. As a learning strategy, one could fix either parameter at the value corresponding to the least constrained model, to ensure that the model is rich enough, and use cross-validation to choose the other parameter. Here the least constrained values are zero weight decay and ten hidden units. Comparing the left panel of Figure 11.7 to Figure 11.8, we see that the test error is less sensitive to the value of the weight

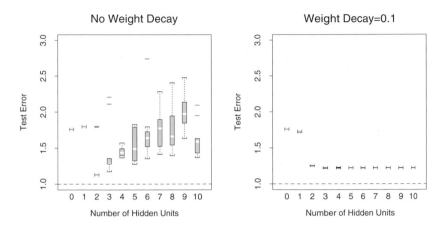

FIGURE 11.7. *Boxplots of test error, for simulated data example, relative to the Bayes error. True function is a sum of two sigmoids. The test error is displayed for ten different starting weights, for a single hidden layer neural network with the number units as indicated. The two panels represent no weight decay (left) and strong weight decay $\lambda = 0.1$ (right).*

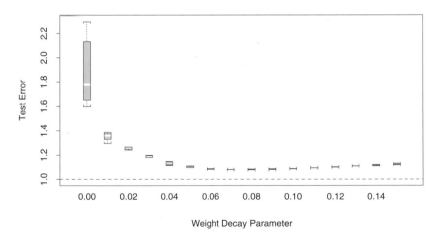

FIGURE 11.8. *Boxplots of test error, for simulated data example. True function is a sum of two sigmoids. The test error is displayed for ten different starting weights, for a single hidden layer neural network with ten hidden units and weight decay parameter value as indicated.*

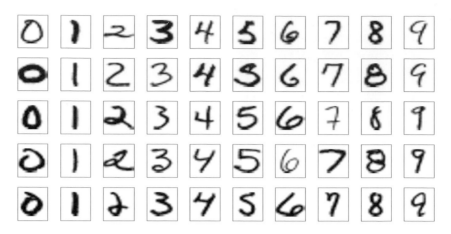

FIGURE 11.9. *Examples of training cases from ZIP code data. Each image is a* 16×16 *8-bit grayscale representation of a handwritten digit.*

decay parameter, and hence cross-validation of this parameter would be preferred.

11.7 Example: ZIP Code Data

This example is a character recognition task: classification of handwritten numerals. This problem captured the attention of the machine learning and neural network community for many years, and has remained a benchmark problem in the field. Figure 11.9 shows some examples of normalized handwritten digits, automatically scanned from envelopes by the U.S. Postal Service. The original scanned digits are binary and of different sizes and orientations; the images shown here have been deslanted and size normalized, resulting in 16×16 grayscale images (Le Cun et al., 1990). These 256 pixel values are used as inputs to the neural network classifier.

A *black box* neural network is not ideally suited to this pattern recognition task, partly because the pixel representation of the images lack certain invariances (such as small rotations of the image). Consequently early attempts with neural networks yielded misclassification rates around 4.5% on various examples of the problem. In this section we show some of the pioneering efforts to handcraft the neural network to overcome some these deficiencies (Le Cun, 1989), which ultimately led to the state of the art in neural network performance(Le Cun et al., 1998)[1].

Although current digit datasets have tens of thousands of training and test examples, the sample size here is deliberately modest in order to em-

---

[1]The figures and tables in this example were recreated from Le Cun (1989).

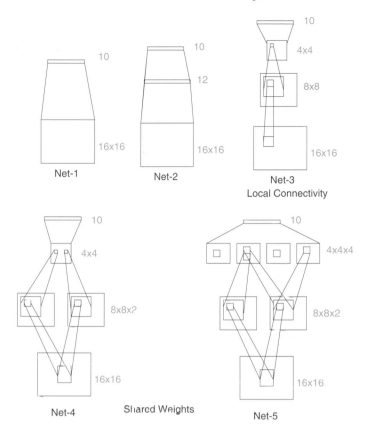

FIGURE 11.10. *Architecture of the five networks used in the ZIP code example.*

phasize the effects. The examples were obtained by scanning some actual hand-drawn digits, and then generating additional images by random horizontal shifts. Details may be found in Le Cun (1989). There are 320 digits in the training set, and 160 in the test set.

Five different networks were fit to the data:

Net-1: No hidden layer, equivalent to multinomial logistic regression.

Net-2: One hidden layer, 12 hidden units fully connected.

Net-3: Two hidden layers locally connected.

Net-4: Two hidden layers, locally connected with weight sharing.

Net-5: Two hidden layers, locally connected, two levels of weight sharing.

These are depicted in Figure 11.10. Net-1 for example has 256 inputs, one each for the 16×16 input pixels, and ten output units for each of the digits 0–9. The predicted value $\hat{f}_k(x)$ represents the estimated probability that an image x has digit class k, for $k = 0, 1, 2, \ldots, 9$.

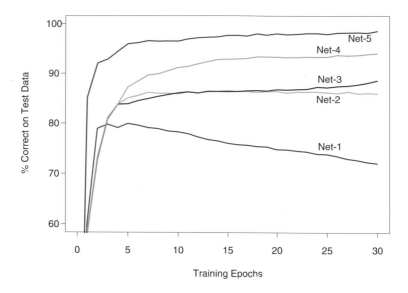

FIGURE 11.11. *Test performance curves, as a function of the number of training epochs, for the five networks of Table 11.1 applied to the ZIP code data. (Le Cun, 1989)*

The networks all have sigmoidal output units, and were all fit with the sum-of-squares error function. The first network has no hidden layer, and hence is nearly equivalent to a linear multinomial regression model (Exercise 11.4). Net-2 is a single hidden layer network with 12 hidden units, of the kind described above.

The training set error for all of the networks was 0%, since in all cases there are more parameters than training observations. The evolution of the test error during the training epochs is shown in Figure 11.11. The linear network (Net-1) starts to overfit fairly quickly, while test performance of the others level off at successively superior values.

The other three networks have additional features which demonstrate the power and flexibility of the neural network paradigm. They introduce constraints on the network, natural for the problem at hand, which allow for more complex connectivity but fewer parameters.

Net-3 uses local connectivity: this means that each hidden unit is connected to only a small patch of units in the layer below. In the first hidden layer (an 8×8 array), each unit takes inputs from a 3×3 patch of the input layer; for units in the first hidden layer that are one unit apart, their receptive fields overlap by one row or column, and hence are two pixels apart. In the second hidden layer, inputs are from a 5×5 patch, and again units that are one unit apart have receptive fields that are two units apart. The weights for all other connections are set to zero. Local connectivity makes each unit responsible for extracting local features from the layer below, and

TABLE 11.1. *Test set performance of five different neural networks on a handwritten digit classification example (Le Cun, 1989).*

	Network Architecture	Links	Weights	% Correct
Net-1:	Single layer network	2570	2570	80.0%
Net-2:	Two layer network	3214	3214	87.0%
Net-3:	Locally connected	1226	1226	88.5%
Net-4:	Constrained network 1	2266	1132	94.0%
Net-5:	Constrained network 2	5194	1060	98.4%

reduces considerably the total number of weights. With many more hidden units than Net-2, Net-3 has fewer links and hence weights (1226 vs. 3214), and achieves similar performance.

Net-4 and Net-5 have local connectivity with shared weights. All units in a local feature map perform the *same* operation on different parts of the image, achieved by sharing the same weights. The first hidden layer of Net-4 has two 8×8 arrays, and each unit takes input from a 3×3 patch just like in Net-3. However, each of the units in a single 8×8 feature map share the same set of nine weights (but have their own bias parameter). This forces the extracted features in different parts of the image to be computed by the same linear functional, and consequently these networks are sometimes known as *convolutional networks*. The second hidden layer of Net-4 has no weight sharing, and is the same as in Net-3. The gradient of the error function R with respect to a shared weight is the sum of the gradients of R with respect to each connection controlled by the weights in question.

Table 11.1 gives the number of links, the number of weights and the optimal test performance for each of the networks. We see that Net-4 has more links but fewer weights than Net-3, and superior test performance. Net-5 has four 4×4 feature maps in the second hidden layer, each unit connected to a 5×5 local patch in the layer below. Weights are shared in each of these feature maps. We see that Net-5 does the best, having errors of only 1.6%, compared to 13% for the "vanilla" network Net-2. The clever design of network Net-5, motivated by the fact that features of handwriting style should appear in more than one part of a digit, was the result of many person years of experimentation. This and similar networks gave better performance on ZIP code problems than any other learning method at that time (early 1990s). This example also shows that neural networks are not a fully automatic tool, as they are sometimes advertised. As with all statistical models, subject matter knowledge can and should be used to improve their performance.

This network was later outperformed by the tangent distance approach (Simard et al., 1993) described in Section 13.3.3, which explicitly incorporates natural affine invariances. At this point the digit recognition datasets become test beds for every new learning procedure, and researchers worked hard to drive down the error rates. As of this writing, the best error rates on a large database (60,000 training, 10,000 test observations), derived from standard NIST[2] databases, were reported to be the following: (Le Cun et al., 1998):

- 1.1% for tangent distance with a 1-nearest neighbor classifier (Section 13.3.3);

- 0.8% for a degree-9 polynomial SVM (Section 12.3);

- 0.8% for *LeNet-5*, a more complex version of the convolutional network described here;

- 0.7% for boosted *LeNet-4*. Boosting is described in Chapter 8. *LeNet-4* is a predecessor of *LeNet-5*.

Le Cun et al. (1998) report a much larger table of performance results, and it is evident that many groups have been working very hard to bring these test error rates down. They report a standard error of 0.1% on the error estimates, which is based on a binomial average with $N = 10,000$ and $p \approx 0.01$. This implies that error rates within 0.1—0.2% of one another are statistically equivalent. Realistically the standard error is even higher, since the test data has been implicitly used in the tuning of the various procedures.

11.8   Discussion

Both projection pursuit regression and neural networks take nonlinear functions of linear combinations ("derived features") of the inputs. This is a powerful and very general approach for regression and classification, and has been shown to compete well with the best learning methods on many problems.

These tools are especially effective in problems with a high signal-to-noise ratio and settings where prediction without interpretation is the goal. They are less effective for problems where the goal is to describe the physical process that generated the data and the roles of individual inputs. Each input enters into the model in many places, in a nonlinear fashion. Some authors (Hinton, 1989) plot a diagram of the estimated weights into each hidden unit, to try to understand the feature that each unit is extracting. This is limited however by the lack of identifiability of the parameter vectors α_m, $m = 1, \ldots, M$. Often there are solutions with α_m spanning the same linear space as the ones found during training, giving predicted values that

---

[2]The National Institute of Standards and Technology maintain large databases, including handwritten character databases; http://www.nist.gov/srd/.

are roughly the same. Some authors suggest carrying out a principal component analysis of these weights, to try to find an interpretable solution. In general, the difficulty of interpreting these models has limited their use in fields like medicine, where interpretation of the model is very important.

There has been a great deal of research on the training of neural networks. Unlike methods like CART and MARS, neural networks are smooth functions of real-valued parameters. This facilitates the development of Bayesian inference for these models. The next sections discusses a successful Bayesian implementation of neural networks.

11.9 Bayesian Neural Nets and the NIPS 2003 Challenge

A classification competition was held in 2003, in which five labeled training datasets were provided to participants. It was organized for a Neural Information Processing Systems (NIPS) workshop. Each of the data sets constituted a two-class classification problems, with different sizes and from a variety of domains (see Table 11.2). Feature measurements for a validation dataset were also available.

Participants developed and applied statistical learning procedures to make predictions on the datasets, and could submit predictions to a website on the validation set for a period of 12 weeks. With this feedback, participants were then asked to submit predictions for a separate test set and they received their results. Finally, the class labels for the validation set were released and participants had one week to train their algorithms on the combined training and validation sets, and submit their final predictions to the competition website. A total of 75 groups participated, with 20 and 16 eventually making submissions on the validation and test sets, respectively.

There was an emphasis on feature extraction in the competition. Artificial "probes" were added to the data: these are noise features with distributions resembling the real features but independent of the class labels. The percentage of probes that were added to each dataset, relative to the total set of features, is shown on Table 11.2. Thus each learning algorithm had to figure out a way of identifying the probes and downweighting or eliminating them.

A number of metrics were used to evaluate the entries, including the percentage correct on the test set, the area under the ROC curve, and a combined score that compared each pair of classifiers head-to-head. The results of the competition are very interesting and are detailed in Guyon et al. (2006). The most notable result: the entries of Neal and Zhang (2006) were the clear overall winners. In the final competition they finished first

TABLE 11.2. *NIPS 2003 challenge data sets. The column labeled p is the number of features. For the* Dorothea *dataset the features are binary. N_{tr}, N_{val} and N_{te} are the number of training, validation and test cases, respectively*

Dataset	Domain	Feature Type	p	Percent Probes	N_{tr}	N_{val}	N_{te}
Arcene	Mass spectrometry	Dense	10,000	30	100	100	700
Dexter	Text classification	Sparse	20,000	50	300	300	2000
Dorothea	Drug discovery	Sparse	100,000	50	800	350	800
Gisette	Digit recognition	Dense	5000	30	6000	1000	6500
Madelon	Artificial	Dense	500	96	2000	600	1800

in three of the five datasets, and were 5th and 7th on the remaining two datasets.

In their winning entries, Neal and Zhang (2006) used a series of pre-processing feature-selection steps, followed by Bayesian neural networks, Dirichlet diffusion trees, and combinations of these methods. Here we focus only on the Bayesian neural network approach, and try to discern which aspects of their approach were important for its success. We rerun their programs and compare the results to boosted neural networks and boosted trees, and other related methods.

11.9.1 Bayes, Boosting and Bagging

Let us first review briefly the Bayesian approach to inference and its application to neural networks. Given training data $\mathbf{X}_{\mathrm{tr}}, \mathbf{y}_{\mathrm{tr}}$, we assume a sampling model with parameters θ; Neal and Zhang (2006) use a two-hidden-layer neural network, with output nodes the class probabilities $\Pr(Y|X, \theta)$ for the binary outcomes. Given a prior distribution $\Pr(\theta)$, the posterior distribution for the parameters is

$$\Pr(\theta|\mathbf{X}_{\mathrm{tr}}, \mathbf{y}_{\mathrm{tr}}) = \frac{\Pr(\theta)\Pr(\mathbf{y}_{\mathrm{tr}}|\mathbf{X}_{\mathrm{tr}}, \theta)}{\int \Pr(\theta)\Pr(\mathbf{y}_{\mathrm{tr}}|\mathbf{X}_{\mathrm{tr}}, \theta)d\theta} \tag{11.19}$$

For a test case with features X_{new}, the predictive distribution for the label Y_{new} is

$$\Pr(Y_{\mathrm{new}}|X_{\mathrm{new}}, \mathbf{X}_{\mathrm{tr}}, \mathbf{y}_{\mathrm{tr}}) = \int \Pr(Y_{\mathrm{new}}|X_{\mathrm{new}}, \theta)\Pr(\theta|\mathbf{X}_{\mathrm{tr}}, \mathbf{y}_{\mathrm{tr}})d\theta \quad (11.20)$$

(c.f. equation 8.24). Since the integral in (11.20) is intractable, sophisticated Markov Chain Monte Carlo (MCMC) methods are used to sample from the posterior distribution $\Pr(Y_{\mathrm{new}}|X_{\mathrm{new}}, \mathbf{X}_{\mathrm{tr}}, \mathbf{y}_{\mathrm{tr}})$. A few hundred values θ are generated and then a simple average of these values estimates the integral. Neal and Zhang (2006) use diffuse Gaussian priors for all of the parameters. The particular MCMC approach that was used is called *hybrid Monte Carlo*, and may be important for the success of the method. It includes an auxiliary momentum vector and implements Hamiltonian dynamics in which the potential function is the target density. This is done to avoid random walk behavior; the successive candidates move across the sample space in larger steps. They tend to be less correlated and hence converge to the target distribution more rapidly.

Neal and Zhang (2006) also tried different forms of pre-processing of the features:

1. univariate screening using t-tests, and

2. automatic relevance determination.

In the latter method (ARD), the weights (coefficients) for the jth feature to each of the first hidden layer units all share a common prior variance σ_j^2, and prior mean zero. The posterior distributions for each variance σ_j^2 are computed, and the features whose posterior variance concentrates on small values are discarded.

There are thus three main features of this approach that could be important for its success:

(a) the feature selection and pre-processing,

(b) the neural network model, and

(c) the Bayesian inference for the model using MCMC.

According to Neal and Zhang (2006), feature screening in (a) is carried out purely for computational efficiency; the MCMC procedure is slow with a large number of features. There is no need to use feature selection to avoid overfitting. The posterior average (11.20) takes care of this automatically.

We would like to understand the reasons for the success of the Bayesian method. In our view, power of modern Bayesian methods does not lie in their use as a formal inference procedure; most people would not believe that the priors in a high-dimensional, complex neural network model are actually correct. Rather the Bayesian/MCMC approach gives an efficient way of sampling the relevant parts of model space, and then averaging the predictions for the high-probability models.

Bagging and boosting are non-Bayesian procedures that have some similarity to MCMC in a Bayesian model. The Bayesian approach fixes the data and perturbs the parameters, according to current estimate of the posterior distribution. Bagging perturbs the data in an i.i.d fashion and then re-estimates the model to give a new set of model parameters. At the end, a simple average of the model predictions from different bagged samples is computed. Boosting is similar to bagging, but fits a model that is additive in the models of each individual base learner, which are learned using non i.i.d. samples. We can write all of these models in the form

$$\hat{f}(\mathbf{x}_{\text{new}}) = \sum_{\ell=1}^{L} w_\ell \mathrm{E}(Y_{\text{new}}|\mathbf{x}_{\text{new}}, \hat{\theta}_\ell) \qquad (11.21)$$

In all cases the $\hat{\theta}_\ell$ are a large collection of model parameters. For the Bayesian model the $w_\ell = 1/L$, and the average estimates the posterior mean (11.21) by sampling θ_ℓ from the posterior distribution. For bagging, $w_\ell = 1/L$ as well, and the $\hat{\theta}_\ell$ are the parameters refit to bootstrap resamples of the training data. For boosting, the weights are all equal to 1, but the $\hat{\theta}_\ell$ are typically chosen in a nonrandom sequential fashion to constantly improve the fit.

11.9.2 Performance Comparisons

Based on the similarities above, we decided to compare Bayesian neural networks to boosted trees, boosted neural networks, random forests and bagged neural networks on the five datasets in Table 11.2. Bagging and boosting of neural networks are not methods that we have previously used in our work. We decided to try them here, because of the success of Bayesian neural networks in this competition, and the good performance of bagging and boosting with trees. We also felt that by bagging and boosting neural nets, we could assess both the choice of model as well as the model search strategy.

Here are the details of the learning methods that were compared:

Bayesian neural nets. The results here are taken from Neal and Zhang (2006), using their Bayesian approach to fitting neural networks. The models had two hidden layers of 20 and 8 units. We re-ran some networks for timing purposes only.

Boosted trees. We used the gbm package (version 1.5-7) in the R language. Tree depth and shrinkage factors varied from dataset to dataset. We consistently bagged 80% of the data at each boosting iteration (the default is 50%). Shrinkage was between 0.001 and 0.1. Tree depth was between 2 and 9.

Boosted neural networks. Since boosting is typically most effective with "weak" learners, we boosted a single hidden layer neural network with two or four units, fit with the nnet package (version 7.2-36) in R.

Random forests. We used the R package randomForest (version 4.5-16) with default settings for the parameters.

Bagged neural networks. We used the same architecture as in the Bayesian neural network above (two hidden layers of 20 and 8 units), fit using both Neal's C language package "Flexible Bayesian Modeling" (2004-11-10 release), and Matlab neural-net toolbox (version 5.1).

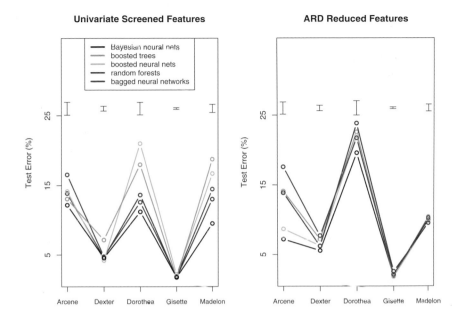

FIGURE 11.12. *Performance of different learning methods on five problems, using both univariate screening of features (top panel) and a reduced feature set from automatic relevance determination. The error bars at the top of each plot have width equal to one standard error of the difference between two error rates. On most of the problems several competitors are within this error bound.*

This analysis was carried out by Nicholas Johnson, and full details may be found in Johnson (2008)[3]. The results are shown in Figure 11.12 and Table 11.3.

The figure and table show Bayesian, boosted and bagged neural networks, boosted trees, and random forests, using both the screened and reduced features sets. The error bars at the top of each plot indicate one standard error of the difference between two error rates. Bayesian neural networks again emerge as the winner, although for some datasets the differences between the test error rates is not statistically significant. Random forests performs the best among the competitors using the selected feature set, while the boosted neural networks perform best with the reduced feature set, and nearly match the Bayesian neural net.

The superiority of boosted neural networks over boosted trees suggest that the neural network model is better suited to these particular problems. Specifically, individual features might not be good predictors here

---

[3]We also thank Isabelle Guyon for help in preparing the results of this section.

TABLE 11.3. *Performance of different methods. Values are average rank of test error across the five problems (low is good), and mean computation time and standard error of the mean, in minutes.*

Method	Screened Features		ARD Reduced Features	
	Average Rank	Average Time	Average Rank	Average Time
Bayesian neural networks	1.5	384(138)	1.6	600(186)
Boosted trees	3.4	3.03(2.5)	4.0	34.1(32.4)
Boosted neural networks	3.8	9.4(8.6)	2.2	35.6(33.5)
Random forests	2.7	1.9(1.7)	3.2	11.2(9.3)
Bagged neural networks	3.6	3.5(1.1)	4.0	6.4(4.4)

and linear combinations of features work better. However the impressive performance of random forests is at odds with this explanation, and came as a surprise to us.

Since the reduced feature sets come from the Bayesian neural network approach, only the methods that use the screened features are legitimate, self-contained procedures. However, this does suggest that better methods for internal feature selection might help the overall performance of boosted neural networks.

The table also shows the approximate training time required for each method. Here the non-Bayesian methods show a clear advantage.

Overall, the superior performance of Bayesian neural networks here may be due to the fact that

(a) the neural network model is well suited to these five problems, and

(b) the MCMC approach provides an efficient way of exploring the important part of the parameter space, and then averaging the resulting models according to their quality.

The Bayesian approach works well for smoothly parametrized models like neural nets; it is not yet clear that it works as well for non-smooth models like trees.

11.10   Computational Considerations

With N observations, p predictors, M hidden units and L training epochs, a neural network fit typically requires $O(NpML)$ operations. There are many packages available for fitting neural networks, probably many more than exist for mainstream statistical methods. Because the available software varies widely in quality, and the learning problem for neural networks is sensitive to issues such as input scaling, such software should be carefully chosen and tested.

Bibliographic Notes

Projection pursuit was proposed by Friedman and Tukey (1974), and specialized to regression by Friedman and Stuetzle (1981). Huber (1985) gives a scholarly overview, and Roosen and Hastie (1994) present a formulation using smoothing splines. The motivation for neural networks dates back to McCulloch and Pitts (1943), Widrow and Hoff (1960) (reprinted in Anderson and Rosenfeld (1988)) and Rosenblatt (1962). Hebb (1949) heavily influenced the development of learning algorithms. The resurgence of neural networks in the mid 1980s was due to Werbos (1974), Parker (1985) and Rumelhart et al. (1986), who proposed the back-propagation algorithm. Today there are many books written on the topic, for a broad range of audiences. For readers of this book, Hertz et al. (1991), Bishop (1995) and Ripley (1996) may be the most informative. Bayesian learning for neural networks is described in Neal (1996). The ZIP code example was taken from Le Cun (1989); see also Le Cun et al. (1990) and Le Cun et al. (1998).

We do not discuss theoretical topics such as approximation properties of neural networks, such as the work of Barron (1993), Girosi et al. (1995) and Jones (1992). Some of these results are summarized by Ripley (1996).

Exercises

Ex. 11.1 Establish the exact correspondence between the projection pursuit regression model (11.1) and the neural network (11.5). In particular, show that the single-layer regression network is equivalent to a PPR model with $g_m(\omega_m^T x) = \beta_m \sigma(\alpha_{0m} + s_m(\omega_m^T x))$, where ω_m is the mth unit vector. Establish a similar equivalence for a classification network.

Ex. 11.2 Consider a neural network for a quantitative outcome as in (11.5), using squared-error loss and identity output function $g_k(t) = t$. Suppose that the weights α_m from the input to hidden layer are nearly zero. Show that the resulting model is nearly linear in the inputs.

Ex. 11.3 Derive the forward and backward propagation equations for the cross-entropy loss function.

Ex. 11.4 Consider a neural network for a K class outcome that uses cross-entropy loss. If the network has no hidden layer, show that the model is equivalent to the multinomial logistic model described in Chapter 4.

Ex. 11.5

(a) Write a program to fit a single hidden layer neural network (ten hidden units) via back-propagation and weight decay.

(b) Apply it to 100 observations from the model

$$Y = \sigma(a_1^T X) + (a_2^T X)^2 + 0.30 \cdot Z,$$

where σ is the sigmoid function, Z is standard normal, $X^T = (X_1, X_2)$, each X_j being independent standard normal, and $a_1 = (3,3), a_2 = (3, -3)$. Generate a test sample of size 1000, and plot the training and test error curves as a function of the number of training epochs, for different values of the weight decay parameter. Discuss the overfitting behavior in each case.

(c) Vary the number of hidden units in the network, from 1 up to 10, and determine the minimum number needed to perform well for this task.

Ex. 11.6 Write a program to carry out projection pursuit regression, using cubic smoothing splines with fixed degrees of freedom. Fit it to the data from the previous exercise, for various values of the smoothing parameter and number of model terms. Find the minimum number of model terms necessary for the model to perform well and compare this to the number of hidden units from the previous exercise.

Ex. 11.7 Fit a neural network to the spam data of Section 9.1.2, and compare the results to those for the additive model given in that chapter. Compare both the classification performance and interpretability of the final model.

12

Support Vector Machines and Flexible Discriminants

12.1 Introduction

In this chapter we describe generalizations of linear decision boundaries for classification. Optimal separating hyperplanes are introduced in Chapter 4 for the case when two classes are linearly separable. Here we cover extensions to the nonseparable case, where the classes overlap. These techniques are then generalized to what is known as the *support vector machine*, which produces nonlinear boundaries by constructing a linear boundary in a large, transformed version of the feature space. The second set of methods generalize Fisher's linear discriminant analysis (LDA). The generalizations include *flexible discriminant analysis* which facilitates construction of nonlinear boundaries in a manner very similar to the support vector machines, *penalized discriminant analysis* for problems such as signal and image classification where the large number of features are highly correlated, and *mixture discriminant analysis* for irregularly shaped classes.

12.2 The Support Vector Classifier

In Chapter 4 we discussed a technique for constructing an *optimal* separating hyperplane between two perfectly separated classes. We review this and generalize to the nonseparable case, where the classes may not be separable by a linear boundary.

T. Hastie et al., *The Elements of Statistical Learning, Second Edition*,
DOI: 10.1007/b94608_12,
© Springer Science+Business Media, LLC 2009

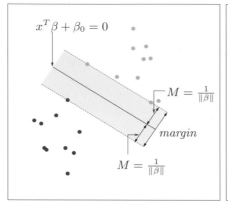 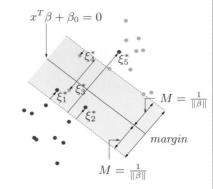

FIGURE 12.1. *Support vector classifiers. The left panel shows the separable case. The decision boundary is the solid line, while broken lines bound the shaded maximal margin of width $2M = 2/\|\beta\|$. The right panel shows the nonseparable (overlap) case. The points labeled ξ_j^* are on the wrong side of their margin by an amount $\xi_j^* = M\xi_j$; points on the correct side have $\xi_j^* = 0$. The margin is maximized subject to a total budget $\sum \xi_i \leq$ constant. Hence $\sum \xi_j^*$ is the total distance of points on the wrong side of their margin.*

Our training data consists of N pairs $(x_1, y_1), (x_2, y_2), \ldots, (x_N, y_N)$, with $x_i \in \mathbb{R}^p$ and $y_i \in \{-1, 1\}$. Define a hyperplane by

$$\{x : f(x) = x^T \beta + \beta_0 = 0\}, \tag{12.1}$$

where β is a unit vector: $\|\beta\| = 1$. A classification rule induced by $f(x)$ is

$$G(x) = \text{sign}[x^T \beta + \beta_0]. \tag{12.2}$$

The geometry of hyperplanes is reviewed in Section 4.5, where we show that $f(x)$ in (12.1) gives the signed distance from a point x to the hyperplane $f(x) = x^T \beta + \beta_0 = 0$. Since the classes are separable, we can find a function $f(x) = x^T \beta + \beta_0$ with $y_i f(x_i) > 0 \ \forall i$. Hence we are able to find the hyperplane that creates the biggest *margin* between the training points for class 1 and -1 (see Figure 12.1). The optimization problem

$$\max_{\beta, \beta_0, \|\beta\|=1} M$$
$$\text{subject to } y_i(x_i^T \beta + \beta_0) \geq M, \ i = 1, \ldots, N, \tag{12.3}$$

captures this concept. The band in the figure is M units away from the hyperplane on either side, and hence $2M$ units wide. It is called the *margin*.

We showed that this problem can be more conveniently rephrased as

$$\min_{\beta, \beta_0} \|\beta\|$$
$$\text{subject to } y_i(x_i^T \beta + \beta_0) \geq 1, \ i = 1, \ldots, N, \tag{12.4}$$

where we have dropped the norm constraint on β. Note that $M = 1/\|\beta\|$. Expression (12.4) is the usual way of writing the support vector criterion for separated data. This is a convex optimization problem (quadratic criterion, linear inequality constraints), and the solution is characterized in Section 4.5.2.

Suppose now that the classes overlap in feature space. One way to deal with the overlap is to still maximize M, but allow for some points to be on the wrong side of the margin. Define the slack variables $\xi = (\xi_1, \xi_2, \ldots, \xi_N)$. There are two natural ways to modify the constraint in (12.3):

$$y_i(x_i^T \beta + \beta_0) \geq M - \xi_i, \tag{12.5}$$

or

$$y_i(x_i^T \beta + \beta_0) \geq M(1 - \xi_i), \tag{12.6}$$

$\forall i$, $\xi_i \geq 0$, $\sum_{i=1}^{N} \xi_i \leq$ constant. The two choices lead to different solutions. The first choice seems more natural, since it measures overlap in actual distance from the margin; the second choice measures the overlap in relative distance, which changes with the width of the margin M. However, the first choice results in a nonconvex optimization problem, while the second is convex; thus (12.6) leads to the "standard" support vector classifier, which we use from here on.

Here is the idea of the formulation. The value ξ_i in the constraint $y_i(x_i^T \beta + \beta_0) \geq M(1 - \xi_i)$ is the proportional amount by which the prediction $f(x_i) = x_i^T \beta + \beta_0$ is on the wrong side of its margin. Hence by bounding the sum $\sum \xi_i$, we bound the total proportional amount by which predictions fall on the wrong side of their margin. Misclassifications occur when $\xi_i > 1$, so bounding $\sum \xi_i$ at a value K say, bounds the total number of training misclassifications at K.

As in (4.48) in Section 4.5.2, we can drop the norm constraint on β, define $M = 1/\|\beta\|$, and write (12.4) in the equivalent form

$$\min \|\beta\| \quad \text{subject to} \quad \begin{cases} y_i(x_i^T \beta + \beta_0) \geq 1 - \xi_i \ \forall i, \\ \xi_i \geq 0, \ \sum \xi_i \leq \text{constant}. \end{cases} \tag{12.7}$$

This is the usual way the support vector classifier is defined for the nonseparable case. However we find confusing the presence of the fixed scale "1" in the constraint $y_i(x_i^T \beta + \beta_0) \geq 1 - \xi_i$, and prefer to start with (12.6). The right panel of Figure 12.1 illustrates this overlapping case.

By the nature of the criterion (12.7), we see that points well inside their class boundary do not play a big role in shaping the boundary. This seems like an attractive property, and one that differentiates it from linear discriminant analysis (Section 4.3). In LDA, the decision boundary is determined by the covariance of the class distributions and the positions of the class centroids. We will see in Section 12.3.3 that logistic regression is more similar to the support vector classifier in this regard.

12.2.1   Computing the Support Vector Classifier

The problem (12.7) is quadratic with linear inequality constraints, hence it is a convex optimization problem. We describe a quadratic programming solution using Lagrange multipliers. Computationally it is convenient to re-express (12.7) in the equivalent form

$$\min_{\beta,\beta_0} \frac{1}{2}\|\beta\|^2 + C\sum_{i=1}^{N}\xi_i \tag{12.8}$$

$$\text{subject to} \quad \xi_i \geq 0, \; y_i(x_i^T\beta + \beta_0) \geq 1 - \xi_i \; \forall i,$$

where the "cost" parameter C replaces the constant in (12.7); the separable case corresponds to $C = \infty$.

The Lagrange (primal) function is

$$L_P = \frac{1}{2}\|\beta\|^2 + C\sum_{i=1}^{N}\xi_i - \sum_{i=1}^{N}\alpha_i[y_i(x_i^T\beta + \beta_0) - (1 - \xi_i)] - \sum_{i=1}^{N}\mu_i\xi_i, \tag{12.9}$$

which we minimize w.r.t β, β_0 and ξ_i. Setting the respective derivatives to zero, we get

$$\beta = \sum_{i=1}^{N}\alpha_i y_i x_i, \tag{12.10}$$

$$0 = \sum_{i=1}^{N}\alpha_i y_i, \tag{12.11}$$

$$\alpha_i = C - \mu_i, \; \forall i, \tag{12.12}$$

as well as the positivity constraints α_i, μ_i, $\xi_i \geq 0 \; \forall i$. By substituting (12.10)–(12.12) into (12.9), we obtain the Lagrangian (Wolfe) dual objective function

$$L_D = \sum_{i=1}^{N}\alpha_i - \frac{1}{2}\sum_{i=1}^{N}\sum_{i'=1}^{N}\alpha_i\alpha_{i'}y_i y_{i'}x_i^T x_{i'}, \tag{12.13}$$

which gives a lower bound on the objective function (12.8) for any feasible point. We maximize L_D subject to $0 \leq \alpha_i \leq C$ and $\sum_{i=1}^{N}\alpha_i y_i = 0$. In addition to (12.10)–(12.12), the Karush–Kuhn–Tucker conditions include the constraints

$$\alpha_i[y_i(x_i^T\beta + \beta_0) - (1 - \xi_i)] = 0, \tag{12.14}$$

$$\mu_i\xi_i = 0, \tag{12.15}$$

$$y_i(x_i^T\beta + \beta_0) - (1 - \xi_i) \geq 0, \tag{12.16}$$

for $i = 1, \ldots, N$. Together these equations (12.10)–(12.16) uniquely characterize the solution to the primal and dual problem.

From (12.10) we see that the solution for β has the form

$$\hat{\beta} = \sum_{i=1}^{N} \hat{\alpha}_i y_i x_i, \qquad (12.17)$$

with nonzero coefficients $\hat{\alpha}_i$ only for those observations i for which the constraints in (12.16) are exactly met (due to (12.14)). These observations are called the *support vectors*, since $\hat{\beta}$ is represented in terms of them alone. Among these support points, some will lie on the edge of the margin ($\hat{\xi}_i = 0$), and hence from (12.15) and (12.12) will be characterized by $0 < \hat{\alpha}_i < C$; the remainder ($\hat{\xi}_i > 0$) have $\hat{\alpha}_i = C$. From (12.14) we can see that any of these margin points ($0 < \hat{\alpha}_i$, $\hat{\xi}_i = 0$) can be used to solve for β_0, and we typically use an average of all the solutions for numerical stability.

Maximizing the dual (12.13) is a simpler convex quadratic programming problem than the primal (12.9), and can be solved with standard techniques (Murray et al., 1981, for example).

Given the solutions $\hat{\beta}_0$ and $\hat{\beta}$, the decision function can be written as

$$\begin{aligned} \hat{G}(x) &= \operatorname{sign}[\hat{f}(x)] \\ &= \operatorname{sign}[x^T \hat{\beta} + \hat{\beta}_0]. \end{aligned} \qquad (12.18)$$

The tuning parameter of this procedure is the cost parameter C.

12.2.2 *Mixture Example (Continued)*

Figure 12.2 shows the support vector boundary for the mixture example of Figure 2.5 on page 21, with two overlapping classes, for two different values of the cost parameter C. The classifiers are rather similar in their performance. Points on the wrong side of the boundary are support vectors. In addition, points on the correct side of the boundary but close to it (in the margin), are also support vectors. The margin is larger for $C = 0.01$ than it is for $C = 10,000$. Hence larger values of C focus attention more on (correctly classified) points near the decision boundary, while smaller values involve data further away. Either way, misclassified points are given weight, no matter how far away. In this example the procedure is not very sensitive to choices of C, because of the rigidity of a linear boundary.

The optimal value for C can be estimated by cross-validation, as discussed in Chapter 7. Interestingly, the leave-one-out cross-validation error can be bounded above by the proportion of support points in the data. The reason is that leaving out an observation that is not a support vector will not change the solution. Hence these observations, being classified correctly by the original boundary, will be classified correctly in the cross-validation process. However this bound tends to be too high, and not generally useful for choosing C (62% and 85%, respectively, in our examples).

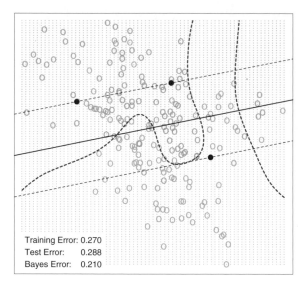

$$C = 10000$$

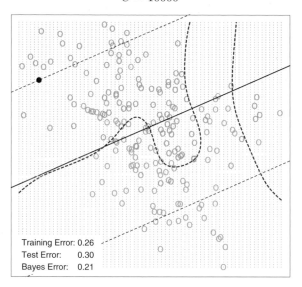

$$C = 0.01$$

FIGURE 12.2. *The linear support vector boundary for the mixture data example with two overlapping classes, for two different values of C. The broken lines indicate the margins, where $f(x) = \pm 1$. The support points ($\alpha_i > 0$) are all the points on the wrong side of their margin. The black solid dots are those support points falling exactly on the margin ($\xi_i = 0$, $\alpha_i > 0$). In the upper panel 62% of the observations are support points, while in the lower panel 85% are. The broken purple curve in the background is the Bayes decision boundary.*

12.3 Support Vector Machines and Kernels

The support vector classifier described so far finds linear boundaries in the input feature space. As with other linear methods, we can make the procedure more flexible by enlarging the feature space using basis expansions such as polynomials or splines (Chapter 5). Generally linear boundaries in the enlarged space achieve better training-class separation, and translate to nonlinear boundaries in the original space. Once the basis functions $h_m(x)$, $m = 1, \ldots, M$ are selected, the procedure is the same as before. We fit the SV classifier using input features $h(x_i) = (h_1(x_i), h_2(x_i), \ldots, h_M(x_i))$, $i = 1, \ldots, N$, and produce the (nonlinear) function $\hat{f}(x) = h(x)^T \hat{\beta} + \hat{\beta}_0$. The classifier is $\hat{G}(x) = \text{sign}(\hat{f}(x))$ as before.

The *support vector machine* classifier is an extension of this idea, where the dimension of the enlarged space is allowed to get very large, infinite in some cases. It might seem that the computations would become prohibitive. It would also seem that with sufficient basis functions, the data would be separable, and overfitting would occur. We first show how the SVM technology deals with these issues. We then see that in fact the SVM classifier is solving a function-fitting problem using a particular criterion and form of regularization, and is part of a much bigger class of problems that includes the smoothing splines of Chapter 5. The reader may wish to consult Section 5.8, which provides background material and overlaps somewhat with the next two sections.

12.3.1 Computing the SVM for Classification

We can represent the optimization problem (12.9) and its solution in a special way that only involves the input features via inner products. We do this directly for the transformed feature vectors $h(x_i)$. We then see that for particular choices of h, these inner products can be computed very cheaply.

The Lagrange dual function (12.13) has the form

$$L_D = \sum_{i=1}^{N} \alpha_i - \frac{1}{2} \sum_{i=1}^{N} \sum_{i'=1}^{N} \alpha_i \alpha_{i'} y_i y_{i'} \langle h(x_i), h(x_{i'}) \rangle. \tag{12.19}$$

From (12.10) we see that the solution function $f(x)$ can be written

$$f(x) = h(x)^T \beta + \beta_0$$
$$= \sum_{i=1}^{N} \alpha_i y_i \langle h(x), h(x_i) \rangle + \beta_0. \tag{12.20}$$

As before, given α_i, β_0 can be determined by solving $y_i f(x_i) = 1$ in (12.20) for any (or all) x_i for which $0 < \alpha_i < C$.

So both (12.19) and (12.20) involve $h(x)$ only through inner products. In fact, we need not specify the transformation $h(x)$ at all, but require only knowledge of the kernel function

$$K(x, x') = \langle h(x), h(x') \rangle \tag{12.21}$$

that computes inner products in the transformed space. K should be a symmetric positive (semi-) definite function; see Section 5.8.1.

Three popular choices for K in the SVM literature are

$$
\begin{aligned}
d\text{th-Degree polynomial: } & K(x, x') = (1 + \langle x, x' \rangle)^d, \\
\text{Radial basis: } & K(x, x') = \exp(-\gamma \|x - x'\|^2), \\
\text{Neural network: } & K(x, x') = \tanh(\kappa_1 \langle x, x' \rangle + \kappa_2).
\end{aligned}
\tag{12.22}
$$

Consider for example a feature space with two inputs X_1 and X_2, and a polynomial kernel of degree 2. Then

$$
\begin{aligned}
K(X, X') &= (1 + \langle X, X' \rangle)^2 \\
&= (1 + X_1 X_1' + X_2 X_2')^2 \\
&= 1 + 2X_1 X_1' + 2X_2 X_2' + (X_1 X_1')^2 + (X_2 X_2')^2 + 2X_1 X_1' X_2 X_2'.
\end{aligned}
\tag{12.23}
$$

Then $M = 6$, and if we choose $h_1(X) = 1$, $h_2(X) = \sqrt{2}X_1$, $h_3(X) = \sqrt{2}X_2$, $h_4(X) = X_1^2$, $h_5(X) = X_2^2$, and $h_6(X) = \sqrt{2}X_1 X_2$, then $K(X, X') = \langle h(X), h(X') \rangle$. From (12.20) we see that the solution can be written

$$\hat{f}(x) = \sum_{i=1}^N \hat{\alpha}_i y_i K(x, x_i) + \hat{\beta}_0. \tag{12.24}$$

The role of the parameter C is clearer in an enlarged feature space, since perfect separation is often achievable there. A large value of C will discourage any positive ξ_i, and lead to an overfit wiggly boundary in the original feature space; a small value of C will encourage a small value of $\|\beta\|$, which in turn causes $f(x)$ and hence the boundary to be smoother. Figure 12.3 show two nonlinear support vector machines applied to the mixture example of Chapter 2. The regularization parameter was chosen in both cases to achieve good test error. The radial basis kernel produces a boundary quite similar to the Bayes optimal boundary for this example; compare Figure 2.5.

In the early literature on support vectors, there were claims that the kernel property of the support vector machine is unique to it and allows one to finesse the curse of dimensionality. Neither of these claims is true, and we go into both of these issues in the next three subsections.

SVM - Degree-4 Polynomial in Feature Space

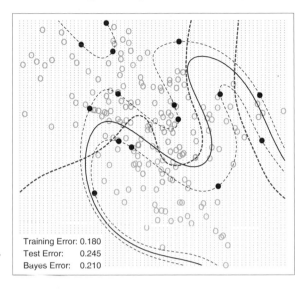

Training Error: 0.180
Test Error: 0.245
Bayes Error: 0.210

SVM - Radial Kernel in Feature Space

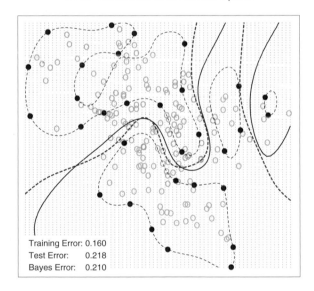

Training Error: 0.160
Test Error: 0.218
Bayes Error: 0.210

FIGURE 12.3. *Two nonlinear SVMs for the mixture data. The upper plot uses a 4th degree polynomial kernel, the lower a radial basis kernel (with $\gamma = 1$). In each case C was tuned to approximately achieve the best test error performance, and $C = 1$ worked well in both cases. The radial basis kernel performs the best (close to Bayes optimal), as might be expected given the data arise from mixtures of Gaussians. The broken purple curve in the background is the Bayes decision boundary.*

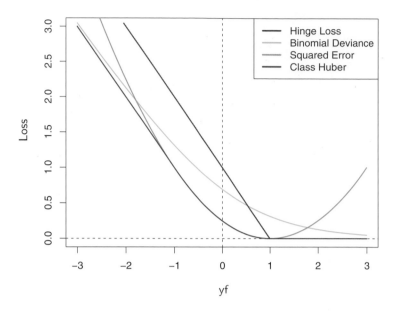

FIGURE 12.4. *The support vector loss function (hinge loss), compared to the negative log-likelihood loss (binomial deviance) for logistic regression, squared-error loss, and a "Huberized" version of the squared hinge loss. All are shown as a function of yf rather than f, because of the symmetry between the $y = +1$ and $y = -1$ case. The deviance and Huber have the same asymptotes as the SVM loss, but are rounded in the interior. All are scaled to have the limiting left-tail slope of -1.*

12.3.2 The SVM as a Penalization Method

With $f(x) = h(x)^T \beta + \beta_0$, consider the optimization problem

$$\min_{\beta_0, \beta} \sum_{i=1}^{N} [1 - y_i f(x_i)]_+ + \frac{\lambda}{2} \|\beta\|^2 \qquad (12.25)$$

where the subscript "+" indicates positive part. This has the form *loss + penalty*, which is a familiar paradigm in function estimation. It is easy to show (Exercise 12.1) that the solution to (12.25), with $\lambda = 1/C$, is the same as that for (12.8).

Examination of the "hinge" loss function $L(y, f) = [1 - yf]_+$ shows that it is reasonable for two-class classification, when compared to other more traditional loss functions. Figure 12.4 compares it to the log-likelihood loss for logistic regression, as well as squared-error loss and a variant thereof. The (negative) log-likelihood or binomial deviance has similar tails as the SVM loss, giving zero penalty to points well inside their margin, and a

TABLE 12.1. *The population minimizers for the different loss functions in Figure 12.4. Logistic regression uses the binomial log-likelihood or deviance. Linear discriminant analysis (Exercise 4.2) uses squared-error loss. The SVM hinge loss estimates the mode of the posterior class probabilities, whereas the others estimate a linear transformation of these probabilities.*

Loss Function	$L[y, f(x)]$	Minimizing Function
Binomial Deviance	$\log[1 + e^{-yf(x)}]$	$f(x) = \log \dfrac{\Pr(Y = +1\|x)}{\Pr(Y = \text{-}1\|x)}$
SVM Hinge Loss	$[1 - yf(x)]_+$	$f(x) = \text{sign}[\Pr(Y = +1\|x) - \frac{1}{2}]$
Squared Error	$[y - f(x)]^2 = [1 - yf(x)]^2$	$f(x) = 2\Pr(Y = +1\|x) - 1$
"Huberised" Square Hinge Loss	$\begin{array}{ll} -4yf(x), & yf(x) < \text{-}1 \\ [1 - yf(x)]_+^2 & \text{otherwise} \end{array}$	$f(x) = 2\Pr(Y = +1\|x) - 1$

linear penalty to points on the wrong side and far away. Squared-error, on the other hand gives a quadratic penalty, and points well inside their own margin have a strong influence on the model as well. The squared hinge loss $L(y, f) = [1 - yf]_+^2$ is like the quadratic, except it is zero for points inside their margin. It still rises quadratically in the left tail, and will be less robust than hinge or deviance to misclassified observations. Recently Rosset and Zhu (2007) proposed a "Huberized" version of the squared hinge loss, which converts smoothly to a linear loss at $yf = -1$.

We can characterize these loss functions in terms of what they are estimating at the population level. We consider minimizing $EL(Y, f(X))$. Table 12.1 summarizes the results. Whereas the hinge loss estimates the classifier $G(x)$ itself, all the others estimate a transformation of the class posterior probabilities. The "Huberized" square hinge loss shares attractive properties of logistic regression (smooth loss function, estimates probabilities), as well as the SVM hinge loss (support points).

Formulation (12.25) casts the SVM as a regularized function estimation problem, where the coefficients of the linear expansion $f(x) = \beta_0 + h(x)^T \beta$ are shrunk toward zero (excluding the constant). If $h(x)$ represents a hierarchical basis having some ordered structure (such as ordered in roughness), then the uniform shrinkage makes more sense if the rougher elements h_j in the vector h have smaller norm.

All the loss-functions in Table 12.1 except squared-error are so called "margin maximizing loss-functions" (Rosset et al., 2004b). This means that if the data are separable, then the limit of $\hat{\beta}_\lambda$ in (12.25) as $\lambda \to 0$ defines the optimal separating hyperplane[1].

12.3.3 Function Estimation and Reproducing Kernels

Here we describe SVMs in terms of function estimation in reproducing kernel Hilbert spaces, where the kernel property abounds. This material is discussed in some detail in Section 5.8. This provides another view of the support vector classifier, and helps to clarify how it works.

Suppose the basis h arises from the (possibly finite) eigen-expansion of a positive definite kernel K,

$$K(x, x') = \sum_{m=1}^{\infty} \phi_m(x)\phi_m(x')\delta_m \qquad (12.26)$$

and $h_m(x) = \sqrt{\delta_m}\phi_m(x)$. Then with $\theta_m = \sqrt{\delta_m}\beta_m$, we can write (12.25) as

$$\min_{\beta_0,\,\theta} \sum_{i=1}^{N} \left[1 - y_i(\beta_0 + \sum_{m=1}^{\infty} \theta_m\phi_m(x_i))\right]_+ + \frac{\lambda}{2} \sum_{m=1}^{\infty} \frac{\theta_m^2}{\delta_m}. \qquad (12.27)$$

Now (12.27) is identical in form to (5.49) on page 169 in Section 5.8, and the theory of reproducing kernel Hilbert spaces described there guarantees a finite-dimensional solution of the form

$$f(x) = \beta_0 + \sum_{i=1}^{N} \alpha_i K(x, x_i). \qquad (12.28)$$

In particular we see there an equivalent version of the optimization criterion (12.19) [Equation (5.67) in Section 5.8.2; see also Wahba et al. (2000)],

$$\min_{\beta_0,\alpha} \sum_{i=1}^{N} (1 - y_i f(x_i))_+ + \frac{\lambda}{2} \alpha^T \mathbf{K}\alpha, \qquad (12.29)$$

where $\mathbf{K}$ is the $N \times N$ matrix of kernel evaluations for all pairs of training features (Exercise 12.2).

These models are quite general, and include, for example, the entire family of smoothing splines, additive and interaction spline models discussed

---

[1]For logistic regression with separable data, $\hat{\beta}_\lambda$ diverges, but $\hat{\beta}_\lambda/\|\hat{\beta}_\lambda\|$ converges to the optimal separating direction.

in Chapters 5 and 9, and in more detail in Wahba (1990) and Hastie and Tibshirani (1990). They can be expressed more generally as

$$\min_{f \in \mathcal{H}} \sum_{i=1}^{N} [1 - y_i f(x_i)]_+ + \lambda J(f), \tag{12.30}$$

where $\mathcal{H}$ is the structured space of functions, and $J(f)$ an appropriate regularizer on that space. For example, suppose $\mathcal{H}$ is the space of additive functions $f(x) = \sum_{j=1}^{p} f_j(x_j)$, and $J(f) = \sum_j \int \{f''_j(x_j)\}^2 dx_j$. Then the solution to (12.30) is an additive cubic spline, and has a kernel representation (12.28) with $K(x, x') = \sum_{j=1}^{p} K_j(x_j, x'_j)$. Each of the K_j is the kernel appropriate for the univariate smoothing spline in x_j (Wahba, 1990).

Conversely this discussion also shows that, for example, *any* of the kernels described in (12.22) above can be used with *any* convex loss function, and will also lead to a finite-dimensional representation of the form (12.28). Figure 12.5 uses the same kernel functions as in Figure 12.3, except using the binomial log-likelihood as a loss function[2]. The fitted function is hence an estimate of the log-odds,

$$\hat{f}(x) = \log \frac{\hat{\Pr}(Y = +1|x)}{\hat{\Pr}(Y = -1|x)}$$

$$- \hat{\beta}_0 + \sum_{i=1}^{N} \hat{\alpha}_i K(x, x_i), \tag{12.31}$$

or conversely we get an estimate of the class probabilities

$$\hat{\Pr}(Y = +1|x) = \frac{1}{1 + e^{-\hat{\beta}_0 - \sum_{i=1}^{N} \hat{\alpha}_i K(x, x_i)}}. \tag{12.32}$$

The fitted models are quite similar in shape and performance. Examples and more details are given in Section 5.8.

It does happen that for SVMs, a sizable fraction of the N values of α_i can be zero (the nonsupport points). In the two examples in Figure 12.3, these fractions are 42% and 45%, respectively. This is a consequence of the piecewise linear nature of the first part of the criterion (12.25). The lower the class overlap (on the training data), the greater this fraction will be. Reducing λ will generally reduce the overlap (allowing a more flexible f). A small number of support points means that $\hat{f}(x)$ can be evaluated more quickly, which is important at lookup time. Of course, reducing the overlap too much can lead to poor generalization.

---

[2] Ji Zhu assisted in the preparation of these examples.

LR - Degree-4 Polynomial in Feature Space

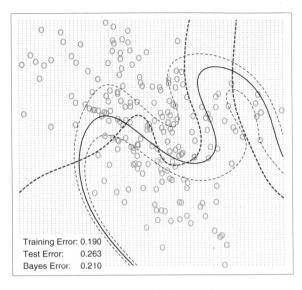

LR - Radial Kernel in Feature Space

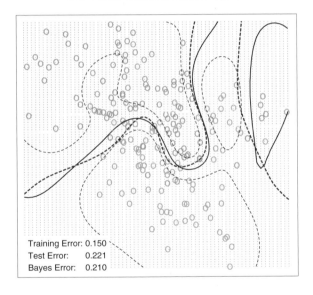

FIGURE 12.5. *The logistic regression versions of the SVM models in Figure 12.3, using the identical kernels and hence penalties, but the log-likelihood loss instead of the SVM loss function. The two broken contours correspond to posterior probabilities of 0.75 and 0.25 for the +1 class (or vice versa). The broken purple curve in the background is the Bayes decision boundary.*

TABLE 12.2. *Skin of the orange: Shown are mean (standard error of the mean) of the test error over 50 simulations. BRUTO fits an additive spline model adaptively, while MARS fits a low-order interaction model adaptively.*

	Method	Test Error (SE)	
		No Noise Features	Six Noise Features
1	SV Classifier	0.450 (0.003)	0.472 (0.003)
2	SVM/poly 2	0.078 (0.003)	0.152 (0.004)
3	SVM/poly 5	0.180 (0.004)	0.370 (0.004)
4	SVM/poly 10	0.230 (0.003)	0.434 (0.002)
5	BRUTO	0.084 (0.003)	0.090 (0.003)
6	MARS	0.156 (0.004)	0.173 (0.005)
	Bayes	0.029	0.029

12.3.4 SVMs and the Curse of Dimensionality

In this section, we address the question of whether SVMs have some edge on the curse of dimensionality. Notice that in expression (12.23) we are not allowed a fully general inner product in the space of powers and products. For example, all terms of the form $2X_j X_j'$ are given equal weight, and the kernel cannot adapt itself to concentrate on subspaces. If the number of features p were large, but the class separation occurred only in the linear subspace spanned by say X_1 and X_2, this kernel would not easily find the structure and would suffer from having many dimensions to search over. One would have to build knowledge about the subspace into the kernel; that is, tell it to ignore all but the first two inputs. If such knowledge were available a priori, much of statistical learning would be made much easier. A major goal of adaptive methods is to discover such structure.

We support these statements with an illustrative example. We generated 100 observations in each of two classes. The first class has four standard normal independent features X_1, X_2, X_3, X_4. The second class also has four standard normal independent features, but conditioned on $9 \leq \sum X_j^2 \leq 16$. This is a relatively easy problem. As a second harder problem, we augmented the features with an additional six standard Gaussian noise features. Hence the second class almost completely surrounds the first, like the skin surrounding the orange, in a four-dimensional subspace. The Bayes error rate for this problem is 0.029 (irrespective of dimension). We generated 1000 test observations to compare different procedures. The average test errors over 50 simulations, with and without noise features, are shown in Table 12.2.

Line 1 uses the support vector classifier in the original feature space. Lines 2–4 refer to the support vector machine with a 2-, 5- and 10-dimensional polynomial kernel. For all support vector procedures, we chose the cost parameter C to minimize the test error, to be as fair as possible to the

Test Error Curves – SVM with Radial Kernel

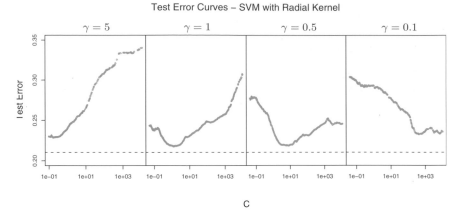

FIGURE 12.6. *Test-error curves as a function of the cost parameter C for the radial-kernel SVM classifier on the mixture data. At the top of each plot is the scale parameter γ for the radial kernel: $K_\gamma(x, y) = \exp\left(-\gamma\|x - y\|^2\right)$. The optimal value for C depends quite strongly on the scale of the kernel. The Bayes error rate is indicated by the broken horizontal lines.*

method. Line 5 fits an additive spline model to the $(-1, +1)$ response by least squares, using the BRUTO algorithm for additive models, described in Hastie and Tibshirani (1990). Line 6 uses MARS (multivariate adaptive regression splines) allowing interaction of all orders, as described in Chapter 9; as such it is comparable with the SVM/poly 10. Both BRUTO and MARS have the ability to ignore redundant variables. Test error was not used to choose the smoothing parameters in either of lines 5 or 6.

In the original feature space, a hyperplane cannot separate the classes, and the support vector classifier (line 1) does poorly. The polynomial support vector machine makes a substantial improvement in test error rate, but is adversely affected by the six noise features. It is also very sensitive to the choice of kernel: the second degree polynomial kernel (line 2) does best, since the true decision boundary is a second-degree polynomial. However, higher-degree polynomial kernels (lines 3 and 4) do much worse. BRUTO performs well, since the boundary is additive. BRUTO and MARS adapt well: their performance does not deteriorate much in the presence of noise.

12.3.5 *A Path Algorithm for the SVM Classifier*

The regularization parameter for the SVM classifier is the cost parameter C, or its inverse λ in (12.25). Common usage is to set C high, leading often to somewhat overfit classifiers.

Figure 12.6 shows the test error on the mixture data as a function of C, using different radial-kernel parameters γ. When $\gamma = 5$ (narrow peaked kernels), the heaviest regularization (small C) is called for. With $\gamma = 1$

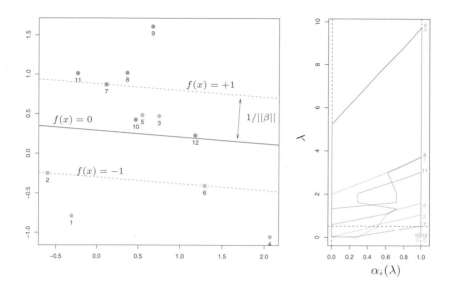

FIGURE 12.7. *A simple example illustrates the SVM path algorithm. (left panel:) This plot illustrates the state of the model at* $\lambda = 1/2$. *The " +1" points are orange, the "−1" blue. The width of the soft margin is* $2/||\beta|| = 2 \times 0.587$. *Two blue points* $\{3,5\}$ *are misclassified, while the two orange points* $\{10,12\}$ *are correctly classified, but on the wrong side of their margin* $f(x) = +1$; *each of these has* $y_i f(x_i) < 1$. *The three square shaped points* $\{2,6,7\}$ *are exactly on their margins. (right panel:) This plot shows the piecewise linear profiles* $\alpha_i(\lambda)$. *The horizontal broken line at* $\lambda = 1/2$ *indicates the state of the* α_i *for the model in the left plot.*

(the value used in Figure 12.3), an intermediate value of C is required. Clearly in situations such as these, we need to determine a good choice for C, perhaps by cross-validation. Here we describe a path algorithm (in the spirit of Section 3.8) for efficiently fitting the entire sequence of SVM models obtained by varying C.

It is convenient to use the loss+penalty formulation (12.25), along with Figure 12.4. This leads to a solution for β at a given value of λ:

$$\beta_\lambda = \frac{1}{\lambda} \sum_{i=1}^{N} \alpha_i y_i x_i. \tag{12.33}$$

The α_i are again Lagrange multipliers, but in this case they all lie in $[0, 1]$.

Figure 12.7 illustrates the setup. It can be shown that the KKT optimality conditions imply that the labeled points (x_i, y_i) fall into three distinct groups:

- Observations correctly classified and outside their margins. They have $y_i f(x_i) > 1$, and Lagrange multipliers $\alpha_i = 0$. Examples are the orange points 8, 9 and 11, and the blue points 1 and 4.

- Observations sitting on their margins with $y_i f(x_i) = 1$, with Lagrange multipliers $\alpha_i \in [0, 1]$. Examples are the orange 7 and the blue 2 and 6.

- Observations inside their margins have $y_i f(x_i) < 1$, with $\alpha_i = 1$. Examples are the blue 3 and 5, and the orange 10 and 12.

The idea for the path algorithm is as follows. Initially λ is large, the margin $1/\|\beta_\lambda\|$ is wide, and all points are inside their margin and have $\alpha_i = 1$. As λ decreases, $1/\|\beta_\lambda\|$ decreases, and the margin gets narrower. Some points will move from inside their margins to outside their margins, and their α_i will change from 1 to 0. By continuity of the $\alpha_i(\lambda)$, these points will *linger* on the margin during this transition. From (12.33) we see that the points with $\alpha_i = 1$ make fixed contributions to $\beta(\lambda)$, and those with $\alpha_i = 0$ make no contribution. So all that changes as λ decreases are the $\alpha_i \in [0, 1]$ of those (small number) of points on the margin. Since all these points have $y_i f(x_i) = 1$, this results in a small set of linear equations that prescribe how $\alpha_i(\lambda)$ and hence β_λ changes during these transitions. This results in piecewise linear paths for each of the $\alpha_i(\lambda)$. The breaks occur when points cross the margin. Figure 12.7 (right panel) shows the $\alpha_i(\lambda)$ profiles for the small example in the left panel.

Although we have described this for linear SVMs, exactly the same idea works for nonlinear models, in which (12.33) is replaced by

$$f_\lambda(x) = \frac{1}{\lambda} \sum_{i=1}^{N} \alpha_i y_i K(x, x_i). \tag{12.34}$$

Details can be found in Hastie et al. (2004). An R package svmpath is available on CRAN for fitting these models.

12.3.6 *Support Vector Machines for Regression*

In this section we show how SVMs can be adapted for regression with a quantitative response, in ways that inherit some of the properties of the SVM classifier. We first discuss the linear regression model

$$f(x) = x^T \beta + \beta_0, \tag{12.35}$$

and then handle nonlinear generalizations. To estimate β, we consider minimization of

$$H(\beta, \beta_0) = \sum_{i=1}^{N} V(y_i - f(x_i)) + \frac{\lambda}{2} \|\beta\|^2, \tag{12.36}$$

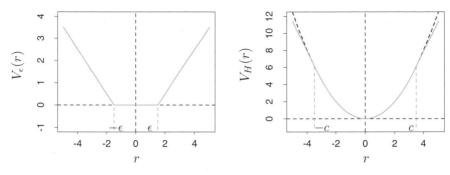

FIGURE 12.8. *The left panel shows the ϵ-insensitive error function used by the support vector regression machine. The right panel shows the error function used in Huber's robust regression (blue curve). Beyond $|c|$, the function changes from quadratic to linear.*

where

$$V_\epsilon(r) = \begin{cases} 0 & \text{if } |r| < \epsilon, \\ |r| - c, & \text{otherwise.} \end{cases} \qquad (12.37)$$

This is an "ϵ-insensitive" error measure, ignoring errors of size less than ϵ (left panel of Figure 12.8). There is a rough analogy with the support vector classification setup, where points on the correct side of the decision boundary and far away from it, are ignored in the optimization. In regression, these "low error" points are the ones with small residuals.

It is interesting to contrast this with error measures used in robust regression in statistics. The most popular, due to Huber (1964), has the form

$$V_H(r) = \begin{cases} r^2/2 & \text{if } |r| \leq c, \\ c|r| - c^2/2, & |r| > c, \end{cases} \qquad (12.38)$$

shown in the right panel of Figure 12.8. This function reduces from quadratic to linear the contributions of observations with absolute residual greater than a prechosen constant c. This makes the fitting less sensitive to outliers. The support vector error measure (12.37) also has linear tails (beyond ϵ), but in addition it flattens the contributions of those cases with small residuals.

If $\hat{\beta}$, $\hat{\beta}_0$ are the minimizers of H, the solution function can be shown to have the form

$$\hat{\beta} = \sum_{i=1}^{N} (\hat{\alpha}_i^* - \hat{\alpha}_i) x_i, \qquad (12.39)$$

$$\hat{f}(x) = \sum_{i=1}^{N} (\hat{\alpha}_i^* - \hat{\alpha}_i) \langle x, x_i \rangle + \beta_0, \qquad (12.40)$$

where $\hat{\alpha}_i, \hat{\alpha}_i^*$ are positive and solve the quadratic programming problem

$$\min_{\alpha_i, \alpha_i^*} \epsilon \sum_{i=1}^{N} (\alpha_i^* + \alpha_i) - \sum_{i=1}^{N} y_i (\alpha_i^* - \alpha_i) + \frac{1}{2} \sum_{i,i'=1}^{N} (\alpha_i^* - \alpha_i)(\alpha_{i'}^* - \alpha_{i'}) \langle x_i, x_{i'} \rangle$$

subject to the constraints

$$0 \le \alpha_i, \ \alpha_i^* \le 1/\lambda,$$

$$\sum_{i=1}^{N} (\alpha_i^* - \alpha_i) = 0, \tag{12.41}$$

$$\alpha_i \alpha_i^* = 0.$$

Due to the nature of these constraints, typically only a subset of the solution values $(\hat{\alpha}_i^* - \hat{\alpha}_i)$ are nonzero, and the associated data values are called the support vectors. As was the case in the classification setting, the solution depends on the input values only through the inner products $\langle x_i, x_{i'} \rangle$. Thus we can generalize the methods to richer spaces by defining an appropriate inner product, for example, one of those defined in (12.22).

Note that there are parameters, ϵ and λ, associated with the criterion (12.36). These seem to play different roles. ϵ is a parameter of the loss function V_ϵ, just like c is for V_H. Note that both V_ϵ and V_H depend on the scale of y and hence r. If we scale our response (and hence use $V_H(r/\sigma)$ and $V_\epsilon(r/\sigma)$ instead), then we might consider using preset values for c and ϵ (the value $c = 1.345$ achieves 95% efficiency for the Gaussian). The quantity λ is a more traditional regularization parameter, and can be estimated for example by cross-validation.

12.3.7 Regression and Kernels

As discussed in Section 12.3.3, this kernel property is not unique to support vector machines. Suppose we consider approximation of the regression function in terms of a set of basis functions $\{h_m(x)\}, m = 1, 2, \ldots, M$:

$$f(x) = \sum_{m=1}^{M} \beta_m h_m(x) + \beta_0. \tag{12.42}$$

To estimate β and β_0 we minimize

$$H(\beta, \beta_0) = \sum_{i=1}^{N} V(y_i - f(x_i)) + \frac{\lambda}{2} \sum \beta_m^2 \tag{12.43}$$

for some general error measure $V(r)$. For any choice of $V(r)$, the solution $\hat{f}(x) = \sum \hat{\beta}_m h_m(x) + \hat{\beta}_0$ has the form

$$\hat{f}(x) = \sum_{i=1}^{N} \hat{a}_i K(x, x_i) \tag{12.44}$$

with $K(x, y) = \sum_{m=1}^{M} h_m(x) h_m(y)$. Notice that this has the same form as both the radial basis function expansion and a regularization estimate, discussed in Chapters 5 and 6.

For concreteness, let's work out the case $V(r) = r^2$. Let $\mathbf{H}$ be the $N \times M$ basis matrix with imth element $h_m(x_i)$, and suppose that $M > N$ is large. For simplicity we assume that $\beta_0 = 0$, or that the constant is absorbed in h; see Exercise 12.3 for an alternative.

We estimate β by minimizing the penalized least squares criterion

$$H(\beta) = (\mathbf{y} - \mathbf{H}\beta)^T (\mathbf{y} - \mathbf{H}\beta) + \lambda \|\beta\|^2. \tag{12.45}$$

The solution is

$$\hat{\mathbf{y}} = \mathbf{H}\hat{\beta} \tag{12.46}$$

with $\hat{\beta}$ determined by

$$-\mathbf{H}^T(\mathbf{y} - \mathbf{H}\hat{\beta}) + \lambda\hat{\beta} = 0. \tag{12.47}$$

From this it appears that we need to evaluate the $M \times M$ matrix of inner products in the transformed space. However, we can premultiply by $\mathbf{H}$ to give

$$\mathbf{H}\hat{\beta} = (\mathbf{H}\mathbf{H}^T + \lambda\mathbf{I})^{-1}\mathbf{H}\mathbf{H}^T\mathbf{y}. \tag{12.48}$$

The $N \times N$ matrix $\mathbf{H}\mathbf{H}^T$ consists of inner products between pairs of observations i, i'; that is, the evaluation of an inner product kernel $\{\mathbf{H}\mathbf{H}^T\}_{i,i'} = K(x_i, x_{i'})$. It is easy to show (12.44) directly in this case, that the predicted values at an arbitrary x satisfy

$$
\begin{aligned}
\hat{f}(x) &= h(x)^T \hat{\beta} \\
&= \sum_{i=1}^{N} \hat{\alpha}_i K(x, x_i),
\end{aligned}
\tag{12.49}
$$

where $\hat{\alpha} = (\mathbf{H}\mathbf{H}^T + \lambda\mathbf{I})^{-1}\mathbf{y}$. As in the support vector machine, we need not specify or evaluate the large set of functions $h_1(x), h_2(x), \ldots, h_M(x)$. Only the inner product kernel $K(x_i, x_{i'})$ need be evaluated, at the N training points for each i, i' and at points x for predictions there. Careful choice of h_m (such as the eigenfunctions of particular, easy-to-evaluate kernels K) means, for example, that $\mathbf{H}\mathbf{H}^T$ can be computed at a cost of $N^2/2$ evaluations of K, rather than the direct cost N^2M.

Note, however, that this property depends on the choice of squared norm $\|\beta\|^2$ in the penalty. It does not hold, for example, for the L_1 norm $|\beta|$, which may lead to a superior model.

12.3.8 *Discussion*

The support vector machine can be extended to multiclass problems, essentially by solving many two-class problems. A classifier is built for each pair of classes, and the final classifier is the one that dominates the most (Kressel, 1999; Friedman, 1996; Hastie and Tibshirani, 1998). Alternatively, one could use the multinomial loss function along with a suitable kernel, as in Section 12.3.3. SVMs have applications in many other supervised and unsupervised learning problems. At the time of this writing, empirical evidence suggests that it performs well in many real learning problems.

Finally, we mention the connection of the support vector machine and structural risk minimization (7.9). Suppose the training points (or their basis expansion) are contained in a sphere of radius R, and let $G(x) = \text{sign}[f(x)] = \text{sign}[\beta^T x + \beta_0]$ as in (12.2). Then one can show that the class of functions $\{G(x), \|\beta\| \leq A\}$ has VC-dimension h satisfying

$$h \leq R^2 A^2. \tag{12.50}$$

If $f(x)$ separates the training data, optimally for $\|\beta\| \leq A$, then with probability at least $1 - \eta$ over training sets (Vapnik, 1996, page 139):

$$\text{Error}_{\text{Test}} \leq 4\frac{h[\log{(2N/h)} + 1] - \log{(\eta/4)}}{N}. \tag{12.51}$$

The support vector classifier was one of the first practical learning procedures for which useful bounds on the VC dimension could be obtained, and hence the SRM program could be carried out. However in the derivation, balls are put around the data points—a process that depends on the observed values of the features. Hence in a strict sense, the VC complexity of the class is not fixed a priori, before seeing the features.

The regularization parameter C controls an upper bound on the VC dimension of the classifier. Following the SRM paradigm, we could choose C by minimizing the upper bound on the test error, given in (12.51). However, it is not clear that this has any advantage over the use of cross-validation for choice of C.

12.4 Generalizing Linear Discriminant Analysis

In Section 4.3 we discussed linear discriminant analysis (LDA), a fundamental tool for classification. For the remainder of this chapter we discuss a class of techniques that produce better classifiers than LDA by directly generalizing LDA.

Some of the virtues of LDA are as follows:

- It is a simple prototype classifier. A new observation is classified to the class with closest centroid. A slight twist is that distance is measured in the Mahalanobis metric, using a pooled covariance estimate.

- LDA is the estimated Bayes classifier if the observations are multivariate Gaussian in each class, with a common covariance matrix. Since this assumption is unlikely to be true, this might not seem to be much of a virtue.

- The decision boundaries created by LDA are linear, leading to decision rules that are simple to describe and implement.

- LDA provides natural low-dimensional views of the data. For example, Figure 12.12 is an informative two-dimensional view of data in 256 dimensions with ten classes.

- Often LDA produces the best classification results, because of its simplicity and low variance. LDA was among the top three classifiers for 7 of the 22 datasets studied in the STATLOG project (Michie et al., 1994)[3].

Unfortunately the simplicity of LDA causes it to fail in a number of situations as well:

- Often linear decision boundaries do not adequately separate the classes. When N is large, it is possible to estimate more complex decision boundaries. Quadratic discriminant analysis (QDA) is often useful here, and allows for quadratic decision boundaries. More generally we would like to be able to model irregular decision boundaries.

- The aforementioned shortcoming of LDA can often be paraphrased by saying that a single prototype per class is insufficient. LDA uses a single prototype (class centroid) plus a common covariance matrix to describe the spread of the data in each class. In many situations, several prototypes are more appropriate.

- At the other end of the spectrum, we may have way too many (correlated) predictors, for example, in the case of digitized analogue signals and images. In this case LDA uses too many parameters, which are estimated with high variance, and its performance suffers. In cases such as this we need to restrict or regularize LDA even further.

In the remainder of this chapter we describe a class of techniques that attend to all these issues by generalizing the LDA model. This is achieved largely by three different ideas.

The first idea is to recast the LDA problem as a linear regression problem. Many techniques exist for generalizing linear regression to more flexible, nonparametric forms of regression. This in turn leads to more flexible forms of discriminant analysis, which we call FDA. In most cases of interest, the

---

[3]This study predated the emergence of SVMs.

regression procedures can be seen to identify an enlarged set of predictors via basis expansions. FDA amounts to LDA in this enlarged space, the same paradigm used in SVMs.

In the case of too many predictors, such as the pixels of a digitized image, we do not want to expand the set: it is already too large. The second idea is to fit an LDA model, but penalize its coefficients to be smooth or otherwise coherent in the spatial domain, that is, as an image. We call this procedure *penalized discriminant analysis* or PDA. With FDA itself, the expanded basis set is often so large that regularization is also required (again as in SVMs). Both of these can be achieved via a suitably regularized regression in the context of the FDA model.

The third idea is to model each class by a mixture of two or more Gaussians with different centroids, but with every component Gaussian, both within and between classes, sharing the same covariance matrix. This allows for more complex decision boundaries, and allows for subspace reduction as in LDA. We call this extension *mixture discriminant analysis* or MDA.

All three of these generalizations use a common framework by exploiting their connection with LDA.

12.5   Flexible Discriminant Analysis

In this section we describe a method for performing LDA using linear regression on derived responses. This in turn leads to nonparametric and flexible alternatives to LDA. As in Chapter 4, we assume we have observations with a quantitative response G falling into one of K classes $\mathcal{G} = \{1, \ldots, K\}$, each having measured features X. Suppose $\theta : \mathcal{G} \mapsto \mathbb{R}^1$ is a function that assigns scores to the classes, such that the transformed class labels are optimally predicted by linear regression on X: If our training sample has the form (g_i, x_i), $i = 1, 2, \ldots, N$, then we solve

$$\min_{\beta, \theta} \sum_{i=1}^{N} \left(\theta(g_i) - x_i^T \beta \right)^2 , \tag{12.52}$$

with restrictions on θ to avoid a trivial solution (mean zero and unit variance over the training data). This produces a one-dimensional separation between the classes.

More generally, we can find up to $L \leq K - 1$ sets of independent scorings for the class labels, $\theta_1, \theta_2, \ldots, \theta_L$, and L corresponding linear maps $\eta_\ell(X) = X^T \beta_\ell$, $\ell = 1, \ldots, L$, chosen to be optimal for multiple regression in $\mathbb{R}^p$. The scores $\theta_\ell(g)$ and the maps β_ℓ are chosen to minimize the average squared residual,

$$ASR = \frac{1}{N} \sum_{\ell=1}^{L} \left[\sum_{i=1}^{N} \left(\theta_\ell(g_i) - x_i^T \beta_\ell \right)^2 \right] . \tag{12.53}$$

The set of scores are assumed to be mutually orthogonal and normalized with respect to an appropriate inner product to prevent trivial zero solutions.

Why are we going down this road? It can be shown that the sequence of discriminant (canonical) vectors ν_ℓ derived in Section 4.3.3 are identical to the sequence β_ℓ up to a constant (Mardia et al., 1979; Hastie et al., 1995). Moreover, the Mahalanobis distance of a test point x to the kth class centroid $\hat{\mu}_k$ is given by

$$\delta_J(x, \hat{\mu}_k) = \sum_{\ell=1}^{K-1} w_\ell(\hat{\eta}_\ell(x) - \bar{\eta}_\ell^k)^2 + D(x), \tag{12.54}$$

where $\bar{\eta}_\ell^k$ is the mean of the $\hat{\eta}_\ell(x_i)$ in the kth class, and $D(x)$ does not depend on k. Here w_ℓ are coordinate weights that are defined in terms of the mean squared residual r_ℓ^2 of the ℓth optimally scored fit

$$w_\ell = \frac{1}{r_\ell^2(1 - r_\ell^2)}. \tag{12.55}$$

In Section 4.3.2 we saw that these canonical distances are all that is needed for classification in the Gaussian setup, with equal covariances in each class. To summarize:

> LDA can be performed by a sequence of linear regressions, followed by classification to the closest class centroid in the space of fits. The analogy applies both to the reduced rank version, or the full rank case when $L = K - 1$.

The real power of this result is in the generalizations that it invites. We can replace the linear regression fits $\eta_\ell(x) = x^T\beta_\ell$ by far more flexible, nonparametric fits, and by analogy achieve a more flexible classifier than LDA. We have in mind generalized additive fits, spline functions, MARS models and the like. In this more general form the regression problems are defined via the criterion

$$ASR(\{\theta_\ell, \eta_\ell\}_{\ell=1}^L) = \frac{1}{N} \sum_{\ell=1}^L \left[\sum_{i=1}^N (\theta_\ell(g_i) - \eta_\ell(x_i))^2 + \lambda J(\eta_\ell) \right], \tag{12.56}$$

where J is a regularizer appropriate for some forms of nonparametric regression, such as smoothing splines, additive splines and lower-order ANOVA spline models. Also included are the classes of functions and associated penalties generated by kernels, as in Section 12.3.3.

Before we describe the computations involved in this generalization, let us consider a very simple example. Suppose we use degree-2 polynomial regression for each η_ℓ. The decision boundaries implied by the (12.54) will be quadratic surfaces, since each of the fitted functions is quadratic, and as

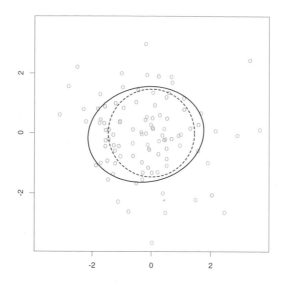

FIGURE 12.9. *The data consist of 50 points generated from each of $N(0, I)$ and $N(0, \frac{9}{4}I)$. The solid black ellipse is the decision boundary found by FDA using degree-two polynomial regression. The dashed purple circle is the Bayes decision boundary.*

in LDA their squares cancel out when comparing distances. We could have achieved *identical* quadratic boundaries in a more conventional way, by augmenting our original predictors with their squares and cross-products. In the enlarged space one performs an LDA, and the linear boundaries in the enlarged space map down to quadratic boundaries in the original space. A classic example is a pair of multivariate Gaussians centered at the origin, one having covariance matrix I, and the other cI for $c > 1$; Figure 12.9 illustrates. The Bayes decision boundary is the sphere $\|x\| = \frac{pc \log c}{2(c-1)}$, which is a linear boundary in the enlarged space.

Many nonparametric regression procedures operate by generating a basis expansion of derived variables, and then performing a linear regression in the enlarged space. The MARS procedure (Chapter 9) is exactly of this form. Smoothing splines and additive spline models generate an extremely large basis set ($N \times p$ basis functions for additive splines), but then perform a penalized regression fit in the enlarged space. SVMs do as well; see also the kernel-based regression example in Section 12.3.7. FDA in this case can be shown to perform a *penalized linear discriminant analysis* in the enlarged space. We elaborate in Section 12.6. Linear boundaries in the enlarged space map down to nonlinear boundaries in the reduced space. This is exactly the same paradigm that is used with support vector machines (Section 12.3).

We illustrate FDA on the speech recognition example used in Chapter 4, with $K = 11$ classes and $p = 10$ predictors. The classes correspond to

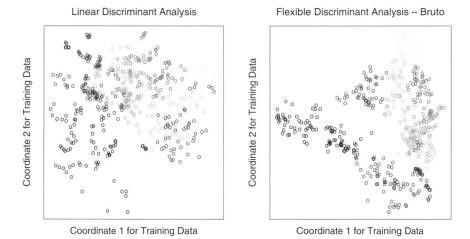

FIGURE 12.10. *The left plot shows the first two LDA canonical variates for the vowel training data. The right plot shows the corresponding projection when FDA/BRUTO is used to fit the model; plotted are the fitted regression functions $\hat{\eta}_1(x_i)$ and $\hat{\eta}_2(x_i)$. Notice the improved separation. The colors represent the eleven different vowel sounds.*

11 vowel sounds, each contained in 11 different words. Here are the words, preceded by the symbols that represent them:

Vowel	Word	Vowel	Word	Vowel	Word	Vowel	Word
i:	heed	O	hod	I	hid	C:	hoard
E	head	U	hood	A	had	u:	who'd
a:	hard	3:	heard	Y	hud		

Each of eight speakers spoke each word six times in the training set, and likewise seven speakers in the test set. The ten predictors are derived from the digitized speech in a rather complicated way, but standard in the speech recognition world. There are thus 528 training observations, and 462 test observations. Figure 12.10 shows two-dimensional projections produced by LDA and FDA. The FDA model used adaptive additive-spline regression functions to model the $\eta_\ell(x)$, and the points plotted in the right plot have coordinates $\hat{\eta}_1(x_i)$ and $\hat{\eta}_2(x_i)$. The routine used in S-PLUS is called `bruto`, hence the heading on the plot and in Table 12.3. We see that flexible modeling has helped to separate the classes in this case. Table 12.3 shows training and test error rates for a number of classification techniques. FDA/MARS refers to Friedman's multivariate adaptive regression splines; degree = 2 means pairwise products are permitted. Notice that for FDA/MARS, the best classification results are obtained in a reduced-rank subspace.

TABLE 12.3. *Vowel recognition data performance results. The results for neural networks are the best among a much larger set, taken from a neural network archive. The notation FDA/BRUTO refers to the regression method used with FDA.*

	Technique	Error Rates	
		Training	Test
(1)	LDA	0.32	0.56
	Softmax	0.48	0.67
(2)	QDA	0.01	0.53
(3)	CART	0.05	0.56
(4)	CART (linear combination splits)	0.05	0.54
(5)	Single-layer perceptron		0.67
(6)	Multi-layer perceptron (88 hidden units)		0.49
(7)	Gaussian node network (528 hidden units)		0.45
(8)	Nearest neighbor		0.44
(9)	FDA/BRUTO	0.06	0.44
	Softmax	0.11	0.50
(10)	FDA/MARS (degree = 1)	0.09	0.45
	Best reduced dimension (=2)	0.18	0.42
	Softmax	0.14	0.48
(11)	FDA/MARS (degree = 2)	0.02	0.42
	Best reduced dimension (=6)	0.13	0.39
	Softmax	0.10	0.50

12.5.1 Computing the FDA Estimates

The computations for the FDA coordinates can be simplified in many important cases, in particular when the nonparametric regression procedure can be represented as a linear operator. We will denote this operator by $\mathbf{S}_\lambda$; that is, $\hat{\mathbf{y}} = \mathbf{S}_\lambda \mathbf{y}$, where $\mathbf{y}$ is the vector of responses and $\hat{\mathbf{y}}$ the vector of fits. Additive splines have this property, if the smoothing parameters are fixed, as does MARS once the basis functions are selected. The subscript λ denotes the entire set of smoothing parameters. In this case optimal scoring is equivalent to a canonical correlation problem, and the solution can be computed by a single eigen-decomposition. This is pursued in Exercise 12.6, and the resulting algorithm is presented here.

We create an $N \times K$ *indicator response matrix* $\mathbf{Y}$ from the responses g_i, such that $y_{ik} = 1$ if $g_i = k$, otherwise $y_{ik} = 0$. For a five-class problem $\mathbf{Y}$ might look like the following:

$$
\begin{array}{cccccc}
 & C_1 & C_2 & C_3 & C_4 & C_5 \\
g_1 = 2 & 0 & 1 & 0 & 0 & 0 \\
g_2 = 1 & 1 & 0 & 0 & 0 & 0 \\
g_3 = 1 & 1 & 0 & 0 & 0 & 0 \\
g_4 = 5 & 0 & 0 & 0 & 0 & 1 \\
g_5 = 4 & 0 & 0 & 0 & 1 & 0 \\
\vdots & & & \vdots & & \\
g_N = 3 & 0 & 0 & 1 & 0 & 0
\end{array}
$$

Here are the computational steps:

1. *Multivariate nonparametric regression.* Fit a multiresponse, adaptive nonparametric regression of $\mathbf{Y}$ on $\mathbf{X}$, giving fitted values $\hat{\mathbf{Y}}$. Let $\mathbf{S}_\lambda$ be the linear operator that fits the final chosen model, and $\eta^*(x)$ be the vector of fitted regression functions.

2. *Optimal scores.* Compute the eigen-decomposition of $\mathbf{Y}^T \hat{\mathbf{Y}} = \mathbf{Y}^T \mathbf{S}_\lambda \mathbf{Y}$, where the eigenvectors $\mathbf{\Theta}$ are normalized: $\mathbf{\Theta}^T \mathbf{D}_\pi \mathbf{\Theta} = \mathbf{I}$. Here $\mathbf{D}_\pi = \mathbf{Y}^T \mathbf{Y}/N$ is a diagonal matrix of the estimated class prior probabilities.

3. *Update* the model from step 1 using the optimal scores: $\eta(x) = \mathbf{\Theta}^T \eta^*(x)$.

The first of the K functions in $\eta(x)$ is the constant function— a trivial solution; the remaining $K-1$ functions are the discriminant functions. The constant function, along with the normalization, causes all the remaining functions to be centered.

Again $\mathbf{S}_\lambda$ can correspond to any regression method. When $\mathbf{S}_\lambda = \mathbf{H}_X$, the linear regression projection operator, then FDA is linear discriminant analysis. The software that we reference in the *Computational Considerations* section on page 455 makes good use of this modularity; the fda function has a method= argument that allows one to supply *any* regression function, as long as it follows some natural conventions. The regression functions we provide allow for polynomial regression, adaptive additive models and MARS. They all efficiently handle multiple responses, so step (1) is a single call to a regression routine. The eigen-decomposition in step (2) simultaneously computes all the optimal scoring functions.

In Section 4.2 we discussed the pitfalls of using linear regression on an indicator response matrix as a method for classification. In particular, severe masking can occur with three or more classes. FDA uses the fits from such a regression in step (1), but then transforms them further to produce useful discriminant functions that are devoid of these pitfalls. Exercise 12.9 takes another view of this phenomenon.

12.6   Penalized Discriminant Analysis

Although FDA is motivated by generalizing optimal scoring, it can also be viewed directly as a form of regularized discriminant analysis. Suppose the regression procedure used in FDA amounts to a linear regression onto a basis expansion $h(X)$, with a quadratic penalty on the coefficients:

$$ASR(\{\theta_\ell, \beta_\ell\}_{\ell=1}^L) = \frac{1}{N} \sum_{\ell=1}^L \left[\sum_{i=1}^N (\theta_\ell(g_i) - h^T(x_i)\beta_\ell)^2 + \lambda \beta_\ell^T \mathbf{\Omega} \beta_\ell \right]. \quad (12.57)$$

The choice of $\mathbf{\Omega}$ depends on the problem. If $\eta_\ell(x) = h(x)\beta_\ell$ is an expansion on spline basis functions, $\mathbf{\Omega}$ might constrain η_ℓ to be smooth over $\mathbb{R}^p$. In the case of additive splines, there are N spline basis functions for each coordinate, resulting in a total of Np basis functions in $h(x)$; $\mathbf{\Omega}$ in this case is $Np \times Np$ and block diagonal.

The steps in FDA can then be viewed as a generalized form of LDA, which we call *penalized discriminant analysis*, or PDA:

- Enlarge the set of predictors X via a basis expansion $h(X)$.

- Use (penalized) LDA in the enlarged space, where the penalized Mahalanobis distance is given by

$$D(x, \mu) = (h(x) - h(\mu))^T (\mathbf{\Sigma}_W + \lambda\mathbf{\Omega})^{-1} (h(x) - h(\mu)), \quad (12.58)$$

  where $\mathbf{\Sigma}_W$ is the within-class covariance matrix of the derived variables $h(x_i)$.

- Decompose the classification subspace using a penalized metric:

$$\max u^T \mathbf{\Sigma}_{\text{Bet}} u \text{ subject to } u^T(\mathbf{\Sigma}_W + \lambda\mathbf{\Omega})u = 1.$$

Loosely speaking, the penalized Mahalanobis distance tends to give less weight to "rough" coordinates, and more weight to "smooth" ones; since the penalty is not diagonal, the same applies to linear combinations that are rough or smooth.

For some classes of problems, the first step, involving the basis expansion, is not needed; we already have far too many (correlated) predictors. A leading example is when the objects to be classified are digitized analog signals:

- the log-periodogram of a fragment of spoken speech, sampled at a set of 256 frequencies; see Figure 5.5 on page 149.

- the grayscale pixel values in a digitized image of a handwritten digit.

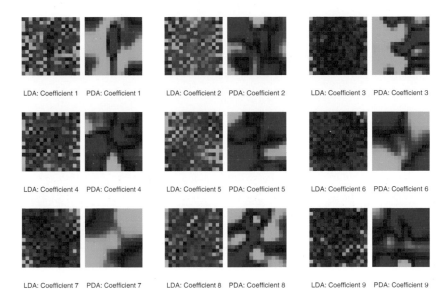

LDA: Coefficient 1   PDA: Coefficient 1   LDA: Coefficient 2   PDA: Coefficient 2   LDA: Coefficient 3   PDA: Coefficient 3

LDA: Coefficient 4   PDA: Coefficient 4   LDA: Coefficient 5   PDA: Coefficient 5   LDA: Coefficient 6   PDA: Coefficient 6

LDA: Coefficient 7   PDA: Coefficient 7   LDA: Coefficient 8   PDA: Coefficient 8   LDA: Coefficient 9   PDA: Coefficient 9

FIGURE 12.11. *The images appear in pairs, and represent the nine discriminant coefficient functions for the digit recognition problem. The left member of each pair is the LDA coefficient, while the right member is the PDA coefficient, regularized to enforce spatial smoothness.*

It is also intuitively clear in these cases why regularization is needed. Take the digitized image as an example. Neighboring pixel values will tend to be correlated, being often almost the same. This implies that the pair of corresponding LDA coefficients for these pixels can be wildly different and opposite in sign, and thus cancel when applied to similar pixel values. Positively correlated predictors lead to noisy, negatively correlated coefficient estimates, and this noise results in unwanted sampling variance. A reasonable strategy is to regularize the *coefficients* to be smooth over the spatial domain, as with images. This is what PDA does. The computations proceed just as for FDA, except that an appropriate penalized regression method is used. Here $h^T(X)\beta_\ell = X\beta_\ell$, and $\boldsymbol{\Omega}$ is chosen so that $\beta_\ell^T \boldsymbol{\Omega} \beta_\ell$ penalizes roughness in β_ℓ when viewed as an image. Figure 1.2 on page 4 shows some examples of handwritten digits. Figure 12.11 shows the discriminant variates using LDA and PDA. Those produced by LDA appear as *salt-and-pepper* images, while those produced by PDA are smooth images. The first smooth image can be seen as the coefficients of a linear contrast functional for separating images with a dark central vertical strip (ones, possibly sevens) from images that are hollow in the middle (zeros, some fours). Figure 12.12 supports this interpretation, and with more difficulty allows an interpretation of the second coordinate. This and other

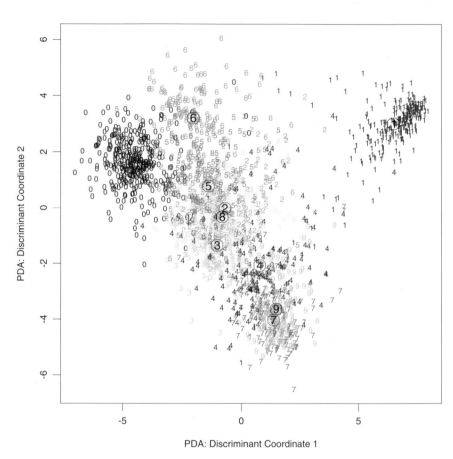

FIGURE 12.12. *The first two penalized canonical variates, evaluated for the test data. The circles indicate the class centroids. The first coordinate contrasts mainly 0's and 1's, while the second contrasts 6's and 7/9's.*

examples are discussed in more detail in Hastie et al. (1995), who also show that the regularization improves the classification performance of LDA on independent test data by a factor of around 25% in the cases they tried.

12.7   Mixture Discriminant Analysis

Linear discriminant analysis can be viewed as a *prototype* classifier. Each class is represented by its centroid, and we classify to the closest using an appropriate metric. In many situations a single prototype is not sufficient to represent inhomogeneous classes, and mixture models are more appropriate. In this section we review Gaussian mixture models and show how they can be generalized via the FDA and PDA methods discussed earlier. A Gaussian mixture model for the kth class has density

$$P(X|G=k) = \sum_{r=1}^{R_k} \pi_{kr} \phi(X; \mu_{kr}, \boldsymbol{\Sigma}), \tag{12.59}$$

where the *mixing proportions* π_{kr} sum to one. This has R_k prototypes for the kth class, and in our specification, the same covariance matrix $\boldsymbol{\Sigma}$ is used as the metric throughout. Given such a model for each class, the class posterior probabilities are given by

$$P(G=k|X=x) = \frac{\sum_{r=1}^{R_k} \pi_{kr} \phi(X; \mu_{kr}, \boldsymbol{\Sigma}) \Pi_k}{\sum_{\ell=1}^{K} \sum_{r=1}^{R_\ell} \pi_{\ell r} \phi(X; \mu_{\ell r}, \boldsymbol{\Sigma}) \Pi_\ell}, \tag{12.60}$$

where Π_k represent the class prior probabilities.

We saw these calculations for the special case of two components in Chapter 8. As in LDA, we estimate the parameters by maximum likelihood, using the joint log-likelihood based on $P(G, X)$:

$$\sum_{k=1}^{K} \sum_{g_i=k} \log \left[\sum_{r=1}^{R_k} \pi_{kr} \phi(x_i; \mu_{kr}, \boldsymbol{\Sigma}) \Pi_k \right]. \tag{12.61}$$

The sum within the log makes this a rather messy optimization problem if tackled directly. The classical and natural method for computing the maximum-likelihood estimates (MLEs) for mixture distributions is the EM algorithm (Dempster et al., 1977), which is known to possess good convergence properties. EM alternates between the two steps:

E-step: Given the current parameters, compute the *responsibility* of sub-class c_{kr} within class k for each of the class-k observations ($g_i = k$):

$$W(c_{kr}|x_i, g_i) = \frac{\pi_{kr}\phi(x_i; \mu_{kr}, \mathbf{\Sigma})}{\sum_{\ell=1}^{R_k} \pi_{k\ell}\phi(x_i; \mu_{k\ell}, \mathbf{\Sigma})}. \qquad (12.62)$$

M-step: Compute the weighted MLEs for the parameters of each of the component Gaussians within each of the classes, using the weights from the E-step.

In the E-step, the algorithm apportions the unit weight of an observation in class k to the various subclasses assigned to that class. If it is close to the centroid of a particular subclass, and far from the others, it will receive a mass close to one for that subclass. On the other hand, observations halfway between two subclasses will get approximately equal weight for both.

In the M-step, an observation in class k is used R_k times, to estimate the parameters in each of the R_k component densities, with a different weight for each. The EM algorithm is studied in detail in Chapter 8. The algorithm requires initialization, which can have an impact, since mixture likelihoods are generally multimodal. Our software (referenced in the *Computational Considerations* on page 455) allows several strategies; here we describe the default. The user supplies the number R_k of subclasses per class. Within class k, a k-means clustering model, with multiple random starts, is fitted to the data. This partitions the observations into R_k disjoint groups, from which an initial weight matrix, consisting of zeros and ones, is created.

Our assumption of an equal component covariance matrix $\mathbf{\Sigma}$ throughout buys an additional simplicity; we can incorporate rank restrictions in the mixture formulation just like in LDA. To understand this, we review a little-known fact about LDA. The rank-L LDA fit (Section 4.3.3) is equivalent to the maximum-likelihood fit of a Gaussian model, where the different mean vectors in each class are confined to a rank-L subspace of $\mathbb{R}^p$ (Exercise 4.8). We can inherit this property for the mixture model, and maximize the log-likelihood (12.61) subject to rank constraints on *all* the $\sum_k R_k$ centroids: $\text{rank}\{\mu_{k\ell}\} = L$.

Again the EM algorithm is available, and the M-step turns out to be a weighted version of LDA, with $R = \sum_{k=1}^{K} R_k$ "classes." Furthermore, we can use optimal scoring as before to solve the weighted LDA problem, which allows us to use a weighted version of FDA or PDA at this stage. One would expect, in addition to an increase in the number of "classes," a similar increase in the number of "observations" in the kth class by a factor of R_k. It turns out that this is not the case if linear operators are used for the optimal scoring regression. The enlarged indicator $\mathbf{Y}$ matrix collapses in this case to a *blurred* response matrix $\mathbf{Z}$, which is intuitively pleasing. For example, suppose there are $K = 3$ classes, and $R_k = 3$ subclasses per class. Then $\mathbf{Z}$ might be

$$
\begin{array}{c}
\begin{array}{ccccccccc}
c_{11} & c_{12} & c_{13} & c_{21} & c_{22} & c_{23} & c_{31} & c_{32} & c_{33}
\end{array}\\
\begin{array}{c}
g_1 = 2\\
g_2 = 1\\
g_3 = 1\\
g_4 = 3\\
g_5 = 2\\
\vdots\\
g_N = 3
\end{array}
\left(
\begin{array}{ccccccccc}
0 & 0 & 0 & 0.3 & 0.5 & 0.2 & 0 & 0 & 0\\
0.9 & 0.1 & 0.0 & 0 & 0 & 0 & 0 & 0 & 0\\
0.1 & 0.8 & 0.1 & 0 & 0 & 0 & 0 & 0 & 0\\
0 & 0 & 0 & 0 & 0 & 0 & 0.5 & 0.4 & 0.1\\
0 & 0 & 0 & 0.7 & 0.1 & 0.2 & 0 & 0 & 0\\
 & & & & \vdots & & & &\\
0 & 0 & 0 & 0 & 0 & 0 & 0.1 & 0.1 & 0.8
\end{array}
\right),
\end{array}
\tag{12.63}
$$

where the entries in a class-k row correspond to $W(c_{kr}|x, g_i)$.

The remaining steps are the same:

$$
\left.
\begin{aligned}
\hat{\mathbf{Z}} &= \mathbf{SZ}\\
\mathbf{Z}^T\hat{\mathbf{Z}} &= \mathbf{\Theta}\mathbf{D}\mathbf{\Theta}^T\\
\text{Update } &\pi s \text{ and } \Pi s
\end{aligned}
\right\}
\quad \text{M-step of MDA.}
$$

These simple modifications add considerable flexibility to the mixture model:

- The dimension reduction step in LDA, FDA or PDA is limited by the number of classes; in particular, for $K = 2$ classes no reduction is possible. MDA substitutes subclasses for classes, and then allows us to look at low-dimensional views of the subspace spanned by these subclass centroids. This subspace will often be an important one for discrimination.

- By using FDA or PDA in the M step, we can adapt even more to particular situations. For example, we can fit MDA models to digitized analog signals and images, with smoothness constraints built in.

Figure 12.13 compares FDA and MDA on the mixture example.

12.7.1 Example: Waveform Data

We now illustrate some of these ideas on a popular simulated example, taken from Breiman et al. (1984, pages 49–55), and used in Hastie and Tibshirani (1996b) and elsewhere. It is a three-class problem with 21 variables, and is considered to be a difficult pattern recognition problem. The predictors are defined by

$$
\begin{aligned}
X_j &= Uh_1(j) + (1 - U)h_2(j) + \epsilon_j \quad &\text{Class 1,}\\
X_j &= Uh_1(j) + (1 - U)h_3(j) + \epsilon_j \quad &\text{Class 2,}\\
X_j &= Uh_2(j) + (1 - U)h_3(j) + \epsilon_j \quad &\text{Class 3,}
\end{aligned}
\tag{12.64}
$$

where $j = 1, 2, \ldots, 21$, U is uniform on $(0, 1)$, ϵ_j are standard normal variates, and the h_ℓ are the shifted triangular waveforms: $h_1(j) = \max(6 -$

FDA / MARS - Degree 2

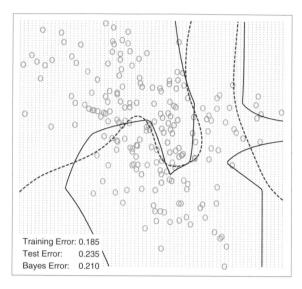

Training Error: 0.185
Test Error: 0.235
Bayes Error: 0.210

MDA - 5 Subclasses per Class

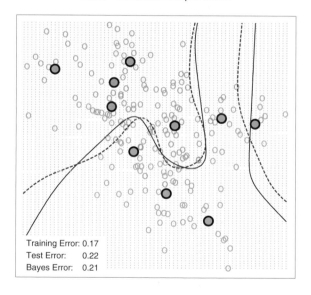

Training Error: 0.17
Test Error: 0.22
Bayes Error: 0.21

FIGURE 12.13. *FDA and MDA on the mixture data. The upper plot uses FDA with MARS as the regression procedure. The lower plot uses MDA with five mixture centers per class (indicated). The MDA solution is close to Bayes optimal, as might be expected given the data arise from mixtures of Gaussians. The broken purple curve in the background is the Bayes decision boundary.*

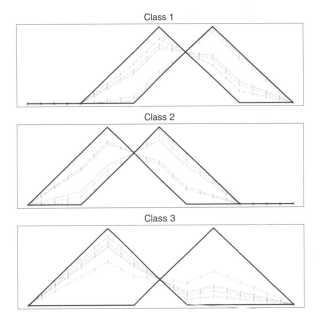

FIGURE 12.14. *Some examples of the waveforms generated from model (12.64) before the Gaussian noise is added.*

$|j - 11|, 0)$, $h_2(j) = h_1(j - 4)$ and $h_3(j) = h_1(j + 4)$. Figure 12.14 shows some example waveforms from each class.

Table 12.4 shows the results of MDA applied to the waveform data, as well as several other methods from this and other chapters. Each training sample has 300 observations, and equal priors were used, so there are roughly 100 observations in each class. We used test samples of size 500. The two MDA models are described in the caption.

Figure 12.15 shows the leading canonical variates for the penalized MDA model, evaluated at the test data. As we might have guessed, the classes appear to lie on the edges of a triangle. This is because the $h_j(i)$ are represented by three points in 21-space, thereby forming vertices of a triangle, and each class is represented as a convex combination of a pair of vertices, and hence lie on an edge. Also it is clear visually that all the information lies in the first two dimensions; the percentage of variance explained by the first two coordinates is 99.8%, and we would lose nothing by truncating the solution there. The Bayes risk for this problem has been estimated to be about 0.14 (Breiman et al., 1984). MDA comes close to the optimal rate, which is not surprising since the structure of the MDA model is similar to the generating model.

TABLE 12.4. *Results for waveform data. The values are averages over ten simulations, with the standard error of the average in parentheses. The five entries above the line are taken from Hastie et al. (1994). The first model below the line is MDA with three subclasses per class. The next line is the same, except that the discriminant coefficients are penalized via a roughness penalty to effectively 4df. The third is the corresponding penalized LDA or PDA model.*

Technique	Error Rates	
	Training	Test
LDA	0.121(0.006)	0.191(0.006)
QDA	0.039(0.004)	0.205(0.006)
CART	0.072(0.003)	0.289(0.004)
FDA/MARS (degree = 1)	0.100(0.006)	0.191(0.006)
FDA/MARS (degree = 2)	0.068(0.004)	0.215(0.002)
MDA (3 subclasses)	0.087(0.005)	0.169(0.006)
MDA (3 subclasses, penalized 4 df)	0.137(0.006)	0.157(0.005)
PDA (penalized 4 df)	0.150(0.005)	0.171(0.005)
Bayes		0.140

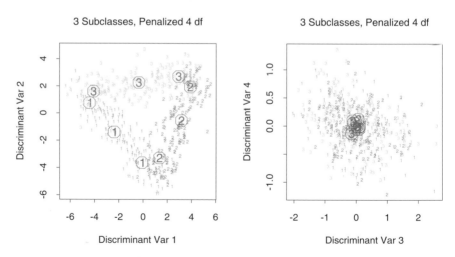

FIGURE 12.15. *Some two-dimensional views of the MDA model fitted to a sample of the waveform model. The points are independent test data, projected on to the leading two canonical coordinates (left panel), and the third and fourth (right panel). The subclass centers are indicated.*

Computational Considerations

With N training cases, p predictors, and m support vectors, the support vector machine requires $m^3 + mN + mpN$ operations, assuming $m \approx N$. They do not scale well with N, although computational shortcuts are available (Platt, 1999). Since these are evolving rapidly, the reader is urged to search the web for the latest technology.

LDA requires $Np^2 + p^3$ operations, as does PDA. The complexity of FDA depends on the regression method used. Many techniques are linear in N, such as additive models and MARS. General splines and kernel-based regression methods will typically require N^3 operations.

Software is available for fitting FDA, PDA and MDA models in the R package mda, which is also available in S-PLUS.

Bibliographic Notes

The theory behind support vector machines is due to Vapnik and is described in Vapnik (1996). There is a burgeoning literature on SVMs; an online bibliography, created and maintained by Alex Smola and Bernhard Schölkopf, can be found at:

http://www.kernel-machines.org.

Our treatment is based on Wahba et al. (2000) and Evgeniou et al. (2000), and the tutorial by Burges (Burges, 1998).

Linear discriminant analysis is due to Fisher (1936) and Rao (1973). The connection with optimal scoring dates back at least to Breiman and Ihaka (1984), and in a simple form to Fisher (1936). There are strong connections with correspondence analysis (Greenacre, 1984). The description of flexible, penalized and mixture discriminant analysis is taken from Hastie et al. (1994), Hastie et al. (1995) and Hastie and Tibshirani (1996b), and all three are summarized in Hastie et al. (2000); see also Ripley (1996).

Exercises

Ex. 12.1 Show that the criteria (12.25) and (12.8) are equivalent.

Ex. 12.2 Show that the solution to (12.29) is the same as the solution to (12.25) for a particular kernel.

Ex. 12.3 Consider a modification to (12.43) where you do not penalize the constant. Formulate the problem, and characterize its solution.

Ex. 12.4 Suppose you perform a reduced-subspace linear discriminant analysis for a K-group problem. You compute the canonical variables of dimension $L \le K - 1$ given by $z = \mathbf{U}^T x$, where $\mathbf{U}$ is the $p \times L$ matrix of discriminant coefficients, and $p > K$ is the dimension of x.

(a) If $L = K - 1$ show that

$$\| z - \bar{z}_k \|^2 - \| z - \bar{z}_{k'} \|^2 = \| x - \bar{x}_k \|_W^2 - \| x - \bar{x}_{k'} \|_W^2 ,$$

where $\|\cdot\|_W$ denotes *Mahalanobis* distance with respect to the covariance $\mathbf{W}$.

(b) If $L < K - 1$, show that the same expression on the left measures the difference in Mahalanobis squared distances for the distributions projected onto the subspace spanned by $\mathbf{U}$.

Ex. 12.5 The data in `phoneme.subset`, available from this book's website

http://www-stat.stanford.edu/ElemStatLearn

consists of digitized log-periodograms for phonemes uttered by 60 speakers, each speaker having produced phonemes from each of five classes. It is appropriate to plot each vector of 256 "features" against the frequencies 0–255.

(a) Produce a separate plot of all the phoneme curves against frequency for each class.

(b) You plan to use a nearest prototype classification scheme to classify the curves into phoneme classes. In particular, you will use a K-means clustering algorithm in each class (`kmeans()` in R), and then classify observations to the class of the closest cluster center. The curves are high-dimensional and you have a rather small sample-size-to-variables ratio. You decide to restrict all the prototypes to be smooth functions of frequency. In particular, you decide to represent each prototype m as $m = B\theta$ where B is a $256 \times J$ matrix of natural spline basis functions with J knots uniformly chosen in $(0, 255)$ and boundary knots at 0 and 255. Describe how to proceed analytically, and in particular, how to avoid costly high-dimensional fitting procedures. (*Hint:* It may help to restrict B to be orthogonal.)

(c) Implement your procedure on the phoneme data, and try it out. Divide the data into a training set and a test set (50-50), making sure that speakers are not split across sets (why?). Use $K = 1, 3, 5, 7$ centers per class, and for each use $J = 5, 10, 15$ knots (taking care to start the K-means procedure at the same starting values for each value of J), and compare the results.

Ex. 12.6 Suppose that the regression procedure used in FDA (Section 12.5.1) is a linear expansion of basis functions $h_m(x)$, $m = 1, \ldots, M$. Let $\mathbf{D}_\pi = \mathbf{Y}^T \mathbf{Y}/N$ be the diagonal matrix of class proportions.

(a) Show that the optimal scoring problem (12.52) can be written in vector notation as

$$\min_{\theta,\beta} \| \mathbf{Y}\theta - \mathbf{H}\beta \|^2 ,\qquad (12.65)$$

where θ is a vector of K real numbers, and $\mathbf{H}$ is the $N \times M$ matrix of evaluations $h_j(x_i)$.

(b) Suppose that the normalization on θ is $\theta^T \mathbf{D}_\pi 1 = 0$ and $\theta^T \mathbf{D}_\pi \theta = 1$. Interpret these normalizations in terms of the original scored $\theta(g_i)$.

(c) Show that, with this normalization, (12.65) can be partially optimized w.r.t. β, and leads to

$$\max_{\theta} \theta^T \mathbf{Y}^T \mathbf{S} \mathbf{Y}\theta,\qquad (12.66)$$

subject to the normalization constraints, where $\mathbf{S}$ is the projection operator corresponding to the basis matrix $\mathbf{H}$.

(d) Suppose that the h_j include the constant function. Show that the largest eigenvalue of $\mathbf{S}$ is 1.

(e) Let Θ be a $K \times K$ matrix of scores (in columns), and suppose the normalization is $\Theta^T \mathbf{D}_\pi \Theta = \mathbf{I}$. Show that the solution to (12.53) is given by the complete set of eigenvectors of $\mathbf{S}$; the first eigenvector is trivial, and takes care of the centering of the scores. The remainder characterize the optimal scoring solution.

Ex. 12.7 Derive the solution to the penalized optimal scoring problem (12.57).

Ex. 12.8 Show that coefficients β_ℓ found by optimal scoring are proportional to the discriminant directions ν_ℓ found by linear discriminant analysis.

Ex. 12.9 Let $\hat{\mathbf{Y}} = \mathbf{X}\hat{\mathbf{B}}$ be the fitted $N \times K$ indicator response matrix after linear regression on the $N \times p$ matrix $\mathbf{X}$, where $p > K$. Consider the reduced features $x_i^* = \hat{\mathbf{B}}^T x_i$. Show that LDA using x_i^* is equivalent to LDA in the original space.

Ex. 12.10 *Kernels and linear discriminant analysis.* Suppose you wish to carry out a linear discriminant analysis (two classes) using a vector of transformations of the input variables $h(x)$. Since $h(x)$ is high-dimensional, you will use a regularized within-class covariance matrix $\mathbf{W}_h + \gamma \mathbf{I}$. Show that the model can be estimated using only the inner products $K(x_i, x_{i'}) = \langle h(x_i), h(x_{i'}) \rangle$. Hence the kernel property of support vector machines is also shared by regularized linear discriminant analysis.

Ex. 12.11 The MDA procedure models each class as a mixture of Gaussians. Hence each mixture center belongs to one and only one class. A more general model allows each mixture center to be shared by all classes. We take the joint density of labels and features to be

$$P(G, X) = \sum_{r=1}^{R} \pi_r P_r(G, X), \tag{12.67}$$

a mixture of joint densities. Furthermore we assume

$$P_r(G, X) = P_r(G)\phi(X; \mu_r, \Sigma). \tag{12.68}$$

This model consists of regions centered at μ_r, and for each there is a class profile $P_r(G)$. The posterior class distribution is given by

$$P(G = k | X = x) = \frac{\sum_{r=1}^{R} \pi_r P_r(G = k)\phi(x; \mu_r, \Sigma)}{\sum_{r=1}^{R} \pi_r \phi(x; \mu_r, \Sigma)}, \tag{12.69}$$

where the denominator is the marginal distribution $P(X)$.

(a) Show that this model (called MDA2) can be viewed as a generalization of MDA since

$$P(X | G = k) = \frac{\sum_{r=1}^{R} \pi_r P_r(G = k)\phi(x; \mu_r, \Sigma)}{\sum_{r=1}^{R} \pi_r P_r(G = k)}, \tag{12.70}$$

where $\pi_{rk} = \pi_r P_r(G = k) / \sum_{r=1}^{R} \pi_r P_r(G = k)$ corresponds to the mixing proportions for the kth class.

(b) Derive the EM algorithm for MDA2.

(c) Show that if the initial weight matrix is constructed as in MDA, involving separate k-means clustering in each class, then the algorithm for MDA2 is identical to the original MDA procedure.

13

Prototype Methods and Nearest-Neighbors

13.1 Introduction

In this chapter we discuss some simple and essentially model-free methods for classification and pattern recognition. Because they are highly unstructured, they typically are not useful for understanding the nature of the relationship between the features and class outcome. However, as *black box* prediction engines, they can be very effective, and are often among the best performers in real data problems. The nearest-neighbor technique can also be used in regression; this was touched on in Chapter 2 and works reasonably well for low-dimensional problems. However, with high-dimensional features, the bias–variance tradeoff does not work as favorably for nearest-neighbor regression as it does for classification.

13.2 Prototype Methods

Throughout this chapter, our training data consists of the N pairs (x_1, g_1), ..., (x_n, g_N) where g_i is a class label taking values in $\{1, 2, \ldots, K\}$. Prototype methods represent the training data by a set of points in feature space. These prototypes are typically not examples from the training sample, except in the case of 1-nearest-neighbor classification discussed later.

Each prototype has an associated class label, and classification of a query point x is made to the class of the closest prototype. "Closest" is usually defined by Euclidean distance in the feature space, after each feature has

T. Hastie et al., *The Elements of Statistical Learning, Second Edition,*
DOI: 10.1007/b94608_13,
© Springer Science+Business Media, LLC 2009

been standardized to have overall mean 0 and variance 1 in the training sample. Euclidean distance is appropriate for quantitative features. We discuss distance measures between qualitative and other kinds of feature values in Chapter 14.

These methods can be very effective if the prototypes are well positioned to capture the distribution of each class. Irregular class boundaries can be represented, with enough prototypes in the right places in feature space. The main challenge is to figure out how many prototypes to use and where to put them. Methods differ according to the number and way in which prototypes are selected.

13.2.1   K-means Clustering

K-means clustering is a method for finding clusters and cluster centers in a set of unlabeled data. One chooses the desired number of cluster centers, say R, and the K-means procedure iteratively moves the centers to minimize the total within cluster variance.[1] Given an initial set of centers, the K-means algorithm alternates the two steps:

- for each center we identify the subset of training points (its cluster) that is closer to it than any other center;

- the means of each feature for the data points in each cluster are computed, and this mean vector becomes the new center for that cluster.

These two steps are iterated until convergence. Typically the initial centers are R randomly chosen observations from the training data. Details of the K-means procedure, as well as generalizations allowing for different variable types and more general distance measures, are given in Chapter 14.

To use K-means clustering for classification of labeled data, the steps are:

- apply K-means clustering to the training data in each class separately, using R prototypes per class;

- assign a class label to each of the $K \times R$ prototypes;

- classify a new feature x to the class of the closest prototype.

Figure 13.1 (upper panel) shows a simulated example with three classes and two features. We used $R = 5$ prototypes per class, and show the classification regions and the decision boundary. Notice that a number of the

---

[1]The "K" in K-means refers to the number of cluster centers. Since we have already reserved K to denote the number of classes, we denote the number of clusters by R.

K-means - 5 Prototypes per Class

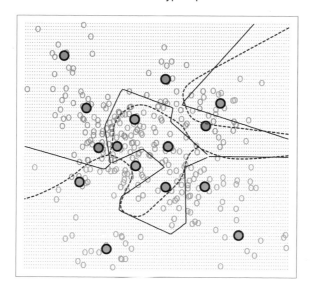

LVQ - 5 Prototypes per Class

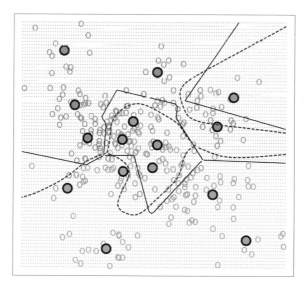

FIGURE 13.1. *Simulated example with three classes and five prototypes per class. The data in each class are generated from a mixture of Gaussians. In the upper panel, the prototypes were found by applying the K-means clustering algorithm separately in each class. In the lower panel, the LVQ algorithm (starting from the K-means solution) moves the prototypes away from the decision boundary. The broken purple curve in the background is the Bayes decision boundary.*

Algorithm 13.1 *Learning Vector Quantization—LVQ.*

---

1. Choose R initial prototypes for each class: $m_1(k), m_2(k), \ldots, m_R(k)$, $k = 1, 2, \ldots, K$, for example, by sampling R training points at random from each class.

2. Sample a training point x_i randomly (with replacement), and let (j, k) index the closest prototype $m_j(k)$ to x_i.

   (a) If $g_i = k$ (i.e., they are in the same class), move the prototype towards the training point:

$$m_j(k) \leftarrow m_j(k) + \epsilon(x_i - m_j(k)),$$

   where ϵ is the *learning rate*.

   (b) If $g_i \neq k$ (i.e., they are in different classes), move the prototype away from the training point:

$$m_j(k) \leftarrow m_j(k) - \epsilon(x_i - m_j(k)).$$

3. Repeat step 2, decreasing the learning rate ϵ with each iteration towards zero.

---

prototypes are near the class boundaries, leading to potential misclassification errors for points near these boundaries. This results from an obvious shortcoming with this method: for each class, the other classes do not have a say in the positioning of the prototypes for that class. A better approach, discussed next, uses all of the data to position all prototypes.

13.2.2 *Learning Vector Quantization*

In this technique due to Kohonen (1989), prototypes are placed strategically with respect to the decision boundaries in an ad-hoc way. LVQ is an *online* algorithm—observations are processed one at a time.

The idea is that the training points attract prototypes of the correct class, and repel other prototypes. When the iterations settle down, prototypes should be close to the training points in their class. The learning rate ϵ is decreased to zero with each iteration, following the guidelines for stochastic approximation learning rates (Section 11.4.)

Figure 13.1 (lower panel) shows the result of LVQ, using the K-means solution as starting values. The prototypes have tended to move away from the decision boundaries, and away from prototypes of competing classes.

The procedure just described is actually called LVQ1. Modifications (LVQ2, LVQ3, etc.) have been proposed, that can sometimes improve performance. A drawback of learning vector quantization methods is the fact that they are defined by algorithms, rather than optimization of some fixed criteria; this makes it difficult to understand their properties.

13.2.3 Gaussian Mixtures

The Gaussian mixture model can also be thought of as a prototype method, similar in spirit to K-means and LVQ. We discuss Gaussian mixtures in some detail in Sections 6.8, 8.5 and 12.7. Each cluster is described in terms of a Gaussian density, which has a centroid (as in K-means), and a covariance matrix. The comparison becomes crisper if we restrict the component Gaussians to have a scalar covariance matrix (Exercise 13.1). The two steps of the alternating EM algorithm are very similar to the two steps in K-means:

- In the E-step, each observation is assigned a *responsibility* or weight for each cluster, based on the likelihood of each of the corresponding Gaussians. Observations close to the center of a cluster will most likely get weight 1 for that cluster, and weight 0 for every other cluster. Observations half-way between two clusters divide their weight accordingly.

- In the M-step, each observation contributes to the weighted means (and covariances) for *every* cluster.

As a consequence, the Gaussian mixture model is often referred to as a *soft* clustering method, while K-means is *hard*.

Similarly, when Gaussian mixture models are used to represent the feature density in each class, it produces smooth posterior probabilities $\hat{p}(x) = \{\hat{p}_1(x), \dots, \hat{p}_K(x)\}$ for classifying x (see (12.60) on page 449.) Often this is interpreted as a soft classification, while in fact the classification rule is $\hat{G}(x) = \arg\max_k \hat{p}_k(x)$. Figure 13.2 compares the results of K-means and Gaussian mixtures on the simulated mixture problem of Chapter 2. We see that although the decision boundaries are roughly similar, those for the mixture model are smoother (although the prototypes are in approximately the same positions.) We also see that while both procedures devote a blue prototype (incorrectly) to a region in the northwest, the Gaussian mixture classifier can ultimately ignore this region, while K-means cannot. LVQ gave very similar results to K-means on this example, and is not shown.

13.3 k-Nearest-Neighbor Classifiers

These classifiers are *memory-based*, and require no model to be fit. Given a query point x_0, we find the k training points $x_{(r)}, r = 1, \dots, k$ closest in distance to x_0, and then classify using majority vote among the k neighbors.

K-means - 5 Prototypes per Class

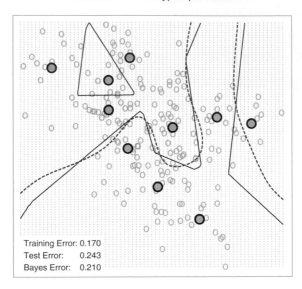

Training Error: 0.170
Test Error: 0.243
Bayes Error: 0.210

Gaussian Mixtures - 5 Subclasses per Class

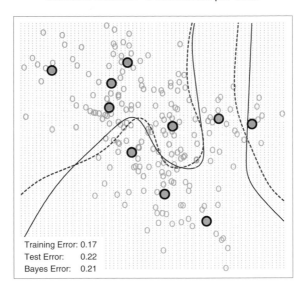

Training Error: 0.17
Test Error: 0.22
Bayes Error: 0.21

FIGURE 13.2. *The upper panel shows the K-means classifier applied to the mixture data example. The decision boundary is piecewise linear. The lower panel shows a Gaussian mixture model with a common covariance for all component Gaussians. The EM algorithm for the mixture model was started at the K-means solution. The broken purple curve in the background is the Bayes decision boundary.*

Ties are broken at random. For simplicity we will assume that the features are real-valued, and we use Euclidean distance in feature space:

$$d_{(i)} = ||x_{(i)} - x_0||. \tag{13.1}$$

Typically we first standardize each of the features to have mean zero and variance 1, since it is possible that they are measured in different units. In Chapter 14 we discuss distance measures appropriate for qualitative and ordinal features, and how to combine them for mixed data. Adaptively chosen distance metrics are discussed later in this chapter.

Despite its simplicity, k-nearest-neighbors has been successful in a large number of classification problems, including handwritten digits, satellite image scenes and EKG patterns. It is often successful where each class has many possible prototypes, and the decision boundary is very irregular. Figure 13.3 (upper panel) shows the decision boundary of a 15-nearest-neighbor classifier applied to the three-class simulated data. The decision boundary is fairly smooth compared to the lower panel, where a 1-nearest-neighbor classifier was used. There is a close relationship between nearest-neighbor and prototype methods: in 1-nearest-neighbor classification, each training point is a prototype.

Figure 13.4 shows the training, test and tenfold cross-validation errors as a function of the neighborhood size, for the two-class mixture problem. Since the tenfold CV errors are averages of ten numbers, we can estimate a standard error.

Because it uses only the training point closest to the query point, the bias of the 1-nearest-neighbor estimate is often low, but the variance is high. A famous result of Cover and Hart (1967) shows that asymptotically the error rate of the 1-nearest-neighbor classifier is never more than twice the Bayes rate. The rough idea of the proof is as follows (using squared-error loss). We assume that the query point coincides with one of the training points, so that the bias is zero. This is true asymptotically if the dimension of the feature space is fixed and the training data fills up the space in a dense fashion. Then the error of the Bayes rule is just the variance of a Bernoulli random variate (the target at the query point), while the error of 1-nearest-neighbor rule is *twice* the variance of a Bernoulli random variate, one contribution each for the training and query targets.

We now give more detail for misclassification loss. At x let k^* be the dominant class, and $p_k(x)$ the true conditional probability for class k. Then

$$\text{Bayes error} = 1 - p_{k^*}(x), \tag{13.2}$$

$$\text{1-nearest-neighbor error} = \sum_{k=1}^{K} p_k(x)(1 - p_k(x)), \tag{13.3}$$

$$\geq 1 - p_{k^*}(x). \tag{13.4}$$

The asymptotic 1-nearest-neighbor error rate is that of a random rule; we pick both the classification and the test point at random with probabili-

15-Nearest Neighbors

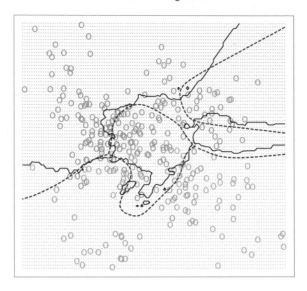

1-Nearest Neighbor

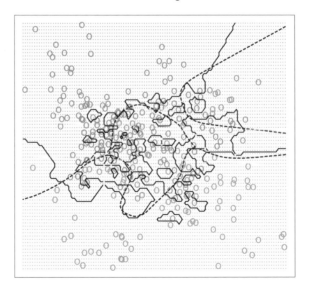

FIGURE 13.3. *k-nearest-neighbor classifiers applied to the simulation data of Figure 13.1. The broken purple curve in the background is the Bayes decision boundary.*

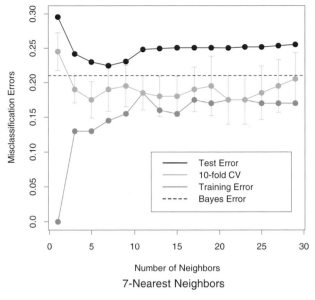

7-Nearest Neighbors

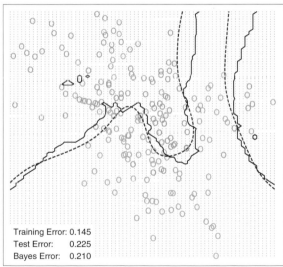

FIGURE 13.4. *k-nearest-neighbors on the two-class mixture data. The upper panel shows the misclassification errors as a function of neighborhood size. Standard error bars are included for 10-fold cross validation. The lower panel shows the decision boundary for 7-nearest-neighbors, which appears to be optimal for minimizing test error. The broken purple curve in the background is the Bayes decision boundary.*

ties $p_k(x)$, $k = 1, \ldots, K$. For $K = 2$ the 1-nearest-neighbor error rate is $2p_{k^*}(x)(1 - p_{k^*}(x)) \leq 2(1 - p_{k^*}(x))$ (twice the Bayes error rate). More generally, one can show (Exercise 13.3)

$$\sum_{k=1}^{K} p_k(x)(1 - p_k(x)) \leq 2(1 - p_{k^*}(x)) - \frac{K}{K-1}(1 - p_{k^*}(x))^2. \qquad (13.5)$$

Many additional results of this kind have been derived; Ripley (1996) summarizes a number of them.

This result can provide a rough idea about the best performance that is possible in a given problem. For example, if the 1-nearest-neighbor rule has a 10% error rate, then asymptotically the Bayes error rate is at least 5%. The kicker here is the asymptotic part, which assumes the bias of the nearest-neighbor rule is zero. In real problems the bias can be substantial. The adaptive nearest-neighbor rules, described later in this chapter, are an attempt to alleviate this bias. For simple nearest-neighbors, the bias and variance characteristics can dictate the optimal number of near neighbors for a given problem. This is illustrated in the next example.

13.3.1 Example: A Comparative Study

We tested the nearest-neighbors, K-means and LVQ classifiers on two simulated problems. There are 10 independent features X_j, each uniformly distributed on $[0, 1]$. The two-class 0-1 target variable is defined as follows:

$$Y = I\left(X_1 > \frac{1}{2}\right); \quad \text{problem 1: "easy",}$$

$$Y = I\left(\text{sign}\left\{\prod_{j=1}^{3}\left(X_j - \frac{1}{2}\right)\right\} > 0\right); \quad \text{problem 2: "difficult."} \qquad (13.6)$$

Hence in the first problem the two classes are separated by the hyperplane $X_1 = 1/2$; in the second problem, the two classes form a checkerboard pattern in the hypercube defined by the first three features. The Bayes error rate is zero in both problems. There were 100 training and 1000 test observations.

Figure 13.5 shows the mean and standard error of the misclassification error for nearest-neighbors, K-means and LVQ over ten realizations, as the tuning parameters are varied. We see that K-means and LVQ give nearly identical results. For the best choices of their tuning parameters, K-means and LVQ outperform nearest-neighbors for the first problem, and they perform similarly for the second problem. Notice that the best value of each tuning parameter is clearly situation dependent. For example 25-nearest-neighbors outperforms 1-nearest-neighbor by a factor of 70% in the

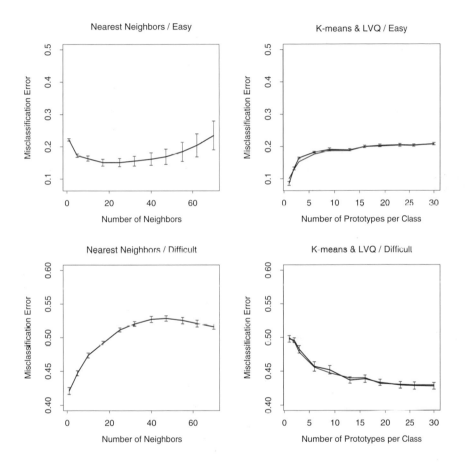

FIGURE 13.5. *Mean ± one standard error of misclassification error for nearest-neighbors, K-means (blue) and LVQ (red) over ten realizations for two simulated problems: "easy" and "difficult," described in the text.*

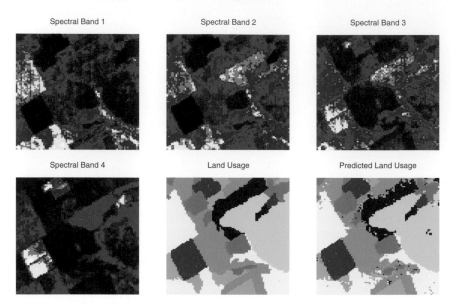

FIGURE 13.6. *The first four panels are LANDSAT images for an agricultural area in four spectral bands, depicted by heatmap shading. The remaining two panels give the actual land usage (color coded) and the predicted land usage using a five-nearest-neighbor rule described in the text.*

first problem, while 1-nearest-neighbor is best in the second problem by a factor of 18%. These results underline the importance of using an objective, data-based method like cross-validation to estimate the best value of a tuning parameter (see Figure 13.4 and Chapter 7).

13.3.2   Example: k-Nearest-Neighbors and Image Scene Classification

The STATLOG project (Michie et al., 1994) used part of a LANDSAT image as a benchmark for classification (82×100 pixels). Figure 13.6 shows four heat-map images, two in the visible spectrum and two in the infrared, for an area of agricultural land in Australia. Each pixel has a class label from the 7-element set $\mathcal{G} = \{$*red soil, cotton, vegetation stubble, mixture, gray soil, damp gray soil, very damp gray soil*$\}$, determined manually by research assistants surveying the area. The lower middle panel shows the actual land usage, shaded by different colors to indicate the classes. The objective is to classify the land usage at a pixel, based on the information in the four spectral bands.

Five-nearest-neighbors produced the predicted map shown in the bottom right panel, and was computed as follows. For each pixel we extracted an 8-neighbor feature map—the pixel itself and its 8 immediate neighbors

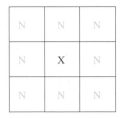

FIGURE 13.7. *A pixel and its 8-neighbor feature map.*

(see Figure 13.7). This is done separately in the four spectral bands, giving $(1+8) \times 4 = 36$ input features per pixel. Then five-nearest-neighbors classification was carried out in this 36-dimensional feature space. The resulting test error rate was about 9.5% (see Figure 13.8). Of all the methods used in the STATLOG project, including LVQ, CART, neural networks, linear discriminant analysis and many others, k-nearest-neighbors performed best on this task. Hence it is likely that the decision boundaries in $\mathbb{R}^{36}$ are quite irregular.

13.3.3 Invariant Metrics and Tangent Distance

In some problems, the training features are invariant under certain natural transformations. The nearest-neighbor classifier can exploit such invariances by incorporating them into the metric used to measure the distances between objects. Here we give an example where this idea was used with great success, and the resulting classifier outperformed all others at the time of its development (Simard et al., 1993).

The problem is handwritten digit recognition, as discussed is Chapter 1 and Section 11.7. The inputs are grayscale images with $16 \times 16 = 256$ pixels; some examples are shown in Figure 13.9. At the top of Figure 13.10, a "3" is shown, in its actual orientation (middle) and rotated 7.5° and 15° in either direction. Such rotations can often occur in real handwriting, and it is obvious to our eye that this "3" is still a "3" after small rotations. Hence we want our nearest-neighbor classifier to consider these two "3"s to be close together (similar). However the 256 grayscale pixel values for a rotated "3" will look quite different from those in the original image, and hence the two objects can be far apart in Euclidean distance in $\mathbb{R}^{256}$.

We wish to remove the effect of rotation in measuring distances between two digits of the same class. Consider the set of pixel values consisting of the original "3" and its rotated versions. This is a one-dimensional curve in $\mathbb{R}^{256}$, depicted by the green curve passing through the "3" in Figure 13.10. Figure 13.11 shows a stylized version of $\mathbb{R}^{256}$, with two images indicated by x_i and $x_{i'}$. These might be two different "3"s, for example. Through each image we have drawn the curve of rotated versions of that image, called

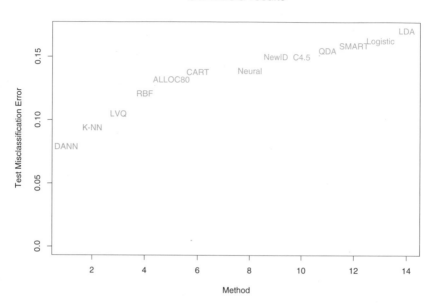

FIGURE 13.8. *Test-error performance for a number of classifiers, as reported by the STATLOG project. The entry DANN is a variant of k-nearest neighbors, using an adaptive metric (Section 13.4.2).*

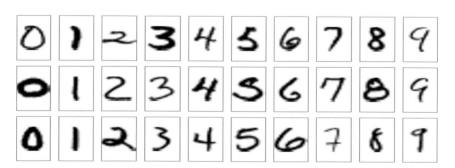

FIGURE 13.9. *Examples of grayscale images of handwritten digits.*

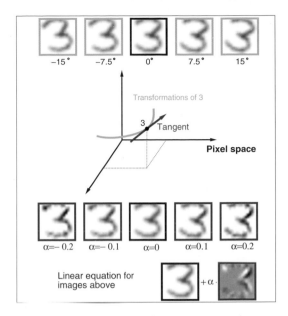

FIGURE 13.10. *The top row shows a "3" in its original orientation (middle) and rotated versions of it. The green curve in the middle of the figure depicts this set of rotated "3" in 256-dimensional space. The red line is the tangent line to the curve at the original image, with some "3"s on this tangent line, and its equation shown at the bottom of the figure.*

invariance manifolds in this context. Now, rather than using the usual Euclidean distance between the two images, we use the shortest distance between the two curves. In other words, the distance between the two images is taken to be the shortest Euclidean distance between any rotated version of first image, and any rotated version of the second image. This distance is called an *invariant metric*.

In principle one could carry out 1-nearest-neighbor classification using this invariant metric. However there are two problems with it. First, it is very difficult to calculate for real images. Second, it allows large transformations that can lead to poor performance. For example a "6" would be considered close to a "9" after a rotation of 180°. We need to restrict attention to small rotations.

The use of *tangent distance* solves both of these problems. As shown in Figure 13.10, we can approximate the invariance manifold of the image "3" by its tangent at the original image. This tangent can be computed by estimating the direction vector from small rotations of the image, or by more sophisticated spatial smoothing methods (Exercise 13.4.) For large rotations, the tangent image no longer looks like a "3," so the problem with large transformations is alleviated.

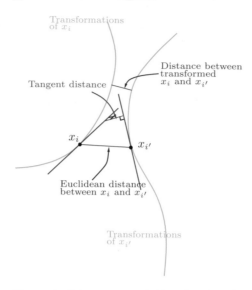

FIGURE 13.11. *Tangent distance computation for two images x_i and $x_{i'}$. Rather than using the Euclidean distance between x_i and $x_{i'}$, or the shortest distance between the two curves, we use the shortest distance between the two tangent lines.*

The idea then is to compute the invariant tangent line for each training image. For a query image to be classified, we compute its invariant tangent line, and find the closest line to it among the lines in the training set. The class (digit) corresponding to this closest line is our predicted class for the query image. In Figure 13.11 the two tangent lines intersect, but this is only because we have been forced to draw a two-dimensional representation of the actual 256-dimensional situation. In $\mathbb{R}^{256}$ the probability of two such lines intersecting is effectively zero.

Now a simpler way to achieve this invariance would be to add into the training set a number of rotated versions of each training image, and then just use a standard nearest-neighbor classifier. This idea is called "hints" in Abu-Mostafa (1995), and works well when the space of invariances is small. So far we have presented a simplified version of the problem. In addition to rotation, there are six other types of transformations under which we would like our classifier to be invariant. There are translation (two directions), scaling (two directions), sheer, and character thickness. Hence the curves and tangent lines in Figures 13.10 and 13.11 are actually 7-dimensional manifolds and hyperplanes. It is infeasible to add transformed versions of each training image to capture all of these possibilities. The tangent manifolds provide an elegant way of capturing the invariances.

Table 13.1 shows the test misclassification error for a problem with 7291 training images and 2007 test digits (the U.S. Postal Services database), for a carefully constructed neural network, and simple 1-nearest-neighbor and

TABLE 13.1. *Test error rates for the handwritten ZIP code problem.*

Method	Error rate
Neural-net	0.049
1-nearest-neighbor/Euclidean distance	0.055
1-nearest-neighbor/tangent distance	0.026

tangent distance 1-nearest-neighbor rules. The tangent distance nearest-neighbor classifier works remarkably well, with test error rates near those for the human eye (this is a notoriously difficult test set). In practice, it turned out that nearest-neighbors are too slow for online classification in this application (see Section 13.5), and neural network classifiers were subsequently developed to mimic it.

13.4   Adaptive Nearest-Neighbor Methods

When nearest-neighbor classification is carried out in a high-dimensional feature space, the nearest neighbors of a point can be very far away, causing bias and degrading the performance of the rule.

To quantify this, consider N data points uniformly distributed in the unit cube $[-\frac{1}{2}, \frac{1}{2}]^p$. Let R be the radius of a 1-nearest-neighborhood centered at the origin. Then

$$\text{median}(R) = v_p^{-1/p} \left(1 - \frac{1}{2}^{1/N} \right)^{1/p}, \tag{13.7}$$

where $v_p r^p$ is the volume of the sphere of radius r in p dimensions. Figure 13.12 shows the median radius for various training sample sizes and dimensions. We see that median radius quickly approaches 0.5, the distance to the edge of the cube.

What can be done about this problem? Consider the two-class situation in Figure 13.13. There are two features, and a nearest-neighborhood at a query point is depicted by the circular region. Implicit in near-neighbor classification is the assumption that the class probabilities are roughly constant in the neighborhood, and hence simple averages give good estimates. However, in this example the class probabilities vary only in the horizontal direction. If we knew this, we would stretch the neighborhood in the vertical direction, as shown by the tall rectangular region. This will reduce the bias of our estimate and leave the variance the same.

In general, this calls for adapting the metric used in nearest-neighbor classification, so that the resulting neighborhoods stretch out in directions for which the class probabilities don't change much. In high-dimensional feature space, the class probabilities might change only a low-dimensional subspace and hence there can be considerable advantage to adapting the metric.

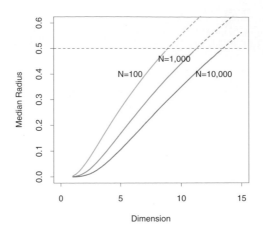

FIGURE 13.12. *Median radius of a 1-nearest-neighborhood, for uniform data with N observations in p dimensions.*

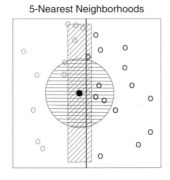

FIGURE 13.13. *The points are uniform in the cube, with the vertical line separating class red and green. The vertical strip denotes the 5-nearest-neighbor region using only the horizontal coordinate to find the nearest-neighbors for the target point (solid dot). The sphere shows the 5-nearest-neighbor region using both coordinates, and we see in this case it has extended into the class-red region (and is dominated by the wrong class in this instance).*

Friedman (1994a) proposed a method in which rectangular neighborhoods are found adaptively by successively carving away edges of a box containing the training data. Here we describe the *discriminant adaptive nearest-neighbor* (DANN) rule of Hastie and Tibshirani (1996a). Earlier, related proposals appear in Short and Fukunaga (1981) and Myles and Hand (1990).

At each query point a neighborhood of say 50 points is formed, and the class distribution among the points is used to decide how to deform the neighborhood—that is, to adapt the metric. The adapted metric is then used in a nearest-neighbor rule at the query point. Thus at each query point a potentially different metric is used.

In Figure 13.13 it is clear that the neighborhood should be stretched in the direction orthogonal to line joining the class centroids. This direction also coincides with the linear discriminant boundary, and is the direction in which the class probabilities change the least. In general this direction of maximum change will not be orthogonal to the line joining the class centroids (see Figure 4.9 on page 116.) Assuming a local discriminant model, the information contained in the local within- and between-class covariance matrices is all that is needed to determine the optimal shape of the neighborhood.

The *discriminant adaptive nearest-neighbor* (DANN) metric at a query point x_0 is defined by

$$D(x, x_0) = (x - x_0)^T \mathbf{\Sigma} (x - x_0), \tag{13.8}$$

where

$$\begin{aligned} \mathbf{\Sigma} &= \mathbf{W}^{-1/2}[\mathbf{W}^{-1/2}\mathbf{B}\mathbf{W}^{-1/2} + \epsilon \mathbf{I}]\mathbf{W}^{-1/2} \\ &= \mathbf{W}^{-1/2}[\mathbf{B}^* + \epsilon \mathbf{I}]\mathbf{W}^{-1/2}. \end{aligned} \tag{13.9}$$

Here $\mathbf{W}$ is the pooled within-class covariance matrix $\sum_{k=1}^{K} \pi_k \mathbf{W}_k$ and $\mathbf{B}$ is the between class covariance matrix $\sum_{k=1}^{K} \pi_k (\bar{x}_k - \bar{x})(\bar{x}_k - \bar{x})^T$, with $\mathbf{W}$ and $\mathbf{B}$ computed using only the 50 nearest neighbors around x_0. After computation of the metric, it is used in a nearest-neighbor rule at x_0.

This complicated formula is actually quite simple in its operation. It first spheres the data with respect to $\mathbf{W}$, and then stretches the neighborhood in the zero-eigenvalue directions of $\mathbf{B}^*$ (the between-matrix for the sphered data). This makes sense, since locally the observed class means do not differ in these directions. The ϵ parameter rounds the neighborhood, from an infinite strip to an ellipsoid, to avoid using points far away from the query point. The value of $\epsilon = 1$ seems to work well in general. Figure 13.14 shows the resulting neighborhoods for a problem where the classes form two concentric circles. Notice how the neighborhoods stretch out orthogonally to the decision boundaries when both classes are present in the neighborhood. In the pure regions with only one class, the neighborhoods remain circular;

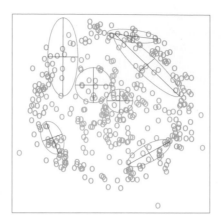

FIGURE 13.14. *Neighborhoods found by the DANN procedure, at various query points (centers of the crosses). There are two classes in the data, with one class surrounding the other. 50 nearest-neighbors were used to estimate the local metrics. Shown are the resulting metrics used to form 15-nearest-neighborhoods.*

in these cases the between matrix $\mathbf{B} = 0$, and the $\boldsymbol{\Sigma}$ in (13.8) is the identity matrix.

13.4.1   Example

Here we generate two-class data in ten dimensions, analogous to the two-dimensional example of Figure 13.14. All ten predictors in class 1 are independent standard normal, conditioned on the squared radius being greater than 22.4 and less than 40, while the predictors in class 2 are independent standard normal without the restriction. There are 250 observations in each class. Hence the first class almost completely surrounds the second class in the full ten-dimensional space.

In this example there are no pure noise variables, the kind that a nearest-neighbor subset selection rule might be able to weed out. At any given point in the feature space, the class discrimination occurs along only one direction. However, this direction changes as we move across the feature space and all variables are important somewhere in the space.

Figure 13.15 shows boxplots of the test error rates over ten realizations, for standard 5-nearest-neighbors, LVQ, and discriminant adaptive 5-nearest-neighbors. We used 50 prototypes per class for LVQ, to make it comparable to 5 nearest-neighbors (since $250/5 = 50$). The adaptive metric significantly reduces the error rate, compared to LVQ or standard nearest-neighbors.

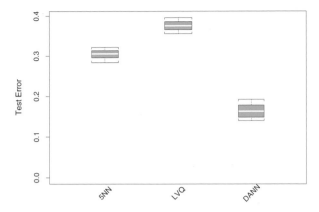

FIGURE 13.15. *Ten-dimensional simulated example: boxplots of the test error rates over ten realizations, for standard 5-nearest-neighbors, LVQ with 50 centers, and discriminant-adaptive 5-nearest-neighbors*

13.4.2 Global Dimension Reduction for Nearest-Neighbors

The discriminant-adaptive nearest-neighbor method carries out local dimension reduction—that is, dimension reduction separately at each query point. In many problems we can also benefit from global dimension reduction, that is, apply a nearest-neighbor rule in some optimally chosen subspace of the original feature space. For example, suppose that the two classes form two nested spheres in four dimensions of feature space, and there are an additional six noise features whose distribution is independent of class. Then we would like to discover the important four-dimensional subspace, and carry out nearest neighbor classification in that reduced subspace. Hastie and Tibshirani (1996a) discuss a variation of the discriminant-adaptive nearest-neighbor method for this purpose. At each training point x_i, the between-centroids sum of squares matrix $\mathbf{B}_i$ is computed, and then these matrices are averaged over all training points:

$$\bar{\mathbf{B}} = \frac{1}{N} \sum_{i=1}^{N} \mathbf{B}_i. \qquad (13.10)$$

Let $e_1, e_2, \ldots, e_p$ be the eigenvectors of the matrix $\bar{\mathbf{B}}$, ordered from largest to smallest eigenvalue θ_k. Then these eigenvectors span the optimal subspaces for global subspace reduction. The derivation is based on the fact that the best rank-L approximation to $\bar{\mathbf{B}}$, $\bar{\mathbf{B}}_{[L]} = \sum_{\ell=1}^{L} \theta_\ell e_\ell e_\ell^T$, solves the least squares problem

$$\min_{\text{rank}(\mathbf{M})=L} \sum_{i=1}^{N} \text{trace}[(\mathbf{B}_i - \mathbf{M})^2]. \qquad (13.11)$$

Since each $\mathbf{B}_i$ contains information on (a) the local discriminant subspace, and (b) the strength of discrimination in that subspace, (13.11) can be seen as a way of finding the best approximating subspace of dimension L to a series of N subspaces by weighted least squares (Exercise 13.5.)

In the four-dimensional sphere example mentioned above and examined in Hastie and Tibshirani (1996a), four of the eigenvalues θ_ℓ turn out to be large (having eigenvectors nearly spanning the interesting subspace), and the remaining six are near zero. Operationally, we project the data into the leading four-dimensional subspace, and then carry out nearest neighbor classification. In the satellite image classification example in Section 13.3.2, the technique labeled DANN in Figure 13.8 used 5-nearest-neighbors in a globally reduced subspace. There are also connections of this technique with the *sliced inverse regression* proposal of Duan and Li (1991). These authors use similar ideas in the regression setting, but do global rather than local computations. They assume and exploit spherical symmetry of the feature distribution to estimate interesting subspaces.

13.5   Computational Considerations

One drawback of nearest-neighbor rules in general is the computational load, both in finding the neighbors and storing the entire training set. With N observations and p predictors, nearest-neighbor classification requires Np operations to find the neighbors per query point. There are fast algorithms for finding nearest-neighbors (Friedman et al., 1975; Friedman et al., 1977) which can reduce this load somewhat. Hastie and Simard (1998) reduce the computations for tangent distance by developing analogs of K-means clustering in the context of this invariant metric.

Reducing the storage requirements is more difficult, and various *editing* and *condensing* procedures have been proposed. The idea is to isolate a subset of the training set that suffices for nearest-neighbor predictions, and throw away the remaining training data. Intuitively, it seems important to keep the training points that are near the decision boundaries and on the correct side of those boundaries, while some points far from the boundaries could be discarded.

The *multi-edit* algorithm of Devijver and Kittler (1982) divides the data cyclically into training and test sets, computing a nearest neighbor rule on the training set and deleting test points that are misclassified. The idea is to keep homogeneous clusters of training observations.

The *condensing* procedure of Hart (1968) goes further, trying to keep only important exterior points of these clusters. Starting with a single randomly chosen observation as the training set, each additional data item is processed one at a time, adding it to the training set only if it is misclassified by a nearest-neighbor rule computed on the current training set.

These procedures are surveyed in Dasarathy (1991) and Ripley (1996). They can also be applied to other learning procedures besides nearest-neighbors. While such methods are sometimes useful, we have not had much practical experience with them, nor have we found any systematic comparison of their performance in the literature.

Bibliographic Notes

The nearest-neighbor method goes back at least to Fix and Hodges (1951). The extensive literature on the topic is reviewed by Dasarathy (1991); Chapter 6 of Ripley (1996) contains a good summary. K-means clustering is due to Lloyd (1957) and MacQueen (1967). Kohonen (1989) introduced learning vector quantization. The tangent distance method is due to Simard et al. (1993). Hastie and Tibshirani (1996a) proposed the discriminant adaptive nearest-neighbor technique.

Exercises

Ex. 13.1 Consider a Gaussian mixture model where the covariance matrices are assumed to be scalar: $\boldsymbol{\Sigma}_r = \sigma\mathbf{I} \ \forall r = 1, \dots, R$, and σ is a fixed parameter. Discuss the analogy between the K-means clustering algorithm and the EM algorithm for fitting this mixture model in detail. Show that in the limit $\sigma \to 0$ the two methods coincide.

Ex. 13.2 Derive formula (13.7) for the median radius of the 1-nearest-neighborhood.

Ex. 13.3 Let E^* be the error rate of the Bayes rule in a K-class problem, where the true class probabilities are given by $p_k(x)$, $k = 1, \dots, K$. Assuming the test point and training point have identical features x, prove (13.5)

$$\sum_{k=1}^{K} p_k(x)(1 - p_k(x)) \le 2(1 - p_{k^*}(x)) - \frac{K}{K-1}(1 - p_{k^*}(x))^2.$$

where $k^* = \arg\max_k p_k(x)$. Hence argue that the error rate of the 1-nearest-neighbor rule converges in L_1, as the size of the training set increases, to a value E_1, bounded above by

$$E^*\left(2 - E^* \frac{K}{K-1}\right). \tag{13.12}$$

[This statement of the theorem of Cover and Hart (1967) is taken from Chapter 6 of Ripley (1996), where a short proof is also given].

Ex. 13.4 Consider an image to be a function $F(x) : \mathbb{R}^2 \mapsto \mathbb{R}^1$ over the two-dimensional spatial domain (paper coordinates). Then $F(c+x_0+\mathbf{A}(x-x_0))$ represents an affine transformation of the image F, where $\mathbf{A}$ is a 2×2 matrix.

1. Decompose $\mathbf{A}$ (via Q-R) in such a way that parameters identifying the four affine transformations (two scale, shear and rotation) are clearly identified.

2. Using the chain rule, show that the derivative of $F(c+x_0+\mathbf{A}(x-x_0))$ w.r.t. each of these parameters can be represented in terms of the two spatial derivatives of F.

3. Using a two-dimensional kernel smoother (Chapter 6), describe how to implement this procedure when the images are quantized to 16×16 pixels.

Ex. 13.5 Let $\mathbf{B}_i, i = 1, 2, \ldots, N$ be square $p \times p$ positive semi-definite matrices and let $\bar{\mathbf{B}} = (1/N)\sum \mathbf{B}_i$. Write the eigen-decomposition of $\bar{\mathbf{B}}$ as $\sum_{\ell=1}^{p} \theta_\ell e_\ell e_\ell^T$ with $\theta_\ell \geq \theta_{\ell-1} \geq \cdots \geq \theta_1$. Show that the best rank-L approximation for the $\mathbf{B}_i$,

$$\min_{\mathrm{rank}(\mathbf{M})=L} \sum_{i=1}^{N} \mathrm{trace}[(\mathbf{B}_i - \mathbf{M})^2],$$

is given by $\bar{\mathbf{B}}_{[L]} = \sum_{\ell=1}^{L} \theta_\ell e_\ell e_\ell^T$. (*Hint:* Write $\sum_{i=1}^{N} \mathrm{trace}[(\mathbf{B}_i - \mathbf{M})^2]$ as

$$\sum_{i=1}^{N} \mathrm{trace}[(\mathbf{B}_i - \bar{\mathbf{B}})^2] + \sum_{i=1}^{N} \mathrm{trace}[(\mathbf{M} - \bar{\mathbf{B}})^2]).$$

Ex. 13.6 Here we consider the problem of *shape averaging*. In particular, $\mathbf{L}_i, \; i = 1, \ldots, M$ are each $N \times 2$ matrices of points in $\mathbb{R}^2$, each sampled from corresponding positions of handwritten (cursive) letters. We seek an *affine invariant average* $\mathbf{V}$, also $N \times 2$, $\mathbf{V}^T\mathbf{V} = I$, of the M letters $\mathbf{L}_i$ with the following property: $\mathbf{V}$ minimizes

$$\sum_{j=1}^{M} \min_{\mathbf{A}_j} \|\mathbf{L}_j - \mathbf{V}\mathbf{A}_j\|^2.$$

Characterize the solution.

This solution can suffer if some of the letters are *big* and dominate the average. An alternative approach is to minimize instead:

$$\sum_{j=1}^{M} \min_{\mathbf{A}_j} \|\mathbf{L}_j \mathbf{A}_j^* - \mathbf{V}\|^2.$$

Derive the solution to this problem. How do the criteria differ? Use the SVD of the $\mathbf{L}_j$ to simplify the comparison of the two approaches.

Ex. 13.7 Consider the application of nearest-neighbors to the "easy" and "hard" problems in the left panel of Figure 13.5.

1. Replicate the results in the left panel of Figure 13.5.

2. Estimate the misclassification errors using fivefold cross-validation, and compare the error rate curves to those in 1.

3. Consider an "AIC-like" penalization of the training set misclassification error. Specifically, add $2t/N$ to the training set misclassification error, where t is the approximate number of parameters N/r, r being the number of nearest-neighbors. Compare plots of the resulting penalized misclassification error to those in 1 and 2. Which method gives a better estimate of the optimal number of nearest-neighbors: cross-validation or AIC?

Ex. 13.8 Generate data in two classes, with two features. These features are all independent Gaussian variates with standard deviation 1. Their mean vectors are $(-1, -1)$ in class 1 and $(1, 1)$ in class 2. To each feature vector apply a random rotation of angle θ, θ chosen uniformly from 0 to 2π. Generate 50 observations from each class to form the training set, and 500 in each class as the test set. Apply four different classifiers:

1. Nearest-neighbors.

2. Nearest-neighbors with hints: ten randomly rotated versions of each data point are added to the training set before applying nearest-neighbors.

3. Invariant metric nearest-neighbors, using Euclidean distance invariant to rotations about the origin.

4. Tangent distance nearest-neighbors.

In each case choose the number of neighbors by tenfold cross-validation. Compare the results.

14
Unsupervised Learning

14.1 Introduction

The previous chapters have been concerned with predicting the values of one or more outputs or response variables $Y = (Y_1, \ldots, Y_m)$ for a given set of input or predictor variables $X^T = (X_1, \ldots, X_p)$. Denote by $x_i^T = (x_{i1}, \ldots, x_{ip})$ the inputs for the ith training case, and let y_i be a response measurement. The predictions are based on the training sample $(x_1, y_1), \ldots, (x_N, y_N)$ of previously solved cases, where the joint values of all of the variables are known. This is called *supervised learning* or "learning with a teacher." Under this metaphor the "student" presents an answer $\hat{y}_i$ for each x_i in the training sample, and the supervisor or "teacher" provides either the correct answer and/or an error associated with the student's answer. This is usually characterized by some loss function $L(y, \hat{y})$, for example, $L(y, \hat{y}) = (y - \hat{y})^2$.

If one supposes that (X, Y) are random variables represented by some joint probability density $\Pr(X, Y)$, then supervised learning can be formally characterized as a density estimation problem where one is concerned with determining properties of the conditional density $\Pr(Y|X)$. Usually the properties of interest are the "location" parameters μ that minimize the expected error at each x,

$$\mu(x) = \operatorname*{argmin}_{\theta} E_{Y|X} L(Y, \theta). \tag{14.1}$$

T. Hastie et al., *The Elements of Statistical Learning, Second Edition*,
DOI: 10.1007/b94608_14,
© Springer Science+Business Media, LLC 2009

Conditioning one has

$$\Pr(X, Y) = \Pr(Y|X) \cdot \Pr(X),$$

where $\Pr(X)$ is the joint marginal density of the X values alone. In supervised learning $\Pr(X)$ is typically of no direct concern. One is interested mainly in the properties of the conditional density $\Pr(Y|X)$. Since Y is often of low dimension (usually one), and only its location $\mu(x)$ is of interest, the problem is greatly simplified. As discussed in the previous chapters, there are many approaches for successfully addressing supervised learning in a variety of contexts.

In this chapter we address *unsupervised learning* or "learning without a teacher." In this case one has a set of N observations $(x_1, x_2, \ldots, x_N)$ of a random p-vector X having joint density $\Pr(X)$. The goal is to directly infer the properties of this probability density without the help of a supervisor or teacher providing correct answers or degree-of-error for each observation. The dimension of X is sometimes much higher than in supervised learning, and the properties of interest are often more complicated than simple location estimates. These factors are somewhat mitigated by the fact that X represents all of the variables under consideration; one is not required to infer how the properties of $\Pr(X)$ change, conditioned on the changing values of another set of variables.

In low-dimensional problems (say $p \leq 3$), there are a variety of effective nonparametric methods for directly estimating the density $\Pr(X)$ itself at all X-values, and representing it graphically (Silverman, 1986, e.g.). Owing to the curse of dimensionality, these methods fail in high dimensions. One must settle for estimating rather crude global models, such as Gaussian mixtures or various simple descriptive statistics that characterize $\Pr(X)$.

Generally, these descriptive statistics attempt to characterize X-values, or collections of such values, where $\Pr(X)$ is relatively large. Principal components, multidimensional scaling, self-organizing maps, and principal curves, for example, attempt to identify low-dimensional manifolds within the X-space that represent high data density. This provides information about the associations among the variables and whether or not they can be considered as functions of a smaller set of "latent" variables. Cluster analysis attempts to find multiple convex regions of the X-space that contain modes of $\Pr(X)$. This can tell whether or not $\Pr(X)$ can be represented by a mixture of simpler densities representing distinct types or classes of observations. Mixture modeling has a similar goal. Association rules attempt to construct simple descriptions (conjunctive rules) that describe regions of high density in the special case of very high dimensional binary-valued data.

With supervised learning there is a clear measure of success, or lack thereof, that can be used to judge adequacy in particular situations and to compare the effectiveness of different methods over various situations.

Lack of success is directly measured by expected loss over the joint distribution $\Pr(X, Y)$. This can be estimated in a variety of ways including cross-validation. In the context of unsupervised learning, there is no such direct measure of success. It is difficult to ascertain the validity of inferences drawn from the output of most unsupervised learning algorithms. One must resort to heuristic arguments not only for motivating the algorithms, as is often the case in supervised learning as well, but also for judgments as to the quality of the results. This uncomfortable situation has led to heavy proliferation of proposed methods, since effectiveness is a matter of opinion and cannot be verified directly.

In this chapter we present those unsupervised learning techniques that are among the most commonly used in practice, and additionally, a few others that are favored by the authors.

14.2 Association Rules

Association rule analysis has emerged as a popular tool for mining commercial data bases. The goal is to find joint values of the variables $X = (X_1, X_2, \ldots, X_p)$ that appear most frequently in the data base. It is most often applied to binary-valued data $X_j \in \{0, 1\}$, where it is referred to as "market basket" analysis. In this context the observations are sales transactions, such as those occurring at the checkout counter of a store. The variables represent all of the items sold in the store. For observation i, each variable X_j is assigned one of two values; $x_{ij} = 1$ if the jth item is purchased as part of the transaction, whereas $x_{ij} = 0$ if it was not purchased. Those variables that frequently have joint values of one represent items that are frequently purchased together. This information can be quite useful for stocking shelves, cross-marketing in sales promotions, catalog design, and consumer segmentation based on buying patterns.

More generally, the basic goal of association rule analysis is to find a collection of prototype X-values $v_1, \ldots, v_L$ for the feature vector X, such that the probability density $\Pr(v_l)$ evaluated at each of those values is relatively large. In this general framework, the problem can be viewed as "mode finding" or "bump hunting." As formulated, this problem is impossibly difficult. A natural estimator for each $\Pr(v_l)$ is the fraction of observations for which $X = v_l$. For problems that involve more than a small number of variables, each of which can assume more than a small number of values, the number of observations for which $X = v_l$ will nearly always be too small for reliable estimation. In order to have a tractable problem, both the goals of the analysis and the generality of the data to which it is applied must be greatly simplified.

The first simplification modifies the goal. Instead of seeking *values* x where $\Pr(x)$ is large, one seeks *regions* of the X-space with high probability content relative to their size or support. Let $\mathcal{S}_j$ represent the set of all possible values of the jth variable (its *support*), and let $s_j \subseteq \mathcal{S}_j$ be a subset of these values. The modified goal can be stated as attempting to find subsets of variable values $s_1, \ldots, s_p$ such that the probability of each of the variables simultaneously assuming a value within its respective subset,

$$\text{Pr}\left[\bigcap_{j=1}^{p}(X_j \in s_j)\right], \tag{14.2}$$

is relatively large. The intersection of subsets $\cap_{j=1}^{p}(X_j \in s_j)$ is called a *conjunctive rule*. For quantitative variables the subsets s_j are contiguous intervals; for categorical variables the subsets are delineated explicitly. Note that if the subset s_j is in fact the entire set of values $s_j = \mathcal{S}_j$, as is often the case, the variable X_j is said *not* to appear in the rule (14.2).

14.2.1 Market Basket Analysis

General approaches to solving (14.2) are discussed in Section 14.2.5. These can be quite useful in many applications. However, they are not feasible for the very large ($p \approx 10^4$, $N \approx 10^8$) commercial data bases to which market basket analysis is often applied. Several further simplifications of (14.2) are required. First, only two types of subsets are considered; either s_j consists of a *single* value of X_j, $s_j = v_{0j}$, or it consists of the entire set of values that X_j can assume, $s_j = \mathcal{S}_j$. This simplifies the problem (14.2) to finding subsets of the integers $\mathcal{J} \subset \{1, \ldots, p\}$, and corresponding values v_{0j}, $j \in \mathcal{J}$, such that

$$\text{Pr}\left[\bigcap_{j \in \mathcal{J}}(X_j = v_{0j})\right] \tag{14.3}$$

is large. Figure 14.1 illustrates this assumption.

One can apply the technique of *dummy variables* to turn (14.3) into a problem involving only binary-valued variables. Here we assume that the support $\mathcal{S}_j$ is finite for each variable X_j. Specifically, a new set of variables $Z_1, \ldots, Z_K$ is created, one such variable for each of the values v_{lj} attainable by each of the original variables $X_1, \ldots, X_p$. The number of dummy variables K is

$$K = \sum_{j=1}^{p} |\mathcal{S}_j|,$$

where $|\mathcal{S}_j|$ is the number of distinct values attainable by X_j. Each dummy variable is assigned the value $Z_k = 1$ if the variable with which it is associated takes on the corresponding value to which Z_k is assigned, and $Z_k = 0$ otherwise. This transforms (14.3) to finding a subset of the integers $\mathcal{K} \subset \{1, \ldots, K\}$ such that

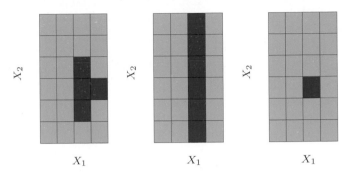

FIGURE 14.1. *Simplifications for association rules. Here there are two inputs* X_1 *and* X_2*, taking four and six distinct values, respectively. The red squares indicate areas of high density. To simplify the computations, we assume that the derived subset corresponds to either a single value of an input or all values. With this assumption we could find either the middle or right pattern, but not the left one.*

$$\Pr\left[\bigcap_{k\in\mathcal{K}}(Z_k=1)\right]=\Pr\left[\prod_{k\in\mathcal{K}}Z_k=1\right] \qquad (14.4)$$

is large. This is the standard formulation of the market basket problem. The set $\mathcal{K}$ is called an "item set." The number of variables Z_k in the item set is called its "size" (note that the size is no bigger than p). The estimated value of (14.4) is taken to be the fraction of observations in the data base for which the conjunction in (14.4) is true:

$$\widehat{\Pr}\left[\prod_{k\in\mathcal{K}}(Z_k=1)\right]=\frac{1}{N}\sum_{i=1}^{N}\prod_{k\in\mathcal{K}}z_{ik}. \qquad (14.5)$$

Here z_{ik} is the value of Z_k for this ith case. This is called the "support" or "prevalence" $T(\mathcal{K})$ of the item set $\mathcal{K}$. An observation i for which $\prod_{k\in\mathcal{K}}z_{ik}=1$ is said to "contain" the item set $\mathcal{K}$.

In association rule mining a lower support bound t is specified, and one seeks *all* item sets $\mathcal{K}_l$ that can be formed from the variables $Z_1,\ldots,Z_K$ with support in the data base greater than this lower bound t

$$\{\mathcal{K}_l|\,T(\mathcal{K}_l)>t\}. \qquad (14.6)$$

14.2.2 The Apriori Algorithm

The solution to this problem (14.6) can be obtained with feasible computation for very large data bases provided the threshold t is adjusted so that (14.6) consists of only a small fraction of all 2^K possible item sets. The "Apriori" algorithm (Agrawal et al., 1995) exploits several aspects of the curse of dimensionality to solve (14.6) with a small number of passes over the data. Specifically, for a given support threshold t:

- The cardinality $|\{\mathcal{K}|\,T(\mathcal{K}) > t\}|$ is relatively small.

- Any item set $\mathcal{L}$ consisting of a subset of the items in $\mathcal{K}$ must have support greater than or equal to that of $\mathcal{K}$, $\mathcal{L} \subseteq \mathcal{K} \Rightarrow T(\mathcal{L}) \geq T(\mathcal{K})$.

The first pass over the data computes the support of all single-item sets. Those whose support is less than the threshold are discarded. The second pass computes the support of all item sets of size two that can be formed from pairs of the single items surviving the first pass. In other words, to generate all frequent itemsets with $|\mathcal{K}| = m$, we need to consider only candidates such that *all* of their m ancestral item sets of size $m - 1$ are frequent. Those size-two item sets with support less than the threshold are discarded. Each successive pass over the data considers only those item sets that can be formed by combining those that survived the previous pass with those retained from the first pass. Passes over the data continue until all candidate rules from the previous pass have support less than the specified threshold. The Apriori algorithm requires only one pass over the data for each value of $|\mathcal{K}|$, which is crucial since we assume the data cannot be fitted into a computer's main memory. If the data are sufficiently sparse (or if the threshold t is high enough), then the process will terminate in reasonable time even for huge data sets.

There are many additional tricks that can be used as part of this strategy to increase speed and convergence (Agrawal et al., 1995). The Apriori algorithm represents one of the major advances in data mining technology.

Each high support item set $\mathcal{K}$ (14.6) returned by the Apriori algorithm is cast into a set of "association rules." The items Z_k, $k \in \mathcal{K}$, are partitioned into two disjoint subsets, $A \cup B = \mathcal{K}$, and written

$$A \Rightarrow B. \tag{14.7}$$

The first item subset A is called the "antecedent" and the second B the "consequent." Association rules are defined to have several properties based on the prevalence of the antecedent and consequent item sets in the data base. The "support" of the rule $T(A \Rightarrow B)$ is the fraction of observations in the union of the antecedent and consequent, which is just the support of the item set $\mathcal{K}$ from which they were derived. It can be viewed as an estimate (14.5) of the probability of simultaneously observing both item sets $\Pr(A \text{ and } B)$ in a randomly selected market basket. The "confidence" or "predictability" $C(A \Rightarrow B)$ of the rule is its support divided by the support of the antecedent

$$C(A \Rightarrow B) = \frac{T(A \Rightarrow B)}{T(A)}, \tag{14.8}$$

which can be viewed as an estimate of $\Pr(B\,|\,A)$. The notation $\Pr(A)$, the probability of an item set A occurring in a basket, is an abbreviation for

$\Pr(\prod_{k \in A} Z_k = 1)$. The "expected confidence" is defined as the support of the consequent $T(B)$, which is an estimate of the unconditional probability $\Pr(B)$. Finally, the "lift" of the rule is defined as the confidence divided by the expected confidence

$$L(A \Rightarrow B) = \frac{C(A \Rightarrow B)}{T(B)}.$$

This is an estimate of the association measure $\Pr(A \text{ and } B)/\Pr(A)\Pr(B)$.

As an example, suppose the item set $\mathcal{K} = \{\texttt{peanut butter, jelly, bread}\}$ and consider the rule $\{\texttt{peanut butter, jelly}\} \Rightarrow \{\texttt{bread}\}$. A support value of 0.03 for this rule means that $\texttt{peanut butter}$, $\texttt{jelly}$, and $\texttt{bread}$ appeared together in 3% of the market baskets. A confidence of 0.82 for this rule implies that when $\texttt{peanut butter}$ and $\texttt{jelly}$ were purchased, 82% of the time $\texttt{bread}$ was also purchased. If bread appeared in 43% of all market baskets then the rule $\{\texttt{peanut butter, jelly}\} \Rightarrow \{\texttt{bread}\}$ would have a lift of 1.95.

The goal of this analysis is to produce association rules (14.7) with both high values of support and confidence (14.8). The Apriori algorithm returns all item sets with high support as defined by the support threshold t (14.6). A confidence threshold c is set, and all rules that can be formed from those item sets (14.6) with confidence greater than this value

$$\{A \Rightarrow B \mid C(A \Rightarrow B) > c\} \tag{14.9}$$

are reported. For each item set $\mathcal{K}$ of size $|\mathcal{K}|$ there are $2^{|\mathcal{K}|-1} - 1$ rules of the form $A \Rightarrow (\mathcal{K} - A)$, $A \subset \mathcal{K}$. Agrawal et al. (1995) present a variant of the Apriori algorithm that can rapidly determine which rules survive the confidence threshold (14.9) from all possible rules that can be formed from the solution item sets (14.6).

The output of the entire analysis is a collection of association rules (14.7) that satisfy the constraints

$$T(A \Rightarrow B) > t \quad \text{and} \quad C(A \Rightarrow B) > c.$$

These are generally stored in a data base that can be queried by the user. Typical requests might be to display the rules in sorted order of confidence, lift or support. More specifically, one might request such a list conditioned on particular items in the antecedent or especially the consequent. For example, a request might be the following:

> *Display all transactions in which ice skates are the consequent that have confidence over* 80% *and support of more than* 2%.

This could provide information on those items (antecedent) that predicate sales of ice skates. Focusing on a particular consequent casts the problem into the framework of supervised learning.

Association rules have become a popular tool for analyzing very large commercial data bases in settings where market basket is relevant. That is when the data can be cast in the form of a multidimensional contingency table. The output is in the form of conjunctive rules (14.4) that are easily understood and interpreted. The Apriori algorithm allows this analysis to be applied to huge data bases, much larger that are amenable to other types of analyses. Association rules are among data mining's biggest successes.

Besides the restrictive form of the data to which they can be applied, association rules have other limitations. Critical to computational feasibility is the support threshold (14.6). The number of solution item sets, their size, and the number of passes required over the data can grow exponentially with decreasing size of this lower bound. Thus, rules with high confidence or lift, but low support, will not be discovered. For example, a high confidence rule such as `vodka` $\Rightarrow$ `caviar` will not be uncovered owing to the low sales volume of the consequent `caviar`.

14.2.3   Example: Market Basket Analysis

We illustrate the use of Apriori on a moderately sized demographics data base. This data set consists of $N = 9409$ questionnaires filled out by shopping mall customers in the San Francisco Bay Area (Impact Resources, Inc., Columbus OH, 1987). Here we use answers to the first 14 questions, relating to demographics, for illustration. These questions are listed in Table 14.1. The data are seen to consist of a mixture of ordinal and (unordered) categorical variables, many of the latter having more than a few values. There are many missing values.

We used a freeware implementation of the Apriori algorithm due to Christian Borgelt[1]. After removing observations with missing values, each ordinal predictor was cut at its median and coded by two dummy variables; each categorical predictor with k categories was coded by k dummy variables. This resulted in a 6876×50 matrix of 6876 observations on 50 dummy variables.

The algorithm found a total of 6288 association rules, involving ≤ 5 predictors, with support of at least 10%. Understanding this large set of rules is itself a challenging data analysis task. We will not attempt this here, but only illustrate in Figure 14.2 the relative frequency of each dummy variable in the data (top) and the association rules (bottom). Prevalent categories tend to appear more often in the rules, for example, the first category in language (English). However, others such as occupation are under-represented, with the exception of the first and fifth level.

Here are three examples of association rules found by the Apriori algorithm:

Association rule 1: Support 25%, confidence 99.7% and lift 1.03.

---

[1]See `http://fuzzy.cs.uni-magdeburg.de/~borgelt`.

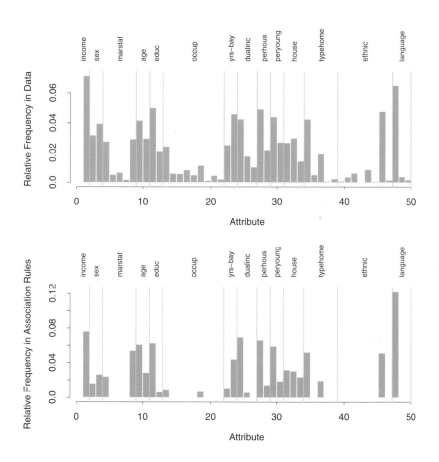

FIGURE 14.2. *Market basket analysis: relative frequency of each dummy variable (coding an input category) in the data (top), and the association rules found by the Apriori algorithm (bottom).*

TABLE 14.1. *Inputs for the demographic data.*

Feature	Demographic	# Values	Type
1	Sex	2	Categorical
2	Marital status	5	Categorical
3	Age	7	Ordinal
4	Education	6	Ordinal
5	Occupation	9	Categorical
6	Income	9	Ordinal
7	Years in Bay Area	5	Ordinal
8	Dual incomes	3	Categorical
9	Number in household	9	Ordinal
10	Number of children	9	Ordinal
11	Householder status	3	Categorical
12	Type of home	5	Categorical
13	Ethnic classification	8	Categorical
14	Language in home	3	Categorical

$$\begin{bmatrix} \text{number in household} & = & 1 \\ \text{number of children} & = & 0 \end{bmatrix}$$
$$\Downarrow$$
$$\text{language in home} = English$$

Association rule 2: Support 13.4%, confidence 80.8%, and lift 2.13.

$$\begin{bmatrix} \text{language in home} & = & English \\ \text{householder status} & = & own \\ \text{occupation} & = & \{professional/managerial\} \end{bmatrix}$$
$$\Downarrow$$
$$\text{income} \geq \$40,000$$

Association rule 3: Support 26.5%, confidence 82.8% and lift 2.15.

$$\begin{bmatrix} \text{language in home} & = & English \\ \text{income} & < & \$40,000 \\ \text{marital status} & = & not\ married \\ \text{number of children} & = & 0 \end{bmatrix}$$
$$\Downarrow$$
$$\text{education} \notin \{college\ graduate,\ graduate\ study\}$$

We chose the first and third rules based on their high support. The second rule is an association rule with a high-income consequent, and could be used to try to target high-income individuals.

As stated above, we created dummy variables for each category of the input predictors, for example, $Z_1 = I(\text{income} < \$40,000)$ and $Z_2 = I(\text{income} \geq \$40,000)$ for below and above the median income. If we were interested only in finding associations with the high-income category, we would include Z_2 but not Z_1. This is often the case in actual market basket problems, where we are interested in finding associations with the presence of a relatively rare item, but not associations with its absence.

14.2.4 Unsupervised as Supervised Learning

Here we discuss a technique for transforming the density estimation problem into one of supervised function approximation. This forms the basis for the generalized association rules described in the next section.

Let $g(x)$ be the unknown data probability density to be estimated, and $g_0(x)$ be a specified probability density function used for reference. For example, $g_0(x)$ might be the uniform density over the range of the variables. Other possibilities are discussed below. The data set $x_1, x_2, \ldots, x_N$ is presumed to be an i.i.d. random sample drawn from $g(x)$. A sample of size N_0 can be drawn from $g_0(x)$ using Monte Carlo methods. Pooling these two data sets, and assigning mass $w = N_0/(N + N_0)$ to those drawn from $g(x)$, and $w_0 = N/(N + N_0)$ to those drawn from $g_0(x)$, results in a random sample drawn from the mixture density $(g(x) + g_0(x))/2$. If one assigns the value $Y = 1$ to each sample point drawn from $g(x)$ and $Y = 0$ those drawn from $g_0(x)$, then

$$\mu(x) = E(Y \mid x) = \frac{g(x)}{g(x) + g_0(x)}$$

$$= \frac{g(x)/g_0(x)}{1 + g(x)/g_0(x)} \tag{14.10}$$

can be estimated by supervised learning using the combined sample

$$(y_1, x_1), (y_2, x_2), \ldots, (y_{N+N_0}, x_{N+N_0}) \tag{14.11}$$

as training data. The resulting estimate $\hat{\mu}(x)$ can be inverted to provide an estimate for $g(x)$

$$\hat{g}(x) = g_0(x) \frac{\hat{\mu}(x)}{1 - \hat{\mu}(x)}. \tag{14.12}$$

Generalized versions of logistic regression (Section 4.4) are especially well suited for this application since the log-odds,

$$f(x) = \log \frac{g(x)}{g_0(x)}, \tag{14.13}$$

are estimated directly. In this case one has

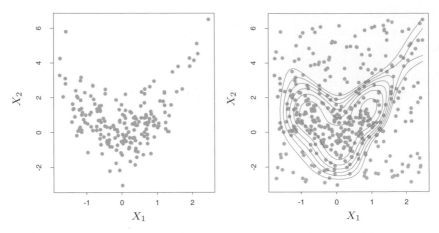

FIGURE 14.3. *Density estimation via classification. (Left panel:) Training set of* 200 *data points. (Right panel:) Training set plus* 200 *reference data points, generated uniformly over the rectangle containing the training data. The training sample was labeled as class 1, and the reference sample class 0, and a semiparametric logistic regression model was fit to the data. Some contours for $\hat{g}(x)$ are shown.*

$$\hat{g}(x) = g_0(x)\, e^{\hat{f}(x)}. \tag{14.14}$$

An example is shown in Figure 14.3. We generated a training set of size 200 shown in the left panel. The right panel shows the reference data (blue) generated uniformly over the rectangle containing the training data. The training sample was labeled as class 1, and the reference sample class 0, and a logistic regression model, using a tensor product of natural splines (Section 5.2.1), was fit to the data. Some probability contours of $\hat{\mu}(x)$ are shown in the right panel; these are also the contours of the density estimate $\hat{g}(x)$, since $\hat{g}(x) = \hat{\mu}(x)/(1 - \hat{\mu}(x))$, is a monotone function. The contours roughly capture the data density.

In principle any reference density can be used for $g_0(x)$ in (14.14). In practice the accuracy of the estimate $\hat{g}(x)$ can depend greatly on particular choices. Good choices will depend on the data density $g(x)$ and the procedure used to estimate (14.10) or (14.13). If accuracy is the goal, $g_0(x)$ should be chosen so that the resulting functions $\mu(x)$ or $f(x)$ are approximated easily by the method being used. However, accuracy is not always the primary goal. Both $\mu(x)$ and $f(x)$ are monotonic functions of the density ratio $g(x)/g_0(x)$. They can thus be viewed as "contrast" statistics that provide information concerning departures of the data density $g(x)$ from the chosen reference density $g_0(x)$. Therefore, in data analytic settings, a choice for $g_0(x)$ is dictated by types of departures that are deemed most interesting in the context of the specific problem at hand. For example, if departures from uniformity are of interest, $g_0(x)$ might be the a uniform density over the range of the variables. If departures from joint normality

are of interest, a good choice for $g_0(x)$ would be a Gaussian distribution
with the same mean vector and covariance matrix as the data. Departures
from independence could be investigated by using

$$g_0(x) = \prod_{j=1}^{p} g_j(x_j), \tag{14.15}$$

where $g_j(x_j)$ is the marginal data density of X_j, the jth coordinate of X.
A sample from this independent density (14.15) is easily generated from the
data itself by applying a different random permutation to the data values
of each of the variables.

As discussed above, unsupervised learning is concerned with revealing
properties of the data density $g(x)$. Each technique focuses on a particular property or set of properties. Although this approach of transforming
the problem to one of supervised learning (14.10)–(14.14) seems to have
been part of the statistics folklore for some time, it does not appear to
have had much impact despite its potential to bring well-developed supervised learning methodology to bear on unsupervised learning problems.
One reason may be that the problem must be enlarged with a simulated
data set generated by Monte Carlo techniques. Since the size of this data
set should be at least as large as the data sample $N_0 \geq N$, the computation and memory requirements of the estimation procedure are at least
doubled. Also, substantial computation may be required to generate the
Monte Carlo sample itself. Although perhaps a deterrent in the past, these
increased computational requirements are becoming much less of a burden
as increased resources become routinely available. We illustrate the use of
supervised learning methods for unsupervised learning in the next section.

14.2.5 *Generalized Association Rules*

The more general problem (14.2) of finding high-density regions in the data
space can be addressed using the supervised learning approach described
above. Although not applicable to the huge data bases for which market
basket analysis is feasible, useful information can be obtained from moderately sized data sets. The problem (14.2) can be formulated as finding
subsets of the integers $\mathcal{J} \subset \{1, 2, \ldots, p\}$ and corresponding value subsets
s_j, $j \in \mathcal{J}$ for the corresponding variables X_j, such that

$$\widehat{\Pr}\left(\bigcap_{j \in \mathcal{J}} (X_j \in s_j)\right) = \frac{1}{N} \sum_{i=1}^{N} I\left(\bigcap_{j \in \mathcal{J}} (x_{ij} \in s_j)\right) \tag{14.16}$$

is large. Following the nomenclature of association rule analysis, $\{(X_j \in s_j)\}_{j \in \mathcal{J}}$ will be called a "generalized" item set. The subsets s_j corresponding to quantitative variables are taken to be contiguous intervals within

their range of values, and subsets for categorical variables can involve more than a single value. The ambitious nature of this formulation precludes a thorough search for all generalized item sets with support (14.16) greater than a specified minimum threshold, as was possible in the more restrictive setting of market basket analysis. Heuristic search methods must be employed, and the most one can hope for is to find a useful collection of such generalized item sets.

Both market basket analysis (14.5) and the generalized formulation (14.16) implicitly reference the uniform probability distribution. One seeks item sets that are more frequent than would be expected if all joint data values $(x_1, x_2, \ldots, x_N)$ were uniformly distributed. This favors the discovery of item sets whose marginal constituents $(X_j \in s_j)$ are *individually* frequent, that is, the quantity

$$\frac{1}{N} \sum_{i=1}^{N} I(x_{ij} \in s_j) \tag{14.17}$$

is large. Conjunctions of frequent subsets (14.17) will tend to appear more often among item sets of high support (14.16) than conjunctions of marginally less frequent subsets. This is why the rule vodka $\Rightarrow$ caviar is not likely to be discovered in spite of a high association (lift); neither item has high marginal support, so that their joint support is especially small. Reference to the uniform distribution can cause highly frequent item sets with low associations among their constituents to dominate the collection of highest support item sets.

Highly frequent subsets s_j are formed as disjunctions of the most frequent X_j-values. Using the product of the variable marginal data densities (14.15) as a reference distribution removes the preference for highly frequent values of the individual variables in the discovered item sets. This is because the density ratio $g(x)/g_0(x)$ is uniform if there are no associations among the variables (complete independence), regardless of the frequency distribution of the individual variable values. Rules like vodka $\Rightarrow$ caviar would have a chance to emerge. It is not clear however, how to incorporate reference distributions other than the uniform into the Apriori algorithm. As explained in Section 14.2.4, it is straightforward to generate a sample from the product density (14.15), given the original data set.

After choosing a reference distribution, and drawing a sample from it as in (14.11), one has a supervised learning problem with a binary-valued output variable $Y \in \{0, 1\}$. The goal is to use this training data to find regions

$$R = \bigcap_{j \in \mathcal{J}} (X_j \in s_j) \tag{14.18}$$

for which the target function $\mu(x) = E(Y \mid x)$ is relatively large. In addition, one might wish to require that the *data* support of these regions

$$T(R) = \int_{x \in R} g(x)\, dx \qquad (14.19)$$

not be too small.

14.2.6 Choice of Supervised Learning Method

The regions (14.18) are defined by conjunctive rules. Hence supervised methods that learn such rules would be most appropriate in this context. The terminal nodes of a CART decision tree are defined by rules precisely of the form (14.18). Applying CART to the pooled data (14.11) will produce a decision tree that attempts to model the target (14.10) over the entire data space by a disjoint set of regions (terminal nodes). Each region is defined by a rule of the form (14.18). Those terminal nodes t with high average y-values

$$\bar{y}_t = \mathrm{ave}(y_i \mid x_i \in t)$$

are candidates for high-support generalized item sets (14.16). The actual (data) support is given by

$$T(R) = \bar{y}_t \cdot \frac{N_t}{N + N_0},$$

where N_t is the number of (pooled) observations within the region represented by the terminal node. By examining the resulting decision tree, one might discover interesting generalized item sets of relatively high-support. These can then be partitioned into antecedents and consequents in a search for generalized association rules of high confidence and/or lift.

Another natural learning method for this purpose is the patient rule induction method PRIM described in Section 9.3. PRIM also produces rules precisely of the form (14.18), but it is especially designed for finding high-support regions that maximize the average target (14.10) value within them, rather than trying to model the target function over the entire data space. It also provides more control over the support/average-target-value tradeoff.

Exercise 14.3 addresses an issue that arises with either of these methods when we generate random data from the product of the marginal distributions.

14.2.7 Example: Market Basket Analysis (Continued)

We illustrate the use of PRIM on the demographics data of Table 14.1.

Three of the high-support generalized item sets emerging from the PRIM analysis were the following:

Item set 1: Support= 24%.

$$
\begin{bmatrix}
\text{marital status} & = & \textit{married} \\
\text{householder status} & = & \textit{own} \\
\text{type of home} & \neq & \textit{apartment}
\end{bmatrix}
$$

Item set 2: Support= 24%.

$$
\begin{bmatrix}
\text{age} & \leq & 24 \\
\text{marital status} & \in & \{\textit{living together-not married, single}\} \\
\text{occupation} & \notin & \{\textit{professional, homemaker, retired}\} \\
\text{householder status} & \in & \{\textit{rent, live with family}\}
\end{bmatrix}
$$

Item set 3: Support= 15%.

$$
\begin{bmatrix}
\text{householder status} & = & \textit{rent} \\
\text{type of home} & \neq & \textit{house} \\
\text{number in household} & \leq & 2 \\
\text{number of children} & = & 0 \\
\text{occupation} & \notin & \{\textit{homemaker, student, unemployed}\} \\
\text{income} & \in & [\$20{,}000, \$150{,}000]
\end{bmatrix}
$$

Generalized association rules derived from these item sets with confidence (14.8) greater than 95% are the following:

Association rule 1: Support 25%, confidence 99.7% and lift 1.35.

$$
\begin{bmatrix}
\text{marital status} & = & \textit{married} \\
\text{householder status} & = & \textit{own}
\end{bmatrix}
$$
$$\Downarrow$$
$$\text{type of home} \neq \textit{apartment}$$

Association rule 2: Support 25%, confidence 98.7% and lift 1.97.

$$
\begin{bmatrix}
\text{age} & \leq & 24 \\
\text{occupation} & \notin & \{\textit{professional, homemaker, retired}\} \\
\text{householder status} & \in & \{\textit{rent, live with family}\}
\end{bmatrix}
$$
$$\Downarrow$$
$$\text{marital status} \in \{\textit{single, living together-not married}\}$$

Association rule 3: Support 25%, confidence 95.9% and lift 2.61.

$$
\begin{bmatrix}
\text{householder status} & = & \textit{own} \\
\text{type of home} & \neq & \textit{apartment}
\end{bmatrix}
$$
$$\Downarrow$$
$$\text{marital status} = \textit{married}$$

Association rule 4: Support 15%, confidence 95.4% and lift 1.50.

$$
\left[
\begin{array}{rcl}
\text{householder status} & = & rent \\
\text{type of home} & \neq & house \\
\text{number in household} & \leq & 2 \\
\text{occupation} & \notin & \{homemaker,\ student,\ unemployed\} \\
\text{income} & \in & [\$20{,}000, \$150{,}000]
\end{array}
\right]
$$

$$\Downarrow$$

$$\text{number of children} = 0$$

There are no great surprises among these particular rules. For the most part they verify intuition. In other contexts where there is less prior information available, unexpected results have a greater chance to emerge. These results do illustrate the type of information generalized association rules can provide, and that the supervised learning approach, coupled with a ruled induction method such as CART or PRIM, can uncover item sets exhibiting high associations among their constituents.

How do these generalized association rules compare to those found earlier by the Apriori algorithm? Since the Apriori procedure gives thousands of rules, it is difficult to compare them. However some general points can be made. The Apriori algorithm is exhaustive—it finds *all* rules with support greater than a specified amount. In contrast, PRIM is a greedy algorithm and is not guaranteed to give an "optimal" set of rules. On the other hand, the Apriori algorithm can deal only with dummy variables and hence could not find some of the above rules. For example, since `type of home` is a categorical input, with a dummy variable for each level, Apriori could not find a rule involving the set

$$\text{type of home} \neq apartment.$$

To find this set, we would have to code a dummy variable for *apartment* versus the other categories of type of home. It will not generally be feasible to precode all such potentially interesting comparisons.

14.3 Cluster Analysis

Cluster analysis, also called data segmentation, has a variety of goals. All relate to grouping or segmenting a collection of objects into subsets or "clusters," such that those within each cluster are more closely related to one another than objects assigned to different clusters. An object can be described by a set of measurements, or by its relation to other objects. In addition, the goal is sometimes to arrange the clusters into a natural hierarchy. This involves successively grouping the clusters themselves so

X_2

X_1

FIGURE 14.4. *Simulated data in the plane, clustered into three classes (repre-*
sented by orange, blue and green) by the K-means clustering algorithm

that at each level of the hierarchy, clusters within the same group are more
similar to each other than those in different groups.

Cluster analysis is also used to form descriptive statistics to ascertain
whether or not the data consists of a set distinct subgroups, each group
representing objects with substantially different properties. This latter goal
requires an assessment of the degree of difference between the objects assigned to the respective clusters.

Central to all of the goals of cluster analysis is the notion of the degree of
similarity (or dissimilarity) between the individual objects being clustered.
A clustering method attempts to group the objects based on the definition
of similarity supplied to it. This can only come from subject matter considerations. The situation is somewhat similar to the specification of a loss or
cost function in prediction problems (supervised learning). There the cost
associated with an inaccurate prediction depends on considerations outside
the data.

Figure 14.4 shows some simulated data clustered into three groups via
the popular K-means algorithm. In this case two of the clusters are not
well separated, so that "segmentation" more accurately describes the part
of this process than "clustering." K-means clustering starts with guesses
for the three cluster centers. Then it alternates the following steps until
convergence:

- for each data point, the closest cluster center (in Euclidean distance)
  is identified;

- each cluster center is replaced by the coordinate-wise average of all data points that are closest to it.

We describe K-means clustering in more detail later, including the problem of how to choose the number of clusters (three in this example). K-means clustering is a *top-down* procedure, while other cluster approaches that we discuss are *bottom-up*. Fundamental to all clustering techniques is the choice of distance or dissimilarity measure between two objects. We first discuss distance measures before describing a variety of algorithms for clustering.

14.3.1   *Proximity Matrices*

Sometimes the data is represented directly in terms of the proximity (alikeness or affinity) between pairs of objects. These can be either *similarities* or *dissimilarities* (difference or lack of affinity). For example, in social science experiments, participants are asked to judge by how much certain objects differ from one another. Dissimilarities can then be computed by averaging over the collection of such judgments. This type of data can be represented by an $N \times N$ matrix $\mathbf{D}$, where N is the number of objects, and each element $d_{ii'}$ records the proximity between the ith and i'th objects. This matrix is then provided as input to the clustering algorithm.

Most algorithms presume a matrix of dissimilarities with nonnegative entries and zero diagonal elements: $d_{ii} = 0$, $i = 1, 2, \ldots, N$. If the original data were collected as similarities, a suitable monotone-decreasing function can be used to convert them to dissimilarities. Also, most algorithms assume symmetric dissimilarity matrices, so if the original matrix $\mathbf{D}$ is not symmetric it must be replaced by $(\mathbf{D} + \mathbf{D}^T)/2$. Subjectively judged dissimilarities are seldom *distances* in the strict sense, since the triangle inequality $d_{ii'} \leq d_{ik} + d_{i'k}$, for all $k \in \{1, \ldots, N\}$ does not hold. Thus, some algorithms that assume distances cannot be used with such data.

14.3.2   *Dissimilarities Based on Attributes*

Most often we have measurements x_{ij} for $i = 1, 2, \ldots, N$, on variables $j = 1, 2, \ldots, p$ (also called *attributes*). Since most of the popular clustering algorithms take a dissimilarity matrix as their input, we must first construct pairwise dissimilarities between the observations. In the most common case, we define a dissimilarity $d_j(x_{ij}, x_{i'j})$ between values of the jth attribute, and then define

$$D(x_i, x_{i'}) = \sum_{j=1}^{p} d_j(x_{ij}, x_{i'j}) \tag{14.20}$$

as the dissimilarity between objects i and i'. By far the most common choice is squared distance

$$d_j(x_{ij}, x_{i'j}) = (x_{ij} - x_{i'j})^2. \tag{14.21}$$

However, other choices are possible, and can lead to potentially different results. For nonquantitative attributes (e.g., categorical data), squared distance may not be appropriate. In addition, it is sometimes desirable to weigh attributes differently rather than giving them equal weight as in (14.20).

We first discuss alternatives in terms of the attribute type:

Quantitative variables. Measurements of this type of variable or attribute are represented by continuous real-valued numbers. It is natural to define the "error" between them as a monotone-increasing function of their absolute difference

$$d(x_i, x_{i'}) = l(|x_i - x_{i'}|).$$

Besides squared-error loss $(x_i - x_{i'})^2$, a common choice is the identity (absolute error). The former places more emphasis on larger differences than smaller ones. Alternatively, clustering can be based on the correlation

$$\rho(x_i, x_{i'}) = \frac{\sum_j (x_{ij} - \bar{x}_i)(x_{i'j} - \bar{x}_{i'})}{\sqrt{\sum_j (x_{ij} - \bar{x}_i)^2 \sum_j (x_{i'j} - \bar{x}_{i'})^2}}, \tag{14.22}$$

with $\bar{x}_i = \sum_j x_{ij}/p$. Note that this is averaged over *variables*, not observations. If the *observations* are first standardized, then $\sum_j (x_{ij} - x_{i'j})^2 \propto 2(1 - \rho(x_i, x_{i'}))$. Hence clustering based on correlation (similarity) is equivalent to that based on squared distance (dissimilarity).

Ordinal variables. The values of this type of variable are often represented as contiguous integers, and the realizable values are considered to be an ordered set. Examples are academic grades (A, B, C, D, F), degree of preference (can't stand, dislike, OK, like, terrific). Rank data are a special kind of ordinal data. Error measures for ordinal variables are generally defined by replacing their M original values with

$$\frac{i - 1/2}{M}, \; i = 1, \ldots, M \tag{14.23}$$

in the prescribed order of their original values. They are then treated as quantitative variables on this scale.

Categorical variables. With unordered categorical (also called nominal) variables, the degree-of-difference between pairs of values must be delineated explicitly. If the variable assumes M distinct values, these can be arranged in a symmetric $M \times M$ matrix with elements $L_{rr'} = L_{r'r}, L_{rr} = 0, L_{rr'} \geq 0$. The most common choice is $L_{rr'} = 1$ for all $r \neq r'$, while unequal losses can be used to emphasize some errors more than others.

14.3.3   Object Dissimilarity

Next we define a procedure for combining the p-individual attribute dissim-ilarities $d_j(x_{ij}, x_{i'j})$, $j = 1, 2, \ldots, p$ into a single overall measure of dissim-ilarity $D(x_i, x_{i'})$ between two objects or observations $(x_i, x_{i'})$ possessing the respective attribute values. This is nearly always done by means of a weighted average (convex combination)

$$D(x_i, x_{i'}) = \sum_{j=1}^{p} w_j \cdot d_j(x_{ij}, x_{i'j}); \quad \sum_{j=1}^{p} w_j = 1. \tag{14.24}$$

Here w_j is a weight assigned to the jth attribute regulating the relative influence of that variable in determining the overall dissimilarity between objects. This choice should be based on subject matter considerations.

It is important to realize that setting the weight w_j to the same value for each variable (say, $w_j = 1 \ \forall \ j$) does *not* necessarily give all attributes equal influence. The influence of the jth attribute X_j on object dissimilarity $D(x_i, x_{i'})$ (14.24) depends upon its relative contribution to the average object dissimilarity measure over all pairs of observations in the data set

$$\bar{D} = \frac{1}{N^2} \sum_{i=1}^{N} \sum_{i'=1}^{N} D(x_i, x_{i'}) = \sum_{j=1}^{p} w_j \cdot \bar{d}_j,$$

with

$$\bar{d}_j = \frac{1}{N^2} \sum_{i=1}^{N} \sum_{i'=1}^{N} d_j(x_{ij}, x_{i'j}) \tag{14.25}$$

being the average dissimilarity on the jth attribute. Thus, the relative in-fluence of the jth variable is $w_j \cdot \bar{d}_j$, and setting $w_j \sim 1/\bar{d}_j$ would give all attributes equal influence in characterizing overall dissimilarity between ob-jects. For example, with p quantitative variables and squared-error distance used for each coordinate, then (14.24) becomes the (weighted) squared Eu-clidean distance

$$D_I(x_i, x_{i'}) = \sum_{j=1}^{p} w_j \cdot (x_{ij} - x_{i'j})^2 \tag{14.26}$$

between pairs of points in an $\mathbb{R}^p$, with the quantitative variables as axes. In this case (14.25) becomes

$$\bar{d}_j = \frac{1}{N^2} \sum_{i=1}^{N} \sum_{i'=1}^{N} (x_{ij} - x_{i'j})^2 = 2 \cdot \mathrm{var}_j, \tag{14.27}$$

where var_j is the sample estimate of $\mathrm{Var}(X_j)$. Thus, the relative impor-tance of each such variable is proportional to its variance over the data

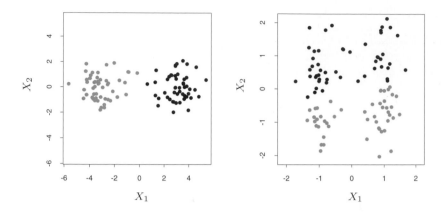

FIGURE 14.5. *Simulated data: on the left, K-means clustering (with K=2) has been applied to the raw data. The two colors indicate the cluster memberships. On the right, the features were first standardized before clustering. This is equivalent to using feature weights $1/[2 \cdot \mathrm{var}(X_j)]$. The standardization has obscured the two well-separated groups. Note that each plot uses the same units in the horizontal and vertical axes.*

set. In general, setting $w_j = 1/\bar{d}_j$ for all attributes, irrespective of type, will cause each one of them to equally influence the overall dissimilarity between pairs of objects $(x_i, x_{i'})$. Although this may seem reasonable, and is often recommended, it can be highly counterproductive. If the goal is to segment the data into groups of similar objects, all attributes may not contribute equally to the (problem-dependent) notion of dissimilarity between objects. Some attribute value differences may reflect greater actual object dissimilarity in the context of the problem domain.

If the goal is to discover natural groupings in the data, some attributes may exhibit more of a grouping tendency than others. Variables that are more relevant in separating the groups should be assigned a higher influence in defining object dissimilarity. Giving all attributes equal influence in this case will tend to obscure the groups to the point where a clustering algorithm cannot uncover them. Figure 14.5 shows an example.

Although simple generic prescriptions for choosing the individual attribute dissimilarities $d_j(x_{ij}, x_{i'j})$ and their weights w_j can be comforting, there is no substitute for careful thought in the context of each individual problem. Specifying an appropriate dissimilarity measure is far more important in obtaining success with clustering than choice of clustering algorithm. This aspect of the problem is emphasized less in the clustering literature than the algorithms themselves, since it depends on domain knowledge specifics and is less amenable to general research.

Finally, often observations have *missing values* in one or more of the attributes. The most common method of incorporating missing values in dissimilarity calculations (14.24) is to omit each observation pair $x_{ij}, x_{i'j}$ having at least one value missing, when computing the dissimilarity between observations x_i and x'_i. This method can fail in the circumstance when both observations have no measured values in common. In this case both observations could be deleted from the analysis. Alternatively, the missing values could be imputed using the mean or median of each attribute over the nonmissing data. For categorical variables, one could consider the value "missing" as just another categorical value, if it were reasonable to consider two objects as being similar if they both have missing values on the same variables.

14.3.4 *Clustering Algorithms*

The goal of cluster analysis is to partition the observations into groups ("clusters") so that the pairwise dissimilarities between those assigned to the same cluster tend to be smaller than those in different clusters. Clustering algorithms fall into three distinct types: combinatorial algorithms, mixture modeling, and mode seeking.

Combinatorial algorithms work directly on the observed data with no direct reference to an underlying probability model. *Mixture modeling* supposes that the data is an *i.i.d* sample from some population described by a probability density function. This density function is characterized by a parameterized model taken to be a mixture of component density functions; each component density describes one of the clusters. This model is then fit to the data by maximum likelihood or corresponding Bayesian approaches. *Mode seekers* ("bump hunters") take a nonparametric perspective, attempting to directly estimate distinct modes of the probability density function. Observations "closest" to each respective mode then define the individual clusters.

Mixture modeling is described in Section 6.8. The PRIM algorithm, discussed in Sections 9.3 and 14.2.5, is an example of mode seeking or "bump hunting." We discuss combinatorial algorithms next.

14.3.5 *Combinatorial Algorithms*

The most popular clustering algorithms directly assign each observation to a group or cluster without regard to a probability model describing the data. Each observation is uniquely labeled by an integer $i \in \{1, \cdots, N\}$. A prespecified number of clusters $K < N$ is postulated, and each one is labeled by an integer $k \in \{1, \ldots, K\}$. Each observation is assigned to one and only one cluster. These assignments can be characterized by a many-to-one mapping, or *encoder* $k = C(i)$, that assigns the ith observation to the kth cluster. One seeks the particular encoder $C^*(i)$ that achieves the required goal (details below), based on the dissimilarities $d(x_i, x_{i'})$ between every pair of observations. These are specified by the user as described above. Generally, the encoder $C(i)$ is explicitly delineated by giving its value (cluster assignment) for each observation i. Thus, the "parameters" of the procedure are the individual cluster assignments for each of the N observations. These are adjusted so as to *minimize* a "loss" function that characterizes the degree to which the clustering goal is *not* met.

One approach is to directly specify a mathematical loss function and attempt to minimize it through some combinatorial optimization algorithm. Since the goal is to assign close points to the same cluster, a natural loss (or "energy") function would be

$$W(C) = \frac{1}{2} \sum_{k=1}^{K} \sum_{C(i)=k} \sum_{C(i')=k} d(x_i, x_{i'}). \tag{14.28}$$

This criterion characterizes the extent to which observations assigned to the same cluster tend to be close to one another. It is sometimes referred to as the "within cluster" point scatter since

$$T = \frac{1}{2} \sum_{i=1}^{N} \sum_{i'=1}^{N} d_{ii'} = \frac{1}{2} \sum_{k=1}^{K} \sum_{C(i)=k} \left(\sum_{C(i')=k} d_{ii'} + \sum_{C(i')\neq k} d_{ii'} \right),$$

or

$$T = W(C) + B(C),$$

where $d_{ii'} = d(x_i, x_{i'})$. Here T is the *total* point scatter, which is a constant given the data, independent of cluster assignment. The quantity

$$B(C) = \frac{1}{2} \sum_{k=1}^{K} \sum_{C(i)=k} \sum_{C(i')\neq k} d_{ii'} \tag{14.29}$$

is the *between-cluster* point scatter. This will tend to be large when observations assigned to different clusters are far apart. Thus one has

$$W(C) = T - B(C)$$

and minimizing $W(C)$ is equivalent to *maximizing* $B(C)$.

Cluster analysis by combinatorial optimization is straightforward in principle. One simply minimizes W or equivalently maximizes B over all possible assignments of the N data points to K clusters. Unfortunately, such optimization by complete enumeration is feasible only for very small data sets. The number of distinct assignments is (Jain and Dubes, 1988)

$$S(N, K) = \frac{1}{K!} \sum_{k=1}^{K} (-1)^{K-k} \binom{K}{k} k^N. \tag{14.30}$$

For example, $S(10, 4) = 34,105$ which is quite feasible. But, $S(N, K)$ grows very rapidly with increasing values of its arguments. Already $S(19, 4) \simeq$

10^{10}, and most clustering problems involve much larger data sets than $N = 19$. For this reason, practical clustering algorithms are able to examine only a very small fraction of all possible encoders $k = C(i)$. The goal is to identify a small subset that is likely to contain the optimal one, or at least a good suboptimal partition.

Such feasible strategies are based on iterative greedy descent. An initial partition is specified. At each iterative step, the cluster assignments are changed in such a way that the value of the criterion is improved from its previous value. Clustering algorithms of this type differ in their prescriptions for modifying the cluster assignments at each iteration. When the prescription is unable to provide an improvement, the algorithm terminates with the current assignments as its solution. Since the assignment of observations to clusters at any iteration is a perturbation of that for the previous iteration, only a very small fraction of all possible assignments (14.30) are examined. However, these algorithms converge to *local* optima which may be highly suboptimal when compared to the global optimum.

14.3.6 K-means

The K-means algorithm is one of the most popular iterative descent clustering methods. It is intended for situations in which all variables are of the quantitative type, and squared Euclidean distance

$$d(x_i, x_{i'}) = \sum_{j=1}^{p}(x_{ij} - x_{i'j})^2 = ||x_i - x_{i'}||^2$$

is chosen as the dissimilarity measure. Note that weighted Euclidean distance can be used by redefining the x_{ij} values (Exercise 14.1).

The within-point scatter (14.28) can be written as

$$\begin{aligned} W(C) &= \frac{1}{2}\sum_{k=1}^{K}\sum_{C(i)=k}\sum_{C(i')=k}||x_i - x_{i'}||^2 \\ &= \sum_{k=1}^{K}N_k\sum_{C(i)=k}||x_i - \bar{x}_k||^2, \end{aligned} \tag{14.31}$$

where $\bar{x}_k = (\bar{x}_{1k}, \ldots, \bar{x}_{pk})$ is the mean vector associated with the kth cluster, and $N_k = \sum_{i=1}^{N}I(C(i) = k)$. Thus, the criterion is minimized by assigning the N observations to the K clusters in such a way that within each cluster the average dissimilarity of the observations from the cluster mean, as defined by the points in that cluster, is minimized.

An iterative descent algorithm for solving

Algorithm 14.1 *K-means Clustering.*

1. For a given cluster assignment C, the total cluster variance (14.33) is minimized with respect to $\{m_1, \ldots, m_K\}$ yielding the means of the currently assigned clusters (14.32).

2. Given a current set of means $\{m_1, \ldots, m_K\}$, (14.33) is minimized by assigning each observation to the closest (current) cluster mean. That is,
$$C(i) = \underset{1 \leq k \leq K}{\operatorname{argmin}} \|x_i - m_k\|^2. \tag{14.34}$$

3. Steps 1 and 2 are iterated until the assignments do not change.

$$C^* = \min_C \sum_{k=1}^{K} N_k \sum_{C(i)=k} \|x_i - \bar{x}_k\|^2$$

can be obtained by noting that for any set of observations S

$$\bar{x}_S = \underset{m}{\operatorname{argmin}} \sum_{i \in S} \|x_i - m\|^2. \tag{14.32}$$

Hence we can obtain C^* by solving the enlarged optimization problem

$$\min_{C, \{m_k\}_1^K} \sum_{k=1}^{K} N_k \sum_{C(i)=k} \|x_i - m_k\|^2. \tag{14.33}$$

This can be minimized by an alternating optimization procedure given in Algorithm 14.1.

Each of steps 1 and 2 reduces the value of the criterion (14.33), so that convergence is assured. However, the result may represent a suboptimal local minimum. The algorithm of Hartigan and Wong (1979) goes further, and ensures that there is no single switch of an observation from one group to another group that will decrease the objective. In addition, one should start the algorithm with many different random choices for the starting means, and choose the solution having smallest value of the objective function.

Figure 14.6 shows some of the K-means iterations for the simulated data of Figure 14.4. The centroids are depicted by "O"s. The straight lines show the partitioning of points, each sector being the set of points closest to each centroid. This partitioning is called the *Voronoi tessellation*. After 20 iterations the procedure has converged.

14.3.7 *Gaussian Mixtures as Soft K-means Clustering*

The K-means clustering procedure is closely related to the EM algorithm for estimating a certain Gaussian mixture model. (Sections 6.8 and 8.5.1).

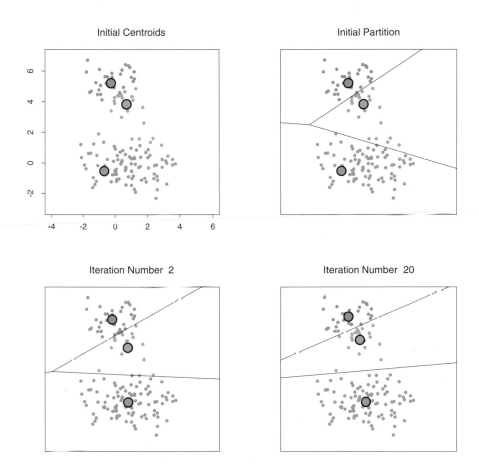

FIGURE 14.6. *Successive iterations of the K-means clustering algorithm for the simulated data of Figure 14.4.*

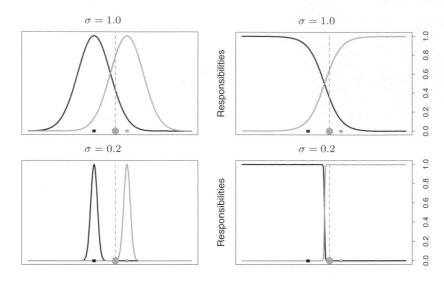

FIGURE 14.7. *(Left panels:) two Gaussian densities $g_0(x)$ and $g_1(x)$ (blue and orange) on the real line, and a single data point (green dot) at $x = 0.5$. The colored squares are plotted at $x = -1.0$ and $x = 1.0$, the means of each density. (Right panels:) the relative densities $g_0(x)/(g_0(x) + g_1(x))$ and $g_1(x)/(g_0(x) + g_1(x))$, called the "responsibilities" of each cluster, for this data point. In the top panels, the Gaussian standard deviation $\sigma = 1.0$; in the bottom panels $\sigma = 0.2$. The EM algorithm uses these responsibilities to make a "soft" assignment of each data point to each of the two clusters. When σ is fairly large, the responsibilities can be near 0.5 (they are 0.36 and 0.64 in the top right panel). As $\sigma \to 0$, the responsibilities $\to 1$, for the cluster center closest to the target point, and 0 for all other clusters. This "hard" assignment is seen in the bottom right panel.*

The E-step of the EM algorithm assigns "responsibilities" for each data point based in its relative density under each mixture component, while the M-step recomputes the component density parameters based on the current responsibilities. Suppose we specify K mixture components, each with a Gaussian density having scalar covariance matrix $\sigma^2 \mathbf{I}$. Then the relative density under each mixture component is a monotone function of the Euclidean distance between the data point and the mixture center. Hence in this setup EM is a "soft" version of K-means clustering, making probabilistic (rather than deterministic) assignments of points to cluster centers. As the variance $\sigma^2 \to 0$, these probabilities become 0 and 1, and the two methods coincide. Details are given in Exercise 14.2. Figure 14.7 illustrates this result for two clusters on the real line.

14.3.8 Example: Human Tumor Microarray Data

We apply K-means clustering to the human tumor microarray data described in Chapter 1. This is an example of high-dimensional clustering.

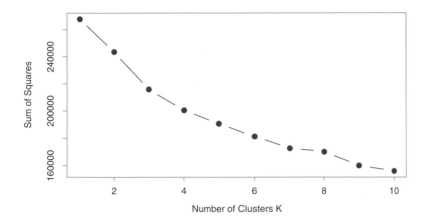

FIGURE 14.8. *Total within-cluster sum of squares for K-means clustering applied to the human tumor microarray data.*

TABLE 14.2. *Human tumor data: number of cancer cases of each type, in each of the three clusters from K-means clustering.*

Cluster	Breast	CNS	Colon	K562	Leukemia	MCF7
1	3	5	0	0	0	0
2	2	0	0	2	6	2
3	2	0	7	0	0	0

Cluster	Melanoma	NSCLC	Ovarian	Prostate	Renal	Unknown
1	1	7	6	2	9	1
2	7	2	0	0	0	0
3	0	0	0	0	0	0

The data are a 6830×64 matrix of real numbers, each representing an expression measurement for a gene (row) and sample (column). Here we cluster the samples, each of which is a vector of length 6830, corresponding to expression values for the 6830 genes. Each sample has a label such as `breast` (for breast cancer), `melanoma`, and so on; we don't use these labels in the clustering, but will examine *posthoc* which labels fall into which clusters.

We applied K-means clustering with K running from 1 to 10, and computed the total within-sum of squares for each clustering, shown in Figure 14.8. Typically one looks for a kink in the sum of squares curve (or its logarithm) to locate the optimal number of clusters (see Section 14.3.11). Here there is no clear indication: for illustration we chose $K = 3$ giving the three clusters shown in Table 14.2.

FIGURE 14.9. *Sir Ronald A. Fisher (1890 − 1962) was one of the founders of modern day statistics, to whom we owe maximum-likelihood, sufficiency, and many other fundamental concepts. The image on the left is a* 1024×1024 *grayscale image at 8 bits per pixel. The center image is the result of* 2 × 2 *block VQ, using* 200 *code vectors, with a compression rate of* 1.9 *bits/pixel. The right image uses only four code vectors, with a compression rate of* 0.50 *bits/pixel*

We see that the procedure is successful at grouping together samples of the same cancer. In fact, the two breast cancers in the second cluster were later found to be misdiagnosed and were melanomas that had metastasized. However, K-means clustering has shortcomings in this application. For one, it does not give a linear ordering of objects within a cluster: we have simply listed them in alphabetic order above. Secondly, as the number of clusters K is changed, the cluster memberships can change in arbitrary ways. That is, with say four clusters, the clusters need not be nested within the three clusters above. For these reasons, hierarchical clustering (described later), is probably preferable for this application.

14.3.9   Vector Quantization

The K-means clustering algorithm represents a key tool in the apparently unrelated area of image and signal compression, particularly in *vector quantization* or VQ (Gersho and Gray, 1992). The left image in Figure 14.9[2] is a digitized photograph of a famous statistician, Sir Ronald Fisher. It consists of 1024×1024 pixels, where each pixel is a grayscale value ranging from 0 to 255, and hence requires 8 bits of storage per pixel. The entire image occupies 1 megabyte of storage. The center image is a VQ-compressed version of the left panel, and requires 0.239 of the storage (at some loss in quality). The right image is compressed even more, and requires only 0.0625 of the storage (at a considerable loss in quality).

The version of VQ implemented here first breaks the image into small blocks, in this case 2×2 blocks of pixels. Each of the 512×512 blocks of four

---

[2]This example was prepared by Maya Gupta.

numbers is regarded as a vector in $\mathbb{R}^4$. A K-means clustering algorithm (also known as Lloyd's algorithm in this context) is run in this space. The center image uses $K = 200$, while the right image $K = 4$. Each of the 512×512 pixel blocks (or points) is approximated by its closest cluster centroid, known as a codeword. The clustering process is called the *encoding* step, and the collection of centroids is called the *codebook*.

To represent the approximated image, we need to supply for each block the identity of the codebook entry that approximates it. This will require $\log_2(K)$ bits per block. We also need to supply the codebook itself, which is $K \times 4$ real numbers (typically negligible). Overall, the storage for the compressed image amounts to $\log_2(K)/(4 \cdot 8)$ of the original (0.239 for $K = 200$, 0.063 for $K = 4$). This is typically expressed as a *rate* in bits per pixel: $\log_2(K)/4$, which are 1.91 and 0.50, respectively. The process of constructing the approximate image from the centroids is called the *decoding* step.

Why do we expect VQ to work at all? The reason is that for typical everyday images like photographs, many of the blocks look the same. In this case there are many almost pure white blocks, and similarly pure gray blocks of various shades. These require only one block each to represent them, and then multiple pointers to that block.

What we have described is known as *lossy* compression, since our images are degraded versions of the original. The degradation or *distortion* is usually measured in terms of mean squared error. In this case $D = 0.89$ for $K = 200$ and $D = 16.95$ for $K = 4$. More generally a rate/distortion curve would be used to assess the tradeoff. One can also perform *lossless* compression using block clustering, and still capitalize on the repeated patterns. If you took the original image and losslessly compressed it, the best you would do is 4.48 bits per pixel.

We claimed above that $\log_2(K)$ bits were needed to identify each of the K codewords in the codebook. This uses a fixed-length code, and is inefficient if some codewords occur many more times than others in the image. Using Shannon coding theory, we know that in general a variable length code will do better, and the rate then becomes $-\sum_{\ell=1}^K p_\ell \log_2(p_\ell)/4$. The term in the numerator is the entropy of the distribution p_ℓ of the codewords in the image. Using variable length coding our rates come down to 1.42 and 0.39, respectively. Finally, there are many generalizations of VQ that have been developed: for example, tree-structured VQ finds the centroids with a top-down, 2-means style algorithm, as alluded to in Section 14.3.12. This allows successive refinement of the compression. Further details may be found in Gersho and Gray (1992).

14.3.10 K-medoids

As discussed above, the K-means algorithm is appropriate when the dissimilarity measure is taken to be squared Euclidean distance $D(x_i, x_{i'})$

Algorithm 14.2 *K-medoids Clustering.*

1. For a given cluster assignment C find the observation in the cluster minimizing total distance to other points in that cluster:

$$i_k^* = \underset{\{i:C(i)=k\}}{\operatorname{argmin}} \sum_{C(i')=k} D(x_i, x_{i'}). \qquad (14.35)$$

Then $m_k = x_{i_k^*}$, $k = 1, 2, \ldots, K$ are the current estimates of the cluster centers.

2. Given a current set of cluster centers $\{m_1, \ldots, m_K\}$, minimize the total error by assigning each observation to the closest (current) cluster center:

$$C(i) = \underset{1 \le k \le K}{\operatorname{argmin}} D(x_i, m_k). \qquad (14.36)$$

3. Iterate steps 1 and 2 until the assignments do not change.

(14.112). This requires all of the variables to be of the quantitative type. In addition, using *squared* Euclidean distance places the highest influence on the largest distances. This causes the procedure to lack robustness against outliers that produce very large distances. These restrictions can be removed at the expense of computation.

The only part of the K-means algorithm that assumes squared Euclidean distance is the minimization step (14.32); the cluster representatives $\{m_1, \ldots, m_K\}$ in (14.33) are taken to be the means of the currently assigned clusters. The algorithm can be generalized for use with arbitrarily defined dissimilarities $D(x_i, x_{i'})$ by replacing this step by an explicit optimization with respect to $\{m_1, \ldots, m_K\}$ in (14.33). In the most common form, centers for each cluster are restricted to be one of the observations assigned to the cluster, as summarized in Algorithm 14.2. This algorithm assumes attribute data, but the approach can also be applied to data described *only* by proximity matrices (Section 14.3.1). There is no need to explicitly compute cluster centers; rather we just keep track of the indices i_k^*.

Solving (14.32) for each provisional cluster k requires an amount of computation proportional to the number of observations assigned to it, whereas for solving (14.35) the computation increases to $O(N_k^2)$. Given a set of cluster "centers," $\{i_1, \ldots, i_K\}$, obtaining the new assignments

$$C(i) = \underset{1 \le k \le K}{\operatorname{argmin}} d_{ii_k^*} \qquad (14.37)$$

requires computation proportional to $K \cdot N$ as before. Thus, K-medoids is far more computationally intensive than K-means.

Alternating between (14.35) and (14.37) represents a particular heuristic search strategy for trying to solve

TABLE 14.3. *Data from a political science survey: values are average pairwise dissimilarities of countries from a questionnaire given to political science students.*

	BEL	BRA	CHI	CUB	EGY	FRA	IND	ISR	USA	USS	YUG
BRA	5.58										
CHI	7.00	6.50									
CUB	7.08	7.00	3.83								
EGY	4.83	5.08	8.17	5.83							
FRA	2.17	5.75	6.67	6.92	4.92						
IND	6.42	5.00	5.58	6.00	4.67	6.42					
ISR	3.42	5.50	6.42	6.42	5.00	3.92	6.17				
USA	2.50	4.92	6.25	7.33	4.50	2.25	6.33	2.75			
USS	6.08	6.67	4.25	2.67	6.00	6.17	6.17	6.92	6.17		
YUG	5.25	6.83	4.50	3.75	5.75	5.42	6.08	5.83	6.67	3.67	
ZAI	4.75	3.00	6.08	6.67	5.00	5.58	4.83	6.17	5.67	6.50	6.92

$$\min_{C,\ \{i_k\}_1^K} \sum_{k-1}^{K} \sum_{C(i)=k} d_{ii_k}. \tag{14.38}$$

Kaufman and Rousseeuw (1990) propose an alternative strategy for directly solving (14.38) that provisionally exchanges each center i_k with an observation that is not currently a center, selecting the exchange that produces the greatest reduction in the value of the criterion (14.38). This is repeated until no advantageous exchanges can be found. Massart et al. (1983) derive a branch-and-bound combinatorial method that finds the global minimum of (14.38) that is practical only for very small data sets.

Example: Country Dissimilarities

This example, taken from Kaufman and Rousseeuw (1990), comes from a study in which political science students were asked to provide pairwise dissimilarity measures for 12 countries: Belgium, Brazil, Chile, Cuba, Egypt, France, India, Israel, United States, Union of Soviet Socialist Republics, Yugoslavia and Zaire. The average dissimilarity scores are given in Table 14.3. We applied 3-medoid clustering to these dissimilarities. Note that K-means clustering could not be applied because we have only distances rather than raw observations. The left panel of Figure 14.10 shows the dissimilarities reordered and blocked according to the 3-medoid clustering. The right panel is a two-dimensional multidimensional scaling plot, with the 3-medoid clusters assignments indicated by colors (multidimensional scaling is discussed in Section 14.8.) Both plots show three well-separated clusters, but the MDS display indicates that "Egypt" falls about halfway between two clusters.

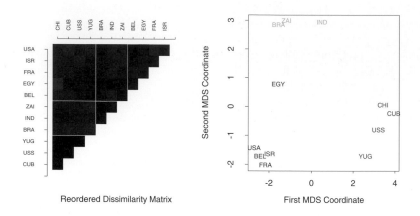

FIGURE 14.10. *Survey of country dissimilarities. (Left panel:) dissimilarities reordered and blocked according to 3-medoid clustering. Heat map is coded from most similar (dark red) to least similar (bright red). (Right panel:) two-dimensional multidimensional scaling plot, with 3-medoid clusters indicated by different colors.*

14.3.11 Practical Issues

In order to apply K-means or K-medoids one must select the number of clusters K^* and an initialization. The latter can be defined by specifying an initial set of centers $\{m_1, \ldots, m_K\}$ or $\{i_1, \ldots, i_K\}$ or an initial encoder $C(i)$. Usually specifying the centers is more convenient. Suggestions range from simple random selection to a deliberate strategy based on forward stepwise assignment. At each step a new center i_k is chosen to minimize the criterion (14.33) or (14.38), given the centers $i_1, \ldots, i_{k-1}$ chosen at the previous steps. This continues for K steps, thereby producing K initial centers with which to begin the optimization algorithm.

A choice for the number of clusters K depends on the goal. For data segmentation K is usually defined as part of the problem. For example, a company may employ K sales people, and the goal is to partition a customer database into K segments, one for each sales person, such that the customers assigned to each one are as similar as possible. Often, however, cluster analysis is used to provide a descriptive statistic for ascertaining the extent to which the observations comprising the data base fall into natural distinct groupings. Here the number of such groups K^* is unknown and one requires that it, as well as the groupings themselves, be estimated from the data.

Data-based methods for estimating K^* typically examine the within-cluster dissimilarity W_K as a function of the number of clusters K. Separate solutions are obtained for $K \in \{1, 2, \ldots, K_{\max}\}$. The corresponding values

$\{W_1, W_2, \ldots, W_{K_{\max}}\}$ generally decrease with increasing K. This will be the case even when the criterion is evaluated on an independent test set, since a large number of cluster centers will tend to fill the feature space densely and thus will be close to all data points. Thus cross-validation techniques, so useful for model selection in supervised learning, cannot be utilized in this context.

The intuition underlying the approach is that if there are actually K^* distinct groupings of the observations (as defined by the dissimilarity measure), then for $K < K^*$ the clusters returned by the algorithm will each contain a subset of the true underlying groups. That is, the solution will not assign observations in the same naturally occurring group to different estimated clusters. To the extent that this is the case, the solution criterion value will tend to decrease substantially with each successive increase in the number of specified clusters, $W_{K+1} \ll W_K$, as the natural groups are successively assigned to separate clusters. For $K > K^*$, one of the estimated clusters must partition at least one of the natural groups into two subgroups. This will tend to provide a smaller decrease in the criterion as K is further increased. Splitting a natural group, within which the observations are all quite close to each other, reduces the criterion less than partitioning the union of two well-separated groups into their proper constituents.

To the extent this scenario is realized, there will be a sharp decrease in successive differences in criterion value, $W_K - W_{K+1}$, at $K = K^*$. That is, $\{W_K - W_{K+1} \mid K < K^*\} \gg \{W_K - W_{K+1} \mid K \geq K^*\}$. An estimate $\hat{K}^*$ for K^* is then obtained by identifying a "kink" in the plot of W_K as a function of K. As with other aspects of clustering procedures, this approach is somewhat heuristic.

The recently proposed *Gap statistic* (Tibshirani et al., 2001b) compares the curve $\log W_K$ to the curve obtained from data uniformly distributed over a rectangle containing the data. It estimates the optimal number of clusters to be the place where the gap between the two curves is largest. Essentially this is an automatic way of locating the aforementioned "kink." It also works reasonably well when the data fall into a single cluster, and in that case will tend to estimate the optimal number of clusters to be one. This is the scenario where most other competing methods fail.

Figure 14.11 shows the result of the Gap statistic applied to simulated data of Figure 14.4. The left panel shows $\log W_K$ for $K = 1, 2, \ldots, 8$ clusters (green curve) and the expected value of $\log W_K$ over 20 simulations from uniform data (blue curve). The right panel shows the gap curve, which is the expected curve minus the observed curve. Shown also are error bars of half-width $s'_K = s_K \sqrt{1 + 1/20}$, where s_K is the standard deviation of $\log W_K$ over the 20 simulations. The Gap curve is maximized at $K = 2$ clusters. If $G(K)$ is the Gap curve at K clusters, the formal rule for estimating K^* is

$$K^* = \operatorname*{argmin}_K \{K | G(K) \geq G(K+1) - s'_{K+1}\}. \tag{14.39}$$

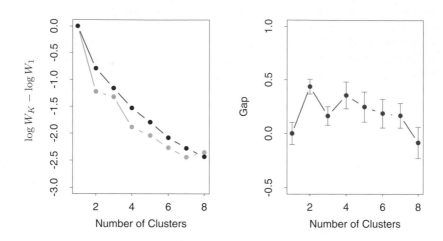

FIGURE 14.11. *(Left panel): observed (green) and expected (blue) values of* $\log W_K$ *for the simulated data of Figure 14.4. Both curves have been translated to equal zero at one cluster. (Right panel): Gap curve, equal to the difference between the observed and expected values of* $\log W_K$. *The Gap estimate* K^* *is the smallest* K *producing a gap within one standard deviation of the gap at* $K+1$; *here* $K^* = 2$.

This gives $K^* = 2$, which looks reasonable from Figure 14.4.

14.3.12   Hierarchical Clustering

The results of applying K-means or K-medoids clustering algorithms depend on the choice for the number of clusters to be searched and a starting configuration assignment. In contrast, hierarchical clustering methods do not require such specifications. Instead, they require the user to specify a measure of dissimilarity between (disjoint) *groups* of observations, based on the pairwise dissimilarities among the observations in the two groups. As the name suggests, they produce hierarchical representations in which the clusters at each level of the hierarchy are created by merging clusters at the next lower level. At the lowest level, each cluster contains a single observation. At the highest level there is only one cluster containing all of the data.

Strategies for hierarchical clustering divide into two basic paradigms: *agglomerative* (bottom-up) and *divisive* (top-down). Agglomerative strategies start at the bottom and at each level recursively merge a selected pair of clusters into a single cluster. This produces a grouping at the next higher level with one less cluster. The pair chosen for merging consist of the two groups with the smallest intergroup dissimilarity. Divisive methods start at the top and at each level recursively split one of the existing clusters at that level into two new clusters. The split is chosen to produce two new groups with the largest between-group dissimilarity. With both paradigms there are $N - 1$ levels in the hierarchy.

Each level of the hierarchy represents a particular grouping of the data into disjoint clusters of observations. The entire hierarchy represents an ordered sequence of such groupings. It is up to the user to decide which level (if any) actually represents a "natural" clustering in the sense that observations within each of its groups are sufficiently more similar to each other than to observations assigned to different groups at that level. The Gap statistic described earlier can be used for this purpose.

Recursive binary splitting/agglomeration can be represented by a rooted binary tree. The nodes of the trees represent groups. The root node represents the entire data set. The N terminal nodes each represent one of the individual observations (singleton clusters). Each nonterminal node ("parent") has two daughter nodes. For divisive clustering the two daughters represent the two groups resulting from the split of the parent; for agglomerative clustering the daughters represent the two groups that were merged to form the parent.

Most agglomerative and some divisive methods (when viewed bottom-up) possess a monotonicity property. That is, the dissimilarity between merged clusters is monotone increasing with the level of the merger. Thus the binary tree can be plotted so that the height of each node is proportional to the value of the intergroup dissimilarity between its two daughters. The terminal nodes representing individual observations are all plotted at zero height. This type of graphical display is called a *dendrogram*.

A dendrogram provides a highly interpretable complete description of the hierarchical clustering in a graphical format. This is one of the main reasons for the popularity of hierarchical clustering methods.

For the microarray data, Figure 14.12 shows the dendrogram resulting from agglomerative clustering with average linkage; agglomerative clustering and this example are discussed in more detail later in this chapter. Cutting the dendrogram horizontally at a particular height partitions the data into disjoint clusters represented by the vertical lines that intersect it. These are the clusters that would be produced by terminating the procedure when the optimal intergroup dissimilarity exceeds that threshold cut value. Groups that merge at high values, relative to the merger values of the subgroups contained within them lower in the tree, are candidates for natural clusters. Note that this may occur at several different levels, indicating a clustering hierarchy: that is, clusters nested within clusters.

Such a dendrogram is often viewed as a graphical summary of the data itself, rather than a description of the results of the algorithm. However, such interpretations should be treated with caution. First, different hierarchical methods (see below), as well as small changes in the data, can lead to quite different dendrograms. Also, such a summary will be valid only to the extent that the pairwise *observation* dissimilarities possess the hierar-

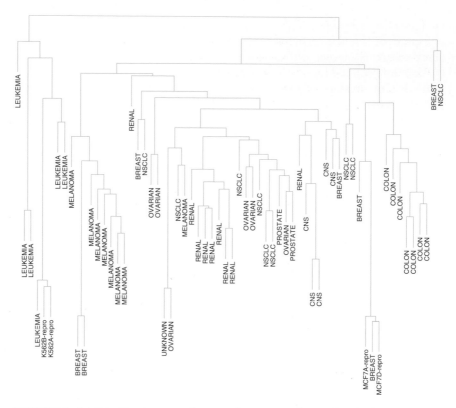

FIGURE 14.12. *Dendrogram from agglomerative hierarchical clustering with average linkage to the human tumor microarray data.*

chical structure produced by the algorithm. Hierarchical methods impose hierarchical structure whether or not such structure actually exists in the data.

The extent to which the hierarchical structure produced by a dendrogram actually represents the data itself can be judged by the *cophenetic correlation coefficient*. This is the correlation between the $N(N-1)/2$ pairwise observation dissimilarities $d_{ii'}$ input to the algorithm and their corresponding *cophenetic* dissimilarities $C_{ii'}$ derived from the dendrogram. The cophenetic dissimilarity $C_{ii'}$ between two observations (i, i') is the intergroup dissimilarity at which observations i and i' are first joined together in the same cluster.

The cophenetic dissimilarity is a very restrictive dissimilarity measure. First, the $C_{ii'}$ over the observations must contain many ties, since only $N-1$ of the total $N(N-1)/2$ values can be distinct. Also these dissimilarities obey the *ultrametric inequality*

$$C_{ii'} \leq \max\{C_{ik}, C_{i'k}\} \tag{14.40}$$

for any three observations (i, i', k). As a geometric example, suppose the data were represented as points in a Euclidean coordinate system. In order for the set of interpoint distances over the data to conform to (14.40), the triangles formed by all triples of points must be isosceles triangles with the unequal length no longer than the length of the two equal sides (Jain and Dubes, 1988). Therefore it is unrealistic to expect general dissimilarities over arbitrary data sets to closely resemble their corresponding cophenetic dissimilarities as calculated from a dendrogram, especially if there are not many tied values. Thus the dendrogram should be viewed mainly as a description of the *clustering* structure of the data as imposed by the particular algorithm employed.

Agglomerative Clustering

Agglomerative clustering algorithms begin with every observation representing a singleton cluster. At each of the $N - 1$ steps the closest two (least dissimilar) clusters are merged into a single cluster, producing one less cluster at the next higher level. Therefore, a measure of dissimilarity between two clusters (groups of observations) must be defined.

Let G and H represent two such groups. The dissimilarity $d(G, H)$ between G and H is computed from the set of pairwise observation dissimilarities $d_{ii'}$ where one member of the pair i is in G and the other i' is in H. *Single linkage* (SL) agglomerative clustering takes the intergroup dissimilarity to be that of the closest (least dissimilar) pair

$$d_{SL}(G, H) = \min_{\substack{i \in G \\ i' \in H}} d_{ii'}. \tag{14.41}$$

This is also often called the *nearest-neighbor* technique. *Complete linkage* (CL) agglomerative clustering (*furthest-neighbor* technique) takes the intergroup dissimilarity to be that of the furthest (most dissimilar) pair

$$d_{CL}(G, H) = \max_{\substack{i \in G \\ i' \in H}} d_{ii'}. \tag{14.42}$$

Group average (GA) clustering uses the average dissimilarity between the groups

$$d_{GA}(G, H) = \frac{1}{N_G N_H} \sum_{i \in G} \sum_{i' \in H} d_{ii'} \tag{14.43}$$

where N_G and N_H are the respective number of observations in each group. Although there have been many other proposals for defining intergroup dissimilarity in the context of agglomerative clustering, the above three are the ones most commonly used. Figure 14.13 shows examples of all three.

If the data dissimilarities $\{d_{ii'}\}$ exhibit a strong clustering tendency, with each of the clusters being compact and well separated from others, then all three methods produce similar results. Clusters are compact if all of the

Average Linkage   Complete Linkage   Single Linkage

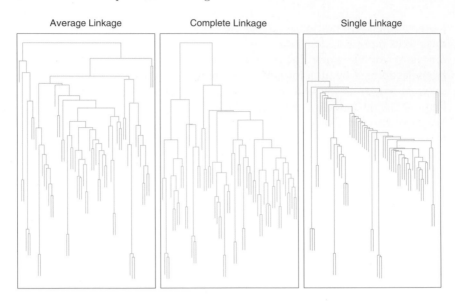

FIGURE 14.13. *Dendrograms from agglomerative hierarchical clustering of human tumor microarray data.*

observations within them are relatively close together (small dissimilarities) as compared with observations in different clusters. To the extent this is not the case, results will differ.

Single linkage (14.41) only requires that a single dissimilarity $d_{ii'}$, $i \in G$ and $i' \in H$, be small for two groups G and H to be considered close together, irrespective of the other observation dissimilarities between the groups. It will therefore have a tendency to combine, at relatively low thresholds, observations linked by a series of close intermediate observations. This phenomenon, referred to as *chaining*, is often considered a defect of the method. The clusters produced by single linkage can violate the "compactness" property that all observations within each cluster tend to be similar to one another, based on the supplied observation dissimilarities $\{d_{ii'}\}$. If we define the *diameter* D_G of a group of observations as the largest dissimilarity among its members

$$D_G = \max_{\substack{i \in G \\ i' \in G}} d_{ii'}, \tag{14.44}$$

then single linkage can produce clusters with very large diameters.

Complete linkage (14.42) represents the opposite extreme. Two groups G and H are considered close only if all of the observations in their union are relatively similar. It will tend to produce compact clusters with small diameters (14.44). However, it can produce clusters that violate the "closeness" property. That is, observations assigned to a cluster can be much closer to members of other clusters than they are to some members of their own cluster.

Group average clustering (14.43) represents a compromise between the two extremes of single and complete linkage. It attempts to produce *relatively* compact clusters that are *relatively* far apart. However, its results depend on the numerical scale on which the observation dissimilarities $d_{ii'}$ are measured. Applying a monotone strictly increasing transformation $h(\cdot)$ to the $d_{ii'}$, $h_{ii'} = h(d_{ii'})$, can change the result produced by (14.43). In contrast, (14.41) and (14.42) depend only on the ordering of the $d_{ii'}$ and are thus invariant to such monotone transformations. This invariance is often used as an argument in favor of single or complete linkage over group average methods.

One can argue that group average clustering has a statistical consistency property violated by single and complete linkage. Assume we have attribute-value data $X^T = (X_1, \ldots, X_p)$ and that each cluster k is a random sample from some population joint density $p_k(x)$. The complete data set is a random sample from a mixture of K such densities. The group average dissimilarity $d_{GA}(G, H)$ (14.43) is an estimate of

$$\int \int d(x, x')\, p_G(x)\, p_H(x')\, dx\, dx', \qquad (14.45)$$

where $d(x, x')$ is the dissimilarity between points x and x' in the space of attribute values. As the sample size N approaches infinity $d_{GA}(G, H)$ (14.43) approaches (14.45), which is a characteristic of the relationship between the two densities $p_G(x)$ and $p_H(x)$. For single linkage, $d_{SL}(G, H)$ (14.41) approaches zero as $N \to \infty$ independent of $p_G(x)$ and $p_H(x)$. For complete linkage, $d_{CL}(G, H)$ (14.42) becomes infinite as $N \to \infty$, again independent of the two densities. Thus, it is not clear what aspects of the population distribution are being estimated by $d_{SL}(G, H)$ and $d_{CL}(G, H)$.

Example: Human Cancer Microarray Data (Continued)

The left panel of Figure 14.13 shows the dendrogram resulting from average linkage agglomerative clustering of the samples (columns) of the microarray data. The middle and right panels show the result using complete and single linkage. Average and complete linkage gave similar results, while single linkage produced unbalanced groups with long thin clusters. We focus on the average linkage clustering.

Like K-means clustering, hierarchical clustering is successful at clustering simple cancers together. However it has other nice features. By cutting off the dendrogram at various heights, different numbers of clusters emerge, and the sets of clusters are nested within one another. Secondly, it gives some partial ordering information about the samples. In Figure 14.14, we have arranged the genes (rows) and samples (columns) of the expression matrix in orderings derived from hierarchical clustering.

Note that if we flip the orientation of the branches of a dendrogram at any merge, the resulting dendrogram is still consistent with the series of hierarchical clustering operations. Hence to determine an ordering of the leaves, we must add a constraint. To produce the row ordering of Figure 14.14, we have used the default rule in S-PLUS: at each merge, the subtree with the tighter cluster is placed to the left (toward the bottom in the rotated dendrogram in the figure.) Individual genes are the tightest clusters possible, and merges involving two individual genes place them in order by their observation number. The same rule was used for the columns. Many other rules are possible—for example, ordering by a multidimensional scaling of the genes; see Section 14.8.

The two-way rearrangement of Figure 14.14 produces an informative picture of the genes and samples. This picture is more informative than the randomly ordered rows and columns of Figure 1.3 of Chapter 1. Furthermore, the dendrograms themselves are useful, as biologists can, for example, interpret the gene clusters in terms of biological processes.

Divisive Clustering

Divisive clustering algorithms begin with the entire data set as a single cluster, and recursively divide one of the existing clusters into two daughter clusters at each iteration in a top-down fashion. This approach has not been studied nearly as extensively as agglomerative methods in the clustering literature. It has been explored somewhat in the engineering literature (Gersho and Gray, 1992) in the context of compression. In the clustering setting, a potential advantage of divisive over agglomerative methods can occur when interest is focused on partitioning the data into a relatively *small* number of clusters.

The divisive paradigm can be employed by recursively applying any of the combinatorial methods such as K-means (Section 14.3.6) or K-medoids (Section 14.3.10), with $K = 2$, to perform the splits at each iteration. However, such an approach would depend on the starting configuration specified at each step. In addition, it would not necessarily produce a splitting sequence that possesses the monotonicity property required for dendrogram representation.

A divisive algorithm that avoids these problems was proposed by Macnaughton Smith et al. (1965). It begins by placing all observations in a single cluster G. It then chooses that observation whose average dissimilarity from all the other observations is largest. This observation forms the first member of a second cluster H. At each successive step that observation in G whose average distance from those in H, minus that for the remaining observations in G is largest, is transferred to H. This continues until the corresponding difference in averages becomes negative. That is, there are no longer any observations in G that are, on average, closer to those in H. The result is a split of the original cluster into two daughter clusters,

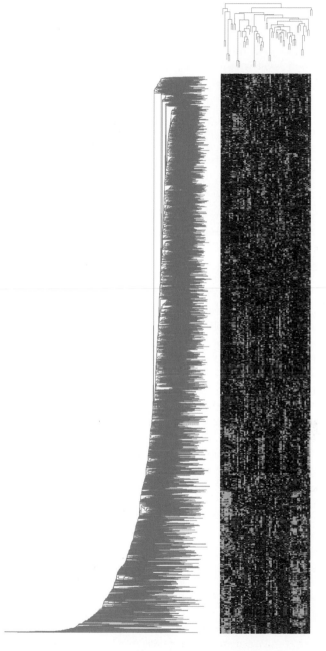

FIGURE 14.14. *DNA microarray data: average linkage hierarchical clustering has been applied independently to the rows (genes) and columns (samples), determining the ordering of the rows and columns (see text). The colors range from bright green (negative, under-expressed) to bright red (positive, over-expressed).*

the observations transferred to H, and those remaining in G. These two clusters represent the second level of the hierarchy. Each successive level is produced by applying this splitting procedure to one of the clusters at the previous level. Kaufman and Rousseeuw (1990) suggest choosing the cluster at each level with the largest diameter (14.44) for splitting. An alternative would be to choose the one with the largest average dissimilarity among its members

$$\bar{d}_G = \frac{1}{N_G{}^2} \sum_{i \in G} \sum_{i' \in G} d_{ii'}.$$

The recursive splitting continues until all clusters either become singletons or all members of each one have zero dissimilarity from one another.

14.4   Self-Organizing Maps

This method can be viewed as a constrained version of K-means clustering, in which the prototypes are encouraged to lie in a one- or two-dimensional manifold in the feature space. The resulting manifold is also referred to as a *constrained topological map*, since the original high-dimensional observations can be mapped down onto the two-dimensional coordinate system. The original SOM algorithm was online—observations are processed one at a time—and later a batch version was proposed. The technique also bears a close relationship to *principal curves and surfaces*, which are discussed in the next section.

We consider a SOM with a two-dimensional rectangular grid of K prototypes $m_j \in \mathbb{R}^p$ (other choices, such as hexagonal grids, can also be used). Each of the K prototypes are parametrized with respect to an integer coordinate pair $\ell_j \in \mathcal{Q}_1 \times \mathcal{Q}_2$. Here $\mathcal{Q}_1 = \{1, 2, \ldots, q_1\}$, similarly $\mathcal{Q}_2$, and $K = q_1 \cdot q_2$. The m_j are initialized, for example, to lie in the two-dimensional principal component plane of the data (next section). We can think of the prototypes as "buttons," "sewn" on the principal component plane in a regular pattern. The SOM procedure tries to bend the plane so that the buttons approximate the data points as well as possible. Once the model is fit, the observations can be mapped down onto the two-dimensional grid.

The observations x_i are processed one at a time. We find the closest prototype m_j to x_i in Euclidean distance in $\mathbb{R}^p$, and then for all neighbors m_k of m_j, move m_k toward x_i via the update

$$m_k \leftarrow m_k + \alpha(x_i - m_k). \tag{14.46}$$

The "neighbors" of m_j are defined to be all m_k such that the distance between ℓ_j and ℓ_k is small. The simplest approach uses Euclidean distance, and "small" is determined by a threshold r. This neighborhood always includes the closest prototype m_j itself.

Notice that distance is defined in the space $\mathcal{Q}_1 \times \mathcal{Q}_2$ of integer topological coordinates of the prototypes, rather than in the feature space $\mathbb{R}^p$. The effect of the update (14.46) is to move the prototypes closer to the data, but also to maintain a smooth two-dimensional spatial relationship between the prototypes.

The performance of the SOM algorithm depends on the learning rate α and the distance threshold r. Typically α is decreased from say 1.0 to 0.0 over a few thousand iterations (one per observation). Similarly r is decreased linearly from starting value R to 1 over a few thousand iterations. We illustrate a method for choosing R in the example below.

We have described the simplest version of the SOM. More sophisticated versions modify the update step according to distance:

$$m_k \leftarrow m_k + \alpha h(\|\ell_j - \ell_k\|)(x_i - m_k), \qquad (14.47)$$

where the *neighborhood function* h gives more weight to prototypes m_k with indices ℓ_k closer to ℓ_j than to those further away.

If we take the distance r small enough so that each neighborhood contains only one point, then the spatial connection between prototypes is lost. In that case one can show that the SOM algorithm is an online version of K-means clustering, and eventually stabilizes at one of the local minima found by K-means. Since the SOM is a constrained version of K-means clustering, it is important to check whether the constraint is reasonable in any given problem. One can do this by computing the reconstruction error $\|x - m_j\|^2$, summed over observations, for both methods. This will necessarily be smaller for K-means, but should not be much smaller if the SOM is a reasonable approximation.

As an illustrative example, we generated 90 data points in three dimensions, near the surface of a half sphere of radius 1. The points were in each of three clusters—red, green, and blue—located near $(0, 1, 0)$, $(0, 0, 1)$ and $(1, 0, 0)$. The data are shown in Figure 14.15

By design, the red cluster was much tighter than the green or blue ones. (Full details of the data generation are given in Exercise 14.5.) A 5×5 grid of prototypes was used, with initial grid size $R = 2$; this meant that about a third of the prototypes were initially in each neighborhood. We did a total of 40 passes through the dataset of 90 observations, and let r and α decrease linearly over the 3600 iterations.

In Figure 14.16 the prototypes are indicated by circles, and the points that project to each prototype are plotted randomly within the corresponding circle. The left panel shows the initial configuration, while the right panel shows the final one. The algorithm has succeeded in separating the clusters; however, the separation of the red cluster indicates that the manifold has folded back on itself (see Figure 14.17). Since the distances in the two-dimensional display are not used, there is little indication in the SOM projection that the red cluster is tighter than the others.

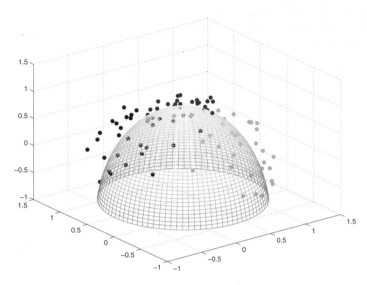

FIGURE 14.15. *Simulated data in three classes, near the surface of a half-sphere.*

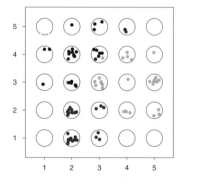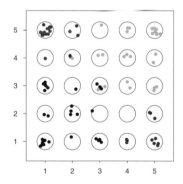

FIGURE 14.16. *Self-organizing map applied to half-sphere data example. Left panel is the initial configuration, right panel the final one. The 5 × 5 grid of prototypes are indicated by circles, and the points that project to each prototype are plotted randomly within the corresponding circle.*

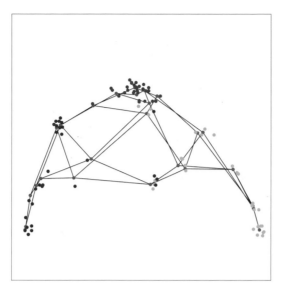

FIGURE 14.17. *Wiremesh representation of the fitted SOM model in* $\mathbb{R}^3$*. The lines represent the horizontal and vertical edges of the topological lattice. The double lines indicate that the surface was folded diagonally back on itself in order to model the red points. The cluster members have been jittered to indicate their color, and the purple points are the node centers.*

Figure 14.18 shows the reconstruction error, equal to the total sum of squares of each data point around its prototype. For comparison we carried out a K-means clustering with 25 centroids, and indicate its reconstruction error by the horizontal line on the graph. We see that the SOM significantly decreases the error, nearly to the level of the K-means solution. This provides evidence that the two-dimensional constraint used by the SOM is reasonable for this particular dataset.

In the batch version of the SOM, we update each m_j via

$$m_j = \frac{\sum w_k x_k}{\sum w_k}. \tag{14.48}$$

The sum is over points x_k that mapped (i.e., were closest to) neighbors m_k of m_j. The weight function may be rectangular, that is, equal to 1 for the neighbors of m_k, or may decrease smoothly with distance $\|\ell_k - \ell_j\|$ as before. If the neighborhood size is chosen small enough so that it consists only of m_k, with rectangular weights, this reduces to the K-means clustering procedure described earlier. It can also be thought of as a discrete version of principal curves and surfaces, described in Section 14.5.

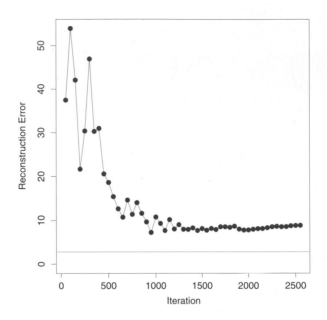

FIGURE 14.18. *Half-sphere data: reconstruction error for the SOM as a function of iteration. Error for k-means clustering is indicated by the horizontal line.*

Example: Document Organization and Retrieval

Document retrieval has gained importance with the rapid development of the Internet and the Web, and SOMs have proved to be useful for organizing and indexing large corpora. This example is taken from the WEBSOM homepage `http://websom.hut.fi/` (Kohonen et al., 2000). Figure 14.19 represents a SOM fit to 12,088 newsgroup `comp.ai.neural-nets` articles. The labels are generated automatically by the WEBSOM software and provide a guide as to the typical content of a node.

In applications such as this, the documents have to be preprocessed in order to create a feature vector. A *term-document* matrix is created, where each row represents a single document. The entries in each row are the relative frequency of each of a predefined set of terms. These terms could be a large set of dictionary entries (50,000 words), or an even larger set of bigrams (word pairs), or subsets of these. These matrices are typically very sparse, and so often some preprocessing is done to reduce the number of features (columns). Sometimes the SVD (next section) is used to reduce the matrix; Kohonen et al. (2000) use a randomized variant thereof. These reduced vectors are then the input to the SOM.

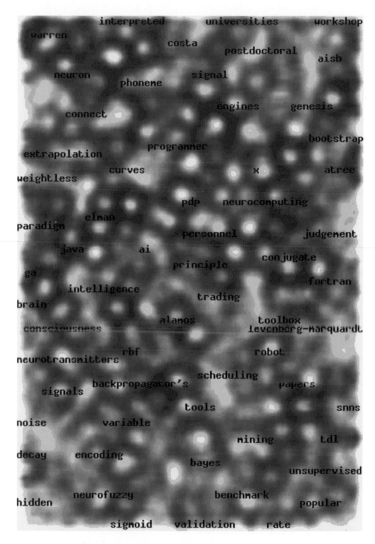

FIGURE 14.19. *Heatmap representation of the SOM model fit to a corpus of 12,088 newsgroup* `comp.ai.neural-nets` *contributions (courtesy WEBSOM homepage). The lighter areas indicate higher-density areas. Populated nodes are automatically labeled according to typical content.*

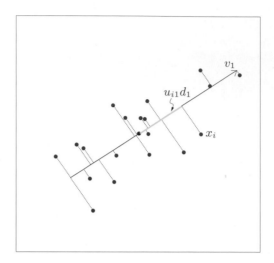

FIGURE 14.20. *The first linear principal component of a set of data. The line minimizes the total squared distance from each point to its orthogonal projection onto the line.*

In this application the authors have developed a "zoom" feature, which allows one to interact with the map in order to get more detail. The final level of zooming retrieves the actual news articles, which can then be read.

14.5 Principal Components, Curves and Surfaces

Principal components are discussed in Sections 3.4.1, where they shed light on the shrinkage mechanism of ridge regression. Principal components are a sequence of projections of the data, mutually uncorrelated and ordered in variance. In the next section we present principal components as linear manifolds approximating a set of N points $x_i \in \mathbb{R}^p$. We then present some nonlinear generalizations in Section 14.5.2. Other recent proposals for nonlinear approximating manifolds are discussed in Section 14.9.

14.5.1 Principal Components

The principal components of a set of data in $\mathbb{R}^p$ provide a sequence of best linear approximations to that data, of all ranks $q \leq p$.

Denote the observations by $x_1, x_2, \ldots, x_N$, and consider the rank-q linear model for representing them

$$f(\lambda) = \mu + \mathbf{V}_q \lambda, \tag{14.49}$$

where μ is a location vector in $\mathbb{R}^p$, $\mathbf{V}_q$ is a $p \times q$ matrix with q orthogonal unit vectors as columns, and λ is a q vector of parameters. This is the parametric representation of an affine hyperplane of rank q. Figures 14.20 and 14.21 illustrate for $q = 1$ and $q = 2$, respectively. Fitting such a model to the data by least squares amounts to minimizing the *reconstruction error*

$$\min_{\mu, \{\lambda_i\}, \mathbf{V}_q} \sum_{i=1}^{N} \|x_i - \mu - \mathbf{V}_q \lambda_i\|^2. \tag{14.50}$$

We can partially optimize for μ and the λ_i (Exercise 14.7) to obtain

$$\hat{\mu} = \bar{x}, \tag{14.51}$$
$$\hat{\lambda}_i = \mathbf{V}_q^T (x_i - \bar{x}). \tag{14.52}$$

This leaves us to find the orthogonal matrix $\mathbf{V}_q$:

$$\min_{\mathbf{V}_q} \sum_{i=1}^{N} \|(x_i - \bar{x}) - \mathbf{V}_q \mathbf{V}_q^T (x_i - \bar{x})\|^2. \tag{14.53}$$

For convenience we assume that $\bar{x} = 0$ (otherwise we simply replace the observations by their centered versions $\tilde{x}_i = x_i - \bar{x}$). The $p \times p$ matrix $\mathbf{H}_q = \mathbf{V}_q \mathbf{V}_q^T$ is a *projection matrix*, and maps each point x_i onto its rank-q reconstruction $\mathbf{H}_q x_i$, the orthogonal projection of x_i onto the subspace spanned by the columns of $\mathbf{V}_q$. The solution can be expressed as follows. Stack the (centered) observations into the rows of an $N \times p$ matrix $\mathbf{X}$. We construct the *singular value decomposition* of $\mathbf{X}$:

$$\mathbf{X} = \mathbf{U} \mathbf{D} \mathbf{V}^T. \tag{14.54}$$

This is a standard decomposition in numerical analysis, and many algorithms exist for its computation (Golub and Van Loan, 1983, for example). Here $\mathbf{U}$ is an $N \times p$ orthogonal matrix ($\mathbf{U}^T \mathbf{U} = \mathbf{I}_p$) whose columns $\mathbf{u}_j$ are called the *left singular vectors*; $\mathbf{V}$ is a $p \times p$ orthogonal matrix ($\mathbf{V}^T \mathbf{V} = \mathbf{I}_p$) with columns v_j called the *right singular vectors*, and $\mathbf{D}$ is a $p \times p$ diagonal matrix, with diagonal elements $d_1 \geq d_2 \geq \cdots \geq d_p \geq 0$ known as the *singular values*. For each rank q, the solution $\mathbf{V}_q$ to (14.53) consists of the first q columns of $\mathbf{V}$. The columns of $\mathbf{UD}$ are called the principal components of $\mathbf{X}$ (see Section 3.5.1). The N optimal $\hat{\lambda}_i$ in (14.52) are given by the first q principal components (the N rows of the $N \times q$ matrix $\mathbf{U}_q \mathbf{D}_q$).

The one-dimensional principal component line in $\mathbb{R}^2$ is illustrated in Figure 14.20. For each data point x_i, there is a closest point on the line, given by $u_{i1} d_1 v_1$. Here v_1 is the direction of the line and $\hat{\lambda}_i = u_{i1} d_1$ measures distance along the line from the origin. Similarly Figure 14.21 shows the

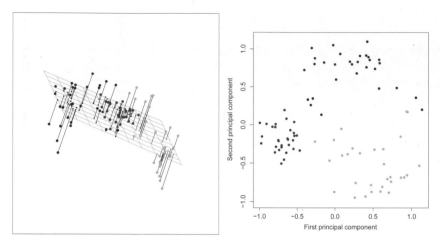

FIGURE 14.21. *The best rank-two linear approximation to the half-sphere data. The right panel shows the projected points with coordinates given by* $\mathbf{U}_2\mathbf{D}_2$, *the first two principal components of the data.*

two-dimensional principal component surface fit to the half-sphere data (left panel). The right panel shows the projection of the data onto the first two principal components. This projection was the basis for the initial configuration for the SOM method shown earlier. The procedure is quite successful at separating the clusters. Since the half-sphere is nonlinear, a nonlinear projection will do a better job, and this is the topic of the next section.

Principal components have many other nice properties, for example, the linear combination $\mathbf{X}v_1$ has the highest variance among all linear combinations of the features; $\mathbf{X}v_2$ has the highest variance among all linear combinations satisfying v_2 orthogonal to v_1, and so on.

Example: Handwritten Digits

Principal components are a useful tool for dimension reduction and compression. We illustrate this feature on the handwritten digits data described in Chapter 1. Figure 14.22 shows a sample of 130 handwritten 3's, each a digitized 16×16 grayscale image, from a total of 658 such 3's. We see considerable variation in writing styles, character thickness and orientation. We consider these images as points x_i in $\mathbb{R}^{256}$, and compute their principal components via the SVD (14.54).

Figure 14.23 shows the first two principal components of these data. For each of these first two principal components $u_{i1}d_1$ and $u_{i2}d_2$, we computed the 5%, 25%, 50%, 75% and 95% quantile points, and used them to define the rectangular grid superimposed on the plot. The circled points indicate

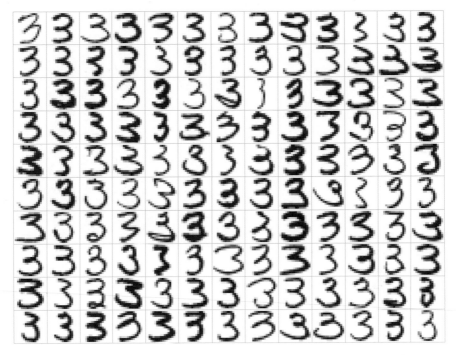

FIGURE 14.22. *A sample of 130 handwritten 3's shows a variety of writing styles.*

those images close to the vertices of the grid, where the distance measure focuses mainly on these projected coordinates, but gives some weight to the components in the orthogonal subspace. The right plot shows the images corresponding to these circled points. This allows us to visualize the nature of the first two principal components. We see that the v_1 (horizontal movement) mainly accounts for the lengthening of the lower tail of the three, while v_2 (vertical movement) accounts for character thickness. In terms of the parametrized model (14.49), this two-component model has the form

$$\hat{f}(\lambda) = \bar{x} + \lambda_1 v_1 + \lambda_2 v_2$$

$$= \boxed{3} + \lambda_1 \cdot \boxed{3} + \lambda_2 \cdot \boxed{3}. \qquad (14.55)$$

Here we have displayed the first two principal component directions, v_1 and v_2, as images. Although there are a possible 256 principal components, approximately 50 account for 90% of the variation in the threes, 12 account for 63%. Figure 14.24 compares the singular values to those obtained for equivalent uncorrelated data, obtained by randomly scrambling each column of $\mathbf{X}$. The pixels in a digitized image are inherently correlated, and since these are all the same digit the correlations are even stronger.

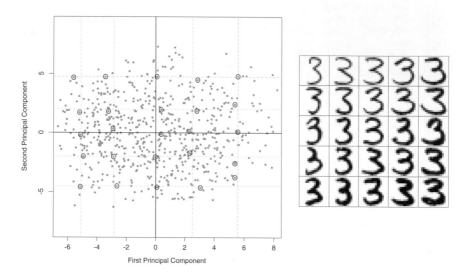

FIGURE 14.23. *(Left panel:) the first two principal components of the hand-written threes. The circled points are the closest projected images to the vertices of a grid, defined by the marginal quantiles of the principal components. (Right panel:) The images corresponding to the circled points. These show the nature of the first two principal components.*

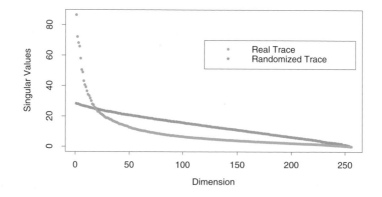

FIGURE 14.24. *The 256 singular values for the digitized threes, compared to those for a randomized version of the data (each column of* **X** *was scrambled).*

A relatively small subset of the principal components serve as excellent lower-dimensional features for representing the high-dimensional data.

Example: Procrustes Transformations and Shape Averaging

FIGURE 14.25. *(Left panel:) Two different digitized handwritten Ss, each represented by 96 corresponding points in* $\mathbb{R}^2$. *The green S has been deliberately rotated and translated for visual effect. (Right panel:) A Procrustes transformation applies a translation and rotation to best match up the two set of points.*

Figure 14.25 represents two sets of points, the orange and green, in the same plot. In this instance these points represent two digitized versions of a handwritten S, extracted from the signature of a subject "Suresh." Figure 14.26 shows the entire signatures from which these were extracted (third and fourth panels). The signatures are recorded dynamically using touch-screen devices, familiar sights in modern supermarkets. There are $N = 96$ points representing each S, which we denote by the $N \times 2$ matrices $\mathbf{X}_1$ and $\mathbf{X}_2$. There is a correspondence between the points—the ith rows of $\mathbf{X}_1$ and $\mathbf{X}_2$ are meant to represent the same positions along the two S's. In the language of morphometrics, these points represent *landmarks* on the two objects. How one finds such corresponding landmarks is in general difficult and subject specific. In this particular case we used *dynamic time warping* of the speed signal along each signature (Hastie et al., 1992), but will not go into details here.

In the right panel we have applied a translation and rotation to the green points so as best to match the orange—a so-called *Procrustes*[3] transformation (Mardia et al., 1979, for example).

Consider the problem

$$\min_{\mu, \mathbf{R}} ||\mathbf{X}_2 - (\mathbf{X}_1 \mathbf{R} + \mathbf{1}\mu^T)||_F, \qquad (14.56)$$

---

[3]Procrustes was an African bandit in Greek mythology, who stretched or squashed his visitors to fit his iron bed (eventually killing them).

with $\mathbf{X}_1$ and $\mathbf{X}_2$ both $N \times p$ matrices of corresponding points, $\mathbf{R}$ an orthonormal $p \times p$ matrix[4], and μ a p-vector of location coordinates. Here $||\mathbf{X}||_F^2 = \text{trace}(\mathbf{X}^T\mathbf{X})$ is the squared *Frobenius* matrix norm.

Let $\bar{x}_1$ and $\bar{x}_2$ be the column mean vectors of the matrices, and $\tilde{\mathbf{X}}_1$ and $\tilde{\mathbf{X}}_2$ be the versions of these matrices with the means removed. Consider the SVD $\tilde{\mathbf{X}}_1^T\tilde{\mathbf{X}}_2 = \mathbf{U}\mathbf{D}\mathbf{V}^T$. Then the solution to (14.56) is given by (Exercise 14.8)

$$\begin{aligned} \hat{\mathbf{R}} &= \mathbf{U}\mathbf{V}^T \\ \hat{\mu} &= \bar{x}_2 - \hat{\mathbf{R}}\bar{x}_1, \end{aligned} \qquad (14.57)$$

and the minimal distances is referred to as the *Procrustes distance*. From the form of the solution, we can center each matrix at its column centroid, and then ignore location completely. Hereafter we assume this is the case.

The *Procrustes distance with scaling* solves a slightly more general problem,

$$\min_{\beta,\mathbf{R}} ||\mathbf{X}_2 - \beta\mathbf{X}_1\mathbf{R}||_F, \qquad (14.58)$$

where $\beta > 0$ is a positive scalar. The solution for $\mathbf{R}$ is as before, with $\hat{\beta} = \text{trace}(D)/||\mathbf{X}_1||_F^2$.

Related to Procrustes distance is the *Procrustes average* of a collection of L shapes, which solves the problem

$$\min_{\{\mathbf{R}_\ell\}_1^L, \mathbf{M}} \sum_{\ell=1}^{L} ||\mathbf{X}_\ell\mathbf{R}_\ell - \mathbf{M}||_F^2; \qquad (14.59)$$

that is, find the shape $\mathbf{M}$ closest in average squared Procrustes distance to all the shapes. This is solved by a simple alternating algorithm:

0. Initialize $\mathbf{M} = \mathbf{X}_1$ (for example).

1. Solve the L Procrustes rotation problems with $\mathbf{M}$ fixed, yielding $\mathbf{X}'_\ell \leftarrow \mathbf{X}\hat{\mathbf{R}}_\ell$.

2. Let $\mathbf{M} \leftarrow \frac{1}{L}\sum_{\ell=1}^{L} \mathbf{X}'_\ell$.

Steps 1. and 2. are repeated until the criterion (14.59) converges.

Figure 14.26 shows a simple example with three shapes. Note that we can only expect a solution up to a rotation; alternatively, we can impose a constraint, such as that $\mathbf{M}$ be upper-triangular, to force uniqueness. One can easily incorporate scaling in the definition (14.59); see Exercise 14.9.

Most generally we can define the *affine-invariant* average of a set of shapes via

---

[4]To simplify matters, we consider only orthogonal matrices which include reflections as well as rotations [the $O(p)$ group]; although reflections are unlikely here, these methods can be restricted further to allow only rotations [$SO(p)$ group].

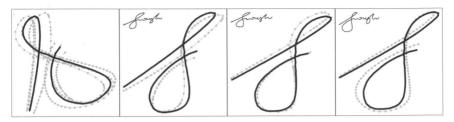

FIGURE 14.26. *The Procrustes average of three versions of the leading S in Suresh's signatures. The left panel shows the* preshape *average, with each of the shapes* $\mathbf{X}'_\ell$ *in preshape space superimposed. The right three panels map the preshape* $\mathbf{M}$ *separately to match each of the original S's.*

$$\min_{\{\mathbf{A}_\ell\}_1^L, \mathbf{M}} \sum_{\ell=1}^L ||\mathbf{X}_\ell \mathbf{A}_\ell - \mathbf{M}||_F^2, \tag{14.60}$$

where the $\mathbf{A}_\ell$ are any $p \times p$ nonsingular matrices. Here we require a standardization, such as $\mathbf{M}^T \mathbf{M} = \mathbf{I}$, to avoid a trivial solution. The solution is attractive, and can be computed without iteration (Exercise 14.10):

1. Let $\mathbf{H}_\ell = \mathbf{X}_\ell (\mathbf{X}_\ell^T \mathbf{X}_\ell)^{-1} \mathbf{X}_\ell^T$ be the rank-p projection matrix defined by $\mathbf{X}_\ell$.

2. $\mathbf{M}$ is the $N \times p$ matrix formed from the p largest eigenvectors of $\bar{\mathbf{H}} = \frac{1}{L} \sum_{\ell=1}^L \mathbf{H}_\ell$.

14.5.2   Principal Curves and Surfaces

Principal curves generalize the principal component line, providing a smooth one-dimensional curved approximation to a set of data points in $\mathbb{R}^p$. A principal surface is more general, providing a curved manifold approximation of dimension 2 or more.

We will first define principal curves for random variables $X \in \mathbb{R}^p$, and then move to the finite data case. Let $f(\lambda)$ be a parameterized smooth curve in $\mathbb{R}^p$. Hence $f(\lambda)$ is a vector function with p coordinates, each a smooth function of the single parameter λ. The parameter λ can be chosen, for example, to be arc-length along the curve from some fixed origin. For each data value x, let $\lambda_f(x)$ define the closest point on the curve to x. Then $f(\lambda)$ is called a principal curve for the distribution of the random vector X if

$$f(\lambda) = \mathrm{E}(X | \lambda_f(X) = \lambda). \tag{14.61}$$

This says $f(\lambda)$ is the average of all data points that project to it, that is, the points for which it is "responsible." This is also known as a *self-consistency* property. Although in practice, continuous multivariate distributes have infinitely many principal curves (Duchamp and Stuetzle, 1996), we are

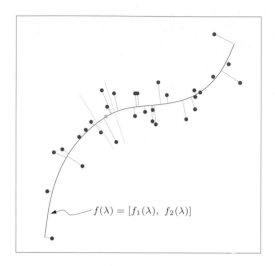

$$f(\lambda) = [f_1(\lambda),\ f_2(\lambda)]$$

FIGURE 14.27. *The principal curve of a set of data. Each point on the curve is the average of all data points that project there.*

interested mainly in the smooth ones. A principal curve is illustrated in Figure 14.27.

Principal points are an interesting related concept. Consider a set of k prototypes and for each point x in the support of a distribution, identify the closest prototype, that is, the prototype that is responsible for it. This induces a partition of the feature space into so-called Voronoi regions. The set of k points that minimize the expected distance from X to its prototype are called the principal points of the distribution. Each principal point is self-consistent, in that it equals the mean of X in its Voronoi region. For example, with $k = 1$, the principal point of a circular normal distribution is the mean vector; with $k = 2$ they are a pair of points symmetrically placed on a ray through the mean vector. Principal points are the distributional analogs of centroids found by K-means clustering. Principal curves can be viewed as $k = \infty$ principal points, but constrained to lie on a smooth curve, in a similar way that a SOM constrains K-means cluster centers to fall on a smooth manifold.

To find a principal curve $f(\lambda)$ of a distribution, we consider its coordinate functions $f(\lambda) = [f_1(\lambda), f_2(\lambda), \ldots, f_p(\lambda)]$ and let $X^T = (X_1, X_2, \ldots, X_p)$. Consider the following alternating steps:

$$
\begin{aligned}
\text{(a)} \quad & \hat{f}_j(\lambda) \leftarrow \mathrm{E}(X_j | \lambda(X) = \lambda); \ j = 1, 2, \ldots, p, \\
\text{(b)} \quad & \hat{\lambda}_f(x) \leftarrow \operatorname{argmin}_{\lambda'} \| x - \hat{f}(\lambda') \|^2.
\end{aligned}
\tag{14.62}
$$

The first equation fixes λ and enforces the self-consistency requirement (14.61). The second equation fixes the curve and finds the closest point on

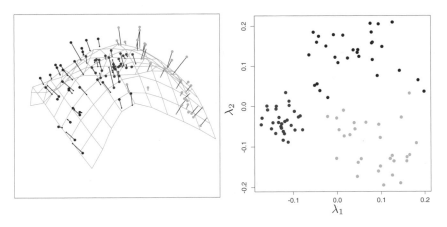

FIGURE 14.28. *Principal surface fit to half-sphere data. (Left panel:) fitted two-dimensional surface. (Right panel:) projections of data points onto the surface, resulting in coordinates* $\hat{\lambda}_1, \hat{\lambda}_2$.

the curve to each data point. With finite data, the principal curve algorithm starts with the linear principal component, and iterates the two steps in (14.62) until convergence. A scatterplot smoother is used to estimate the conditional expectations in step (a) by smoothing each X_j as a function of the arc-length $\hat{\lambda}(X)$, and the projection in (b) is done for each of the observed data points. Proving convergence in general is difficult, but one can show that if a linear least squares fit is used for the scatterplot smoothing, then the procedure converges to the first linear principal component, and is equivalent to the *power method* for finding the largest eigenvector of a matrix.

Principal surfaces have exactly the same form as principal curves, but are of higher dimension. The mostly commonly used is the two-dimensional principal surface, with coordinate functions

$$f(\lambda_1, \lambda_2) = [f_1(\lambda_1, \lambda_2), \ldots, f_p(\lambda_1, \lambda_2)].$$

The estimates in step (a) above are obtained from two-dimensional surface smoothers. Principal surfaces of dimension greater than two are rarely used, since the visualization aspect is less attractive, as is smoothing in high dimensions.

Figure 14.28 shows the result of a principal surface fit to the half-sphere data. Plotted are the data points as a function of the estimated nonlinear coordinates $\hat{\lambda}_1(x_i), \hat{\lambda}_2(x_i)$. The class separation is evident.

Principal surfaces are very similar to self-organizing maps. If we use a kernel surface smoother to estimate each coordinate function $f_j(\lambda_1, \lambda_2)$, this has the same form as the batch version of SOMs (14.48). The SOM weights w_k are just the weights in the kernel. There is a difference, however:

the principal surface estimates a separate prototype $f(\lambda_1(x_i), \lambda_2(x_i))$ for each data point x_i, while the SOM shares a smaller number of prototypes for all data points. As a result, the SOM and principal surface will agree only as the number of SOM prototypes grows very large.

There also is a conceptual difference between the two. Principal surfaces provide a smooth parameterization of the entire manifold in terms of its coordinate functions, while SOMs are discrete and produce only the estimated prototypes for approximating the data. The smooth parameterization in principal surfaces preserves distance locally: in Figure 14.28 it reveals that the red cluster is tighter than the green or blue clusters. In simple examples the estimates coordinate functions themselves can be informative: see Exercise 14.13.

14.5.3 Spectral Clustering

Traditional clustering methods like K-means use a spherical or elliptical metric to group data points. Hence they will not work well when the clusters are non-convex, such as the concentric circles in the top left panel of Figure 14.29. Spectral clustering is a generalization of standard clustering methods, and is designed for these situations. It has close connections with the local multidimensional-scaling techniques (Section 14.9) that generalize MDS.

The starting point is a $N \times N$ matrix of pairwise similarities $s_{ii'} \geq 0$ between all observation pairs. We represent the observations in an undirected *similarity graph* $G = \langle V, E \rangle$. The N vertices v_i represent the observations, and pairs of vertices are connected by an edge if their similarity is positive (or exceeds some threshold). The edges are weighted by the $s_{ii'}$. Clustering is now rephrased as a graph-partition problem, where we identify connected components with clusters. We wish to partition the graph, such that edges between different groups have low weight, and within a group have high weight. The idea in spectral clustering is to construct similarity graphs that represent the local neighborhood relationships between observations.

To make things more concrete, consider a set of N points $x_i \in \mathbb{R}^p$, and let $d_{ii'}$ be the Euclidean distance between x_i and $x_{i'}$. We will use as similarity matrix the radial-kernel gram matrix; that is, $s_{ii'} = \exp(-d_{ii'}^2/c)$, where $c > 0$ is a scale parameter.

There are many ways to define a similarity matrix and its associated similarity graph that reflect local behavior. The most popular is the *mutual K-nearest-neighbor graph*. Define $\mathcal{N}_K$ to be the symmetric set of nearby pairs of points; specifically a pair (i, i') is in $\mathcal{N}_K$ if point i is among the K-nearest neighbors of i', or vice-versa. Then we connect all symmetric nearest neighbors, and give them edge weight $w_{ii'} = s_{ii'}$; otherwise the edge weight is zero. Equivalently we set to zero all the pairwise similarities not in $\mathcal{N}_K$, and draw the graph for this modified similarity matrix.

Alternatively, a fully connected graph includes all pairwise edges with weights $w_{ii'} = s_{ii'}$, and the local behavior is controlled by the scale parameter c.

The matrix of edge weights $\mathbf{W} = \{w_{ii'}\}$ from a similarity graph is called the *adjacency matrix*. The *degree* of vertex i is $g_i = \sum_{i'} w_{ii'}$, the sum of the weights of the edges connected to it. Let $\mathbf{G}$ be a diagonal matrix with diagonal elements g_i.

Finally, the *graph Laplacian* is defined by

$$\mathbf{L} = \mathbf{G} - \mathbf{W} \tag{14.63}$$

This is called the *unnormalized graph Laplacian*; a number of normalized versions have been proposed—these standardize the Laplacian with respect to the node degrees g_i, for example, $\tilde{\mathbf{L}} = \mathbf{I} - \mathbf{G}^{-1}\mathbf{W}$.

Spectral clustering finds the m eigenvectors $\mathbf{Z}_{N \times m}$ corresponding to the m *smallest* eigenvalues of $\mathbf{L}$ (ignoring the trivial constant eigenvector). Using a standard method like K-means, we then cluster the rows of $\mathbf{Z}$ to yield a clustering of the original data points.

An example is presented in Figure 14.29. The top left panel shows 450 simulated data points in three circular clusters indicated by the colors. K-means clustering would clearly have difficulty identifying the outer clusters. We applied spectral clustering using a 10-nearest neighbor similarity graph, and display the eigenvector corresponding to the second and third smallest eigenvalue of the graph Laplacian in the lower left. The 15 smallest eigenvalues are shown in the top right panel. The two eigenvectors shown have identified the three clusters, and a scatterplot of the rows of the eigenvector matrix $\mathbf{Y}$ in the bottom right clearly separates the clusters. A procedure such as K-means clustering applied to these transformed points would easily identify the three groups.

Why does spectral clustering work? For any vector $\mathbf{f}$ we have

$$
\begin{aligned}
\mathbf{f}^T \mathbf{L} \mathbf{f} &= \sum_{i=1}^{N} g_i f_i^2 - \sum_{i=1}^{N} \sum_{i'=1}^{N} f_i f_{i'} w_{ii'} \\
&= \frac{1}{2} \sum_{i=1}^{N} \sum_{i'=1}^{N} w_{ii'} (f_i - f_{i'})^2.
\end{aligned}
\tag{14.64}
$$

Formula 14.64 suggests that a small value of $\mathbf{f}^T \mathbf{L} \mathbf{f}$ will be achieved if pairs of points with large adjacencies have coordinates f_i and $f_{i'}$ close together.

Since $\mathbf{1}^T \mathbf{L} \mathbf{1} = 0$ for any graph, the constant vector is a trivial eigenvector with eigenvalue zero. Not so obvious is the fact that if the graph is connected[5], it is the *only* zero eigenvector (Exercise 14.21). Generalizing this argument, it is easy to show that for a graph with m connected components,

---

[5] A graph is connected if any two nodes can be reached via a path of connected nodes.

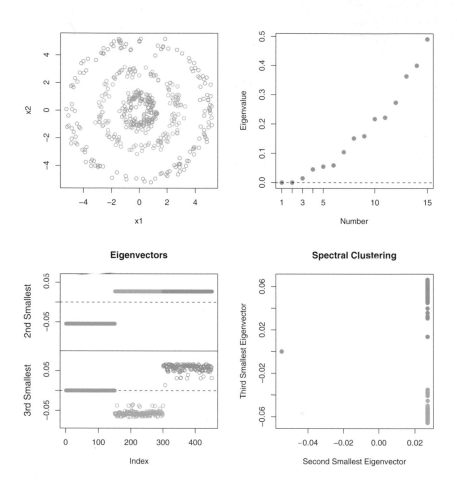

FIGURE 14.29. *Toy example illustrating spectral clustering. Data in top left are 450 points falling in three concentric clusters of 150 points each. The points are uniformly distributed in angle, with radius 1, 2.8 and 5 in the three groups, and Gaussian noise with standard deviation 0.25 added to each point. Using a k = 10 nearest-neighbor similarity graph, the eigenvector corresponding to the second and third smallest eigenvalues of* **L** *are shown in the bottom left; the smallest eigenvector is constant. The data points are colored in the same way as in the top left. The 15 smallest eigenvalues are shown in the top right panel. The coordinates of the 2nd and 3rd eigenvectors (the 450 rows of* **Z***) are plotted in the bottom right panel. Spectral clustering does standard (e.g., K-means) clustering of these points and will easily recover the three original clusters.*

the nodes can be reordered so that $\mathbf{L}$ is block diagonal with a block for each connected component. Then $\mathbf{L}$ has m eigenvectors of eigenvalue zero, and the eigenspace of eigenvalue zero is spanned by the indicator vectors of the connected components. In practice one has strong and weak connections, so zero eigenvalues are approximated by small eigenvalues.

Spectral clustering is an interesting approach for finding non-convex clusters. When a normalized graph Laplacian is used, there is another way to view this method. Defining $\mathbf{P} = \mathbf{G}^{-1}\mathbf{W}$, we consider a random walk on the graph with transition probability matrix $\mathbf{P}$. Then spectral clustering yields groups of nodes such that the random walk seldom transitions from one group to another.

There are a number of issues that one must deal with in applying spectral clustering in practice. We must choose the type of similarity graph—eg. fully connected or nearest neighbors, and associated parameters such as the number of nearest of neighbors k or the scale parameter of the kernel c. We must also choose the number of eigenvectors to extract from $\mathbf{L}$ and finally, as with all clustering methods, the number of clusters. In the toy example of Figure 14.29 we obtained good results for $k \in [5, 200]$, the value 200 corresponding to a fully connected graph. With $k < 5$ the results deteriorated. Looking at the top-right panel of Figure 14.29, we see no strong separation between the smallest three eigenvalues and the rest. Hence it is not clear how many eigenvectors to select.

14.5.4 Kernel Principal Components

Spectral clustering is related to *kernel principal components*, a non-linear version of linear principal components. Standard linear principal components (PCA) are obtained from the eigenvectors of the covariance matrix, and give directions in which the data have maximal variance. Kernel PCA (Schölkopf et al., 1999) expand the scope of PCA, mimicking what we would obtain if we were to expand the features by non-linear transformations, and then apply PCA in this transformed feature space.

We show in Section 18.5.2 that the principal components variables $\mathbf{Z}$ of a data matrix $\mathbf{X}$ can be computed from the inner-product (gram) matrix $\mathbf{K} = \mathbf{X}\mathbf{X}^T$. In detail, we compute the eigen-decomposition of the double-centered version of the gram matrix

$$\widetilde{\mathbf{K}} = (\mathbf{I} - \mathbf{M})\mathbf{K}(\mathbf{I} - \mathbf{M}) = \mathbf{U}\mathbf{D}^2\mathbf{U}^T, \qquad (14.65)$$

with $\mathbf{M} = \mathbf{1}\mathbf{1}^T/N$, and then $\mathbf{Z} = \mathbf{U}\mathbf{D}$. Exercise 18.15 shows how to compute the projections of new observations in this space.

Kernel PCA simply mimics this procedure, interpreting the kernel matrix $\mathbf{K} = \{K(x_i, x_{i'})\}$ as an inner-product matrix of the implicit features $\langle \phi(x_i), \phi(x_{i'}) \rangle$ and finding its eigenvectors. The elements of the mth component $\mathbf{z}_m$ (mth column of $\mathbf{Z}$) can be written (up to centering) as $z_{im} = \sum_{j=1}^{N} \alpha_{jm} K(x_i, x_j)$, where $\alpha_{jm} = u_{jm}/d_m$ (Exercise 14.16).

We can gain more insight into kernel PCA by viewing the $\mathbf{z}_m$ as sample evaluations of principal component *functions* $g_m \in \mathcal{H}_K$, with $\mathcal{H}_K$ the reproducing kernel Hilbert space generated by K (see Section 5.8.1). The first principal component function g_1 solves

$$\max_{g_1 \in \mathcal{H}_K} \text{Var}_{\mathcal{T}} g_1(X) \text{ subject to } ||g_1||_{\mathcal{H}_K} = 1 \qquad (14.66)$$

Here $\text{Var}_{\mathcal{T}}$ refers to the sample variance over training data $\mathcal{T}$. The norm constraint $||g_1||_{\mathcal{H}_K} = 1$ controls the size and roughness of the function g_1, as dictated by the kernel K. As in the regression case it can be shown that the solution to (14.66) is finite dimensional with representation $g_1(x) = \sum_{j=1}^{N} c_j K(x, x_j)$. Exercise 14.17 shows that the solution is defined by $\hat{c}_j = \alpha_{j1}$, $j = 1, \ldots, N$ above. The second principal component function is defined in a similar way, with the additional constraint that $\langle g_1, g_2 \rangle_{\mathcal{H}_K} = 0$, and so on.[6]

Schölkopf et al. (1999) demonstrate the use of kernel principal components as features for handwritten-digit classification, and show that they can improve the performance of a classifier when these are used instead of linear principal components.

Note that if we use the radial kernel

$$K(x, x') \quad = \exp(-||x - x'||^2/c), \qquad (14.67)$$

then the kernel matrix $\mathbf{K}$ has the same form as the similarity matrix $\mathbf{S}$ in spectral clustering. The matrix of edge weights $\mathbf{W}$ is a localized version of $\mathbf{K}$, setting to zero all similarities for pairs of points that are not nearest neighbors.

Kernel PCA finds the eigenvectors corresponding to the largest eigenvalues of $\widetilde{\mathbf{K}}$; this is equivalent to finding the eigenvectors corresponding to the *smallest* eigenvalues of

$$\mathbf{I} - \widetilde{\mathbf{K}}. \qquad (14.68)$$

This is almost the same as the Laplacian (14.63), the differences being the centering of $\widetilde{\mathbf{K}}$ and the fact that $\mathbf{G}$ has the degrees of the nodes along the diagonal.

Figure 14.30 examines the performance of kernel principal components in the toy example of Figure 14.29. In the upper left panel we used the radial kernel with $c = 2$, the same value that was used in spectral clustering. This does not separate the groups, but with $c = 10$ (upper right panel), the first component separates the groups well. In the lower-left panel we applied kernel PCA using the nearest-neighbor radial kernel $\mathbf{W}$ from spectral clustering. In the lower right panel we use the kernel matrix itself as the

---

[6]This section benefited from helpful discussions with Jonathan Taylor.

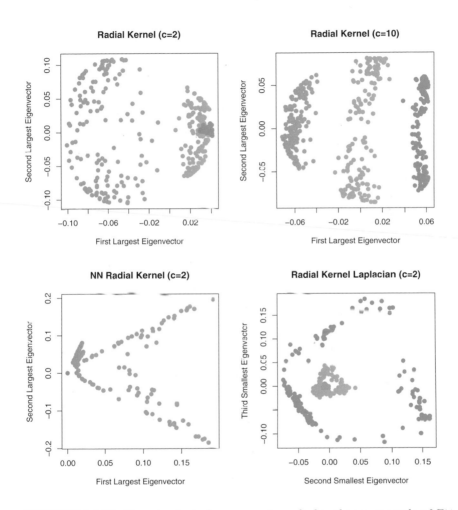

FIGURE 14.30. *Kernel principal components applied to the toy example of Figure 14.29, using different kernels. (Top left:) Radial kernel (14.67) with c = 2. (Top right:) Radial kernel with c = 10. (Bottom left): Nearest neighbor radial kernel* **W** *from spectral clustering. (Bottom right:) Spectral clustering with Laplacian constructed from the radial kernel.*

similarity matrix for constructing the Laplacian (14.63) in spectral clustering. In neither case do the projections separate the two groups. Adjusting c did not help either.

In this toy example, we see that kernel PCA is quite sensitive to the scale and nature of the kernel. We also see that the nearest-neighbor truncation of the kernel is important for the success of spectral clustering.

14.5.5 Sparse Principal Components

We often interpret principal components by examining the direction vectors v_j, also known as *loadings*, to see which variables play a role. We did this with the image loadings in (14.55). Often this interpretation is made easier if the loadings are sparse. In this section we briefly discuss some methods for deriving principal components with sparse loadings. They are all based on lasso (L_1) penalties.

We start with an $N \times p$ data matrix $\mathbf{X}$, with centered columns. The proposed methods focus on either the maximum-variance property of principal components, or the minimum reconstruction error. The *SCoTLASS* procedure of Joliffe et al. (2003) takes the first approach, by solving

$$\max v^T (\mathbf{X}^T \mathbf{X}) v, \text{ subject to } \sum_{j=1}^{p} |v_j| \leq t, \, v^T v = 1. \tag{14.69}$$

The absolute-value constraint encourages some of the loadings to be zero and hence v to be sparse. Further sparse principal components are found in the same way, by forcing the kth component to be orthogonal to the first $k-1$ components. Unfortunately this problem is not convex and the computations are difficult.

Zou et al. (2006) start instead with the regression/reconstruction property of PCA, similar to the approach in Section 14.5.1. Let x_i be the ith row of $\mathbf{X}$. For a single component, their *sparse principal component* technique solves

$$\min_{\theta, v} \sum_{i=1}^{N} ||x_i - \theta v^T x_i||_2^2 + \lambda ||v||_2^2 + \lambda_1 ||v||_1 \tag{14.70}$$

$$\text{subject to } ||\theta||_2 = 1.$$

Let's examine this formulation in more detail.

- If both λ and λ_1 are zero and $N > p$, it is easy to show that $v = \theta$ and is the largest principal component direction.

- When $p \gg N$ the solution is not necessarily unique unless $\lambda > 0$. For any $\lambda > 0$ and $\lambda_1 = 0$ the solution for v is proportional to the largest principal component direction.

- The second penalty on v encourages sparseness of the loadings.

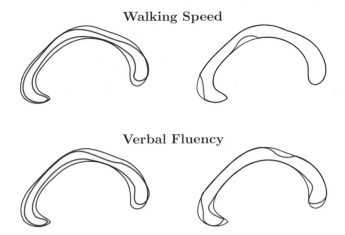

FIGURE 14.31. *Standard and sparse principal components from a study of the corpus callosum variation. The shape variations corresponding to significant principal components (red curves) are overlaid on the mean CC shape (black curves).*

For multiple components, the sparse principal components procedures minimizes

$$\sum_{i=1}^{N} ||x_i - \boldsymbol{\Theta} \mathbf{V}^T x_i||^2 + \lambda \sum_{k=1}^{K} ||v_k||_2^2 + \sum_{k=1}^{K} \lambda_{1k} ||v_k||_1, \qquad (14.71)$$

subject to $\boldsymbol{\Theta}^T \boldsymbol{\Theta} = \mathbf{I}_K$. Here $\mathbf{V}$ is a $p \times K$ matrix with columns v_k and $\boldsymbol{\Theta}$ is also $p \times K$.

Criterion (14.71) is not jointly convex in $\mathbf{V}$ and $\boldsymbol{\Theta}$, but it is convex in each parameter with the other parameter fixed[7]. Minimization over $\mathbf{V}$ with $\boldsymbol{\Theta}$ fixed is equivalent to K elastic net problems (Section 18.4) and can be done efficiently. On the other hand, minimization over $\boldsymbol{\Theta}$ with $\mathbf{V}$ fixed is a version of the Procrustes problem (14.56), and is solved by a simple SVD calculation (Exercise 14.12). These steps are alternated until convergence.

Figure 14.31 shows an example of sparse principal components analysis using (14.71), taken from Sjöstrand et al. (2007). Here the shape of the mid-sagittal cross-section of the corpus callosum (CC) is related to various clinical parameters in a study involving 569 elderly persons[8]. In this exam-

---

[7]Note that the usual principal component criterion, for example (14.50), is not jointly convex in the parameters either. Nevertheless, the solution is well defined and an efficient algorithm is available.

[8]We thank Rasmus Larsen and Karl Sjöstrand for suggesting this application, and supplying us with the postscript figures reproduced here.

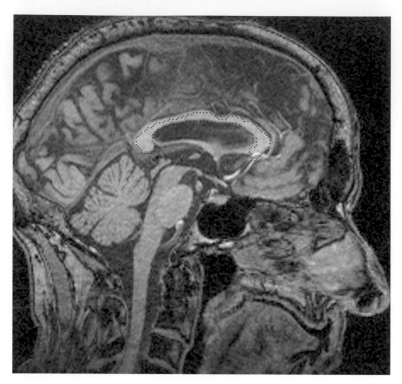

FIGURE 14.32. *An example of a mid-saggital brain slice, with the corpus collosum annotated with landmarks.*

ple PCA is applied to *shape* data, and is a popular tool in morphometrics. For such applications, a number of landmarks are identified along the circumference of the shape; an example is given in Figure 14.32. These are aligned by Procrustes analysis to allow for rotations, and in this case scaling as well (see Section 14.5.1). The features used for PCA are the sequence of coordinate pairs for each landmark, unpacked into a single vector.

In this analysis, both standard and sparse principal components were computed, and components that were significantly associated with various clinical parameters were identified. In the figure, the shape variations corresponding to significant principal components (red curves) are overlaid on the mean CC shape (black curves). Low walking speed relates to CCs that are thinner (displaying atrophy) in regions connecting the motor control and cognitive centers of the brain. Low verbal fluency relates to CCs that are thinner in regions connecting auditory/visual/cognitive centers. The sparse principal components procedure gives a more parsimonious, and potentially more informative picture of the important differences.

14.6   Non-negative Matrix Factorization

Non-negative matrix factorization (Lee and Seung, 1999) is a recent alternative approach to principal components analysis, in which the data and components are assumed to be non-negative. It is useful for modeling non-negative data such as images.

The $N \times p$ data matrix $\mathbf{X}$ is approximated by

$$\mathbf{X} \approx \mathbf{WH} \tag{14.72}$$

where $\mathbf{W}$ is $N \times r$ and $\mathbf{H}$ is $r \times p$, $r \leq \max(N, p)$. We assume that $x_{ij}, w_{ik}, h_{kj} \geq 0$.

The matrices $\mathbf{W}$ and $\mathbf{H}$ are found by maximizing

$$L(\mathbf{W}, \mathbf{H}) = \sum_{i=1}^{N} \sum_{j=1}^{p} [x_{ij} \log(\mathbf{WH})_{ij} - (\mathbf{WH})_{ij}]. \tag{14.73}$$

This is the log-likelihood from a model in which x_{ij} has a Poisson distribution with mean $(\mathbf{WH})_{ij}$—quite reasonable for positive data.

The following alternating algorithm (Lee and Seung, 2001) converges to a local maximum of $L(\mathbf{W}, \mathbf{H})$:

$$
\begin{aligned}
w_{ik} &\leftarrow w_{ik} \frac{\sum_{j=1}^{p} h_{kj} x_{ij} / (\mathbf{WH})_{ij}}{\sum_{j=1}^{p} h_{kj}} \\
h_{kj} &\leftarrow h_{kj} \frac{\sum_{i=1}^{N} w_{ik} x_{ij} / (\mathbf{WH})_{ij}}{\sum_{i=1}^{N} w_{ik}}
\end{aligned}
\tag{14.74}
$$

This algorithm can be derived as a minorization procedure for maximizing $L(\mathbf{W}, \mathbf{H})$ (Exercise 14.23) and is also related to the iterative-proportional-scaling algorithm for log-linear models (Exercise 14.24).

Figure 14.33 shows an example taken from Lee and Seung (1999)[9], comparing non-negative matrix factorization (NMF), vector quantization (VQ, equivalent to k-means clustering) and principal components analysis (PCA). The three learning methods were applied to a database of $N = 2,429$ facial images, each consisting of 19×19 pixels, resulting in a $2,429 \times 381$ matrix $\mathbf{X}$. As shown in the 7×7 array of montages (each a 19×19 image), each method has learned a set of $r = 49$ basis images. Positive values are illustrated with black pixels and negative values with red pixels. A particular instance of a face, shown at top right, is approximated by a linear superposition of basis images. The coefficients of the linear superposition are shown next to each montage, in a 7×7 array[10], and the resulting superpositions are shown to the right of the equality sign. The authors point

---

[9]We thank Sebastian Seung for providing this image.

[10]These 7×7 arrangements allow for a compact display, and have no structural significance.

out that unlike VQ and PCA, NMF learns to represent faces with a set of basis images resembling parts of faces.

Donoho and Stodden (2004) point out a potentially serious problem with non-negative matrix factorization. Even in situations where $\mathbf{X} = \mathbf{WH}$ holds exactly, the decomposition may not be unique. Figure 14.34 illustrates the problem. The data points lie in $p = 2$ dimensions, and there is "open space" between the data and the coordinate axes. We can choose the basis vectors h_1 and h_2 anywhere in this open space, and represent each data point exactly with a nonnegative linear combination of these vectors. This non-uniqueness means that the solution found by the above algorithm depends on the starting values, and it would seem to hamper the interpretability of the factorization. Despite this interpretational drawback, the non-negative matrix factorization and its applications has attracted a lot of interest.

14.6.1 Archetypal Analysis

This method, due to Cutler and Breiman (1994), approximates data points by prototypes that are themselves linear combinations of data points. In this sense it has a similar flavor to K-means clustering. However, rather than approximating each data point by a single nearby prototype, archetypal analysis approximates each data point by a convex combination of a collection of prototypes. The use of a convex combination forces the prototypes to lie on the convex hull of the data cloud. In this sense, the prototypes are "pure.", or "archetypal."

As in (14.72), the $N \times p$ data matrix $\mathbf{X}$ is modeled as

$$\mathbf{X} \approx \mathbf{WH} \tag{14.75}$$

where $\mathbf{W}$ is $N \times r$ and $\mathbf{H}$ is $r \times p$. We assume that $w_{ik} \geq 0$ and $\sum_{k=1}^{r} w_{ik} = 1 \ \forall i$. Hence the N data points (rows of $\mathbf{X}$) in p-dimensional space are represented by convex combinations of the r archetypes (rows of $\mathbf{H}$). We also assume that

$$\mathbf{H} = \mathbf{BX} \tag{14.76}$$

where $\mathbf{B}$ is $r \times N$ with $b_{ki} \geq 0$ and $\sum_{i=1}^{N} b_{ki} = 1 \ \forall k$. Thus the archetypes themselves are convex combinations of the data points. Using both (14.75) and (14.76) we minimize

$$\begin{aligned} J(\mathbf{W}, \mathbf{B}) &= ||\mathbf{X} - \mathbf{WH}||^2 \\ &= ||\mathbf{X} - \mathbf{WBX}||^2 \end{aligned} \tag{14.77}$$

over the weights $\mathbf{W}$ and $\mathbf{B}$. This function is minimized in an alternating fashion, with each separate minimization involving a convex optimization. The overall problem is not convex however, and so the algorithm converges to a local minimum of the criterion.

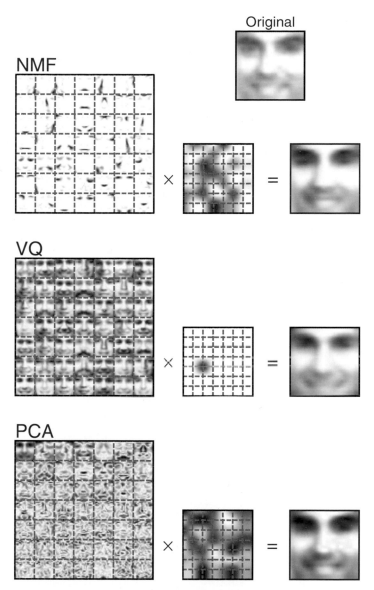

FIGURE 14.33. *Non-negative matrix factorization (NMF), vector quantization (VQ, equivalent to k-means clustering) and principal components analysis (PCA) applied to a database of facial images. Details are given in the text. Unlike VQ and PCA, NMF learns to represent faces with a set of basis images resembling parts of faces.*

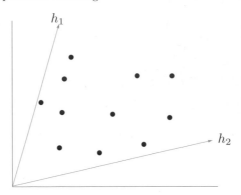

FIGURE 14.34. *Non-uniqueness of the non-negative matrix factorization. There are 11 data points in two dimensions. Any choice of the basis vectors h_1 and h_2 in the open space between the coordinate axes and data, gives an exact reconstruction of the data.*

Figure 14.35 shows an example with simulated data in two dimensions. The top panel displays the results of archetypal analysis, while the bottom panel shows the results from K-means clustering. In order to best reconstruct the data from *convex* combinations of the prototypes, it pays to locate the prototypes on the convex hull of the data. This is seen in the top panels of Figure 14.35 and is the case in general, as proven by Cutler and Breiman (1994). K-means clustering, shown in the bottom panels, chooses prototypes in the middle of the data cloud.

We can think of K-means clustering as a special case of the archetypal model, in which each row of $\mathbf{W}$ has a single one and the rest of the entries are zero.

Notice also that the archetypal model (14.75) has the same general form as the non-negative matrix factorization model (14.72). However, the two models are applied in different settings, and have somewhat different goals. Non-negative matrix factorization aims to approximate the columns of the data matrix $\mathbf{X}$, and the main output of interest are the columns of $\mathbf{W}$ representing the primary non-negative components in the data. Archetypal analysis focuses instead on the approximation of the rows of $\mathbf{X}$ using the rows of $\mathbf{H}$, which represent the archetypal data points. Non-negative matrix factorization also assumes that $r \le p$. With $r = p$, we can get an exact reconstruction simply choosing $\mathbf{W}$ to be the data $\mathbf{X}$ with columns scaled so that they sum to 1. In contrast, archetypal analysis requires $r \le N$, but allows $r > p$. In Figure 14.35, for example, $p = 2, N = 50$ while $r = 2, 4$ or 8. The additional constraint (14.76) implies that the archetypal approximation will not be perfect, even if $r > p$.

Figure 14.36 shows the results of archetypal analysis applied to the database of 3's displayed in Figure 14.22. The three rows in Figure 14.36 are the resulting archetypes from three runs, specifying two, three and four

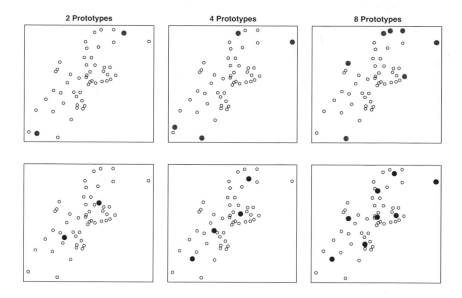

FIGURE 14.35. *Archetypal analysis (top panels) and K-means clustering (bottom panels) applied to 50 data points drawn from a bivariate Gaussian distribution. The colored points show the positions of the prototypes in each case.*

archetypes, respectively. As expected, the algorithm has produced *extreme* 3's both in size and shape.

14.7 Independent Component Analysis and Exploratory Projection Pursuit

Multivariate data are often viewed as multiple indirect measurements arising from an underlying source, which typically cannot be directly measured. Examples include the following:

- Educational and psychological tests use the answers to questionnaires to measure the underlying intelligence and other mental abilities of subjects.

- EEG brain scans measure the neuronal activity in various parts of the brain indirectly via electromagnetic signals recorded at sensors placed at various positions on the head.

- The trading prices of stocks change constantly over time, and reflect various unmeasured factors such as market confidence, external in-

FIGURE 14.36. *Archetypal analysis applied to the database of digitized 3's. The rows in the figure show the resulting archetypes from three runs, specifying two, three and four archetypes, respectively.*

fluences, and other driving forces that may be hard to identify or measure.

Factor analysis is a classical technique developed in the statistical literature that aims to identify these latent sources. Factor analysis models are typically wed to Gaussian distributions, which has to some extent hindered their usefulness. More recently, independent component analysis has emerged as a strong competitor to factor analysis, and as we will see, relies on the non-Gaussian nature of the underlying sources for its success.

14.7.1 Latent Variables and Factor Analysis

The singular-value decomposition $\mathbf{X} = \mathbf{U}\mathbf{D}\mathbf{V}^T$ (14.54) has a latent variable representation. Writing $\mathbf{S} = \sqrt{N}\mathbf{U}$ and $\mathbf{A}^T = \mathbf{D}\mathbf{V}^T/\sqrt{N}$, we have $\mathbf{X} = \mathbf{S}\mathbf{A}^T$, and hence each of the columns of $\mathbf{X}$ is a linear combination of the columns of $\mathbf{S}$. Now since $\mathbf{U}$ is orthogonal, and assuming as before that the columns of $\mathbf{X}$ (and hence $\mathbf{U}$) each have mean zero, this implies that the columns of $\mathbf{S}$ have zero mean, are uncorrelated and have unit variance. In terms of random variables, we can interpret the SVD, or the corresponding principal component analysis (PCA) as an estimate of a latent variable model

$$
\begin{aligned}
X_1 &= a_{11}S_1 + a_{12}S_2 + \cdots + a_{1p}S_p \\
X_2 &= a_{21}S_1 + a_{22}S_2 + \cdots + a_{2p}S_p \\
&\vdots \qquad\qquad \vdots \\
X_p &= a_{p1}S_1 + a_{p2}S_2 + \cdots + a_{pp}S_p,
\end{aligned}
\tag{14.78}
$$

or simply $X = \mathbf{A}S$. The correlated X_j are each represented as a linear expansion in the uncorrelated, unit variance variables S_ℓ. This is not too satisfactory, though, because given any orthogonal $p \times p$ matrix $\mathbf{R}$, we can write

$$
\begin{aligned}
X &= \mathbf{A}S \\
&= \mathbf{A}\mathbf{R}^T \mathbf{R} S \\
&= \mathbf{A}^* S^*,
\end{aligned}
\tag{14.79}
$$

and $\mathrm{Cov}(S^*) = \mathbf{R}\,\mathrm{Cov}(S)\,\mathbf{R}^T = \mathbf{I}$. Hence there are many such decompositions, and it is therefore impossible to identify any particular latent variables as unique underlying sources. The SVD decomposition does have the property that any rank $q < p$ truncated decomposition approximates $\mathbf{X}$ in an optimal way.

The classical *factor analysis* model, developed primarily by researchers in psychometrics, alleviates these problems to some extent; see, for example, Mardia et al. (1979). With $q < p$, a factor analysis model has the form

$$
\begin{aligned}
X_1 &= a_{11}S_1 + \cdots + a_{1q}S_q + \varepsilon_1 \\
X_2 &= a_{21}S_1 + \cdots + a_{2q}S_q + \varepsilon_2 \\
&\vdots \qquad\qquad \vdots \\
X_p &= a_{p1}S_1 + \cdots + a_{pq}S_q + \varepsilon_p,
\end{aligned}
\tag{14.80}
$$

or $X = \mathbf{A}S + \varepsilon$. Here S is a vector of $q < p$ underlying latent variables or factors, $\mathbf{A}$ is a $p \times q$ matrix of factor *loadings*, and the ε_j are uncorrelated zero-mean disturbances. The idea is that the latent variables S_ℓ are common sources of variation amongst the X_j, and account for their correlation structure, while the uncorrelated ε_j are unique to each X_j and pick up the remaining unaccounted variation. Typically the S_ℓ and the ε_j are modeled as Gaussian random variables, and the model is fit by maximum likelihood. The parameters all reside in the covariance matrix

$$
\boldsymbol{\Sigma} = \mathbf{A}\mathbf{A}^T + \mathbf{D}_\varepsilon,
\tag{14.81}
$$

where $\mathbf{D}_\varepsilon = \mathrm{diag}[\mathrm{Var}(\varepsilon_1), \ldots, \mathrm{Var}(\varepsilon_p)]$. The S_ℓ being Gaussian and uncorrelated makes them statistically independent random variables. Thus a battery of educational test scores would be thought to be driven by the independent underlying factors such as *intelligence*, *drive* and so on. The columns of $\mathbf{A}$ are referred to as the *factor loadings*, and are used to name and interpret the factors.

Unfortunately the identifiability issue (14.79) remains, since $\mathbf{A}$ and $\mathbf{A}\mathbf{R}^T$ are equivalent in (14.81) for any $q \times q$ orthogonal $\mathbf{R}$. This leaves a certain subjectivity in the use of factor analysis, since the user can search for rotated versions of the factors that are more easily interpretable. This aspect has left many analysts skeptical of factor analysis, and may account for its lack of popularity in contemporary statistics. Although we will not go into details here, the SVD plays a key role in the estimation of (14.81). For example, if the $\mathrm{Var}(\varepsilon_j)$ are all assumed to be equal, the leading q components of the SVD identify the subspace determined by $\mathbf{A}$.

Because of the separate disturbances ε_j for each X_j, factor analysis can be seen to be modeling the correlation structure of the X_j rather than the covariance structure. This can be easily seen by standardizing the covariance structure in (14.81) (Exercise 14.14). This is an important distinction between factor analysis and PCA, although not central to the discussion here. Exercise 14.15 discusses a simple example where the solutions from factor analysis and PCA differ dramatically because of this distinction.

14.7.2 Independent Component Analysis

The independent component analysis (ICA) model has exactly the same form as (14.78), except the S_ℓ are assumed to be *statistically independent* rather than uncorrelated. Intuitively, lack of correlation determines the second-degree cross-moments (covariances) of a multivariate distribution, while in general statistical independence determines all of the cross-moments. These extra moment conditions allow us to identify the elements of $\mathbf{A}$ uniquely. Since the multivariate Gaussian distribution is determined by its second moments alone, it is the exception, and any Gaussian independent components can be determined only up to a rotation, as before. Hence identifiability problems in (14.78) and (14.80) can be avoided if we assume that the S_ℓ are independent and *non-Gaussian*.

Here we will discuss the full p-component model as in (14.78), where the S_ℓ are independent with unit variance; ICA versions of the factor analysis model (14.80) exist as well. Our treatment is based on the survey article by Hyvärinen and Oja (2000).

We wish to recover the mixing matrix $\mathbf{A}$ in $X = \mathbf{A}S$. Without loss of generality, we can assume that X has already been *whitened* to have $\mathrm{Cov}(X) = \mathbf{I}$; this is typically achieved via the SVD described above. This in turn implies that $\mathbf{A}$ is orthogonal, since S also has covariance $\mathbf{I}$. So solving the ICA problem amounts to finding an orthogonal $\mathbf{A}$ such that the components of the vector random variable $S = \mathbf{A}^T X$ are independent (and non-Gaussian).

Figure 14.37 shows the power of ICA in separating two mixed signals. This is an example of the classical *cocktail party problem*, where different microphones X_j pick up mixtures of different independent sources S_ℓ (music, speech from different speakers, etc.). ICA is able to perform *blind*

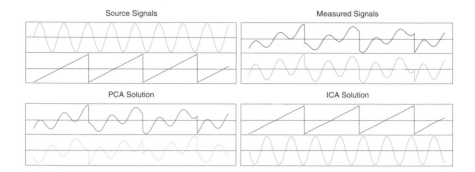

FIGURE 14.37. *Illustration of ICA vs. PCA on artificial time-series data. The upper left panel shows the two source signals, measured at 1000 uniformly spaced time points. The upper right panel shows the observed mixed signals. The lower two panels show the principal components and independent component solutions.*

source separation, by exploiting the independence and non-Gaussianity of the original sources.

Many of the popular approaches to ICA are based on entropy. The differential entropy H of a random variable Y with density $g(y)$ is given by

$$H(Y) = -\int g(y) \log g(y) dy. \tag{14.82}$$

A well-known result in information theory says that among all random variables with equal variance, Gaussian variables have the maximum entropy. Finally, the *mutual information* $I(Y)$ between the components of the random vector Y is a natural measure of dependence:

$$I(Y) = \sum_{j=1}^{p} H(Y_j) - H(Y). \tag{14.83}$$

The quantity $I(Y)$ is called the *Kullback–Leibler distance* between the density $g(y)$ of Y and its independence version $\prod_{j=1}^{p} g_j(y_j)$, where $g_j(y_j)$ is the marginal density of Y_j. Now if X has covariance $\mathbf{I}$, and $Y = \mathbf{A}^T X$ with $\mathbf{A}$ orthogonal, then it is easy to show that

$$I(Y) = \sum_{j=1}^{p} H(Y_j) - H(X) - \log|\det \mathbf{A}| \tag{14.84}$$

$$= \sum_{j=1}^{p} H(Y_j) - H(X). \tag{14.85}$$

Finding an $\mathbf{A}$ to minimize $I(Y) = I(\mathbf{A}^T X)$ looks for the orthogonal transformation that leads to the most independence between its components. In

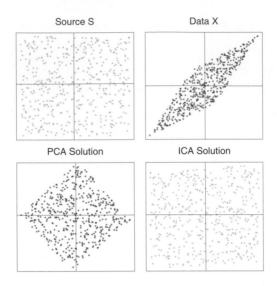

FIGURE 14.38. *Mixtures of independent uniform random variables. The upper left panel shows 500 realizations from the two independent uniform sources, the upper right panel their mixed versions. The lower two panels show the PCA and ICA solutions, respectively.*

light of (14.84) this is equivalent to minimizing the sum of the entropies of the separate components of Y, which in turn amounts to maximizing their departures from Gaussianity.

For convenience, rather than using the entropy $H(Y_j)$, Hyvärinen and Oja (2000) use the *negentropy* measure $J(Y_j)$ defined by

$$J(Y_j) = H(Z_j) - H(Y_j), \qquad (14.86)$$

where Z_j is a Gaussian random variable with the same variance as Y_j. Negentropy is non-negative, and measures the departure of Y_j from Gaussianity. They propose simple approximations to negentropy which can be computed and optimized on data. The ICA solutions shown in Figures 14.37–14.39 use the approximation

$$J(Y_j) \approx [EG(Y_j) - EG(Z_j)]^2, \qquad (14.87)$$

where $G(u) = \frac{1}{a} \log \cosh(au)$ for $1 \leq a \leq 2$. When applied to a sample of x_i, the expectations are replaced by data averages. This is one of the options in the `FastICA` software provided by these authors. More classical (and less robust) measures are based on fourth moments, and hence look for departures from the Gaussian via kurtosis. See Hyvärinen and Oja (2000) for more details. In Section 14.7.4 we describe their approximate Newton algorithm for finding the optimal directions.

In summary then, ICA applied to multivariate data looks for a sequence of orthogonal projections such that the projected data look as far from

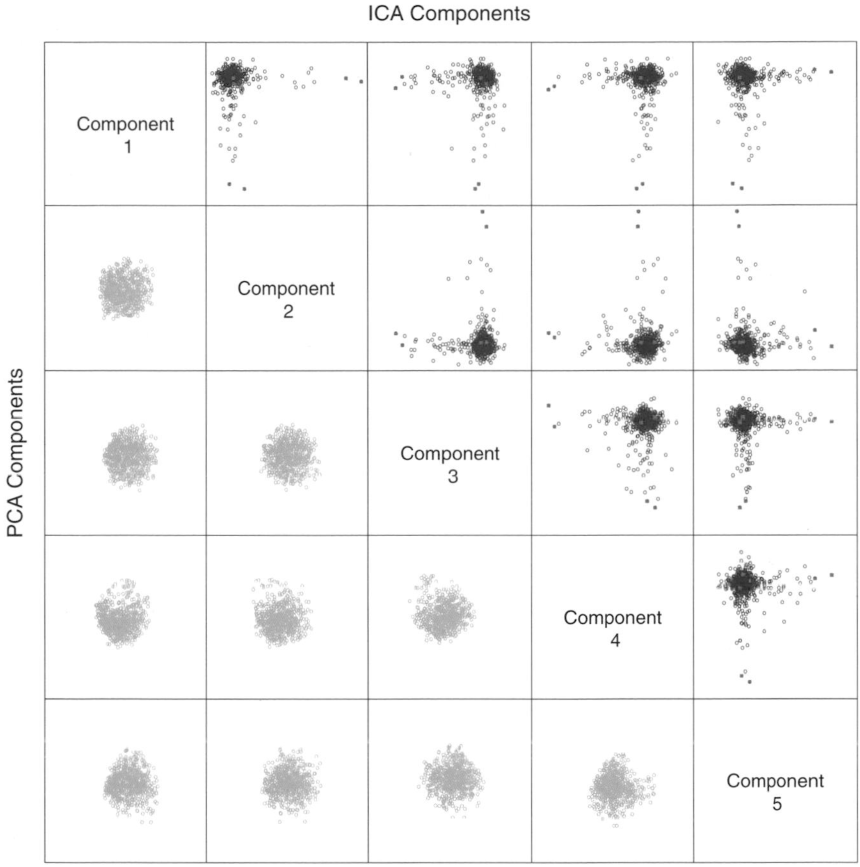

ICA Components

FIGURE 14.39. *A comparison of the first five ICA components computed using* FastICA *(above diagonal) with the first five PCA components(below diagonal). Each component is standardized to have unit variance.*

Gaussian as possible. With pre-whitened data, this amounts to looking for components that are as independent as possible.

ICA starts from essentially a factor analysis solution, and looks for rotations that lead to independent components. From this point of view, ICA is just another factor rotation method, along with the traditional "varimax" and "quartimax" methods used in psychometrics.

Example: Handwritten Digits

We revisit the handwritten threes analyzed by PCA in Section 14.5.1. Figure 14.39 compares the first five (standardized) principal components with the first five ICA components, all shown in the same standardized units. Note that each plot is a two-dimensional projection from a 256-dimensional

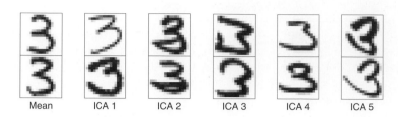

FIGURE 14.40. *The highlighted digits from Figure 14.39. By comparing with the mean digits, we see the nature of the ICA component.*

space. While the PCA components all appear to have joint Gaussian distributions, the ICA components have long-tailed distributions. This is not too surprising, since PCA focuses on variance, while ICA specifically looks for non-Gaussian distributions. All the components have been standardized, so we do not see the decreasing variances of the principal components.

For each ICA component we have highlighted two of the extreme digits, as well as a pair of central digits and displayed them in Figure 14.40. This illustrates the nature of each of the components. For example, ICA component five picks up the long sweeping tailed threes.

Example: EEG Time Courses

ICA has become an important tool in the study of brain dynamics—the example we present here uses ICA to untangle the components of signals in multi-channel electroencephalographic (EEG) data (Onton and Makeig, 2006).

Subjects wear a cap embedded with a lattice of 100 EEG electrodes, which record brain activity at different locations on the scalp. Figure 14.41[11] (top panel) shows 15 seconds of output from a subset of nine of these electrodes from a subject performing a standard "two-back" learning task over a 30 minute period. The subject is presented with a letter (B, H, J, C, F, or K) at roughly 1500-ms intervals, and responds by pressing one of two buttons to indicate whether the letter presented is the same or different from that presented two steps back. Depending on the answer, the subject earns or loses points, and occasionally earns bonus or loses penalty points. The time-course data show spatial correlation in the EEG signals—the signals of nearby sensors look very similar.

The key assumption here is that signals recorded at each scalp electrode are a mixture of independent potentials arising from different cortical ac-

---

[11]Reprinted from *Progress in Brain Research*, Vol. 159, Julie Onton and Scott Makeig, "Information based modeling of event-related brain dynamics," Page 106 , Copyright (2006), with permission from Elsevier. We thank Julie Onton and Scott Makeig for supplying an electronic version of the image.

tivities, as well as non-cortical artifact domains; see the reference for a detailed overview of ICA in this domain.

The lower part of Figure 14.41 shows a selection of ICA components. The colored images represent the estimated unmixing coefficient vectors $\hat{a}_j$ as heatmap images superimposed on the scalp, indicating the location of activity. The corresponding time-courses show the activity of the learned ICA components.

For example, the subject blinked after each performance feedback signal (colored vertical lines), which accounts for the location and artifact signal in IC1 and IC3. IC12 is an artifact associated with the cardiac pulse. IC4 and IC7 account for frontal theta-band activities, and appear after a stretch of correct performance. See Onton and Makeig (2006) for a more detailed discussion of this example, and the use of ICA in EEG modeling.

14.7.3   Exploratory Projection Pursuit

Friedman and Tukey (1974) proposed *exploratory projection pursuit,* a graphical exploration technique for visualizing high-dimensional data. Their view was that most low (one- or two-dimensional) projections of high-dimensional data look Gaussian. Interesting structure, such as clusters or long tails, would be revealed by non-Gaussian projections. They proposed a number of *projection indices* for optimization, each focusing on a different departure from Gaussianity. Since their initial proposal, a variety of improvements have been suggested (Huber, 1985; Friedman, 1987), and a variety of indices, including entropy, are implemented in the interactive graphics package Xgobi (Swayne et al., 1991, now called GGobi). These projection indices are exactly of the same form as $J(Y_j)$ above, where $Y_j = a_j^T X$, a normalized linear combination of the components of X. In fact, some of the approximations and substitutions for cross-entropy coincide with indices proposed for projection pursuit. Typically with projection pursuit, the directions a_j are not constrained to be orthogonal. Friedman (1987) transforms the data to look Gaussian in the chosen projection, and then searches for subsequent directions. Despite their different origins, ICA and exploratory projection pursuit are quite similar, at least in the representation described here.

14.7.4   A Direct Approach to ICA

Independent components have by definition a joint product density

$$f_S(s) = \prod_{j=1}^{p} f_j(s_j), \tag{14.88}$$

so here we present an approach that estimates this density directly using generalized additive models (Section 9.1). Full details can be found in

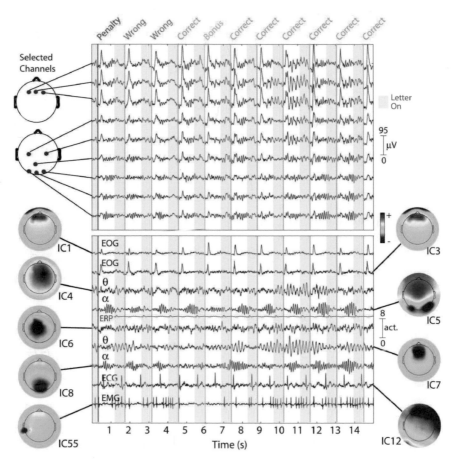

FIGURE 14.41. *Fifteen seconds of EEG data (of 1917 seconds) at nine (of 100) scalp channels (top panel), as well as nine ICA components (lower panel). While nearby electrodes record nearly identical mixtures of brain and non-brain activity, ICA components are temporally distinct. The colored scalps represent the ICA unmixing coefficients $\hat{a}_j$ as a heatmap, showing brain or scalp location of the source.*

Hastie and Tibshirani (2003), and the method is implemented in the R package `ProDenICA`, available from CRAN.

In the spirit of representing departures from Gaussianity, we represent each f_j as

$$f_j(s_j) = \phi(s_j)e^{g_j(s_j)}, \qquad (14.89)$$

a *tilted* Gaussian density. Here ϕ is the standard Gaussian density, and g_j satisfies the normalization conditions required of a density. Assuming as before that X is pre-whitened, the log-likelihood for the observed data $X = \mathbf{A}S$ is

$$\ell(\mathbf{A}, \{g_j\}_1^p; \mathbf{X}) = \sum_{i=1}^{N} \sum_{j=1}^{p} \left[\log \phi_j(a_j^T x_i) + g_j(a_j^T x_i) \right], \qquad (14.90)$$

which we wish to maximize subject to the constraints that $\mathbf{A}$ is orthogonal and that the g_j result in densities in (14.89). Without imposing any further restrictions on g_j, the model (14.90) is over-parametrized, so we instead maximize a regularized version

$$\sum_{j=1}^{p} \left[\frac{1}{N} \sum_{i=1}^{N} \left[\log \phi(a_j^T x_i) + g_j(a_j^T x_i) \right] - \int \phi(t)e^{g_j(t)} dt - \lambda_j \int \{g_j'''(t)\}^2(t) dt \right].$$
$$(14.91)$$

We have subtracted two penalty terms (for each j) in (14.91), inspired by Silverman (1986, Section 5.4.4):

- The first enforces the density constraint $\int \phi(t)e^{\hat{g}_j(t)} dt = 1$ on any solution $\hat{g}_j$.

- The second is a roughness penalty, which guarantees that the solution $\hat{g}_j$ is a quartic-spline with knots at the observed values of $s_{ij} = a_j^T x_i$.

It can further be shown that the solution densities $\hat{f}_j = \phi e^{\hat{g}_j}$ each have mean zero and variance one (Exercise 14.18). As we increase λ_j, these solutions approach the standard Gaussian ϕ.

Algorithm 14.3 *Product Density ICA Algorithm:* `ProDenICA`

---

1. Initialize $\mathbf{A}$ (random Gaussian matrix followed by orthogonalization).

2. Alternate until convergence of $\mathbf{A}$:

   (a) Given $\mathbf{A}$, optimize (14.91) w.r.t. g_j (separately for each j).

   (b) Given g_j, $j = 1, \ldots, p$, perform one step of a fixed point algorithm towards finding the optimal $\mathbf{A}$.

We fit the functions g_j and directions a_j by optimizing (14.91) in an alternating fashion, as described in Algorithm 14.3.

Step 2(a) amounts to a semi-parametric density estimation, which can be solved using a novel application of generalized additive models. For convenience we extract one of the p separate problems,

$$\frac{1}{N}\sum_{i=1}^{N}[\log \phi(s_i) + g(s_i)] - \int \phi(t)e^{g(t)}dt - \lambda \int \{g'''(t)\}^2(t)dt. \quad (14.92)$$

Although the second integral in (14.92) leads to a smoothing spline, the first integral is problematic, and requires an approximation. We construct a fine grid of L values s_ℓ^* in increments Δ covering the observed values s_i, and count the number of s_i in the resulting bins:

$$y_\ell^* = \frac{\#s_i \in (s_\ell^* - \Delta/2, s_\ell^* + \Delta/2)}{N}. \quad (14.93)$$

Typically we pick L to be 1000, which is more than adequate. We can then approximate (14.92) by

$$\sum_{\ell=1}^{L}\left\{y_\ell^*\left[\log(\phi(s_\ell^*)) + g(s_\ell^*)\right] - \Delta\phi(s_\ell^*)e^{g(s_\ell^*)}\right\} - \lambda \int g'''^2(s)ds. \quad (14.94)$$

This last expression can be seen to be proportional to a penalized Poisson log-likelihood with response y_ℓ^*/Δ and penalty parameter λ/Δ, and mean $\mu(s) = \phi(s)e^{g(s)}$. This is a *generalized additive spline model* (Hastie and Tibshirani, 1990; Efron and Tibshirani, 1996), with an *offset* term $\log \phi(s)$, and can be fit using a Newton algorithm in $O(L)$ operations. Although a quartic spline is called for, we find in practice that a cubic spline is adequate. We have p tuning parameters λ_j to set; in practice we make them all the same, and specify the amount of smoothing via the effective degrees-of-freedom $\mathrm{df}(\lambda)$. Our software uses 5df as a default value.

Step 2(b) in Algorithm 14.3 requires optimizing (14.91) with respect to $\mathbf{A}$, holding the $\hat{g}_j$ fixed. Only the first terms in the sum involve $\mathbf{A}$, and since $\mathbf{A}$ is orthogonal, the collection of terms involving ϕ do not depend on $\mathbf{A}$ (Exercise 14.19). Hence we need to maximize

$$C(\mathbf{A}) = \frac{1}{N}\sum_{j=1}^{p}\sum_{i=1}^{N}\hat{g}_j(a_j^T x_i) \quad (14.95)$$

$$= \sum_{j=1}^{p}C_j(a_j)$$

$C(\mathbf{A})$ is a log-likelihood ratio between the fitted density and a Gaussian, and can be seen as an estimate of negentropy (14.86), with each $\hat{g}_j$ a contrast function as in (14.87). The fixed point update in step 2(b) is a modified Newton step (Exercise 14.20)

1. For each j update

$$a_j \leftarrow \mathrm{E}\left\{X\hat{g}_j'(a_j^T X) - \mathrm{E}[\hat{g}_j''(a_j^T X)]a_j\right\},\qquad(14.96)$$

where E represents expectation w.r.t the sample x_i. Since $\hat{g}_j$ is a fitted quartic (or cubic) spline, the first and second derivatives are readily available.

2. Orthogonalize $\mathbf{A}$ using the symmetric square-root transformation $(\mathbf{A}\mathbf{A}^T)^{-\frac{1}{2}}\mathbf{A}$. If $\mathbf{A} = \mathbf{U}\mathbf{D}\mathbf{V}^T$ is the SVD of $\mathbf{A}$, it is easy to show that this leads to the update $\mathbf{A} \leftarrow \mathbf{U}\mathbf{V}^T$.

Our `ProDenICA` algorithm works as well as `FastICA` on the artificial time series data of Figure 14.37, the mixture of uniforms data of Figure 14.38, and the digit data in Figure 14.39.

Example: Simulations

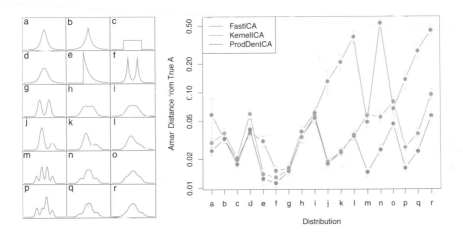

FIGURE 14.42. *The left panel shows 18 distributions used for comparisons. These include the "t", uniform, exponential, mixtures of exponentials, symmetric and asymmetric Gaussian mixtures. The right panel shows (on the log scale) the average Amari metric for each method and each distribution, based on 30 simulations in $\mathbb{R}^2$ for each distribution.*

Figure 14.42 shows the results of a simulation comparing `ProDenICA` to `FastICA`, and another semi-parametric competitor `KernelICA` (Bach and Jordan, 2002). The left panel shows the 18 distributions used as a basis of comparison. For each distribution, we generated a pair of independent components ($N = 1024$), and a random mixing matrix in $\mathbb{R}^2$ with condition number between 1 and 2. We used our R implementations of `FastICA`, using the negentropy criterion (14.87), and `ProDenICA`. For `KernelICA` we used the authors MATLAB code.[12] Since the search criteria are nonconvex, we used five random starts for each method. Each of the algorithms delivers an orthogonal mixing matrix $\mathbf{A}$ (the data were *pre-whitened*), which is available for comparison with the generating orthogonalized mixing matrix $\mathbf{A}_0$. We used the Amari metric (Bach and Jordan, 2002) as a measure of the closeness of the two frames:

$$d(\mathbf{A}_0, \mathbf{A}) = \frac{1}{2p}\sum_{i=1}^{p}\left(\frac{\sum_{j=1}^{p}|r_{ij}|}{\max_j |r_{ij}|} - 1\right) + \frac{1}{2p}\sum_{j=1}^{p}\left(\frac{\sum_{i=1}^{p}|r_{ij}|}{\max_i |r_{ij}|} - 1\right), \quad (14.97)$$

where $r_{ij} = (\mathbf{A}_o\mathbf{A}^{-1})_{ij}$. The right panel in Figure 14.42 compares the averages (on the log scale) of the Amari metric between the truth and the estimated mixing matrices. *ProDenICA* is competitive with *FastICA* and *KernelICA* in all situations, and dominates most of the mixture simulations.

14.8   Multidimensional Scaling

Both self-organizing maps and principal curves and surfaces map data points in $\mathbb{R}^p$ to a lower-dimensional manifold. Multidimensional scaling (MDS) has a similar goal, but approaches the problem in a somewhat different way.

We start with observations $x_1, x_2, \ldots, x_N \in \mathbb{R}^p$, and let d_{ij} be the distance between observations i and j. Often we choose Euclidean distance $d_{ij} = ||x_i - x_j||$, but other distances may be used. Further, in some applications we may not even have available the data points x_i, but only have some *dissimilarity* measure d_{ij} (see Section 14.3.10). For example, in a wine tasting experiment, d_{ij} might be a measure of how different a subject judged wines i and j, and the subject provides such a measure for all pairs of wines i, j. MDS requires only the dissimilarities d_{ij}, in contrast to the SOM and principal curves and surfaces which need the data points x_i.

Multidimensional scaling seeks values $z_1, z_2, \ldots, z_N \in \mathbb{R}^k$ to minimize the so-called *stress function*[13]

$$S_M(z_1, z_2, \ldots, z_N) = \sum_{i\neq i'}(d_{ii'} - ||z_i - z_{i'}||)^2. \quad (14.98)$$

This is known as *least squares* or *Kruskal–Shephard* scaling. The idea is to find a lower-dimensional representation of the data that preserves the pairwise distances as well as possible. Notice that the approximation is

---

[12]Francis Bach kindly supplied this code, and helped us set up the simulations.
[13]Some authors define stress as the square-root of S_M; since it does not affect the optimization, we leave it squared to make comparisons with other criteria simpler.

in terms of the distances rather than squared distances (which results in slightly messier algebra). A gradient descent algorithm is used to minimize S_M.

A variation on least squares scaling is the so-called *Sammon mapping* which minimizes

$$S_{Sm}(z_1, z_2, \ldots, z_N) = \sum_{i \neq i'} \frac{(d_{ii'} - ||z_i - z_{i'}||)^2}{d_{ii'}}. \qquad (14.99)$$

Here more emphasis is put on preserving smaller pairwise distances.

In *classical scaling*, we instead start with similarities $s_{ii'}$: often we use the centered inner product $s_{ii'} = \langle x_i - \bar{x}, x_{i'} - \bar{x} \rangle$. The problem then is to minimize

$$S_C(z_1, z_2, \ldots, z_N) = \sum_{i,i'} (s_{ii'} - \langle z_i - \bar{z}, z_{i'} - \bar{z} \rangle)^2 \qquad (14.100)$$

over $z_1, z_2, \ldots, z_N \in \mathbb{R}^k$. This is attractive because there is an explicit solution in terms of eigenvectors: see Exercise 14.11. If we have distances rather than inner-products, we can convert them to centered inner-products if the distances are *Euclidean*;[14] see (18.31) on page 671 in Chapter 18. If the similarities are in fact centered inner-products, classical scaling is exactly equivalent to principal components, an inherently linear dimension-reduction technique. Classical scaling is not equivalent to least squares scaling; the loss functions are different, and the mapping can be nonlinear.

Least squares and classical scaling are referred to as *metric* scaling methods, in the sense that the actual dissimilarities or similarities are approximated. *Shephard–Kruskal nonmetric scaling* effectively uses only ranks. Nonmetric scaling seeks to minimize the stress function

$$S_{NM}(z_1, z_2, \ldots, z_N) = \frac{\sum_{i \neq i'} \left[||z_i - z_{i'}|| - \theta(d_{ii'})\right]^2}{\sum_{i \neq i'} ||z_i - z_{i'}||^2} \qquad (14.101)$$

over the z_i and an arbitrary increasing function θ. With θ fixed, we minimize over z_i by gradient descent. With the z_i fixed, the method of isotonic regression is used to find the best monotonic approximation $\theta(d_{ii'})$ to $||z_i - z_{i'}||$. These steps are iterated until the solutions stabilize.

Like the self-organizing map and principal surfaces, multidimensional scaling represents high-dimensional data in a low-dimensional coordinate system. Principal surfaces and SOMs go a step further, and approximate the original data by a low-dimensional manifold, parametrized in the low dimensional coordinate system. In a principal surface and SOM, points

---

[14]An $N \times N$ distance matrix is Euclidean if the entries represent pairwise Euclidean distances between N points in some dimensional space.

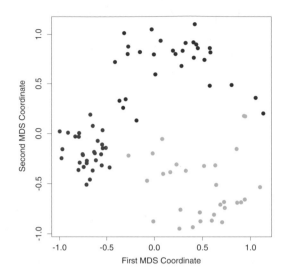

FIGURE 14.43. *First two coordinates for half-sphere data, from classical multidimensional scaling.*

close together in the original feature space should map close together on the manifold, but points far apart in feature space might also map close together. This is less likely in multidimensional scaling since it explicitly tries to preserve all pairwise distances.

Figure 14.43 shows the first two MDS coordinates from classical scaling for the half-sphere example. There is clear separation of the clusters, and the tighter nature of the red cluster is apparent.

14.9 Nonlinear Dimension Reduction and Local Multidimensional Scaling

Several methods have been recently proposed for nonlinear dimension reduction, similar in spirit to principal surfaces. The idea is that the data lie close to an intrinsically low-dimensional nonlinear manifold embedded in a high-dimensional space. These methods can be thought of as "flattening" the manifold, and hence reducing the data to a set of low-dimensional coordinates that represent their relative positions in the manifold. They are useful for problems where signal-to-noise ratio is very high (e.g., physical systems), and are probably not as useful for observational data with lower signal-to-noise ratios.

The basic goal is illustrated in the left panel of Figure 14.44. The data lie near a parabola with substantial curvature. Classical MDS does not pre-

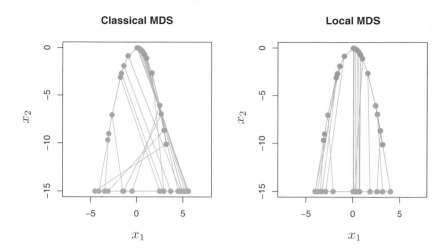

FIGURE 14.44. *The orange points show data lying on a parabola, while the blue points shows multidimensional scaling representations in one dimension. Classical multidimensional scaling (left panel) does not preserve the ordering of the points along the curve, because it judges points on opposite ends of the curve to be close together. In contrast, local multidimensional scaling (right panel) does a good job of preserving the ordering of the points along the curve.*

serve the ordering of the points along the curve, because it judges points on opposite ends of the curve to be close together. The right panel shows the results of *local multi-dimensional scaling*, one of the three methods for non-linear multi-dimensional scaling that we discuss below. These methods use only the coordinates of the points in p dimensions, and have no other information about the manifold. Local MDS has done a good job of preserving the ordering of the points along the curve.

We now briefly describe three new approaches to nonlinear dimension reduction and manifold mapping.

Isometric feature mapping (ISOMAP) (Tenenbaum et al., 2000) constructs a graph to approximate the geodesic distance between points along the manifold. Specifically, for each data point we find its neighbors—points within some small Euclidean distance of that point. We construct a graph with an edge between any two neighboring points. The geodesic distance between any two points is then approximated by the shortest path between points on the graph. Finally, classical scaling is applied to the graph distances, to produce a low-dimensional mapping.

Local linear embedding (Roweis and Saul, 2000) takes a very different approach, trying to preserve the local affine structure of the high-dimensional data. Each data point is approximated by a linear combination of neighboring points. Then a lower dimensional representation is constructed that

best preserves these local approximations. The details are interesting, so
we give them here.

1. For each data point x_i in p dimensions, we find its K-nearest neighbors $\mathcal{N}(i)$ in Euclidean distance.

2. We approximate each point by an affine mixture of the points in its neighborhood:

$$\min_{W_{ik}} ||x_i - \sum_{k \in \mathcal{N}(i)} w_{ik} x_k||^2 \tag{14.102}$$

over weights w_{ik} satisfying $w_{ik} = 0$, $k \notin \mathcal{N}(i)$, $\sum_{k=1}^{N} w_{ik} = 1$. w_{ik} is the contribution of point k to the reconstruction of point i. Note that for a hope of a unique solution, we must have $K < p$.

3. Finally, we find points y_i in a space of dimension $d < p$ to minimize

$$\sum_{i=1}^{N} ||y_i - \sum_{k=1}^{N} w_{ik} y_k||^2 \tag{14.103}$$

with w_{ik} fixed.

In step 3, we minimize

$$\text{tr}[(\mathbf{Y} - \mathbf{WY})^T (\mathbf{Y} - \mathbf{WY})] = \text{tr}[\mathbf{Y}^T (\mathbf{I} - \mathbf{W})^T (\mathbf{I} - \mathbf{W})\mathbf{Y}] \tag{14.104}$$

where $\mathbf{W}$ is $N \times N$; $\mathbf{Y}$ is $N \times d$, for some small $d < p$. The solutions $\hat{\mathbf{Y}}$ are the trailing eigenvectors of $\mathbf{M} = (\mathbf{I} - \mathbf{W})^T (\mathbf{I} - \mathbf{W})$. Since $\mathbf{1}$ is a trivial eigenvector with eigenvalue 0, we discard it and keep the next d. This has the side effect that $\mathbf{1}^T \mathbf{Y} = 0$, and hence the embedding coordinates are mean centered.

Local MDS (Chen and Buja, 2008) takes the simplest and arguably the most direct approach. We define $\mathcal{N}$ to be the symmetric set of nearby pairs of points; specifically a pair (i, i') is in $\mathcal{N}$ if point i is among the K-nearest neighbors of i', or vice-versa. Then we construct the stress function

$$S_L(z_1, z_2, \ldots, z_N) = \sum_{(i,i') \in \mathcal{N}} (d_{ii'} - ||z_i - z_{i'}||)^2$$
$$+ \sum_{(i,i') \notin \mathcal{N}} w \cdot (D - ||z_i - z_{i'}||)^2. \tag{14.105}$$

Here D is some large constant and w is a weight. The idea is that points that are not neighbors are considered to be very far apart; such pairs are given a small weight w so that they don't dominate the overall stress function. To simplify the expression, we take $w \sim 1/D$, and let $D \to \infty$. Expanding (14.105), this gives

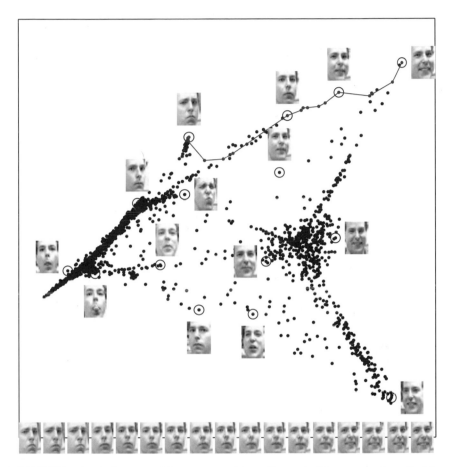

FIGURE 14.45. *Images of faces mapped into the embedding space described by the first two coordinates of LLE. Next to the circled points, representative faces are shown in different parts of the space. The images at the bottom of the plot correspond to points along the top right path (linked by solid line), and illustrate one particular mode of variability in pose and expression.*

$$S_L(z_1, z_2, \ldots, z_N) = \sum_{(i,i') \in \mathcal{N}} (d_{ii'} - ||z_i - z_{i'}||)^2 - \tau \sum_{(i,i') \notin \mathcal{N}} ||z_i - z_{i'}||,$$

$$(14.106)$$

where $\tau = 2wD$. The first term in (14.106) tries to preserve local structure in the data, while the second term encourages the representations $z_i, z_{i'}$ for pairs (i, i') that are non-neighbors to be farther apart. Local MDS minimizes the stress function (14.106) over z_i, for fixed values of the number of neighbors K and the tuning parameter τ.

The right panel of Figure 14.44 shows the result of local MDS, using $k = 2$ neighbors and $\tau = 0.01$. We used coordinate descent with multiple starting values to find a good minimum of the (nonconvex) stress function (14.106). The ordering of the points along the curve has been largely preserved,

Figure 14.45 shows a more interesting application of one of these methods (LLE)[15]. The data consist of 1965 photographs, digitized as 20×28 grayscale images. The result of the first two-coordinates of LLE are shown and reveal some variability in pose and expression. Similar pictures were produced by local MDS.

In experiments reported in Chen and Buja (2008), local MDS shows superior performance, as compared to ISOMAP and LLE. They also demonstrate the usefulness of local MDS for graph layout. There are also close connections between the methods discussed here, spectral clustering (Section 14.5.3) and kernel PCA (Section 14.5.4).

14.10   The Google PageRank Algorithm

In this section we give a brief description of the original *PageRank* algorithm used by the Google search engine, an interesting recent application of unsupervised learning methods.

We suppose that we have N web pages and wish to rank them in terms of importance. For example, the N pages might all contain a string match to "statistical learning" and we might wish to rank the pages in terms of their likely relevance to a websurfer.

The *PageRank* algorithm considers a webpage to be important if many other webpages point to it. However the linking webpages that point to a given page are not treated equally: the algorithm also takes into account both the importance (*PageRank*) of the linking pages and the number of outgoing links that they have. Linking pages with higher *PageRank* are given more weight, while pages with more outgoing links are given less weight. These ideas lead to a recursive definition for *PageRank*, detailed next.

---

[15]Sam Roweis and Lawrence Saul kindly provided this figure.

Let $L_{ij} = 1$ if page j points to page i, and zero otherwise. Let $c_j = \sum_{i=1}^{N} L_{ij}$ equal the number of pages pointed to by page j (number of out-links). Then the Google *PageRanks* p_i are defined by the recursive relationship

$$p_i = (1 - d) + d \sum_{j=1}^{N} \left(\frac{L_{ij}}{c_j} \right) p_j \tag{14.107}$$

where d is a positive constant (apparently set to 0.85).

The idea is that the importance of page i is the sum of the importances of pages that point to that page. The sums are weighted by $1/c_j$, that is, each page distributes a total vote of 1 to other pages. The constant d ensures that each page gets a *PageRank* of at least $1 - d$. In matrix notation

$$\mathbf{p} = (1 - d)\mathbf{e} + d \cdot \mathbf{L} \mathbf{D}_c^{-1} \mathbf{p} \tag{14.108}$$

where $\mathbf{e}$ is a vector of N ones and $\mathbf{D}_c = \mathrm{diag}(\mathbf{c})$ is a diagonal matrix with diagonal elements c_j. Introducing the normalization $\mathbf{e}^T \mathbf{p} = N$ (i.e., the average *PageRank* is 1), we can write (14.108) as

$$\begin{aligned} \mathbf{p} &= \left[(1 - d)\mathbf{e}\mathbf{e}^T / N + d\mathbf{L}\mathbf{D}_c^{-1} \right] \mathbf{p} \\ &= \mathbf{A}\mathbf{p} \end{aligned} \tag{14.109}$$

where the matrix $\mathbf{A}$ is the expression in square braces.

Exploiting a connection with Markov chains (see below), it can be shown that the matrix $\mathbf{A}$ has a real eigenvalue equal to one, and one is its largest eigenvalue. This means that we can find $\hat{\mathbf{p}}$ by the power method: starting with some $\mathbf{p} = \mathbf{p}_0$ we iterate

$$\mathbf{p}_k \leftarrow \mathbf{A}\mathbf{p}_{k-1}; \quad \mathbf{p}_k \leftarrow N \frac{\mathbf{p}_k}{\mathbf{e}^T \mathbf{p}_k}. \tag{14.110}$$

The fixed points $\hat{\mathbf{p}}$ are the desired *PageRanks*.

In the original paper of Page et al. (1998), the authors considered *PageRank* as a model of user behavior, where a random web surfer clicks on links at random, without regard to content. The surfer does a random walk on the web, choosing among available outgoing links at random. The factor $1 - d$ is the probability that he does not click on a link, but jumps instead to a random webpage.

Some descriptions of *PageRank* have $(1 - d)/N$ as the first term in definition (14.107), which would better coincide with the random surfer interpretation. Then the page rank solution (divided by N) is the stationary distribution of an irreducible, aperiodic Markov chain over the N webpages.

Definition (14.107) also corresponds to an irreducible, aperiodic Markov chain, with different transition probabilities than those from the $(1-d)/N$ version. Viewing *PageRank* as a Markov chain makes clear why the matrix $\mathbf{A}$ has a maximal real eigenvalue of 1. Since $\mathbf{A}$ has positive entries with

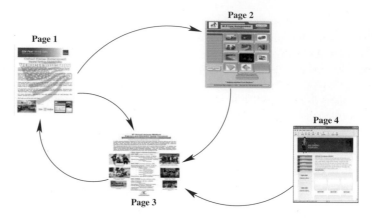

FIGURE 14.46. *PageRank algorithm: example of a small network*

each column summing to one, Markov chain theory tells us that it has a unique eigenvector with eigenvalue one, corresponding to the stationary distribution of the chain (Bremaud, 1999).

A small network is shown for illustration in Figure 14.46. The link matrix is

$$
\mathbf{L} = \begin{pmatrix} 0 & 0 & 1 & 0 \\ 1 & 0 & 0 & 0 \\ 1 & 1 & 0 & 1 \\ 0 & 0 & 0 & 0 \end{pmatrix}
\tag{14.111}
$$

and the number of outlinks is $\mathbf{c} = (2, 1, 1, 1)$.

The *PageRank* solution is $\hat{\mathbf{p}} = (1.49, 0.78, 1.58, 0.15)$. Notice that page 4 has no incoming links, and hence gets the minimum *PageRank* of 0.15.

Bibliographic Notes

There are many books on clustering, including Hartigan (1975), Gordon (1999) and Kaufman and Rousseeuw (1990). K-means clustering goes back at least to Lloyd (1957), Forgy (1965), Jancey (1966) and MacQueen (1967). Applications in engineering, especially in image compression via vector quantization, can be found in Gersho and Gray (1992). The k-medoid procedure is described in Kaufman and Rousseeuw (1990). Association rules are outlined in Agrawal et al. (1995). The self-organizing map was proposed by Kohonen (1989) and Kohonen (1990); Kohonen et al. (2000) give a more recent account. Principal components analysis and multidimensional scaling are described in standard books on multivariate analysis, for example, Mardia et al. (1979). Buja et al. (2008) have implemented a powerful environment called Ggvis for multidimensional scaling, and the user manual contains a lucid overview of the subject. Figures 14.17, 14.21 (left panel) and 14.28 (left panel) were produced in Xgobi, a multidimensional data visualization package by the same authors. GGobi is a more recent implementation (Cook and Swayne, 2007). Goodall (1991) gives a technical overview of Procrustes methods in statistics, and Ramsay and Silverman (1997) discuss the shape registration problem. Principal curves and surfaces were proposed in Hastie (1984) and Hastie and Stuetzle (1989). The idea of principal points was formulated in Flury (1990), Tarpey and Flury (1996) give an exposition of the general concept of self-consistency. An excellent tutorial on spectral clustering can be found in von Luxburg (2007); this was the main source for Section 14.5.3. Luxborg credits Donath and Hoffman (1973) and Fiedler (1973) with the earliest work on the subject. A history of spectral clustering my be found in Spielman and Teng (1996). Independent component analysis was proposed by Comon (1994), with subsequent developments by Bell and Sejnowski (1995); our treatment in Section 14.7 is based on Hyvärinen and Oja (2000). Projection pursuit was proposed by Friedman and Tukey (1974), and is discussed in detail in Huber (1985). A dynamic projection pursuit algorithm is implemented in GGobi.

Exercises

Ex. 14.1 *Weights for clustering.* Show that weighted Euclidean distance

$$d_e^{(w)}(x_i, x_{i'}) = \frac{\sum_{l=1}^p w_l (x_{il} - x_{i'l})^2}{\sum_{l=1}^p w_l}$$

satisfies

$$d_e^{(w)}(x_i, x_{i'}) = d_e(z_i, z_{i'}) - \sum_{l=1}^p (z_{il} - z_{i'l})^2, \qquad (14.112)$$

where

$$z_{il} = x_{il} \cdot \left(\frac{w_l}{\sum_{l=1}^p w_l} \right)^{1/2}. \qquad (14.113)$$

Thus weighted Euclidean distance based on x is equivalent to unweighted Euclidean distance based on z.

Ex. 14.2 Consider a mixture model density in p-dimensional feature space,

$$g(x) = \sum_{k=1}^K \pi_k g_k(x), \qquad (14.114)$$

where $g_k = N(\mu_k, \mathbf{L} \cdot \sigma^2)$ and $\pi_k \geq 0 \ \forall k$ with $\sum_k \pi_k = 1$. Here $\{\mu_k, \pi_k\}, k = 1, \ldots, K$ and σ^2 are unknown parameters.

Suppose we have data $x_1, x_2, \ldots, x_N \sim g(x)$ and we wish to fit the mixture model.

1. Write down the log-likelihood of the data

2. Derive an EM algorithm for computing the maximum likelihood estimates (see Section 8.1).

3. Show that if σ has a known value in the mixture model and we take $\sigma \to 0$, then in a sense this EM algorithm coincides with K-means clustering.

Ex. 14.3 In Section 14.2.6 we discuss the use of CART or PRIM for constructing generalized association rules. Show that a problem occurs with either of these methods when we generate the random data from the product-marginal distribution; i.e., by randomly permuting the values for each of the variables. Propose ways to overcome this problem.

Ex. 14.4 Cluster the demographic data of Table 14.1 using a classification tree. Specifically, generate a reference sample of the same size of the training set, by randomly permuting the values within each feature. Build a classification tree to the training sample (class 1) and the reference sample (class 0) and describe the terminal nodes having highest estimated class 1 probability. Compare the results to the PRIM results near Table 14.1 and also to the results of K-means clustering applied to the same data.

Ex. 14.5 Generate data with three features, with 30 data points in each of three classes as follows:

$$
\begin{aligned}
\theta_1 &= U(-\pi/8, \pi/8) \\
\phi_1 &= U(0, 2\pi) \\
x_1 &= \sin(\theta_1)\cos(\phi_1) + W_{11} \\
y_1 &= \sin(\theta_1)\sin(\phi_1) + W_{12} \\
z_1 &= \cos(\theta_1) + W_{13}
\end{aligned}
$$

$$
\begin{aligned}
\theta_2 &= U(\pi/2 - \pi/4, \pi/2 + \pi/4) \\
\phi_2 &= U(-\pi/4, \pi/4) \\
x_2 &= \sin(\theta_2)\cos(\phi_2) + W_{21} \\
y_2 &= \sin(\theta_2)\sin(\phi_2) + W_{22} \\
z_2 &= \cos(\theta_2) + W_{23}
\end{aligned}
$$

$$
\begin{aligned}
\theta_3 &= U(\pi/2 - \pi/4, \pi/2 + \pi/4) \\
\phi_3 &= U(\pi/2 - \pi/4, \pi/2 + \pi/4) \\
x_3 &= \sin(\theta_3)\cos(\phi_3) + W_{31} \\
y_3 &= \sin(\theta_3)\sin(\phi_3) + W_{32} \\
z_3 &= \cos(\theta_3) + W_{33}
\end{aligned}
$$

Here $U(a, b)$ indicates a uniform variate on the range $[a, b]$ and W_{jk} are independent normal variates with standard deviation 0.6. Hence the data lie near the surface of a sphere in three clusters centered at $(1, 0, 0)$, $(0, 1, 0)$ and $(0, 0, 1)$.

Write a program to fit a SOM to these data, using the learning rates given in the text. Carry out a K-means clustering of the same data, and compare the results to those in the text.

Ex. 14.6 Write programs to implement K-means clustering and a self-organizing map (SOM), with the prototype lying on a two-dimensional grid. Apply them to the columns of the human tumor microarray data, using $K = 2, 5, 10, 20$ centroids for both. Demonstrate that as the size of the SOM neighborhood is taken to be smaller and smaller, the SOM solution becomes more similar to the K-means solution.

Ex. 14.7 Derive (14.51) and (14.52) in Section 14.5.1. Show that $\hat{\mu}$ is not unique, and characterize the family of equivalent solutions.

Ex. 14.8 Derive the solution (14.57) to the Procrustes problem (14.56). Derive also the solution to the Procrustes problem with scaling (14.58).

Ex. 14.9 Write an algorithm to solve

$$\min_{\{\beta_\ell, \mathbf{R}_\ell\}_1^L, \mathbf{M}} \sum_{\ell=1}^{L} ||\mathbf{X}_\ell \mathbf{R}_\ell - \mathbf{M}||_F^2. \qquad (14.115)$$

Apply it to the three S's, and compare the results to those shown in Figure 14.26.

Ex. 14.10 Derive the solution to the affine-invariant average problem (14.60). Apply it to the three S's, and compare the results to those computed in Exercise 14.9.

Ex. 14.11 *Classical multidimensional scaling.* Let $\mathbf{S}$ be the centered inner product matrix with elements $\langle x_i - \bar{x}, x_j - \bar{x} \rangle$. Let $\lambda_1 > \lambda_2 > \cdots > \lambda_k$ be the k largest eigenvalues of $\mathbf{S}$, with associated eigenvectors $\mathbf{E}_k = (e_1, e_2, \ldots, e_k)$. Let $\mathbf{D}_k$ be a diagonal matrix with diagonal entries $\sqrt{\lambda_1}$, $\sqrt{\lambda_2}$, $\ldots$, $\sqrt{\lambda_k}$. Show that the solutions z_i to the classical scaling problem (14.100) are the *rows* of $\mathbf{E}_k \mathbf{D}_k$.

Ex. 14.12 Consider the sparse PCA criterion (14.71).

1. Show that with $\boldsymbol{\Theta}$ fixed, solving for $\mathbf{V}$ amounts to K separate elastic-net regression problems, with responses the K elements of $\boldsymbol{\Theta}^T x_i$.

2. Show that with $\mathbf{V}$ fixed, solving for $\boldsymbol{\Theta}$ amounts to a reduced-rank version of the Procrustes problem, which reduces to

$$\max_{\boldsymbol{\Theta}} \text{trace}(\boldsymbol{\Theta}^T \mathbf{M}) \text{ subject to } \boldsymbol{\Theta}^T \boldsymbol{\Theta} = \mathbf{I}_K, \qquad (14.116)$$

   where $\mathbf{M}$ and $\boldsymbol{\Theta}$ are both $p \times K$ with $K \leq p$. If $\mathbf{M} = \mathbf{UDQ}^T$ is the SVD of $\mathbf{M}$, show that the optimal $\boldsymbol{\Theta} = \mathbf{UQ}^T$.

Ex. 14.13 Generate 200 data points with three features, lying close to a
helix. In detail, define $X_1 = \cos(s) + 0.1 \cdot Z_1, X_2 = \sin(s) + 0.1 \cdot Z_2, X_3 = s + 0.1 \cdot Z_3$ where s takes on 200 equally spaced values between 0 and 2π, and Z_1, Z_2, Z_3 are independent and have standard Gaussian distributions.

(a) Fit a principal curve to the data and plot the estimated coordinate
    functions. Compare them to the underlying functions $\cos(s), \sin(s)$
    and s.

(b) Fit a self-organizing map to the same data, and see if you can discover
    the helical shape of the original point cloud.

Ex. 14.14 Pre- and post-multiply equation (14.81) by a diagonal matrix
containing the inverse variances of the X_j. Hence obtain an equivalent
decomposition for the correlation matrix, in the sense that a simple scaling
is applied to the matrix $\mathbf{A}$.

Ex. 14.15 Generate 200 observations of three variates X_1, X_2, X_3 according
to

$$
\begin{aligned}
X_1 &\sim Z_1 \\
X_2 &= X_1 + 0.001 \cdot Z_2 \\
X_3 &= 10 \cdot Z_3
\end{aligned}
\tag{14.117}
$$

where Z_1, Z_2, Z_3 are independent standard normal variates. Compute the
leading principal component and factor analysis directions. Hence show
that the leading principal component aligns itself in the maximal variance
direction X_3, while the leading factor essentially ignores the uncorrelated
component X_3, and picks up the correlated component $X_2 + X_1$ (Geoffrey
Hinton, personal communication).

Ex. 14.16 Consider the kernel principal component procedure outlined in
Section 14.5.4. Argue that the number M of principal components is equal
to the rank of $\mathbf{K}$, which is the number of non-zero elements in $\mathbf{D}$. Show
that the mth component $\mathbf{z}_m$ (mth column of $\mathbf{Z}$) can be written (up to
centering) as $z_{im} = \sum_{j=1}^N \alpha_{jm} K(x_i, x_j)$, where $\alpha_{jm} = u_{jm}/d_m$. Show that
the mapping of a new observation x_0 to the mth component is given by
$z_{0m} = \sum_{j=1}^N \alpha_{jm} K(x_0, x_j)$.

Ex. 14.17 Show that with $g_1(x) = \sum_{j=1}^N c_j K(x, x_j)$, the solution to (14.66)
is given by $\hat{c}_j = u_{j1}/d_1$, where $\mathbf{u}_1$ is the first column of $\mathbf{U}$ in (14.65), and
d_1 the first diagonal element of $\mathbf{D}$. Show that the second and subsequent
principal component functions are defined in a similar manner (*hint*: see
Section 5.8.1.)

Ex. 14.18 Consider the regularized log-likelihood for the density estimation
problem arising in ICA,

$$\frac{1}{N}\sum_{i=1}^{N}[\log\phi(s_i)+g(s_i)]-\int\phi(t)e^{g(t)}\,dt-\lambda\int\{g'''(t)\}^2(t)dt. \quad (14.118)$$

The solution $\hat g$ is a quartic smoothing spline, and can be written as $\hat g(s)=\hat q(s)+\hat q_\perp(s)$, where q is a quadratic function (in the null space of the penalty). Let $q(s)=\theta_0+\theta_1 s+\theta_2 s^2$. By examining the stationarity conditions for $\hat\theta_k$, $k=1,2,3$, show that the solution $\hat f=\phi e^{\hat g}$ is a density, and has mean zero and variance one. If we used a second-derivative penalty $\int\{g''(t)\}^2(t)dt$ instead, what simple modification could we make to the problem to maintain the three moment conditions?

Ex. 14.19 If $\mathbf{A}$ is $p\times p$ orthogonal, show that the first term in (14.91) on page 567

$$\sum_{j=1}^{p}\sum_{i=1}^{N}\log\phi(a_j^T x_i),$$

with a_j the jth column of $\mathbf{A}$, does not depend on $\mathbf{A}$.

Ex. 14.20 *Fixed point algorithm for ICA* (Hyvärinen et al., 2001). Consider maximizing $C(a)=E\{g(a^T X)\}$ with respect to a, with $||a||=1$ and $\mathrm{Cov}(X)=I$. Use a Lagrange multiplier to enforce the norm constraint, and write down the first two derivatives of the modified criterion. Use the approximation

$$E\{XX^T g''(a^T X)\}\approx E\{XX^T\}E\{g''(a^T X)\}$$

to show that the Newton update can be written as the fixed-point update (14.96).

Ex. 14.21 Consider an undirected graph with non-negative edge weights $w_{ii'}$ and graph Laplacian $\mathbf{L}$. Suppose there are m connected components $A_1,A_2,\ldots,A_m$ in the graph. Show that there are m eigenvectors of $\mathbf{L}$ corresponding to eigenvalue zero, and the indicator vectors of these components $I_{A_1},I_{A_2},\ldots,I_{A_m}$ span the zero eigenspace.

Ex. 14.22

(a) Show that definition (14.108) implies that the sum of the *PageRanks* p_i is N, the number of web pages.

(b) Write a program to compute the *PageRank* solutions by the power method using formulation (14.107). Apply it to the network of Figure 14.47.

Ex. 14.23 *Algorithm for non-negative matrix factorization* (Wu and Lange, 2007). A function $g(x,y)$ to said to *minorize* a function $f(x)$ if

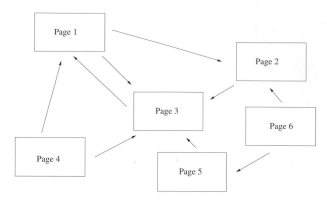

FIGURE 14.47. *Example of a small network.*

$$g(x,y) \leq f(x), \quad g(x,x) = f(x) \qquad (14.119)$$

for all x, y in the domain. This is useful for maximizing $f(x)$ since it is easy to show that $f(x)$ is nondecreasing under the update

$$x^{s+1} = \operatorname{argmax}_x g(x, x^s) \qquad (14.120)$$

There are analogous definitions for *majorization*, for minimizing a function $f(x)$. The resulting algorithms are known as *MM* algorithms, for "minorize-maximize" or "majorize-minimize" (Lange, 2004). It also can be shown that the EM algorithm (8.5) is an example of an MM algorithm: see Section 8.5.3 and Exercise 8.2 for details.

(a) Consider maximization of the function $L(\mathbf{W}, \mathbf{H})$ in (14.73), written here without the matrix notation

$$L(\mathbf{W}, \mathbf{H}) = \sum_{i=1}^{N} \sum_{j=1}^{p} \left[x_{ij} \log \left(\sum_{k=1}^{r} w_{ik} h_{kj} \right) - \sum_{k=1}^{r} w_{ik} h_{kj} \right].$$

Using the concavity of $\log(x)$, show that for any set of r values $y_k \geq 0$ and $0 \leq c_k \leq 1$ with $\sum_{k=1}^{r} c_k = 1$,

$$\log \left(\sum_{k=1}^{r} y_k \right) \geq \sum_{k=1}^{r} c_k \log(y_k / c_k)$$

Hence

$$\log \left(\sum_{k=1}^{r} w_{ik} h_{kj} \right) \geq \sum_{k=1}^{r} \frac{a_{ikj}^s}{b_{ij}^s} \log \left(\frac{b_{ij}^s}{a_{ikj}^s} w_{ik} h_{kj} \right),$$

where

$$a_{ikj}^s = w_{ik}^s h_{kj}^s \text{ and } b_{ij}^s = \sum_{k=1}^{r} w_{ik}^s h_{kj}^s,$$

and s indicates the current iteration.

(b) Hence show that, ignoring constants, the function

$$g(\mathbf{W}, \mathbf{H} \mid \mathbf{W}^s, \mathbf{H}^s) = \sum_{i=1}^{N} \sum_{j=1}^{p} \sum_{k=1}^{r} x_{ij} \frac{a_{ikj}^s}{b_{ij}^s} \left(\log w_{ik} + \log h_{kj} \right)$$

$$- \sum_{i=1}^{N} \sum_{j=1}^{p} \sum_{k=1}^{r} w_{ik} h_{kj}$$

minorizes $L(\mathbf{W}, \mathbf{H})$.

(c) Set the partial derivatives of $g(\mathbf{W}, \mathbf{H} \mid \mathbf{W}^s, \mathbf{H}^s)$ to zero and hence derive the updating steps (14.74).

Ex. 14.24 Consider the non-negative matrix factorization (14.72) in the rank one case $(r = 1)$.

(a) Show that the updates (14.74) reduce to

$$w_i \leftarrow w_i \frac{\sum_{j=1}^{p} x_{ij}}{\sum_{j=1}^{p} w_i h_j}$$

$$h_j \leftarrow h_j \frac{\sum_{i=1}^{N} x_{ij}}{\sum_{i=1}^{N} w_i h_j} \qquad (14.121)$$

where $w_i = w_{i1}$, $h_j = h_{1j}$. This is an example of the *iterative proportional scaling* procedure, applied to the independence model for a two way contingency table (Fienberg, 1977, for example).

(b) Show that the final iterates have the explicit form

$$w_i = c \cdot \frac{\sum_{j=1}^{p} x_{ij}}{\sum_{i=1}^{N} \sum_{j=1}^{p} x_{ij}}, \qquad h_k = \frac{1}{c} \cdot \frac{\sum_{i=1}^{N} x_{ik}}{\sum_{i=1}^{N} \sum_{j=1}^{p} x_{ij}} \qquad (14.122)$$

for any constant $c > 0$. These are equivalent to the usual row and column estimates for a two-way independence model.

Ex. 14.25 Fit a non-negative matrix factorization model to the collection of two's in the digits database. Use 25 basis elements, and compare with a 24- component (plus mean) PCA model. In both cases display the $\mathbf{W}$ and $\mathbf{H}$ matrices as in Figure 14.33.

15
Random Forests

15.1 Introduction

Bagging or *bootstrap aggregation* (section 8.7) is a technique for reducing the variance of an estimated prediction function. Bagging seems to work especially well for high-variance, low-bias procedures, such as trees. For regression, we simply fit the same regression tree many times to bootstrap-sampled versions of the training data, and average the result. For classification, a *committee* of trees each cast a vote for the predicted class.

Boosting in Chapter 10 was initially proposed as a committee method as well, although unlike bagging, the committee of *weak learners* evolves over time, and the members cast a weighted vote. Boosting appears to dominate bagging on most problems, and became the preferred choice.

Random forests (Breiman, 2001) is a substantial modification of bagging that builds a large collection of *de-correlated* trees, and then averages them. On many problems the performance of random forests is very similar to boosting, and they are simpler to train and tune. As a consequence, random forests are popular, and are implemented in a variety of packages.

15.2 Definition of Random Forests

The essential idea in bagging (Section 8.7) is to average many noisy but approximately unbiased models, and hence reduce the variance. Trees are ideal candidates for bagging, since they can capture complex interaction

T. Hastie et al., *The Elements of Statistical Learning, Second Edition,*
DOI: 10.1007/b94608_15,
© Springer Science+Business Media, LLC 2009

Algorithm 15.1 *Random Forest for Regression or Classification.*

1. For $b = 1$ to B:

   (a) Draw a bootstrap sample $\mathbf{Z}^*$ of size N from the training data.

   (b) Grow a random-forest tree T_b to the bootstrapped data, by recursively repeating the following steps for each terminal node of the tree, until the minimum node size n_{min} is reached.

      i. Select m variables at random from the p variables.

      ii. Pick the best variable/split-point among the m.

      iii. Split the node into two daughter nodes.

2. Output the ensemble of trees $\{T_b\}_1^B$.

To make a prediction at a new point x:

Regression: $\hat{f}_{\text{rf}}^B(x) = \frac{1}{B}\sum_{b=1}^{B} T_b(x)$.

Classification: Let $\hat{C}_b(x)$ be the class prediction of the bth random-forest tree. Then $\hat{C}_{\text{rf}}^B(x) = $ *majority vote* $\{\hat{C}_b(x)\}_1^B$.

---

structures in the data, and if grown sufficiently deep, have relatively low bias. Since trees are notoriously noisy, they benefit greatly from the averaging. Moreover, since each tree generated in bagging is identically distributed (i.d.), the expectation of an average of B such trees is the same as the expectation of any one of them. This means the bias of bagged trees is the same as that of the individual (bootstrap) trees, and the only hope of improvement is through variance reduction. This is in contrast to boosting, where the trees are grown in an adaptive way to remove bias, and hence are not i.d.

An average of B i.i.d. random variables, each with variance σ^2, has variance $\frac{1}{B}\sigma^2$. If the variables are simply i.d. (identically distributed, but not necessarily independent) with positive pairwise correlation ρ, the variance of the average is (Exercise 15.1)

$$\rho\sigma^2 + \frac{1-\rho}{B}\sigma^2. \tag{15.1}$$

As B increases, the second term disappears, but the first remains, and hence the size of the correlation of pairs of bagged trees limits the benefits of averaging. The idea in random forests (Algorithm 15.1) is to improve the variance reduction of bagging by reducing the correlation between the trees, without increasing the variance too much. This is achieved in the tree-growing process through random selection of the input variables.

Specifically, when growing a tree on a bootstrapped dataset:

Before each split, select $m \leq p$ of the input variables at random as candidates for splitting.

Typically values for m are $\sqrt{p}$ or even as low as 1.

After B such trees $\{T(x; \Theta_b)\}_1^B$ are grown, the random forest (regression) predictor is

$$\hat{f}_{\text{rf}}^B(x) = \frac{1}{B} \sum_{b=1}^{B} T(x; \Theta_b). \qquad (15.2)$$

As in Section 10.9 (page 356), Θ_b characterizes the bth random forest tree in terms of split variables, cutpoints at each node, and terminal-node values. Intuitively, reducing m will reduce the correlation between any pair of trees in the ensemble, and hence by (15.1) reduce the variance of the average.

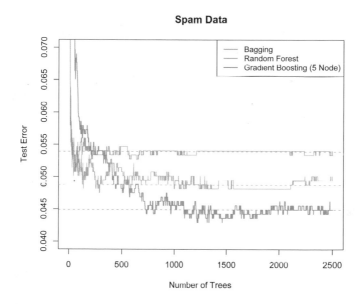

FIGURE 15.1. *Bagging, random forest, and gradient boosting, applied to the spam data. For boosting, 5-node trees were used, and the number of trees were chosen by 10-fold cross-validation (2500 trees). Each "step" in the figure corresponds to a change in a single misclassification (in a test set of 1536).*

Not all estimators can be improved by shaking up the data like this. It seems that highly nonlinear estimators, such as trees, benefit the most. For bootstrapped trees, ρ is typically small (0.05 or lower is typical; see Figure 15.9), while σ^2 is not much larger than the variance for the original tree. On the other hand, bagging does not change *linear* estimates, such as the sample mean (hence its variance either); the pairwise correlation between bootstrapped means is about 50% (Exercise 15.4).

Random forests are popular. Leo Breiman's[1] collaborator Adele Cutler maintains a random forest website[2] where the software is freely available, with more than 3000 downloads reported by 2002. There is a `randomForest` package in R, maintained by Andy Liaw, available from the `CRAN` website.

The authors make grand claims about the success of random forests: "most accurate," "most interpretable," and the like. In our experience random forests do remarkably well, with very little tuning required. A random forest classifier achieves 4.88% misclassification error on the `spam` test data, which compares well with all other methods, and is not significantly worse than gradient boosting at 4.5%. Bagging achieves 5.4% which is significantly worse than either (using the McNemar test outlined in Exercise 10.6), so it appears on this example the additional randomization helps.

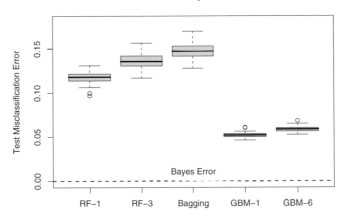

FIGURE 15.2. *The results of* 50 *simulations from the "nested spheres" model in* $\mathbb{R}^{10}$. *The Bayes decision boundary is the surface of a sphere (additive). "RF-3" refers to a random forest with* $m = 3$, *and "GBM-6" a gradient boosted model with interaction order six; similarly for "RF-1" and "GBM-1." The training sets were of size* 2000, *and the test sets* 10,000.

Figure 15.1 shows the test-error progression on 2500 trees for the three methods. In this case there is some evidence that gradient boosting has started to overfit, although 10-fold cross-validation chose all 2500 trees.

---

[1] Sadly, Leo Breiman died in July, 2005.
[2] http://www.math.usu.edu/~adele/forests/

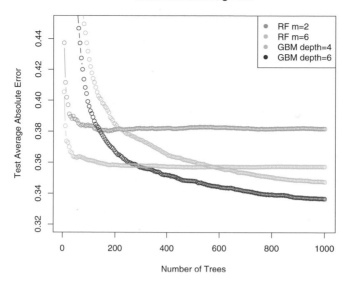

California Housing Data

FIGURE 15.3. *Random forests compared to gradient boosting on the California housing data. The curves represent mean absolute error on the test data as a function of the number of trees in the models. Two random forests are shown, with $m = 2$ and $m = 6$. The two gradient boosted models use a shrinkage parameter $\nu = 0.05$ in (10.41), and have interaction depths of 4 and 6. The boosted models outperform random forests.*

Figure 15.2 shows the results of a simulation[3] comparing random forests to gradient boosting on the *nested spheres* problem [Equation (10.2) in Chapter 10]. Boosting easily outperforms random forests here. Notice that smaller m is better here, although part of the reason could be that the true decision boundary is additive.

Figure 15.3 compares random forests to boosting (with shrinkage) in a regression problem, using the California housing data (Section 10.14.1). Two strong features that emerge are

- Random forests stabilize at about 200 trees, while at 1000 trees boosting continues to improve. Boosting is slowed down by the shrinkage, as well as the fact that the trees are much smaller.

- Boosting outperforms random forests here. At 1000 terms, the weaker boosting model (GBM depth 4) has a smaller error than the stronger

---

[3]Details: The random forests were fit using the R package `randomForest 4.5-11`, with 500 trees. The gradient boosting models were fit using R package `gbm 1.5`, with shrinkage parameter set to 0.05, and 2000 trees.

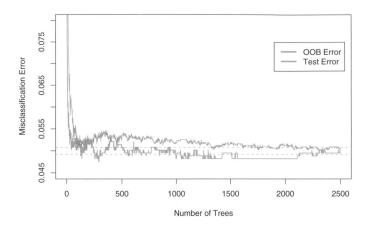

FIGURE 15.4. OOB *error computed on the* **spam** *training data, compared to the test error computed on the test set.*

random forest (RF $m = 6$); a Wilcoxon test on the mean differences in absolute errors has a p-value of 0.007. For larger m the random forests performed no better.

15.3 Details of Random Forests

We have glossed over the distinction between random forests for classification versus regression. When used for classification, a random forest obtains a class vote from each tree, and then classifies using majority vote (see Section 8.7 on bagging for a similar discussion). When used for regression, the predictions from each tree at a target point x are simply averaged, as in (15.2). In addition, the inventors make the following recommendations:

- For classification, the default value for m is $\lfloor \sqrt{p} \rfloor$ and the minimum node size is one.

- For regression, the default value for m is $\lfloor p/3 \rfloor$ and the minimum node size is five.

In practice the best values for these parameters will depend on the problem, and they should be treated as tuning parameters. In Figure 15.3 $m = 6$ performs much better than the default value $\lfloor 8/3 \rfloor = 2$.

15.3.1 Out of Bag Samples

An important feature of random forests is its use of *out-of-bag* (OOB) samples:

For each observation $z_i = (x_i, y_i)$, construct its random forest predictor by averaging only *those trees corresponding to boot-strap samples in which z_i did not* appear.

An OOB error estimate is almost identical to that obtained by N-fold cross-validation; see Exercise 15.2. Hence unlike many other nonlinear estimators, random forests can be fit in one sequence, with cross-validation being performed along the way. Once the OOB error stabilizes, the training can be terminated.

Figure 15.4 shows the OOB misclassification error for the spam data, compared to the test error. Although 2500 trees are averaged here, it appears from the plot that about 200 would be sufficient.

15.3.2 Variable Importance

Variable importance plots can be constructed for random forests in exactly the same way as they were for gradient-boosted models (Section 10.13). At each split in each tree, the improvement in the split-criterion is the importance measure attributed to the splitting variable, and is accumulated over all the trees in the forest separately for each variable. The left plot of Figure 15.5 shows the variable importances computed in this way for the spam data; compare with the corresponding Figure 10.6 on page 354 for gradient boosting. Boosting ignores some variables completely, while the random forest does not. The candidate split-variable selection increases the chance that any single variable gets included in a random forest, while no such selection occurs with boosting.

Random forests also use the OOB samples to construct a different *variable-importance* measure, apparently to measure the prediction strength of each variable. When the bth tree is grown, the OOB samples are passed down the tree, and the prediction accuracy is recorded. Then the values for the jth variable are randomly permuted in the OOB samples, and the accuracy is again computed. The decrease in accuracy as a result of this permuting is averaged over all trees, and is used as a measure of the importance of variable j in the random forest. These are expressed as a percent of the maximum in the right plot in Figure 15.5. Although the rankings of the two methods are similar, the importances in the right plot are more uniform over the variables. The randomization effectively voids the effect of a variable, much like setting a coefficient to zero in a linear model (Exercise 15.7). This does not measure the effect on prediction were this variable not available, because if the model was refitted without the variable, other variables could be used as surrogates.

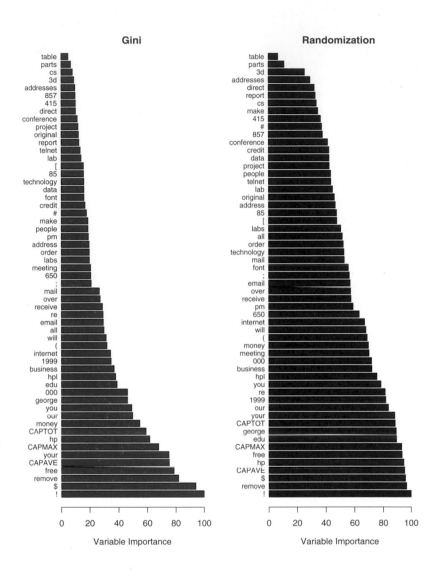

FIGURE 15.5. *Variable importance plots for a classification random forest grown on the* **spam** *data. The left plot bases the importance on the Gini splitting index, as in gradient boosting. The rankings compare well with the rankings produced by gradient boosting (Figure 10.6 on page 354). The right plot uses* OOB *randomization to compute variable importances, and tends to spread the importances more uniformly.*

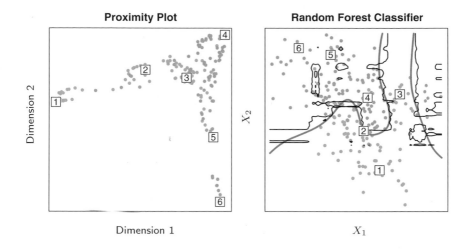

FIGURE 15.6. *(Left): Proximity plot for a random forest classifier grown to the mixture data. (Right): Decision boundary and training data for random forest on mixture data. Six points have been identified in each plot.*

15.3.3 Proximity Plots

One of the advertised outputs of a random forest is a *proximity plot*. Figure 15.6 shows a proximity plot for the mixture data defined in Section 2.3.3 in Chapter 2. In growing a random forest, an $N \times N$ proximity matrix is accumulated for the training data. For every tree, any pair of OOB observations sharing a terminal node has their proximity increased by one. This proximity matrix is then represented in two dimensions using multidimensional scaling (Section 14.8). The idea is that even though the data may be high-dimensional, involving mixed variables, etc., the proximity plot gives an indication of which observations are effectively close together in the eyes of the random forest classifier.

Proximity plots for random forests often look very similar, irrespective of the data, which casts doubt on their utility. They tend to have a star shape, one arm per class, which is more pronounced the better the classification performance.

Since the mixture data are two-dimensional, we can map points from the proximity plot to the original coordinates, and get a better understanding of what they represent. It seems that points in pure regions class-wise map to the extremities of the star, while points nearer the decision boundaries map nearer the center. This is not surprising when we consider the construction of the proximity matrices. Neighboring points in pure regions will often end up sharing a bucket, since when a terminal node is pure, it is no longer split by a random forest tree-growing algorithm. On the other hand, pairs of points that are close but belong to different classes will sometimes share a terminal node, but not always.

15.3.4 Random Forests and Overfitting

When the number of variables is large, but the fraction of relevant variables small, random forests are likely to perform poorly with small m. At each split the chance can be small that the relevant variables will be selected. Figure 15.7 shows the results of a simulation that supports this claim. Details are given in the figure caption and Exercise 15.3. At the top of each pair we see the hyper-geometric probability that a relevant variable will be selected at any split by a random forest tree (in this simulation, the relevant variables are all equal in stature). As this probability gets small, the gap between boosting and random forests increases. When the number of relevant variables increases, the performance of random forests is surprisingly robust to an increase in the number of noise variables. For example, with 6 relevant and 100 noise variables, the probability of a relevant variable being selected at any split is 0.46, assuming $m = \sqrt{(6 + 100)} \approx 10$. According to Figure 15.7, this does not hurt the performance of random forests compared with boosting. This robustness is largely due to the relative insensitivity of misclassification cost to the bias and variance of the probability estimates in each tree. We consider random forests for regression in the next section.

Another claim is that random forests "cannot overfit" the data. It is certainly true that increasing B does not cause the random forest sequence to overfit; like bagging, the random forest estimate (15.2) approximates the expectation

$$\hat{f}_{\mathrm{rf}}(x) = \mathrm{E}_{\Theta} T(x; \Theta) = \lim_{B \to \infty} \hat{f}(x)_{\mathrm{rf}}^{B} \tag{15.3}$$

with an average over B realizations of Θ. The distribution of Θ here is conditional on the training data. However, *this limit can overfit the data*; the average of fully grown trees can result in too rich a model, and incur unnecessary variance. Segal (2004) demonstrates small gains in performance by controlling the depths of the individual trees grown in random forests. Our experience is that using full-grown trees seldom costs much, and results in one less tuning parameter.

Figure 15.8 shows the modest effect of depth control in a simple regression example. Classifiers are less sensitive to variance, and this effect of overfitting is seldom seen with random-forest classification.

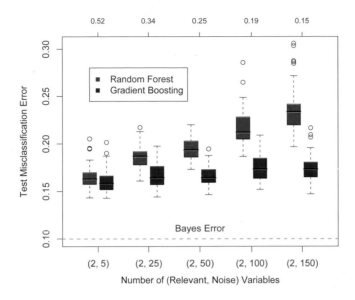

FIGURE 15.7. *A comparison of random forests and gradient boosting on problems with increasing numbers of noise variables. In each case the true decision boundary depends on two variables, and an increasing number of noise variables are included. Random forests uses its default value* $m = \sqrt{p}$. *At the top of each pair is the probability that one of the relevant variables is chosen at any split. The results are based on* 50 *simulations for each pair, with a training sample of* 300, *and a test sample of* 500. *See Exercise 15.3.*

15.4   Analysis of Random Forests

In this section we analyze the mechanisms at play with the additional randomization employed by random forests. For this discussion we focus on regression and squared error loss, since this gets at the main points, and bias and variance are more complex with 0–1 loss (see Section 7.3.1). Furthermore, even in the case of a classification problem, we can consider the random-forest average as an estimate of the class posterior probabilities, for which bias and variance are appropriate descriptors.

15.4.1   *Variance and the De-Correlation Effect*

The limiting form ($B \to \infty$) of the random forest regression estimator is

$$\hat{f}_{\mathrm{rf}}(x) = \mathrm{E}_{\Theta|\mathbf{Z}} T(x; \Theta(\mathbf{Z})), \qquad (15.4)$$

where we have made explicit the dependence on the training data $\mathbf{Z}$. Here we consider estimation at a single target point x. From (15.1) we see that

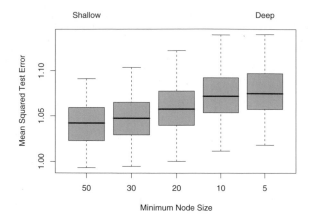

FIGURE 15.8. *The effect of* tree size *on the error in random forest regression. In this example, the true surface was additive in two of the 12 variables, plus additive unit-variance Gaussian noise. Tree depth is controlled here by the minimum node size; the smaller the minimum node size, the deeper the trees.*

$$\text{Var}\hat{f}_{\text{rf}}(x) = \rho(x)\sigma^2(x). \tag{15.5}$$

Here

- $\rho(x)$ is the *sampling* correlation between any pair of trees used in the averaging:

$$\rho(x) = \text{corr}[T(x; \Theta_1(\mathbf{Z})), T(x; \Theta_2(\mathbf{Z}))], \tag{15.6}$$

  where $\Theta_1(\mathbf{Z})$ and $\Theta_2(\mathbf{Z})$ are a randomly drawn pair of random forest trees grown to the randomly sampled $\mathbf{Z}$;

- $\sigma^2(x)$ is the sampling variance of any single randomly drawn tree,

$$\sigma^2(x) = \text{Var}\, T(x; \Theta(\mathbf{Z})). \tag{15.7}$$

It is easy to confuse $\rho(x)$ with the average correlation between fitted trees in a *given* random-forest ensemble; that is, think of the fitted trees as N-vectors, and compute the average pairwise correlation between these vectors, conditioned on the data. This is *not* the case; this conditional correlation is not directly relevant in the averaging process, and the dependence on x in $\rho(x)$ warns us of the distinction. Rather, $\rho(x)$ is the theoretical correlation between a pair of random-forest trees evaluated at x, induced by repeatedly making training sample draws $\mathbf{Z}$ from the population, and then drawing a pair of random forest trees. In statistical jargon, this is the correlation induced by the *sampling distribution* of $\mathbf{Z}$ and Θ.

More precisely, the variability averaged over in the calculations in (15.6) and (15.7) is both

- conditional on **Z**: due to the bootstrap sampling and feature sampling at each split, and

- a result of the sampling variability of **Z** itself.

In fact, the conditional covariance of a pair of tree fits at x is zero, because the bootstrap and feature sampling is i.i.d; see Exercise 15.5.

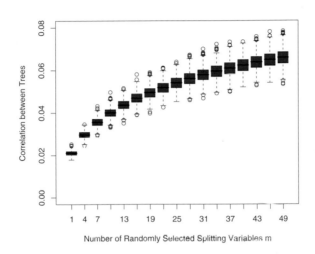

FIGURE 15.9. *Correlations between pairs of trees drawn by a random-forest regression algorithm, as a function of m. The boxplots represent the correlations at* 600 *randomly chosen prediction points x.*

The following demonstrations are based on a simulation model

$$Y = \frac{1}{\sqrt{50}} \sum_{j=1}^{50} X_j + \varepsilon, \qquad (15.8)$$

with all the X_j and ε iid Gaussian. We use 500 training sets of size 100, and a single set of test locations of size 600. Since regression trees are nonlinear in **Z**, the patterns we see below will differ somewhat depending on the structure of the model.

Figure 15.9 shows how the correlation (15.6) between pairs of trees decreases as m decreases: pairs of tree predictions at x for different training sets **Z** are likely to be less similar if they do not use the same splitting variables.

In the left panel of Figure 15.10 we consider the variances of single tree predictors, $\mathrm{Var}T(x; \Theta(\mathbf{Z}))$ (averaged over 600 prediction points x drawn randomly from our simulation model). This is the total variance, and can be decomposed into two parts using standard conditional variance arguments (see Exercise 15.5):

$$\text{Var}_{\Theta,\mathbf{Z}}T(x;\Theta(\mathbf{Z})) \;=\; \text{Var}_{\mathbf{Z}}\text{E}_{\Theta|\mathbf{Z}}T(x;\Theta(\mathbf{Z})) \;+\; \text{E}_{\mathbf{Z}}\text{Var}_{\Theta|\mathbf{Z}}T(x;\Theta(\mathbf{Z}))$$

$$\textit{Total Variance} \;=\; \text{Var}_{\mathbf{Z}}\hat{f}_{\text{rf}}(x) \;+\; \textit{within-}\mathbf{Z}\ \textit{Variance}$$

(15.9)

The second term is the within-$\mathbf{Z}$ variance—a result of the randomization, which increases as m decreases. The first term is in fact the sampling variance of the random forest ensemble (shown in the right panel), which decreases as m decreases. The variance of the individual trees does not change appreciably over much of the range of m, hence in light of (15.5), the variance of the ensemble is dramatically lower than this tree variance.

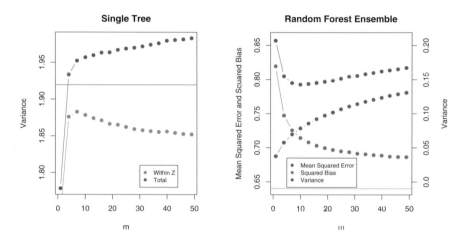

FIGURE 15.10. *Simulation results. The left panel shows the average variance of a single random forest tree, as a function of m. "Within $\mathbf{Z}$" refers to the average within-sample contribution to the variance, resulting from the bootstrap sampling and split-variable sampling (15.9). "Total" includes the sampling variability of $\mathbf{Z}$. The horizontal line is the average variance of a single fully grown tree (without bootstrap sampling). The right panel shows the average mean-squared error, squared bias and variance of the ensemble, as a function of m. Note that the variance axis is on the right (same scale, different level). The horizontal line is the average squared-bias of a fully grown tree.*

15.4.2 Bias

As in bagging, the bias of a random forest is the same as the bias of any of the individual sampled trees $T(x;\Theta(\mathbf{Z}))$:

$$\mathrm{Bias}(x) = \mu(x) - \mathrm{E}_{\mathbf{Z}}\hat{f}_{\mathrm{rf}}(x)$$
$$= \mu(x) - \mathrm{E}_{\mathbf{Z}}\mathrm{E}_{\Theta|\mathbf{Z}}T(x;\Theta(\mathbf{Z})). \qquad (15.10)$$

This is also typically greater (in absolute terms) than the bias of an unpruned tree grown to $\mathbf{Z}$, since the randomization and reduced sample space impose restrictions. Hence the improvements in prediction obtained by bagging or random forests are *solely a result of variance reduction*.

Any discussion of bias depends on the unknown true function. Figure 15.10 (right panel) shows the squared bias for our additive model simulation (estimated from the 500 realizations). Although for different models the shape and rate of the bias curves may differ, the general trend is that as m decreases, the bias increases. Shown in the figure is the mean-squared error, and we see a classical bias-variance trade-off in the choice of m. For all m the squared bias of the random forest is greater than that for a single tree (horizontal line).

These patterns suggest a similarity with ridge regression (Section 3.4.1). Ridge regression is useful (in linear models) when one has a large number of variables with similarly sized coefficients; ridge shrinks their coefficients toward zero, and those of strongly correlated variables toward each other. Although the size of the training sample might not permit all the variables to be in the model, this regularization via ridge stabilizes the model and allows all the variables to have their say (albeit diminished). Random forests with small m perform a similar averaging. Each of the relevant variables get their turn to be the primary split, and the ensemble averaging reduces the contribution of any individual variable. Since this simulation example (15.8) is based on a linear model in all the variables, ridge regression achieves a lower mean-squared error (about 0.45 with $\mathrm{df}(\lambda_{\mathrm{opt}}) \approx 29$).

15.4.3 Adaptive Nearest Neighbors

The random forest classifier has much in common with the k-nearest neighbor classifier (Section 13.3); in fact a weighted version thereof. Since each tree is grown to maximal size, for a particular Θ^*, $T(x;\Theta^*(\mathbf{Z}))$ is the response value for one of the training samples[4]. The tree-growing algorithm finds an "optimal" path to that observation, choosing the most informative predictors from those at its disposal. The averaging process assigns weights to these training responses, which ultimately vote for the prediction. Hence via the random-forest voting mechanism, those observations *close* to the target point get assigned weights—an equivalent kernel—which combine to form the classification decision.

Figure 15.11 demonstrates the similarity between the decision boundary of 3-nearest neighbors and random forests on the mixture data.

---

[4]We gloss over the fact that pure nodes are not split further, and hence there can be more than one observation in a terminal node

Random Forest Classifier

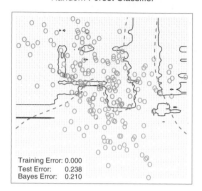

3–Nearest Neighbors

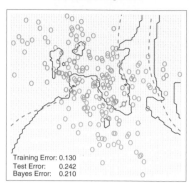

FIGURE 15.11. *Random forests versus 3-NN on the mixture data. The axis-oriented nature of the individual trees in a random forest lead to decision regions with an axis-oriented flavor.*

Bibliographic Notes

Random forests as described here were introduced by Breiman (2001), although many of the ideas had cropped up earlier in the literature in different forms. Notably Ho (1995) introduced the term "random forest," and used a consensus of trees grown in random subspaces of the features. The idea of using stochastic perturbation and averaging to avoid overfitting was introduced by Kleinberg (1990), and later in Kleinberg (1996). Amit and Geman (1997) used randomized trees grown on image features for image classification problems. Breiman (1996a) introduced bagging, a precursor to his version of random forests. Dietterich (2000b) also proposed an improvement on bagging using additional randomization. His approach was to rank the top 20 candidate splits at each node, and then select from the list at random. He showed through simulations and real examples that this additional randomization improved over the performance of bagging. Friedman and Hall (2007) showed that sub-sampling (without replacement) is an effective alternative to bagging. They showed that growing and averaging trees on samples of size $N/2$ is approximately equivalent (in terms bias/variance considerations) to bagging, while using smaller fractions of N reduces the variance even further (through decorrelation).

There are several free software implementations of random forests. In this chapter we used the `randomForest` package in R, maintained by Andy Liaw, available from the `CRAN` website. This allows both split-variable selection, as well as sub-sampling. Adele Cutler maintains a random forest website `http://www.math.usu.edu/~adele/forests/` where (as of August 2008) the software written by Leo Breiman and Adele Cutler is freely available. Their code, and the name "random forests", is exclusively licensed to Salford Systems for commercial release. The Weka machine learning archive http://www.cs.waikato.ac.nz/ml/weka/ at Waikato University, New Zealand, offers a free java implementation of random forests.

Exercises

Ex. 15.1 Derive the variance formula (15.1). This appears to fail if ρ is negative; diagnose the problem in this case.

Ex. 15.2 Show that as the number of bootstrap samples B gets large, the OOB error estimate for a random forest approaches its N-fold CV error estimate, and that in the limit, the identity is exact.

Ex. 15.3 Consider the simulation model used in Figure 15.7 (Mease and Wyner, 2008). Binary observations are generated with probabilities

$$\Pr(Y = 1|X) = q + (1 - 2q) \cdot I \left[\sum_{j=1}^{J} X_j > J/2 \right], \qquad (15.11)$$

where $X \sim U[0, 1]^p$, $0 \le q \le \frac{1}{2}$, and $J \le p$ is some predefined (even) number. Describe this probability surface, and give the Bayes error rate.

Ex. 15.4 Suppose x_i, $i = 1, \ldots, N$ are iid (μ, σ^2). Let $\bar{x}_1^*$ and $\bar{x}_2^*$ be two bootstrap realizations of the sample mean. Show that the sampling correlation $\mathrm{corr}(\bar{x}_1^*, \bar{x}_2^*) = \frac{n}{2n-1} \approx 50\%$. Along the way, derive $\mathrm{var}(\bar{x}_1^*)$ and the variance of the bagged mean $\bar{x}_{bag}$. Here $\bar{x}$ is a *linear* statistic; bagging produces no reduction in variance for linear statistics.

Ex. 15.5 Show that the sampling correlation between a pair of random-forest trees at a point x is given by

$$\rho(x) = \frac{\mathrm{Var}_{\mathbf{Z}}[\mathrm{E}_{\Theta|\mathbf{Z}} T(x; \Theta(\mathbf{Z}))]}{\mathrm{Var}_{\mathbf{Z}}[\mathrm{E}_{\Theta|\mathbf{Z}} T(x; \Theta(\mathbf{Z}))] + \mathrm{E}_{\mathbf{Z}} \mathrm{Var}_{\Theta|\mathbf{Z}}[T(x; \Theta(\mathbf{Z}))]}. \qquad (15.12)$$

The term in the numerator is $\mathrm{Var}_{\mathbf{Z}}[\hat{f}_{\mathrm{rf}}(x)]$, and the second term in the denominator is the expected conditional variance due to the randomization in random forests.

Ex. 15.6 Fit a series of random-forest classifiers to the spam data, to explore the sensitivity to the parameter m. Plot both the OOB error as well as the test error against a suitably chosen range of values for m.

Ex. 15.7 Suppose we fit a linear regression model to N observations with response y_i and predictors $x_{i1}, \ldots, x_{ip}$. Assume that all variables are standardized to have mean zero and standard deviation one. Let RSS be the mean-squared residual on the training data, and $\hat{\beta}$ the estimated coefficient. Denote by RSS_j^* the mean-squared residual on the training data using the same $\hat{\beta}$, but with the N values for the jth variable randomly permuted before the predictions are calculated. Show that

$$\mathrm{E}_P[RSS_j^* - RSS] = 2\hat{\beta}_j^2, \tag{15.13}$$

where E_P denotes expectation with respect to the permutation distribution. Argue that this is approximately true when the evaluations are done using an independent test set.

16
Ensemble Learning

16.1 Introduction

The idea of ensemble learning is to build a prediction model by combining the strengths of a collection of simpler base models. We have already seen a number of examples that fall into this category.

Bagging in Section 8.7 and random forests in Chapter 15 are ensemble methods for classification, where a *committee* of trees each cast a vote for the predicted class. Boosting in Chapter 10 was initially proposed as a committee method as well, although unlike random forests, the committee of *weak learners* evolves over time, and the members cast a weighted vote. Stacking (Section 8.8) is a novel approach to combining the strengths of a number of fitted models. In fact one could characterize any dictionary method, such as regression splines, as an ensemble method, with the basis functions serving the role of weak learners.

Bayesian methods for nonparametric regression can also be viewed as ensemble methods: a large number of candidate models are averaged with respect to the posterior distribution of their parameter settings (e.g. (Neal and Zhang, 2006)).

Ensemble learning can be broken down into two tasks: developing a population of base learners from the training data, and then combining them to form the composite predictor. In this chapter we discuss boosting technology that goes a step further; it builds an ensemble model by conducting a regularized and supervised search in a high-dimensional space of weak learners.

T. Hastie et al., *The Elements of Statistical Learning, Second Edition,*
DOI: 10.1007/b94608_16,
© Springer Science+Business Media, LLC 2009

An early example of a learning ensemble is a method designed for multi-class classification using *error-correcting output codes* (Dieterich and Bakiri, 1995, ECOC). Consider the 10-class digit classification problem, and the coding matrix $\mathbf{C}$ given in Table 16.1.

TABLE 16.1. *Part of a 15-bit error-correcting coding matrix $\mathbf{C}$ for the 10-class digit classification problem. Each column defines a two-class classification problem.*

Digit	C_1	C_2	C_3	C_4	C_5	C_6	$\cdots$	C_{15}
0	1	1	0	0	0	0	$\cdots$	1
1	0	0	1	1	1	1	$\cdots$	0
2	1	0	0	1	0	0	$\cdots$	1
$\vdots$	$\vdots$	$\vdots$	$\vdots$	$\vdots$	$\vdots$	$\vdots$	$\cdots$	$\vdots$
8	1	1	0	1	0	1	$\cdots$	1
9	0	1	1	1	0	0	$\cdots$	0

Note that the ℓth column of the coding matrix C_ℓ defines a two-class variable that merges all the original classes into two groups. The method works as follows:

1. Learn a separate classifier for each of the $L = 15$ two class problems defined by the columns of the coding matrix.

2. At a test point x, let $\hat{p}_\ell(x)$ be the predicted probability of a one for the ℓth response.

3. Define $\delta_k(x) = \sum_{\ell=1}^{L} |C_{k\ell} - \hat{p}_\ell(x)|$, the discriminant function for the kth class, where $C_{k\ell}$ is the entry for row k and column ℓ in Table 16.1.

Each row of $\mathbf{C}$ is a binary code for representing that class. The rows have more bits than is necessary, and the idea is that the redundant "error-correcting" bits allow for some inaccuracies, and can improve performance. In fact, the full code matrix $\mathbf{C}$ above has a minimum Hamming distance[1] of 7 between any pair of rows. Note that even the indicator response coding (Section 4.2) is redundant, since 10 classes require only $\lceil \log_2 10 = 4$ bits for their unique representation. Dieterich and Bakiri (1995) showed impressive improvements in performance for a variety of multiclass problems when classification trees were used as the base classifier.

James and Hastie (1998) analyzed the ECOC approach, and showed that random code assignment worked as well as the optimally constructed error-correcting codes. They also argued that the main benefit of the coding was in variance reduction (as in bagging and random forests), because the different coded problems resulted in different trees, and the decoding step (3) above has a similar effect as averaging.

---

[1] The Hamming distance between two vectors is the number of mismatches between corresponding entries.

16.2   Boosting and Regularization Paths

In Section 10.12.2 of the first edition of this book, we suggested an analogy between the sequence of models produced by a gradient boosting algorithm and regularized model fitting in high-dimensional feature spaces. This was primarily motivated by observing the close connection between a boosted version of linear regression and the lasso (Section 3.4.2). These connections have been pursued by us and others, and here we present our current thinking in this area. We start with the original motivation, which fits more naturally in this chapter on ensemble learning.

16.2.1   Penalized Regression

Intuition for the success of the shrinkage strategy (10.41) of gradient boosting (page 364 in Chapter 10) can be obtained by drawing analogies with penalized linear regression with a large basis expansion. Consider the dictionary of all possible J-terminal node regression trees $\mathcal{T} = \{T_k\}$ that could be realized on the training data as basis functions in $\mathbb{R}^p$. The linear model is

$$f(x) = \sum_{k=1}^{K} \alpha_k T_k(x), \qquad (16.1)$$

where $K = \text{card}(\mathcal{T})$. Suppose the coefficients are to be estimated by least squares. Since the number of such trees is likely to be much larger than even the largest training data sets, some form of regularization is required. Let $\hat{\alpha}(\lambda)$ solve

$$\min_{\alpha} \left\{ \sum_{i=1}^{N} \left(y_i - \sum_{k=1}^{K} \alpha_k T_k(x_i) \right)^2 + \lambda \cdot J(\alpha) \right\}, \qquad (16.2)$$

$J(\alpha)$ is a function of the coefficients that generally penalizes larger values. Examples are

$$J(\alpha) = \sum_{k=1}^{K} |\alpha_k|^2 \qquad \text{ridge regression,} \qquad (16.3)$$

$$J(\alpha) = \sum_{k=1}^{K} |\alpha_k| \qquad \text{lasso,} \qquad (16.4)$$

$$(16.5)$$

both covered in Section 3.4. As discussed there, the solution to the lasso problem with moderate to large λ tends to be sparse; many of the $\hat{\alpha}_k(\lambda) = 0$. That is, only a small fraction of all possible trees enter the model (16.1).

Algorithm 16.1 *Forward Stagewise Linear Regression.*

1. Initialize $\breve{\alpha}_k = 0$, $k = 1, \dots, K$. Set $\varepsilon > 0$ to some small constant, and M large.

2. For $m = 1$ to M:

   (a) $(\beta^*, k^*) = \arg\min_{\beta,k} \sum_{i=1}^{N} \left(y_i - \sum_{l=1}^{K} \breve{\alpha}_l T_l(x_i) - \beta T_k(x_i) \right)^2$.

   (b) $\breve{\alpha}_{k^*} \leftarrow \breve{\alpha}_{k^*} + \varepsilon \cdot \mathrm{sign}(\beta^*)$.

3. Output $f_M(x) = \sum_{k=1}^{K} \breve{\alpha}_k T_k(x)$.

This seems reasonable since it is likely that only a small fraction of all possible trees will be relevant in approximating any particular target function. However, the relevant subset will be different for different targets. Those coefficients that are not set to zero are shrunk by the lasso in that their absolute values are smaller than their corresponding least squares values[2]: $|\hat{\alpha}_k(\lambda)| < |\hat{\alpha}_k(0)|$. As λ increases, the coefficients all shrink, each one ultimately becoming zero.

Owing to the very large number of basis functions T_k, directly solving (16.2) with the lasso penalty (16.4) is not possible. However, a feasible forward stagewise strategy exists that closely approximates the effect of the lasso, and is very similar to boosting and the forward stagewise Algorithm 10.2. Algorithm 16.1 gives the details. Although phrased in terms of tree basis functions T_k, the algorithm can be used with any set of basis functions. Initially all coefficients are zero in line 1; this corresponds to $\lambda = \infty$ in (16.2). At each successive step, the tree T_{k^*} is selected that best fits the current residuals in line 2(a). Its corresponding coefficient $\breve{\alpha}_{k^*}$ is then incremented or decremented by an infinitesimal amount in 2(b), while all other coefficients $\breve{\alpha}_k$, $k \neq k^*$ are left unchanged. In principle, this process could be iterated until either all the residuals are zero, or $\beta^* = 0$. The latter case can occur if $K < N$, and at that point the coefficient values represent a least squares solution. This corresponds to $\lambda = 0$ in (16.2).

After applying Algorithm 16.1 with $M < \infty$ iterations, many of the coefficients will be zero, namely, those that have yet to be incremented. The others will tend to have absolute values smaller than their corresponding least squares solution values, $|\breve{\alpha}_k(M)| < |\hat{\alpha}_k(0)|$. Therefore this M-iteration solution qualitatively resembles the lasso, with M inversely related to λ.

Figure 16.1 shows an example, using the prostate data studied in Chapter 3. Here, instead of using trees $T_k(X)$ as basis functions, we use the origi-

---

[2]If $K > N$, there is in general no unique "least squares value," since infinitely many solutions will exist that fit the data perfectly. We can pick the minimum L_1-norm solution amongst these, which is the unique lasso solution.

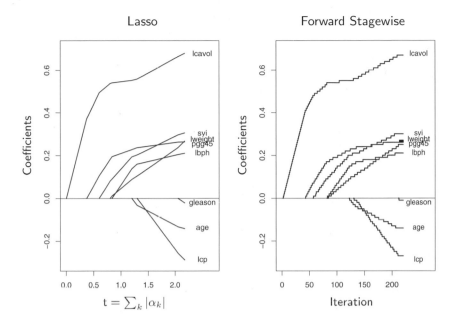

FIGURE 16.1. *Profiles of estimated coefficients from linear regression, for the prostate data studied in Chapter 3. The left panel shows the results from the lasso, for different values of the bound parameter $t = \sum_k |\alpha_k|$. The right panel shows the results of the stagewise linear regression Algorithm 16.1, using $M = 220$ consecutive steps of size $\varepsilon = .01$.*

nal variables X_k themselves; that is, a multiple linear regression model. The left panel displays the profiles of estimated coefficients from the lasso, for different values of the bound parameter $t = \sum_k |\alpha_k|$. The right panel shows the results of the stagewise Algorithm 16.1, with $M = 250$ and $\varepsilon = 0.01$. [The left and right panels of Figure 16.1 are the same as Figure 3.10 and the left panel of Figure 3.19, respectively.] The similarity between the two graphs is striking.

In some situations the resemblance is more than qualitative. For example, if all of the basis functions T_k are mutually uncorrelated, then as $\varepsilon \downarrow 0$, $M \uparrow$ such that $M\epsilon \to t$, Algorithm 16.1 yields exactly the same solution as the lasso for bound parameter $t = \sum_k |\alpha_k|$ (and likewise for all solutions along the path). Of course, tree-based regressors are not uncorrelated. However, the solution sets are also identical if the coefficients $\hat{\alpha}_k(\lambda)$ are all monotone functions of λ. This is often the case when the correlation between the variables is low. When the $\hat{\alpha}_k(\lambda)$ are not monotone in λ, then the solution sets are not identical. The solution sets for Algorithm 16.1 tend to change less rapidly with changing values of the regularization parameter than those of the lasso.

Efron et al. (2004) make the connections more precise, by characterizing the exact solution paths in the ε-limiting case. They show that the coefficient paths are piece-wise linear functions, both for the lasso and forward stagewise. This facilitates efficient algorithms which allow the entire paths to be computed with the same cost as a single least-squares fit. This *least angle regression* algorithm is described in more detail in Section 3.8.1.

Hastie et al. (2007) show that this infinitesimal forward stagewise algorithm (FS_0) fits a monotone version of the lasso, which optimally reduces at each step the loss function for a given increase in the *arc length* of the coefficient path (see Sections 16.2.3 and 3.8.1). The arc-length for the $\epsilon > 0$ case is $M\epsilon$, and hence proportional to the number of steps.

Tree boosting (Algorithm 10.3) with shrinkage (10.41) closely resembles Algorithm 16.1, with the learning rate parameter ν corresponding to ε. For squared error loss, the only difference is that the optimal tree to be selected at each iteration T_{k*} is approximated by the standard top-down greedy tree-induction algorithm. For other loss functions, such as the exponential loss of AdaBoost and the binomial deviance, Rosset et al. (2004a) show similar results to what we see here. Thus, one can view tree boosting with shrinkage as a form of monotone ill-posed regression on all possible (J-terminal node) trees, with the lasso penalty (16.4) as a regularizer. We return to this topic in Section 16.2.3.

The choice of no shrinkage [$\nu = 1$ in equation (10.41)] is analogous to forward-stepwise regression, and its more aggressive cousin best-subset selection, which penalizes the *number* of non zero coefficients $J(\alpha) = \sum_k |\alpha_k|^0$. With a small fraction of dominant variables, best subset approaches often work well. But with a moderate fraction of strong variables, it is well known that subset selection can be excessively greedy (Copas, 1983), often yielding poor results when compared to less aggressive strategies such as the lasso or ridge regression. The dramatic improvements often seen when shrinkage is used with boosting are yet another confirmation of this approach.

16.2.2 The "Bet on Sparsity" Principle

As shown in the previous section, boosting's forward stagewise strategy with shrinkage approximately minimizes the same loss function with a lasso-style L_1 penalty. The model is built up slowly, searching through "model space" and adding shrunken basis functions derived from important predictors. In contrast, the L_2 penalty is computationally much easier to deal with, as shown in Section 12.3.7. With the basis functions and L_2 penalty chosen to match a particular positive-definite kernel, one can solve the corresponding optimization problem without explicitly searching over individual basis functions.

However, the sometimes superior performance of boosting over procedures such as the support vector machine may be largely due to the implicit use of the L_1 versus L_2 penalty. The shrinkage resulting from the

L_1 penalty is better suited to *sparse* situations, where there are few basis functions with nonzero coefficients (among all possible choices).

We can strengthen this argument through a simple example, taken from Friedman et al. (2004). Suppose we have $10,000$ data points and our model is a linear combination of a million trees. If the true population coefficients of these trees arose from a Gaussian distribution, then we know that in a Bayesian sense the best predictor is ridge regression (Exercise 3.6). That is, we should use an L_2 rather than an L_1 penalty when fitting the coefficients. On the other hand, if there are only a small number (e.g., 1000) coefficients that are nonzero, the lasso (L_1 penalty) will work better. We think of this as a *sparse* scenario, while the first case (Gaussian coefficients) is *dense*. Note however that in the dense scenario, although the L_2 penalty is best, neither method does very well since there is too little data from which to estimate such a large number of nonzero coefficients. This is the curse of dimensionality taking its toll. In a sparse setting, we can potentially do well with the L_1 penalty, since the number of nonzero coefficients is small. The L_2 penalty fails again.

In other words, use of the L_1 penalty follows what we call the "bet on sparsity" principle for high-dimensional problems:

> *Use a procedure that does well in sparse problems, since no procedure does well in dense problems.*

These comments need some qualification:

- For any given application, the degree of sparseness/denseness depends on the unknown true target function, and the chosen dictionary $\mathcal{T}$.

- The notion of sparse versus dense is relative to the size of the training data set and/or the noise-to-signal ratio (NSR). Larger training sets allow us to estimate coefficients with smaller standard errors. Likewise in situations with small NSR, we can identify more nonzero coefficients with a given sample size than in situations where the NSR is larger.

- The size of the dictionary plays a role as well. Increasing the size of the dictionary may lead to a sparser representation for our function, but the search problem becomes more difficult leading to higher variance.

Figure 16.2 illustrates these points in the context of linear models using simulation. We compare ridge regression and lasso, both for classification and regression problems. Each run has 50 observations with 300 independent Gaussian predictors. In the top row all 300 coefficients are nonzero, generated from a Gaussian distribution. In the middle row, only 10 are nonzero and generated from a Gaussian, and the last row has 30 non zero Gaussian coefficients. For regression, standard Gaussian noise is

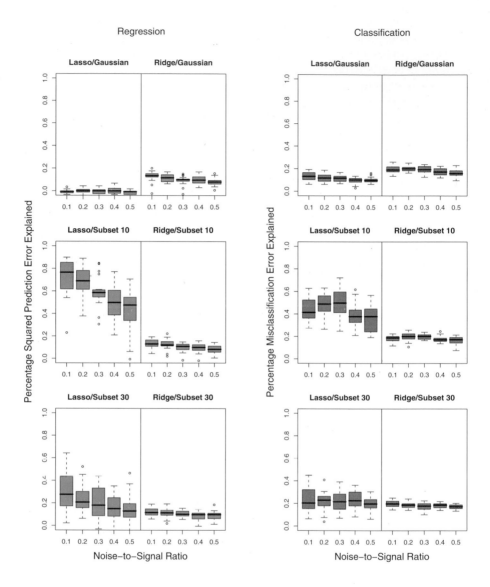

FIGURE 16.2. *Simulations that show the superiority of the L_1 (lasso) penalty over L_2 (ridge) in regression and classification. Each run has 50 observations with 300 independent Gaussian predictors. In the top row all 300 coefficients are nonzero, generated from a Gaussian distribution. In the middle row, only 10 are nonzero, and the last row has 30 nonzero. Gaussian errors are added to the linear predictor $\eta(X)$ for the regression problems, and binary responses generated via the inverse-logit transform for the classification problems. Scaling of $\eta(X)$ resulted in the noise-to-signal ratios shown. Lasso is used in the left sub-columns, ridge in the right. We report the optimal percentage of error explained on test data (relative to the error of a constant model), displayed as boxplots over 20 realizations for each combination. In the only situation where ridge beats lasso (top row), neither do well.*

added to the linear predictor $\eta(X) = X^T\beta$ to produce a continuous response. For classification the linear predictor is transformed via the inverse-logit to a probability, and a binary response is generated. Five different noise-to-signal ratios are presented, obtained by scaling $\eta(X)$ prior to generating the response. In both cases this is defined to be $NSR = \mathrm{Var}(Y|\eta(X))/\mathrm{Var}(\eta(X))$. Both the ridge regression and lasso coefficient paths were fit using a series of 50 values of λ corresponding to a range of df from 1 to 50 (see Chapter 3 for details). The models were evaluated on a large test set (infinite for Gaussian, 5000 for binary), and in each case the value for λ was chosen to minimize the test-set error. We report percentage variance explained for the regression problems, and percentage misclassification error explained for the classification problems (relative to a baseline error of 0.5). There are 20 simulation runs for each scenario.

Note that for the classification problems, we are using squared-error loss to fit the binary response. Note also that we do not using the training data to select λ, but rather are reporting the best possible behavior for each method in the different scenarios. The L_2 penalty performs poorly everywhere. The Lasso performs reasonably well in the only two situations where it can (sparse coefficients). As expected the performance gets worse as the NSR increases (less so for classification), and as the model becomes denser. The differences are less marked for classification than for regression.

These empirical results are supported by a large body of theoretical results (Donoho and Johnstone, 1994; Donoho and Elad, 2003; Donoho, 2006b; Candes and Tao, 2007) that support the superiority of L_1 estimation in sparse settings.

16.2.3 *Regularization Paths, Over-fitting and Margins*

It has often been observed that boosting "does not overfit," or more astutely is "slow to overfit." Part of the explanation for this phenomenon was made earlier for random forests — misclassification error is less sensitive to variance than is mean-squared error, and classification is the major focus in the boosting community. In this section we show that the regularization paths of boosted models are "well behaved," and that for certain loss functions they have an appealing limiting form.

Figure 16.3 shows the coefficient paths for lasso and infinitesimal forward stagewise (FS_0) in a simulated regression setting. The data consists of a dictionary of 1000 Gaussian variables, strongly correlated ($\rho = 0.95$) within blocks of 20, but uncorrelated between blocks. The generating model has nonzero coefficients for 50 variables, one drawn from each block, and the coefficient values are drawn from a standard Gaussian. Finally, Gaussian noise is added, with a noise-to-signal ratio of 0.72 (Exercise 16.1.) The FS_0 algorithm is a limiting form of algorithm 16.1, where the step size ε is shrunk to zero (Section 3.8.1). The grouping of the variables is intended to mimic the correlations of nearby trees, and with the forward-stagewise

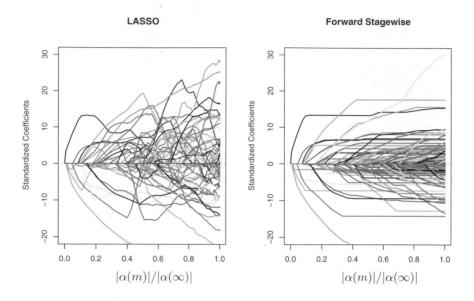

FIGURE 16.3. *Comparison of lasso and infinitesimal forward stagewise paths on simulated regression data. The number of samples is* 60 *and the number of variables is* 1000. *The forward-stagewise paths fluctuate less than those of lasso in the final stages of the algorithms.*

algorithm, this setup is intended as an idealized version of gradient boosting with shrinkage. For both these algorithms, the coefficient paths can be computed exactly, since they are piecewise linear (see the LARS algorithm in Section 3.8.1).

Here the coefficient profiles are similar only in the early stages of the paths. For the later stages, the forward stagewise paths tend to be monotone and smoother, while those for the lasso fluctuate widely. This is due to the strong correlations among subsets of the variables   lasso suffers somewhat from the multi-collinearity problem (Exercise 3.28).

The performance of the two models is rather similar (Figure 16.4), and they achieve about the same minimum. In the later stages forward stagewise takes longer to overfit, a likely consequence of the smoother paths.

Hastie et al. (2007) show that FS_0 solves a *monotone* version of the lasso problem for squared error loss. Let $\mathcal{T}^a = \mathcal{T} \cup \{-\mathcal{T}\}$ be the augmented dictionary obtained by including a negative copy of every basis element in $\mathcal{T}$. We consider models $f(x) = \sum_{T_k \in \mathcal{T}^a} \alpha_k T_k(x)$ with non-negative coefficients $\alpha_k \geq 0$. In this expanded space, the lasso coefficient paths are positive, while those of FS_0 are monotone nondecreasing.

The monotone lasso path is characterized by a differential equation

$$\frac{\partial \alpha}{\partial \ell} = \rho^{ml}(\alpha(\ell)), \qquad (16.6)$$

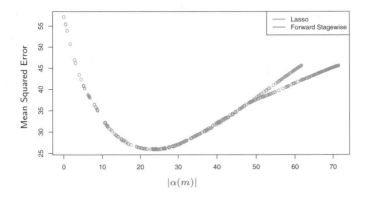

FIGURE 16.4. *Mean squared error for lasso and infinitesimal forward stagewise on the simulated data. Despite the difference in the coefficient paths, the two models perform similarly over the critical part of the regularization path. In the right tail, lasso appears to overfit more rapidly.*

with initial condition $\alpha(0) = 0$, where ℓ is the L_1 arc-length of the path $\alpha(\ell)$ (Exercise 16.2). The monotone lasso move direction (velocity vector) $\rho^{ml}(\alpha(\ell))$ decreases the loss at the optimal quadratic rate per unit increase in the L_1 arc-length of the path. Since $\rho_k^{ml}(\alpha(\ell)) \geq 0 \ \forall k, \ell$, the solution paths are monotone.

The lasso can similarly be characterized as the solution to a differential equation as in (16.6), except that the move directions decrease the loss optimally per unit increase in the L_1 norm of the path. As a consequence, they are not necessarily positive, and hence the lasso paths need not be monotone.

In this augmented dictionary, restricting the coefficients to be positive is natural, since it avoids an obvious ambiguity. It also ties in more naturally with tree boosting—we always find trees positively correlated with the current residual.

There have been suggestions that boosting performs well (for two-class classification) because it exhibits maximal-margin properties, much like the support-vector machines of Chapters 4.5.2 and 12. Schapire et al. (1998) define the *normalized L_1 margin* of a fitted model $f(x) = \sum_k \alpha_k T_k(x)$ as

$$m(f) = \min_i \frac{y_i f(x_i)}{\sum_{k=1}^{K} |\alpha_k|}. \tag{16.7}$$

Here the minimum is taken over the training sample, and $y_i \in \{-1, +1\}$. Unlike the L_2 margin (4.40) of support vector machines, the L_1 margin $m(f)$ measures the distance to the closest training point in L_∞ units (maximum coordinate distance).

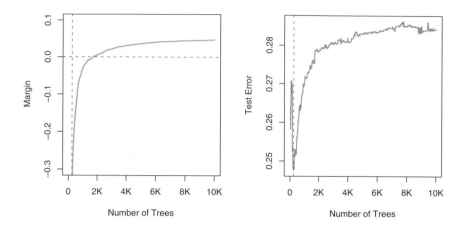

FIGURE 16.5. *The left panel shows the L_1 margin $m(f)$ for the Adaboost classifier on the mixture data, as a function of the number of 4-node trees. The model was fit using the R package* **gbm**, *with a shrinkage factor of* 0.02. *After* 10,000 *trees, $m(f)$ has settled down. Note that when the margin crosses zero, the training error becomes zero. The right panel shows the test error, which is minimized at* 240 *trees. In this case, Adaboost overfits dramatically if run to convergence.*

Schapire et al. (1998) prove that with separable data, Adaboost increases $m(f)$ with each iteration, converging to a margin-symmetric solution. Rätsch and Warmuth (2002) prove the asymptotic convergence of Adaboost with shrinkage to a L_1-margin-maximizing solution. Rosset et al. (2004a) consider regularized models of the form (16.2) for general loss functions. They show that as $\lambda \downarrow 0$, for particular loss functions the solution converges to a margin-maximizing configuration. In particular they show this to be the case for the exponential loss of Adaboost, as well as binomial deviance.

Collecting together the results of this section, we reach the following summary for boosted classifiers:

> *The sequence of boosted classifiers form an L_1-regularized monotone path to a margin-maximizing solution.*

Of course the margin-maximizing end of the path can be a very poor, overfit solution, as it is in the example in Figure 16.5. Early stopping amounts to picking a point along the path, and should be done with the aid of a validation dataset.

16.3   Learning Ensembles

The insights learned from the previous sections can be harnessed to produce a more effective and efficient ensemble model. Again we consider functions of the form

$$f(x) = \alpha_0 + \sum_{T_k \in \mathcal{T}} \alpha_k T_k(x), \qquad (16.8)$$

where $\mathcal{T}$ is a dictionary of basis functions, typically trees. For gradient boosting and random forests, $|\mathcal{T}|$ is very large, and it is quite typical for the final model to involve many thousands of trees. In the previous section we argue that gradient boosting with shrinkage fits an L_1 regularized monotone path in this space of trees.

Friedman and Popescu (2003) propose a hybrid approach which breaks this process down into two stages:

- A finite dictionary $\mathcal{T}_L = \{T_1(x), T_2(x), \dots, T_M(x)\}$ of basis functions is induced from the training data;

- A family of functions $f_\lambda(x)$ is built by fitting a lasso path in this dictionary:

$$\alpha(\lambda) = \arg\min_\alpha \sum_{i=1}^N L[y_i, \alpha_0 + \sum_{m-1}^M \alpha_m T_m(x_i)] + \lambda \sum_{m=1}^M |\alpha_m|. \quad (16.9)$$

In its simplest form this model could be seen as a way of post-processing boosting or random forests, taking for $\mathcal{T}_L$ the collection of trees produced by the gradient boosting or random forest algorithms. By fitting the lasso path to these trees, we would typically use a much reduced set, which would save in computations and storage for future predictions. In the next section we describe modifications of this prescription that reduce the correlations in the ensemble $\mathcal{T}_L$, and improve the performance of the lasso post processor.

As an initial illustration, we apply this procedure to a random forest ensemble grown on the spam data.

Figure 16.6 shows that a lasso post-processing offers modest improvement over the random forest (blue curve), and reduces the forest to about 40 trees, rather than the original 1000. The post-processed performance matches that of gradient boosting. The orange curves represent a modified version of random forests, designed to reduce the correlations between trees even more. Here a random sub-sample (without replacement) of 5% of the training sample is used to grow each tree, and the trees are restricted to be shallow (about six terminal nodes). The post-processing offers more dramatic improvements here, and the training costs are reduced by a factor of about 100. However, the performance of the post-processed model falls somewhat short of the blue curves.

16.3.1 *Learning a Good Ensemble*

Not all ensembles $\mathcal{T}_L$ will perform well with post-processing. In terms of basis functions, we want a collection that covers the space well in places

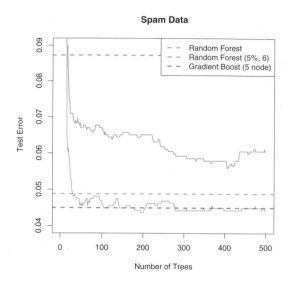

FIGURE 16.6. *Application of the lasso post-processing (16.9) to the spam data. The horizontal blue line is the test error of a random forest fit to the spam data, using 1000 trees grown to maximum depth (with m = 7; see Algorithm 15.1). The jagged blue curve is the test error after post-processing the first 500 trees using the lasso, as a function of the number of trees with nonzero coefficients. The orange curve/line use a modified form of random forest, where a random draw of 5% of the data are used to grow each tree, and the trees are forced to be shallow (typically six terminal nodes). Here the post-processing offers much greater improvement over the random forest that generated the ensemble.*

where they are needed, and are sufficiently different from each other for the post-processor to be effective.

Friedman and Popescu (2003) gain insights from numerical quadrature and importance sampling. They view the unknown function as an integral

$$f(x) = \int \beta(\gamma) b(x; \gamma) d\gamma, \qquad (16.10)$$

where $\gamma \in \Gamma$ indexes the basis functions $b(x; \gamma)$. For example, if the basis functions are trees, then γ indexes the splitting variables, the split-points and the values in the terminal nodes. Numerical quadrature amounts to finding a set of M evaluation points $\gamma_m \in \Gamma$ and corresponding weights α_m so that $f_M(x) = \alpha_0 + \sum_{m=1}^{M} \alpha_m b(x; \gamma_m)$ approximates $f(x)$ well over the domain of x. Importance sampling amounts to sampling γ at random, but giving more weight to relevant regions of the space Γ. Friedman and Popescu (2003) suggest a measure of (lack of) relevance that uses the loss function (16.9):

$$Q(\gamma) = \min_{c_0, c_1} \sum_{i=1}^{N} L(y_i, c_0 + c_1 b(x_i; \gamma)), \qquad (16.11)$$

evaluated on the training data.

If a single basis function were to be selected (e.g., a tree), it would be the global minimizer $\gamma^* = \arg\min_{\gamma \in \Gamma} Q(\gamma)$. Introducing randomness in the selection of γ would necessarily produce less optimal values with $Q(\gamma) \geq Q(\gamma^*)$. They propose a natural measure of the characteristic *width* σ of the sampling scheme $\mathcal{S}$,

$$\sigma = \mathrm{E}_{\mathcal{S}}[Q(\gamma) - Q(\gamma^*)]. \qquad (16.12)$$

- σ too narrow suggests too many of the $b(x; \gamma_m)$ look alike, and similar to $b(x; \gamma^*)$;

- σ too wide implies a large spread in the $b(x; \gamma_m)$, but possibly consisting of many irrelevant cases.

Friedman and Popescu (2003) use sub-sampling as a mechanism for introducing randomness, leading to their ensemble-generation algorithm 16.2.

Algorithm 16.2 *ISLE Ensemble Generation.*

---

1. $f_0(x) = \arg\min_c \sum_{i=1}^{N} L(y_i, c)$

2. For $m = 1$ to M do

    (a) $\gamma_m = \arg\min_\gamma \sum_{i \in S_m(\eta)} L(y_i, f_{m-1}(x_i) + b(x_i; \gamma))$

    (b) $f_m(x) = f_{m-1}(x) + \nu b(x; \gamma_m)$

3. $\mathcal{T}_{ISLE} = \{b(x; \gamma_1), b(x; \gamma_2), \ldots, b(x; \gamma_M)\}$.

$S_m(\eta)$ refers to a subsample of $N \cdot \eta$ ($\eta \in (0, 1]$) of the training observations, typically *without* replacement. Their simulations suggest picking $\eta \leq \frac{1}{2}$, and for large N picking $\eta \sim 1/\sqrt{N}$. Reducing η increases the randomness, and hence the width σ. The parameter $\nu \in [0, 1]$ introduces *memory* into the randomization process; the larger ν, the more the procedure avoids $b(x; \gamma)$ similar to those found before. A number of familiar randomization schemes are special cases of Algorithm 16.2:

Bagging has $\eta = 1$, but samples with replacement, and has $\nu = 0$. Friedman and Hall (2007) argue that sampling without replacement with $\eta = 1/2$ is equivalent to sampling with replacement with $\eta = 1$, and the former is much more efficient.

Random forest sampling is similar, with more randomness introduced by the selection of the splitting variable. Reducing $\eta < 1/2$ in algorithm 16.2 has a similar effect to reducing m in random forests, but does not suffer from the potential biases discussed in Section 15.4.2.

Gradient boosting with shrinkage (10.41) uses $\eta = 1$, but typically does not produce sufficient width σ.

Stochastic gradient boosting (Friedman, 1999) follows the recipe exactly.

The authors recommend values $\nu = 0.1$ and $\eta \leq \frac{1}{2}$, and call their combined procedure (ensemble generation and post processing) *Importance sampled learning ensemble* (ISLE).

Figure 16.7 shows the performance of an ISLE on the spam data. It does

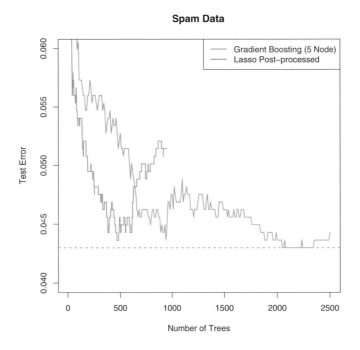

Spam Data

FIGURE 16.7. *Importance sampling learning ensemble (ISLE) fit to the spam data. Here we used $\eta = 1/2$, $\nu = 0.05$, and trees with five terminal nodes. The lasso post-processed ensemble does not improve the prediction error in this case, but it reduces the number of trees by a factor of five.*

not improve the predictive performance, but is able to produce a more parsimonious model. Note that in practice the post-processing includes the selection of the regularization parameter λ in (16.9), which would be chosen by cross-validation. Here we simply demonstrate the effects of post-processing by showing the entire path on the test data.

Figure 16.8 shows various ISLEs on a regression example. The generating

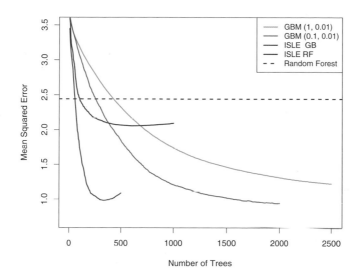

FIGURE 16.8. *Demonstration of ensemble methods on a regression simulation example. The notation GBM (0.1, 0.01) refers to a gradient boosted model, with parameters (η, ν). We report mean-squared error from the true (known) function. Note that the sub-sampled GBM model (green) outperforms the full GBM model (orange). The lasso post-processed version achieves similar error. The random forest is outperformed by its post-processed version, but both fall short of the other models.*

function is

$$f(X) = 10 \cdot \prod_{j=1}^{5} e^{-2X_j^2} + \sum_{j=6}^{35} X_j, \qquad (16.13)$$

where $X \sim U[0, 1]^{100}$ (the last 65 elements are noise variables). The response $Y = f(X) + \varepsilon$ where $\varepsilon \sim N(0, \sigma^2)$; we chose $\sigma = 1.3$ resulting in a signal-to-noise ratio of approximately 2. We used a training sample of size 1000, and estimated the mean squared error $E(\hat{f}(X) - f(X))^2$ by averaging over a test set of 500 samples. The sub-sampled GBM curve (light blue) is an instance of *stochastic gradient boosting* (Friedman, 1999) discussed in Section 10.12, and it outperforms gradient boosting on this example.

16.3.2 *Rule Ensembles*

Here we describe a modification of the tree-ensemble method that focuses on individual rules (Friedman and Popescu, 2003). We encountered rules in Section 9.3 in the discussion of the PRIM method. The idea is to enlarge an ensemble of trees by constructing a set of rules from each of the trees in the collection.

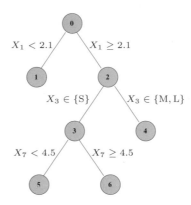

FIGURE 16.9. *A typical tree in an ensemble, from which rules can be derived.*

Figure 16.9 depicts a small tree, with numbered nodes. The following rules can be derived from this tree:

$$
\begin{aligned}
R_1(X) &= I(X_1 < 2.1) \\
R_2(X) &= I(X_1 \geq 2.1) \\
R_3(X) &= I(X_1 \geq 2.1) \cdot I(X_3 \in \{S\}) \\
R_4(X) &= I(X_1 \geq 2.1) \cdot I(X_3 \in \{M, L\}) \\
R_5(X) &= I(X_1 \geq 2.1) \cdot I(X_3 \in \{S\}) \cdot I(X_7 < 4.5) \\
R_6(X) &= I(X_1 \geq 2.1) \cdot I(X_3 \in \{S\}) \cdot I(X_7 \geq 4.5)
\end{aligned}
\tag{16.14}
$$

A linear expansion in rules 1, 4, 5 and 6 is equivalent to the tree itself (Exercise 16.3); hence (16.14) is an *over-complete* basis for the tree.

For each tree T_m in an ensemble $\mathcal{T}$, we can construct its mini-ensemble of rules $\mathcal{T}_{RULE}^m$, and then combine them all to form a larger ensemble

$$
\mathcal{T}_{\text{RULE}} = \bigcup_{m=1}^{M} \mathcal{T}_{\text{RULE}}^m.
\tag{16.15}
$$

This is then treated like any other ensemble, and post-processed via the lasso or similar regularized procedure.

There are several advantages to this approach of deriving rules from the more complex trees:

- The space of models is enlarged, and can lead to improved performance.

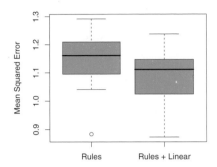

FIGURE 16.10. *Mean squared error for rule ensembles, using* 20 *realizations of the simulation example (16.13).*

- Rules are easier to interpret than trees, so there is the potential for a simplified model.

- It is often natural to augment $\mathcal{T}_{RULE}$ by including each variable X_j separately as well, thus allowing the ensemble to model linear functions well.

Friedman and Popescu (2008) demonstrate the power of this procedure on a number of illustrative examples, including the simulation example (16.13). Figure 16.10 shows boxplots of the mean-squared error from the true model for twenty realizations from this model. The models were all fit using the `Rulefit` software, available on the ESL homepage[3], which runs in an automatic mode.

On the same training set as used in Figure 16.8, the rule based model achieved a mean-squared error of 1.06. Although slightly worse than the best achieved in that figure, the results are not comparable because cross-validation was used here to select the final model.

Bibliographic Notes

As noted in the introduction, many of the new methods in machine learning have been dubbed "ensemble" methods. These include neural networks boosting, bagging and random forests; Dietterich (2000a) gives a survey of tree-based ensemble methods. Neural networks (Chapter 11) are perhaps more deserving of the name, since they simultaneously learn the parameters

---

[3]ESL homepage: `www-stat.stanford.edu/ElemStatLearn`

of the hidden units (basis functions), along with how to combine them. Bishop (2006) discusses neural networks in some detail, along with the Bayesian perspective (MacKay, 1992; Neal, 1996). Support vector machines (Chapter 12) can also be regarded as an ensemble method; they perform L_2 regularized model fitting in high-dimensional feature spaces. Boosting and lasso exploit sparsity through L_1 regularization to overcome the high-dimensionality, while SVMs rely on the "kernel trick" characteristic of L_2 regularization.

C5.0 (Quinlan, 2004) is a commercial tree and rule generation package, with some goals in common with `Rulefit`.

There is a vast and varied literature often referred to as "combining classifiers" which abounds in ad-hoc schemes for mixing methods of different types to achieve better performance. For a principled approach, see Kittler et al. (1998).

Exercises

Ex. 16.1 Describe exactly how to generate the block correlated data used in the simulation in Section 16.2.3.

Ex. 16.2 Let $\alpha(t) \in \mathbb{R}^p$ be a piecewise-differentiable and continuous coefficient profile, with $\alpha(0) = 0$. The L_1 arc-length of α from time 0 to t is defined by

$$\Lambda(t) = \int_0^t |\dot{\alpha}(t)|_1 dt. \tag{16.16}$$

Show that $\Lambda(t) > |\alpha(t)|_1$, with equality iff $\alpha(t)$ is monotone.

Ex. 16.3 Show that fitting a linear regression model using rules 1, 4, 5 and 6 in equation (16.14) gives the same fit as the regression tree corresponding to this tree. Show the same is true for classification, if a logistic regression model is fit.

Ex. 16.4 Program and run the simulation study described in Figure 16.2.

17
Undirected Graphical Models

17.1 Introduction

A graph consists of a set of vertices (nodes), along with a set of edges joining some pairs of the vertices. In graphical models, each vertex represents a random variable, and the graph gives a visual way of understanding the joint distribution of the entire set of random variables. They can be useful for either unsupervised or supervised learning. In an *undirected graph*, the edges have no directional arrows. We restrict our discussion to undirected graphical models, also known as *Markov random fields* or *Markov networks*. In these graphs, the absence of an edge between two vertices has a special meaning: the corresponding random variables are conditionally independent, given the other variables.

Figure 17.1 shows an example of a graphical model for a flow-cytometry dataset with $p = 11$ proteins measured on $N = 7466$ cells, from Sachs et al. (2005). Each vertex in the graph corresponds to the real-valued expression level of a protein. The network structure was estimated assuming a multivariate Gaussian distribution, using the graphical lasso procedure discussed later in this chapter.

Sparse graphs have a relatively small number of edges, and are convenient for interpretation. They are useful in a variety of domains, including genomics and proteomics, where they provide rough models of cell pathways. Much work has been done in defining and understanding the structure of graphical models; see the Bibliographic Notes for references.

T. Hastie et al., *The Elements of Statistical Learning, Second Edition*,
DOI: 10.1007/b94608_17,
© Springer Science+Business Media, LLC 2009

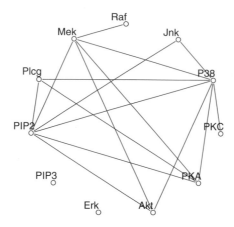

FIGURE 17.1. *Example of a sparse undirected graph, estimated from a flow cytometry dataset, with $p = 11$ proteins measured on $N = 7466$ cells. The network structure was estimated using the graphical lasso procedure discussed in this chapter.*

As we will see, the edges in a graph are parametrized by values or *potentials* that encode the strength of the conditional dependence between the random variables at the corresponding vertices. The main challenges in working with graphical models are model selection (choosing the structure of the graph), estimation of the edge parameters from data, and computation of marginal vertex probabilities and expectations, from their joint distribution. The last two tasks are sometimes called *learning* and *inference* in the computer science literature.

We do not attempt a comprehensive treatment of this interesting area. Instead, we introduce some basic concepts, and then discuss a few simple methods for estimation of the parameters and structure of undirected graphical models; methods that relate to the techniques already discussed in this book. The estimation approaches that we present for continuous and discrete-valued vertices are different, so we treat them separately. Sections 17.3.1 and 17.3.2 may be of particular interest, as they describe new, regression-based procedures for estimating graphical models.

There is a large and active literature on *directed graphical models* or *Bayesian networks*; these are graphical models in which the edges have directional arrows (but no directed cycles). Directed graphical models represent probability distributions that can be factored into products of conditional distributions, and have the potential for causal interpretations. We refer the reader to Wasserman (2004) for a brief overview of both undirected and directed graphs; the next section follows closely his Chapter 18.

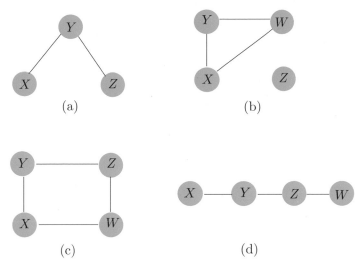

FIGURE 17.2. *Examples of undirected graphical models or Markov networks. Each node or vertex represents a random variable, and the lack of an edge between two nodes indicates conditional independence. For example, in graph (a), X and Z are conditionally independent, given Y. In graph (b), Z is independent of each of X, Y, and W.*

A longer list of useful references is given in the Bibliographic Notes on page 645.

17.2   Markov Graphs and Their Properties

In this section we discuss the basic properties of graphs as models for the joint distribution of a set of random variables. We defer discussion of (a) parametrization and estimation of the edge parameters from data, and (b) estimation of the topology of a graph, to later sections.

Figure 17.2 shows four examples of undirected graphs. A graph $\mathcal{G}$ consists of a pair (V, E), where V is a set of vertices and E the set of edges (defined by pairs of vertices). Two vertices X and Y are called *adjacent* if there is a edge joining them; this is denoted by $X \sim Y$. A *path* $X_1, X_2, \ldots, X_n$ is a set of vertices that are joined, that is $X_{i-1} \sim X_i$ for $i = 2, \ldots, n$. A *complete graph* is a graph with every pair of vertices joined by an edge. A *subgraph* $U \in V$ is a subset of vertices together with their edges. For example, (X, Y, Z) in Figure 17.2(a) form a path but not a complete graph.

Suppose that we have a graph $\mathcal{G}$ whose vertex set V represents a set of random variables having joint distribution P. In a Markov graph $\mathcal{G}$, the absence of an edge implies that the corresponding random variables are conditionally independent given the variables at the other vertices. This is expressed with the following notation:

$$\text{No edge joining } X \text{ and } Y \Longleftrightarrow X \perp Y|\text{rest} \qquad (17.1)$$

where "rest" refers to all of the other vertices in the graph. For example in Figure 17.2(a) $X \perp Z|Y$. These are known as the *pairwise Markov independencies* of $\mathcal{G}$.

If A, B and C are subgraphs, then C is said to *separate* A and B if every path between A and B intersects a node in C. For example, Y separates X and Z in Figures 17.2(a) and (d), and Z separates Y and W in (d). In Figure 17.2(b) Z is not connected to X, Y, W so we say that the two sets are separated by the empty set. In Figure 17.2(c), $C = \{X, Z\}$ separates Y and W.

Separators have the nice property that they break the graph into conditionally independent pieces. Specifically, in a Markov graph $\mathcal{G}$ with subgraphs A, B and C,

$$\text{if } C \text{ separates } A \text{ and } B \text{ then } A \perp B|C. \qquad (17.2)$$

These are known as the *global Markov properties* of $\mathcal{G}$. It turns out that the pairwise and global Markov properties of a graph are equivalent (for graphs with positive distributions). That is, the set of graphs with associated probability distributions that satisfy the pairwise Markov independencies and global Markov assumptions are the same. This result is useful for inferring global independence relations from simple pairwise properties. For example in Figure 17.2(d) $X \perp Z|\{Y, W\}$ since it is a Markov graph and there is no link joining X and Z. But Y also separates X from Z and W and hence by the global Markov assumption we conclude that $X \perp Z|Y$ and $X \perp W|Y$. Similarly we have $Y \perp W|Z$.

The global Markov property allows us to decompose graphs into smaller more manageable pieces and thus leads to essential simplifications in computation and interpretation. For this purpose we separate the graph into cliques. A *clique* is a complete subgraph— a set of vertices that are all adjacent to one another; it is called *maximal* if it is a clique and no other vertices can be added to it and still yield a clique. The maximal cliques for the graphs of Figure 17.2 are

(a) $\{X, Y\}, \{Y, Z\}$,

(b) $\{X, Y, W\}, \{Z\}$,

(c) $\{X, Y\}, \{Y, Z\}, \{Z, W\}, \{X, W\}$, and

(d) $\{X, Y\}, \{Y, Z\}, \{Z, W\}$.

Although the following applies to both continuous and discrete distributions, much of the development has been for the latter. A probability density function f over a Markov graph $\mathcal{G}$ can be can represented as

$$f(x) = \frac{1}{Z} \prod_{C \in \mathcal{C}} \psi_C(x_C) \tag{17.3}$$

where $\mathcal{C}$ is the set of maximal cliques, and the positive functions $\psi_C(\cdot)$ are called *clique potentials*. These are not in general density functions[1], but rather are affinities that capture the dependence in X_C by scoring certain instances x_C higher than others. The quantity

$$Z = \sum_{x \in \mathcal{X}} \prod_{C \in \mathcal{C}} \psi_C(x_C) \tag{17.4}$$

is the normalizing constant, also known as the *partition* function. Alternatively, the representation (17.3) implies a graph with independence properties defined by the cliques in the product. This result holds for Markov networks $\mathcal{G}$ with positive distributions, and is known as the *Hammersley-Clifford* theorem (Hammersley and Clifford, 1971; Clifford, 1990).

Many of the methods for estimation and computation on graphs first decompose the graph into its maximal cliques. Relevant quantities are computed in the individual cliques and then accumulated across the entire graph. A prominent example is the *join tree* or *junction tree* algorithm for computing marginal and low order probabilities from the joint distribution on a graph. Details can be found in Pearl (1986), Lauritzen and Spiegelhalter (1988), Pearl (1988), Shenoy and Shafer (1988), Jensen et al. (1990), or Koller and Friedman (2007).

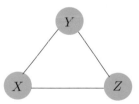

FIGURE 17.3. *A complete graph does not uniquely specify the higher-order dependence structure in the joint distribution of the variables.*

A graphical model does not always uniquely specify the higher-order dependence structure of a joint probability distribution. Consider the complete three-node graph in Figure 17.3. It could represent the dependence structure of either of the following distributions:

$$\begin{aligned} f^{(2)}(x,y,z) &= \tfrac{1}{Z}\psi(x,y)\psi(x,z)\psi(y,z); \\ f^{(3)}(x,y,z) &= \tfrac{1}{Z}\psi(x,y,z). \end{aligned} \tag{17.5}$$

The first specifies only second order dependence (and can be represented with fewer parameters). Graphical models for discrete data are a special

---

[1]If the cliques are separated, then the potentials can be densities, but this is in general not the case.

case of *loglinear models for multiway contingency tables* (Bishop et al., 1975, e.g.); in that language $f^{(2)}$ is referred to as the "no second-order interaction" model.

For the remainder of this chapter we focus on *pairwise Markov graphs* (Koller and Friedman, 2007). Here there is a potential function for each edge (pair of variables as in $f^{(2)}$ above), and at most second–order interactions are represented. These are more parsimonious in terms of parameters, easier to work with, and give the minimal complexity implied by the graph structure. The models for both continuous and discrete data are functions of only the pairwise marginal distributions of the variables represented in the edge set.

17.3   Undirected Graphical Models for Continuous Variables

Here we consider Markov networks where all the variables are continuous. The Gaussian distribution is almost always used for such graphical models, because of its convenient analytical properties. We assume that the observations have a multivariate Gaussian distribution with mean μ and covariance matrix $\boldsymbol{\Sigma}$. Since the Gaussian distribution represents at most second-order relationships, it automatically encodes a pairwise Markov graph. The graph in Figure 17.1 is an example of a Gaussian graphical model.

The Gaussian distribution has the property that all conditional distributions are also Gaussian. The inverse covariance matrix $\boldsymbol{\Sigma}^{-1}$ contains information about the *partial covariances* between the variables; that is, the covariances between pairs i and j, conditioned on all other variables. In particular, if the ijth component of $\boldsymbol{\Theta} = \boldsymbol{\Sigma}^{-1}$ is zero, then variables i and j are conditionally independent, given the other variables (Exercise 17.3).

It is instructive to examine the conditional distribution of one variable versus the rest, where the role of $\boldsymbol{\Theta}$ is explicit. Suppose we partition $X = (Z, Y)$ where $Z = (X_1, \ldots, X_{p-1})$ consists of the first $p - 1$ variables and $Y = X_p$ is the last. Then we have the conditional distribution of Y give Z (Mardia et al., 1979, e.g.)

$$Y|Z = z \sim N\left(\mu_Y + (z - \mu_Z)^T \boldsymbol{\Sigma}_{ZZ}^{-1}\sigma_{ZY},\ \sigma_{YY} - \sigma_{ZY}^T\boldsymbol{\Sigma}_{ZZ}^{-1}\sigma_{ZY}\right), \quad (17.6)$$

where we have partitioned $\boldsymbol{\Sigma}$ as

$$\boldsymbol{\Sigma} = \begin{pmatrix} \boldsymbol{\Sigma}_{ZZ} & \sigma_{ZY} \\ \sigma_{ZY}^T & \sigma_{YY} \end{pmatrix}. \tag{17.7}$$

The conditional mean in (17.6) has exactly the same form as the population multiple linear regression of Y on Z, with regression coefficient $\beta = \boldsymbol{\Sigma}_{ZZ}^{-1}\sigma_{ZY}$ [see (2.16) on page 19]. If we partition $\boldsymbol{\Theta}$ in the same way, since $\boldsymbol{\Sigma}\boldsymbol{\Theta} = \mathbf{I}$ standard formulas for partitioned inverses give

$$\theta_{ZY} = -\theta_{YY} \cdot \boldsymbol{\Sigma}_{ZZ}^{-1}\sigma_{ZY}, \tag{17.8}$$

where $1/\theta_{YY} = \sigma_{YY} - \sigma_{ZY}^T\boldsymbol{\Sigma}_{ZZ}^{-1}\sigma_{ZY} > 0$. Hence

$$
\begin{aligned}
\beta &= \boldsymbol{\Sigma}_{ZZ}^{-1}\sigma_{ZY} \\
&= -\theta_{ZY}/\theta_{YY}.
\end{aligned}
\tag{17.9}
$$

We have learned two things here:

- The dependence of Y on Z in (17.6) is in the mean term alone. Here we see explicitly that zero elements in β and hence θ_{ZY} mean that the corresponding elements of Z are conditionally independent of Y, given the rest.

- We can learn about this dependence structure through multiple linear regression.

Thus $\boldsymbol{\Theta}$ captures all the second-order information (both structural and quantitative) needed to describe the conditional distribution of each node given the rest, and is the so-called "natural" parameter for the Gaussian graphical model[2].

Another (different) kind of graphical model is the *covariance graph* or *relevance network*, in which vertices are connected by bidirectional edges if the covariance (rather than the partial covariance) between the corresponding variables is nonzero. These are popular in genomics, see especially Butte et al. (2000). The negative log-likelihood from these models is not convex, making the computations more challenging (Chaudhuri et al., 2007).

17.3.1 Estimation of the Parameters when the Graph Structure is Known

Given some realizations of X, we would like to estimate the parameters of an undirected graph that approximates their joint distribution. Suppose first that the graph is complete (fully connected). We assume that we have N multivariate normal realizations x_i, $i = 1, \ldots, N$ with population mean μ and covariance $\boldsymbol{\Sigma}$. Let

$$\mathbf{S} = \frac{1}{N}\sum_{i=1}^{N}(x_i - \bar{x})(x_i - \bar{x})^T \tag{17.10}$$

be the empirical covariance matrix, with $\bar{x}$ the sample mean vector. Ignoring constants, the log-likelihood of the data can be written as

---

[2]The distribution arising from a Gaussian graphical model is a Wishart distribution. This is a member of the exponential family, with canonical or "natural" parameter $\boldsymbol{\Theta} = \boldsymbol{\Sigma}^{-1}$. Indeed, the partially maximized log-likelihood (17.11) is (up to constants) the Wishart log-likelihood.

$$\ell(\mathbf{\Theta}) = \log \det \mathbf{\Theta} - \text{trace}(\mathbf{S}\mathbf{\Theta}). \tag{17.11}$$

In (17.11) we have partially maximized with respect to the mean parameter μ. The quantity $-\ell(\mathbf{\Theta})$ is a convex function of $\mathbf{\Theta}$. It is easy to show that the maximum likelihood estimate of $\mathbf{\Sigma}$ is simply $\mathbf{S}$.

Now to make the graph more useful (especially in high-dimensional settings) let's assume that some of the edges are missing; for example, the edge between PIP3 and Erk is one of several missing in Figure 17.1. As we have seen, for the Gaussian distribution this implies that the corresponding entries of $\mathbf{\Theta} = \mathbf{\Sigma}^{-1}$ are zero. Hence we now would like to maximize (17.11) under the constraints that some pre-defined subset of the parameters are zero. This is an equality-constrained convex optimization problem, and a number of methods have been proposed for solving it, in particular the iterative proportional fitting procedure (Speed and Kiiveri, 1986). This and other methods are summarized for example in Whittaker (1990) and Lauritzen (1996). These methods exploit the simplifications that arise from decomposing the graph into its maximal cliques, as described in the previous section. Here we outline a simple alternate approach, that exploits the sparsity in a different way. The fruits of this approach will become apparent later when we discuss the problem of estimation of the graph structure.

The idea is based on linear regression, as inspired by (17.6) and (17.9). In particular, suppose that we want to estimate the edge parameters θ_{ij} for the vertices that are joined to a given vertex i, restricting those that are not joined to be zero. Then it would seem that the linear regression of the node i values on the other relevant vertices might provide a reasonable estimate. But this ignores the dependence structure among the predictors in this regression. It turns out that if instead we use our current (model-based) estimate of the cross-product matrix of the predictors when we perform our regressions, this gives the correct solutions and solves the constrained maximum-likelihood problem exactly. We now give details.

To constrain the log-likelihood (17.11), we add Lagrange constants for all missing edges

$$\ell_C(\mathbf{\Theta}) = \log \det \mathbf{\Theta} - \text{trace}(\mathbf{S}\mathbf{\Theta}) - \sum_{(j,k)\notin E} \gamma_{jk}\theta_{jk}. \tag{17.12}$$

The gradient equation for maximizing (17.12) can be written as

$$\mathbf{\Theta}^{-1} - \mathbf{S} - \mathbf{\Gamma} = \mathbf{0}, \tag{17.13}$$

using the fact that the derivative of $\log \det \mathbf{\Theta}$ equals $\mathbf{\Theta}^{-1}$ (Boyd and Vandenberghe, 2004, for example, page 641). $\mathbf{\Gamma}$ is a matrix of Lagrange parameters with nonzero values for all pairs with edges absent.

We will show how we can use regression to solve for $\mathbf{\Theta}$ and its inverse $\mathbf{W} = \mathbf{\Theta}^{-1}$ one row and column at a time. For simplicity let's focus on the last row and column. Then the upper right block of equation (17.13) can be written as

$$w_{12} - s_{12} - \gamma_{12} = 0. \tag{17.14}$$

Here we have partitioned the matrices into two parts as in (17.7): part 1 being the first $p-1$ rows and columns, and part 2 the pth row and column. With $\mathbf{W}$ and its inverse $\boldsymbol{\Theta}$ partitioned in a similar fashion, we have

$$\begin{pmatrix} \mathbf{W}_{11} & w_{12} \\ w_{12}^T & w_{22} \end{pmatrix} \begin{pmatrix} \boldsymbol{\Theta}_{11} & \theta_{12} \\ \theta_{12}^T & \theta_{22} \end{pmatrix} = \begin{pmatrix} \mathbf{I} & 0 \\ 0^T & 1 \end{pmatrix}. \tag{17.15}$$

This implies

$$
\begin{aligned}
w_{12} &= -\mathbf{W}_{11}\theta_{12}/\theta_{22} & (17.16) \\
&= \mathbf{W}_{11}\beta & (17.17)
\end{aligned}
$$

where $\beta = -\theta_{12}/\theta_{22}$ as in (17.9). Now substituting (17.17) into (17.14) gives

$$\mathbf{W}_{11}\beta - s_{12} - \gamma_{12} = 0. \tag{17.18}$$

These can be interpreted as the $p-1$ estimating equations for the constrained regression of X_p on the other predictors, except that the observed mean cross-products matrix $\mathbf{S}_{11}$ is replaced by $\mathbf{W}_{11}$, the current estimated covariance matrix from the model.

Now we can solve (17.18) by simple subset regression. Suppose there are $p-q$ nonzero elements in γ_{12}—i.e., $p-q$ edges constrained to be zero. These $p-q$ rows carry no information and can be removed. Furthermore we can reduce β to β^* by removing its $p-q$ zero elements, yielding the reduced $q \times q$ system of equations

$$\mathbf{W}_{11}^* \beta^* - s_{12}^* = 0, \tag{17.19}$$

with solution $\hat{\beta}^* = \mathbf{W}_{11}^{*-1} s_{12}^*$. This is padded with $p-q$ zeros to give $\hat{\beta}$.

Although it appears from (17.16) that we only recover the elements θ_{12} up to a scale factor $1/\theta_{22}$, it is easy to show that

$$\frac{1}{\theta_{22}} = w_{22} - w_{12}^T\beta \tag{17.20}$$

(using partitioned inverse formulas). Also $w_{22} = s_{22}$, since the diagonal of $\boldsymbol{\Gamma}$ in (17.13) is zero.

This leads to the simple iterative procedure given in Algorithm 17.1 for estimating both $\hat{\mathbf{W}}$ and its inverse $\hat{\boldsymbol{\Theta}}$, subject to the constraints of the missing edges.

Note that this algorithm makes conceptual sense. The graph estimation problem is not p separate regression problems, but rather p coupled problems. The use of the common $\mathbf{W}$ in step (b), in place of the observed cross-products matrix, couples the problems together in the appropriate fashion. Surprisingly, we were not able to find this procedure in the literature. However it is related to the covariance selection procedures of

---

Algorithm 17.1 *A Modified Regression Algorithm for Estimation of an Undirected Gaussian Graphical Model with Known Structure.*

---

1. Initialize $\mathbf{W} = \mathbf{S}$.

2. Repeat for $j = 1, 2, \ldots, p, 1, \ldots$ until convergence:

   (a) Partition the matrix $\mathbf{W}$ into part 1: all but the jth row and column, and part 2: the jth row and column.

   (b) Solve $\mathbf{W}_{11}^* \beta^* - s_{12}^* = 0$ for the unconstrained edge parameters β^*, using the reduced system of equations as in (17.19). Obtain $\hat{\beta}$ by padding $\hat{\beta}^*$ with zeros in the appropriate positions.

   (c) Update $w_{12} = \mathbf{W}_{11}\hat{\beta}$

3. In the final cycle (for each j) solve for $\hat{\theta}_{12} = -\hat{\beta} \cdot \hat{\theta}_{22}$, with $1/\hat{\theta}_{22} = s_{22} - w_{12}^T \hat{\beta}$.

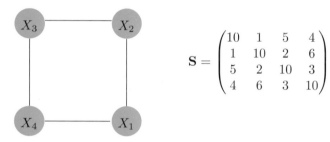

$$\mathbf{S} = \begin{pmatrix} 10 & 1 & 5 & 4 \\ 1 & 10 & 2 & 6 \\ 5 & 2 & 10 & 3 \\ 4 & 6 & 3 & 10 \end{pmatrix}$$

FIGURE 17.4. *A simple graph for illustration, along with the empirical covariance matrix.*

Dempster (1972), and is similar in flavor to the iterative conditional fitting procedure for covariance graphs, proposed by Chaudhuri et al. (2007).

Here is a little example, borrowed from Whittaker (1990). Suppose that our model is as depicted in Figure 17.4, along with its empirical covariance matrix $\mathbf{S}$. We apply algorithm (17.1) to this problem; for example, in the modified regression for variable 1 in step (b), variable 3 is left out. The procedure quickly converged to the solutions:

$$\hat{\Sigma} = \begin{pmatrix} 10.00 & 1.00 & 1.31 & 4.00 \\ 1.00 & 10.00 & 2.00 & 0.87 \\ 1.31 & 2.00 & 10.00 & 3.00 \\ 4.00 & 0.87 & 3.00 & 10.00 \end{pmatrix}, \hat{\Sigma}^{-1} = \begin{pmatrix} 0.12 & -0.01 & 0.00 & -0.05 \\ -0.01 & 0.11 & -0.02 & 0.00 \\ 0.00 & -0.02 & 0.11 & -0.03 \\ -0.05 & 0.00 & -0.03 & 0.13 \end{pmatrix}.$$

Note the zeroes in $\hat{\Sigma}^{-1}$, corresponding to the missing edges (1,3) and (2,4). Note also that the corresponding elements in $\hat{\Sigma}$ are the only elements different from $\mathbf{S}$. The estimation of $\hat{\Sigma}$ is an example of what is sometimes called the positive definite "completion" of $\mathbf{S}$.

17.3.2 Estimation of the Graph Structure

In most cases we do not know which edges to omit from our graph, and so would like to try to discover this from the data itself. In recent years a number of authors have proposed the use of L_1 (lasso) regularization for this purpose.

Meinshausen and Bühlmann (2006) take a simple approach to the problem: rather than trying to fully estimate $\mathbf{\Sigma}$ or $\mathbf{\Theta} = \mathbf{\Sigma}^{-1}$, they only estimate which components of θ_{ij} are nonzero. To do this, they fit a lasso regression using each variable as the response and the others as predictors. The component θ_{ij} is then estimated to be nonzero if either the estimated coefficient of variable i on j is nonzero, OR the estimated coefficient of variable j on i is nonzero (alternatively they use an AND rule). They show that asymptotically this procedure consistently estimates the set of nonzero elements of $\mathbf{\Theta}$.

We can take a more systematic approach with the lasso penalty, following the development of the previous section. Consider maximizing the penalized log-likelihood

$$\log \det \mathbf{\Theta} - \text{trace}(\mathbf{S\Theta}) - \lambda ||\mathbf{\Theta}||_1, \tag{17.21}$$

where $||\mathbf{\Theta}||_1$ is the L_1 norm— the sum of the absolute values of the elements of $\mathbf{\Sigma}^{-1}$, and we have ignored constants. The negative of this penalized likelihood is a convex function of $\mathbf{\Theta}$.

It turns out that one can adapt the lasso to give the exact maximizer of the penalized log-likelihood. In particular, we simply replace the modified regression step (b) in Algorithm 17.1 by a modified lasso step. Here are the details.

The analog of the gradient equation (17.13) is now

$$\mathbf{\Theta}^{-1} - \mathbf{S} - \lambda \cdot \text{Sign}(\mathbf{\Theta}) = \mathbf{0}. \tag{17.22}$$

Here we use *sub-gradient* notation, with $\text{Sign}(\theta_{jk}) = \text{sign}(\theta_{jk})$ if $\theta_{jk} \neq 0$, else $\text{Sign}(\theta_{jk}) \subset [-1, 1]$ if $\theta_{jk} = 0$. Continuing the development in the previous section, we reach the analog of (17.18)

$$\mathbf{W}_{11}\beta - s_{12} + \lambda \cdot \text{Sign}(\beta) = 0 \tag{17.23}$$

(recall that β and θ_{12} have opposite signs). We will now see that this system is exactly equivalent to the estimating equations for a lasso regression.

Consider the usual regression setup with outcome variables $\mathbf{y}$ and predictor matrix $\mathbf{Z}$. There the lasso minimizes

$$\tfrac{1}{2}(\mathbf{y} - \mathbf{Z}\beta)^T(\mathbf{y} - \mathbf{Z}\beta) + \lambda \cdot ||\beta||_1 \tag{17.24}$$

[see (3.52) on page 68; here we have added a factor $\tfrac{1}{2}$ for convenience]. The gradient of this expression is

---

Algorithm 17.2 *Graphical Lasso.*

---

1. Initialize $\mathbf{W} = \mathbf{S} + \lambda\mathbf{I}$. The diagonal of $\mathbf{W}$ remains unchanged in what follows.

2. Repeat for $j = 1, 2, \ldots p, 1, 2, \ldots p, \ldots$ until convergence:

   (a) Partition the matrix $\mathbf{W}$ into part 1: all but the jth row and column, and part 2: the jth row and column.

   (b) Solve the estimating equations $\mathbf{W}_{11}\beta - s_{12} + \lambda \cdot \mathrm{Sign}(\beta) = 0$ using the cyclical coordinate-descent algorithm (17.26) for the modified lasso.

   (c) Update $w_{12} = \mathbf{W}_{11}\hat{\beta}$

3. In the final cycle (for each j) solve for $\hat{\theta}_{12} = -\hat{\beta} \cdot \hat{\theta}_{22}$, with $1/\hat{\theta}_{22} = w_{22} - w_{12}^T\hat{\beta}$.

$$\mathbf{Z}^T\mathbf{Z}\beta - \mathbf{Z}^T\mathbf{y} + \lambda \cdot \mathrm{Sign}(\beta) = 0 \qquad (17.25)$$

So up to a factor $1/N$, $\mathbf{Z}^T\mathbf{y}$ is the analog of s_{12}, and we replace $\mathbf{Z}^T\mathbf{Z}$ by $\mathbf{W}_{11}$, the estimated cross-product matrix from our current model.

The resulting procedure is called the *graphical lasso*, proposed by Friedman et al. (2008b) building on the work of Banerjee et al. (2008). It is summarized in Algorithm 17.2.

Friedman et al. (2008b) use the pathwise coordinate descent method (Section 3.8.6) to solve the modified lasso problem at each stage. Here are the details of pathwise coordinate descent for the graphical lasso algorithm. Letting $\mathbf{V} = \mathbf{W}_{11}$, the update has the form

$$\hat{\beta}_j \leftarrow S\left(s_{12j} - \sum_{k \neq j} V_{kj}\hat{\beta}_k, \lambda\right)/V_{jj} \qquad (17.26)$$

for $j = 1, 2, \ldots, p - 1, 1, 2, \ldots, p - 1, \ldots$, where S is the soft-threshold operator:

$$S(x, t) = \mathrm{sign}(x)(|x| - t)_+. \qquad (17.27)$$

The procedure cycles through the predictors until convergence.

It is easy to show that the diagonal elements w_{jj} of the solution matrix $\mathbf{W}$ are simply $s_{jj} + \lambda$, and these are fixed in step 1 of Algorithm 17.2[3].

The graphical lasso algorithm is extremely fast, and can solve a moderately sparse problem with 1000 nodes in less than a minute. It is easy to modify the algorithm to have edge-specific penalty parameters λ_{jk}; since

---

[3]An alternative formulation of the problem (17.21) can be posed, where we don't penalize the diagonal of $\boldsymbol{\Theta}$. Then the diagonal elements w_{jj} of the solution matrix are s_{jj}, and the rest of the algorithm is unchanged.

$\lambda_{jk} = \infty$ will force $\hat{\theta}_{jk}$ to be zero, this algorithm subsumes Algorithm 17.1.
By casting the sparse inverse-covariance problem as a series of regressions,
one can also quickly compute and examine the solution paths as a function
of the penalty parameter λ. More details can be found in Friedman et al.
(2008b).

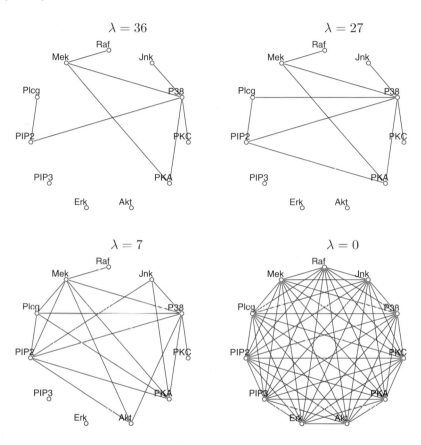

FIGURE 17.5. *Four different graphical-lasso solutions for the flow-cytometry
data.*

Figure 17.1 shows the result of applying the graphical lasso to the flow-cytometry dataset. Here the lasso penalty parameter λ was set at 14. In
practice it is informative to examine the different sets of graphs that are
obtained as λ is varied. Figure 17.5 shows four different solutions. The
graph becomes more sparse as the penalty parameter is increased.

Finally note that the values at some of the nodes in a graphical model can
be unobserved; that is, missing or hidden. If only some values are missing
at a node, the EM algorithm can be used to impute the missing values

(Exercise 17.9). However, sometimes the entire node is hidden or *latent*. In the Gaussian model, if a node has all missing values, due to linearity one can simply average over the missing nodes to yield another Gaussian model over the observed nodes. Hence the inclusion of hidden nodes does not enrich the resulting model for the observed nodes; in fact, it imposes additional structure on its covariance matrix. However in the discrete model (described next) the inherent nonlinearities make hidden units a powerful way of expanding the model.

17.4 Undirected Graphical Models for Discrete Variables

Undirected Markov networks with all discrete variables are popular, and in particular pairwise Markov networks with binary variables being the most common. They are sometimes called *Ising models* in the statistical mechanics literature, and *Boltzmann machines* in the machine learning literature, where the vertices are referred to as "nodes" or "units" and are binary-valued.

In addition, the values at each node can be observed ("visible") or unobserved ("hidden"). The nodes are often organized in layers, similar to a neural network. Boltzmann machines are useful both for unsupervised and supervised learning, especially for structured input data such as images, but have been hampered by computational difficulties. Figure 17.6 shows a restricted Boltzmann machine (discussed later), in which some variables are hidden, and only some pairs of nodes are connected. We first consider the simpler case in which all p nodes are visible with edge pairs (j, k) enumerated in E.

Denoting the binary valued variable at node j by X_j, the Ising model for their joint probabilities is given by

$$p(X, \boldsymbol{\Theta}) = \exp \Big[\sum_{(j,k) \in E} \theta_{jk} X_j X_k - \Phi(\boldsymbol{\Theta}) \Big] \text{ for } X \in \mathcal{X}, \qquad (17.28)$$

with $\mathcal{X} = \{0, 1\}^p$. As with the Gaussian model of the previous section, only pairwise interactions are modeled. The Ising model was developed in statistical mechanics, and is now used more generally to model the joint effects of pairwise interactions. $\Phi(\boldsymbol{\Theta})$ is the log of the partition function, and is defined by

$$\Phi(\boldsymbol{\Theta}) = \log \sum_{x \in \mathcal{X}} \Big[\exp \Big(\sum_{(j,k) \in E} \theta_{jk} x_j x_k \Big) \Big]. \qquad (17.29)$$

The partition function ensures that the probabilities add to one over the sample space. The terms $\theta_{jk} X_j X_k$ represent a particular parametrization of the (log) potential functions (17.5), and for technical reasons requires a *constant* node $X_0 \equiv 1$ to be included (Exercise 17.10), with "edges" to all the other nodes. In the statistics literature, this model is equivalent to a first-order-interaction Poisson log-linear model for multiway tables of counts (Bishop et al., 1975; McCullagh and Nelder, 1989; Agresti, 2002).

The Ising model implies a logistic form for each node conditional on the others (exercise 17.11):

$$\Pr(X_j = 1 | X_{-j} = x_{-j}) = \frac{1}{1 + \exp(-\theta_{j0} - \sum_{(j,k) \in E} \theta_{jk} x_k)}, \quad (17.30)$$

where X_{-j} denotes all of the nodes except j. Hence the parameter θ_{jk} measures the dependence of X_j on X_k, conditional on the other nodes.

17.4.1 *Estimation of the Parameters when the Graph Structure is Known*

Given some data from this model, how can we estimate the parameters? Suppose we have observations $x_i = (x_{i1}, x_{i2}, \ldots, x_{ip}) \in \{0, 1\}^p$, $i = 1, \ldots, N$. The log-likelihood is

$$
\begin{aligned}
\ell(\mathbf{\Theta}) \quad &= \quad \sum_{i=1}^{N} \log \Pr_{\mathbf{\Theta}}(X_i = x_i) \\
&= \quad \sum_{i=1}^{N} \left[\sum_{(j,k) \in E} \theta_{jk} x_{ij} x_{ik} - \Phi(\mathbf{\Theta}) \right] \quad (17.31)
\end{aligned}
$$

The gradient of the log-likelihood is

$$\frac{\partial \ell(\mathbf{\Theta})}{\partial \theta_{jk}} = \sum_{i=1}^{N} x_{ij} x_{ik} - N \frac{\partial \Phi(\mathbf{\Theta})}{\partial \theta_{jk}} \quad (17.32)$$

and

$$
\begin{aligned}
\frac{\partial \Phi(\mathbf{\Theta})}{\partial \theta_{jk}} \quad &= \quad \sum_{x \in \mathcal{X}} x_j x_k \cdot p(x, \mathbf{\Theta}) \\
&= \quad \mathrm{E}_{\mathbf{\Theta}}(X_j X_k) \quad (17.33)
\end{aligned}
$$

Setting the gradient to zero gives

$$\hat{\mathrm{E}}(X_j X_k) - \mathrm{E}_{\mathbf{\Theta}}(X_j X_k) = 0 \quad (17.34)$$

where we have defined

$$\hat{\mathrm{E}}(X_j X_k) = \frac{1}{N} \sum_{i=1}^{N} x_{ij} x_{ik}, \qquad (17.35)$$

the expectation taken with respect to the empirical distribution of the data. Looking at (17.34), we see that the maximum likelihood estimates simply match the estimated inner products between the nodes to their observed inner products. This is a standard form for the score (gradient) equation for exponential family models, in which sufficient statistics are set equal to their expectations under the model.

To find the maximum likelihood estimates, we can use gradient search or Newton methods. However the computation of $E_{\Theta}(X_j X_k)$ involves enumeration of $p(X, \Theta)$ over 2^{p-2} of the $|\mathcal{X}| = 2^p$ possible values of X, and is not generally feasible for large p (e.g., larger than about 30). For smaller p, a number of standard statistical approaches are available:

Poisson log-linear modeling, where we treat the problem as a large regression problem (Exercise 17.12). The response vector **y** is the vector of 2^p counts in each of the cells of the multiway tabulation of the data[4]. The predictor matrix **Z** has 2^p rows and up to $1 + p + p^2$ columns that characterize each of the cells, although this number depends on the sparsity of the graph. The computational cost is essentially that of a regression problem of this size, which is $O(p^4 2^p)$ and is manageable for $p < 20$. The Newton updates are typically computed by iteratively reweighted least squares, and the number of steps is usually in the single digits. See Agresti (2002) and McCullagh and Nelder (1989) for details. Standard software (such as the R package `glm`) can be used to fit this model.

Gradient descent requires at most $O(p^2 2^{p-2})$ computations to compute the gradient, but may require many more gradient steps than the second–order Newton methods. Nevertheless, it can handle slightly larger problems with $p \leq 30$. These computations can be reduced by exploiting the special clique structure in sparse graphs, using the junction-tree algorithm. Details are not given here.

Iterative proportional fitting (IPF) performs cyclical coordinate descent on the gradient equations (17.34). At each step a parameter is updated so that its gradient equation is exactly zero. This is done in a cyclical fashion until all the gradients are zero. One complete cycle costs the same as a gradient evaluation, but may be more efficient. Jiroušek and Přeučil (1995) implement an efficient version of IPF, using junction trees.

---

[4]Each of the cell counts is treated as an independent Poisson variable. We get the multinomial model corresponding to (17.28) by conditioning on the total count N (which is also Poisson under this framework).

When p is large (> 30) other approaches have been used to approximate the gradient.

- The mean field approximation (Peterson and Anderson, 1987) estimates $E_{\Theta}(X_j X_k)$ by $E_{\Theta}(X_j)E_{\Theta}(X_j)$, and replaces the input variables by their means, leading to a set of nonlinear equations for the parameters θ_{jk}.

- To obtain near-exact solutions, Gibbs sampling (Section 8.6) is used to approximate $E_{\Theta}(X_j X_k)$ by successively sampling from the estimated model probabilities $\mathrm{Pr}_{\Theta}(X_j|X_{-j})$ (see e.g. Ripley (1996)).

We have not discussed *decomposable models*, for which the maximum likelihood estimates can be found in closed form without any iteration whatsoever. These models arise, for example, in *trees*: special graphs with tree-structured topology. When computational tractability is a concern, trees represent a useful class of models and they sidestep the computational concerns raised in this section. For details, see for example Chapter 12 of Whittaker (1990).

17.4.2 Hidden Nodes

We can increase the complexity of a discrete Markov network by including latent or hidden nodes. Suppose that a subset of the variables $X_{\mathcal{H}}$ are unobserved or "hidden", and the remainder $X_{\mathcal{V}}$ are observed or "visible." Then the log-likelihood of the observed data is

$$
\begin{aligned}
\ell(\mathbf{\Theta}) &= \sum_{i=1}^{N} \log[\mathrm{Pr}_{\Theta}(X_{\mathcal{V}} = x_{i\mathcal{V}})] \\
&= \sum_{i=1}^{N} \Big[\log \sum_{x_{\mathcal{H}} \subset \mathcal{X}_{\mathcal{H}}} \exp \sum_{(j,k)\in E} (\theta_{jk}x_{ij}x_{ik} - \Phi(\mathbf{\Theta}))\Big]. \quad (17.36)
\end{aligned}
$$

The sum over $x_{\mathcal{H}}$ means that we are summing over all possible $\{0,1\}$ values for the hidden units. The gradient works out to be

$$
\frac{d\ell(\mathbf{\Theta})}{d\theta_{jk}} = \hat{E}_{\mathcal{V}}E_{\Theta}(X_j X_k | X_{\mathcal{V}}) - E_{\Theta}(X_j X_k) \quad (17.37)
$$

The first term is an empirical average of $X_j X_k$ if both are visible; if one or both are hidden, they are first imputed given the visible data, and then averaged over the hidden variables. The second term is the unconditional expectation of $X_j X_k$.

The inner expectation in the first term can be evaluated using basic rules of conditional expectation and properties of Bernoulli random variables. In detail, for observation i

$$E_\Theta(X_j X_k | X_\mathcal{V} = x_{i\mathcal{V}}) = \begin{cases} x_{ij}x_{ik} & \text{if } j,k \in \mathcal{V} \\ x_{ij}\text{Pr}_\Theta(X_k = 1 | X_\mathcal{V} = x_{i\mathcal{V}}) & \text{if } j \in \mathcal{V}, k \in \mathcal{H} \\ \text{Pr}_\Theta(X_j = 1, X_k = 1 | X_\mathcal{V} = x_{i\mathcal{V}}) & \text{if } j,k \in \mathcal{H}. \end{cases}$$

(17.38)

Now two separate runs of Gibbs sampling are required; the first to estimate $E_\Theta(X_j X_k)$ by sampling from the model as above, and the second to estimate $E_\Theta(X_j X_k | X_\mathcal{V} = x_{i\mathcal{V}})$. In this latter run, the visible units are fixed ("clamped") at their observed values and only the hidden variables are sampled. Gibbs sampling must be done for each observation in the training set, at each stage of the gradient search. As a result this procedure can be very slow, even for moderate-sized models. In Section 17.4.4 we consider further model restrictions to make these computations manageable.

17.4.3 Estimation of the Graph Structure

The use of a lasso penalty with binary pairwise Markov networks has been suggested by Lee et al. (2007) and Wainwright et al. (2007). The first authors investigate a conjugate gradient procedure for exact maximization of a penalized log-likelihood. The bottleneck is the computation of $E_\Theta(X_j X_k)$ in the gradient; exact computation via the junction tree algorithm is manageable for sparse graphs but becomes unwieldy for dense graphs.

The second authors propose an approximate solution, analogous to the Meinshausen and Bühlmann (2006) approach for the Gaussian graphical model. They fit an L_1-penalized logistic regression model to each node as a function of the other nodes, and then symmetrize the edge parameter estimates in some fashion. For example if $\tilde\theta_{jk}$ is the estimate of the j-k edge parameter from the logistic model for outcome node j, the "min" symmetrization sets $\hat\theta_{jk}$ to either $\tilde\theta_{jk}$ or $\tilde\theta_{kj}$, whichever is smallest in absolute value. The "max" criterion is defined similarly. They show that under certain conditions either approximation estimates the nonzero edges correctly as the sample size goes to infinity. Hoefling and Tibshirani (2008) extend the graphical lasso to discrete Markov networks, obtaining a procedure which is somewhat faster than conjugate gradients, but still must deal with computation of $E_\Theta(X_j X_k)$. They also compare the exact and approximate solutions in an extensive simulation study and find the "min" or "max" approximations are only slightly less accurate than the exact procedure, both for estimating the nonzero edges and for estimating the actual values of the edge parameters, and are *much* faster. Furthermore, they can handle denser graphs because they never need to compute the quantities $E_\Theta(X_j X_k)$.

Finally, we point out a key difference between the Gaussian and binary models. In the Gaussian case, both Σ and its inverse will often be of interest, and the graphical lasso procedure delivers estimates for both of these quantities. However, the approximation of Meinshausen and Bühlmann (2006) for Gaussian graphical models, analogous to the Wainwright et al. (2007)

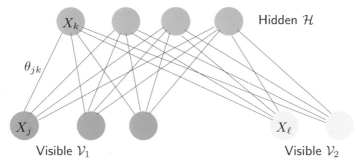

FIGURE 17.6. *A restricted Boltzmann machine (RBM) in which there are no connections between nodes in the same layer. The visible units are subdivided to allow the RBM to model the joint density of feature $\mathcal{V}_1$ and their labels $\mathcal{V}_2$.*

approximation for the binary case, only yields an estimate of $\boldsymbol{\Sigma}^{-1}$. In contrast, in the Markov model for binary data, $\boldsymbol{\Theta}$ is the object of interest, and its inverse is not of interest. The approximate method of Wainwright et al. (2007) estimates $\boldsymbol{\Theta}$ efficiently and hence is an attractive solution for the binary problem.

17.4.4 Restricted Boltzmann Machines

In this section we consider a particular architecture for graphical models inspired by neural networks, where the units are organized in layers. A restricted Boltzmann machine (RBM) consists of one layer of visible units and one layer of hidden units with no connections within each layer. It is much simpler to compute the conditional expectations (as in (17.37) and (17.38)) if the connections between hidden units are removed [5]. Figure 17.6 shows an example; the visible layer is divided into input variables $\mathcal{V}_1$ and output variables $\mathcal{V}_2$, and there is a hidden layer $\mathcal{H}$. We denote such a network by

$$\mathcal{V}_1 \leftrightarrow \mathcal{H} \leftrightarrow \mathcal{V}_2. \tag{17.39}$$

For example, $\mathcal{V}_1$ could be the binary pixels of an image of a handwritten digit, and $\mathcal{V}_2$ could have 10 units, one for each of the observed class labels 0-9.

The restricted form of this model simplifies the Gibbs sampling for estimating the expectations in (17.37), since the variables in each layer are independent of one another, given the variables in the other layers. Hence they can be sampled together, using the conditional probabilities given by expression (17.30).

The resulting model is less general than a Boltzmann machine, but is still useful; for example it can learn to extract interesting features from images.

---

[5] We thank Geoffrey Hinton for assistance in the preparation of the material on RBMs.

By alternately sampling the variables in each layer of the RBM shown in Figure 17.6, it is possible to generate samples from the joint density model. If the $\mathcal{V}_1$ part of the visible layer is clamped at a particular feature vector during the alternating sampling, it is possible to sample from the distribution over labels given $\mathcal{V}_1$. Alternatively classification of test items can also be achieved by comparing the unnormalized joint densities of each label category with the observed features. We do not need to compute the partition function as it is the same for all of these combinations.

As noted the restricted Boltzmann machine has the same generic form as a single hidden layer neural network (Section 11.3). The edges in the latter model are directed, the hidden units are usually real-valued, and the fitting criterion is different. The neural network minimizes the error (cross-entropy) between the targets and their model predictions, conditional on the input features. In contrast, the restricted Boltzmann machine maximizes the log-likelihood for the joint distribution of all visible units—that is, the features and targets. It can extract information from the input features that is useful for predicting the labels, but, unlike supervised learning methods, it may also use some of its hidden units to model structure in the feature vectors that is not immediately relevant for predicting the labels. These features may turn out to be useful, however, when combined with features derived from other hidden layers.

Unfortunately, Gibbs sampling in a restricted Boltzmann machine can be very slow, as it can take a long time to reach stationarity. As the network weights get larger, the chain mixes more slowly and we need to run more steps to get the unconditional estimates. Hinton (2002) noticed empirically that learning still works well if we estimate the second expectation in (17.37) by starting the Markov chain at the data and only running for a few steps (instead of to convergence). He calls this *contrastive divergence*: we sample $\mathcal{H}$ given $\mathcal{V}_1, \mathcal{V}_2$, then $\mathcal{V}_1, \mathcal{V}_2$ given $\mathcal{H}$ and finally $\mathcal{H}$ given $\mathcal{V}_1, \mathcal{V}_2$ again. The idea is that when the parameters are far from the solution, it may be wasteful to iterate the Gibbs sampler to stationarity, as just a single iteration will reveal a good direction for moving the estimates.

We now give an example to illustrate the use of an RBM. Using contrastive divergence, it is possible to train an RBM to recognize hand-written digits from the MNIST dataset (LeCun et al., 1998). With 2000 hidden units, 784 visible units for representing binary pixel intensities and one 10-way multinomial visible unit for representing labels, the RBM achieves an error rate of 1.9% on the test set. This is a little higher than the 1.4% achieved by a support vector machine and comparable to the error rate achieved by a neural network trained with backpropagation. The error rate of the RBM, however, can be reduced to 1.25% by replacing the 784 pixel intensities by 500 features that are produced from the images without using any label information. First, an RBM with 784 visible units and 500 hidden units is trained, using contrastive divergence, to model the set of images. Then the hidden states of the first RBM are used as data for training a

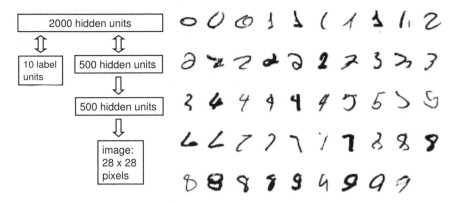

FIGURE 17.7. *Example of a restricted Boltzmann machine for handwritten digit classification. The network is depicted in the schematic on the left. Displayed on the right are some difficult test images that the model classifies correctly.*

second RBM that has 500 visible units and 500 hidden units. Finally, the hidden states of the second RBM are used as the features for training an RBM with 2000 hidden units as a joint density model. The details and justification for learning features in this greedy, layer-by-layer way are described in Hinton et al. (2006). Figure 17.7 gives a representation of the composite model that is learned in this way and also shows some examples of the types of distortion that it can cope with.

Bibliographic Notes

Much work has been done in defining and understanding the structure of graphical models. Comprehensive treatments of graphical models can be found in Whittaker (1990), Lauritzen (1996), Cox and Wermuth (1996), Edwards (2000), Pearl (2000), Anderson (2003), Jordan (2004), and Koller and Friedman (2007). Wasserman (2004) gives a brief introduction, and Chapter 8 of Bishop (2006) gives a more detailed overview. Boltzmann machines were proposed in Ackley et al. (1985). Ripley (1996) has a detailed chapter on topics in graphical models that relate to machine learning. We found this particularly useful for its discussion of Boltzmann machines.

Exercises

Ex. 17.1 For the Markov graph of Figure 17.8, list all of the implied conditional independence relations and find the maximal cliques.

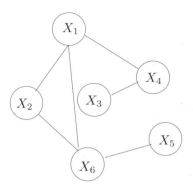

FIGURE 17.8.

Ex. 17.2 Consider random variables X_1, X_2, X_3, X_4. In each of the following cases draw a graph that has the given independence relations:

(a) $X_1 \perp X_3 | X_2$ and $X_2 \perp X_4 | X_3$.

(b) $X_1 \perp X_4 | X_2, X_3$ and $X_2 \perp X_4 | X_1, X_3$.

(c) $X_1 \perp X_4 | X_2, X_3$, $X_1 \perp X_3 | X_2, X_4$ and $X_3 \perp X_4 | X_1, X_2$.

Ex. 17.3 Let $\boldsymbol{\Sigma}$ be the covariance matrix of a set of p variables X. Consider the partial covariance matrix $\boldsymbol{\Sigma}_{a.b} = \boldsymbol{\Sigma}_{aa} - \boldsymbol{\Sigma}_{ab}\boldsymbol{\Sigma}_{bb}^{-1}\boldsymbol{\Sigma}_{ba}$ between the two subsets of variables $X_a = (X_1, X_2)$ consisting of the first two, and X_b the rest. This is the covariance matrix between these two variables, after linear adjustment for all the rest. In the Gaussian distribution, this is the covariance matrix of the conditional distribution of $X_a | X_b$. The partial correlation coefficient $\rho_{jk|\text{rest}}$ between the pair X_a conditional on the rest X_b, is simply computed from this partial covariance. Define $\boldsymbol{\Theta} = \boldsymbol{\Sigma}^{-1}$.

1. Show that $\boldsymbol{\Sigma}_{a.b} = \boldsymbol{\Theta}_{aa}^{-1}$.

2. Show that if any off-diagonal element of $\boldsymbol{\Theta}$ is zero, then the partial correlation coefficient between the corresponding variables is zero.

3. Show that if we treat $\boldsymbol{\Theta}$ as if it were a covariance matrix, and compute the corresponding "correlation" matrix

$$\mathbf{R} = \operatorname{diag}(\boldsymbol{\Theta})^{-1/2} \cdot \boldsymbol{\Theta} \cdot \operatorname{diag}(\boldsymbol{\Theta})^{-1/2}, \qquad (17.40)$$

   then $r_{jk} = -\rho_{jk|\text{rest}}$

Ex. 17.4 Denote by

$$f(X_1 | X_2, X_3, \dots, X_p)$$

the conditional density of X_1 given $X_2, \dots, X_p$. If

$$f(X_1 | X_2, X_3, \dots, X_p) = f(X_1 | X_3, \dots, X_p),$$

show that $X_1 \perp X_2 | X_3, \dots, X_p$.

Ex. 17.5 Consider the setup in Section 17.3.1 with no missing edges. Show that

$$\mathbf{S}_{11}\beta - s_{12} = 0$$

are the estimating equations for the multiple regression coefficients of the last variable on the rest.

Ex. 17.6 *Recovery of* $\hat{\boldsymbol{\Theta}} = \hat{\boldsymbol{\Sigma}}^{-1}$ *from Algorithm 17.1.* Use expression (17.16) to derive the standard partitioned inverse expressions

$$\theta_{12} = -\mathbf{W}_{11}^{-1} w_{12} \theta_{22} \qquad (17.41)$$
$$\theta_{22} = 1/(w_{22} - w_{12}^T \mathbf{W}_{11}^{-1} w_{12}). \qquad (17.42)$$

Since $\hat{\beta} = \mathbf{W}_{11}^{-1} w_{12}$, show that $\hat{\theta}_{22} = 1/(w_{22} - w_{12}^T \hat{\beta})$ and $\hat{\theta}_{12} = -\hat{\beta}\hat{\theta}_{22}$. Thus $\hat{\theta}_{12}$ is a simply rescaling of $\hat{\beta}$ by $-\hat{\theta}_{22}$.

Ex. 17.7 Write a program to implement the modified regression procedure in Algorithm 17.1 for fitting the Gaussian graphical model with pre-specified edges missing. Test it on the flow cytometry data from the book website, using the graph of Figure 17.1.

Ex. 17.8

(a) Write a program to fit the lasso using the coordinate descent procedure (17.26). Compare its results to those from the `lars` program or some other convex optimizer, to check that it is working correctly.

(b) Using the program from (a), write code to implement the graphical lasso (Algorithm 17.2). Apply it to the flow cytometry data from the book website. Vary the regularization parameter and examine the resulting networks.

Ex. 17.9 Suppose that we have a Gaussian graphical model in which some or all of the data at some vertices are missing.

(a) Consider the EM algorithm for a dataset of N i.i.d. multivariate observations $x_i \in \mathbb{R}^p$ with mean μ and covariance matrix $\boldsymbol{\Sigma}$. For each sample i, let o_i and m_i index the predictors that are observed and missing, respectively. Show that in the E step, the observations are imputed from the current estimates of μ and $\boldsymbol{\Sigma}$:

$$\hat{x}_{i,m_i} = \mathrm{E}(x_{i,m_i}|x_{i,o_i}, \theta) = \hat{\mu}_{m_i} + \hat{\Sigma}_{m_i,o_i} \hat{\Sigma}_{o_i,o_i}^{-1}(x_{i,o_i} - \hat{\mu}_{o_i})$$
$$(17.43)$$

while in the M step, μ and $\boldsymbol{\Sigma}$ are re-estimated from the empirical mean and (modified) covariance of the imputed data:

$$\hat{\mu}_j = \sum_{i=1}^{N} \hat{x}_{ij}/N$$

$$\hat{\Sigma}_{jj'} \; = \; \sum_{i=1}^{N} [(\hat{x}_{ij} - \hat{\mu}_j)(\hat{x}_{ij'} - \hat{\mu}_{j'}) + c_{i,jj'}]/N \qquad (17.44)$$

where $c_{i,jj'} = \hat{\Sigma}_{jj'}$ if $j, j' \in m_i$ and zero otherwise. Explain the reason for the correction term $c_{i,jj'}$ (Little and Rubin, 2002).

(b) Implement the EM algorithm for the Gaussian graphical model using the modified regression procedure from Exercise 17.7 for the M-step.

(c) For the flow cytometry data on the book website, set the data for the last protein Jnk in the first 1000 observations to missing, fit the model of Figure 17.1, and compare the predicted values to the actual values for Jnk. Compare the results to those obtained from a regression of Jnk on the other vertices with edges to Jnk in Figure 17.1, using only the non-missing data.

Ex. 17.10 Using a simple binary graphical model with just two variables, show why it is essential to include a constant node $X_0 \equiv 1$ in the model.

Ex. 17.11 Show that the Ising model (17.28) for the joint probabilities in a discrete graphical model implies that the conditional distributions have the logistic form (17.30).

Ex. 17.12 Consider a Poisson regression problem with p binary variables x_{ij}, $j = 1, \ldots, p$ and response variable y_i which measures the number of observations with predictor $x_i \in \{0, 1\}^p$. The design is balanced, in that all $n = 2^p$ possible combinations are measured. We assume a log-linear model for the Poisson mean in each cell

$$\log \mu(X) = \theta_{00} + \sum_{(j,k) \in E} x_{ij} x_{ik} \theta_{jk}, \qquad (17.45)$$

using the same notation as in Section 17.4.1 (including the constant variable $x_{i0} = 1 \forall i$). We assume the response is distributed as

$$\Pr(Y = y | X = x) = \frac{e^{-\mu(x)} \mu(x)^y}{y!}. \qquad (17.46)$$

Write down the conditional log-likelihood for the observed responses y_i, and compute the gradient.

(a) Show that the gradient equation for θ_{00} computes the partition function (17.29).

(b) Show that the gradient equations for the remainder of the parameters are equivalent to the gradient (17.34).

18
High-Dimensional Problems: $p \gg N$

18.1 When p is Much Bigger than N

In this chapter we discuss prediction problems in which the number of features p is much larger than the number of observations N, often written $p \gg N$. Such problems have become of increasing importance, especially in genomics and other areas of computational biology. We will see that high variance and overfitting are a major concern in this setting. As a result, simple, highly regularized approaches often become the methods of choice. The first part of the chapter focuses on prediction in both the classification and regression settings, while the second part discusses the more basic problem of feature selection and assessment.

To get us started, Figure 18.1 summarizes a small simulation study that demonstrates the "less fitting is better" principle that applies when $p \gg N$. For each of $N = 100$ samples, we generated p standard Gaussian features X with pairwise correlation 0.2. The outcome Y was generated according to a linear model

$$Y = \sum_{j=1}^{p} X_j \beta_j + \sigma\varepsilon \qquad (18.1)$$

where ε was generated from a standard Gaussian distribution. For each dataset, the set of coefficients β_j were also generated from a standard Gaussian distribution. We investigated three cases: $p = 20, 100,$ and 1000. The standard deviation σ was chosen in each case so that the signal-to-noise ratio $\mathrm{Var}[\mathrm{E}(Y|X)]/\sigma^2$ equaled 2. As a result, the number of significant uni-

T. Hastie et al., *The Elements of Statistical Learning, Second Edition,*
DOI: 10.1007/b94608_18,
© Springer Science+Business Media, LLC 2009

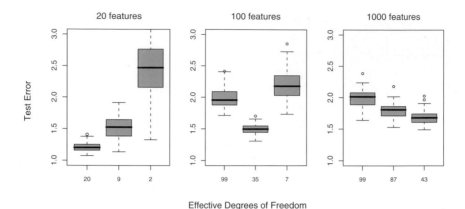

FIGURE 18.1. *Test-error results for simulation experiments. Shown are box-plots of the relative test errors over* 100 *simulations, for three different values of p, the number of features. The relative error is the test error divided by the Bayes error, σ^2. From left to right, results are shown for ridge regression with three different values of the regularization parameter λ:* 0.001, 100 *and* 1000. *The (average) effective degrees of freedom in the fit is indicated below each plot.*

variate regression coefficients[1] was 9, 33 and 331, respectively, averaged over the 100 simulation runs. The $p = 1000$ case is designed to mimic the kind of data that we might see in a high-dimensional genomic or proteomic dataset, for example.

We fit a ridge regression to the data, with three different values for the regularization parameter λ: 0.001, 100, and 1000. When $\lambda = 0.001$, this is nearly the same as least squares regression, with a little regularization just to ensure that the problem is non-singular when $p > N$. Figure 18.1 shows boxplots of the relative test error achieved by the different estimators in each scenario. The corresponding average degrees of freedom used in each ridge-regression fit is indicated (computed using formula (3.50) on page 68[2]). The degrees of freedom is a more interpretable parameter than λ. We see that ridge regression with $\lambda = 0.001$ (20 df) wins when $p = 20$; $\lambda = 100$ (35 df) wins when $p = 100$, and $\lambda = 1000$ (43 df) wins when $p = 1000$.

Here is an explanation for these results. When $p = 20$, we fit all the way and we can identify as many of the significant coefficients as possible with

---

[1]We call a regression coefficient significant if $|\hat{\beta}_j / \widehat{se}_j| \geq 2$, where $\hat{\beta}_j$ is the estimated (univariate) coefficient and $\widehat{se}_j$ is its estimated standard error.

[2]For a fixed value of the regularization parameter λ, the degrees of freedom depends on the observed predictor values in each simulation. Hence we compute the average degrees of freedom over simulations.

low bias. When $p = 100$, we can identify some non-zero coefficients using moderate shrinkage. Finally, when $p = 1000$, even though there are many nonzero coefficients, we don't have a hope for finding them and we need to shrink all the way down. As evidence of this, let $t_j = \widehat{\beta}_j / \widehat{se}_j$, where $\widehat{\beta}_j$ is the ridge regression estimate and $\widehat{se}_j$ its estimated standard error. Then using the optimal ridge parameter in each of the three cases, the median value of $|t_j|$ was 2.0, 0.6 and 0.2, and the average number of $|t_j|$ values exceeding 2 was equal to 9.8, 1.2 and 0.0.

Ridge regression with $\lambda = 0.001$ successfully exploits the correlation in the features when $p < N$, but cannot do so when $p \gg N$. In the latter case there is not enough information in the relatively small number of samples to efficiently estimate the high-dimensional covariance matrix. In that case, more regularization leads to superior prediction performance.

Thus it is not surprising that the analysis of high-dimensional data requires either modification of procedures designed for the $N > p$ scenario, or entirely new procedures. In this chapter we discuss examples of both kinds of approaches for high dimensional classification and regression; these methods tend to regularize quite heavily, using scientific contextual knowledge to suggest the appropriate form for this regularization. The chapter ends with a discussion of feature selection and multiple testing.

18.2 Diagonal Linear Discriminant Analysis and Nearest Shrunken Centroids

Gene expression arrays are an important new technology in biology, and are discussed in Chapters 1 and 14. The data in our next example form a matrix of 2308 genes (columns) and 63 samples (rows), from a set of microarray experiments. Each expression value is a log-ratio $\log(R/G)$. R is the amount of gene-specific RNA in the target sample that hybridizes to a particular (gene-specific) spot on the microarray, and G is the corresponding amount of RNA from a reference sample. The samples arose from small, round blue-cell tumors (SRBCT) found in children, and are classified into four major types: BL (Burkitt lymphoma), EWS (Ewing's sarcoma), NB (neuroblastoma), and RMS (rhabdomyosarcoma). There is an additional test data set of 20 observations. We will not go into the scientific background here.

Since $p \gg N$, we cannot fit a full linear discriminant analysis (LDA) to the data; some sort of regularization is needed. The method we describe here is similar to the methods of Section 4.3.1, but with important modifications that achieve feature selection. The simplest form of regularization assumes that the features are independent within each class, that is, the within-class covariance matrix is diagonal. Despite the fact that features will rarely be independent within a class, when $p \gg N$ we don't have enough data to estimate their dependencies. The assumption of independence greatly reduces the number of parameters in the model and often results in an effective and interpretable classifier.

Thus we consider the *diagonal-covariance* LDA rule for classifying the classes. The *discriminant score* [see (4.12) on page 110] for class k is

$$\delta_k(x^*) = -\sum_{j=1}^{p} \frac{(x_j^* - \bar{x}_{kj})^2}{s_j^2} + 2\log \pi_k. \tag{18.2}$$

Here $x^* = (x_1^*, x_2^*, \ldots, x_p^*)^T$ is a vector of expression values for a test observation, s_j is the pooled within-class standard deviation of the jth gene, and $\bar{x}_{kj} = \sum_{i \in C_k} x_{ij}/N_k$ is the mean of the N_k values for gene j in class k, with C_k being the index set for class k. We call $\tilde{x}_k = (\bar{x}_{k1}, \bar{x}_{k2}, \ldots \bar{x}_{kp})^T$ the *centroid* of class k. The first part of (18.2) is simply the (negative) standardized squared distance of x^* to the kth centroid. The second part is a correction based on the class *prior probability* π_k, where $\sum_{k=1}^{K} \pi_k = 1$. The classification rule is then

$$C(x^*) = \ell \text{ if } \delta_\ell(x^*) = \max_k \delta_k(x^*). \tag{18.3}$$

We see that the diagonal LDA classifier is equivalent to a nearest centroid classifier after appropriate standardization. It is also a special case of the naive-Bayes classifier, as described in Section 6.6.3. It assumes that the features in each class have independent Gaussian distributions with the same variance.

The diagonal LDA classifier is often effective in high dimensional settings. It is also called the "independence rule" in Bickel and Levina (2004), who demonstrate theoretically that it will often outperform standard linear discriminant analysis in high-dimensional problems. Here the diagonal LDA classifier yielded five misclassification errors for the 20 test samples. One drawback of the diagonal LDA classifier is that it uses all of the features (genes), and hence is not convenient for interpretation. With further regularization we can do better—both in terms of test error and interpretability.

We would like to regularize in a way that automatically drops out features that are not contributing to the class predictions. We can do this by shrinking the classwise mean toward the overall mean, for each feature separately. The result is a regularized version of the nearest centroid classifier, or equivalently a regularized version of the diagonal-covariance form of LDA. We call the procedure *nearest shrunken centroids* (NSC).

The shrinkage procedure is defined as follows. Let

$$d_{kj} = \frac{\bar{x}_{kj} - \bar{x}_j}{m_k(s_j + s_0)}, \tag{18.4}$$

where $\bar{x}_j$ is the overall mean for gene j, $m_k^2 = 1/N_k - 1/N$ and s_0 is a small positive constant, typically chosen to be the median of the s_j values.

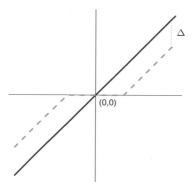

FIGURE 18.2. *Soft thresholding function* $\text{sign}(x)(|x| - \Delta)_+$ *is shown in orange, along with the* $45°$ *line in red.*

This constant guards against large d_{kj} values that arise from expression values near zero. With constant within-class variance σ^2, the variance of the contrast $\bar{x}_{kj} - \bar{x}_j$ in the numerator is $m_k^2\sigma^2$, and hence the form of the standardization in the denominator. We shrink the d_{kj} toward zero using soft thresholding

$$d'_{kj} = \text{sign}(d_{kj})(|d_{kj}| - \Delta)_+ ; \qquad (18.5)$$

see Figure 18.2. Here Δ is a parameter to be determined; we used 10-fold cross-validation in the example (see the top panel of Figure 18.4). Each d_{kj} is reduced by an amount Δ in absolute value, and is set to zero if its value is less than zero. The soft-thresholding function is shown in Figure 18.2; the same thresholding is applied to wavelet coefficients in Section 5.9. An alternative is to use hard thresholding

$$d'_{kj} = d_{kj} \cdot I(|d_{kj}| \geq \Delta); \qquad (18.6)$$

we prefer soft-thresholding, as it is a smoother operation and typically works better. The shrunken versions of $\bar{x}_{kj}$ are then obtained by reversing the transformation in (18.4):

$$\bar{x}'_{kj} = \bar{x}_j + m_k(s_j + s_0)d'_{kj}. \qquad (18.7)$$

We then use the shrunken centroids $\bar{x}'_{kj}$ in place of the original $\bar{x}_{kj}$ in the discriminant score (18.2). The estimator (18.7) can also be viewed as a lasso-style estimator for the class means (Exercise 18.2).

Notice that only the genes that have a nonzero d'_{kj} for at least one of the classes play a role in the classification rule, and hence the vast majority of genes can often be discarded. In this example, all but 43 genes were discarded, leaving a small interpretable set of genes that characterize each class. Figure 18.3 represents the genes in a heatmap.

Figure 18.4 (top panel) demonstrates the effectiveness of the shrinkage. With no shrinkage we make 5/20 errors on the test data, and several errors on the training and CV data. The shrunken centroids achieve zero test errors for a fairly broad band of values for Δ. The bottom panel of Figure 18.4 shows the four centroids for the SRBCT data (gray), relative to the overall centroid. The blue bars are shrunken versions of these centroids, obtained by soft-thresholding the gray bars, using $\Delta = 4.3$. The discriminant scores (18.2) can be used to construct class probability estimates:

$$\hat{p}_k(x^*) = \frac{e^{\frac{1}{2}\delta_k(x^*)}}{\sum_{\ell=1}^{K} e^{\frac{1}{2}\delta_\ell(x^*)}}. \tag{18.8}$$

These can be used to rate the classifications, or to decide not to classify a particular sample at all.

Note that other forms of feature selection can be used in this setting, including hard thresholding. Fan and Fan (2008) show theoretically the importance of carrying out some kind of feature selection with diagonal linear discriminant analysis in high-dimensional problems.

18.3 Linear Classifiers with Quadratic Regularization

Ramaswamy et al. (2001) present a more difficult microarray classification problem, involving a training set of 144 patients with 14 different types of cancer, and a test set of 54 patients. Gene expression measurements were available for $16,063$ genes.

Table 18.1 shows the prediction results from eight different classification methods. The data from each patient was first standardized to have mean 0 and variance 1; this seems to improve prediction accuracy overall this example, suggesting that the "shape" of each gene-expression profile is important, rather than the absolute expression levels. In each case, the

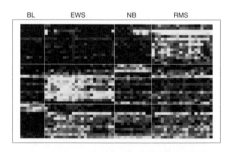

FIGURE 18.3. *Heat-map of the chosen 43 genes. Within each of the horizontal partitions, we have ordered the genes by hierarchical clustering, and similarly for the samples within each vertical partition. Yellow represents over- and blue under-expression.*

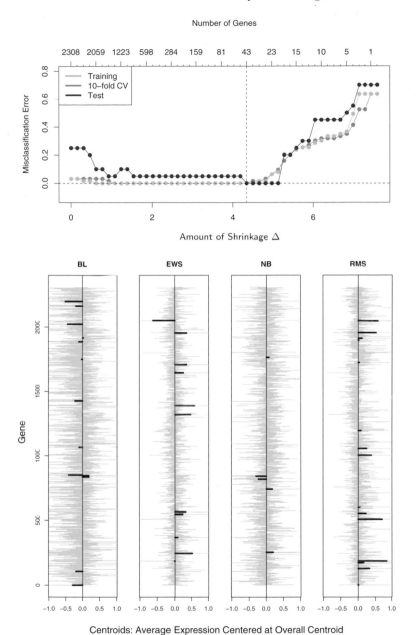

FIGURE 18.4. *(Top): Error curves for the SRBCT data. Shown are the training, 10-fold cross-validation, and test misclassification errors as the threshold parameter Δ is varied. The value $\Delta = 4.34$ is chosen by CV, resulting in a subset of 43 selected genes. (Bottom): Four centroids profiles d_{kj} for the SRBCT data (gray), relative to the overall centroid. Each centroid has 2308 components, and we see considerable noise. The blue bars are shrunken versions d'_{kj} of these centroids, obtained by soft-thresholding the gray bars, using $\Delta = 4.3$.*

TABLE 18.1. *Prediction results for microarray data with 14 cancer classes. Method 1 is described in Section 18.2. Methods 2, 3 and 6 are discussed in Section 18.3, while 4, 7 and 8 are discussed in Section 18.4. Method 5 is described in Section 13.3. The elastic-net penalized multinomial does the best on the test data, but the standard error of each test-error estimate is about 3, so such comparisons are inconclusive.*

Methods	CV errors (SE) Out of 144	Test errors Out of 54	Number of Genes Used
1. Nearest shrunken centroids	35 (5.0)	17	6,520
2. L_2-penalized discriminant analysis	25 (4.1)	12	16,063
3. Support vector classifier	26 (4.2)	14	16,063
4. Lasso regression (one vs all)	30.7 (1.8)	12.5	1,429
5. k-nearest neighbors	41 (4.6)	26	16,063
6. L_2-penalized multinomial	26 (4.2)	15	16,063
7. L_1-penalized multinomial	17 (2.8)	13	269
8. Elastic-net penalized multinomial	22 (3.7)	11.8	384

regularization parameter has been chosen to minimize the cross-validation error, and the test error at that value of the parameter is shown. When more than one value of the regularization parameter yields the minimal cross-validation error, the average test error at these values is reported.

RDA (regularized discriminant analysis), regularized multinomial logistic regression, and the support vector machine are more complex methods that try to exploit multivariate information in the data. We describe each in turn, as well as a variety of regularization methods, including both L_1 and L_2 and some in between.

18.3.1 *Regularized Discriminant Analysis*

Regularized discriminant analysis (RDA) is described in Section 4.3.1. Linear discriminant analysis involves the inversion of a $p \times p$ within-covariance matrix. When $p \gg N$, this matrix can be huge, has rank at most $N < p$, and hence is singular. RDA overcomes the singularity issues by regularizing the within-covariance estimate $\hat{\mathbf{\Sigma}}$. Here we use a version of RDA that shrinks $\hat{\mathbf{\Sigma}}$ towards its diagonal:

$$\hat{\mathbf{\Sigma}}(\gamma) = \gamma\hat{\mathbf{\Sigma}} + (1 - \gamma)\mathrm{diag}(\hat{\mathbf{\Sigma}}), \text{ with } \gamma \in [0, 1]. \tag{18.9}$$

Note that $\gamma = 0$ corresponds to diagonal LDA, which is the "no shrinkage" version of nearest shrunken centroids. The form of shrinkage in (18.9) is much like ridge regression (Section 3.4.1), which shrinks the total covariance matrix of the features towards a diagonal (scalar) matrix. In fact, viewing linear discriminant analysis as linear regression with optimal scoring of the categorical response (see (12.57) in Section 12.6), the equivalence becomes more precise.

The computational burden of inverting this large $p \times p$ matrix is overcome using the methods discussed in Section 18.3.5. The value of γ was chosen by cross-validation in line 2 of Table 18.1; all values of $\gamma \in (0.002, 0.550)$ gave the same CV and test error. Further development of RDA, including shrinkage of the centroids in addition to the covariance matrix, can be found in Guo et al. (2006).

18.3.2 Logistic Regression with Quadratic Regularization

Logistic regression (Section 4.4) can be modified in a similar way, to deal with the $p \gg N$ case. With K classes, we use a symmetric version of the multiclass logistic model (4.17) on page 119:

$$\Pr(G = k | X = x) = \frac{\exp(\beta_{k0} + x^T \beta_k)}{\sum_{\ell=1}^{K} \exp(\beta_{\ell 0} + x^T \beta_\ell)}. \tag{18.10}$$

This has K coefficient vectors of log-odds parameters $\beta_1, \beta_2, \ldots, \beta_K$. We regularize the fitting by maximizing the penalized log-likelihood

$$\max_{\{\beta_{0k}, \beta_k\}_1^K} \left[\sum_{i=1}^{N} \log \Pr(g_i | x_i) - \frac{\lambda}{2} \sum_{k-1}^{K} ||\beta_k||_2^2 \right]. \tag{18.11}$$

This regularization automatically resolves the redundancy in the parametrization, and forces $\sum_{k=1}^{K} \hat{\beta}_{kj} = 0$, $j = 1, \ldots, p$ (Exercise 18.3). Note that the constant terms β_{k0} are not regularized (and so one should be set to zero). The resulting optimization problem is convex, and can be solved by a Newton algorithm or other numerical techniques. Details are given in Zhu and Hastie (2004). Friedman et al. (2010) provide software for computing the regularization path for the two- and multiclass logistic regression models. Table 18.1, line 6 reports the results for the multiclass logistic regression model, referred to there as "multinomial". It can be shown (Rosset et al., 2004a) that for separable data, as $\lambda \rightarrow 0$, the regularized (two-class) logistic regression estimate (renormalized) converges to the maximal margin classifier (Section 12.2). This gives an attractive alternative to the support-vector machine, discussed next, especially in the multiclass case.

18.3.3 The Support Vector Classifier

The support vector classifier is described for the two-class case in Section 12.2. When $p > N$, it is especially attractive because in general the classes are perfectly separable by a hyperplane unless there are identical feature vectors in different classes. Without any regularization the support vector classifier finds the separating hyperplane with the largest margin; that is, the hyperplane yielding the biggest gap between the classes in the training data. Somewhat surprisingly, when $p \gg N$ the unregularized support vector classifier often works about as well as the best regularized version. Overfitting often does not seem to be a problem, partly because of the insensitivity of misclassification loss.

There are many different methods for generalizing the two-class support-vector classifier to $K > 2$ classes. In the "one versus one" (OVO) approach, we compute all $\binom{K}{2}$ pairwise classifiers. For each test point, the predicted class is the one that wins the most pairwise contests. In the "one versus all" (OVA) approach, each class is compared to all of the others in K two-class comparisons. To classify a test point, we compute the confidences (signed distance from the hyperplane) for each of the K classifiers. The winner is the class with the highest confidence. Finally, Vapnik (1998) and Weston and Watkins (1999) suggested (somewhat complex) multiclass criteria which generalize the two-class criterion (12.7).

Tibshirani and Hastie (2007) propose the *margin tree* classifier, in which support-vector classifiers are used in a binary tree, much as in CART (Chapter 9). The classes are organized in a hierarchical manner, which can be useful for classifying patients into different cancer types, for example.

Line 3 of Table 18.1 shows the results for the support vector classifier using the OVA method; Ramaswamy et al. (2001) reported (and we confirmed) that this approach worked best for this problem. The errors are very similar to those in line 6, as we might expect from the comments at the end of the previous section. The error rates are insensitive to the choice of C [the regularization parameter in (12.8) on page 420], for values of $C > 0.001$. Since $p > N$, the support vector hyperplane can perfectly separate the training data by setting $C = \infty$.

18.3.4 *Feature Selection*

Feature selection is an important scientific requirement for a classifier when p is large. Neither discriminant analysis, logistic regression, nor the support-vector classifier perform feature selection automatically, because all use quadratic regularization. All features have nonzero weights in both models. Ad-hoc methods for feature selection have been proposed, for example, removing genes with small coefficients, and refitting the classifier. This is done in a backward stepwise manner, starting with the smallest weights and moving on to larger weights. This is known as *recursive feature elimination* (Guyon et al., 2002). It was not successful in this example; Ramaswamy et al. (2001) report, for example, that the accuracy of the support-vector classifier starts to degrade as the number of genes is reduced from the full set of $16,063$. This is rather remarkable, as the number of training samples is only 144. We do not have an explanation for this behavior.

All three methods discussed in this section (RDA, LR and SVM) can be modified to fit nonlinear decision boundaries using kernels. Usually the motivation for such an approach is to increase the model complexity. With $p \gg N$ the models are already sufficiently complex and overfitting is always a danger. Yet despite the high dimensionality, radial kernels (Section 12.3.3) sometimes deliver superior results in these high dimensional problems. The radial kernel tends to dampen inner products between points far away from each other, which in turn leads to robustness to outliers. This occurs often in high dimensions, and may explain the positive results. We tried a radial kernel with the SVM in Table 18.1, but in this case the performance was inferior.

18.3.5 Computational Shortcuts When $p \gg N$

The computational techniques discussed in this section apply to any method that fits a linear model with quadratic regularization on the coefficients. That includes all the methods discussed in this section, and many more. When $p > N$, the computations can be carried out in an N-dimensional space, rather than p, via the singular value decomposition introduced in Section 14.5. Here is the geometric intuition: just like two points in three-dimensional space always lie on a line, N points in p-dimensional space lie in an $(N-1)$-dimensional affine subspace.

Given the $N \times p$ data matrix $\mathbf{X}$, let

$$\mathbf{X} \quad = \quad \mathbf{U}\mathbf{D}\mathbf{V}^T \qquad\qquad (18.12)$$

$$= \quad \mathbf{R}\mathbf{V}^T \qquad\qquad (18.13)$$

be the singular-value decomposition (SVD) of $\mathbf{X}$; that is, $\mathbf{V}$ is $p \times N$ with orthonormal columns, $\mathbf{U}$ is $N \times N$ orthogonal, and $\mathbf{D}$ a diagonal matrix with elements $d_1 \geq d_2 \geq d_N \geq 0$. The matrix $\mathbf{R}$ is $N \times N$, with rows r_i^T.

As a simple example, let's first consider the estimates from a ridge regression:

$$\hat{\beta} = (\mathbf{X}^T\mathbf{X} + \lambda\mathbf{I})^{-1}\mathbf{X}^T\mathbf{y}. \qquad\qquad (18.14)$$

Replacing $\mathbf{X}$ by $\mathbf{R}\mathbf{V}^T$ and after some further manipulations, this can be shown to equal

$$\hat{\beta} \quad = \quad \mathbf{V}(\mathbf{R}^T\mathbf{R} + \lambda\mathbf{I})^{-1}\mathbf{R}^T\mathbf{y} \qquad\qquad (18.15)$$

(Exercise 18.4). Thus $\hat{\beta} = \mathbf{V}\hat{\theta}$, where $\hat{\theta}$ is the ridge-regression estimate using the N observations (r_i, y_i), $i = 1, 2, \ldots, N$. In other words, we can simply reduce the data matrix from $\mathbf{X}$ to $\mathbf{R}$, and work with the rows of $\mathbf{R}$. This trick reduces the computational cost from $O(p^3)$ to $O(pN^2)$ when $p > N$.

These results can be generalized to *all* models that are linear in the parameters and have quadratic penalties. Consider any supervised learning problem where we use a linear function $f(X) = \beta_0 + X^T \beta$ to model a parameter in the conditional distribution of $Y|X$. We fit the parameters β by minimizing some loss function $\sum_{i=1}^{N} L(y_i, f(x_i))$ over the data with a quadratic penalty on β. Logistic regression is a useful example to have in mind. Then we have the following simple theorem:

Let $f^*(r_i) = \theta_0 + r_i^T \theta$ with r_i defined in (18.13), and consider the pair of optimization problems:

$$(\hat{\beta}_0, \hat{\beta}) = \arg\min_{\beta_0, \beta \in \mathbb{R}^p} \sum_{i=1}^{N} L(y_i, \beta_0 + x_i^T \beta) + \lambda \beta^T \beta; \qquad (18.16)$$

$$(\hat{\theta}_0, \hat{\theta}) = \arg\min_{\theta_0, \theta \in \mathbb{R}^N} \sum_{i=1}^{N} L(y_i, \theta_0 + r_i^T \theta) + \lambda \theta^T \theta. \qquad (18.17)$$

Then the $\hat{\beta}_0 = \hat{\theta}_0$, and $\hat{\beta} = \mathbf{V}\hat{\theta}$.

The theorem says that we can simply replace the p vectors x_i by the N-vectors r_i, and perform our penalized fit as before, but with far fewer predictors. The N-vector solution $\hat{\theta}$ is then transformed back to the p-vector solution via a simple matrix multiplication. This result is part of the statistics folklore, and deserves to be known more widely—see Hastie and Tibshirani (2004) for further details.

Geometrically, we are rotating the features to a coordinate system in which all but the first N coordinates are zero. Such rotations are allowed since the quadratic penalty is invariant under rotations, and linear models are equivariant.

This result can be applied to many of the learning methods discussed in this chapter, such as regularized (multiclass) logistic regression, linear discriminant analysis (Exercise 18.6), and support vector machines. It also applies to neural networks with quadratic regularization (Section 11.5.2). Note, however, that it does not apply to methods such as the lasso, which uses nonquadratic (L_1) penalties on the coefficients.

Typically we use cross-validation to select the parameter λ. It can be seen (Exercise 18.12) that we only need to construct $\mathbf{R}$ once, on the original data, and use it as the data for each of the CV folds.

The support vector "kernel trick" of Section 12.3.7 exploits the same reduction used in this section, in a slightly different context. Suppose we have at our disposal the $N \times N$ gram (inner-product) matrix $\mathbf{K} = \mathbf{X}\mathbf{X}^T$. From (18.12) we have $\mathbf{K} = \mathbf{U}\mathbf{D}^2\mathbf{U}^T$, and so $\mathbf{K}$ captures the same information as $\mathbf{R}$. Exercise 18.13 shows how we can exploit the ideas in this section to fit a ridged logistic regression with $\mathbf{K}$ using its SVD.

18.4 Linear Classifiers with L_1 Regularization

The methods of Section 18.3 use an L_2 penalty to regularize their parameters, just as in ridge regression. All of the estimated coefficients are nonzero, and hence no feature selection is performed. In this section we discuss methods that use L_1 penalties instead, and hence provide automatic feature selection.

Recall the lasso of Section 3.4.2,

$$\min_{\beta} \frac{1}{2} \sum_{i=1}^{N} \left(y_i - \beta_0 - \sum_{j=1}^{p} x_{ij}\beta_j \right)^2 + \lambda \sum_{j=1}^{p} |\beta_j|, \qquad (18.18)$$

which we have written in the Lagrange form (3.52). As discussed there, the use of the L_1 penalty causes a subset of the solution coefficients $\hat{\beta}_j$ to be exactly zero, for a sufficiently large value of the tuning parameter λ.

In Section 3.8.1 we discussed the LARS algorithm, an efficient procedure for computing the lasso solution for all λ. When $p > N$ (as in this chapter), as λ approaches zero, the lasso fits the training data exactly. In fact, by convex duality one can show that when $p > N$ the number of non-zero coefficients is at most N for all values of λ (Rosset and Zhu, 2007, for example). Thus the lasso provides a (severe) form of feature selection.

Lasso regression can be applied to a two-class classification problem by coding the outcome ± 1, and applying a cutoff (usually 0) to the predictions. For more than two classes, there are many possible approaches, including the OVA and OVO methods discussed in Section 18.3.3. We tried the OVA-approach on the cancer data in Section 18.3. The results are shown in line (4) of Table 18.1. Its performance is among the best.

A more natural approach for classification problems is to use the lasso penalty to regularize logistic regression. Several implementations have been proposed in the literature, including path algorithms similar to LARS (Park and Hastie, 2007). Because the paths are piecewise smooth but nonlinear, exact methods are slower than the LARS algorithm, and are less feasible when p is large.

Friedman et al. (2010) provide very fast algorithms for fitting L_1-penalized logistic and multinomial regression models. They use the symmetric multinomial logistic regression model as in (18.10) in Section 18.3.2, and maximize the penalized log-likelihood

$$\max_{\{\beta_{0k}, \beta_k \in \mathbb{R}^p\}_1^K} \left[\sum_{i=1}^{N} \log \Pr(g_i|x_i) - \lambda \sum_{k=1}^{K} \sum_{j=1}^{p} |\beta_{kj}| \right]; \qquad (18.19)$$

compare with (18.11). Their algorithm computes the exact solution at a pre-chosen sequence of values for λ by cyclical coordinate descent (Section 3.8.6), and exploits the fact that solutions are sparse when $p \gg N$, as well as the fact that solutions for neighboring values of λ tend to be very similar. This method was used in line (7) of Table 18.1, with the overall tuning parameter λ chosen by cross-validation. The performance was similar to that of the best methods, except here the automatic feature selection chose 269 genes altogether. A similar approach is used in Genkin et al. (2007); although they present their model from a Bayesian point of view, they in fact compute the posterior mode, which solves the penalized maximum-likelihood problem.

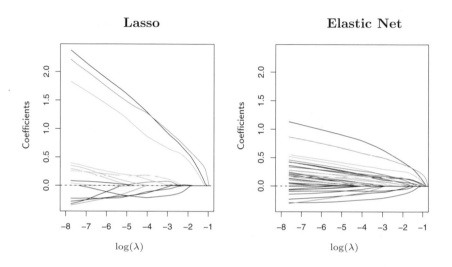

FIGURE 18.5. *Regularized logistic regression paths for the leukemia data. The left panel is the lasso path, the right panel the elastic-net path with $\alpha = 0.8$. At the ends of the path (extreme left), there are 19 nonzero coefficients for the lasso, and 39 for the elastic net. The averaging effect of the elastic net results in more non-zero coefficients than the lasso, but with smaller magnitudes.*

In genomic applications, there are often strong correlations among the variables; genes tend to operate in molecular pathways. The lasso penalty is somewhat indifferent to the choice among a set of strong but correlated variables (Exercise 3.28). The ridge penalty, on the other hand, tends to shrink the coefficients of correlated variables toward each other (Exercise 3.29 on page 99). The *elastic net* penalty (Zou and Hastie, 2005) is a compromise, and has the form

$$\sum_{j=1}^{p} \left(\alpha |\beta_j| + (1 - \alpha)\beta_j^2 \right). \qquad (18.20)$$

The second term encourages highly correlated features to be averaged, while the first term encourages a sparse solution in the coefficients of these averaged features. The elastic net penalty can be used with any linear model, in particular for regression or classification.

Hence the multinomial problem above with elastic-net penalty becomes

$$\max_{\{\beta_{0k}, \beta_k \in \mathbb{R}^p\}_1^K} \left[\sum_{i=1}^{N} \log \Pr(g_i|x_i) - \lambda \sum_{k=1}^{K} \sum_{j=1}^{p} \left(\alpha|\beta_{kj}| + (1-\alpha)\beta_{kj}^2 \right) \right].$$

$$(18.21)$$

The parameter α determines the mix of the penalties, and is often pre-chosen on qualitative grounds. The elastic net can yield more that N non-zero coefficients when $p > N$, a potential advantage over the lasso. Line (8) in Table 18.1 uses this model, with α and λ chosen by cross-validation. We used a sequence of 20 values of α between 0.05 and 1.0, and a 100 values of λ uniform on the log scale covering the entire range. Values of $\alpha \in [0.75, 0.80]$ gave the minimum CV error, with values of $\lambda < 0.001$ for all tied solutions. Although it has the lowest test error among all methods, the margin is small and not significant. Interestingly, when CV is performed separately for each value of α, a minimum test error of 8.8 is achieved at $\alpha = 0.10$, but this is not the value chosen in the two-dimensional CV.

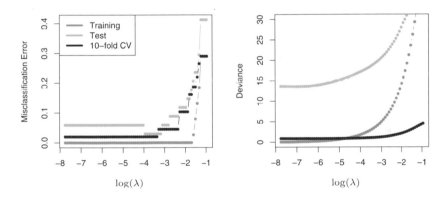

FIGURE 18.6. *Training, test, and 10-fold cross validation curves for lasso logistic regression on the leukemia data. The left panel shows misclassification errors, the right panel shows deviance.*

Figure 18.5 shows the lasso and elastic-net coefficient paths on the two-class leukemia data (Golub et al., 1999). There are 7129 gene-expression measurements on 38 samples, 27 of them in class ALL (acute lymphocytic leukemia), and 11 in class AML (acute myelogenous leukemia). There is also a test set with 34 samples (20, 14). Since the data are linearly separable, the solution is undefined at $\lambda = 0$ (Exercise 18.11), and degrades for very small values of λ. Hence the paths have been truncated as the fitted probabilities approach 0 and 1. There are 19 non-zero coefficients in the left plot, and 39 in the right. Figure 18.6 (left panel) shows the misclassification errors for the lasso logistic regression on the training and test data, as well as for 10-fold cross-validation on the training data. The right panel uses binomial deviance to measure errors, and is much smoother. The small sample sizes lead to considerable sampling variance in these curves, even though individual curves are relatively smooth (see, for example, Figure 7.1 on page 220). Both of these plots suggest that the limiting solution $\lambda \downarrow 0$ is adequate, leading to 3/34 misclassifications in the test set. The corresponding figures for the elastic net are qualitatively similar and are not shown.

For $p \gg N$, the limiting coefficients diverge for all regularized logistic regression models, so in practical software implementations a minimum value for $\lambda > 0$ is either explicitly or implicitly set. However, renormalized versions of the coefficients converge, and these limiting solutions can be thought of as interesting alternatives to the linear optimal separating hyperplane (SVM). With $\alpha = 0$ the limiting solution coincides with the SVM (see end of Section 18.3.2), but all the 7129 genes are selected. With $\alpha = 1$, the limiting solution coincides with an L_1 separating hyperplane (Rosset et al., 2004a), and includes at most 38 genes. As α decreases from 1, the elastic-net solutions include more genes in the separating hyperplane.

18.4.1 *Application of Lasso to Protein Mass Spectroscopy*

Protein mass spectrometry has become a popular technology for analyzing the proteins in blood, and can be used to diagnose a disease or understand the processes underlying it.

For each blood serum sample i, we observe the intensity x_{ij} for many *time of flight* values t_j. This intensity is related to the number of particles observed to take approximately t_j time to pass from the emitter to the detector during a cycle of operation of the machine. The time of flight has a known relationship to the mass over charge ratio (m/z) of the constituent proteins in the blood. Hence the identification of a peak in the spectrum at a certain t_j tells us that there is a protein with a corresponding mass and charge. The identity of this protein can then be determined by other means.

Figure 18.7 shows an example taken from Adam et al. (2003). It shows the average spectra for healthy patients and those with prostate cancer. There are 16,898 m/z sites in total, ranging in value from 2000 to 40,000. The full dataset consists of 157 healthy patients and 167 with cancer, and the goal is to find m/z sites that discriminate between the two groups. This is an example of *functional* data; the predictors can be viewed as a function of m/z. There has been much interest in this problem in the past few years; see e.g. Petricoin et al. (2002).

The data were first standardized (baseline subtraction and normalization), and we restricted attention to m/z values between 2000 and 40,000 (spectra outside of this range were not of interest). We then applied near-

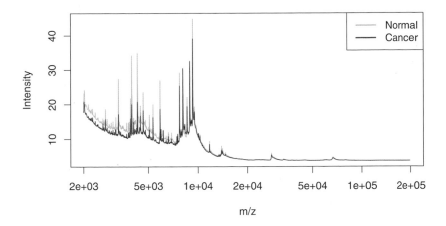

FIGURE 18.7. *Protein mass spectrometry data: average profiles from normal and prostate cancer patients.*

est shrunken centroids and lasso regression to the data, with the results for both methods shown in Table 18.2.

By fitting harder to the data, the lasso achieves a considerably lower test error rate. However, it may not provide a scientifically useful solution. Ideally, protein mass spectrometry resolves a biological sample into its constituent proteins, and these should appear as peaks in the spectra. The lasso doesn't treat peaks in any special way, so not surprisingly only some of the non-zero lasso weights were situated near peaks in the spectra. Furthermore, the same protein may yield a peak at slightly different m/z values in different spectra. In order to identify common peaks, some kind of m/z *warping* is needed from sample to sample.

To address this, we applied a standard peak-extraction algorithm to each spectrum, yielding a total of 5178 peaks in the 217 training spectra. Our idea was to pool the collection of peaks from all patients, and hence construct a set of common peaks. For this purpose, we applied hierarchical clustering to the positions of these peaks along the log m/z axis. We cut the resulting dendrogram horizontally at height $\log(0.005)^3$, and computed averages of the peak positions in each resulting cluster. This process yielded 728 common clusters and their corresponding peak centers.

Given these 728 common peaks, we determined which of these were present in each individual spectrum, and if present, the height of the peak. A peak height of zero was assigned if that peak was not found. This produced a 217×728 matrix of peak heights as features, which was used in a lasso regression. We scored the test spectra for the same 728 peaks.

---

[3] Use of the value 0.005 means that peaks with positions less than 0.5% apart are considered the same peak, a fairly common assumption.

TABLE 18.2. *Results for the prostate data example. The standard deviation for the test errors is about 4.5.*

Method	Test Errors/108	Number of Sites
1. Nearest shrunken centroids	34	459
2. Lasso	22	113
3. Lasso on peaks	28	35

The prediction results for this application of the lasso to the peaks are shown in the last line of Table 18.2: it does fairly well, but not as well as the lasso on the raw spectra. However, the fitted model may be more useful to the biologist as it yields 35 peak positions for further study. On the other hand, the results suggest that there may be useful discriminatory information between the peaks of the spectra, and the positions of the lasso sites from line (2) of the table also deserve further examination.

18.4.2   The Fused Lasso for Functional Data

In the previous example, the features had a natural order, determined by the mass-to-charge ratio m/z. More generally, we may have functional features $x_i(t)$ that are ordered according to some index variable t. We have already discussed several approaches for exploiting such structure.

We can represent $x_i(t)$ by their coefficients in a basis of functions in t, such as splines, wavelets or Fourier bases, and then apply a regression using these coefficients as predictors. Equivalently, one can instead represent the coefficients of the original features in these bases. These approaches are described in Section 5.3.

In the classification setting, we discuss the analogous approach of penalized discriminant analysis in Section 12.6. This uses a penalty that explicitly controls the resulting smoothness of the coefficient vector.

The above methods tend to smooth the coefficients uniformly. Here we present a more adaptive strategy that modifies the lasso penalty to take into account the ordering of the features. The *fused lasso* (Tibshirani et al., 2005) solves

$$\min_{\beta \in \mathbb{R}^p} \left\{ \sum_{i=1}^{N}(y_i - \beta_0 - \sum_{j=1}^{p} x_{ij}\beta_j)^2 + \lambda_1 \sum_{j=1}^{p}|\beta_j| + \lambda_2 \sum_{j=1}^{p-1}|\beta_{j+1} - \beta_j| \right\}. \quad (18.22)$$

This criterion is strictly convex in β, so a unique solution exists. The first penalty encourages the solution to be sparse, while the second encourages it to be smooth in the index j.

The difference penalty in (18.22) assumes an uniformly spaced index j. If instead the underlying index variable t has nonuniform values t_j, a natural generalization of (18.22) would be based on divided differences

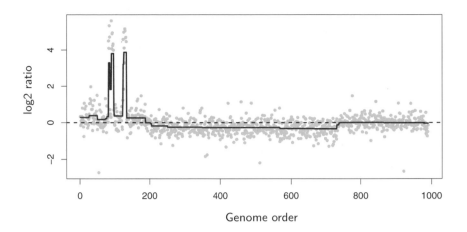

FIGURE 18.8. *Fused lasso applied to CGH data. Each point represents the copy-number of a gene in a tumor sample, relative to that of a control (on the log base-2 scale).*

$$\lambda_2 \sum_{j=1}^{p-1} \frac{|\beta_{j+1} - \beta_j|}{|t_{j+1} - t_j|}. \tag{18.23}$$

This amounts to having a penalty modifier for each of the terms in the series.

A particularly useful special case arises when the predictor matrix $\mathbf{X} = \mathbf{I}_N$, the $N \times N$ identity matrix. This is a special case of the fused lasso, used to approximate a sequence $\{y_i\}_1^N$. The *fused lasso signal approximator* solves

$$\min_{\beta \in \mathbb{R}^N} \left\{ \sum_{i=1}^{N} (y_i - \beta_0 - \beta_i)^2 + \lambda_1 \sum_{i=1}^{N} |\beta_i| + \lambda_2 \sum_{i=1}^{N-1} |\beta_{i+1} - \beta_i| \right\}. \tag{18.24}$$

Figure 18.8 shows an example taken from Tibshirani and Wang (2007). The data in the panel come from a Comparative Genomic Hybridization (CGH) array, measuring the approximate log (base-two) ratio of the number of copies of each gene in a tumor sample, as compared to a normal sample. The horizontal axis represents the chromosomal location of each gene. The idea is that in cancer cells, genes are often amplified (duplicated) or deleted, and it is of interest to detect these events. Furthermore, these events tend to occur in contiguous regions. The smoothed signal estimate from the fused lasso signal approximator is shown in dark red (with appropriately chosen values for λ_1 and λ_2). The significantly nonzero regions can be used to detect locations of gains and losses of genes in the tumor.

There is also a two-dimensional version of the fused lasso, in which the parameters are laid out in a grid of pixels, and a penalty is applied to the first differences to the left, right, above and below the target pixel. This can be useful for denoising or classifying images. Friedman et al. (2007) develop fast generalized coordinate descent algorithms for the one- and two-dimensional fused lasso.

18.5 Classification When Features are Unavailable

In some applications the objects under study are more abstract in nature, and it is not obvious how to define a feature vector. As long as we can fill in an $N \times N$ *proximity* matrix of similarities between pairs of objects in our database, it turns out we can put to use many of the classifiers in our arsenal by interpreting the proximities as inner-products. Protein structures fall into this category, and we explore an example in Section 18.5.1 below.

In other applications, such as document classification, feature vectors are available but can be extremely high-dimensional. Here we may not wish to compute with such high-dimensional data, but rather store the inner-products between pairs of documents. Often these inner-products can be approximated by sampling techniques.

Pairwise distances serve a similar purpose, because they can be turned into centered inner-products. Proximity matrices are discussed in more detail in Chapter 14.

18.5.1 *Example: String Kernels and Protein Classification*

An important problem in computational biology is to classify proteins into functional and structural classes based on their sequence similarities. Protein molecules are strings of amino acids, differing in both length and composition. In the example we consider, the lengths vary between 75–160 amino-acid molecules, each of which can be one of 20 different types, labeled using letters. Here are two examples, of length 110 and 153, respectively:

IPTSALVKETLALLSTHRTLLIANETLRIPVPVHKNHQLCTEEIFQGIGTLESQTVQGGTV
ERLFKNLSLIKKYIDGQKKKCGEERRRVNQFLDY**LQE**FLGVMNTEWI

PHRRDLCSRSIWLARKIRSDLTALTESYVKHQGLWSELTEAER**LQE**NLQAYRTFHVLLA
RLLEDQQVHFTPTEGDFHQAIHTLLLQVAAFAYQIEELMILLEYKIPRNEADGMLFEKK
LWGLKV**LQE**LSQWTVRSIHDLRFISSHQTGIP

There have been many proposals for measuring the similarity between a pair of protein molecules. Here we focus on a measure based on the count of matching substrings (Leslie et al., 2004), such as the **LQE** above.

To construct our features, we count the number of times that a given sequence of length m occurs in our string, and we compute this number for all possible sequences of length m. Formally, for a string x, we define a feature map

$$\Phi_m(x) = \{\phi_a(x)\}_{a \in \mathcal{A}_m} \qquad (18.25)$$

where $\mathcal{A}_m$ is the set of subsequences of length m, and $\phi_a(x)$ is the number of times that "a" occurs in our string x. Using this, we define the inner product

$$K_m(x_1, x_2) = \langle \Phi_m(x_1), \Phi_m(x_2) \rangle, \qquad (18.26)$$

which measures the similarity between the two strings x_1, x_2. This can be used to drive, for example, a support vector classifier for classifying strings into different protein classes.

Now the number of possible sequences a is $|\mathcal{A}_m| = 20^m$, which can be very large for moderate m, and the vast majority of the subsequences do not match the strings in our training set. It turns out that we can compute the $N \times N$ inner-product matrix or *string kernel* $\mathbf{K}_m$ (18.26) efficiently using tree-structures, without actually computing the individual vectors. This methodology, and the data to follow, come from Leslie et al. (2004).[4]

The data consist of 1708 proteins in two classes— negative (1663) and positive (45). The two examples above, which we will call "x_1" and "x_2", are from this set. We have marked the occurrences of subsequence **LQE**, which appears in both proteins. There are 20^3 possible subsequences, so $\Phi_3(x)$ will be a vector of length 8000. For this example $\phi_{LQE}(x_1) = 1$ and $\phi_{LQE}(x_2) = 2$.

Using software from Leslie et al. (2004), we computed the string kernel for $m = 4$, which was then used in a support vector classifier to find the maximal margin solution in this $20^4 = 160,000$-dimensional feature space. We used 10-fold cross-validation to compute the SVM predictions on all of the training data. The orange curve in Figure 18.9 shows the cross-validated ROC curve for the support vector classifier, computed by varying the cut-point on the real-valued predictions from the cross-validated support vector classifier. The area under the curve is 0.84. Leslie et al. (2004) show that the string kernel method is competitive with, but perhaps not as accurate as, more specialized methods for protein string matching.

Many other classifiers can be computed using only the information in the kernel matrix; some details are given in the next section. The results for the nearest centroid classifier (green), and distance-weighted one-nearest neighbors (blue) are shown in Figure 18.9. Their performance is similar to that of the support vector classifier.

---

[4]We thank Christina Leslie for her help and for providing the data, which is available on our book website.

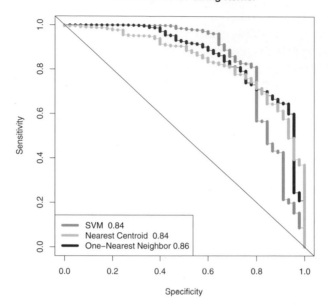

FIGURE 18.9. *Cross-validated ROC curves for protein example using the string kernel. The numbers next to each method in the legend give the area under the curve, an overall measure of accuracy. The SVM achieves better sensitivities than the other two, which achieve better specificities.*

18.5.2 *Classification and Other Models Using Inner-Product Kernels and Pairwise Distances*

There are a number of other classifiers, besides the support-vector machine, that can be implemented using only inner-product matrices. This also implies they can be "kernelized" like the SVM.

An obvious example is nearest-neighbor classification, since we can transform pairwise inner-products to pairwise distances:

$$||x_i - x_{i'}||^2 = \langle x_i, x_i \rangle + \langle x_{i'}, x_{i'} \rangle - 2\langle x_i, x_{i'} \rangle. \qquad (18.27)$$

A variation of 1-NN classification is used in Figure 18.9, which produces a continuous discriminant score needed to construct a ROC curve. This distance-weighted 1-NN makes use of the distance of a test points to the closest member of each class; see Exercise 18.14.

Nearest-centroid classification follows easily as well. For training pairs (x_i, g_i), $i = 1, \ldots, N$, a test point x_0, and class centroids $\bar{x}_k$, $k = 1, \ldots, K$ we can write

$$||x_0 - \bar{x}_k||^2 = \langle x_0, x_0 \rangle - \frac{2}{N_k} \sum_{g_i = k} \langle x_0, x_i \rangle + \frac{1}{N_k^2} \sum_{g_i = k} \sum_{g_{i'} = k} \langle x_i, x_{i'} \rangle, \quad (18.28)$$

Hence we can compute the distance of the test point to each of the centroids, and perform nearest centroid classification. This also implies that methods like K-means clustering can also be implemented, using only the inner products of the data points.

Logistic and multinomial regression with quadratic regularization can also be implemented with inner-product kernels; see Section 12.3.3 and Exercise 18.13. Exercise 12.10 derives linear discriminant analysis using an inner-product kernel.

Principal components can be computed using inner-product kernels as well; since this is frequently useful, we give some details. Suppose first that we have a centered data matrix $\mathbf{X}$, and let $\mathbf{X} = \mathbf{U}\mathbf{D}\mathbf{V}^T$ be its SVD (18.12). Then $\mathbf{Z} = \mathbf{U}\mathbf{D}$ is the matrix of principal component variables (see Section 14.5.1). But if $\mathbf{K} = \mathbf{X}\mathbf{X}^T$, then it follows that $\mathbf{K} = \mathbf{U}\mathbf{D}^2\mathbf{U}^T$, and hence we can compute $\mathbf{Z}$ from the eigen decomposition of $\mathbf{K}$. If $\mathbf{X}$ is *not* centered, then we can center it using $\tilde{\mathbf{X}} = (\mathbf{I} - \mathbf{M})\mathbf{X}$, where $\mathbf{M} = \frac{1}{N}\mathbf{1}\mathbf{1}^T$ is the mean operator. Thus we compute the eigenvectors of the *double-centered* kernel $(\mathbf{I} - \mathbf{M})\mathbf{K}(\mathbf{I} - \mathbf{M})$ for the principal components from an uncentered inner-product matrix. Exercise 18.15 explores this further, and Section 14.5.4 discusses in more detail kernel PCA for general kernels, such as the radial kernel used in SVMs.

If instead we had available only the pairwise (squared) Euclidean distances between observations,

$$\Delta_{ii'}^2 = ||x_i - x_{i'}||^2, \tag{18.29}$$

it turns out we can do all of the above as well. The trick is to convert the pairwise distances to centered inner-products, and then proceed as before. We write

$$\Delta_{ii'}^2 = ||x_i - \bar{x}||^2 + ||x_{i'} - \bar{x}||^2 - 2\langle x_i - \bar{x}, x_{i'} - \bar{x}\rangle. \tag{18.30}$$

Defining $\mathbf{B} = \{-\Delta_{ii'}^2/2\}$, we double center $\mathbf{B}$:

$$\tilde{\mathbf{K}} = (\mathbf{I} - \mathbf{M})\mathbf{B}(\mathbf{I} - \mathbf{M}); \tag{18.31}$$

it is easy to check that $\tilde{K}_{ii'} = \langle x_i - \bar{x}, x_{i'} - \bar{x}\rangle$, the centered inner-product matrix.

Distances and inner-products also allow us to compute the medoid in each class—the observation with smallest average distance to other observations in that class. This can be used for classification (closest medoids), as well as to drive k-medoids clustering (Section 14.3.10). With abstract data objects like proteins, medoids have a practical advantage over means. The medoid is one of the training examples, and can be displayed. We tried closest medoids in the example in the next section (see Table 18.3), and its performance is disappointing.

It is useful to consider what we *cannot* do with inner-product kernels and distances:

TABLE 18.3. *Cross-validated error rates for the abstracts example. The nearest shrunken centroids ended up using no-shrinkage, but does use a word-by-word standardization (section 18.2). This standardization gives it a distinct advantage over the other methods.*

	Method	CV Error (SE)
1.	Nearest shrunken centroids	0.17 (0.05)
2.	SVM	0.23 (0.06)
3.	Nearest medoids	0.65 (0.07)
4.	1-NN	0.44 (0.07)
5.	Nearest centroids	0.29 (0.07)

- We cannot standardize the variables; standardization significantly improves performance in the example in the next section.

- We cannot assess directly the contributions of individual variables. In particular, we cannot perform individual t-tests, fit the nearest shrunken centroids model, or fit any model that uses the lasso penalty.

- We cannot separate the good variables from the noise: all variables get an equal say. If, as is often the case, the ratio of relevant to irrelevant variables is small, methods that use kernels are not likely to work as well as methods that do feature selection.

18.5.3 Example: Abstracts Classification

This somewhat whimsical example serves to illustrate a limitation of kernel approaches. We collected the abstracts from 48 papers, 16 each from Bradley Efron (BE), Trevor Hastie and Rob Tibshirani (HT) (frequent co-authors), and Jerome Friedman (JF). We extracted all unique words from these abstracts, and defined features x_{ij} to be the number of times word j appears in abstract i. This is the so-called *bag of words* representation. Quotations, parentheses and special characters were first removed from the abstracts, and all characters were converted to lower case. We also removed the word "we", which could unfairly discriminate HT abstracts from the others.

There were 4492 total words, of which $p = 1310$ were unique. We sought to classify the documents into BE, HT or JF on the basis of the features x_{ij}. Although it is artificial, this example allows us to assess the possible degradation in performance if information specific to the raw features is not used.

We first applied the nearest shrunken centroid classifier to the data, using 10-fold cross-validation. It essentially chose no shrinkage, and so used all the features; see the first line of Table 18.3. The error rate is 17%; the number of features can be reduced to about 500 without much loss in accuracy.

Note that the nearest shrunken classifier requires the raw feature matrix **X** in order to standardize the features individually. Figure 18.10 shows the

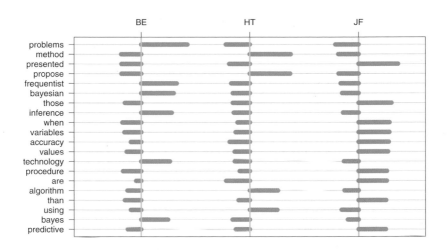

FIGURE 18.10. *Abstracts example: top 20 scores from nearest shrunken centroids. Each score is the standardized difference in frequency for the word in the given class (BE, HT or JF) versus all classes. Thus a positive score (to the right of the vertical grey zero lines) indicates a higher frequency in that class; a negative score indicates a lower relative frequency.*

top 20 discriminating words, with a positive score indicating that a word appears more in that class than in the other classes.

Some of these terms make sense: for example "frequentist" and "Bayesian" reflect Efron's greater emphasis on statistical inference. However, many others are surprising, and reflect personal writing styles: for example, Friedman's use of "presented" and HT's use of "propose".

We then applied the support vector classifier with linear kernel and no regularization, using the "all pairs" (OVO) method to handle the three classes (regularization of the SVM did not improve its performance). The result is shown in Table 18.3. It does somewhat worse than the nearest shrunken centroid classifier.

As mentioned, the first line of Table 18.3 represents nearest shrunken centroids (with no shrinkage). Denote by s_j the pooled within-class standard deviation for feature j, and s_0 the median of the s_j values. Then line (1) also corresponds to nearest centroid classification, after first standardizing each feature by $s_j + s_0$ [recall (18.4) on page 652].

Line (3) shows that the performance of nearest medoids is very poor, something which surprised us. It is perhaps due to the small sample sizes and high dimensions, with medoids having much higher variance than means. The performance of the one-nearest neighbor classifier is also poor.

The performance of the nearest centroid classifier is also shown in Table 18.3 in line (5): it is better than nearest medoids, but worse than that of nearest shrunken centroids, even with no shrinkage. The difference seems to be the standardization of each feature that is done in nearest shrunken centroids. This standardization is important here, and requires access to the individual feature values. Nearest centroids uses a spherical metric, and relies on the fact that the features are in similar units. The support vector machine estimates a linear combination of the features and can better deal with unstandardized features.

18.6 High-Dimensional Regression: Supervised Principal Components

In this section we describe a simple approach to regression and generalized regression that is especially useful when $p \gg N$. We illustrate the method on another microarray data example. The data is taken from Rosenwald et al. (2002) and consists of 240 samples from patients with diffuse large B-cell lymphoma (DLBCL), with gene expression measurements for 7399 genes. The outcome is survival time, either observed or right censored. We randomly divided the lymphoma samples into a training set of size 160 and a test set of size 80.

Although supervised principal components is useful for linear regression, its most interesting applications may be in survival studies, which is the focus of this example.

We have not yet discussed regression with censored survival data in this book; it represents a generalized form of regression in which the outcome variable (survival time) is only partly observed for some individuals. Suppose for example we carry out a medical study that lasts for 365 days, and for simplicity all subjects are recruited on day one. We might observe one individual to die 200 days after the start of the study. Another individual might still be alive at 365 days when the study ends. This individual is said to be "right censored" at 365 days. We know only that he or she lived *at least* 365 days. Although we do not know how long past 365 days the individual actually lived, the censored observation is still informative. This is illustrated in Figure 18.11. Figure 18.12 shows the survival curve estimated by the Kaplan–Meier method for the 80 patients in the test set. See for example Kalbfleisch and Prentice (1980) for a description of the Kaplan–Meier method.

Our objective in this example is to find a set of features (genes) that can predict the survival of an independent set of patients. This could be

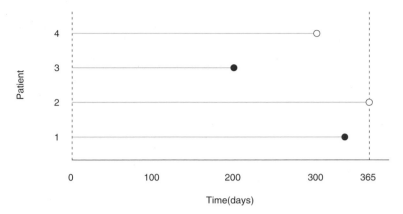

FIGURE 18.11. *Censored survival data. For illustration there are four patients. The first and third patients die before the study ends. The second patient is alive at the end of the study (365 days), while the fourth patient is lost to follow-up before the study ends. For example, this patient might have moved out of the country. The survival times for patients two and four are said to be "censored."*

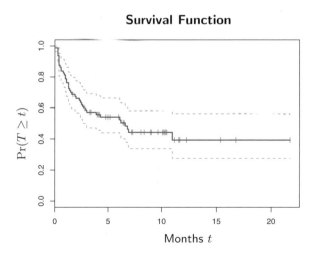

FIGURE 18.12. *Lymphoma data. The Kaplan–Meier estimate of the survival function for the 80 patients in the test set, along with one-standard-error curves. The curve estimates the probability of surviving past t months. The ticks indicate censored observations.*

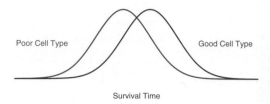

FIGURE 18.13. *Underlying conceptual model for supervised principal components. There are two cell types, and patients with the good cell type live longer on the average. Supervised principal components estimate the cell type, by averaging the expression of genes that reflect it.*

useful as a prognostic indicator to aid in choosing treatments, or to help understand the biological basis for the disease.

The underlying conceptual model for supervised principal components is shown in Figure 18.13. We imagine that there are two cell types, and patients with the good cell type live longer on the average. However there is considerable overlap in the two sets of survival times. We might think of survival time as a "noisy surrogate" for cell type. A fully supervised approach would give the most weight to those genes having the strongest relationship with survival. These genes are partially, but not perfectly, related to cell type. If we could instead discover the underlying cell types of the patients, often reflected by a sizable signature of genes acting together in pathways, then we might do a better job of predicting patient survival.

Although the cell type in Figure 18.13 is discrete, it is useful to imagine a continuous cell type, define by some linear combination of the features. We will estimate the cell type as a continuous quantity, and then discretize it for display and interpretation.

How can we find the linear combination that defines the important underlying cell types? Principal components analysis (Section 14.5) is an effective method for finding linear combinations of features that exhibit large variation in a dataset. But what we seek here are linear combinations with both high variance *and* significant correlation with the outcome. The lower right panel of Figure 18.14 shows the result of applying standard principal components in this example; the leading component does not correlate strongly with survival (details are given in the figure caption).

Hence we want to encourage principal component analysis to find linear combinations of features that have high correlation with the outcome. To do this, we restrict attention to features which by themselves have a sizable correlation with the outcome. This is summarized in the *supervised principal components* Algorithm 18.1, and illustrated in Figure 18.14.

The details in steps (1) and (2b) will depend on the type of outcome variable. For a standard regression problem, we use the univariate linear least squares coefficients in step (1) and a linear least squares model in

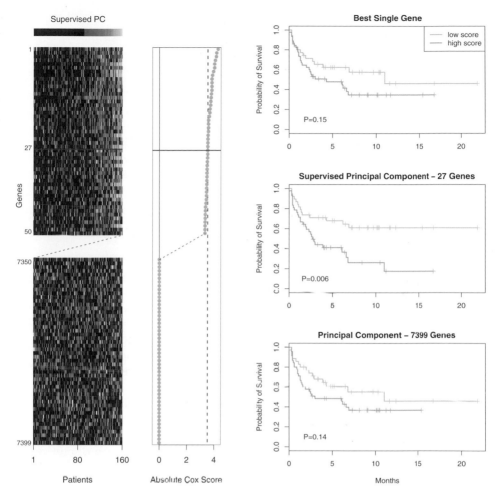

FIGURE 18.14. *Supervised principal components on the lymphoma data. The left panel shows a heatmap of a subset of the gene-expression training data. The rows are ordered by the magnitude of the univariate Cox-score, shown in the middle vertical column. The top 50 and bottom 50 genes are shown. The supervised principal component uses the top 27 genes (chosen by 10-fold CV). It is represented by the bar at the top of the heatmap, and is used to order the columns of the expression matrix. In addition, each row is multiplied by the sign of the Cox-score. The middle panel on the right shows the survival curves on the test data when we create a low and high group by splitting this supervised PC at zero (training data mean). The curves are well separated, as indicated by the p-value for the log-rank test. The top panel does the same, using the top-scoring gene on the training data. The curves are somewhat separated, but not significantly. The bottom panel uses the first principal component on all the genes, and the separation is also poor. Each of the top genes can be interpreted as noisy surrogates for a latent underlying cell-type characteristic, and supervised principal components uses them all to estimate this latent factor.*

Algorithm 18.1 *Supervised Principal Components.*

---

1. Compute the standardized univariate regression coefficients for the outcome as a function of each feature separately.

2. For each value of the threshold θ from the list $0 \le \theta_1 < \theta_2 < \cdots < \theta_K$:

   (a) Form a reduced data matrix consisting of only those features whose univariate coefficient exceeds θ in absolute value, and compute the first m principal components of this matrix.

   (b) Use these principal components in a regression model to predict the outcome.

3. Pick θ (and m) by cross-validation.

---

step (2b). For survival problems, Cox's proportional hazards regression model is widely used; hence we use the score test from this model in step (1) and the multivariate Cox model in step (2b). The details are not essential for understanding the basic method; they may be found in Bair et al. (2006).

Figure 18.14 shows the results of supervised principal components in this example. We used a Cox-score cutoff of 3.53, yielding 27 genes, where the value 3.53 was found through 10-fold cross-validation. We then computed the first principal component ($m = 1$) using just this subset of the data, as well as its value for each of the test observations. We included this as a quantitative predictor in a Cox regression model, and its likelihood-ratio significance was $p = 0.005$. When dichotomized (using the mean score on the training data as a threshold), it clearly separates the patients in the test set into low and high risk groups (middle-right panel of Figure 18.14, $p = 0.006$).

The top-right panel of Figure 18.14 uses the top scoring gene (dichotomized) alone as a predictor of survival. It is not significant on the test set. Likewise, the lower-right panel shows the dichotomized principal component using all the training data, which is also not significant.

Our procedure allows $m > 1$ principal components in step (2a). However, the supervision in step (1) encourages the principal components to align with the outcome, and thus in most cases only the first or first few components tend to be useful for prediction. In the mathematical development below, we consider only the first component, but extensions to more than one component can be derived in a similar way.

18.6.1 Connection to Latent-Variable Modeling

A formal connection between supervised principal components and the underlying cell type model (Figure 18.13) can be seen through a latent variable model for the data. Suppose we have a response variable Y which is related to an underlying latent variable U by a linear model

$$Y = \beta_0 + \beta_1 U + \varepsilon. \tag{18.32}$$

In addition, we have measurements on a set of features X_j indexed by $j \in \mathcal{P}$ (for pathway), for which

$$X_j = \alpha_{0j} + \alpha_{1j}U + \epsilon_j, \quad j \in \mathcal{P}. \tag{18.33}$$

The errors ε and ϵ_j are assumed to have mean zero and are independent of all other random variables in their respective models.

We also have many additional features X_k, $k \notin \mathcal{P}$ which are independent of U. We would like to identify $\mathcal{P}$, estimate U, and hence fit the prediction model (18.32). This is a special case of a latent-structure model, or single-component factor-analysis model (Mardia et al., 1979, see also Section 14.7). The latent factor U is a continuous version of the cell type conceptualized in Figure 18.13.

The supervised principal component algorithm can be seen as a method for fitting this model:

- The screening step (1) estimates the set $\mathcal{P}$.

- Given $\widehat{\mathcal{P}}$, the largest principal component in step (2a) estimates the latent factor U.

- Finally, the regression fit in step (2b) estimates the coefficient in model (18.32).

Step (1) is natural, since on average the regression coefficient is nonzero only if α_{1j} is non-zero. Hence this step should select the features $j \in \mathcal{P}$. Step (2a) is natural if we assume that the errors ϵ_j have a Gaussian distribution, with the same variance. In this case the principal component is the maximum likelihood estimate for the single factor model (Mardia et al., 1979). The regression in (2b) is an obvious final step.

Suppose there are a total of p features, with p_1 features in the relevant set $\mathcal{P}$. Then if p and p_1 grow but p_1 is small relative to p, one can show (under reasonable conditions) that the leading supervised principal component is consistent for the underlying latent factor. The usual leading principal component may not be consistent, since it can be contaminated by the presence of a large number of "noise" features.

Finally, suppose that the threshold used in step (1) of the supervised principal component procedure yields a large number of features for computation of the principal component. Then for interpretational purposes, as well as for practical uses, we would like some way of finding a reduced a set of features that approximates the model. Pre-conditioning (Section 18.6.3) is one way of doing this.

18.6.2 Relationship with Partial Least Squares

Supervised principal components is closely related to partial least squares regression (Section 3.5.2). Bair et al. (2006) found that the key to the good performance of supervised principal components was the filtering out of noisy features in step (2a). Partial least squares (Section 3.5.2) downweights noisy features, but does not throw them away; as a result a large number of noisy features can contaminate the predictions. However, a modification of the partial least squares procedure has been proposed that has a similar flavor to supervised principal components [Brown et al. (1991),Nadler and Coifman (2005), for example]. We select the features as in steps (1) and (2a) of supervised principal components, but then apply PLS (rather than principal components) to these features. For our current discussion, we call this "thresholded PLS."

Thresholded PLS can be viewed as a noisy version of supervised principal components, and hence we might not expect it to work as well in practice. Assume the variables are all standardized. The first PLS variate has the form

$$\mathbf{z} = \sum_{j \in \mathcal{P}} \langle \mathbf{y}, \mathbf{x}_j \rangle \mathbf{x}_j, \qquad (18.34)$$

and can be thought of as an estimate of the latent factor U in model (18.33). In contrast, the supervised principal components direction $\hat{\mathbf{u}}$ satisfies

$$\hat{\mathbf{u}} = \frac{1}{d^2} \sum_{j \in \mathcal{P}} \langle \hat{\mathbf{u}}, \mathbf{x}_j \rangle \mathbf{x}_j, \qquad (18.35)$$

where d is the leading singular value of $\mathbf{X}_{\mathcal{P}}$. This follows from the definition of the leading principal component. Hence thresholded PLS uses weights which are the inner product of $\mathbf{y}$ with each of the features, while supervised principal components uses the features to derive a "self-consistent" estimate $\hat{\mathbf{u}}$. Since many features contribute to the estimate $\hat{\mathbf{u}}$, rather than just the single outcome $\mathbf{y}$, we can expect $\hat{\mathbf{u}}$ to be less noisy than $\mathbf{z}$. In fact, if there are p_1 features in the set $\mathcal{P}$, and N, p and p_1 go to infinity with $p_1/N \to 0$, then it can be shown using the techniques in Bair et al. (2006) that

$$\begin{aligned} \mathbf{z} &= \mathbf{u} + O_p(1) \\ \hat{\mathbf{u}} &= \mathbf{u} + O_p(\sqrt{p_1/N}), \end{aligned} \qquad (18.36)$$

where $\mathbf{u}$ is the true (unobservable) latent variable in the model (18.32), (18.33).

We now present a simulation example to compare the methods numerically. There are $N = 100$ samples and $p = 5000$ genes. We generated the data as follows:

FIGURE 18.15. *Heatmap of the outcome (left column) and first 500 genes from a realization from model (18.37). The genes are in the columns, and the samples are in the rows.*

$$
\begin{aligned}
x_{ij} &= \begin{cases} 3 + \epsilon_{ij} & \text{if } i \le 50, \\ 4 + \epsilon_{ij} & \text{if } i > 50 \end{cases} & j = 1, \dots, 50 \\[2mm]
x_{ij} &= \begin{cases} 1.5 + \epsilon_{ij} & \text{if } 1 \le i \le 25 \text{ or } 51 \le i \le 75 \\ 5.5 + \epsilon_{ij} & \text{if } 26 \le i \le 50 \text{ or } 76 \le i \le 100 \end{cases} & j = 51, \dots, 250 \\[2mm]
x_{ij} &= \epsilon_{ij} & j = 251, \dots, 5000 \\[1mm]
y_i &= 2 \cdot \tfrac{1}{50} \sum_{j=1}^{50} x_{ij} + \varepsilon_i &
\end{aligned}
$$

$$(18.37)$$

where ϵ_{ij} and ε_i are independent normal random variables with mean 0 and standard deviations 1 and 1.5, respectively. Thus in the first 50 genes, there is an average difference of 1 unit between samples 1–50 and 51–100, and this difference correlates with the outcome y. The next 200 genes have a large average difference of 4 units between samples (1–25, 51–75) and (26–50, 76–100), but this difference is uncorrelated with the outcome. The rest of the genes are noise. Figure 18.15 shows a heatmap of a typical realization, with the outcome at the left, and the first 500 genes to the right.

We generated 100 simulations from this model, and summarize the test error results in Figure 18.16. The test errors of principal components and partial least squares are shown at the right of the plot; both are badly affected by the noisy features in the data. Supervised principal components and thresholded PLS work best over a wide range of the number of selected features, with the former showing consistently lower test errors.

While this example seems "tailor-made" for supervised principal components, its good performance seems to hold in other simulated and real datasets (Bair et al., 2006).

18.6.3 Pre-Conditioning for Feature Selection

Supervised principal components can yield lower test errors than competing methods, as shown in Figure 18.16. However, it does not always produce a sparse model involving only a small number of features (genes). Even if the thresholding in Step (1) of the algorithm yields a relatively small number

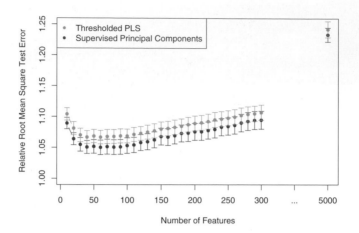

FIGURE 18.16. *Root mean squared test error ($\pm$ one standard error), for supervised principal components and thresholded PLS on* 100 *realizations from model (18.37). All methods use one component, and the errors are relative to the noise standard deviation (the Bayes error is* 1.0*). For both methods, different values for the filtering threshold were tried and the number of features retained is shown on the horizontal axis. The extreme right points correspond to regular principal components and partial least squares, using all the genes.*

of features, it may be that some of the omitted features have sizable inner products with the supervised principal component (and could act as a good surrogate). In addition, highly correlated features will tend to be chosen together, and there may be great deal of redundancy in the set of selected features.

The lasso (Sections 18.4 and 3.4.2), on the other hand, produces a sparse model from the data. How do the test errors of the two methods compare on the simulated example of the last section? Figure 18.17 shows the test errors for one realization from model (18.37) for the lasso, supervised principal components, and the pre-conditioned lasso (described below).

We see that supervised principal components (orange curve) reaches its lowest error when about 50 features are included in the model, which is the correct number for the simulation. Although a linear model in the first 50 features is optimal, the lasso (green) is adversely affected by the large number of noisy features, and starts overfitting when far fewer are in the model.

Can we get the low test error of supervised principal components along with the sparsity of the lasso? This is the goal of *pre-conditioning* (Paul et al., 2008). In this approach, one first computes the supervised principal component predictor $\hat{y}_i$ for each observation in the training set (with the

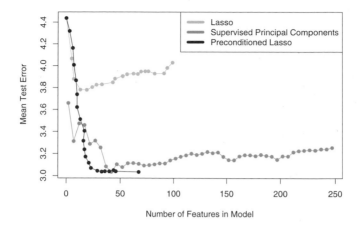

FIGURE 18.17. *Test errors for the lasso, supervised principal components, and pre-conditioned lasso, for one realization from model (18.37). Each model is indexed by the number of non-zero features. The supervised principal component path is truncated at 250 features. The lasso self truncates at 100, the sample size (see Section 18.4). In this case, the pre-conditioned lasso achieves the lowest error with about 25 features.*

threshold selected by cross-validation). Then we apply the lasso with $\hat{y}_i$ as the outcome variable, in place of the usual outcome y_i. All features are used in the lasso fit, not just those that were retained in the thresholding step in supervised principal components. The idea is that by first denoising the outcome variable, the lasso should not be as adversely affected by the large number of noise features. Figure 18.17 shows that pre-conditioning (purple curve) has been successful here, yielding much lower test error than the usual lasso, and as low (in this case) as for supervised principal components. It also can achieve this using less features. The usual lasso, applied to the raw outcome, starts to overfit more quickly than the pre-conditioned version. Overfitting is not a problem, since the outcome variable has been denoised. We usually select the tuning parameter for the pre-conditioned lasso on more subjective grounds, like parsimony.

Pre-conditioning can be applied in a variety of settings, using initial estimates other than supervised principal components and post-processors other than the lasso. More details may be found in Paul et al. (2008).

18.7 Feature Assessment and the Multiple-Testing Problem

In the first part of this chapter we discuss prediction models in the $p \gg N$ setting. Here we consider the more basic problem of assessing the significance of each of the p features. Consider the protein mass spectrometry example of Section 18.4.1. In that problem, the scientist might not be interested in predicting whether a given patient has prostate cancer. Rather the goal might be to identify proteins whose abundance differs between normal and cancer samples, in order to enhance understanding of the disease and suggest targets for drug development. Thus our goal is to assess the significance of individual features. This assessment is usually done without the use of a multivariate predictive model like those in the first part of this chapter. The feature assessment problem moves our focus from prediction to the traditional statistical topic of *multiple hypothesis testing*. For the remainder of this chapter we will use M instead of p to denote the number of features, since we will frequently be referring to *p-values*.

TABLE 18.4. *Subset of the* 12,625 *genes from microarray study of radiation sensitivity. There are a total of* 44 *samples in the normal group and* 14 *in the radiation sensitive group; we only show three samples from each group.*

	Normal				Radiation Sensitive			
Gene 1	7.85	29.74	29.50	...	17.20	-50.75	-18.89	...
Gene 2	15.44	2.70	19.37	...	6.57	-7.41	79.18	...
Gene 3	-1.79	15.52	-3.13	...	-8.32	12.64	4.75	...
Gene 4	-11.74	22.35	-36.11	...	-52.17	7.24	-2.32	...
$\vdots$	$\vdots$	$\vdots$	$\vdots$	$\vdots$	$\vdots$	$\vdots$	$\vdots$	$\vdots$
Gene 12,625	-14.09	32.77	57.78	...	-32.84	24.09	-101.44	...

Consider, for example, the microarray data in Table 18.4, taken from a study on the sensitivity of cancer patients to ionizing radiation treatment (Rieger et al., 2004). Each row consists of the expression of genes in 58 patient samples: 44 samples were from patients with a normal reaction, and 14 from patients who had a severe reaction to radiation. The measurements were made on oligo-nucleotide microarrays. The object of the experiment was to find genes whose expression was different in the radiation sensitive group of patients. There are $M = 12,625$ genes altogether; the table shows the data for some of the genes and samples for illustration.

To identify informative genes, we construct a two-sample t-statistic for each gene.

$$t_j = \frac{\bar{x}_{2j} - \bar{x}_{1j}}{se_j}, \tag{18.38}$$

where $\bar{x}_{kj} = \sum_{i \in C_\ell} x_{ij}/N_\ell$. Here C_ℓ are the indices of the N_ℓ samples in group ℓ, where $\ell = 1$ is the normal group and $\ell = 2$ is the sensitive group. The quantity se_j is the pooled within-group standard error for gene j:

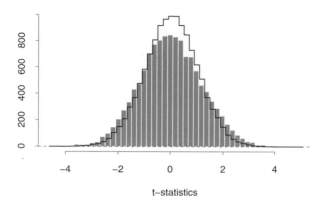

FIGURE 18.18. *Radiation sensitivity microarray example. A histogram of the 12,625 t-statistics comparing the radiation-sensitive versus insensitive groups. Overlaid in blue is the histogram of the t-statistics from 1000 permutations of the sample labels.*

$$\text{se}_j = \hat{\sigma}_j \sqrt{\tfrac{1}{N_1} + \tfrac{1}{N_2}}; \quad \hat{\sigma}_j^2 = \tfrac{1}{N_1+N_2-2}\left(\sum_{i\in C_1}(x_{ij}-\bar{x}_{1j})^2 + \sum_{i\in C_2}(x_{ij}-\bar{x}_{2j})^2\right).$$

$$(18.39)$$

A histogram of the 12,625 t-statistics is shown in orange in Figure 18.18, ranging in value from -4.7 to 5.0. If the t_j values were normally distributed we could consider any value greater than two in absolute value to be significantly large. This would correspond to a significance level of about 5%. Here there are 1189 genes with $|t_j| \geq 2$. However with 12,625 genes we would expect many large values to occur by chance, even if the grouping is unrelated to any gene. For example, if the genes were independent (which they are surely not), the number of falsely significant genes would have a binomial distribution with mean $12,625 \cdot 0.05 = 631.3$ and standard deviation 24.5; the actual 1189 is way out of range.

How do we assess the results for all 12,625 genes? This is called the *multiple testing* problem. We can start as above by computing a p-value for each gene. This can be done using the theoretical t-distribution probabilities, which assumes the features are normally distributed. An attractive alternative approach is to use the permutation distribution, since it avoids assumptions about the distribution of the data. We compute (in principle) all $K = \binom{58}{14}$ permutations of the sample labels, and for each permutation k compute the t-statistics t_j^k. Then the p-value for gene j is

$$p_j = \frac{1}{K} \sum_{k=1}^{K} I(|t_j^k| > |t_j|). \qquad (18.40)$$

Of course, $\binom{58}{14}$ is a large number (around 10^{13}) and so we can't enumerate all of the possible permutations. Instead we take a random sample of the possible permutations; here we took a random sample of $K = 1000$ permutations.

To exploit the fact that the genes are similar (e.g., measured on the same scale), we can instead pool the results for all genes in computing the p-values.

$$p_j = \frac{1}{MK} \sum_{j'=1}^{M} \sum_{k=1}^{K} I(|t_{j'}^k| > |t_j|). \qquad (18.41)$$

This also gives more granular p-values than does (18.40), since there many more values in the pooled null distribution than there are in each individual null distribution.

Using this set of p-values, we would like to test the hypotheses:

$$
\begin{aligned}
H_{0j} &= \quad \text{treatment has no effect on gene } j \\
&\qquad\qquad\qquad versus \qquad\qquad\qquad\qquad\qquad (18.42) \\
H_{1j} &= \quad \text{treatment has an effect on gene } j
\end{aligned}
$$

for all $j = 1, 2, \ldots, M$. We reject H_{0j} at level α if $p_j < \alpha$. This test has type-I error equal to α; that is, the probability of falsely rejecting H_{0j} is α.

Now with many tests to consider, it is not clear what we should use as an overall measure of error. Let A_j be the event that H_{0j} is falsely rejected; by definition $\Pr(A_j) = \alpha$. The *family-wise error rate* (FWER) is the probability of at least one false rejection, and is a commonly used overall measure of error. In detail, if $A = \cup_{j=1}^{M} A_j$ is the event of at least one false rejection, then the FWER is $\Pr(A)$. Generally $\Pr(A) \gg \alpha$ for large M, and depends on the correlation between the tests. If the tests are independent each with type-I error rate α, then the family-wise error rate of the collection of tests is $(1 - (1 - \alpha)^M)$. On the other hand, if the tests have positive dependence, that is $\Pr(A_j|A_k) > \Pr(A_j)$, then the FWER will be less than $(1 - (1 - \alpha)^M)$. Positive dependence between tests often occurs in practice, in particular in genomic studies.

One of the simplest approaches to multiple testing is the *Bonferroni* method. It makes each individual test more stringent, in order to make the FWER equal to at most α: we reject H_{0j} if $p_j < \alpha/M$. It is easy to show that the resulting FWER is $\leq \alpha$ (Exercise 18.16). The Bonferroni method can be useful if M is relatively small, but for large M it is too conservative, that is, it calls too few genes significant.

In our example, if we test at level say $\alpha = 0.05$, then we must use the threshold $0.05/12,625 = 3.9 \times 10^{-6}$. None of the $12,625$ genes had a p-value this small.

There are variations to this approach that adjust the individual p-values to achieve an FWER of at most α, with some approaches avoiding the assumption of independence; see, e.g., Dudoit et al. (2002b).

18.7.1 The False Discovery Rate

A different approach to multiple testing does not try to control the FWER, but focuses instead on the proportion of falsely significant genes. As we will see, this approach has a strong practical appeal.

Table 18.5 summarizes the theoretical outcomes of M hypothesis tests. Note that the family-wise error rate is $\Pr(V \geq 1)$. Here we instead focus

TABLE 18.5. *Possible outcomes from M hypothesis tests. Note that V is the number of false-positive tests; the type-I error rate is $\mathrm{E}(V)/M_0$. The type-II error rate is $\mathrm{E}(T)/M_1$, and the power is $1 - \mathrm{E}(T)/M_1$.*

	Called Not Significant	Called Significant	Total
H_0 True	U	V	M_0
H_0 False	T	S	M_1
Total	$M - R$	R	M

on the *false discovery rate*

$$\mathrm{FDR} = \mathrm{E}(V/R). \tag{18.43}$$

In the microarray setting, this is the expected proportion of genes that are incorrectly called significant, among the R genes that are called significant. The expectation is taken over the population from which the data are generated. Benjamini and Hochberg (1995) first proposed the notion of false discovery rate, and gave a testing procedure (Algorithm 18.2) whose FDR is bounded by a user-defined level α. The Benjamini–Hochberg (BH) procedure is based on p-values; these can be obtained from an asymptotic approximation to the test statistic (e.g., Gaussian), or a permutation distribution, as is done here.

If the hypotheses are independent, Benjamini and Hochberg (1995) show that regardless of how many null hypotheses are true and regardless of the distribution of the p-values when the null hypothesis is false, this procedure has the property

$$\mathrm{FDR} \leq \frac{M_0}{M}\alpha \leq \alpha. \tag{18.45}$$

For illustration we chose $\alpha = 0.15$. Figure 18.19 shows a plot of the ordered p-values $p_{(j)}$, and the line with slope $0.15/12625$.

Algorithm 18.2 *Benjamini–Hochberg (BH) Method.*

---

1. Fix the false discovery rate α and let $p_{(1)} \le p_{(2)} \le \cdots \le p_{(M)}$ denote the ordered p-values

2. Define

$$L = \max\left\{j : p_{(j)} < \alpha \cdot \frac{j}{M}\right\}. \qquad (18.44)$$

3. Reject all hypotheses H_{0j} for which $p_j \le p_{(L)}$, the BH rejection threshold.

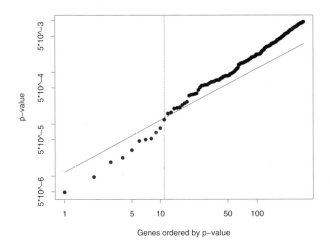

FIGURE 18.19. *Microarray example continued. Shown is a plot of the ordered p-values $p_{(j)}$ and the line $0.15 \cdot (j/12,625)$, for the Benjamini–Hochberg method. The largest j for which the p-value $p_{(j)}$ falls below the line, gives the BH threshold. Here this occurs at $j = 11$, indicated by the vertical line. Thus the BH method calls significant the 11 genes (in red) with smallest p-values.*

Algorithm 18.3 *The Plug-in Estimate of the False Discovery Rate.*

---

1. Create K permutations of the data, producing t-statistics t_j^k for features $j = 1, 2, \ldots, M$ and permutations $k = 1, 2, \ldots, K$.

2. For a range of values of the cut-point C, let

$$R_{\text{obs}} = \sum_{j=1}^{M} I(|t_j| > C), \quad \widehat{\text{E}(V)} = \frac{1}{K} \sum_{j=1}^{M} \sum_{k=1}^{K} I(|t_j^k| > C). \quad (18.46)$$

3. Estimate the FDR by $\widehat{\text{FDR}} = \widehat{\text{E}(V)}/R_{\text{obs}}$.

Starting at the left and moving right, the BH method finds the last time that the p-values fall below the line. This occurs at $j = 11$, so we reject the 11 genes with smallest p-values. Note that the cutoff occurs at the 11th smallest p-value, 0.00012, and the 11th largest of the values $|t_j|$ is 4.101 Thus we reject the 11 genes with $|t_j| \geq 4.101$.

From our brief description, it is not clear how the BH procedure works; that is, why the corresponding FDR is at most 0.15, the value used for α. Indeed, the proof of this fact is quite complicated (Benjamini and Hochberg, 1995).

A more direct way to proceed is a *plug-in* approach. Rather than starting with a value for α, we fix a cut-point for our t-statistics, say the value 4.101 that appeared above. The number of observed values $|t_j|$ equal or greater than 4.101 is 11. The total number of permutation values $|t_j^k|$ equal or greater than 4.101 is 1518, for an average of $1518/1000 = 1.518$ per permutation. Thus a direct estimate of the false discovery rate is $\widehat{\text{FDR}} = 1.518/11 \approx 14\%$. Note that 14% is approximately equal to the value of $\alpha = 0.15$ used above (the difference is due to discreteness). This procedure is summarized in Algorithm 18.3. To recap:

The plug-in estimate of FDR of Algorithm 18.3 is equivalent to the BH procedure of Algorithm 18.2, using the permutation p-values (18.40).

This correspondence between the BH method and the plug-in estimate is not a coincidence. Exercise 18.17 shows that they are equivalent in general. Note that this procedure makes no reference to p-values at all, but rather works directly with the test statistics.

The plug-in estimate is based on the approximation

$$\text{E}(V/R) \approx \frac{\text{E}(V)}{\text{E}(R)}, \quad (18.47)$$

and in general $\widehat{\text{FDR}}$ is a consistent estimate of FDR (Storey, 2002; Storey et al., 2004). Note that the numerator $\widehat{\text{E}(V)}$ actually estimates $(M/M_0)\text{E}(V)$, since the permutation distribution uses M rather M_0 null hypotheses. Hence if an estimate of M_0 is available, a better estimate of FDR can be obtained from $(\hat{M}_0/M) \cdot \widehat{\text{FDR}}$. Exercise 18.19 shows a way to estimate M_0. The most conservative (upwardly biased) estimate of FDR uses $M_0 = M$. Equivalently, an estimate of M_0 can be used to improve the BH method, through relation (18.45).

The reader might be surprised that we chose a value as large as 0.15 for α, the FDR bound. We must remember that the FDR is not the same as type-I error, for which 0.05 is the customary choice. For the scientist, the false discovery rate is the expected proportion of false positive genes among the list of genes that the statistician tells him are significant. Microarray experiments with FDRs as high as 0.15 might still be useful, especially if they are exploratory in nature.

18.7.2 Asymmetric Cutpoints and the SAM Procedure

In the testing methods described above, we used the absolute value of the test statistic t_j, and hence applied the same cut-points to both positive and negative values of the statistic. In some experiments, it might happen that most or all of the differentially expressed genes change in the positive direction (or all in the negative direction). For this situation it is advantageous to derive separate cut-points for the two cases.

The *significance analysis of microarrays* (SAM) approach offers a way of doing this. The basis of the SAM method is shown in Figure 18.20. On the vertical axis we have plotted the ordered test statistics $t_{(1)} \leq t_{(2)} \leq \cdots \leq t_{(M)}$, while the horizontal axis shows the expected order statistics from the permutations of the data: $\tilde{t}_{(j)} = (1/K) \sum_{k=1}^{K} t_{(j)}^k$, where $t_{(1)}^k \leq t_{(2)}^k \leq \cdots \leq t_{(M)}^k$ are the ordered test statistics from permutation k.

Two lines are drawn, parallel to the 45° line, Δ units away. Starting at the origin and moving to the right, we find the first place that the genes leave the band. This defines the upper cutpoint $\mathtt{C_{hi}}$ and all genes beyond that point are called significant (marked red). Similarly we find the lower cutpoint $\mathtt{C_{low}}$ for genes in the bottom left corner. Thus each value of the tuning parameter Δ defines upper and lower cutpoints, and the plug-in estimate $\widehat{\text{FDR}}$ for each of these cutpoints is estimated as before. Typically a range of values of Δ and associated $\widehat{\text{FDR}}$ values are computed, from which a particular pair are chosen on subjective grounds.

The advantage of the SAM approach lies in the possible asymmetry of the cutpoints. In the example of Figure 18.20, with $\Delta = 0.71$ we obtain 11 significant genes; they are all in the upper right. The data points in the bottom left never leave the band, and hence $\mathtt{C_{low}} = -\infty$. Hence for this value of Δ, no genes are called significant on the left (negative) side. We do not impose symmetry on the cutpoints, as was done in Section 18.7.1, as there is no reason to assume similar behavior at the two ends.

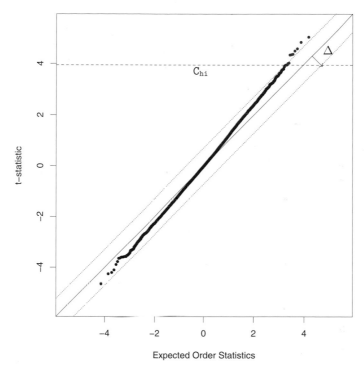

FIGURE 18.20. *SAM plot for the radiation sensitivity microarray data. On the vertical axis we have plotted the ordered test statistics, while the horizontal axis shows the expected order statistics of the test statistics from permutations of the data. Two lines are drawn, parallel to the 45° line, Δ units away from it. Starting at the origin and moving to the right, we find the first place that the genes leave the band. This defines the upper cut-point C_{hi} and all genes beyond that point are called significant (marked in red). Similarly we define a lower cutpoint C_{low}. For the particular value of $\Delta = 0.71$ in the plot, no genes are called significant in the bottom left.*

There is some similarity between this approach and the asymmetry possible with likelihood-ratio tests. Suppose we have a log-likelihood $\ell_0(t_j)$ under the null-hypothesis of no effect, and a log-likelihood $\ell(t_j)$ under the alternative. Then a likelihood ratio test amounts to rejecting the null-hypothesis if

$$\ell(t_j) - \ell_0(t_j) > \Delta, \qquad (18.48)$$

for some Δ. Depending on the likelihoods, and particularly their relative values, this can result in a different threshold for t_j than for $-t_j$. The SAM procedure rejects the null-hypothesis if

$$|t_{(j)} - \tilde{t}_{(j)}| > \Delta \qquad (18.49)$$

Again, the threshold for each $t_{(j)}$ depends on the corresponding value of the null value $\tilde{t}_{(j)}$.

18.7.3 A Bayesian Interpretation of the FDR

There is an interesting Bayesian view of the FDR, developed in Storey (2002) and Efron and Tibshirani (2002). First we need to define the *positive false discovery rate* (pFDR) as

$$\text{pFDR} = \mathrm{E}\left[\frac{V}{R} \middle| R > 0 \right]. \qquad (18.50)$$

The additional term *positive* refers to the fact that we are only interested in estimating an error rate where positive findings have occurred. It is this slightly modified version of the FDR that has a clean Bayesian interpretation. Note that the usual FDR [expression (18.43)] is not defined if $\Pr(R = 0) > 0$.

Let Γ be a rejection region for a single test; in the example above we used $\Gamma = (-\infty, -4.10) \cup (4.10, \infty)$. Suppose that M identical simple hypothesis tests are performed with the i.i.d. statistics $t_1, \ldots, t_M$ and rejection region Γ. We define a random variable Z_j which equals 0 if the jth null hypothesis is true, and 1 otherwise. We assume that each pair (t_j, Z_j) are i.i.d random variables with

$$t_j | Z_j \sim (1 - Z_j) \cdot F_0 + Z_j \cdot F_1 \qquad (18.51)$$

for some distributions F_0 and F_1. This says that each test statistic t_j comes from one of two distributions: F_0 if the null hypothesis is true, and F_1 otherwise. Letting $\Pr(Z_j = 0) = \pi_0$, marginally we have:

$$t_j \sim \pi_0 \cdot F_0 + (1 - \pi_0) \cdot F_1. \qquad (18.52)$$

Then it can be shown (Efron et al., 2001; Storey, 2002) that

$$\mathrm{pFDR}(\Gamma) = \Pr(Z_j = 0 | t_j \in \Gamma). \tag{18.53}$$

Hence under the mixture model (18.51), the pFDR is the posterior probability that the null hypothesis it true, given that test statistic falls in the rejection region for the test; that is, given that we reject the null hypothesis (Exercise 18.20).

The false discovery rate provides a measure of accuracy for tests based on an entire rejection region, such as $|t_j| \geq 2$. But if the FDR of such a test is say 10%, then a gene with say $t_j = 5$ will be more significant than a gene with $t_j = 2$. Thus it is of interest to derive a local (gene-specific) version of the FDR. The *q-value* (Storey, 2003) of a test statistic t_j is defined to be the smallest FDR over all rejection regions that reject t_j. That is, for symmetric rejection regions, the q-value for $t_j = 2$ is defined to be the FDR for the rejection region $\Gamma = \{-(\infty, -2) \cup (2, \infty)\}$. Thus the q-value for $t_j = 5$ will be smaller than that for $t_j = 2$, reflecting the fact that $t_j = 5$ is more significant than $t_j = 2$. The *local false discovery rate* (Efron and Tibshirani, 2002) at $t = t_0$ is defined to be

$$\Pr(Z_j = 0 | t_j = t_0). \tag{18.54}$$

This is the (positive) FDR for an infinitesimal rejection region surrounding the value $t_j = t_0$.

18.8 Bibliographic Notes

Many references were given at specific points in this chapter; we give some additional ones here. Dudoit et al. (2002a) give an overview and comparison of discrimination methods for gene expression data. Levina (2002) does some mathematical analysis comparing diagonal LDA to full LDA, as $p, N \to \infty$ with $p > N$. She shows that with reasonable assumptions diagonal LDA has a lower asymptotic error rate than full LDA. Tibshirani et al. (2001a) and Tibshirani et al. (2003) proposed the nearest shrunken-centroid classifier. Zhu and Hastie (2004) study regularized logistic regression. High-dimensional regression and the lasso are very active areas of research, and many references are given in Section 3.8.5. The fused lasso was proposed by Tibshirani et al. (2005), while Zou and Hastie (2005) introduced the elastic net. Supervised principal components is discussed in Bair and Tibshirani (2004) and Bair et al. (2006). For an introduction to the analysis of censored survival data, see Kalbfleisch and Prentice (1980).

Microarray technology has led to a flurry of statistical research: see for example the books by Speed (2003), Parmigiani et al. (2003), Simon et al. (2004), and Lee (2004).

The false discovery rate was proposed by Benjamini and Hochberg (1995), and studied and generalized in subsequent papers by these authors and many others. A partial list of papers on FDR may be found on Yoav Benjamini's homepage. Some more recent papers include Efron and Tibshirani (2002), Storey (2002), Genovese and Wasserman (2004), Storey and Tibshirani (2003) and Benjamini and Yekutieli (2005). Dudoit et al. (2002b) review methods for identifying differentially expressed genes in microarray studies.

Exercises

Ex. 18.1 For a coefficient estimate $\hat{\beta}_j$, let $\hat{\beta}_j / ||\hat{\beta}_j||_2$ be the normalized version. Show that as $\lambda \to \infty$, the normalized ridge-regression estimates converge to the renormalized partial-least-squares one-component estimates.

Ex. 18.2 *Nearest shrunken centroids and the lasso.* Consider a (naive Bayes) Gaussian model for classification in which the features $j = 1, 2, \ldots, p$ are assumed to be independent within each class $k = 1, 2, \ldots, K$. With observations $i = 1, 2, \ldots, N$ and C_k equal to the set of indices of the N_k observations in class k, we observe $x_{ij} \sim N(\mu_j + \mu_{jk}, \sigma_j^2)$ for $i \in C_k$ with $\sum_{k=1}^{K} \mu_{jk} = 0$. Set $\hat{\sigma}_j^2 = s_j^2$, the pooled within-class variance for feature j, and consider the lasso-style minimization problem

$$\min_{\{\mu_j, \mu_{jk}\}} \left\{ \frac{1}{2} \sum_{j=1}^{p} \sum_{k=1}^{K} \sum_{i \in C_k} \frac{(x_{ij} - \mu_j - \mu_{jk})^2}{s_j^2} + \lambda \sqrt{N_k} \sum_{j=1}^{p} \sum_{k=1}^{K} \frac{|\mu_{jk}|}{s_j} \right\}. \quad (18.55)$$

Show that the solution is equivalent to the nearest shrunken centroid estimator (18.7), with s_0 set to zero, and m_k^2 equal to $1/N_k$ instead of $1/N_k - 1/N$ as before.

Ex. 18.3 Show that the fitted coefficients for the regularized multiclass logistic regression problem (18.10) satisfy $\sum_{k=1}^{K} \hat{\beta}_{kj} = 0$, $j = 1, \ldots, p$. What about the $\hat{\beta}_{k0}$? Discuss issues with these constant parameters, and how they can be resolved.

Ex. 18.4 Derive the computational formula (18.15) for ridge regression. [*Hint:* Use the first derivative of the penalized sum-of-squares criterion to show that if $\lambda > 0$, then $\hat{\beta} = \mathbf{X}^T s$ for some $s \in \mathbb{R}^N$.]

Ex. 18.5 Prove the theorem (18.16)–(18.17) in Section 18.3.5, by decomposing β and the rows of $\mathbf{X}$ into their projections into the column space of $\mathbf{V}$ and its complement in $\mathbb{R}^p$.

Ex. 18.6 Show how the theorem in Section 18.3.5 can be applied to regularized discriminant analysis [Equations 4.14 and (18.9)].

Ex. 18.7 Consider a linear regression problem where $p \gg N$, and assume the rank of $\mathbf{X}$ is N. Let the SVD of $\mathbf{X} = \mathbf{U}\mathbf{D}\mathbf{V}^T = \mathbf{R}\mathbf{V}^T$, where $\mathbf{R}$ is $N \times N$ nonsingular, and $\mathbf{V}$ is $p \times N$ with orthonormal columns.

(a) Show that there are infinitely many least-squares solutions all with zero residuals.

(b) Show that the ridge-regression estimate for β can be written

$$\hat{\beta}_\lambda = \mathbf{V}(\mathbf{R}^T\mathbf{R} + \lambda\mathbf{I})^{-1}\mathbf{R}^T\mathbf{y} \qquad (18.56)$$

(c) Show that when $\lambda = 0$, the solution $\hat{\beta}_0 = \mathbf{V}\mathbf{D}^{-1}\mathbf{U}^T\mathbf{y}$ has residuals all equal to zero, and is unique in that it has the smallest Euclidean norm amongst all zero-residual solutions.

Ex. 18.8 *Data Piling*. Exercise 4.2 shows that the two-class LDA solution can be obtained by a linear regression of a binary response vector $\mathbf{y}$ consisting of -1s and $+1$s. The prediction $\hat{\beta}^T x$ for any x is (up to a scale and shift) the LDA score $\delta(x)$. Suppose now that $p \gg N$.

(a) Consider the linear regression model $f(x) = \alpha + \beta^T x$ fit to a binary response $Y \in \{-1, +1\}$. Using Exercise 18.7, show that there are infinitely many directions defined by $\hat{\beta}$ in $\mathbb{R}^p$ onto which the data project to *exactly two* points, one for each class. These are known as *data piling* directions (Ahn and Marron, 2005).

(b) Show that the distance between the projected points is $2/||\hat{\beta}||$, and hence these directions define separating hyperplanes with that margin.

(c) Argue that there is a single *maximal data piling* direction for which this distance is largest, and is defined by $\hat{\beta}_0 = \mathbf{V}\mathbf{D}^{-1}\mathbf{U}^T\mathbf{y} = \mathbf{X}^-\mathbf{y}$, where $\mathbf{X} = \mathbf{U}\mathbf{D}\mathbf{V}^T$ is the SVD of $\mathbf{X}$.

Ex. 18.9 Compare the data piling direction of Exercise 18.8 to the direction of the optimal separating hyperplane (Section 4.5.2) qualitatively. Which makes the widest margin, and why? Use a small simulation to demonstrate the difference.

Ex. 18.10 When $p \gg N$, linear discriminant analysis (see Section 4.3) is degenerate because the within-class covariance matrix $\mathbf{W}$ is singular. One version of regularized discriminant analysis (4.14) replaces $\mathbf{W}$ by a ridged version $\mathbf{W} + \lambda\mathbf{I}$, leading to a regularized discriminant function $\delta_\lambda(x) = x^T(\mathbf{W} + \lambda\mathbf{I})^{-1}(\bar{x}_1 - \bar{x}_{-1})$. Show that $\delta_0(x) = \lim_{\lambda \downarrow 0} \delta_\lambda(x)$ corresponds to the maximal data piling direction defined in Exercise 18.8.

Ex. 18.11 Suppose you have a sample of N pairs (x_i, y_i), with y_i binary and $x_i \in \mathbb{R}^1$. Suppose also that the two classes are separable; e.g., for each pair i, i' with $y_i = 0$ and $y_{i'} = 1$, $x_{i'} - x_i \geq C$ for some $C > 0$. You wish to fit a linear logistic regression model $\text{logitPr}(Y = 1|X) = \alpha + \beta X$ by maximum-likelihood. Show that $\hat{\beta}$ is undefined.

Ex. 18.12 Suppose we wish to select the ridge parameter λ by 10-fold cross-validation in a $p \gg N$ situation (for any linear model). We wish to use the computational shortcuts described in Section 18.3.5. Show that we need only to reduce the $N \times p$ matrix $\mathbf{X}$ to the $N \times N$ matrix $\mathbf{R}$ *once*, and can use it in all the cross-validation runs.

Ex. 18.13 Suppose our $p > N$ predictors are presented as an $N \times N$ inner-product matrix $\mathbf{K} = \mathbf{X}\mathbf{X}^T$, and we wish to fit the equivalent of a linear logistic regression model in the original features with quadratic regularization. Our predictions are also to be made using inner products; a new x_0 is presented as $k_0 = \mathbf{X}x_0$. Let $\mathbf{K} = \mathbf{U}\mathbf{D}^2\mathbf{U}^T$ be the eigen-decomposition of $\mathbf{K}$. Show that the predictions are given by $\hat{f}_0 = k_0^T \hat{\alpha}$, where

(a) $\hat{\alpha} = \mathbf{U}\mathbf{D}^{-1}\hat{\beta}$, and

(b) $\hat{\beta}$ is the ridged logistic regression estimate with input matrix $\mathbf{R} = \mathbf{U}\mathbf{D}$.

Argue that the same approach can be used for any appropriate kernel matrix $\mathbf{K}$.

Ex. 18.14 *Distance weighted 1-NN classification.* Consider the 1-nearest-neighbor method (Section 13.3) in a two-class classification problem. Let $d_+(x_0)$ be the shortest distance to a training observation in class $+1$, and likewise $d_-(x_0)$ the shortest distance for class -1. Let N_- be the number of samples in class -1, N_+ the number in class $+1$, and $N = N_- + N_+$.

(a) Show that
$$\delta(x_0) = \log \frac{d_-(x_0)}{d_+(x_0)} \qquad (18.57)$$

can be viewed as a nonparametric discriminant function corresponding to 1-NN classification. [*Hint*: Show that $\hat{f}_+(x_0) = \frac{1}{N_+ d_+(x_0)}$ can be viewed as a nonparametric estimate of the density in class $+1$ at x_0].

(b) How would you modify this function to introduce class prior probabilities π_+ and π_- different from the sample-priors N_+/N and N_-/N?

(c) How would you generalize this approach for K-NN classification?

Ex. 18.15 *Kernel PCA.* In Section 18.5.2 we show how to compute the principal component variables $\mathbf{Z}$ from an uncentered inner-product matrix $\mathbf{K}$. We compute the eigen-decomposition $(\mathbf{I} - \mathbf{M})\mathbf{K}(\mathbf{I} - \mathbf{M}) = \mathbf{U}\mathbf{D}^2\mathbf{U}^T$, with $\mathbf{M} = \mathbf{1}\mathbf{1}^T/N$, and then $\mathbf{Z} = \mathbf{U}\mathbf{D}$. Suppose we have the inner-product vector $\mathbf{k}_0$, containing the N inner-products between a new point x_0 and each of the x_i in our training set. Show that the (centered) projections of x_0 onto the principal-component directions are given by

$$\mathbf{z}_0 = \mathbf{D}^{-1}\mathbf{U}^T(\mathbf{I} - \mathbf{M})\left[\mathbf{k}_0 - \mathbf{K}\mathbf{1}/N\right]. \qquad (18.58)$$

Ex. 18.16 *Bonferroni method for multiple comparisons.* Suppose we are in a multiple-testing scenario with null hypotheses H_{0j}, $j = 1, 2, \ldots, M$, and corresponding p-values p_j, $i = 1, 2, \ldots, M$. Let A be the event that at least one null hypothesis is falsely rejected, and let A_j be the event that the jth null hypothesis is falsely rejected. Suppose that we use the Bonferroni method, rejecting the jth null hypothesis if $p_j < \alpha/M$.

(a) Show that $\Pr(A) \leq \alpha$. [*Hint:* $\Pr(A_j \cup A_{j'}) = \Pr(A_j) + \Pr(A_{j'}) - \Pr(A_j \cap A_{j'})$]

(b) If the hypotheses $H_{0j}, j = 1, 2, \ldots, M$, are independent, then $\Pr(A) = 1 - \Pr(A^C) = 1 - \prod_{j=1}^{M}\Pr(A_j^C) = 1 - (1 - \alpha/M)^M$. Use this to show that $\Pr(A) \approx \alpha$ in this case.

Ex. 18.17 *Equivalence between Benjamini–Hochberg and plug-in methods.*

(a) In the notation of Algorithm 18.2, show that for rejection threshold $p_0 = p_{(L)}$, a proportion of at most p_0 of the permuted values t_j^k exceed $|T|_{(L)}$ where $|T|_{(L)}$ is the Lth largest value among the $|t_j|$. Hence show that the plug-in FDR estimate $\widehat{\mathrm{FDR}}$ is less than or equal to $p_0 \cdot M/L = \alpha$.

(b) Show that the cut-point $|T|_{(L+1)}$ produces a test with estimated FDR greater than α.

Ex. 18.18 Use result (18.53) to show that

$$\mathrm{pFDR} = \frac{\pi_0 \cdot \{\text{Type I error of }\Gamma\}}{\pi_0 \cdot \{\text{Type I error of }\Gamma\} + \pi_1\{\text{Power of }\Gamma\}} \qquad (18.59)$$

(Storey, 2003).

Ex. 18.19 Consider the data in Table 18.4 of Section (18.7), available from the book website.

(a) Using a symmetric two-sided rejection region based on the t-statistic, compute the plug-in estimate of the FDR for various values of the cut-point.

(b) Carry out the BH procedure for various FDR levels α and show the equivalence of your results, with those from part (a).

(c) Let $(q_{.25}, q_{.75})$ be the quartiles of the t-statistics from the permuted datasets. Let $\hat{\pi}_0 = \{\#t_j \in (q_{.25}, q_{.75})\}/(.5M)$, and set $\hat{\pi}_0 = \min(\hat{\pi}_0, 1)$. Multiply the FDR estimates from (a) by $\hat{\pi}_0$ and examine the results.

(d) Give a motivation for the estimate in part (c).

(Storey, 2003)

Ex. 18.20 *Proof of result (18.53)*. Write

$$\text{pFDR} = E\left(\frac{V}{R}\bigg|R > 0\right) \tag{18.60}$$

$$= \sum_{k=1}^{M} E\left[\frac{V}{R}\bigg|R = k\right]\Pr(R = k|R > 0) \tag{18.61}$$

Use the fact that given $R = k$, V is a binomial random variable, with k trials and probability of success $\Pr(H = 0|T \in \Gamma)$, to complete the proof.

References

Abu-Mostafa, Y. (1995). Hints, *Neural Computation* **7**: 639–671.

Ackley, D. H., Hinton, G. and Sejnowski, T. (1985). A learning algorithm for Boltzmann machines, *Trends in Cognitive Sciences* **9**: 147–169.

Adam, B.-L., Qu, Y., Davis, J. W., Ward, M. D., Clements, M. A., Cazares, L. H., Semmes, O. J., Schellhammer, P. F., Yasui, Y., Feng, Z. and Wright, G. (2003). Serum protein fingerprinting coupled with a pattern-matching algorithm distinguishes prostate cancer from benign prostate hyperplasia and healthy mean, *Cancer Research* **63**(10): 3609–3614.

Agrawal, R., Mannila, H., Srikant, R., Toivonen, H. and Verkamo, A. I. (1995). Fast discovery of association rules, *Advances in Knowledge Discovery and Data Mining*, AAAI/MIT Press, Cambridge, MA.

Agresti, A. (1996). *An Introduction to Categorical Data Analysis*, Wiley, New York.

Agresti, A. (2002). *Categorical Data Analysis (2nd Ed.)*, Wiley, New York.

Ahn, J. and Marron, J. (2005). The direction of maximal data piling in high dimensional space, *Technical report*, Statistics Department, University of North Carolina, Chapel Hill.

Akaike, H. (1973). Information theory and an extension of the maximum likelihood principle, *Second International Symposium on Information Theory*, pp. 267–281.

T. Hastie et al., *The Elements of Statistical Learning, Second Edition*,
DOI: 10.1007/b94608,
© Springer Science+Business Media, LLC 2009

Allen, D. (1974). The relationship between variable selection and data augmentation and a method of prediction, *Technometrics* **16**: 125–7.

Ambroise, C. and McLachlan, G. (2002). Selection bias in gene extraction on the basis of microarray gene-expression data, *Proceedings of the National Academy of Sciences* **99**: 6562–6566.

Amit, Y. and Geman, D. (1997). Shape quantization and recognition with randomized trees, *Neural Computation* **9**: 1545–1588.

Anderson, J. and Rosenfeld, E. (eds) (1988). *Neurocomputing: Foundations of Research*, MIT Press, Cambridge, MA.

Anderson, T. (2003). *An Introduction to Multivariate Statistical Analysis, 3rd ed.*, Wiley, New York.

Bach, F. and Jordan, M. (2002). Kernel independent component analysis, *Journal of Machine Learning Research* **3**: 1–48.

Bair, E. and Tibshirani, R. (2004). Semi-supervised methods to predict patient survival from gene expression data, *PLOS Biology* **2**: 511–522.

Bair, E., Hastie, T., Paul, D. and Tibshirani, R. (2006). Prediction by supervised principal components, *Journal of the American Statistical Association* **101**: 119–137.

Bakin, S. (1999). Adaptive regression and model selection in data mining problems, *Technical report*, PhD. thesis, Australian National University, Canberra.

Banerjee, O., Ghaoui, L. E. and d'Aspremont, A. (2008). Model selection through sparse maximum likelihood estimation for multivariate gaussian or binary data, *Journal of Machine Learning Research* **9**: 485–516.

Barron, A. (1993). Universal approximation bounds for superpositions of a sigmoid function, *IEEE Transactions on Information Theory* **39**: 930–945.

Bartlett, P. and Traskin, M. (2007). Adaboost is consistent, *in* B. Schölkopf, J. Platt and T. Hoffman (eds), *Advances in Neural Information Processing Systems 19*, MIT Press, Cambridge, MA, pp. 105–112.

Becker, R., Cleveland, W. and Shyu, M. (1996). The visual design and control of trellis display, *Journal of Computational and Graphical Statistics* **5**: 123–155.

Bell, A. and Sejnowski, T. (1995). An information-maximization approach to blind separation and blind deconvolution, *Neural Computation* **7**: 1129–1159.

Bellman, R. E. (1961). *Adaptive Control Processes*, Princeton University Press.

Benjamini, Y. and Hochberg, Y. (1995). Controlling the false discovery rate: a practical and powerful approach to multiple testing, *Journal of the Royal Statistical Society Series B.* **85**: 289–300.

Benjamini, Y. and Yekutieli, Y. (2005). False discovery rate controlling confidence intervals for selected parameters, *Journal of the American Statistical Association* **100**: 71–80.

Bickel, P. and Levina, E. (2004). Some theory for Fisher's linear discriminant function,"Naive Bayes", and some alternatives when there are many more variables than observations, *Bernoulli* **10**: 989–1010.

Bickel, P. J., Ritov, Y. and Tsybakov, A. (2008). Simultaneous analysis of lasso and Dantzig selector, *Annals of Statistics.* to appear.

Bishop, C. (1995). *Neural Networks for Pattern Recognition*, Clarendon Press, Oxford.

Bishop, C. (2006). *Pattern Recognition and Machine Learning*, Springer, New York.

Bishop, Y., Fienberg, S. and Holland, P. (1975). *Discrete Multivariate Analysis*, MIT Press, Cambridge, MA.

Boyd, S. and Vandenberghe, L. (2004). *Convex Optimization*, Cambridge University Press.

Breiman, L. (1992). The little bootstrap and other methods for dimensionality selection in regression: X-fixed prediction error, *Journal of the American Statistical Association* **87**: 738–754.

Breiman, L. (1996a). Bagging predictors, *Machine Learning* **26**: 123–140.

Breiman, L. (1996b). Stacked regressions, *Machine Learning* **24**: 51–64.

Breiman, L. (1998). Arcing classifiers (with discussion), *Annals of Statistics* **26**: 801–849.

Breiman, L. (1999). Prediction games and arcing algorithms, *Neural Computation* **11**(7): 1493–1517.

Breiman, L. (2001). Random forests, *Machine Learning* **45**: 5–32.

Breiman, L. and Friedman, J. (1997). Predicting multivariate responses in multiple linear regression (with discussion), *Journal of the Royal Statistical Society Series B.* **59**: 3–37.

Breiman, L. and Ihaka, R. (1984). Nonlinear discriminant analysis via scaling and ACE, *Technical report*, University of California, Berkeley.

Breiman, L. and Spector, P. (1992). Submodel selection and evaluation in regression: the X-random case, *International Statistical Review* **60**: 291–319.

Breiman, L., Friedman, J., Olshen, R. and Stone, C. (1984). *Classification and Regression Trees*, Wadsworth, New York.

Bremaud, P. (1999). *Markov Chains: Gibbs Fields, Monte Carlo Simulation, and Queues*, Springer, New York.

Brown, P., Spiegelman, C. and Denham, M. (1991). Chemometrics and spectral frequency selection, *Transactions of the Royal Society of London Series A.* **337**: 311–322.

Bruce, A. and Gao, H. (1996). *Applied Wavelet Analysis with S-PLUS*, Springer, New York.

Bühlmann, P. and Hothorn, T. (2007). Boosting algorithms: regularization, prediction and model fitting (with discussion), *Statistical Science* **22**(4): 477–505.

Buja, A., Hastie, T. and Tibshirani, R. (1989). Linear smoothers and additive models (with discussion), *Annals of Statistics* **17**: 453–555.

Buja, A., Swayne, D., Littman, M., Hofmann, H. and Chen, L. (2008). Data vizualization with multidimensional scaling, *Journal of Computational and Graphical Statistics.* to appear.

Bunea, F., Tsybakov, A. and Wegkamp, M. (2007). Sparsity oracle inequalities for the lasso, *Electronic Journal of Statistics* **1**: 169–194.

Burges, C. (1998). A tutorial on support vector machines for pattern recognition, *Knowledge Discovery and Data Mining* **2**(2): 121–167.

Butte, A., Tamayo, P., Slonim, D., Golub, T. and Kohane, I. (2000). Discovering functional relationships between RNA expression and chemotherapeutic susceptibility using relevance networks, *Proceedings of the National Academy of Sciences* pp. 12182–12186.

Candes, E. (2006). Compressive sampling, *Proceedings of the International Congress of Mathematicians*, European Mathematical Society, Madrid, Spain.

Candes, E. and Tao, T. (2007). The Dantzig selector: Statistical estimation when p is much larger than n, *Annals of Statistics* **35**(6): 2313–2351.

Chambers, J. and Hastie, T. (1991). *Statistical Models in S*, Wadsworth/Brooks Cole, Pacific Grove, CA.

Chaudhuri, S., Drton, M. and Richardson, T. S. (2007). Estimation of a covariance matrix with zeros, *Biometrika* **94**(1): 1–18.

Chen, L. and Buja, A. (2008). Local multidimensional scaling for nonlinear dimension reduction, graph drawing and proximity analysis, *Journal of the American Statistical Association*.

Chen, S. S., Donoho, D. and Saunders, M. (1998). Atomic decomposition by basis pursuit, *SIAM Journal on Scientific Computing* **20**(1): 33–61.

Cherkassky, V. and Ma, Y. (2003). Comparison of model selection for regression, *Neural computation* **15**(7): 1691–1714.

Cherkassky, V. and Mulier, F. (2007). *Learning from Data (2nd Edition)*, Wiley, New York.

Chui, C. (1992). *An Introduction to Wavelets*, Academic Press, London.

Clifford, P. (1990). Markov random fields in statistics, *in* G. R. Grimmett and D. J. A. Welsh (eds), *Disorder in Physical Systems. A Volume in Honour of John M. Hammersley*, Clarendon Press, Oxford, pp. 19–32.

Comon, P. (1994). Independent component analysis—a new concept?, *Signal Processing* **36**: 287–314.

Cook, D. and Swayne, D. (2007). *Interactive and Dynamic Graphics for Data Analysis; with R and GGobi*, Springer, New York. With contributions from A. Buja, D. Temple Lang, H. Hofmann, H. Wickham and M. Lawrence.

Cook, N. (2007). Use and misuse of the receiver operating characteristic curve in risk prediction, *Circulation* **116**(6): 928–35.

Copas, J. B. (1983). Regression, prediction and shrinkage (with discussion), *Journal of the Royal Statistical Society, Series B, Methodological* **45**: 311–354.

Cover, T. and Hart, P. (1967). Nearest neighbor pattern classification, *IEEE Transactions on Information Theory* **IT-11**: 21–27.

Cover, T. and Thomas, J. (1991). *Elements of Information Theory*, Wiley, New York.

Cox, D. and Hinkley, D. (1974). *Theoretical Statistics*, Chapman and Hall, London.

Cox, D. and Wermuth, N. (1996). *Multivariate Dependencies: Models, Analysis and Interpretation*, Chapman and Hall, London.

Cressie, N. (1993). *Statistics for Spatial Data (Revised Edition)*, Wiley-Interscience, New York.

Csiszar, I. and Tusnády, G. (1984). Information geometry and alternating minimization procedures, *Statistics & Decisions Supplement Issue* **1**: 205–237.

Cutler, A. and Breiman, L. (1994). Archetypal analysis, *Technometrics* **36**(4): 338–347.

Dasarathy, B. (1991). *Nearest Neighbor Pattern Classification Techniques*, IEEE Computer Society Press, Los Alamitos, CA.

Daubechies, I. (1992). *Ten Lectures on Wavelets*, Society for Industrial and Applied Mathematics, Philadelphia, PA.

Daubechies, I., Defrise, M. and De Mol, C. (2004). An iterative thresholding algorithm for linear inverse problems with a sparsity constraint, *Communications on Pure and Applied Mathematics* **57**: 1413–1457.

de Boor, C. (1978). *A Practical Guide to Splines*, Springer, New York.

Dempster, A. (1972). Covariance selection, *Biometrics* **28**: 157–175.

Dempster, A., Laird, N. and Rubin, D. (1977). Maximum likelihood from incomplete data via the EM algorithm (with discussion), *Journal of the Royal Statistical Society Series B* **39**: 1–38.

Devijver, P. and Kittler, J. (1982). *Pattern Recognition: A Statistical Approach*, Prentice-Hall, Englewood Cliffs, N.J.

Dietterich, T. (2000a). Ensemble methods in machine learning, *Lecture Notes in Computer Science* **1857**: 1–15.

Dietterich, T. (2000b). An experimental comparison of three methods for constructing ensembles of decision trees: bagging, boosting, and randomization, *Machine Learning* **40**(2): 139–157.

Dietterich, T. and Bakiri, G. (1995). Solving multiclass learning problems via error-correcting output codes, *Journal of Artificial Intelligence Research* **2**: 263–286.

Donath, W. E. and Hoffman, A. J. (1973). Lower bounds for the partitioning of graphs, *IBM Journal of Research and Development* pp. 420–425.

Donoho, D. (2006a). Compressed sensing, *IEEE Transactions on Information Theory* **52**(4): 1289–1306.

Donoho, D. (2006b). For most large underdetermined systems of equations, the minimal ℓ^1-norm solution is the sparsest solution, *Communications on Pure and Applied Mathematics* **59**: 797–829.

Donoho, D. and Elad, M. (2003). Optimally sparse representation from overcomplete dictionaries via ℓ^1-norm minimization, *Proceedings of the National Academy of Sciences* **100**: 2197–2202.

Donoho, D. and Johnstone, I. (1994). Ideal spatial adaptation by wavelet shrinkage, *Biometrika* **81**: 425–455.

Donoho, D. and Stodden, V. (2004). When does non-negative matrix factorization give a correct decomposition into parts?, *in* S. Thrun, L. Saul and B. Schölkopf (eds), *Advances in Neural Information Processing Systems 16*, MIT Press, Cambridge, MA.

Duan, N. and Li, K.-C. (1991). Slicing regression: a link-free regression method, *Annals of Statistics* **19**: 505–530.

Duchamp, T. and Stuetzle, W. (1996). Extremal properties of principal curves in the plane, *Annals of Statistics* **24**: 1511–1520.

Duda, R., Hart, P. and Stork, D. (2000). *Pattern Classification (2nd Edition)*, Wiley, New York.

Dudoit, S., Fridlyand, J. and Speed, T. (2002a). Comparison of discrimination methods for the classification of tumors using gene expression data, *Journal of the American Statistical Association* **97**(457): 77–87.

Dudoit, S., Yang, Y., Callow, M. and Speed, T. (2002b). Statistical methods for identifying differentially expressed genes in replicated cDNA microarray experiments, *Statistica Sinica* pp. 111–139.

Edwards, D. (2000). *Introduction to Graphical Modelling, 2nd Edition*, Springer, New York.

Efron, B. (1975). The efficiency of logistic regression compared to normal discriminant analysis, *Journal of the American Statistical Association* **70**: 892–898.

Efron, B. (1979). Bootstrap methods: another look at the jackknife, *Annals of Statistics* **7**: 1–26.

Efron, B. (1983). Estimating the error rate of a prediction rule: some improvements on cross-validation, *Journal of the American Statistical Association* **78**: 316–331.

Efron, B. (1986). How biased is the apparent error rate of a prediction rule?, *Journal of the American Statistical Association* **81**: 461–70.

706   References

Efron, B. and Tibshirani, R. (1991). Statistical analysis in the computer age, *Science* **253**: 390–395.

Efron, B. and Tibshirani, R. (1993). *An Introduction to the Bootstrap*, Chapman and Hall, London.

Efron, B. and Tibshirani, R. (1996). Using specially designed exponential families for density estimation, *Annals of Statistics* **24**(6): 2431–2461.

Efron, B. and Tibshirani, R. (1997). Improvements on cross-validation: the 632+ bootstrap: method, *Journal of the American Statistical Association* **92**: 548–560.

Efron, B. and Tibshirani, R. (2002). Microarrays, empirical Bayes methods, and false discovery rates, *Genetic Epidemiology* **1**: 70–86.

Efron, B., Hastie, T. and Tibshirani, R. (2007). Discussion of "Dantzig selector" by Candes and Tao, *Annals of Statistics* **35**(6): 2358–2364.

Efron, B., Hastie, T., Johnstone, I. and Tibshirani, R. (2004). Least angle regression (with discussion), *Annals of Statistics* **32**(2): 407–499.

Efron, B., Tibshirani, R., Storey, J. and Tusher, V. (2001). Empirical Bayes analysis of a microarray experiment, *Journal of the American Statistical Association* **96**: 1151–1160.

Evgeniou, T., Pontil, M. and Poggio, T. (2000). Regularization networks and support vector machines, *Advances in Computational Mathematics* **13**(1): 1–50.

Fan, J. and Fan, Y. (2008). High dimensional classification using features annealed independence rules, *Annals of Statistics*. to appear.

Fan, J. and Gijbels, I. (1996). *Local Polynomial Modelling and Its Applications*, Chapman and Hall, London.

Fan, J. and Li, R. (2005). Variable selection via nonconcave penalized likelihood and its oracle properties, *Journal of the American Statistical Association* **96**: 1348–1360.

Fiedler, M. (1973). Algebraic connectivity of graphs, *Czechoslovak Mathematics Journal* **23**(98): 298–305.

Fienberg, S. (1977). *The Analysis of Cross-Classified Categorical Data*, MIT Press, Cambridge.

Fisher, R. A. (1936). The use of multiple measurements in taxonomic problems, *Eugen.* **7**: 179–188.

Fisher, W. (1958). On grouping for maximum homogeniety, *Journal of the American Statistical Association* **53**(284): 789–798.

Fix, E. and Hodges, J. (1951). Discriminatory analysis—nonparametric discrimination: Consistency properties, *Technical Report 21-49-004,4*, U.S. Air Force, School of Aviation Medicine, Randolph Field, TX.

Flury, B. (1990). Principal points, *Biometrika* **77**: 33–41.

Forgy, E. (1965). Cluster analysis of multivariate data: efficiency vs. interpretability of classifications, *Biometrics* **21**: 768–769.

Frank, I. and Friedman, J. (1993). A statistical view of some chemometrics regression tools (with discussion), *Technometrics* **35**(2): 109–148.

Freund, Y. (1995). Boosting a weak learning algorithm by majority, *Information and Computation* **121**(2): 256–285.

Freund, Y. and Schapire, R. (1996a). Experiments with a new boosting algorithm, *Machine Learning: Proceedings of the Thirteenth International Conference*, Morgan Kauffman, San Francisco, pp. 148–156.

Freund, Y. and Schapire, R. (1996b). Game theory, on-line prediction and boosting, *Proceedings of the Ninth Annual Conference on Computational Learning Theory*, Desenzano del Garda, Italy, pp. 325–332.

Freund, Y. and Schapire, R. (1997). A decision-theoretic generalization of online learning and an application to boosting, *Journal of Computer and System Sciences* **55**: 119–139.

Friedman, J. (1987). Exploratory projection pursuit, *Journal of the American Statistical Association* **82**: 249–266.

Friedman, J. (1989). Regularized discriminant analysis, *Journal of the American Statistical Association* **84**: 165–175.

Friedman, J. (1991). Multivariate adaptive regression splines (with discussion), *Annals of Statistics* **19**(1): 1–141.

Friedman, J. (1994a). Flexible metric nearest-neighbor classification, *Technical report*, Stanford University.

Friedman, J. (1994b). An overview of predictive learning and function approximation, *in* V. Cherkassky, J. Friedman and H. Wechsler (eds), *From Statistics to Neural Networks*, Vol. 136 of *NATO ISI Series F*, Springer, New York.

Friedman, J. (1996). Another approach to polychotomous classification, *Technical report*, Stanford University.

708   References

Friedman, J. (1997). On bias, variance, 0-1 loss and the curse of dimensionality, *Journal of Data Mining and Knowledge Discovery* **1**: 55–77.

Friedman, J. (1999). Stochastic gradient boosting, *Technical report*, Stanford University.

Friedman, J. (2001). Greedy function approximation: A gradient boosting machine, *Annals of Statistics* **29**(5): 1189–1232.

Friedman, J. and Fisher, N. (1999). Bump hunting in high dimensional data, *Statistics and Computing* **9**: 123–143.

Friedman, J. and Hall, P. (2007). On bagging and nonlinear estimation, *Journal of Statistical Planning and Inference* **137**: 669–683.

Friedman, J. and Popescu, B. (2003). Importance sampled learning ensembles, *Technical report*, Stanford University, Department of Statistics.

Friedman, J. and Popescu, B. (2008). Predictive learning via rule ensembles, *Annals of Applied Statistics, to appear*.

Friedman, J. and Silverman, B. (1989). Flexible parsimonious smoothing and additive modelling (with discussion), *Technometrics* **31**: 3–39.

Friedman, J. and Stuetzle, W. (1981). Projection pursuit regression, *Journal of the American Statistical Association* **76**: 817–823.

Friedman, J. and Tukey, J. (1974). A projection pursuit algorithm for exploratory data analysis, *IEEE Transactions on Computers, Series C* **23**: 881–889.

Friedman, J., Baskett, F. and Shustek, L. (1975). An algorithm for finding nearest neighbors, *IEEE Transactions on Computers* **24**: 1000–1006.

Friedman, J., Bentley, J. and Finkel, R. (1977). An algorthm for finding best matches in logarithmic expected time, *ACM Transactions on Mathematical Software* **3**: 209–226.

Friedman, J., Hastie, T. and Tibshirani, R. (2000). Additive logistic regression: a statistical view of boosting (with discussion), *Annals of Statistics* **28**: 337–307.

Friedman, J., Hastie, T. and Tibshirani, R. (2008a). Response to "Mease and Wyner: Evidence contrary to the statistical view of boosting", *Journal of Machine Learning Research* **9**: 175–180.

Friedman, J., Hastie, T. and Tibshirani, R. (2008b). Sparse inverse covariance estimation with the graphical lasso, *Biostatistics* **9**: 432–441.

Friedman, J., Hastie, T. and Tibshirani, R. (2010). Regularization paths for generalized linear models via coordinate descent, *Journal of Statistical Software* **33**(1): 1–22.

Friedman, J., Hastie, T., Hoefling, H. and Tibshirani, R. (2007). Pathwise coordinate optimization, *Annals of Applied Statistics* **2**(1): 302–332.

Friedman, J., Hastie, T., Rosset, S., Tibshirani, R. and Zhu, J. (2004). Discussion of three boosting papers by Jiang, Lugosi and Vayatis, and Zhang, *Annals of Statistics* **32**: 102–107.

Friedman, J., Stuetzle, W. and Schroeder, A. (1984). Projection pursuit density estimation, *Journal of the American Statistical Association* **79**: 599–608.

Fu, W. (1998). Penalized regressions: the bridge vs. the lasso, *Journal of Computational and Graphical Statistics* **7**(3): 397–416.

Furnival, G. and Wilson, R. (1974). Regression by leaps and bounds, *Technometrics* **16**: 499–511.

Gelfand, A. and Smith, A. (1990). Sampling based approaches to calculating marginal densities, *Journal of the American Statistical Association* **85**: 398–409.

Gelman, A., Carlin, J., Stern, H. and Rubin, D. (1995). *Bayesian Data Analysis*, CRC Press, Boca Raton, FL.

Geman, S. and Geman, D. (1984). Stochastic relaxation, Gibbs distributions and the Bayesian restoration of images, *IEEE Transactions on Pattern Analysis and Machine Intelligence* **6**: 721–741.

Genkin, A., Lewis, D. and Madigan, D. (2007). Large-scale Bayesian logistic regression for text categorization, *Technometrics* **49**(3): 291–304.

Genovese, C. and Wasserman, L. (2004). A stochastic process approach to false discovery rates, *Annals of Statistics* **32**(3): 1035–1061.

Gersho, A. and Gray, R. (1992). *Vector Quantization and Signal Compression*, Kluwer Academic Publishers, Boston, MA.

Girosi, F., Jones, M. and Poggio, T. (1995). Regularization theory and neural network architectures, *Neural Computation* **7**: 219–269.

Golub, G. and Van Loan, C. (1983). *Matrix Computations*, Johns Hopkins University Press, Baltimore.

Golub, G., Heath, M. and Wahba, G. (1979). Generalized cross-validation as a method for choosing a good ridge parameter, *Technometrics* **21**: 215–224.

Golub, T., Slonim, D., Tamayo, P., Huard, C., Gaasenbeek, M., Mesirov, J., Coller, H., Loh, M., Downing, J., Caligiuri, M., Bloomfield, C. and Lander, E. (1999). Molecular classification of cancer: Class discovery and class prediction by gene expression monitoring, *Science* **286**: 531–536.

Goodall, C. (1991). Procrustes methods in the statistical analysis of shape, *Journal of the Royal Statistical Society, Series B* **53**: 285–321.

Gordon, A. (1999). *Classification (2nd edition)*, Chapman and Hall/CRC Press, London.

Green, P. and Silverman, B. (1994). *Nonparametric Regression and Generalized Linear Models: A Roughness Penalty Approach*, Chapman and Hall, London.

Greenacre, M. (1984). *Theory and Applications of Correspondence Analysis*, Academic Press, New York.

Greenshtein, E. and Ritov, Y. (2004). Persistence in high-dimensional linear predictor selection and the virtue of overparametrization, *Bernoulli* **10**: 971–988.

Guo, Y., Hastie, T. and Tibshirani, R. (2006). Regularized linear discriminant analysis and its application in microarrays, *Biostatistics* **8**: 86–100.

Guyon, I., Gunn, S., Nikravesh, M. and Zadeh, L. (eds) (2006). *Feature Extraction, Foundations and Applications*, Springer, New York.

Guyon, I., Weston, J., Barnhill, S. and Vapnik, V. (2002). Gene selection for cancer classification using support vector machines, *Machine Learning* **46**: 389–422.

Hall, P. (1992). *The Bootstrap and Edgeworth Expansion*, Springer, New York.

Hammersley, J. M. and Clifford, P. (1971). Markov field on finite graphs and lattices, unpublished.

Hand, D. (1981). *Discrimination and Classification*, Wiley, Chichester.

Hanley, J. and McNeil, B. (1982). The meaning and use of the area under a receiver operating characteristic (roc) curve, *Radiology* **143**: 29–36.

Hart, P. (1968). The condensed nearest-neighbor rule, *IEEE Transactions on Information Theory* **14**: 515–516.

Hartigan, J. A. (1975). *Clustering Algorithms*, Wiley, New York.

Hartigan, J. A. and Wong, M. A. (1979). [(Algorithm AS 136] A *k*-means clustering algorithm (AS R39: 81v30 p355-356), *Applied Statistics* **28**: 100–108.

Hastie, T. (1984). *Principal Curves and Surfaces*, PhD thesis, Stanford University.

Hastie, T. and Herman, A. (1990). An analysis of gestational age, neonatal size and neonatal death using nonparametric logistic regression, *Journal of Clinical Epidemiology* **43**: 1179–90.

Hastie, T. and Simard, P. (1998). Models and metrics for handwritten digit recognition, *Statistical Science* **13**: 54–65.

Hastie, T. and Stuetzle, W. (1989). Principal curves, *Journal of the American Statistical Association* **84**(406): 502–516.

Hastie, T. and Tibshirani, R. (1987). Nonparametric logistic and proportional odds regression, *Applied Statistics* **36**: 260–276.

Hastie, T. and Tibshirani, R. (1990). *Generalized Additive Models*, Chapman and Hall, London.

Hastie, T. and Tibshirani, R. (1996a). Discriminant adaptive nearest-neighbor classification, *IEEE Pattern Recognition and Machine Intelligence* **18**: 607–616.

Hastie, T. and Tibshirani, R. (1996b). Discriminant analysis by Gaussian mixtures, *Journal of the Royal Statistical Society Series B.* **58**: 155–176.

Hastie, T. and Tibshirani, R. (1998). Classification by pairwise coupling, *Annals of Statistics* **26**(2): 451–471.

Hastie, T. and Tibshirani, R. (2003). Independent components analysis through product density estimation, *in* S. T. S. Becker and K. Obermayer (eds), *Advances in Neural Information Processing Systems 15*, MIT Press, Cambridge, MA, pp. 649–656.

Hastie, T. and Tibshirani, R. (2004). Efficient quadratic regularization for expression arrays, *Biostatistics* **5**(3): 329–340.

Hastie, T. and Zhu, J. (2006). Discussion of "Support vector machines with applications" by Javier Moguerza and Alberto Munoz, *Statistical Science* **21**(3): 352–357.

Hastie, T., Botha, J. and Schnitzler, C. (1989). Regression with an ordered categorical response, *Statistics in Medicine* **43**: 884–889.

Hastie, T., Buja, A. and Tibshirani, R. (1995). Penalized discriminant analysis, *Annals of Statistics* **23**: 73–102.

Hastie, T., Kishon, E., Clark, M. and Fan, J. (1992). A model for signature verification, *Technical report*, AT&T Bell Laboratories. http://www-stat.stanford.edu/~hastie/Papers/signature.pdf.

Hastie, T., Rosset, S., Tibshirani, R. and Zhu, J. (2004). The entire regularization path for the support vector machine, *Journal of Machine Learning Research* **5**: 1391–1415.

Hastie, T., Taylor, J., Tibshirani, R. and Walther, G. (2007). Forward stagewise regression and the monotone lasso, *Electronic Journal of Statistics* **1**: 1–29.

Hastie, T., Tibshirani, R. and Buja, A. (1994). Flexible discriminant analysis by optimal scoring, *Journal of the American Statistical Association* **89**: 1255–1270.

Hastie, T., Tibshirani, R. and Buja, A. (2000). Flexible discriminant and mixture models, *in* J. Kay and M. Titterington (eds), *Statistics and Artificial Neural Networks*, Oxford University Press.

Hastie, T., Tibshirani, R. and Friedman, J. (2003). A note on "Comparison of model selection for regression" by Cherkassky and Ma, *Neural computation* **15**(7): 1477–1480.

Hathaway, R. J. (1986). Another interpretation of the EM algorithm for mixture distributions, *Statistics & Probability Letters* **4**: 53–56.

Hebb, D. (1949). *The Organization of Behavior*, Wiley, New York.

Hertz, J., Krogh, A. and Palmer, R. (1991). *Introduction to the Theory of Neural Computation*, Addison Wesley, Redwood City, CA.

Hinton, G. (1989). Connectionist learning procedures, *Artificial Intelligence* **40**: 185–234.

Hinton, G. (2002). Training products of experts by minimizing contrastive divergence, *Neural Computation* **14**: 1771–1800.

Hinton, G., Osindero, S. and Teh, Y.-W. (2006). A fast learning algorithm for deep belief nets, *Neural Computation* **18**: 1527–1554.

Ho, T. K. (1995). Random decision forests, *in* M. Kavavaugh and P. Storms (eds), *Proc. Third International Conference on Document Analysis and Recognition*, Vol. 1, IEEE Computer Society Press, New York, pp. 278–282.

Hoefling, H. and Tibshirani, R. (2008). Estimation of sparse Markov networks using modified logistic regression and the lasso, submitted.

Hoerl, A. E. and Kennard, R. (1970). Ridge regression: biased estimation for nonorthogonal problems, *Technometrics* **12**: 55–67.

Hothorn, T. and Bühlmann, P. (2006). Model-based boosting in high dimensions, *Bioinformatics* **22**(22): 2828–2829.

Huber, P. (1964). Robust estimation of a location parameter, *Annals of Mathematical Statistics* **53**: 73–101.

Huber, P. (1985). Projection pursuit, *Annals of Statistics* **13**: 435–475.

Hunter, D. and Lange, K. (2004). A tutorial on MM algorithms, *The American Statistician* **58**(1): 30–37.

Hyvärinen, A. and Oja, E. (2000). Independent component analysis: algorithms and applications, *Neural Networks* **13**: 411–430.

Hyvärinen, A., Karhunen, J. and Oja, E. (2001). *Independent Component Analysis*, Wiley, New York.

Izenman, A. (1975). Reduced-rank regression for the multivariate linear model, *Journal of Multivariate Analysis* **5**: 248–264.

Jacobs, R., Jordan, M., Nowlan, S. and Hinton, G. (1991). Adaptive mixtures of local experts, *Neural computation* **3**: 79–87.

Jain, A. and Dubes, R. (1988). *Algorithms for Clustering Data*, Prentice-Hall, Englewood Cliffs, N.J.

James, G. and Hastie, T. (1998). The error coding method and PICTs, *Journal of Computational and Graphical Statistics* **7**(3): 377–387.

Jancey, R. (1966). Multidimensional group analysis, *Australian Journal of Botany* **14**: 127–130.

Jensen, F. V., Lauritzen, S. and Olesen, K. G. (1990). Bayesian updating in recursive graphical models by local computation, *Computational Statistics Quarterly* **4**: 269–282.

Jiang, W. (2004). Process consistency for Adaboost, *Annals of Statistics* **32**(1): 13–29.

Jiroušek, R. and Přeučil, S. (1995). On the effective implementation of the iterative proportional fitting procedure, *Computational Statistics and Data Analysis* **19**: 177–189.

Johnson, N. (2008). A study of the NIPS feature selection challenge, Submitted.

Joliffe, I. T., Trendafilov, N. T. and Uddin, M. (2003). A modified principal component technique based on the lasso, *Journal of Computational and Graphical Statistics* **12**: 531–547.

Jones, L. (1992). A simple lemma on greedy approximation in Hilbert space and convergence rates for projection pursuit regression and neural network training, *Annals of Statistics* **20**: 608–613.

Jordan, M. (2004). Graphical models, *Statistical Science (Special Issue on Bayesian Statistics)* **19**: 140–155.

Jordan, M. and Jacobs, R. (1994). Hierachical mixtures of experts and the EM algorithm, *Neural Computation* **6**: 181–214.

Kalbfleisch, J. and Prentice, R. (1980). *The Statistical Analysis of Failure Time Data*, Wiley, New York.

Kaufman, L. and Rousseeuw, P. (1990). *Finding Groups in Data: An Introduction to Cluster Analysis*, Wiley, New York.

Kearns, M. and Vazirani, U. (1994). *An Introduction to Computational Learning Theory*, MIT Press, Cambridge, MA.

Kittler, J., Hatef, M., Duin, R. and Matas, J. (1998). On combining classifiers, *IEEE Transaction on Pattern Analysis and Machine Intelligence* **20**(3): 226–239.

Kleinberg, E. M. (1990). Stochastic discrimination, *Annals of Mathematical Artificial Intelligence* **1**: 207–239.

Klcinberg, E. M. (1996). An overtraining-resistant stochastic modeling method for pattern recognition, *Annals of Statistics* **24**: 2319–2349.

Knight, K. and Fu, W. (2000). Asymptotics for lasso-type estimators, *Annals of Statistics* **28**(5): 1356–1378.

Koh, K., Kim, S.-J. and Boyd, S. (2007). An interior-point method for large-scale L1-regularized logistic regression, *Journal of Machine Learning Research* **8**: 1519–1555.

Kohavi, R. (1995). A study of cross-validation and bootstrap for accuracy estimation and model selection, *International Joint Conference on Artificial Intelligence (IJCAI)*, Morgan Kaufmann, pp. 1137–1143.

Kohonen, T. (1989). *Self-Organization and Associative Memory (3rd edition)*, Springer, Berlin.

Kohonen, T. (1990). The self-organizing map, *Proceedings of the IEEE* **78**: 1464–1479.

Kohonen, T., Kaski, S., Lagus, K., Salojärvi, J., Paatero, A. and Saarela, A. (2000). Self-organization of a massive document collection, *IEEE Transactions on Neural Networks* **11**(3): 574–585. Special Issue on Neural Networks for Data Mining and Knowledge Discovery.

Koller, D. and Friedman, N. (2007). *Structured Probabilistic Models*, Stanford Bookstore Custom Publishing. (Unpublished Draft).

Kressel, U. (1999). Pairwise classification and support vector machines, *in* B. Schölkopf, C. Burges and A. Smola (eds), *Advances in Kernel Methods - Support Vector Learning*, MIT Press, Cambridge, MA., pp. 255–268.

Lambert, D. (1992). Zero-inflated Poisson regression, with an application to defects in manufacturing, *Technometrics* **34**(1): 1–14.

Lange, K. (2004). *Optimization*, Springer, New York.

Lauritzen, S. (1996). *Graphical Models*, Oxford University Press.

Lauritzen, S. and Spiegelhalter, D. (1988). Local computations with probabilities on graphical structures and their application to expert systems, *J. Royal Statistical Society B.* **50**: 157–224.

Lawson, C. and Hansen, R. (1974). *Solving Least Squares Problems*, Prentice-Hall, Englewood Cliffs, NJ.

Le Cun, Y. (1989). Generalization and network design strategies, *Technical Report CRG-TR-89-4*, Department of Computer Science, Univ. of Toronto.

Le Cun, Y., Boser, B., Denker, J., Henderson, D., Howard, R., Hubbard, W. and Jackel, L. (1990). Handwritten digit recognition with a back-propogation network, *in* D. Touretzky (ed.), *Advances in Neural Information Processing Systems*, Vol. 2, Morgan Kaufman, Denver, CO, pp. 386–404.

Le Cun, Y., Bottou, L., Bengio, Y. and Haffner, P. (1998). Gradient-based learning applied to document recognition, *Proceedings of the IEEE* **86**(11): 2278–2324.

Leathwick, J., Elith, J., Francis, M., Hastie, T. and Taylor, P. (2006). Variation in demersal fish species richness in the oceans surrounding new zealand: an analysis using boosted regression trees, *Marine Ecology Progress Series* **77**: 802–813.

Leathwick, J., Rowe, D., Richardson, J., Elith, J. and Hastie, T. (2005). Using multivariate adaptive regression splines to predict the distributions of New Zealand's freshwater diadromous fish, *Freshwater Biology* **50**: 2034–2051.

Leblanc, M. and Tibshirani, R. (1996). Combining estimates in regression and classification, *Journal of the American Statistical Association* **91**: 1641–1650.

LeCun, Y., Bottou, L., Bengio, Y. and Haffner, P. (1998). Gradient-based learning applied to document recognition, *Proceedings of the IEEE* **86**(11): 2278–2324.

Lee, D. and Seung, H. (1999). Learning the parts of objects by non-negative matrix factorization, *Nature* **401**: 788.

Lee, D. and Seung, H. (2001). Algorithms for non-negative matrix factorization, *Advances in Neural Information Processing Systems, (NIPS 2001)*, Vol. 13, Morgan Kaufman, Denver., pp. 556–562.

Lee, M.-L. (2004). *Analysis of Microarray Gene Expression Data*, Kluwer Academic Publishers.

Lee, S.-I., Ganapathi, V. and Koller, D. (2007). Efficient structure learning of markov networks using l_1-regularization, *in* B. Schölkopf, J. Platt and T. Hoffman (eds), *Advances in Neural Information Processing Systems 19*, MIT Press, Cambridge, MA, pp. 817–824.

Leslie, C., Eskin, E., Cohen, A., Weston, J. and Noble, W. S. (2004). Mismatch string kernels for discriminative protein classification, *Bioinformatics* **20**(4): 467–476.

Levina, E. (2002). *Statistical issues in texture analysis*, PhD thesis, Department. of Statistics, University of California, Berkeley.

Lin, H., McCulloch, C., Turnbull, B., Slate, E. and Clark, L. (2000). A latent class mixed model for analyzing biomarker trajectories in longitudinal data with irregularly scheduled observations, *Statistics in Medicine* **19**: 1303–1318.

Lin, Y. and Zhang, H. (2006). Component selection and smoothing in smoothing spline analysis of variance models, *Annals of Statistics* **34**: 2272–2297.

Little, R. and Rubin, D. (2002). *Statistical Analysis with Missing Data (2nd Edition)*, Wiley, New York.

Lloyd, S. (1957). Least squares quantization in PCM., *Technical report*, Bell Laboratories. Published in 1982 in IEEE Transactions on Information Theory **28** 128-137.

Loader, C. (1999). *Local Regression and Likelihood*, Springer, New York.

Loh, W. and Vanichsetakul, N. (1988). Tree structured classification via generalized discriminant analysis, *Journal of the American Statistical Association* **83**: 715–728.

Lugosi, G. and Vayatis, N. (2004). On the bayes-risk consistency of regularized boosting methods, *Annals of Statistics* **32**(1): 30–55.

Macnaughton Smith, P., Williams, W., Dale, M. and Mockett, L. (1965). Dissimilarity analysis: a new technique of hierarchical subdivision, *Nature* **202**: 1034–1035.

MacKay, D. (1992). A practical Bayesian framework for backpropagation neural networks, *Neural Computation* **4**: 448–472.

MacQueen, J. (1967). Some methods for classification and analysis of multivariate observations, *Proceedings of the Fifth Berkeley Symposium on Mathematical Statistics and Probability, eds. L.M. LeCam and J. Neyman*, University of California Press, pp. 281–297.

Madigan, D. and Raftery, A. (1994). Model selection and accounting for model uncertainty using Occam's window, *Journal of the American Statistical Association* **89**: 1535–46.

Mardia, K., Kent, J. and Bibby, J. (1979). *Multivariate Analysis*, Academic Press.

Mason, L., Baxter, J., Bartlett, P. and Frean, M. (2000). Boosting algorithms as gradient descent, **12**: 512–518.

Massart, D., Plastria, F. and Kaufman, L. (1983). Non-hierarchical clustering with MASLOC, *The Journal of the Pattern Recognition Society* **16**: 507–516.

McCullagh, P. and Nelder, J. (1989). *Generalized Linear Models*, Chapman and Hall, London.

McCulloch, W. and Pitts, W. (1943). A logical calculus of the ideas imminent in nervous activity, *Bulletin of Mathematical Biophysics* **5**: 115–133. Reprinted in Anderson and Rosenfeld (1988), pp 96-104.

McLachlan, G. (1992). *Discriminant Analysis and Statistical Pattern Recognition*, Wiley, New York.

Mease, D. and Wyner, A. (2008). Evidence contrary to the statistical view of boosting (with discussion), *Journal of Machine Learning Research* **9**: 131–156.

Meinshausen, N. (2007). Relaxed lasso, *Computational Statistics and Data Analysis* **52**(1): 374–393.

Meinshausen, N. and Bühlmann, P. (2006). High-dimensional graphs and variable selection with the lasso, *Annals of Statistics* **34**: 1436–1462.

Meir, R. and Rätsch, G. (2003). An introduction to boosting and leveraging, *in* S. Mendelson and A. Smola (eds), *Lecture notes in Computer Science*, Advanced Lectures in Machine Learning, Springer, New York.

Michie, D., Spiegelhalter, D. and Taylor, C. (eds) (1994). *Machine Learning, Neural and Statistical Classification*, Ellis Horwood Series in Artificial Intelligence, Ellis Horwood.

Morgan, J. N. and Sonquist, J. A. (1963). Problems in the analysis of survey data, and a proposal, *Journal of the American Statistical Association* **58**: 415–434.

Murray, W., Gill, P. and Wright, M. (1981). *Practical Optimization*, Academic Press.

Myles, J. and Hand, D. (1990). The multiclass metric problem in nearest neighbor classification, *Pattern Recognition* **23**: 1291–1297.

Nadler, B. and Coifman, R. R. (2005). An exact asymptotic formula for the error in CLS and in PLS: The importance of dimensional reduction in multivariate calibration, *Journal of Chemometrics* **102**: 107–118.

Neal, R. (1996). *Bayesian Learning for Neural Networks*, Springer, New York.

Neal, R. and Hinton, G. (1998). *A view of the EM algorithm that justifies incremental, sparse, and other variants; in Learning in Graphical Models, M. Jordan (ed.)*, Dordrecht: Kluwer Academic Publishers, Boston, MA., pp. 355–368.

Neal, R. and Zhang, J. (2006). High dimensional classification with bayesian neural networks and dirichlet diffusion trees, *in* I. Guyon, S. Gunn, M. Nikravesh and L. Zadeh (eds), *Feature Extraction, Foundations and Applications*, Springer, New York, pp. 265–296.

Onton, J. and Makeig, S. (2006). Information-based modeling of event-related brain dynamics, *in* Neuper and Klimesch (eds), *Progress in Brain Research*, Vol. 159, Elsevier, pp. 99–120.

Osborne, M., Presnell, B. and Turlach, B. (2000a). A new approach to variable selection in least squares problems, *IMA Journal of Numerical Analysis* **20**: 389–404.

Osborne, M., Presnell, B. and Turlach, B. (2000b). On the lasso and its dual, *Journal of Computational and Graphical Statistics* **9**: 319–337.

Pace, R. K. and Barry, R. (1997). Sparse spatial autoregressions, *Statistics and Probability Letters* **33**: 291–297.

Page, L., Brin, S., Motwani, R. and Winograd, T. (1998). The pagerank citation ranking: bringing order to the web, *Technical report*, Stanford Digital Library Technologies Project. http://citeseer.ist.psu.edu/page98pagerank.html.

Park, M. Y. and Hastie, T. (2007). l_1-regularization path algorithm for generalized linear models, *Journal of the Royal Statistical Society Series B* **69**: 659–677.

Parker, D. (1985). Learning logic, *Technical Report TR-87*, Cambridge MA: MIT Center for Research in Computational Economics and Management Science.

Parmigiani, G., Garett, E. S., Irizarry, R. A. and Zeger, S. L. (eds) (2003). *The Analysis of Gene Expression Data*, Springer, New York.

Paul, D., Bair, E., Hastie, T. and Tibshirani, R. (2008). "Pre-conditioning" for feature selection and regression in high-dimensional problems, *Annals of Statistics* **36**(4): 1595–1618.

Pearl, J. (1986). On evidential reasoning in a hierarchy of hypotheses, *Artificial Intelligence* **28**: 9–15.

Pearl, J. (1988). *Probabilistic reasoning in intelligent systems: networks of plausible inference*, Morgan Kaufmann, San Francisco, CA.

Pearl, J. (2000). *Causality: Models, Reasoning and Inference*, Cambridge University Press.

Peterson and Anderson, J. R. (1987). A mean field theory learning algorithm for neural networks, *Complex Systems* **1**: 995–1019.

Petricoin, E. F., Ardekani, A. M., Hitt, B. A., Levine, P. J., Fusaro, V., Steinberg, S. M., Mills, G. B., Simone, C., Fishman, D. A., Kohn, E. and Liotta, L. A. (2002). Use of proteomic patterns in serum to identify ovarian cancer, *Lancet* **359**: 572–577.

Platt, J. (1999). *Fast Training of Support Vector Machines using Sequential Minimal Optimization; in Advances in Kernel Methods—Support Vector Learning, B. Schölkopf and C. J. C. Burges and A. J. Smola (eds)*, MIT Press, Cambridge, MA., pp. 185–208.

Quinlan, R. (1993). *C4.5: Programs for Machine Learning*, Morgan Kaufmann, San Mateo.

Quinlan, R. (2004). C5.0, www.rulequest.com.

Ramaswamy, S., Tamayo, P., Rifkin, R., Mukherjee, S., Yeang, C., Angelo, M., Ladd, C., Reich, M., Latulippe, E., Mesirov, J., Poggio, T., Gerald, W., Loda, M., Lander, E. and Golub, T. (2001). Multiclass cancer diagnosis using tumor gene expression signature, *PNAS* **98**: 15149–15154.

Ramsay, J. and Silverman, B. (1997). *Functional Data Analysis*, Springer, New York.

Rao, C. R. (1973). *Linear Statistical Inference and Its Applications*, Wiley, New York.

Rätsch, G. and Warmuth, M. (2002). Maximizing the margin with boosting, *Proceedings of the 15th Annual Conference on Computational Learning Theory*, pp. 334–350.

Ravikumar, P., Liu, H., Lafferty, J. and Wasserman, L. (2008). Spam: Sparse additive models, *in* J. Platt, D. Koller, Y. Singer and S. Roweis (eds), *Advances in Neural Information Processing Systems 20*, MIT Press, Cambridge, MA, pp. 1201–1208.

Ridgeway, G. (1999). The state of boosting, *Computing Science and Statistics* **31**: 172–181.

Rieger, K., Hong, W., Tusher, V., Tang, J., Tibshirani, R. and Chu, G. (2004). Toxicity from radiation therapy associated with abnormal transcriptional responses to DNA damage, *Proceedings of the National Academy of Sciences* **101**: 6634–6640.

Ripley, B. D. (1996). *Pattern Recognition and Neural Networks*, Cambridge University Press.

Rissanen, J. (1983). A universal prior for integers and estimation by minimum description length, *Annals of Statistics* **11**: 416–431.

Robbins, H. and Munro, S. (1951). A stochastic approximation method, *Annals of Mathematical Statistics* **22**: 400–407.

Roosen, C. and Hastie, T. (1994). Automatic smoothing spline projection pursuit, *Journal of Computational and Graphical Statistics* **3**: 235–248.

Rosenblatt, F. (1958). The perceptron: a probabilistic model for information storage and organization in the brain, *Psychological Review* **65**: 386–408.

Rosenblatt, F. (1962). *Principles of Neurodynamics: Perceptrons and the Theory of Brain Mechanisms*, Spartan, Washington, D.C.

Rosenwald, A., Wright, G., Chan, W. C., Connors, J. M., Campo, E., Fisher, R. I., Gascoyne, R. D., Muller-Hermelink, H. K., Smeland, E. B. and Staudt, L. M. (2002). The use of molecular profiling to predict survival after chemotherapy for diffuse large b-cell lymphoma, *The New England Journal of Medicine* **346**: 1937–1947.

Rosset, S. and Zhu, J. (2007). Piecewise linear regularized solution paths, *Annals of Statistics* **35**(3): 1012–1030.

Rosset, S., Zhu, J. and Hastie, T. (2004a). Boosting as a regularized path to a maximum margin classifier, *Journal of Machine Learning Research* **5**: 941–973.

Rosset, S., Zhu, J. and Hastie, T. (2004b). Margin maximizing loss functions, *in* S. Thrun, L. Saul and B. Schölkopf (eds), *Advances in Neural Information Processing Systems 16*, MIT Press, Cambridge, MA.

Rousseauw, J., du Plessis, J., Benade, A., Jordaan, P., Kotze, J., Jooste, P. and Ferreira, J. (1983). Coronary risk factor screening in three rural communities, *South African Medical Journal* **64**: 430–436.

Roweis, S. T. and Saul, L. K. (2000). Locally linear embedding, *Science* **290**: 2323–2326.

Rumelhart, D., Hinton, G. and Williams, R. (1986). Learning internal representations by error propagation, *in* D. Rumelhart and J. McClelland (eds), *Parallel Distributed Processing: Explorations in the Microstructure of Cognition*, The MIT Press, Cambridge, MA., pp. 318–362.

Sachs, K., Perez, O., Pe'er, D., Lauffenburger, D. and Nolan, G. (2005). Causal protein-signaling networks derived from multiparameter single-cell data, *Science* **308**: 523–529.

Schapire, R. (1990). The strength of weak learnability, *Machine Learning* **5**(2): 197–227.

Schapire, R. (2002). The boosting approach to machine learning: an overview, *in* D. Denison, M. Hansen, C. Holmes, B. Mallick and B. Yu (eds), *MSRI workshop on Nonlinear Estimation and Classification*, Springer, New York.

Schapire, R. and Singer, Y. (1999). Improved boosting algorithms using confidence-rated predictions, *Machine Learning* **37**(3): 297–336.

Schapire, R., Freund, Y., Bartlett, P. and Lee, W. (1998). Boosting the margin: a new explanation for the effectiveness of voting methods, *Annals of Statistics* **26**(5): 1651–1686.

Schölkopf, B., Smola, A. and Müller, K.-R. (1999). Kernel principal component analysis, *in* B. Schölkopf, C. Burges and A. Smola (eds), *Advances in Kernel Methods—Support Vector Learning*, MIT Press, Cambridge, MA, USA, pp. 327–352.

Schwarz, G. (1978). Estimating the dimension of a model, *Annals of Statistics* **6**(2): 461–464.

Scott, D. (1992). *Multivariate Density Estimation: Theory, Practice, and Visualization*, Wiley, New York.

Seber, G. (1984). *Multivariate Observations*, Wiley, New York.

Segal, M. (2004). Machine learning benchmarks and random forest regression, *Technical report*, eScholarship Repository, University of California. http://repositories.edlib.org/cbmb/bench_rf_regn.

Shao, J. (1996). Bootstrap model selection, *Journal of the American Statistical Association* **91**: 655–665.

Shenoy, P. and Shafer, G. (1988). An axiomatic framework for Bayesian and belief-function propagation, *AAAI Workshop on Uncertainty in AI*, North-Holland, pp. 307–314.

Short, R. and Fukunaga, K. (1981). The optimal distance measure for nearest neighbor classification, *IEEE Transactions on Information Theory* **27**: 622–627.

Silverman, B. (1986). *Density Estimation for Statistics and Data Analysis*, Chapman and Hall, London.

Silvey, S. (1975). *Statistical Inference*, Chapman and Hall, London.

Simard, P., Cun, Y. L. and Denker, J. (1993). Efficient pattern recognition using a new transformation distance, *Advances in Neural Information Processing Systems*, Morgan Kaufman, San Mateo, CA, pp. 50–58.

Simon, R. M., Korn, E. L., McShane, L. M., Radmacher, M. D., Wright, G. and Zhao, Y. (2004). *Design and Analysis of DNA Microarray Investigations*, Springer, New York.

Sjöstrand, K., Rostrup, E., Ryberg, C., Larsen, R., Studholme, C., Baezner, H., Ferro, J., Fazekas, F., Pantoni, L., Inzitari, D. and Waldemar, G. (2007). Sparse decomposition and modeling of anatomical shape variation, *IEEE Transactions on Medical Imaging* **26**(12): 1625–1635.

Speed, T. and Kiiveri, H. T. (1986). Gaussian Markov distributions over finite graphs, *Annals of Statistics* **14**: 138–150.

Speed, T. (ed.) (2003). *Statistical Analysis of Gene Expression Microarray Data*, Chapman and Hall, London.

Spiegelhalter, D., Best, N., Gilks, W. and Inskip, H. (1996). Hepatitis B: a case study in MCMC methods, *in* W. Gilks, S. Richardson and D. Spegelhalter (eds), *Markov Chain Monte Carlo in Practice*, Interdisciplinary Statistics, Chapman and Hall, London, pp. 21–43.

Spielman, D. A. and Teng, S.-H. (1996). Spectral partitioning works: Planar graphs and finite element meshes, *IEEE Symposium on Foundations of Computer Science*, pp. 96–105.

Stamey, T., Kabalin, J., McNeal, J., Johnstone, I., Freiha, F., Redwine, E. and Yang, N. (1989). Prostate specific antigen in the diagnosis and treatment of adenocarcinoma of the prostate II radical prostatectomy treated patients, *Journal of Urology* **16**: 1076–1083.

Stone, C., Hansen, M., Kooperberg, C. and Truong, Y. (1997). Polynomial splines and their tensor products (with discussion), *Annals of Statistics* **25**(4): 1371–1470.

Stone, M. (1974). Cross-validatory choice and assessment of statistical predictions, *Journal of the Royal Statistical Society Series B* **36**: 111–147.

Stone, M. (1977). An asymptotic equivalence of choice of model by cross-validation and Akaike's criterion, *Journal of the Royal Statistical Society Series B.* **39**: 44 7.

Stone, M. and Brooks, R. J. (1990). Continuum regression: cross-validated sequentially constructed prediction embracing ordinary least squares, partial least squares and principal components regression (Corr: V54 p906-907), *Journal of the Royal Statistical Society, Series B* **52**: 237–269.

Storey, J. (2002). A direct approach to false discovery rates, *Journal of the Royal Statistical Society B.* **64**(3): 479–498.

Storey, J. (2003). The positive false discovery rate: A Bayesian interpretation and the q-value, *Annals of Statistics* **31**: 2013–2025.

Storey, J. and Tibshirani, R. (2003). Statistical significance for genomewide studies, *Proceedings of the National Academy of Sciences* **100-**: 9440–9445.

Storey, J., Taylor, J. and Siegmund, D. (2004). Strong control, conservative point estimation, and simultaneous conservative consistency of false discovery rates: A unified approach., *Journal of the Royal Statistical Society, Series B* **66**: 187–205.

Surowiecki, J. (2004). *The Wisdom of Crowds: Why the Many are Smarter than the Few and How Collective Wisdom Shapes Business, Economics, Societies and Nations.*, Little, Brown.

Swayne, D., Cook, D. and Buja, A. (1991). Xgobi: Interactive dynamic graphics in the X window system with a link to S, *ASA Proceedings of Section on Statistical Graphics*, pp. 1–8.

Tanner, M. and Wong, W. (1987). The calculation of posterior distributions by data augmentation (with discussion), *Journal of the American Statistical Association* **82**: 528–550.

Tarpey, T. and Flury, B. (1996). Self-consistency: A fundamental concept in statistics, *Statistical Science* **11**: 229–243.

Tenenbaum, J. B., de Silva, V. and Langford, J. C. (2000). A global geometric framework for nonlinear dimensionality reduction, *Science* **290**: 2319–2323.

Tibshirani, R. (1996). Regression shrinkage and selection via the lasso, *Journal of the Royal Statistical Society, Series B* **58**: 267–288.

Tibshirani, R. and Hastie, T. (2007). Margin trees for high-dimensional classification, *Journal of Machine Learning Research* **8**: 637–652.

Tibshirani, R. and Knight, K. (1999). Model search and inference by bootstrap "bumping, *Journal of Computational and Graphical Statistics* **8**: 671–686.

Tibshirani, R. and Wang, P. (2007). Spatial smoothing and hot spot detection for CGH data using the fused lasso, *Biostatistics* **9**: 18–29.

Tibshirani, R., Hastie, T., Narasimhan, B. and Chu, G. (2001a). Diagnosis of multiple cancer types by shrunken centroids of gene expression, *Proceedings of the National Academy of Sciences* **99**: 6567–6572.

Tibshirani, R., Hastie, T., Narasimhan, B. and Chu, G. (2003). Class prediction by nearest shrunken centroids, with applications to DNA microarrays, *Statistical Science* **18**(1): 104–117.

Tibshirani, R., Saunders, M., Rosset, S., Zhu, J. and Knight, K. (2005). Sparsity and smoothness via the fused lasso, *Journal of the Royal Statistical Society, Series B* **67**: 91–108.

Tibshirani, R., Walther, G. and Hastie, T. (2001b). Estimating the number of clusters in a dataset via the gap statistic, *Journal of the Royal Statistical Society, Series B.* **32**(2): 411–423.

Tropp, J. (2004). Greed is good: algorithmic results for sparse approximation, *IEEE Transactions on Information Theory* **50**: 2231– 2242.

Tropp, J. (2006). Just relax: convex programming methods for identifying sparse signals in noise, *IEEE Transactions on Information Theory* **52**: 1030–1051.

Valiant, L. G. (1984). A theory of the learnable, *Communications of the ACM* **27**: 1134–1142.

van der Merwe, A. and Zidek, J. (1980). Multivariate regression analysis and canonical variates, *The Canadian Journal of Statistics* **8**: 27–39.

Vapnik, V. (1996). *The Nature of Statistical Learning Theory*, Springer, New York.

Vapnik, V. (1998). *Statistical Learning Theory*, Wiley, New York.

Vidakovic, B. (1999). *Statistical Modeling by Wavelets*, Wiley, New York.

von Luxburg, U. (2007). A tutorial on spectral clustering, *Statistics and Computing* **17**(4): 395–416.

Wahba, G. (1980). Spline bases, regularization, and generalized cross-validation for solving approximation problems with large quantities of noisy data, *Proceedings of the International Conference on Approximation theory in Honour of George Lorenz*, Academic Press, Austin, Texas, pp. 905–912.

Wahba, G. (1990). *Spline Models for Observational Data*, SIAM, Philadelphia.

Wahba, G., Lin, Y. and Zhang, H. (2000). GACV for support vector machines, *in* A. Smola, P. Bartlett, B. Schölkopf and D. Schuurmans (eds), *Advances in Large Margin Classifiers*, MIT Press, Cambridge, MA., pp. 297–311.

Wainwright, M. (2006). Sharp thresholds for noisy and high-dimensional recovery of sparsity using ℓ_1-constrained quadratic programming, *Technical report*, Department of Statistics, University of California, Berkeley.

Wainwright, M. J., Ravikumar, P. and Lafferty, J. D. (2007). High-dimensional graphical model selection using ℓ_1-regularized logistic regression, *in* B. Schölkopf, J. Platt and T. Hoffman (eds), *Advances in Neural Information Processing Systems 19*, MIT Press, Cambridge, MA, pp. 1465–1472.

Wasserman, L. (2004). *All of Statistics: a Concise Course in Statistical Inference*, Springer, New York.

Weisberg, S. (1980). *Applied Linear Regression*, Wiley, New York.

Werbos, P. (1974). *Beyond Regression*, PhD thesis, Harvard University.

Weston, J. and Watkins, C. (1999). Multiclass support vector machines, *in* M. Verleysen (ed.), *Proceedings of ESANN99*, D. Facto Press, Brussels.

Whittaker, J. (1990). *Graphical Models in Applied Multivariate Statistics*, Wiley, Chichester.

Wickerhauser, M. (1994). *Adapted Wavelet Analysis from Theory to Software*, A.K. Peters Ltd, Natick, MA.

Widrow, B. and Hoff, M. (1960). Adaptive switching circuits, *IRE WESCON Convention record*, Vol. 4. pp 96-104; Reprinted in Andersen and Rosenfeld (1988).

Wold, H. (1975). Soft modelling by latent variables: the nonlinear iterative partial least squares (NIPALS) approach, *Perspectives in Probability and Statistics, In Honor of M. S. Bartlett*, pp. 117–144.

Wolpert, D. (1992). Stacked generalization, *Neural Networks* **5**: 241–259.

Wu, T. and Lange, K. (2007). The MM alternative to EM, unpublished.

Wu, T. and Lange, K. (2008). Coordinate descent procedures for lasso penalized regression, *Annals of Applied Statistics* **2**(1): 224–244.

Yee, T. and Wild, C. (1996). Vector generalized additive models, *Journal of the Royal Statistical Society, Series B*. **58**: 481–493.

Yuan, M. and Lin, Y. (2007). Model selection and estimation in regression with grouped variables, *Journal of the Royal Statistical Society, Series B* **68**(1): 49–67.

Zhang, P. (1993). Model selection via multifold cross-validation, *Annals of Statistics* **21**: 299–311.

Zhang, T. and Yu, B. (2005). Boosting with early stopping: convergence and consistency, *Annals of Statistics* **33**: 1538–1579.

Zhao, P. and Yu, B. (2006). On model selection consistency of lasso, *Journal of Machine Learning Research* **7**: 2541–2563.

Zhao, P., Rocha, G. and Yu, B. (2008). The composite absolute penalties for grouped and hierarchichal variable selection, *Annals of Statistics*. (to appear).

Zhu, J. and Hastie, T. (2004). Classification of gene microarrays by penalized logistic regression, *Biostatistics* **5**(2): 427–443.

Zhu, J., Zou, H., Rosset, S. and Hastie, T. (2005). Multiclass adaboost, Unpublished.

Zou, H. (2006). The adaptive lasso and its oracle properties, *Journal of the American Statistical Association* **101**: 1418–1429.

Zou, H. and Hastie, T. (2005). Regularization and variable selection via the elastic net, *Journal of the Royal Statistical Society Series B.* **67**(2): 301–320.

Zou, H., Hastie, T. and Tibshirani, R. (2006). Sparse principal component analysis, *Journal of Computational and Graphical Statistics* **15**(2): 265–28.

Zou, H., Hastie, T. and Tibshirani, R. (2007). On the degrees of freedom of the lasso, *Annals of Statistics* **35**(5): 2173–2192.

Author Index

Abu-Mostafa, Y. 95, 474
Ackley, D. H. 645
Adam, B.-L. 664
Agrawal, R. 489–491, 578
Agresti, A. 385, 638, 640
Ahn, J. 695
Akaike, H. 257
Allen, D. 257
Ambroise, C. 247
Amit, Y. 602
Anderson, J. R. 641
Anderson, T. 645
Angelo, M. 654, 658
Ardekani, A. M. 664

Bach, F. 569
Baezner, H. 551
Bair, E. 676, 679–683, 693
Bakin, S. 90
Bakiri, G. 605, 606
Banerjee, O. 636
Barnhill, S. 658
Barron, A. 415
Barry, R. 371
Bartlett, P. 384, 615

Baskett, F. 480
Baxter, J. 384
Becker, R. 369
Bell, A. 578
Bellman, R. E. 22
Benade, A. 122
Bengio, Y. 404, 407, 408, 414, 644
Benjamini, Y. 687, 689, 693
Bentley, J. 480
Best, N. 292
Bibby, J. 94, 135, 441, 539, 559,
    578, 630, 679
Bickel, P. 652
Bickel, P. J. 89
Bishop, C. 38, 233, 414, 623, 645
Bishop, Y. 629, 638
Bloomfield, C. 663
Boser, B. 404, 414
Botha, J. 334
Bottou, L. 404, 407, 408, 414, 644
Boyd, S. 125, 632
Breiman, L. 85, 243, 251, 257, 292,
    308, 310, 334, 339, 367,
    384, 451, 453, 455, 554,
    587, 602

T. Hastie et al., *The Elements of Statistical Learning, Second Edition,*
DOI: 10.1007/b94608,
© Springer Science+Business Media, LLC 2009

Index

T. Hastie et al., *The Elements of Statistical Learning, Second Edition,*
DOI: 10.1007/b94608,
© Springer Science+Business Media, LLC 2009

Printed by Wilco bv, the Netherlands